Disaster Resilience and Green Growth

In recent years, it has become clear that environmental changes and disaster risks are closely linked. Climate change, changes in land use, and the depletion of natural resources make communities and infrastructure more vulnerable. This has created a need for comprehensive and collaborative solutions that bring together different fields. Effective management of disaster risks and adaptation to climate change requires coordinated efforts across infrastructure, social systems, and ecological resilience. Factors related to the environment and human activities greatly influence risk exposure. Therefore, risk management must consider these elements. Bridging fields such as environmental science, engineering, economics, and policy can lead to effective disaster resilience.

Traditional economic growth models often overlook environmental sustainability, resulting in more resource depletion and vulnerability. In response, new green growth strategies focus on sustainability, including nature-based solutions and climate actions to promote circular economies and low-carbon development, enhancing the approach of disaster resilience. Current policy discussions emphasize the importance of building resilient systems in infrastructure, industry, and public health as part of the sustainability agenda post-2030. Furthermore, new energy technologies present both opportunities and risks, requiring strategies to manage their impacts effectively. Chemical and industrial safety risks in addition to natural hazard induced technological disasters as complex emergencies are also a growing concerns in disaster management and sustainability in the emerging world.

The scope of book series relates all the SDG in implicit ways but is focused particularly on SDG 2, 3, 8, 9, 11, 12, 13 and 15. The book series "Disaster Resilience and Green Growth" aims to share research and insights on these topics surrounding safety and resilience of human life and society in the changing world, fostering innovation and knowledge-sharing for a sustainable future.

Y. D. Imran Khan • Saurabh Purohit
Jibrin Mohammed Jibrin
Madhav Dwarkanath Gholkar
Editors

Land Restoration Through Ecosystem-Based Approach

Contexts from Drylands of the Global South

Springer

Editors
Y. D. Imran Khan
WOTR Centre for Resilience Studies (W-CReS)
Watershed Organisation Trust (WOTR)
Pune, Maharashtra, India

Saurabh Purohit
WOTR Centre for Resilience Studies (W-CReS)
Watershed Organisation Trust (WOTR)
Pune, Maharashtra, India

Jibrin Mohammed Jibrin
Centre of Excellence in Dryland Agriculture
Bayero University
Kano, Nigeria

Madhav Dwarkanath Gholkar
WOTR Centre for Resilience Studies (W-CReS)
Watershed Organisation Trust (WOTR)
Pune, Maharashtra, India

ISSN 2662-4885 ISSN 2662-4893 (electronic)
Disaster Resilience and Green Growth
ISBN 978-981-95-0388-9 ISBN 978-981-95-0389-6 (eBook)
https://doi.org/10.1007/978-981-95-0389-6

This Springer imprint is published by the registered company Springer Nature Singapore Pte Ltd.
The registered company address is: 152 Beach Road, #21-01/04 Gateway East, Singapore 189721, Singapore

This book is dedicated to Fr. Hermann Bacher (1924-2021), the visionary Jesuit pioneer of large-scale participatory watershed management in India, whose relentless efforts transformed the lives of millions in the country's drought-prone and rain-fed landscapes. He was a true champion of community-led conservation and water stewardship. His legacy continues to inspire generations in the fight against land degradation.

With his unwavering commitment to rural development, he recognised early on that true watershed restoration could not be achieved through top-down approaches but by empowering communities as stewards of their own natural resources. Beginning in the 1960s, his ground-breaking work laid the foundation for integrated, community-led watershed development, emphasising soil and water conservation, afforestation, and sustainable agriculture. As the initiator and key architect of the Indo-German Watershed Development Program (IGWDP)—a large-scale bilateral project between the Governments of Germany and India—and

Co-founder of the Watershed Organisation Trust (WOTR)—an NGO he set up to support and expand the IGWDP, he institutionalised a self-governing, ecosystem restoration and climate-resilient water management model that has inspired generations of developmental practitioners and policymakers.

His participatory and multi-stakeholder collaborative approach helped restore degraded lands, revive groundwater, and build climate-resilient livelihoods in some of the country's most water-stressed regions, especially in the Indian state of Maharashtra. Through his tireless efforts at mobilising communities to harvest the rain wherever it falls across the watersheds and landscapes they live, villages once plagued by drought and poverty have witnessed renewed prosperity, environmental revival, and social empowerment, particularly among women and marginalised communities. His model has not only reshaped Maharashtra's watershed development policies but has also been replicated across India. His legacy is the now widely accepted understanding that when communities come together to regenerate their landscapes and capture the rains, are empowered to drive, implement and manage the effort, are adequately supported with the required resources, and held accountable for desired outcomes, drylands can bloom and set a community on the path to prosperity, resilience, and sustainable living.

This book honours his unparalleled contribution to restoring drylands, reviving ecosystems, enhancing water security, and empowering rural communities. It is a tribute to a man whose life's mission was to ensure that every village had access to the most fundamental of resources—water, the lifeline of life and prosperity.

Foreword

Land degradation is one of our time's most pressing environmental challenges, with profound consequences for biodiversity, livelihoods, and climate resilience. Nowhere is this challenge more pronounced than in the drylands of the Global South, where millions depend on fragile ecosystems for sustenance and where the impacts of climate change, unsustainable land use, and socio-economic pressures intersect. Addressing these concerns requires a holistic and innovative approach that moves beyond conventional restoration strategies and embraces the principles of ecosystem-based approaches.

Land Restoration Through Ecosystem-Based Approach: Contexts from Drylands of the Global South is a timely and insightful contribution to the growing body of knowledge on sustainable land management. This book brings together scholars, practitioners, and policymakers to explore the potential of nature-based solutions for restoring degraded landscapes. The authors illustrate how ecosystem-based approaches can foster ecological recovery, enhance community resilience, and promote sustainable development in dry regions through diverse case studies, empirical analyses, and theoretical perspectives. One of the key strengths of this book is its interdisciplinary lens. Land degradation is not merely an ecological problem—it is deeply intertwined with socio-economic, political, and cultural dimensions. The contributors highlight the complex interactions that shape land restoration efforts by integrating perspectives from ecology, agronomy, hydrology, and social sciences. From agroforestry and sustainable grazing practices to water conservation and community-led initiatives, the book showcases a wide array of restoration strategies that harness the power of nature to heal degraded ecosystems.

Another significant aspect of this work is its emphasis on localized knowledge and community participation. Too often, restoration efforts fail because they do not account for the lived realities of local communities. The case studies presented in this volume underscore the importance of engaging indigenous knowledge systems, traditional land management practices, and participatory governance models. The Global South is home to rich and diverse cultural traditions that have long adapted to arid and semi-arid environments. Leveraging these practices while integrating modern scientific insights can lead to more effective and equitable restoration outcomes.

Moreover, this book serves as a crucial reminder of the role of policy and governance in enabling or hindering restoration efforts. Effective land restoration cannot occur in isolation—it requires supportive institutional frameworks, coherent policies, and cross-sectoral collaboration. The authors provide valuable insights into policy interventions facilitating ecosystem-based restoration and the challenges of land tenure insecurity, inadequate funding, and conflicting land-use priorities. Their analysis offers actionable recommendations for policymakers, donors, and development agencies seeking to scale up restoration initiatives.

As the world faces accelerating environmental change, the lessons from this book are more relevant than ever. Restoring degraded drylands is a matter of ecological necessity and a pathway to achieving broader global goals, including food security, poverty reduction, and climate resilience. The ecosystem-based approaches discussed in this volume offer hope and practical solutions for regenerating landscapes while ensuring the well-being of communities that depend on them.

I congratulate the editors and contributors for their outstanding work compiled in this important book. I hope this book will inspire researchers, practitioners, and policymakers to advance ecosystem-based restoration efforts and recognize the transformative potential of working with nature. In doing so, we can move towards a more sustainable and resilient future for drylands and those who call them home.

Oluwole A. Fatunbi
Director of Research and Innovation
Forum for Agricultural Research in Africa (FARA),
Accra, Ghana

Preface

Land degradation in drylands is a pressing challenge across many regions of the Global South, threatening livelihoods, food security, and biodiversity. Driven by climate change, unsustainable land-use practices, and socio-economic pressures, these fragile ecosystems require urgent and innovative restoration strategies. While conventional restoration approaches often rely on engineering-based interventions, ecosystem-based approaches offer a sustainable, cost-effective, and ecologically sound alternative. These nature-based solutions restore degraded landscapes and enhance biodiversity, improve soil health, increase water retention, and provide long-term economic benefits for local communities. However, the successful implementation of ecosystem-based restoration requires a deep understanding of ecological processes, community engagement, policy integration, and innovative financial mechanisms.

This book, *Land Restoration Through Ecosystem-Based Approach: Contexts from Drylands of the Global South*, presents a comprehensive analysis of ecosystem-based solutions tailored to the specific ecological, cultural, and economic contexts of drylands. Despite the growing recognition of ecosystem-based restoration, there is still a significant gap in practical knowledge, regional case studies, and policy frameworks tailored to the Global South's unique socio-economic and ecological contexts. This book aims to bridge that gap by providing a comprehensive and interdisciplinary resource for practitioners, policymakers, researchers, and development professionals working on land restoration in these regions. This volume brings together insights from researchers, practitioners, and policymakers working on dryland restoration across the Global South, providing a comprehensive framework for ecosystem-based restoration in these regions.

The chapters in this book delve into key themes such as sustainable land management, nature-based restoration strategies, technological advancements, and participatory governance in restoration efforts. Special emphasis is placed on the role of local communities, institutions, and policymakers in driving sustainable land management practices that balance ecological restoration with economic development.

This book is primarily aimed at researchers and practitioners dedicated to confronting the intricate and escalating challenges of land degradation in dryland regions. However, its relevance extends far beyond academia. It is equally valuable for government agencies, non-governmental organisations, and local communities striving to implement practical solutions that enhance resilience to climate-induced stresses, such as droughts, desertification, and erratic rainfall. Additionally, it is an essential resource for students and educators seeking to expand their knowledge of ecosystem-based approaches and their application in restoring and sustainably managing dryland ecosystems.

Through this book, we hope to inspire actionable change and contribute to the global movement towards resilient, restored, and thriving dryland ecosystems in the developing world.

Pune, Maharashtra, India — Y. D. Imran Khan
Pune, Maharashtra, India — Saurabh Purohit
Kano, Nigeria — Jibrin Mohammed Jibrin
Pune, Maharashtra, India — Madhav Dwarkanath Gholkar

Message from Crispino Lobo

Land, water and climate provide the basis of life and are the foundations of our communities and civilisations. On their bounty we build our prosperity, and when they can no longer provide us with the resources we depend on, our economies and civilisations falter and decline. History is replete with sophisticated civilisations that collapsed and disappeared due to environmental degradation and climate change, notable examples of which are the Old Egyptian Kingdom (4500 years ago), the Harappan (3000 years ago), Mayan (AD 900) and Angkor Wat (circa 1400s) civilisations. When one considers that it takes between 500 and 1000 years to create one inch of soil and that most of our food comes from a layer of soil 6 to 40 inches in depth, it is not hard to understand why entire societies crumble when severe, high frequency or prolonged natural (and man-made) disasters irretrievably erode soils and degrade lands. Just as we consider water as life, soil is the foundation that sustains us and climate, the interlocutor—all three of which are interwoven into our lives and shape our destinies.

It is indeed a fitting tribute to these three pillars of our existence that this book compiles and shares technologies which can help communities across the world to sustainably manage their water and soil resources while adapting to weather vagaries and climate change. Well-deserved kudos to the editors and authors!

Following the dedication of this book to the memory of Fr. Hermann Bacher, who conceived and initiated the Indo-German Watershed Development Program (IGWDP), a large-scale community-led watershed development programme that began in Maharashtra and whose success significantly contributed to and influenced the national discourse and practice of watershed development in India,[1] I would like

[1] Fr. Bacher conceived the IGWDP in 1989; it became operational in 1992 and came to an end in 2014. Its impact led the Government of India to set up the Watershed Development Fund (WDF) at NABARD in 1999, which has continued the up-scaling of the IGWDP paradigm of watershed development across the country. The IGWDP had as its partners the German Ministry of Development Cooperation (represented by the KfW and the GIZ), the Government of India (represented by NABARD), the State Government of Maharashtra and WOTR (Watershed Organisation Trust), an NGO.

to share some insights into what likely made this programme the huge success it was, aimed at practitioners and policymakers.[2]

1. Large-scale projects involving natural resources management are complex and politically sensitive. These can only be successfully implemented if the most significant stakeholders, at all social and administrative levels, share a basic consensus on the need for such a project, are involved in its design and are agreed upon the expected outcomes. This helps create the political, institutional, social and community "buy-in" and will to achieve these desired outcomes, secure the necessary resources, find a way to overcoming conflicts (and they will be many given the often divergent interest of various stakeholders) and staying the course over the medium-to-long term, this latter being especially important given that natural resources management projects (NRM) in dryland areas generally have a 5-12-year period to become sustainable.
2. The design of such programmes must be such that the community and their participatively elected/nominated representatives must be in the "driver's seat" deciding every aspect of the project and being responsible and accountable for its implementation and management, during and in the post-project period. This is necessary to create the required ownership, without which a project will not likely achieve its intended goals. The reason being that unless the majority of stakeholders feel that they will benefit from the project and their interests are assured, they will not likely cooperate and willingly shoulder the responsibilities that such projects impose on communities such as controlling free grazing in treated areas, making voluntary contributions, permitting work on own lands that are contiguous to lands under treatment, etc.
3. Projects must be implemented in a technically sound manner in order to deliver the expected results and become sustainable[3]. This would imply that required resources—financial, human, institutional and community "buy-in"—are secured before implementation begins or can be achieved in the near term. NRM projects require the trust of the community and stakeholders to be successful. The knowledge or belief that required resources are available or have been provisioned for helps keep enthusiasm high, mitigates conflicts and sustains the momentum of the project.
4. A key resource is the availability of quality human resources at the implementation level, often a key constraint. Adequate attention needs to be paid to building the capacities of project functionaries, especially at the village level. Apart from providing technical and managerial support during project implementation, this

[2] The author of this Message, Crispino Lobo, has had the privilege of working with Fr. Bacher from the very initiation of the IGWDP and was its first (and only) Program Coordinator for 12 years until 2001.

[3] Ideally, watershed projects adopt a "ridge-to-valley" approach where one begins treating lands in the higher slopes before coming to those below with emphasis on in situ soil and water conservation measures (including vegetative grasses, plantations, afforestation) followed by water harvesting structures along the water courses. Area treatment measures and water harvesting works undertaken should be robust, interlinked and should meet sound technical standards.

cohort of trained personnel can provide locally vital knowledge and skills to help maintain structures created in the post-project period.

5. Sustainability should be part of project design from its very beginning. Apart from building up the human, organisational and institutional capabilities of villagers to implement and manage the project, a key requirement is fostering a change in the mindset, perspectives and attitudes of the villagers. Specifically, they must be viewed as "project holders" or drivers of the project; not as beneficiaries or recipients of largesse. This is crucial if the project is to be "owned" by the villagers, for only then will sustainability of the project become a realistic possibility.
6. Together with facilitating villagers to plan project measures, a tried and tested way of putting communities in the "driver's seat" (project holders) is to devolve project funds to a representative body of the villagers together with a clearly and publicly communicated activities, transparency and accountability plan accompanied by mechanisms that hold this body publicly accountable on a regular basis. The underlying principle being that discharging responsibilities must also be accompanied by devolution and exercising of authority, fund utilisation and accountability to related stakeholders. The insight being that only then will a community (and stakeholders) acquire a strong and enduring stake in sustaining what they believe they have achieved, rather than in what they have just contributed to or been part of.
7. At the programme level, there is a need to establish a point of coordination and networking to keep all stakeholders on board and engaged at least during key periods of programme and project execution. In the Indo-German Watershed Development Program (IGWDP), this was the Office of the Program Coordinator[4] whose task was to promote collaboration between the various stakeholders at all levels, facilitate engagement of related government institutions and other resource agencies, identify potential project villages and NGOs/ facilitating organisations, lead efforts to develop a strategy and pedagogy of action[5] and capacity building of field functionaries and villagers, share insights and learnings from the ground to government agencies and policy channels in order to contribute to better programme design and implementation, and secure the necessary resources and enablements (permissions, public funds, technical support) so that the programme could be successfully implemented and result in scalar impact. Case in point is the fact that despite the IGWDP having ended in 2014, the watershed program continues to be up-scaled with Indian public funds, now covering an area of 2.7 million hectares across 28 states, involving an outlay of

[4] Crispino Lobo was the first Program Coordinator and the Watershed Organisation Trust (WOTR) which he co-founded served as the institutional base of the Office of the Program Coordinator delivering the services mentioned above. WOTR was the Capacity Building Agency of the IGWDP (www.wotr.org)

[5] It was called the Gender-Oriented Participatory Operational Pedagogy (GO-POP)

approximately USD 330 million, as on February 2025.[6] The success and impact of the IGWDP on communities (and the environment) in dryland regions led the Government of India to set up a self-financing fund at NABARD with a mandate to scale up the approach to dryland regions across the country.

When community-led watershed development is done in the right way, the social, economic and financial returns can be substantial. A study conducted of one of the treated watershed villages by the World Resources Institute (WRI) and WOTR showed that the estimated net agricultural income had increased about three times from 1998 to 2012 while the benefit-cost ratio (BC) ranged from 2.28 to 3.76.[7] Other studies conducted by WOTR in 15 drought-prone villages in which watershed development was undertaken showed that within a short span of 4-5 years, area under water bodies had increased by 20%, seasonally irrigated area by 25% and gross cultivated area by 80%. Importantly, annual average household income had increased by 2.5 times, leading to a marked reduction in indebted households by 47% and an increase in construction of cement houses by 89%. Prior to the watershed effort, only 39 households were able to afford LPG (natural gas) for cooking; this figure rose to 592 households (an increase of 1418%)[8]. Significantly, distress migration has stopped and reverse or in-migration has occurred in most villages; water shortages have greatly reduced and school attendance, especially of girls, has gone up significantly. Hunger and debilitating poverty are now a distant memory and most of the families are well on the pathway out of poverty.

The way the IGWDP was designed and implemented, involving multiple stakeholders across sectors and levels, could provide a blueprint, suitably modified, for scalar expansion in other parts of the world where land degradation and water scarcity are posing existential threats to communities, societies and regions. An overview of the salient and innovative features of the IGWDP is available in a chapter written by the author titled "Soil Protection through Watershed Development: A Grassroots Perspective", published in *Soil Law and Governance in India* by Springer, Switzerland.[9]

The chapters in this volume offer a smorgasbord of experiences and perspectives on the concept and practice of ecological restoration in different geographies and contexts. It is hoped that this conspectus will help inspire policymakers,

---

[6] https://www.nabard.org/contentsearch.aspx?AID=95&Key=watershed+development+fund, accessed on 12.3.2025

[7] https://wotr-website-publications.s3.ap-south-1.amazonaws.com/18_wsd_in_india.pdf. The village is Kumbharwadi, whose watershed development was funded by the IGWDP and implemented by WOTR.

[8] WOTR (2020), "Regeneration of Natural Resources through Integrated and Participatory Watershed Development in Akole and Sangamner Block of Ahmednagar District- Year 2019-20: Impact Assessment Report".

[9] Lobo, C. (2023). Soil Protection Through Watershed Development: A Grassroots Perspective. In: Desai, B.H. (eds) Soil Law and Governance in India. International Yearbook of Soil Law and Policy. Springer, Cham. https://doi.org/10.1007/978-3-031-32360-7_10

practitioners and researchers to work together to mobilise communities to restore and sustainably manage the ecosystems they live in to better their livelihoods, the local economy and overall quality of life.

–Crispino Lobo[10]

[10] Crispino Lobo, Co-Founder and Managing Trustee of the Watershed Organization Trust (WOTR), has 35+ years of experience tackling water challenges, environmental degradation and the impacts of climate change on agriculture and livelihoods in rain-dependent and dryland regions of rural India and internationally. He has co-founded four non-profits, one of which is the Watershed Organisation Trust (WOTR). WOTR is dedicated to transforming the lives of millions of poor villagers across India through participatory watershed development and ecosystem restoration, climate-resilient agriculture, integrated and efficient water management, sustainable livelihoods and climate change adaptation, with a special emphasis on building the resilience of vulnerable communities, farmers and women. He was the Program Coordinator of the large-scale, highly successful bilaterally funded Indo-German Watershed Development Program (IGWDP) in Maharashtra, India. He has been a member of several government advisory committees at both the state and federal levels. He has contributed to setting up two significant national public funds to promote participatory watershed development and climate change adaptation in the country. An alumnus of Harvard University and the Gokhale Institute of Politics and Economics, Crispino has 5 academic degrees and has authored/co-authored/helmed 12 books besides several articles/chapters in national and international publications/books.

Acknowledgement

The journey of this book, *Land Restoration Through Ecosystem-Based Approach: Contexts from Drylands of the Global South*, has been enlightening and humbling. Its completion would not have been possible without the invaluable contributions and unwavering support of the Watershed Organisation Trust (WOTR), the WOTR Centre for Resilience Studies (W-CReS), Pune, and the Bayero University, Kano (BUK), Nigeria.

We are immensely grateful to Crispino Lobo, Co-Founder of WOTR; Dr. Marcella D'Souza, Director of the WOTR Centre for Resilience Studies (W-CReS); Prakash Keskar, Executive Director of WOTR; and all WOTRians for supporting this initiative. Their commitment to environmental sustainability and resilience-building in dryland regions has facilitated the research and discussions underpinning this book.

We extend our heartfelt gratitude to Professor Aliyu Salisu Barau, Dean of the Faculty of Earth and Environmental Sciences at BUK, for his invaluable support. We also deeply appreciate the contributions of the staff at the Centre for Dryland Agriculture (CDA), particularly Dr. Muhammad Nuraddeen Danjuma, Dr. Adnan Aminu Adnan, and Professor Murtala Muhammad Badamasi, whose insights and encouragement were instrumental to this work.

Our sincere appreciation also goes to Dr. Abdurrazak Ibrahim of the Forum for Agricultural Research in Africa (FARA) for his generous efforts in widely disseminating our call for abstracts across the FARA network in Africa and Brazil, helping us reach a broader academic and professional community.

We express our profound gratitude to the researchers, practitioners, and policymakers whose unwavering dedication to land restoration in the Global South has provided the foundation for this work. Their efforts on the ground—often amid immense ecological and socio-economic challenges—continue to inspire and inform the restoration movement. A heartfelt appreciation is extended to local communities, indigenous groups, farmers, and the dedicated stewards of the land. Their lived experiences and voices form the very foundation of this book, reflecting the essence of community-driven restoration efforts.

We also extend our deepest gratitude to the Series Editors and their exceptional team, whose unwavering dedication to reviewing, refining, and structuring the content has been instrumental in shaping this book. Their meticulous attention to detail and commitment to ensuring the coherence and clarity of each chapter have greatly elevated the quality of this publication. We also profoundly thank the Springer Nature team for their invaluable guidance, patience, and professionalism throughout the publication process. Their steadfast support has made this journey smoother and infinitely more rewarding.

We are very grateful to our families, mentors, and friends whose unwavering support made this book possible. To our families who shared their time, the mentors who guided with wisdom, and the friends who offered motivation—their contributions, though often behind the scenes, have been invaluable. This book is as much a reflection of their support as it is of the knowledge within its pages.

Finally, special acknowledgement goes out to all those working tirelessly to heal the Earth's degraded lands, ensuring a sustainable and resilient future for generations to come. May it serve as a catalyst for knowledge-sharing, collaboration, and action in the global effort to restore our planet's drylands.

Y. D. Imran Khan
Saurabh Purohit
Jibrin Mohammed Jibrin
Madhav Dwarkanath Gholkar

Contents

1 **Land Restoration Using Ecosystem Approaches in the Countries of the Global South: Prologue** 1
Y. D. Imran Khan, Saurabh Purohit, Jibrin Mohammed Jibrin, and Madhav Dwarkanath Gholkar

Part I Restoration Dynamics in Dryland Ecosystems

2 **Climate Change and Ecosystem-Based Approach: Adapting to Climate Change in Dryland Ecosystems; Mitigation and Resilience Through Ecosystem-Based Solutions** 9
Akshay Kumar Sagar, Rajiv Kumar Srivastava, and A. Chakraborty

3 **Nature-Based Solutions (NbS) for Restoration and Management of Dryland Ecosystems** 33
Subrata Gorain, Ayushman Malakar, and Suman Dutta

4 **Why Less Visual and Undervalued Lands Need Urgent Restoration: Perspectives from India's Semi-arid Rangelands** 55
Dada R. Dadas, Faraz Rupani, and Dipak Zade

5 **Restoration of Tropical Drylands: Lessons from Tamil Nadu, India** 75
R. J. Ranjit Daniels

6 **Study of the Impact of Irrigation on Groundwater Quality in the Jam River Basin, Nashik and Ahilyanagar Districts, Maharashtra, India** 85
U. G. Mhaske, Resham Bhalla, Chetan A. Patil, Pravin M. Nalawade, and G. D. Mhaske

Part II Nature-Based Restoration Strategies for Climate Resilient Ecosystems

7 **Ecosystem-Based Approaches to Natural Resource Management: Adapting to Climate Change in Arid Regions** 105
Archana Verma, Dheeraj Singh, M. B. Noor Mohamed, K. Shiran, A. Keerthika, N. V. Saresh, and Navjot Kaler

8 **Fostering the Future of Drylands: Balancing Productivity and Resilience Through Nature-Based Solutions** 129
M. Saritha, Shrvan Kumar, Anand Kumar Naorem, Om Prakash Meena, Kamlesh Kumar Meena, and Dheeraj Singh

9 **Adapting to Climate Change in Dryland Ecosystems; Mitigation and Resilience Through Ecosystem-Based Adaptation** 167
Chandan Kumar, Dheeraj Singh, Om Prakash Meena, and Anand Kumar Naorem

10 **Soil-Based Dryland Restoration Practices in Southern Africa: A Systematic Review** 185
Lazarus Chapungu and Kelvin Zhanda

11 **Nature-Based Disaster Risk Management Approach in the Sudano-Sahelian Zone of Cameroon** 209
Fréderic Saha and Thierry Gaitan Seutchueng Tchuenga

12 **Avian Diversity and Feeding Guild of Inland Wetland in a Semiarid Region: Implications for Wetland Conservation and Restoration** 229
Saurabh Purohit, Krishanmurti, Manendra Singh, Omkar M. Hande, Ashwini M. Wadhu, and Y. D. Imran Khan

13 **Reviving Groundwater Ecosystem Health: Industrial Effluent Impacts on Water Quality and Agricultural Sustainability in Solapur's Chincholi Industrial Zone** 247
Shrikant Mukate, Dipak Panaskar, Vasant Wagh, Yogesh Deshpande, and Uday Sahu

Part III Sustainable Land Management and Innovative Agricultural Practices

14 **Soybean Farmers' Ecosystem-Based Approach in Mitigating Climate Change in Savannah Agroecological Zones in Northern Ghana** 277
Selorm Omega, Henry De-Graft Acquah, and Selorm Akaba

15 **Mitigating Climate Change Through Sustainable Farming: An Organic Approach for Dryland Resilience in Maharashtra** 327
Sandhya Sreekumar, Prachi Nautiyal, Sheetal Walvekar, Ganesh Deshmukh, K. Gireesan, Enosha Arshad, and Swati Nirala

16 **Scaling Up of Vermicomposting in the Selected Landscapes of Madhya Pradesh and Chhattisgarh for Sustainable Agriculture and Land Productivity** 343
R. S. Rawat, A. K. Thakur, and Shilpa Gautam

17 **Plant Breeding Strategies for Sustainable Agriculture in Climate Change Context: A Literature Review** 359
Ali Yetgin, Rajiv Kumar Srivastava, Siddharth Panda, and Sweta Rath

Part IV Cutting-Edge Innovations, Digital Tools, and Co-creation Strategies

18 **Utilizing Decision Support Tools to Optimize Resource Allocation for Different Categories of Farmers in Northern Ghana** 383
Dilys S. MacCarthy, Samuel G. K. Adiku, Bright S. Freduah, Edward E. Onumah, Collins N. O. Larkai, Pierre S. Traore, and Helen Greatrex

19 **Promoting Ecosystem-Based Adaptation to Climate Change in a Fragile Ecological Context of the North Region of Cameroon: A contribution of Innovation Niche Approach** 399
Thierry Gaitan Seutchueng Tchuenga, Émile Blaise Siewe Pougoue, and Fréderic Saha

20 **Soil and Water Conservation Technologies for Ecosystem Restoration in Drylands** 425
Stephen Mureithi, Samuel Mwendwa, and Yonathan Ayalew

21 **MaxEnt's Forecast of Climate Change's Effect on Habitat Suitability of *Anacardium occidentale* L. in Togo (West Africa)** 485
Kossi-Messan Jacques Agboka, Dodiomon Soro, Komi Agboka, and Wouyo Atakpama

22 **Integrating Remote Sensing and Land-Use Change to Identify Potential Wetland Restoration Areas: A Multiyear Analysis Using Landsat Data** 501
Jintu moni Bhuyan, Manju Tiwari, and Nitish Tiwari

23 **Multi-stakeholder Participatory Approaches for Sustainable Dryland Restoration in Sub-Saharan Africa** 519
Leonard Chitongo and Kelvin Zhanda

24 **Indigenous Ecological Knowledge and Climate Change Adaptation** 545
Samson Olusoji Afolabi, Emmanuel Babatunde Ajala, Innocent Onu Alenyi, and Olubunmi Abosede Odusanya

Part V Adaptive Policies, Institutional Governance, and Long-Term Sustainability Strategies

25 **Mainstreaming Circular Ecosystem-Based Approach into National Climate Adaptation Strategies: Enhancing Resilience and Sustainability of Crop-Livestock Systems in Africa's Drylands** 571
Mujeyi Angeline, Musemwa Lovemore, and Munyati Vincent

26 **Restoring Dryland Ecosystems: Local Knowledge, Native Species and Tree Restoration Policies and Practices in Northern Nigeria** 595
Nicola Favretto, Lindsay C. Stringer, Aliyu S. Barau, Jibrin Mohammed Jibrin, Saminu Ado, Ibrahim B. Yakubu, and Martin Dallimer

27 **Understanding Government Policies on Climate-Smart Agriculture in West Africa: Insights from Burkina Faso** 621
Eveline Sawadogo Compaoré, Roman Hinz, Reginald Tang Guuroh, Jewel Andoh, Constantin Gueswindé Compaoré, Sibiri Birba, and Farid Traoré

28 **The Potential and Policy Options for Mainstreaming Ecosystem-Based Adaptation in Africa's Drylands** 639
Leonard Omullo Orondo

29 **Probing into Nexus Between Industrial Development and Dryland Degradation in the Global South** 667
K. Vishaal, Aamir Ahmad Teeli, Hanh Pham Thi, and Showkat Anwar Bhat

30 **Restoring Drylands: Policy Pathways, Governance, and Stakeholder Synergy for Ecosystem Revival** 681
Deepanjana Saha

Part VI Epilogue

31 **Lessons Learned and the Way Forward for Land Restoration Through Ecosystem-Based Approach in the Drylands** 695
Y. D. Imran Khan, Saurabh Purohit, Jibrin Mohammed Jibrin, and Madhav Dwarkanath Gholkar

Editors and Contributors

About the Editors

Y. D. Imran Khan currently working as a Program Manager and Thematic Lead (Ecology) at the WOTR Centre for Resilience Studies (W-CReS), Watershed Organisation Trust (WOTR) in Pune, India. He holds a Ph.D. in Zoology (with a specialisation in Ecology) from Bangalore University, Bengaluru, and an M.Sc. in Zoology from Sri Krishnadevaraya University, Anantapur. Before joining WOTR, Dr. Khan worked at the Institute for Social and Economic Change (ISEC) in Bengaluru, the National Institute of Rural Development and Panchayati Raj (NIRD&PR) in Hyderabad, and the Tata Institute of Social Sciences (TISS) in Mumbai. He has held various positions, including Research Assistant, Junior Research Fellow, Senior Research Fellow, Research Associate, and Training Manager and Senior Researcher in the field of Natural Resource Management and Rural Development. Dr. Khan's areas of interest include conservation biology, socio-ecology, rural livelihoods, and watershed Ecology. He is dedicated to developing an integrated interdisciplinary approach that supports sustainable landscape development. Dr. Khan collaborates with interdisciplinary and transdisciplinary teams of scholars from various national and international organisations to integrate ecological and social science approaches in assessing the impact of policy and behavioural changes.

Saurabh Purohit is currently working as a Senior Researcher (Ecology) at the WOTR Centre for Resilient Ecosystem (W-CReS), Watershed Organisation Trust (WOTR), Pune. He has over 10 years of research expertise in natural resource management from the Western Himalayan region to the foothills of the Western Ghats. He earned both a Master's and a PhD degree in Forestry from the Forest Research Institute (Deemed to be University), India. His research was conducted at the Indian Institute of Remote Sensing (IIRS), which is affiliated with the Indian Space Research Organisation (ISRO) in Dehradun. In his current role, Dr. Purohit is working on using nature-based solutions to increase the resilience of rural communities in the semi-arid landscape of India. His specialisation includes a wide range of subjects, including nature-based solutions, carbon sequestration, sustainable forest practices, ecological niche modelling, remote sensing, GIS, and natural resource management.

Jibrin Mohammed Jibrin is a Professor of Soil Science with over 30 years of experience in agriculture and rural development. His public service started in 1998 as an Extension Specialist with the National Agricultural Extension and Research Liaison Services in Nigeria. He is currently the Director of the World Bank-funded Africa Centre of Excellence in Dryland Agriculture (CDA), President of the Soil Science Society of Nigeria, and President of the Food for West Africa (Food4WA) Network. Professor Jibrin has published several scientific articles in national and international peer-reviewed journals and has contributed to book chapters and edited conference proceedings. His research focuses on plant nutrient management in semi-arid environments, simulating the effects of future climates and soil management on crop yields in semi-arid areas, and natural resource management in drylands.

Madhav Dwarkanath Gholkar a Thematic Lead (Climate Resilient Agriculture) at WOTR Centre for Resilience Studies (W-CReS), possesses over 16 years of professional expertise in climate-resilient agricultural practices, GHG emission assessment, energy and carbon exchanges, valuation of ecosystem services, knowledge transfer, and capacity building of stakeholders for ecosystem-based adaptation. He holds a Master's degree in Agriculture Systems and Engineering from the Asian Institute of Technology (AIT), Bangkok, and a Bachelor's degree in Agricultural Engineering from Mahatma Phule Krishi Vidyapeeth (MPKV), Rahuri, Maharashtra, India. For the past 9 years, he has been associated with WOTR, primarily focusing on implementation and applied research in sustainable agriculture and allied sectors. Before joining WOTR, he worked at the Space Applications Centre (SAC, ISRO), specialising in remote sensing (RS) and GIS applications in agriculture, and at GVK Bio-Sciences Pvt. Ltd., where he focused on environmental fate modelling of agrochemicals. He has also presented his work at national and international events, including UNCCD COP14 and the ADB conference.

Contributors

Henry De-Graft Acquah Department of Agricultural Economics and Extension, School of Agriculture, College of Agriculture and Natural Sciences, University of Cape Coast, Cape Coast, Ghana

Samuel G. K. Adiku Department of Soil Science, School of Agriculture, College of Basic and Applied Sciences, University of Ghana, Kpong, Ghana

Saminu Ado Department of Environmental Management, Bayero University Kano, Kano, Nigeria

Samson Olusoji Afolabi Glorious Vision University (Formerly, Samuel Adegboyega University), Ogwa, Nigeria

Komi Agboka Ecole Supérieure d'Agronomie (ESA), Université de Lomé, Lomé, Togo

West African Science Service Centre on Climate Change and Adapted Land Use (WASCAL), Université de Lomé, Lomé, Togo

Kossi-Messan Jacques Agboka Ecole Supérieure d'Agronomie (ESA), Université de Lomé, Lomé, Togo

WASCAL-Graduate Research Program on Climate Change and Biodiversity, Université Felix Houphouet-Boigny, Abidjan, Côte d'Ivoire

Emmanuel Babatunde Ajala Glorious Vision University (Formerly, Samuel Adegboyega University), Ogwa, Nigeria

Selorm Akaba Department of Agricultural Economics and Extension, School of Agriculture, College of Agriculture and Natural Sciences, University of Cape Coast, Cape Coast, Ghana

Innocent Onu Alenyi National Agency for the Great Green Wall,, Abuja, Nigeria

Jewel Andoh CSIR-Forestry Research Institute of Ghana,, Kumasi, Ghana

Mujeyi Angeline Department of Agricultural Economics, Education and Extension (AEEE), Faculty of Agriculture and Environmental Sciences, Astra Campus, Bindura University of Science Education (BUSE), Bindura, Zimbabwe

Enosha Arshad Department of Economics and Public Policy, MIT-World Peace University, Pune, Maharashtra, India

Wouyo Atakpama Laboratoire de Botanique et d'Ecologie Végétale (LBEV), Faculté des Sciences, Université de Lomé, Lomé, Togo

Yonathan Ayalew World Food Programme (WFP) Climate and Resilience Service, Rome, Italy

Aliyu S. Barau Department of Environmental Management, Bayero University Kano, Kano, Nigeria

Resham Bhalla Department of Zoology, LVH Arts, Science and Commerce College, Nashik, Maharashtra, India

Showkat Anwar Bhat Population Research Centre, Central University of Kashmir, Srinagar, Jammu and Kashmir, India

Jintu moni Bhuyan Forestry and Ecology, Indian Institute of Remote Sensing, ISRO, Dehradun, Dehradun, Uttarakhand, India

Sibiri Birba Institute for Environment and Agricultural Research, Ouagadougou, Burkina Faso

A. Chakraborty Centre for Ocean, River, Atmosphere and Land Sciences, Indian Institute of Technology, Kharagpur, West Bengal, India

Lazarus Chapungu Institute for Corporate Citizenship, University of South Africa, Pretoria, South Africa

Leonard Chitongo Department of Biological and Environmental Sciences, Faculty of Natural Sciences, Walter Sisulu University, Mthatha, South Africa

Constantin Gueswindé Compaoré Nazi Boni University, Bobo-Dioulasso, Burkina Faso

Eveline Sawadogo Compaoré Institute for Environment and Agricultural Research, Ouagadougou, Burkina Faso

Dada R. Dadas WOTR Centre for Resilience Studies (W-CReS), Watershed Organisation Trust (WOTR), Pune, Maharashtra, India

Martin Dallimer Centre for Environmental Policy, Imperial College London, London, UK

R. J. Ranjit Daniels Trustee & Ecologist, Care Earth Trust,, Chennai, Tamil Nadu, India

Ganesh Deshmukh Department of Economics and Public Policy, Dr. Vishwanath Karad MIT World Peace University (MIT WPU), Pune, Maharashtra, India

Yogesh Deshpande School of Open Learning, SPPU University, Pune, Maharashtra, India

Suman Dutta Faculty Center for Integrated Rural Development and Management, Ramakrishna Mission Vivekananda Educational and Research Institute, Howrah, West Bengal, India

Nicola Favretto Department of Environment and Geography, University of York, York, UK

Bright S. Freduah Soil and Irrigation Research Centre, School of Agriculture, College of Basic and Applied Sciences, University of Ghana, Kpong, Ghana

Shilpa Gautam Biodiversity and Climate Change Division, Indian Council of Forestry Research and Education, Dehradun, Uttarakhand, India

Madhav Dwarkanath Gholkar WOTR Centre for Resilience Studies (W-CReS), Watershed Organisation Trust (WOTR), Pune, Maharashtra, India

K. Gireesan Department of Public Administration, Rajiv Gandhi National Institute of Youth Development (Institute of National Importance), Ministry of Youth Affairs and Sports, Government of India, Regional Centre, Chandigarh, Punjab, India

Subrata Gorain Department of Agricultural Economics, Palli Siksha Bhavana (Institute of Agriculture), Visva-Bharati (Central University), Sriniketan,, West Bengal, India

Helen Greatrex Department of Geography, Department of Statistics and the Institute for Computational and Data Sciences, The Pennsylvania State University, University Park, PA, USA

Reginald Tang Guuroh CSIR-Forestry Research Institute of Ghana, Kumasi, Ghana

Omkar M. Hande WOTR Centre for Resilience Studies (W-CReS), Watershed Organisation Trust (WOTR), Pune, Maharashtra, India

Roman Hinz Kassel Institute for Sustainability, University of Kassel, Kassel, Germany

Jibrin Mohammed Jibrin Department of Environmental Management, Centre for Dryland Agriculture (CDA), Bayero University, Kano, Nigeria

Navjot Kaler Dr Y S Parmar UHF, Solan, Himachal Pradesh, India

A. Keerthika ICAR-Central Arid Zone Research Institute, Jodhpur, Rajasthan, India

Y. D. Imran Khan The WOTR-Centre for Resilience Studies (W-CReS), Watershed Organisation Trust (WOTR), Pune, Maharashtra, India

Krishanmurti WOTR Centre for Resilience Studies (W-CReS), Watershed Organisation Trust (WOTR), Pune, Maharashtra, India

Chandan Kumar ICAR-CAZRI, KVK, Pali-Marwar, Rajasthan, India

Shrvan Kumar ICAR-Central Arid Zone Research Institute,, Jodhpur, Rajasthan, India

Collins N. O. Larkai Department of Agriculture Economics and Agribusiness, School of Agriculture, College of Basic and Applied Sciences, University of Ghana, Kpong, Ghana

Musemwa Lovemore Department of Agricultural Economics, Education and Extension (AEEE), Faculty of Agriculture and Environmental Sciences, Astra Campus, Bindura University of Science Education (BUSE), Bindura, Zimbabwe

Dilys S. MacCarthy Soil and Irrigation Research Centre, School of Agriculture, College of Basic and Applied Sciences, University of Ghana, Kpong, Ghana

Ayushman Malakar ICFRE-Institute of Forest Productivity, Lalgutwa, Ranchi, Jharkhand, India

Kamlesh Kumar Meena ICAR-Central Arid Zone Research Institute, Jodhpur, Rajasthan, India

Om Prakash Meena ICAR-Central Arid Zone Research Institute, Jodhpur, Rajasthan, India

G. D. Mhaske Department of Environmental Science, KRT Arts, BH Commerce & AM Science (KTHM) College, Nashik, Maharashtra, India

U. G. Mhaske Department of Physics, HPT Arts and RYK Science College, Nashik, Maharashtra, India

M. B. Noor Mohamed ICAR-Central Arid Zone Research Institute, Jodhpur, Rajasthan, India

Shrikant Mukate Sustainable Sapiens, Pune, Maharashtra, India

Stephen Mureithi Department of Land Resource Management and Agricultural Technology (LARMAT), University of Nairobi, Kangemi-Nairobi, Kenya

Samuel Mwendwa Department of Land Resource Management and Agricultural Technology (LARMAT), University of Nairobi, Nairobi, Kenya

Pravin M. Nalawade Department of Environmental Science, KRT Arts, BH Commerce & AM Science (KTHM) College, Nashik, Maharashtra, India

Anand kumar Naorem ICAR-Central Arid Zone Research Institute, Jodhpur, Rajasthan, India

Prachi Nautiyal Department of Economics and Public Policy, Dr. Vishwanath Karad MIT World Peace University (MIT WPU), Pune, Maharashtra, India

Swati Nirala Department of Economics and Public Policy, Dr. Vishwanath Karad MIT World Peace University (MIT WPU), Pune, Maharashtra, India

Olubunmi Abosede Odusanya Olabisi Onabanjo University, Ago Iwoye, Ogun State, Nigeria

Selorm Omega Department of Agricultural Economics and Extension, School of Agriculture, College of Agriculture and Natural Sciences, University of Cape Coast, Cape Coast, Ghana

Edward E. Onumah Department of Agriculture Economics and Agribusiness, School of Agriculture, College of Basic and Applied Sciences, University of Ghana, Accra, Ghana

Leonard Omullo Orondo Landscapes Integrity Solutions (LIS), Nairobi, Kenya

Dipak Panaskar School of Earth Sciences, SRTM University, Nanded, Maharashtra, India

Siddharth Panda Department of Genetics and Plant Breeding, Siksha 'O' Anusandhan, Bhubaneswar, Odisha, India

Chetan A. Patil Department of Environmental Science, KRT Arts, BH Commerce & AM Science (KTHM) College, Nashik, Maharashtra, India

Émile Blaise Siewe Pougoue National Polytechnic School, Department of Agriculture, Livestock and Derived Products (AGEPD), Bio-resources and Food Technology Laboratory, University of Maroua, Maroua, Cameroon

Saurabh Purohit WOTR Centre for Resilience Studies (W-CReS), Watershed Organisation Trust (WOTR), Pune, Maharashtra, India

Sweta Rath Department of Agronomy, Siksha 'O' Anusandhan, Bhubaneswar, Odisha, India

R. S. Rawat Biodiversity and Climate Change Division, Indian Council of Forestry Research and Education, Dehradun, Uttarakhand, India

Faraz Rupani WOTR-Centre for Resilience Studies (W-CReS), Watershed Organisation Trust (WOTR), Pune, Maharashtra, India

Akshay Kumar Sagar Centre for Ocean, River, Atmosphere and Land Sciences, Indian Institute of Technology, Kharagpur, West Bengal, India

Deepanjana Saha Conservation Genetics lab, Ashoka Trust for Research in Ecology and the Environment (ATREE), Bengaluru, Karnataka, India

Manipal Academy of Higher Education (MAHE), Manipal, Karnataka, India

Fréderic Saha Faculty of Arts, Letters and Human Sciences, The University of Yaoundé 1, Yaoundé, Cameroon

CIFOR-ICRAF, ReSI-NoC Project, Garoua, Cameroon

Uday Sahu Department of Geology, Toshniwal College of Arts, Commerce and Science, Hingoli, Maharashtra, India

N. V. Saresh ARS, Keshwana, Agriculture University, Jodhpur, Rajasthan, India

M. Saritha ICAR-Central Arid Zone Research Institute, Jodhpur, Rajasthan, India

K. Shiran ICAR-Central Arid Zone Research Institute,, Jodhpur, Rajasthan, India

Dheeraj Singh ICAR-Central Arid Zone Research Institute, Jodhpur, Rajasthan, India

ICAR-CAZRI, Jodhpur, Rajasthan, India

Manendra Singh WOTR Centre for Resilience Studies (W-CReS), Watershed Organisation Trust (WOTR), Pune, Maharashtra, India

Dodiomon Soro Laboratoire des Milieux Naturels et Conservation de la Biodiversité-UFR Biosciences-UFHB (Côte d'Ivoire), Abidjan, Côte d'Ivoire

Sandhya Sreekumar Department of Economics and Public Policy, Dr. Vishwanath Karad MIT World Peace University (MIT WPU), Pune, Maharashtra, India

Rajiv Kumar Srivastava Biological and Agricultural Engineering Department, Texas A&M University, College Station, TX, USA

Lindsay C. Stringer Department of Environment and Geography, University of York, York, UK

Thierry Gaitan Seutchueng Tchuenga Faculty of Arts, Letters and Humanities, University of Yaoundé 1, Yaoundé, Cameroon

CIFOR-ICRAF, Project ReSI-NoC, Garoua, Cameroon

Agricultural Research Institute for Development, Yaoundé, Cameroon

Aamir Ahmad Teeli Department of Economics, Central University of Tamil Nadu, Thiruvarur, Tamil Nadu, India

A. K. Thakur Biodiversity and Climate Change Division, Indian Council of Forestry Research and Education, Dehradun, Uttarakhand, India

Hanh Pham Thi Head of Tourism Department, Faculty of Business Administration, Vietnam Women's Academy, Ha Noi, Vietnam

Manju Tiwari SVCAET&RS,2 Indira Gandhi Krishi Vishwavidyalaya, Raipur, Chhattisgarh, India

Nitish Tiwari Deparetment of Agronomy, Indira Gandhi Krishi Vishwavidyalaya, Raipur, Chhattisgarh, India

Farid Traoré Institute for Environment and Agricultural Research, Ouagadougou, Burkina Faso

Pierre S. Traore International Crops Research Institute for the Semi-arid Tropics (ICRISAT), Bamako, Mali

Manobi Africa PLC, Port-Louis, Mauritius

Archana Verma ICAR-Central Arid Zone Research Institute, Jodhpur, Rajasthan, India

Munyati Vincent Department of Agricultural Economics, Education and Extension (AEEE), Faculty of Agriculture and Environmental Sciences, Astra Campus, Bindura University of Science Education (BUSE), Bindura, Zimbabwe

K. Vishaal Department of Economics, Central University of Tamil Nadu, Thiruvarur, Tamil Nadu, India

Ashwini M. Wadhu WOTR Centre for Resilience Studies (W-CReS), Watershed Organisation Trust (WOTR), Pune, Maharashtra, India

Vasant Wagh School of Earth Sciences, SRTM University, Nanded, Maharashtra, India

Sheetal Walvekar Department of Economics and Public Policy, Dr. Vishwanath Karad MIT World Peace University (MIT WPU), Pune, Maharashtra, India

Ibrahim B. Yakubu Department of Environmental Management, Bayero University Kano, Kano, Nigeria

Ali Yetgin Toros Agri Industry and Trade Co. Inc., Research and Development Center, Mersin, Turkey

Department of Biotechnology, Institute of Nature and Applied Sciences, Cukurova University, Adana, Turkey

Dipak Zade WOTR-Centre for Resilience Studies (W-CReS), Watershed Organisation Trust (WOTR), Pune, Maharashtra, India

Kelvin Zhanda Department of Architecture and Real Estate, University of Zimbabwe, Harare, Zimbabwe

Chapter 1
Land Restoration Using Ecosystem Approaches in the Countries of the Global South: Prologue

Y. D. Imran Khan, Saurabh Purohit, Jibrin Mohammed Jibrin, and Madhav Dwarkanath Gholkar

1.1 Introduction and Contextual Background

Drylands encompass land areas characterised by a mean annual precipitation to mean annual potential evapotranspiration ratio (known as the aridity index) below 0.65 (UNEP 1992). The aridity index defines four primary dryland categories: hyper-arid (aridity index <0.05), arid (0.05 ≤ aridity index <0.20), semi-arid (0.20 ≤ aridity index <0.50) and dry sub-humid (0.50 ≤ aridity index <0.65). Drylands cover about 42% of the earth's terrestrial surface, encompass diverse and unique regions globally (Lewin et al. 2024) and support nearly 2.5 billion people (Eldridge et al. 2024; Huang et al. 2017), making them some of the most fragile yet vital ecosystems of the planet (Fig. 1.1).

Drylands strongly influence global interannual variability in carbon and water cycling due to their substantial heterogeneity over space and time (Ahlström et al. 2015; Kannenberg et al. 2024). They harbour a remarkable diversity of unique and endemic plant and animal species shaped by extensive adaptations to extreme environmental conditions across varied and heterogeneous habitats (Maestre et al. 2024; Zhang et al. 2023). The concept of "Global South" has emerged to replace the colonial-era terminologies like "The Third World" and "Periphery". Adopted by the developed countries in Europe and North America, pointing to the planet's

Y. D. I. Khan (✉) · S. Purohit · M. D. Gholkar
WOTR Centre for Resilience Studies (W-CReS), Watershed Organisation Trust (WOTR), Pune, Maharashtra, India
e-mail: ikeducationkhan08@gmail.com; saurabhpurohit10@gmail.com; madhav.gholkar@wotr.org.in

J. Mohammed Jibrin
Centre for Dryland Agriculture (CDA), Bayero University, Kano, Nigeria
e-mail: jibrin@buk.edu.ng

Y. D. I. Khan et al. (eds.), *Land Restoration Through Ecosystem-Based Approach*, Disaster Resilience and Green Growth, https://doi.org/10.1007/978-981-95-0389-6_1

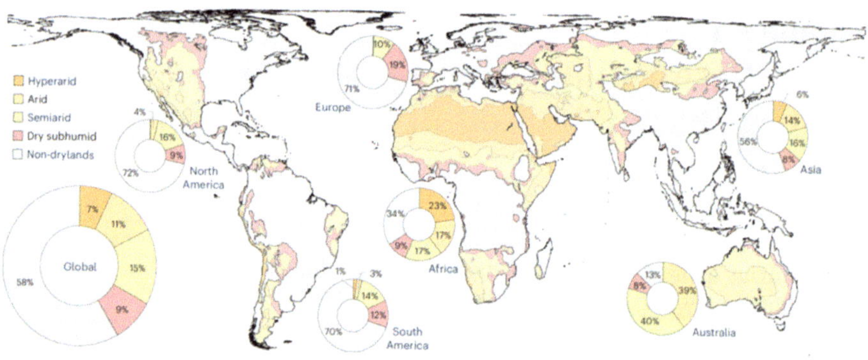

Fig. 1.1 Global drylands and subtypes (UNEP-WCMC). (Source: Lewin et al. 2024)

low-income and often politically or culturally marginalised countries (Dados and Connell 2012).

The area broadly refers to the regions of Latin America, Asia, Africa and Oceania. The "Global South" idea emphasises geopolitical relations of power among more equal nations. The drylands are mainly concentrated in different countries of the global south, especially in Africa and Asia, which combined comprise 64% of global drylands and the largest areas of hyperarid and arid subtypes (Lewin et al. 2024). Over the years, the drylands of Global South have suffered substantial losses of productivity and biodiversity, increasing the severity and frequency of droughts, food insecurity, poverty, conflicts, emigration and social disintegration (Fu et al. 2024; Li et al. 2024; Varela Pérez et al. 2022), which are exacerbated by climate change and unsustainable land-use practices.

For decades, the measures to address land degradation in the drylands have been towards the use of hard or grey infrastructure—such as rainwater harvesting structures, large-scale irrigation projects or small water infrastructures (SWIs) like deep wells and borewells (Piemontese et al. 2024). While these measures may have provided short-term relief, many have failed to address the root causes of land degradation and, in some cases, are not climate resilient. It has become clear that a different approach is needed—one that aligns with nature rather than imposing control over it.

1.2 Rethinking Land Restoration: An Ecosystem-Based Approach

The ecosystem-based approach is a strategy for the integrated management of land, water and living resources that promotes conservation and sustainable use in an equitable way (Sukhwani et al. 2021). This approach applies scientific methods to study biological organisation at various levels, emphasising the essential structures, processes, functions and interactions between organisms and their environment (Ma and Jiang 2023). It recognises that humans, with their cultural diversity, are integral to many ecosystems. Ecosystem-based approaches address the crucial links between climate change, biodiversity, ecosystem services and sustainable resource management, and thus, they have the potential to simultaneously contribute to the adaptation and mitigation of climate change (Munang et al. 2013). These approaches also maintain existing carbon stocks, regulate water flow and storage, maintain and increase resilience, reduce the vulnerability of ecosystems and people, help adapt to climate change impacts, improve biodiversity conservation and livelihood opportunities and provide health and recreational benefits (Naumann et al. 2011; Nautiyal et al. 2016). These approaches can be broadly divided into ecosystem-based approaches for adaptation (EbA) and ecosystem-based approaches for mitigation (EbM). EbA uses biodiversity and ecosystem services as part of an overall adaptation strategy to help people adapt to the adverse effects of climate change (CBD 2009; Harvey et al. 2017; Reid 2016). EbM uses ecosystems for its carbon storage and sequestration service to aid climate change mitigation. Emissions reductions are achieved through creating, restoring and managing ecosystems (e.g. forest restoration, peat conservation) (Margottini et al. 2015). Unlike conventional land rehabilitation methods, which often overlook socio-cultural and ecological nuances, EbA recognises that dryland ecosystems are dynamic and deeply interconnected with human livelihoods. It integrates agroecology, rangeland management, regenerative agriculture, watershed conservation and Indigenous stewardship practices to create ecologically viable and socio-economically inclusive solutions.

1.3 Why This Book?

The Global South hosts some of the most climate-vulnerable yet knowledge-rich communities, where generations of pastoralists, smallholder farmers and indigenous groups have adapted to arid conditions through resourcefulness and ecological wisdom (Hussainzad and Gou 2024; Jones-Garcia and Krishna 2021; Taylor et al. 2023). However, their practices are often undervalued or dismissed in favour of industrialised solutions. This book seeks to bridge that gap by bringing forward case studies, evidence-based insights and community-driven strategies demonstrating ecosystem-based approaches' effectiveness in restoring drylands. The book provides insights into:

- Land restoration dynamics in dryland ecosystems
- Nature-based restoration strategies for climate-resilient dryland ecosystems
- Sustainable land management and innovative agricultural practices for dryland ecosystems
- Advanced innovations, digital technologies and collaborative strategies for dryland restoration
- Adaptive policies, institutional governance and long-term sustainability strategies for sustainable restoration of drylands

1.4 A Call to Action

The restoration of drylands is not just an ecological necessity—it is a moral and economic imperative for ensuring food security, mitigating climate change and safeguarding the well-being of millions of people. By embracing an ecosystem-based approach, we can move towards a future where degraded lands are revived, biodiversity thrives and communities regain control over their natural heritage.

This book is an invitation—to reimagine restoration beyond engineering solutions, recognise the power of nature and respect the knowledge of those who have lived in harmony with drylands for centuries. It is a call to act with urgency and wisdom, ensuring that the landscapes of the Global South do not turn into wastelands, but instead, become models of resilience and regeneration.

References

Ahlström A, Raupach MR, Schurgers G, Smith B, Arneth A, Jung M, Reichstein M, Canadell JG, Friedlingstein P, Jain AK, Kato E, Poulter B, Sitch S, Stocker BD, Viovy N, Wang YP, Wiltshire A, Zaehle S, Zeng N (2015) The dominant role of semi-arid ecosystems in the trend and variability of the land CO2 sink. Science 348:895–899. https://doi.org/10.1126/science.aaa1668

CBD (2009) Connecting biodiversity and climate change mitigation and adaptation: report of the second ad hoc technical expert group on biodiversity and climate change. https://doi.org/10.13140/RG.2.2.12351.10409

Dados N, Connell R (2012) The global south. Contexts 11:12–13. https://doi.org/10.1177/1536504212436479

Eldridge DJ, Wang C, Liu Y, Ding J, Li Y, Wu X, Li C (2024) Nature's contribution to people in drylands. Cambridge Prism Drylands 1:e2. https://doi.org/10.1017/dry.2024.2

Fu B, Stafford-Smith M, Fu C, Liu Y, Wang Y, Wu B, Yu X, Lu N, Ojima DS (2024) The global-DEP: a research programme to promote sustainability of dryland social-ecological systems. In: Fu B, Stafford-Smith M (eds) BT – dryland social-ecological systems in changing environments. Springer Nature Singapore, Singapore, pp 1–20. https://doi.org/10.1007/978-981-99-9375-8_1

Harvey CA, Martínez-Rodríguez MR et al (2017) The use of ecosystem-based adaptation practices by smallholder farmers in Central America. Agric Ecosyst Environ 246:279–290. https://doi.org/10.1016/j.agee.2017.04.018

Huang J, Li Y, Fu C, Chen F, Fu Q, Dai A, Shinoda M, Ma Z, Guo W, Li Z, Zhang L, Liu Y, Yu H, He Y, Xie Y, Guan X, Ji M, Lin L, Wang S, Yan H, Wang G (2017) Dryland climate change: recent progress and challenges. Rev Geophys 55:719–778. https://doi.org/10.1002/2016RG000550

Hussainzad EA, Gou Z (2024) Climate risk and vulnerability assessment in informal settlements of the global south: a critical review. Land 13. https://doi.org/10.3390/land13091357

Jones-Garcia E, Krishna VV (2021) Farmer adoption of sustainable intensification technologies in the maise systems of the global south. A review. Agron Sustain Dev 41:8. https://doi.org/10.1007/s13593-020-00658-9

Kannenberg SA, Anderegg WRL, Barnes ML, Dannenberg MP, Knapp AK (2024) Dominant role of soil moisture in mediating carbon and water fluxes in dryland ecosystems. Nat Geosci 17:38–43. https://doi.org/10.1038/s41561-023-01351-8

Lewin A, Murali G, Rachmilevitch S, Roll U (2024) Global evaluation of current and future threats to drylands and their vertebrate biodiversity. Nat Ecol Evol 8:1448–1458. https://doi.org/10.1038/s41559-024-02450-4

Li T, Singh RK, Cui L, Xu Z, Pandey R, Liu Y, Cui X, Liu Y, Fava F, Yang Y, Wang Y (2024) Managing multiple stressors for sustainable livelihoods in dryland ecosystems: insights and entry points for resource management. L Degrad Dev 35:968–984. https://doi.org/10.1002/ldr.4964

Ma Y, Jiang Y (2023) Ecosystem-based adaptation to address urbanisation and climate change challenges: the case of China's sponge city initiative. Clim Pol 23:268–284. https://doi.org/10.1080/14693062.2022.2131503

Maestre FT, Biancari L, Chen N, Corrochano-Monsalve M, Jenerette GD, Nelson C, Shilula KN, Shpilkina Y (2024) Research needs on the biodiversity–ecosystem functioning relationship in drylands. NPJ Biodivers 3:12. https://doi.org/10.1038/s44185-024-00046-6

Margottini C, Fidolini F, Iadanza C, Trigila A, Ubelmann Y (2015) The conservation of the Shahr-e-Zohak archaeological site (Central Afghanistan): geomorphological processes and ecosystem-based mitigation. Geomorphology 239:73–90. https://doi.org/10.1016/j.geomorph.2014.12.047

Munang R, Thiaw I, Alverson K, Mumba M, Liu J, Rivington M (2013) Climate change and ecosystem-based adaptation: a new pragmatic approach to buffering climate change impacts. Curr Opin Environ Sustain 5:67–71. https://doi.org/10.1016/j.cosust.2012.12.001

Naumann S, Anzaldua G, Berry P, Burch S, Davis M, Frelih-Larsen A, Gerdes H, Sanders M (2011) Assessment of the potential of ecosystem-based approaches to climate change adaptation and mitigation in Europe. Final Rep. to Eur. Comm. DG Environ

Nautiyal S, Schaldah R, Raju KV, Kaechele H, Pritchard B, Rao KS (2016) Climate change challenge (3C) and social-economic-ecological interface-building – exploring potential adaptation strategies for bio-resource conservation and livelihood development: epilogue. Springer International Publishing, pp 631–639

Piemontese L, Terzi S, Di Baldassarre G, Menestrey Schwieger DA, Castelli G, Bresci E (2024) Over-reliance on water infrastructure can hinder climate resilience in pastoral drylands. Nat Clim Chang 14:267–274. https://doi.org/10.1038/s41558-024-01929-z

Reid H (2016) Ecosystem- and community-based adaptation: learning from community-based natural resource management. Clim Dev 8:4–9. https://doi.org/10.1080/17565529.2015.1034233

Sukhwani V, Thapa K, Shaw R, Deshkar S, Mitra BK, Yan W (2021) Ecosystem-based approaches for water stress management – lessons from Nagpur metropolitan area, India. In: Mukherjee M, Shaw R (eds) BT – ecosystem-based disaster and climate resilience: integration of blue-green infrastructure in sustainable development. Springer Singapore, Singapore, pp 389–410. https://doi.org/10.1007/978-981-16-4815-1_17

Taylor A, Methner N, Barkai KR, McClure A, Jack C, New M, Ziervogel G (2023) Operationalising climate-resilient development pathways in the global south. Curr Opin Environ Sustain 64:101328. https://doi.org/10.1016/j.cosust.2023.101328

UNEP (1992) World atlas of desertification. Edward Arnold

Varela Pérez P, Greiner BE, von Cossel M (2022) Socio-economic and environmental implications of bioenergy crop cultivation on marginal African drylands and key principles for a sustainable development. Earth 3:652. https://doi.org/10.3390/earth3020038

Zhang Y, Tariq A, Hughes AC, Hong D, Wei F, Sun H, Sardans J, Peñuelas J, Perry G, Qiao J, Kurban A, Jia X, Raimondo D, Pan B, Yang W, Zhang D, Li W, Ahmed Z, Beierkuhnlein C, Lazkov G, Toderich K, Karryeva S, Dehkonov D, Hisoriev H, Dimeyeva L, Milko D, Soule A, Suska-Malawska M, Saparmuradov J, Bekzod A, Allin P, Dieye S, Cissse B, Whibesilassie W, Ma K (2023) Challenges and solutions to biodiversity conservation in arid lands. Sci Total Environ 857:159695. https://doi.org/10.1016/j.scitotenv.2022.159695

Part I
Restoration Dynamics in Dryland Ecosystems

Chapter 2
Climate Change and Ecosystem-Based Approach: Adapting to Climate Change in Dryland Ecosystems; Mitigation and Resilience Through Ecosystem-Based Solutions

Akshay Kumar Sagar, Rajiv Kumar Srivastava, and A. Chakraborty

Abstract Climate change presents significant challenges to dryland ecosystems, which are particularly vulnerable to fluctuations in temperature and precipitation patterns. These regions face heightened risks such as desertification, declining agricultural productivity, and biodiversity loss. This chapter explores the role of ecosystem-based approaches in strengthening resilience and mitigating climate change impacts in dryland ecosystems. This study identifies vulnerable areas and proposes targeted interventions by integrating advanced methods such as rainfall mapping and soil moisture budget analysis. Key ecosystem-based approaches—including reforestation, wetland restoration, and sustainable land management practices—offer promising strategies for adaptation and mitigation. These approaches enhance biodiversity and ecosystem services and support local communities in building resilience against climate-related stresses. Through case studies and practical examples, this chapter highlights the critical importance of leveraging natural processes and data-driven strategies to foster sustainable development in dryland ecosystems.

Keywords Climate change · Ecosystem-based approach · Dryland ecosystems · Soil moisture · Resilience

A. K. Sagar (✉) · A. Chakraborty
Centre for Ocean, River, Atmosphere and Land Sciences, Indian Institute of Technology, Kharagpur, West Bengal, India

R. K. Srivastava
Biological and Agricultural Engineering Department, Texas A&M University, College Station, TX, USA

Y. D. I. Khan et al. (eds.), *Land Restoration Through Ecosystem-Based Approach*, Disaster Resilience and Green Growth, https://doi.org/10.1007/978-981-95-0389-6_2

2.1 Introduction

Climate change is reshaping ecosystems globally, with profound implications for both natural systems and human societies (Sadhukhan et al. 2024). Nowhere are these effects more pronounced than in dryland ecosystems, which cover over 40% of the Earth's terrestrial surface and are home to more than 2 billion people (Osborne et al. 2022). Drylands include arid, semiarid, and dry sub-humid zones, where water scarcity is a fundamental challenge due to low and erratic rainfall, high evaporation rates, and often limited soil water retention capacity (Koster et al. 2010). As climate change accelerates, the pressures on these fragile ecosystems mount, making drylands particularly vulnerable to degradation, biodiversity loss, and reduced agricultural productivity (Sharma et al. 2024). These risks call for innovative and adaptive management strategies that mitigate the impacts of climate change and enhance the resilience of dryland ecosystems. This chapter explores ecosystem-based approaches as a viable pathway to address these challenges by fostering resilience and sustainable development in dryland regions.

2.1.1 The Impact of Climate Change on Dryland Ecosystems

Climate change intensifies several environmental stressors affecting drylands (Sharma et al. 2024). Increasing temperatures, altered precipitation patterns, and the growing frequency of extreme weather events—such as droughts, heatwaves, and sandstorms—exacerbate the natural challenges of drylands (Hartzell et al. 2024). In many regions, temperatures are projected to increase by more than the global average, leading to greater water loss through evaporation and a decrease in soil moisture. Drought conditions, in particular, result in reduced moisture retention, significantly impacting agricultural productivity and local ecosystems (Lorenz et al. 2010).

In conjunction with these warming trends, drylands are expected to experience more irregular rainfall patterns, where periods of intense rainfall are followed by extended dry spells (Srivastava et al. 2024). Such erratic precipitation can lead to soil erosion, loss of soil fertility, and reduced water availability for agriculture and natural vegetation (Grillakis 2019). This fluctuation in precipitation is a primary driver of changes in soil moisture, and it impacts the water balance, further accelerating dryland degradation (Stringer et al. 2021).

One of the most visible impacts of climate change in drylands is desertification, a process whereby fertile land progressively loses its productivity and turns into barren, unproductive deserts (Stringer et al. 2021). Desertification threatens biodiversity and compromises the livelihoods of millions who depend on these lands for grazing, farming, and water resources (Jain et al. 2024). Additionally, dryland degradation disrupts carbon storage in soils, releasing stored carbon into the atmosphere and contributing to the cycle of climate change (Trenberth 2011). This phenomenon underscores the importance of addressing dryland resilience as an environmental concern and a critical component of global climate mitigation efforts.

2.1.2 *The Importance of Ecosystem-Based Approaches for Adaptation and Mitigation*

Ecosystem-based approaches, often termed nature-based solutions, emphasize using natural processes and ecosystem services to adapt to and mitigate the impacts of climate change (Seddon et al. 2020). Unlike conventional engineering solutions, ecosystem-based approaches focus on enhancing the inherent resilience of ecosystems through sustainable practices that support biodiversity, water management, and soil health (McVittie et al. 2018). In the context of drylands, these strategies include reforestation, wetland restoration, agroforestry, and sustainable land management practices that align with the unique ecological characteristics and needs of arid and semiarid regions (Seddon et al. 2020).

The appeal of ecosystem-based approaches lies in their multifunctional benefits. For instance, reforestation and afforestation sequester carbon, provide shade, reduce soil erosion, and promote groundwater recharge (Locatelli et al. 2015). Similarly, restoring wetlands and riparian zones can enhance water storage, regulate local temperatures, and support biodiversity, creating buffers against climate-induced stresses (Weiskopf et al. 2020). These nature-based interventions are also cost-effective, particularly in resource-limited dryland areas, where extensive infrastructure projects may not be viable (Irvine et al. 2016). Furthermore, ecosystem-based approaches foster community involvement and stewardship, as they often leverage local knowledge and practices, making them socially sustainable in addition to being environmentally beneficial.

2.1.3 *Objectives and Scope of the Chapter*

The primary objective of this chapter is to explore the potential of ecosystem-based approaches to enhance resilience and support climate adaptation in dryland ecosystems. It comprehensively analyzes how these approaches can mitigate climate change impacts, conserve biodiversity, and improve water and soil resources. Advanced techniques like rainfall mapping and soil moisture budget analysis are valuable tools for managing water resources in dryland areas, providing a data-driven basis for implementing ecosystem-based solutions by identifying regions most vulnerable to water stress and erosion. The chapter also evaluates ecosystem-based interventions through practical case studies, highlighting best practices and potential challenges. Emphasizing the role of community participation and local knowledge, it recognizes that the success of these strategies depends on the active engagement and support of local inhabitants.

2.2 Ecosystem-Based Approaches for Climate Adaptation and Mitigation

Ecosystem-based approaches, also known as nature-based solutions, offer a promising pathway to enhance resilience and support sustainable development in dryland regions (Seddon et al. 2020). These approaches leverage natural processes to restore ecosystem health, improve water and soil resources, and support biodiversity, making them suitable for adapting to climate change in resource-limited environments.

2.2.1 Key Ecosystem-Based Strategies in Drylands

2.2.1.1 Reforestation and Afforestation

Planting trees in degraded or deforested areas can stabilize soil, reduce erosion, and increase carbon sequestration. In drylands, specific species adapted to low water conditions—such as acacias—enhance soil moisture retention and provide shade, thereby reducing evaporation rates. This approach mitigates climate impacts and contributes to restoring local biodiversity (Sang et al. 2013).

2.2.1.2 Agroforestry

Integrating trees with crops or livestock in a shared system offers multiple benefits, including improved soil structure, enhanced water retention, and protection from wind erosion (Altieri and Nicholls 2017). Agroforestry systems, such as using shade trees for coffee or cocoa crops, promote biodiversity while helping regulate local climates and soil moisture (Mortimer et al. 2018).

2.2.1.3 Wetland and Riparian Restoration

Wetlands, even those adapted to arid environments, are critical for water storage and habitat provision (Weiskopf et al. 2020). Restoring these areas increases local water availability, supports biodiversity, and buffers against temperature extremes (Irvine et al. 2016). These benefits make wetlands and riparian zones important for enhancing resilience in dryland ecosystems (Weiskopf et al. 2020).

2.2.1.4 Sustainable Land Management

Conservation tillage, rotational grazing, and soil mulching help maintain soil health and reduce degradation (Verhulst et al. 2010). Conservation tillage minimizes soil disturbance, improving soil structure and enhancing water infiltration (Verhulst et al. 2010). Meanwhile, rotational grazing prevents overgrazing and allows vegetation to recover, supporting soil structure and reducing erosion (Siddique et al. 2024; Teague and Kreuter 2020).

2.2.2 *Advanced Technologies Supporting Ecosystem-Based Approaches*

Advanced techniques, such as remote sensing and land surface modeling, enable precise measurements of soil moisture variability, providing valuable data to inform management strategies and forecast drought conditions (Rasheed et al. 2022). Table 2.1 showcases advanced technologies that align with ecosystem-based approaches by enhancing data-driven decision-making and resource management in dryland ecosystems. Each technology provides specific advantages, from real-time monitoring with remote sensing to predictive insights using machine learning models, enabling targeted interventions for soil moisture retention, water conservation, and drought response.

Table 2.1 Advanced technologies supporting ecosystem-based approaches in dryland ecosystems

Technology	Description	Applications in drylands	Benefits
Remote sensing Mohanty et al. (2017)	Use of satellite and drone imagery to monitor environmental changes in real time	Mapping soil moisture, vegetation cover, and rainfall distribution	High spatial and temporal coverage; cost-effective
Soil moisture sensors (Rasheed et al. 2022)	In situ sensors that measure soil moisture at different depths and time intervals	Monitoring soil moisture for agricultural planning and drought response	Precision data for improved water management
Geographic information systems (GIS) (Petropoulos et al. 2015)	Spatial analysis tools for mapping and visualizing ecosystem data	Identifying vulnerable areas and planning interventions	Improved data integration and analysis
Machine learning models (Verma et al. 2023)	Algorithms to predict soil moisture and climate trends based on historical data	Forecasting droughts and supporting early warning systems	Enhanced predictive capacity for resource planning
Rainwater harvesting systems (Kloos and Renaud 2016)	Technology for collecting and storing rainwater during rainy seasons	Supplementing water supply in drought-prone areas	Increases water availability and soil moisture
Water budget modeling (Amatya et al. 2016)	Computational models to calculate precipitation, runoff, and evaporation	Assessing water availability and identifying soil moisture deficits	Data-driven insights for sustainable water use

2.2.3 *The Role of Community and Policy Support*

The success of ecosystem-based approaches depends on active community involvement and supportive policy frameworks (Reid et al. 2018). Engaging local populations helps ensure these approaches align with traditional practices and meet community needs. Furthermore, policy support is essential to incentivize ecosystem-based approaches, such as providing subsidies for reforestation or creating protected areas to prevent overgrazing and deforestation (Sanz et al. 2017).

2.3 Climate and Hydrological Characteristics of Drylands

Dryland climates are marked by extremes in temperature and precipitation (Lorenz et al. 2010). High temperatures, especially in arid regions, contribute to accelerated evaporation, which reduces soil moisture availability and exacerbates water scarcity (Berg and Sheffield 2018). Precipitation in drylands is limited and sporadic, often coming in short, intense bursts that can lead to flash floods and soil erosion (Table 2.1). This rainfall pattern leads to erratic water availability, making it difficult to sustain consistent agricultural practices without supplemental irrigation or specialized conservation techniques (Srivastav et al. 2021).

Table 2.2 clearly shows that the drylands are particularly sensitive to climate change, intensifying the existing stresses on water availability, soil quality, and vegetation cover (Berg and Sheffield 2018). Projected increases in temperature and alterations in precipitation patterns due to climate change will likely worsen drought conditions and expand the extent of desertification (Trenberth 2011; Sagar and Chakraborty 2024). This process leads to significant loss of productive land, declining agricultural yields, and displacement of communities reliant on these lands for their livelihoods. Increased desertification also impacts global carbon storage, as degraded drylands release stored carbon into the atmosphere, further accelerating climate change (Lorenz et al. 2010).

Table 2.2 Impacts of climate variables on soil moisture dynamics in drylands

Climate variable	Impact on soil moisture	Mechanism	References
Temperature	Increased temperature reduces soil moisture	Accelerates evaporation and increases water demand	Lorenz et al. (2010)
Precipitation	Directly affects soil moisture levels	Rainfall replenishes soil moisture reserves	Trenberth (2011)
Relative humidity	Low RH accelerates soil drying	Reduces evaporation when high; increases it when low	Grünzweig et al. (2022)
Wind speed	High wind speeds reduce soil moisture	Enhances evaporation rates	Koster et al. (2010)
Solar radiation	High solar radiation decreases soil moisture	Increases soil and surface temperature	Mondal et al. (2024)

2.4 Soil Moisture Budget and Climate Change Implications

Soil moisture is essential in defining local and regional water balance and affecting agricultural productivity, hydrological cycles, and climatic dynamics (Lorenz et al. 2010). It is also an essential element in the soil-plant-atmosphere continuum, influencing surface energy and water fluxes (Koster et al. 2010). Precise comprehension and measurement of soil moisture are crucial for water resource management, drought prediction, and assessing the possible effects of climate change on hydrological systems (Mondal et al. 2024).

The soil moisture budget represents the balance between moisture inputs, precipitation, and outputs, like evaporation and runoff (Amatya et al. 2016). In drylands, maintaining this balance is crucial for sustaining vegetation and managing water resources (Stringer et al. 2021). However, climate change impacts, including increased temperatures and altered precipitation patterns, disrupt this equilibrium, leading to more frequent and severe soil moisture deficits (Trenberth 2011).

Total precipitation, relative humidity (RH), specific humidity (SH), temperatures at various atmospheric levels, evaporation, and runoff are the principal factors influencing soil moisture dynamics (Table 2.2). Precipitation is the main input to the soil moisture system, replenishing moisture levels that may have been reduced by evaporation or plant transpiration. Total precipitation levels vary significantly across regions and seasons, influenced by air circulation patterns, local topography, and climatic conditions. High precipitation generally accumulates soil moisture, while prolonged dry periods reduce soil moisture and cause drought conditions.

Relative and specific humidity are key meteorological variables that affect evaporation rates from the soil surface (Grünzweig et al. 2022). High relative humidity reduces evaporation potential, while low relative humidity increases it, leading to greater soil moisture loss (Grünzweig et al. 2022). Specific humidity, which measures the absolute water content in the atmosphere, also influences moisture transport within the boundary layer (Petropoulos et al. 2015). These humidity variables, along with temperature, air pressure, and local weather patterns, are crucial for understanding moisture fluxes at the soil surface. The temperature at 2 meters is closely linked to surface heat fluxes and plays a vital role in regulating evaporation and transpiration rates and precipitation (Oleson et al. 2013). As air temperature rises, the potential for increased evaporation also grows, resulting in more significant soil moisture loss without sufficient precipitation (Oleson et al. 2013).

Evaporation, the release of water from soil and plant surfaces into the atmosphere, is influenced by temperature, humidity, wind speed, and solar radiation (Srivastava et al. 2020). It is a significant output in soil moisture budgets, and accurate measurement is essential for understanding soil moisture responses to changing climatic conditions (Seneviratne et al. 2010). Advances in quantification and modeling, including remote sensing and land surface models, have improved assessment accuracy (Berg et al. 2014). Runoff, the water flows across the land surface, occurs when soil is saturated or infiltration rates are exceeded due to heavy precipitation (Srivastava and Imtiyaz 2016). Factors like soil texture, land cover, slope, and precipitation intensity influence runoff volume. Understanding runoff is crucial for assessing flood risk and developing sustainable land management strategies to reduce surface runoff and enhance soil water retention (Srivastava and Imtiyaz 2016).

2.5 Case Study: Spatiotemporal Analysis of Soil Moisture Over India Using a Soil Moisture Budgeting Approach

This case study investigates the factors affecting the soil moisture budget, focusing on the relationships among precipitation, temperature, humidity, evaporation, and runoff. It aims to measure the impact of climatic conditions on soil moisture dynamics and evaluate their temporal and spatial variability. By combining observational data from meteorological stations with satellite-derived remote sensing products, the study elucidates the geographical and temporal variations in soil moisture and their correlation with atmospheric factors. Additionally, it examines how changes in climate variables, such as temperature and precipitation, may alter the soil moisture budget and affect water availability across different regions. The increasing frequency and severity of extreme weather events, including droughts and floods, underscore the need for precise soil moisture evaluations to manage water resources effectively (Trenberth 2011).

2.5.1 Data Sources and Tools

To investigate the soil moisture budget and other atmospheric variables, this study used data from the Copernicus Climate Data Store (CDS), specifically the ERA5 reanalysis dataset provided by the European Centre for Medium-Range Weather Forecasts (ECMWF). ERA5 offers high-resolution, hourly climate data with a spatial resolution of approximately 31 km, integrating satellite and ground-based observations for accuracy and continuity (Hersbach et al. 2020). The dataset covers key parameters influencing the soil moisture budget, including evaporation, precipitation, relative and specific humidity, temperature, and runoff (Table 2.3).

The study period spans from 2006 to 2020, allowing for a comprehensive examination of trends and variability over a significant timeframe. ERA5's high temporal resolution enabled monthly and annual data aggregation, facilitating detailed temporal analyses of these variables.

Table 2.3 Variables and definitions of the datasets used from ERA 5

Variable	S.I. unit	Definition
Evaporation	m	This measure shows how much water, including transpiration from vegetation, evaporates off the earth's surface and turns into vapor in the atmosphere
Relative humidity	%	When water vapor condenses into liquid water or deposits into ice, this parameter shows the water vapor pressure as a proportion of air saturation
Specific humidity	kg^{-1}	When dry air, water vapor, cloud liquid, cloud ice, rain, and snow are combined, the parameter measures the mass of water vapor per kilogram of moist air
Saturated soil moisture	%	The amount of water in soil when all its pore spaces are filled with water is known as SSM
Total precipitation	m	Water that has accumulated on the surface of the earth, including rain and snow
Temperature (2 m)	°C	This parameter is the temperature of the air two meters above the surface of land, sea, or inland waters
Runoff	m	Some water is retained in the soil by rainfall, snowmelt, or soil depth. If not, the water flows away either below the ground (sub-surface runoff) or above it (surface runoff), and the total of these two is known as runoff

2.5.2 *Data Processing*

Data processing was performed in several stages to ensure data quality and usability:

1. *Data Cleaning and Preprocessing*: Initial data cleaning involved handling missing values for certain regions, using interpolation techniques to maintain data continuity across the study area and avoid potential biases.
2. *Temporal Aggregation*: Monthly data were aggregated to annual averages for each year within the 2006–2020 period. This aggregation provides a view of yearly trends in soil moisture and related parameters, highlighting seasonal and interannual variability.
3. *Calculation of Soil Moisture Budget*: The soil moisture budget was calculated using the following formula.

$$\text{Water Budget} = \text{Precipitation}\,?\left(\text{Evaporation} + \text{Runoff}\right) \tag{2.1}$$

This formula captures the net moisture balance in the soil by accounting for the primary inputs (precipitation) and outputs (evaporation and runoff). The resulting values indicate areas experiencing soil moisture surpluses or deficits, critical for assessing vulnerability in dryland ecosystems (Singh et al. 2020).

2.5.3 Temporal and Spatial Analysis

The temporal and spatial analysis involved evaluating soil moisture and climate variables across time and geographic regions, providing a nuanced understanding of drylands' variability.

- *Temporal Analysis*: To examine long-term trends, monthly and annual averages of climate variables (e.g., precipitation, evaporation, and temperature) were calculated. Temporal analysis provided insights into how climate change has influenced soil moisture levels over the 15-year study period, capturing consistent trends and anomalies. This was especially useful for identifying years with significant deviations from average soil moisture levels indicative of extreme climate events or shifting weather patterns (Hirschi et al. 2011).
- *Spatial Analysis*: The study also included a spatial analysis of soil moisture distribution, using regional maps to visualize variations across dryland regions. This analysis highlighted geographic differences in soil moisture availability and other key parameters, identifying zones with high evaporation rates, low humidity, or significant runoff. Spatial mapping of these variables was conducted using geographic information system (GIS) tools to represent spatial patterns and variability accurately.

2.5.4 Visualization and Interpretation

To better understand the soil moisture dynamics in dryland regions, this study produced several visualizations:

- *Regional Maps*: Regional maps were generated to display the spatial distribution of evaporation, precipitation, humidity, temperature, and soil moisture across the study area. These maps provided a visual overview of soil moisture levels, helping to identify areas with high water stress or vulnerability to desertification.
- *Trend Graphs*: Graphical representations of annual averages and trends were created for key variables. These graphs illustrated year-on-year changes, allowing for easy comparison of the impact of different climate variables on soil moisture over time.

The combination of temporal and spatial visualizations offers a comprehensive view of how climatic factors influence soil moisture dynamics in drylands, aiding in the identification of trends that may inform ecosystem-based adaptation strategies.

2.5.5 *Limitations*

While the ERA5 dataset provides high-resolution climate data, certain limitations must be considered. The study's reliance on reanalysis data, which combines observations and model outputs, means that some areas may experience data gaps or estimation inaccuracies, particularly in remote or sparsely populated regions with fewer observational data points (Zandler et al. 2020). Additionally, the spatial resolution of 31 km, while detailed, may not capture micro-scale variations in soil moisture within heterogeneous landscapes. These limitations were considered in interpreting results, and caution was applied when generalizing findings across diverse dryland regions.

2.6 Results and Discussion

This section presents the findings from the spatial and temporal analysis of soil moisture and related climate variables across the study period (2006–2020) for dryland regions. By examining key parameters such as evaporation, precipitation, humidity, temperature, and the overall water budget, this analysis identifies trends and regional patterns that provide insight into the effects of climate change on soil moisture dynamics in dryland ecosystems.

2.6.1 *Spatial Distribution of Climate and Land Parameters*

2.6.1.1 Evaporation

The spatial distribution of evaporation rates across the Indian dryland regions for the 2006–2020 period is depicted in Fig. 2.1. Higher evaporation rates are observed in the western parts of India, the Himalayan region, and northeastern India. These areas experience greater evaporation due to their higher temperatures and variable moisture availability, which leads to accelerated water loss from soil surfaces. In contrast, regions near the coast exhibit lower evaporation rates due to higher humidity and consistent rainfall, which helps maintain soil moisture levels. This pattern aligns with findings by Lorenz et al. (2010), who highlighted that evaporation rates are sensitive to both temperature and humidity levels, especially in semiarid and arid environments.

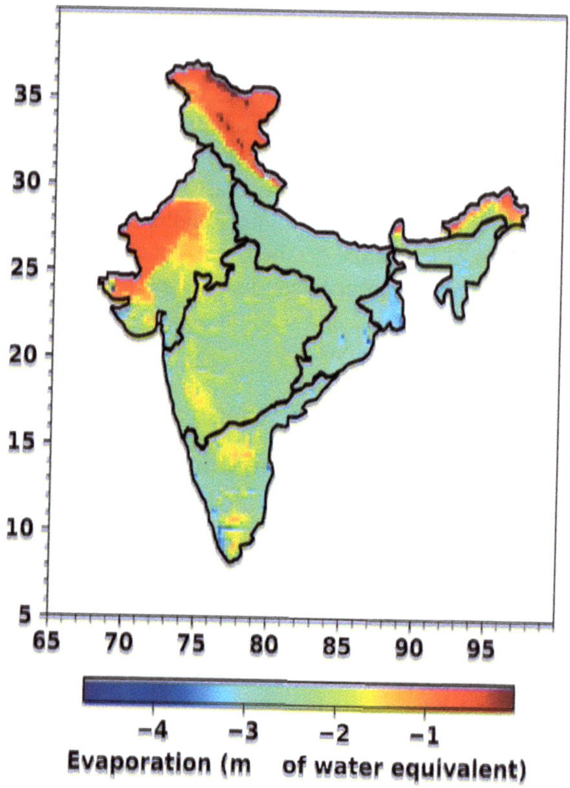

Fig. 2.1 Overall mean of evaporation for the period of 2006–2020 over the Indian region

2.6.1.2 Evaporation-Precipitation (E-P) Balance

The spatial distribution of the evaporation-precipitation balance (Fig. 2.2) mirrors the evaporation patterns, with significant E-P changes noted in the northern, western, and northeastern regions of India. Areas with high evaporation rates tend to have negative E-P values, indicating a deficit in soil moisture and higher susceptibility to drought conditions. Conversely, coastal and central regions with positive E-P values have a better moisture balance, which supports vegetation growth and agricultural activities. These findings support previous studies by Trenberth (2011) on the critical role of E-P balance in determining soil moisture availability.

2.6.1.3 Relative Humidity (RH)

The spatial variation in relative humidity across the study area is shown in Fig. 2.3. The western, central, and northeastern parts of India show lower RH levels, which correlate with reduced soil moisture and increased evaporation. In contrast, northern and coastal regions have higher RH, promoting moisture retention in soils. Lorenz et al. (2010) emphasize that lower relative humidity accelerates soil moisture depletion, especially in regions with limited rainfall.

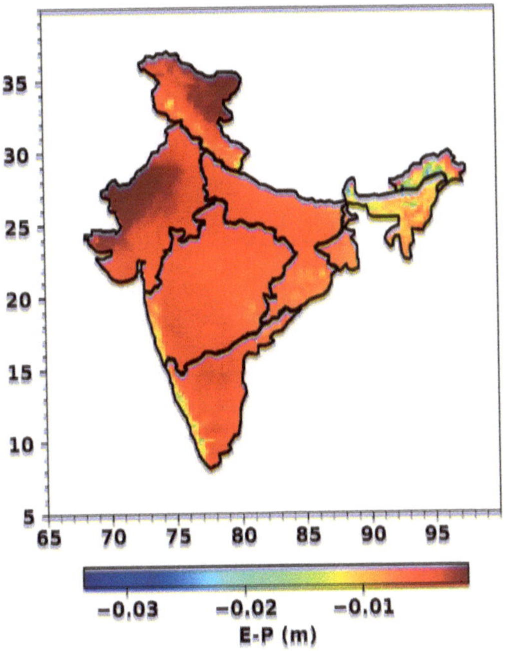

Fig. 2.2 Overall mean of evaporation-precipitation for the period of 2006–2020 over the Indian region

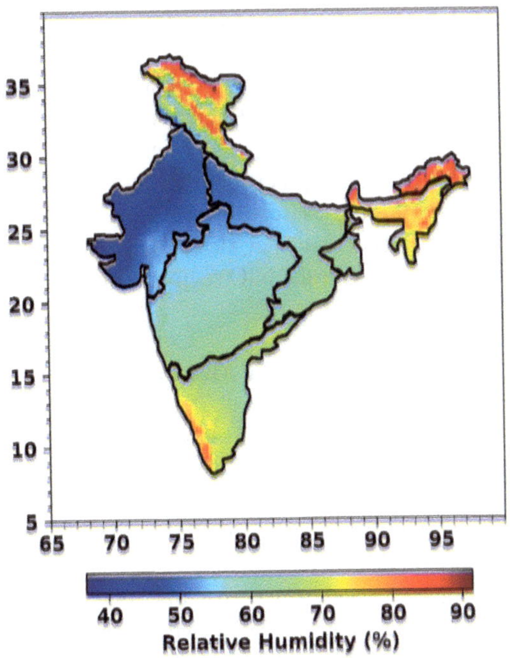

Fig. 2.3 Overall mean of relative humidity for the period of 2006–2020 over the Indian region

Fig. 2.4 Overall mean of specific humidity for the period of 2006–2020 over the Indian region

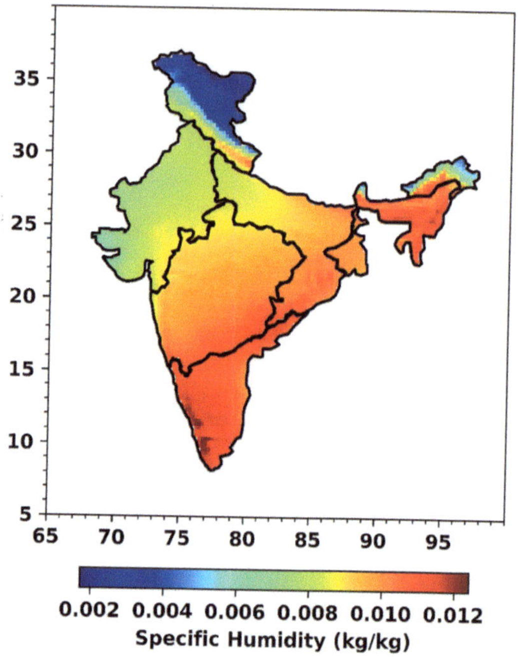

2.6.1.4 Specific Humidity (SH)

Specific humidity is inversely related to relative humidity patterns, with higher SH observed in areas where RH is low (Fig. 2.4). For instance, the western and central-northeastern regions show elevated SH levels, a characteristic of dry climates. This inverse relationship between SH and RH impacts soil moisture, as low RH combined with high SH can increase the demand for soil water, further accelerating moisture loss (Koster et al. 2010, 2018). Higher SH and RH values also help in cyclogenesis over the NIO region (Kumar et al. 2024, Sagar et al. 2025a, b, c).

2.6.1.5 Saturation Soil Moisture (SSM)

The spatial distribution of saturation soil moisture across India is shown in Fig. 2.5. Northern and northeastern regions display high SSM levels, while the western, central, and coastal areas show lower SSM. These variations are associated with differences in soil texture, rainfall patterns, and temperature. High SSM in northern regions indicates greater water-holding capacity, which benefits agriculture and vegetation resilience, whereas low SSM in western drylands signals limited capacity for water retention, increasing vulnerability to droughts (Mondal and Mishra 2024; Ganeshi et al. 2023).

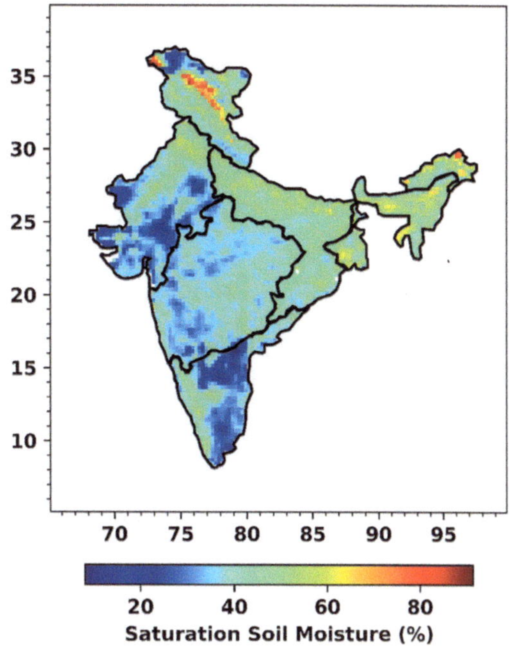

Fig. 2.5 Overall mean of saturation soil moisture for the period of 2006–2020 over the Indian region

2.6.1.6 Total Precipitation

The spatial distribution of total precipitation over the study area (Fig. 2.6) reflects the variability in rainfall patterns across dryland regions. Areas near the western coast and northeastern India receive the highest rainfall, which supports high soil moisture and reduces the need for irrigation. Conversely, regions with minimal precipitation face greater risks of desertification, soil erosion, and reduced agricultural productivity. Trenberth (2011) has noted that precipitation is a primary input in the soil moisture budget and plays a critical role in maintaining dryland resilience.

2.6.1.7 Temperature at 2 m

The spatial distribution of near-surface temperature, with most of India experiencing temperatures above 28 °C, except for the hilly and northern regions (Fig. 2.7). High temperatures in dryland regions lead to increased evaporation rates, stressing water resources and accelerating soil moisture loss. These findings are consistent with those of Hirschi et al. (2011), who discussed how elevated temperatures increase soil moisture deficits, especially in areas with limited rainfall. Variations in greenhouse gas level over the region contribute to higher temperature (Sagar 2025f, Sagar et al. 2025d, e).

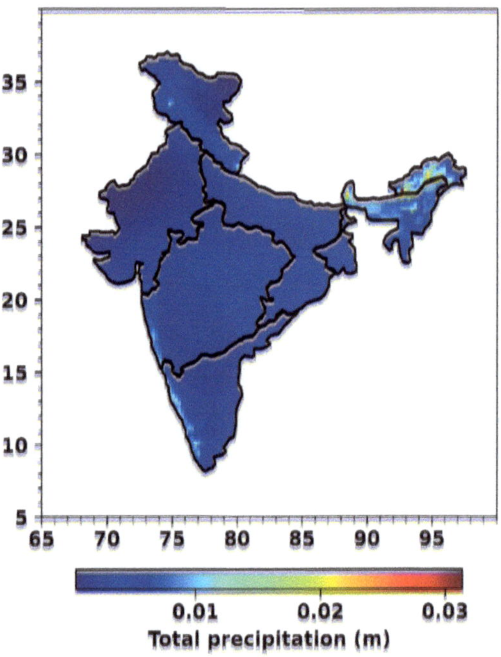

Fig. 2.6 Overall mean of total precipitation for the period of 2006–2020 over the Indian region

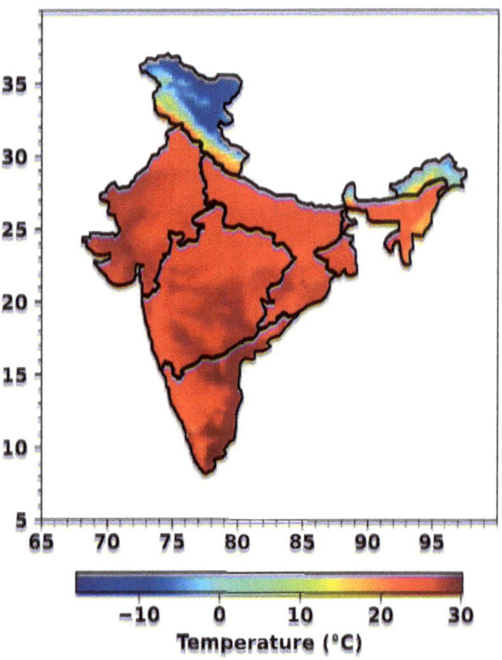

Fig. 2.7 Overall mean of 2 m temperature for the period of 2006–2020 over the Indian region

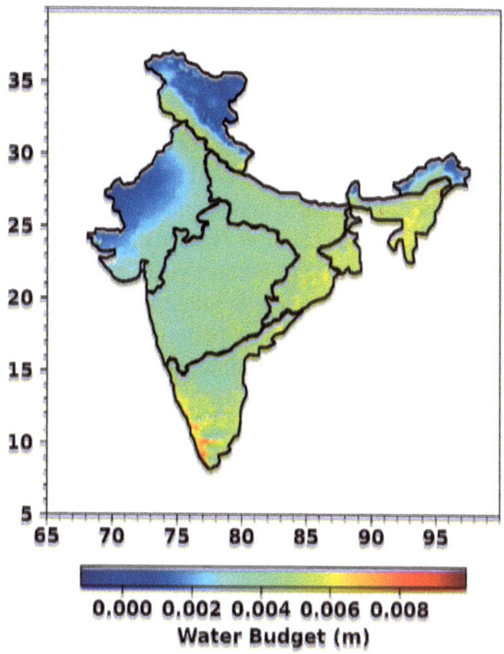

Fig. 2.8 Overall mean of water budget for the period of 2006–2020 over the Indian region

2.6.1.8 Water Budget

The water budget for the study period is displayed in Fig. 2.8, calculated using Eq. 2.1. The Western Ghats and northeastern parts of India, with their higher precipitation and moderate evaporation, show a positive water budget, meaning soil moisture levels are sustained over time. In contrast, western and hilly areas exhibit negative water budgets due to low precipitation and high evaporation, posing challenges for soil moisture retention and increasing the risk of drought (Sharma and Choudhury (2021). These trends align with Moyers et al. (2023), who highlighted the importance of positive water budgets for soil moisture sustainability in dryland ecosystems.

2.6.2 *Temporal Changes and Trends (2006–2020)*

The temporal analysis revealed distinct year-on-year variations in climate parameters and soil moisture levels:

- *Annual Variation in Soil Moisture*: Soil moisture levels fluctuated significantly over the study period, with notable deficits during years marked by drought or extreme heat events. These fluctuations underscore the sensitivity of soil moisture to short-term climatic changes, as noted by Mondal and Mishra (2024), who emphasized that annual variability in soil moisture is primarily driven by precipitation and temperature changes.

- *Trend of Increasing Temperatures*: The consistent increase in 2 m temperatures throughout the study period correlates with higher evaporation rates and reduced soil moisture availability. This trend is consistent with findings from Oleson et al. (2013), who observed that rising temperatures increase soil moisture stress, particularly in arid and semiarid regions.
- *Shifts in Precipitation Patterns*: The analysis also identified shifts in precipitation patterns, with certain years experiencing lower-than-average rainfall, contributing to soil moisture deficits. Dirmeyer (2003) demonstrated that changes in seasonal precipitation patterns impact soil moisture storage, reducing water availability for vegetation and exacerbating drought conditions in dryland areas.

2.6.3 Implications for Dryland Ecosystem Resilience

The spatial and temporal variability in soil moisture and related climate variables underscores the challenges facing dryland ecosystems under climate change (Koster et al. 2018). Regions with negative water budgets, low RH, and high evaporation rates are at heightened risk of desertification, biodiversity loss, and agricultural decline. By identifying these vulnerable areas, the study highlights the critical need for ecosystem-based approaches that enhance resilience through water conservation, soil restoration, and sustainable land management practices.

The analysis of soil moisture and climate variables across dryland regions has demonstrated the significant influence of climate change on water availability and soil health. These findings support the implementation of targeted ecosystem-based interventions, such as reforestation and sustainable land management, to mitigate the effects of climate change and improve resilience in dryland ecosystems. Further research is needed to explore the impacts of extreme weather events and to refine soil moisture models to improve climate adaptation strategies for drylands.

2.7 Implications for Sustainable Development

Dryland ecosystems face unique challenges due to climate change, and their vulnerability has significant implications for sustainable development. These regions support large human populations, agriculture, and biodiversity and play a crucial role in global food security and environmental stability. However, the increasing severity of climate-induced stresses—such as desertification, soil erosion, and water scarcity—threatens the sustainability of these ecosystems.

2.7.1 *Importance of Soil Moisture Management for Climate Resilience*

Soil moisture is a vital resource for dryland agriculture, which is highly dependent on limited and erratic rainfall. The findings from this study underscore the importance of managing soil moisture as a key component of resilience in dryland areas. Effective soil moisture management can reduce vulnerability to drought, enhance crop yields, and support livelihoods that depend on agriculture. Practices like conservation tillage, mulching, and agroforestry can increase soil moisture retention, thereby reducing the need for irrigation and making agricultural practices more sustainable. Kloos and Renaud (2016) highlight that improved soil moisture management in drylands has the potential to enhance water availability, reduce greenhouse gas emissions, and improve ecosystem health.

Table 2.4 provides a summary of various ecosystem-based interventions implemented in different regions of India, highlighting their objectives and outcomes. For example, reforestation in the Aravalli Hills of Rajasthan aims to combat soil erosion and improve soil health, reducing soil loss and increasing carbon sequestration. Similarly, agroforestry in Tamil Nadu seeks to increase crop resilience to drought, leading to improved crop yields and enhanced soil fertility. Each entry in the table includes the specific intervention, the region where it was implemented, its primary objective, the observed outcomes or impacts, and the reference to the study or report documenting these results. This table helps illustrate the practical applications and benefits of ecosystem-based approaches in various dryland regions.

Table 2.4 Examples of ecosystem-based interventions in Indian drylands

Intervention	Region	Objective	Outcome/impact	References
Reforestation (Aravalli Hills)	Rajasthan	Combat soil erosion and improve soil health	Reduced soil loss, increased carbon sequestration	Gupta et al. (2024)
Agroforestry	Tamil Nadu	Increase crop resilience to drought	Improved crop yields, enhanced soil fertility	Keprate et al. (2024)
Water harvesting (Jal Swavlamban)	Rajasthan	Enhance water availability	Increased groundwater levels, improved crop irrigation	Saxena and Brighu (2022)
Check dams	Gujarat	Reduce runoff and improve groundwater recharge	Improved local water table, reduced soil erosion	Mozzi et al. (2021)
Grassland restoration	Maharashtra	Improve grazing quality and reduce soil loss	Increased soil carbon, better livestock productivity	Malaviya et al. (2018)

2.7.2 Contributions to Sustainable Development Goals (SDGs)

The ecosystem-based approaches explored in this study align closely with several United Nations Sustainable Development Goals (SDGs), particularly those focused on climate action (SDG 13), life on land (SDG 15), and clean water and sanitation (SDG 6) (Diz 2019). Enhancing soil moisture management and adopting sustainable land practices contribute directly to these goals by addressing water scarcity, improving land productivity, and reducing degradation (Sharma and Choudhury 2021). Reforestation, sustainable agriculture, and wetland restoration initiatives can enhance biodiversity, support carbon sequestration, and promote resilience to climate change impacts (Gupta et al. 2024).

Additionally, dryland regions are often home to marginalized and vulnerable communities. Sustainable development in these areas supports SDG 1 (No Poverty) and SDG 2 (Zero Hunger) by improving food security, creating employment opportunities, and promoting climate-resilient agriculture (Lucatello and Huber-Sannwald 2020). When local communities are involved in planning and implementing ecosystem-based solutions, sustainable practices' social and economic benefits are amplified, helping to build long-term resilience in these communities (Dirmeyer 2003).

2.7.3 Policy Implications for Resource Management in Drylands

The study's findings have significant policy implications, highlighting the need for integrated, climate-sensitive policies that support sustainable dryland management. Policymakers should prioritize:

- *Investment in Ecosystem-Based Solutions*: Allocating resources to reforestation, wetland restoration, and agroforestry projects can improve soil health and water availability, addressing core vulnerabilities in dryland ecosystems (Lucatello and Huber-Sannwald 2020).
- *Incentives for Sustainable Agriculture*: Providing subsidies for sustainable practices, like conservation tillage and rotational grazing, can encourage farmers to adopt soil-friendly practices that increase resilience to climate impacts (Diz 2019).
- *Community Engagement*: Policies should emphasize the involvement of local communities in resource management, recognizing the value of traditional knowledge in sustainable practices (Dirmeyer 2003).

These policy measures can help protect dryland ecosystems while promoting sustainable livelihoods, strengthening the resilience of both people and the environment.

2.8 Conclusion

This chapter has highlighted the critical role of ecosystem-based approaches in enhancing resilience and supporting sustainable development in dryland ecosystems. As climate change intensifies, drylands face increasing threats to soil health, water availability, and agricultural productivity, underscoring the urgent need for adaptive management strategies. By focusing on natural processes, ecosystem-based approaches—such as reforestation, agroforestry, and sustainable land management—offer effective and sustainable solutions for climate adaptation.

The analysis of soil moisture dynamics across dryland regions demonstrates the complex interplay between precipitation, evaporation, temperature, and soil moisture, which collectively impact water availability and ecosystem resilience. The study's findings reinforce the importance of soil moisture management, not only as a tool for supporting agriculture but also as a critical element in maintaining ecological balance and mitigating the effects of drought and desertification.

Furthermore, aligning ecosystem-based approaches with global sustainable development goals highlights the potential of these strategies to enhance climate resilience, food security, and community well-being. Effective soil and water management in dryland regions can mitigate climate impacts while supporting biodiversity and improving rural livelihoods. Therefore, the challenges faced by dryland ecosystems demand a holistic approach that integrates scientific insights, community participation, and policy support. Continued investment in ecosystem-based solutions is essential to foster resilience, mitigate climate risks, and promote sustainable development in these vulnerable landscapes. Through targeted interventions and collaboration at multiple levels, dryland ecosystems can become resilient landscapes that support both people and the planet.

Acknowledgments We want to express our sincere gratitude to the IIT Kharagpur for their invaluable support and assistance in facilitating this study. Their resources, expertise, and encouragement have been instrumental in the successful completion of our research. We deeply appreciate their contribution and collaboration.

References

Altieri MA, Nicholls CI (2017) The adaptation and mitigation potential of traditional agriculture in a changing climate. Clim Chang 140:33–45

Amatya DM, Irmak S, Gowda P, Sun G, Nettles JE, Douglas-Mankin KR (2016) Ecosystem evapotranspiration: challenges in measurements, estimates, and modeling. Trans ASABE 59(2):555–560

Berg A, Sheffield J (2018) Climate change and drought: the soil moisture perspective. Curr Clim Chang Rep 4:180–191

Berg A, Lintner BR, Findell KL, Malyshev S, Loikith PC, Gentine P (2014) Impact of soil moisture–atmosphere interactions on surface temperature distribution. J Clim 27(21):7976–7993

Dirmeyer PA (2003) The role of the land surface background state in climate predictability. J Hydrometeorol 4(3):599–610

Diz D (2019) The ecosystem approach as a frame for SDG 14 implementation. Ocean Yearbook Online 33(1):187–206

Ganeshi NG, Mujumdar M, Takaya Y, Goswami MM, Singh BB, Krishnan R, Terao T (2023) Soil moisture revamps the temperature extremes in a warming climate over India. Npj climate and atmospheric. Science 6(1):12

Grillakis MG (2019) Increase in severe and extreme soil moisture droughts for Europe under climate change. Sci Total Environ 660:1245–1255

Grünzweig JM, De Boeck HJ, Rey A, Santos MJ, Adam O, Bahn M et al (2022) Dryland mechanisms could widely control ecosystem functioning in a drier and warmer world. Nat Ecol Evolut 6(8):1064–1076

Gupta SK, Kanga S, Meraj G, Singh SK, Singh S, Sajan B et al (2024) Optimizing land use for climate mitigation using nature based solution (NBS) strategy: a study on afforestation potential and carbon sequestration in Rajasthan, India. Discov Geosci 2(1):36

Hartzell J Altman M, Srivastava R, Bhandari P, Aryee JNA (2024) Oceanic Oscillations and Congo River Basin Climatology (1.0). Climatematch Impact Scholars Program 2023–2024 (CISP), Virtual. Zenodo. https://doi.org/10.5281/zenodo.12192494

Hersbach H, Bell B, Berrisford P, Hirahara S, Horányi A, Muñoz-Sabater J et al (2020) The ERA5 global reanalysis. Q J R Meteorol Soc 146(730):1999–2049

Hirschi M, Seneviratne SI, Alexandrov V, Boberg F, Boroneant C, Christensen OB et al (2011) Observational evidence for soil-moisture impact on hot extremes in southeastern Europe. Nat Geosci 4(1):17–21

Irvine K, Castello L, Junqueira A, Moulton T (2016) Linking ecology with social development for tropical aquatic conservation. Aquat Conserv Mar Freshwat Ecosyst 26(5):917–941

Jain S, Srivastava A, Khadke L, Chatterjee U, Elbeltagi A (2024) Global-scale water security and desertification management amidst climate change. Environ Sci Pollut Res 31:1–25

Keprate A, Bhardwaj DR, Sharma P, Verma K, Abbas G, Sharma V et al (2024) Climate resilient agroforestry systems for sustainable land use and livelihood. In: Transforming agricultural management for a sustainable future: climate change and machine learning perspectives. Cham, Springer Nature, pp 141–161

Kloos J, Renaud FG (2016) Overview of ecosystem-based approaches to drought risk reduction targeting small-scale farmers in sub-Saharan Africa. In: Ecosystem-based disaster risk reduction and adaptation in practice, pp 199–226

Koster RD, Mahanama SPP, Yamada TJ, Balsamo G, Berg AA, Boisserie M et al (2010) Contribution of land surface initialization to subseasonal forecast skill: first results from a multi-model experiment. Geophys Res Lett 37(2)

Koster RD, Crow WT, Reichle RH, Mahanama SP (2018) Estimating basin-scale water budgets with SMAP soil moisture data. Water Resour Res 54(7):4228–4244

Kumar A, Chakraborty A, Sagar AK (2024) Analysis of the Bay of Bengal cyclone and its rejuvenation in the Arabian Sea after passing over peninsular India. Geomatics, Natural Hazards and Risk 15(1):2433110.

Locatelli B, Pavageau C, Pramova E, Di Gregorio M (2015) Integrating climate change mitigation and adaptation in agriculture and forestry: opportunities and trade-offs. Wiley Interdiscip Rev Clim Chang 6(6):585–598

Lorenz R, Jaeger EB, Seneviratne SI (2010) Persistence of heat waves and its link to soil moisture memory. Geophys Res Lett 37(9)

Lucatello S, Huber-Sannwald E (2020) Sustainable development goals and drylands: addressing the interconnection. In: Stewardship of future drylands and climate change in the global south: challenges and opportunities for the agenda, vol 2030, pp 27–40

Malaviya DR, Roy AK, Kaushal P (2018) Rangelands/grasslands of India: current status and future prospects. Grasslands of the World:235–252

McVittie A, Cole L, Wreford A, Sgobbi A, Yordi B (2018) Ecosystem-based solutions for disaster risk reduction: lessons from European applications of ecosystem-based adaptation measures. Int J Disaster Risk Reduct 32:42–54

Mohanty BP, Cosh MH, Lakshmi V, Montzka C (2017) Soil moisture remote sensing: state-of-the-science. Vadose Zone J 16(1):1–9

Mondal S, Mishra A (2024) Quantifying the precipitation, evapotranspiration, and soil moisture network's interaction over global land surface hydrological cycle. Water Resour Res 60(2):e2023WR034861

Mondal SK, An SI, Min SK, Jiang T, Su B (2024) Enhanced soil moisture–temperature coupling could exacerbate drought under net-negative emissions. NPJ Clim Atmos Sci 7(1):265

Mortimer R, Saj S, David C (2018) Supporting and regulating ecosystem services in cacao agroforestry systems. Agrofor Syst 92(6):1639–1657

Moyers K, Sabie R, Waring E, Preciado J, Naughton CC, Harmon T, Safeeq M, Torres-Rua A, Fernald A, Viers JH (2023) A decade of data-driven water budgets: synthesis and bibliometric review. Water Resour Res 59(11):e2022WR034310

Mozzi G, Pavelic P, Alam MF, Stefan C, Villholth KG (2021) Hydrologic assessment of check dam performances in semi-arid areas: a case study from Gujarat, India. Frontiers in Water 3:628955

Oleson KW, Lawrence DM, Bonan GB, Fisher RA, Lawrence PJ, Muszala SP (2013) Technical description of version 4.5 of the Community Land Model (CLM). Technical description of version 4.5 of the Community Land Model (CLM)(2013) NCAR/TN-503+ STR, 503.

Osborne BB, Bestelmeyer BT, Currier CM, Homyak PM, Throop HL, Young K, Reed SC (2022) The consequences of climate change for dryland biogeochemistry. New Phytologist 236(1):15–20

Petropoulos GP, Ireland G, Barrett B (2015) Surface soil moisture retrievals from remote sensing: current status, products & future trends. Phys Chem Earth 83:36–56

Rasheed MW, Tang J, Sarwar A, Shah S, Saddique N, Khan MU et al (2022) Soil moisture measuring techniques and factors affecting the moisture dynamics: a comprehensive review. Sustainability 14(18):11538

Reid H, Bourne A, Muller H, Podvin K, Scorgie S, Orindi V (2018) A framework for assessing the effectiveness of ecosystem-based approaches to adaptation. In: Resilience. Elsevier, pp 207–216

Sadhukhan B, Srivastava RK, Chakraborty A, Panda RK (2024) Scientific evidence supporting the progression of climate change–induced drought through history. In: Integrated drought management, vol 1. CRC Press, pp 525–540

Sagar AK, Chakraborty A (2024) Spatial changes of driving parameters affecting cyclonic activity over the North Indian Ocean from 1960 to 2020. J Atmos Sci Res 7(4):13–22. https://doi.org/10.30564/jasr.v7i4.6926

Sagar AK, Chakraborty A, Behera SK, Kumar A (2025a) Study of accumulated cyclone energy and associated drivers afecting tropical cyclone activities over the North Indian Ocean during the preand post monsoon season. Nat Haz 1-20

Sagar AK, Chakraborty A, Kumar A, Kumar P (2025b) Identifying robust precursor regions of driving parameters afecting the Accumulated Cyclone Energy in the North Indian Ocean using the Causal Efect Network. Results Earth Sci 3:100078

Sagar AK, Chakraborty A (2025c) Climate Response of the Seven Homogeneous Monsoon Rainfall Regions of India Using the SOLAR Geoengineering Simulation Datasets. Aerosol Science and Engineering, 1–14

Sagar AK (2025f) Analysis of Spatial and Temporal Changes in Methane Gas Levels Over the Indian Region. In Mitigation and Adaptation Strategies Against Climate Change in Natural Systems (pp. 289–299). Cham: Springer Nature Switzerland

Sagar AK, Chakraborty A, Srivastava RK (2025e) Nitrogen Dioxide Trends: A Global Perspective with Regional Insights from India. In Mitigation and Adaptation Strategies Against Climate Change in Natural Systems (pp. 321–340). Cham: Springer Nature Switzerland

Sagar AK, Chakraborty A, Srivastava RK (2025d) Global Atmospheric Trends and Climate Change: A Comprehensive Analysis of Greenhouse Gases and Temperature (2003–2020). In Mitigation and Adaptation Strategies Against Climate Change in Natural Systems (pp. 301–320). Cham: Springer Nature Switzerland

Sang PM, Lamb D, Bonner M, Schmidt S (2013) Carbon sequestration and soil fertility of tropical tree plantations and secondary forest established on degraded land. Plant Soil 362:187–200

Sanz MJ, De Vente J, Chotte JL, Bernoux M, Kust G, Ruiz I. et al (2017) Sustainable land management contribution to successful land-based climate change adaptation and mitigation: a report of the science-policy interface. In Bonn, Germany: United Nations Convention to Combat Desertification (UNCCD)

Saxena K, Brighu U (2022) Jal Swavlamban: water Independence: a case study of semiarid state of Rajasthan in India. In: The Palgrave handbook of climate resilient societies. Springer International Publishing, Cham, pp 321–346. https://doi.org/10.1007/978-3-030-42462-6_76

Seddon N, Chausson A, Berry P, Girardin CA, Smith A, Turner B (2020) Understanding the value and limits of nature-based solutions to climate change and other global challenges. Philos Trans R Soc B 375(1794):20190120

Seneviratne SI, Corti T, Davin EL, Hirschi M, Jaeger EB, Lehner I et al (2010) Investigating soil moisture–climate interactions in a changing climate: a review. Earth Sci Rev 99(3–4):125–161

Sharma A, Choudhury M (2021) Changing climate and depleting water resources in the mountains with a case study from the Himalayas. In: Water conservation in the era of global climate change. Elsevier, pp 411–436

Sharma APM, Jhajharia D, Yurembam GS, Srivastava RK (2024) Drought preparedness under changing climate with special reference to Manipur. In: Climate change impacts on soil-plant-atmosphere continuum. Springer Nature, Singapore, pp 701–711

Siddique KH, Bolan N, Rehman A, Farooq M (2024) Enhancing crop productivity for recarbonizing soil. Soil Tillage Res 235:105863

Singh A, Gaurav K, Meena GK, Kumar S (2020) Estimation of soil moisture applying modified Dubois model to Sentinel-1; a regional study from Central India. Remote Sens 12(14):2266

Srivastav AL, Dhyani R, Ranjan M, Madhav S, Sillanpää M (2021) Climate-resilient strategies for sustainable management of water resources and agriculture. Environ Sci Pollut Res 28(31):41576–41595

Srivastava RK, Imtiyaz M (2016) Testing of coupled SCS curve number model for estimating runoff and sediment yield for eleven watersheds. J Geol Soc India 88:627–636

Srivastava RK, Panda RK, Chakraborty A, Halder D (2020) Quantitative estimation of water use efficiency and evapotranspiration under varying nitrogen levels and sowing dates for rainfed and irrigated maize. Theor Appl Climatol 139(3):1385–1400

Srivastava RK, Sadhukhan B, Chakraborty A, Panda RK (2024) Bias correction and trend analysis of temperature and rainfall in eastern India. In: Proceedings of the National Academy of Sciences, India section a: physical sciences, vol 94, pp 1–16

Stringer LC, Mirzabaev A, Benjaminsen TA, Harris RM, Jafari M, Lissner TK et al (2021) Climate change impacts on water security in global drylands. One Earth 4(6):851–864

Teague R, Kreuter U (2020) Managing grazing to restore soil health, ecosystem function, and ecosystem services. Front Sustain Food Syst 4:534187

Trenberth KE (2011) Changes in precipitation with climate change. Clim Res 47(1–2):123–138

Verhulst N, Govaerts B, Verachtert E, Castellanos-Navarrete A, Mezzalama M, Wall PC et al (2010) Conservation agriculture, improving soil quality for sustainable production systems. In: Advances in soil science: food security and soil quality, pp 137–208

Verma VK, Sur K, Prakash C (2023) Ecotope-based diversity monitoring of wetland using infused machine learning technique. Water Conserv Sci Eng 8(1):38

Weiskopf SR, Rubenstein MA, Crozier LG, Gaichas S, Griffis R, Halofsky JE et al (2020) Climate change effects on biodiversity, ecosystems, ecosystem services, and natural resource management in the United States. Sci Total Environ 733:137782

Zandler H, Senftl T, Vanselow KA (2020) Reanalysis datasets outperform other gridded climate products in vegetation change analysis in peripheral conservation areas of Central Asia. Sci Rep 10(1):22446

Chapter 3
Nature-Based Solutions (NbS) for Restoration and Management of Dryland Ecosystems

Subrata Gorain, Ayushman Malakar, and Suman Dutta

Abstract Drylands, encompassing over 40% of the Earth's surface and home to more than 2 billion people, face significant challenges due to environmental degradation, desertification, and climate change. The management and restoration of these ecosystems are crucial for sustaining agriculture, ensuring human welfare, and conserving biodiversity. Nature-based solutions (NbS) offer a promising approach by leveraging natural processes to enhance biodiversity, ecosystem services, and climate resilience in dryland regions. Key strategies within NbS include agroforestry, which integrates trees with crops and livestock to improve soil fertility and moisture retention, and water harvesting techniques, such as contour bunds and check dams, that facilitate collecting and storing rainfall for agricultural use. Additionally, initiatives focused on vegetation restoration, particularly with drought-resistant species, contribute to soil sstabilization and the recovery of ecosystem functions. Sustainable grazing practices further support ecosystem health by preventing overgrazing and fostering landscape regeneration. The economic implications of NbS are substantial; these approaches not only enhance ecosystem services like carbon sequestration and water regulation but are often more cost-effective than traditional engineering solutions. Furthermore, NbS promote long-term economic resilience by mitigating climate change impacts, supporting diverse livelihoods, and generating employment in restoration activities. While challenges remain, such as securing sustainable funding and addressing land tenure issues, the potential of NbS to transform drylands is immense. By fostering resilience and restoring ecosystem

S. Gorain
Department of Agricultural Economics, Palli Siksha Bhavana (Institute of Agriculture), Visva-Bharati (Central University), Sriniketan, West Bengal, India

A. Malakar
ICFRE-Institute of Forest Productivity, Lalgutwa, Ranchi, Jharkhand, India

S. Dutta (✉)
Faculty Center for Integrated Rural Development and Management, Ramakrishna Mission Vivekananda Educational and Research Institute, Howrah, West Bengal, India

Y. D. I. Khan et al. (eds.), *Land Restoration Through Ecosystem-Based Approach*, Disaster Resilience and Green Growth, https://doi.org/10.1007/978-981-95-0389-6_3

integrity, NbS present a viable pathway to combat environmental degradation and support local communities in dryland areas.

Keywords Nature-based solutions · Dryland restoration · Ecosystem services · Agroforestry · Water harvesting · Biodiversity conservation

3.1 Introduction

Ecosystem restoration and management in drylands are paramount, as these regions face increasing environmental degradation that threatens biodiversity, agriculture, and human livelihoods (Wale and Dejenie 2013). Drylands, which make up more than 40% of the planet's surface and are inhabited by over 2 billion people, are extremely vulnerable to unsustainable land-use practices, desertification, and climate change (Lopez Porras et al. 2020). These regions are especially vulnerable to land degradation, biodiversity loss, and diminishing agricultural output because of their delicate ecosystems and harsh climate. In response, a potential strategy to deal with these intricate problems is nature-based solutions (NbS) (Gooden and Pritzlaff 2021). The concept of NbS originated from the "ecosystem approach" methodology of the Convention on Biological Diversity (CBD) in the late 2000s (Alikhanova and Bull 2023). Subsequently, the International Union for Conservation of Nature (IUCN) and the European Commission (EC) redefined the definition of NbS. Although these definitions share a common objective of tackling significant societal issues through the efficient utilization of ecosystems and their services, they emphasize different aspects. (Liu et al. 2021). The IUCN's definition was formulated from a global viewpoint and highlights the importance of a well-managed or restored ecosystem as central to any NbS, whereas the EC definition focuses more on implementing solutions that not only utilize nature but are also inspired and supported by it (Debele et al. 2019). Nonetheless, the formal definition of NbS by IUCN is widely referenced around the world, describing it as "*Actions to protect, sustainably manage, and restore natural or modified ecosystems, that address societal challenges effectively and adaptively, simultaneously providing human well-being and biodiversity benefits*" (IUCN 2020). NbS offer long-term, sustainable solutions that are advantageous to society and the environment by utilizing ecosystem functions and natural processes (Bagwan 2023). NbS can improve ecosystem services, boost resilience to climate change, and promote biodiversity by supporting practices like agroforestry, sustainable grazing, water collection, and revegetation of degraded lands (Sharma et al. 2024). The sustainability of agricultural systems in arid regions depends on factors including soil health, moisture retention, and soil erosion prevention, which are all enhanced by agroforestry—the integration of trees with crops and cattle.

The principles underlying NbS in dryland restoration focus on ecosystem integrity, sustainability, resilience, and the active involvement of local communities (Bagwan 2023). The natural processes of dryland ecosystems, which offer vital services like carbon sequestration, erosion control, and groundwater recharge, can be restored by implementing strategies that enhance soil health, water availability, and vegetation cover (Li et al. 2021). Additionally, these ecosystems grow more tolerant to the severe weather patterns of arid regions, including heat waves and droughts (Owusu et al. 2022). Another fundamental idea is sustainability, since NbS places a strong emphasis on long-term ecological health, which enables restored ecosystems to more effectively adjust to shifting environmental conditions (Seddon et al. 2021). Crucially, NbS enhance livelihoods, which also affects nearby communities (Seddon 2022). For instance, agroforestry increases food security and diversifies farmers' revenue streams by improving crop yields and soil fertility while simultaneously producing feed for animals (Barman et al. 2024). Furthermore, sustainable grazing methods that avoid overgrazing and encourage vegetation regrowth assist community agricultural requirements while also aiding in ecosystem restoration (Guo and Chen 2024). Integrating Indigenous knowledge, encouraging community involvement, and fostering inclusive collaboration are all necessary for the successful implementation of NbS (Seddon 2022). Participation from the local community guarantees that restoration initiatives are customized to particular dryland habitats and satisfy the distinct socioeconomic requirements of these areas (Osaka et al. 2021). As a result, NbS provide a comprehensive, community-based method of dryland restoration that tackles socioeconomic and ecological issues while promoting resilience and sustainability over the long run.

Despite the growing recognition of NbS as effective strategies for addressing environmental degradation in drylands, several research gaps remain (Bagwan 2023). One major knowledge gap is incomplete knowledge of the long-term ecological and socioeconomic effects of NbS in many dryland habitats. Although research has shown the advantages of techniques like water harvesting, sustainable grazing, and agroforestry, thorough, long-term evaluations are required to measure their efficacy in various soil types and climate zones (Sharma et al. 2024). Furthermore, little is known about the precise processes by which NbS enhance resistance to extreme weather events like heat waves and droughts, especially when considering the magnitude and severity of these occurrences (Seddon et al. 2021). The socioeconomic integration of NbS in drylands is another area of disparity. Research on the most effective ways to integrate traditional land-use practices and local government systems into contemporary NbS frameworks is lacking, despite the fact that NbS strongly emphasize community involvement and the use of indigenous knowledge (Gooden and Pritzlaff 2021). Furthermore, more research is needed to determine whether NbS might produce economic co-benefits like improved food security and employment development, especially when it comes to how scalable these solutions are for bigger populations (Bagwan 2023). Lastly, even though NbS have the potential to draw in green investments and advance sustainable development, further study is required to create precise rules and rewards for both public and private sector participation, guaranteeing that NbS investments are efficient and fair (Maes and Jacobs 2017). By filling these gaps, NbS implementation in drylands would be improved, and more resilient and sustainable ecosystems would be supported (Nelson et al. 2020). With an emphasis on ecosystem restoration, biodiversity protection, and climate change resilience, this chapter aims to investigate the potential of NbS in tackling environmental deterioration in drylands. Key NbS activities like agroforestry, water harvesting, and sustainable grazing will be examined, with an emphasis on the ecological and socioeconomic advantages of each. Along with identifying research needs, the chapter offers insights into community integration, long-term efficacy, and the possibility of green investment in dryland restoration projects.

3.2 Principles of Nature-Based Solutions in Dryland Restoration

3.2.1 Definition

As the idea of NbS emerged in the late 2000s, it was reformulated by IUCN as an "*umbrella concept, which covered ecosystem-based management and issue-specific ecosystem related approaches, infrastructure-related approaches and ecosystem protection approaches*" (Cohen-Shacham et al. 2019). This IUCN definition has also received formal recognition recently from the United Nations (UN) during the fifth session of the United Nations Environment Assembly, where a motion on NbS for promoting sustainable development was adopted (UNEP 2022). NbS are an integrated approach that encompasses various ecosystem-based concepts to address societal challenges. NbS refer to actions that work with and enhance nature to help address societal challenges. These actions are combined in a way that provides both human well-being and biodiversity benefits. These approaches may address more than one challenge simultaneously. NbS prioritize collaboration with nature instead of opposition, acknowledging that ecosystems are essential for sustainable development. By integrating these diverse methodologies, NbS offer a framework for cohesive solutions that can yield several advantages to human well-being. The concepts under NbS can be categorized into restorative, issue-specific, climate adaptation, infrastructure, management, and protection (Cohen-Shacham et al. 2019). The IUCN and its Commission on Ecosystem Management (CEM) established the NbS definitional framework for a comprehensive understanding of NbS and facilitating its operationalization across different organizations and contexts. The framework comprises eight principles that illustrate the tactics for the concerns NbS seek to tackle (IUCN 2016). Many of the concepts are interconnected, and in some situations, they may even be depending on one another. NbS frameworks are continuously evolving as our understanding of ecosystem-based techniques and the ways in which they might be applied to address social concerns increases.

3.2.2 *Principle of NbS to Embrace Nature Conservation Norms*

NbS a holistic approach to tackling social concerns while protecting and improving natural ecosystems. These solutions are based on conservation principles; however, it is crucial to emphasize that not all conservation activities are NbS (Cohen-Shacham et al. 2016). NbS's adaptability allows them to be used as independent treatments or in combination with technical and engineering solutions, frequently complementing and strengthening traditional techniques (Nesshöver et al. 2017). A significant characteristic of NbS is its context-specificity, which draws on a complex tapestry of traditional, local, and scientific knowledge. This method guarantees that solutions are adapted to the specific ecological and cultural settings where they are implemented, resulting in a more holistic and participatory decision-making process (Raymond et al. 2017). Furthermore, NbS prioritize fair benefit-sharing across stakeholders, from local communities to national organizations, while emphasizing openness and wide engagement in its design and implementation (Kabisch et al. 2016). NbS implementation aims to preserve and improve the biological and cultural variety, acknowledging the fundamental importance of ecosystem complexity and its role in delivering resilient services in the face of environmental change (Seddon et al. 2020). These solutions are often implemented at the landscape level, involving several ecosystems and sometimes crossing governmental lines. This broad-scale strategy enables more comprehensive and integrated natural resource management (Keesstra et al. 2018). NbS recognize possible trade-offs between short-term economic advantages and long-term ecological services. This requires careful stakeholder analysis and discussion to balance current development requirements with long-term ecological advantages (Maginnis et al. 2004). To have a broad impact, NbS must be incorporated into broader policies and comprehensive activities rather than executed in isolation (Faivre et al. 2017). Working in harmony with nature, NbS provide a framework for delivering various benefits across diverse sectors, solving various social concerns while also increasing ecosystem health and resilience (Maes and Jacobs 2017).

3.2.2.1 Ecosystem Integrity in NbS for Dryland Restoration

NbS have emerged as an innovative strategy and promising solution to address the challenging issues that dryland ecosystems pose. Fundamentally, these solutions are based on ecosystem integrity, which is the restoration of the basic elements defining and maintaining these special surroundings. Developing ecological metrics to assess and monitor ecosystem health is one of the primary approaches to maintaining the ecosystem integrity in dryland. Moving beyond traditional meteorological indices, Chenoweth et al. (2023) developed ecologically relevant moisture and temperature metrics that provide insights into dryland plant growth dynamics. Combining qualitative observations with quantitative data for effective land management is also important to effectively implement NbS in drylands (Herrick et al. 2010). Additionally, restoration of soil microbial diversity can enhance nutrient cycling and improve plant growth, thereby contributing to overall ecosystem integrity (Maestre et al. 2017). This is crucial in drylands, where soil degradation can severely impact the ecosystem services. Furthermore, Zhan et al. (2022) pointed out the importance of understanding spatial dynamics in dryland ecosystems to inform restoration efforts and enhance ecosystem integrity.

Dryland ecosystems are particularly sensitive to deterioration and are marked by restricted water supply. Three main linked elements—soil health, water availability, and plant cover—emphasize the restoration of ecological integrity in these regions. Through bettering these elements, NbS hope to revive the capacity of the ecosystem to offer necessary functions and increase its resistance to environmental pressures (Reynolds et al. 2007). Thus, the basis of the integrity of a dryland ecosystem is soil condition. Using native plant species or bioengineering structures can help to enhance soil structure and fertility using NbS methods greatly. These treatments improve the soil's water retention capacity and nutrient cycling, therefore fostering a more friendly habitat for microbial activity and plant development (Yan et al. 2015). Another primary focus of NbS is water availability, a crucial limiting element in drylands. Rainwater collecting, contour bunding, and natural water catchment restoration all help to lower surface runoff and boost water infiltration (Binyam and Desale 2015). These steps help restore long-term groundwater supplies and increase immediate water availability for plants (Glendenning et al. 2012). Maintaining ecological integrity depends much on vegetation cover, the obvious expression of ecosystem health. NbS techniques often entail reintroducing native plant species that are suited to local circumstances. These plants provide good microclimates, aid in stabilizing the ground, and lower erosion. Furthermore, they are the main agents of carbon sequestration, helping slow global warming (Lal 2004). Multiple advantages follow from the restoration of ecological integrity by NbS. While better soil stability lowers erosion and desertification hazards, enhanced carbon sequestration capability helps offset climate change's effects. Furthermore, restoring natural water cycles helps groundwater recharge, benefiting the ecology and human populations depending on these water supplies (Maestre et al. 2012). We need to remember that NbS's effectiveness in restoring dryland ecosystem integrity depends on a thorough awareness of local biological processes and social settings. Every dryland ecosystem is different; restoration initiatives must be catered to these particular circumstances to succeed (Reed et al. 2015).

3.2.2.2 Sustainability and Resilience of NbS in Dryland Ecosystems

In the framework of dryland ecosystems, NbS are rather important for fostering sustainability and improving resistance against environmental pressures. Characterized by low water supply and sharp temperature changes, these arid and semi-arid areas provide special difficulties that require flexible and strong responses (Reynolds et al. 2007). NbS's capacity to interact with rather than against natural processes explains its sustainability in drylands. These ideas build a self-sustaining cycle of environmental improvement by improving the activities of ecosystems. Restoring native flora, for example, increases soil stability and improves water retention, fostering a more friendly habitat for better plant development (Maestre et al. 2012; Raghavendra et al. 2020; Kimmell et al. 2023). A fundamental quality of effective NbS resilience is the ability of an ecosystem to tolerate and recover from disruptions while preserving its vital purposes. This resilience is especially important given the frequent environmental stresses like extended droughts and strong heat waves in dryland settings (Bestelmeyer et al. 2015). NbS's primary means of improving resilience is via supporting biodiversity. Different species react differently to environmental stresses; hence, diverse ecosystems are more suited to change with the times. This ecological redundancy guarantees that certain species may preserve vital ecosystem activities even if some suffer under different circumstances (Hooper et al. 2005). Another important component of NbS resilience in drylands is water management. Rainwater collecting and better irrigation techniques not only assist in replenishing groundwater aquifers but also aid in saving limited water supplies. This method builds a buffer against drought, improving the ecosystem's survival capacity in trying circumstances (Vohland and Barry 2009). Crucially, the resilience encouraged by NbS spans societal and ecological systems. NbS can increase the adaptive ability of human populations by including nearby communities in the design and execution of these solutions. This participative method guarantees that treatments are financially feasible and culturally suitable, raising their chances of long-term success (Reed et al. 2015). Furthermore, its capacity to offer several ecological functions concurrently is directly related to NbS's sustainability in drylands. Through sustainable harvesting methods, reforestation projects can, for instance, sequester carbon, lower soil erosion, and give habitat for wildlife—all while perhaps providing financial gains. This multifarious nature helps NbS to be more resilient against shifting priorities or environmental situations (Chazdon 2008). Understanding how these treatments behave over time and under different environmental pressures needs long-term monitoring and study (Aronson et al. 2010).

3.2.2.3 Co-benefits and Inclusive Collaboration in NbS for Dryland Restoration

Using NbS in dryland environments results in a wide range of co-benefits that solve environmental and socioeconomic issues for nearby populations. Improving food security and livelihood possibilities is one of the most important co-benefits. For example, a mainstay of many NbS projects, agroforestry techniques perfectly illustrate its multifarious use (Powell et al. 2015). Agroforestry systems combine shrubs and trees into agricultural settings to establish a symbiotic interaction between crops, cattle, and woody perennials. Using higher organic matter intake and nitrogen fixation, this method raises soil fertility, hence improving crop yields. These systems yield feed for cattle, lumber, non-timber forest products (NTFPs), and other materials, diversifying revenue streams for nearby residences (Mbow et al. 2014). The soil enhancements advocated through NbS provide long-term benefits in agricultural lands. Improved water retention via enhanced soil structure and higher organic matter content helps lower droughts' effect and strengthen general ecosystem resilience. Not only does this enhanced water availability help nearby natural habitats but also regions under cultivation, therefore aiding more general efforts at biodiversity preservation (Lal 2004). NbS's economic co-benefits generate green employment in sustainable land management and restoration. By offering sustainable livelihoods within local communities, these job possibilities can assist in lowering rural-urban migration—a significant problem in many dryland areas (Nkonya et al. 2016).

The active involvement and participation of local populations greatly determines the effectiveness and sustainability of NbS in dryland environments. This inclusive approach guarantees that actions are socially acceptable, economically feasible, and ecologically sensible. Including indigenous traditional knowledge (ITK) in NbS design and application is vital. Developed over millennia of dryland existence, ITK usually consists of complex awareness of local ecosystems, climatic patterns, and sustainable resource management techniques. Using this information improves the efficacy and cultural fit of NbS treatments (Berkes et al. 2000). This co-creation of knowledge improves communities' adaptive ability, allowing them to react more successfully to evolving environmental situations (Tengö et al. 2014). In NbS implementation, cooperative techniques help local stakeholders feel more responsible, improving the possibility of long-term success. Stronger and context-appropriate solutions can result from participatory planning initiatives involving local officials, community people, and scientific specialists. Early in the process, this multi-stakeholder participation also helps to spot and resolve possible trade-offs or disputes (Reed et al. 2015). Furthermore, inclusive NbS approaches can assist in solving socioeconomic imbalances in local areas. NbS can help guarantee more fair results by ensuring that underprivileged groups—including women and indigenous people—have a say in decision-making procedures. This is especially crucial in dryland areas where a lack of resources could aggravate already existing social conflicts (Djoudi et al. 2016).

3.3 Ecosystem Restoration in Drylands Using NbS

Dryland ecosystems cover around 41% of the Earth's land surface and support over 2 billion people worldwide (Millennium Ecosystem Assessment 2005). Land deterioration, desertification, and the effects of climate change are serious problems for these areas. NbS give local communities and the environment several advantages and a viable way to repair and sustainably manage these delicate ecosystems. NbS call for several ecosystem restoration strategies in drylands, including sustainable land management, water collecting techniques, and reforestation measures (Fig. 3.1).

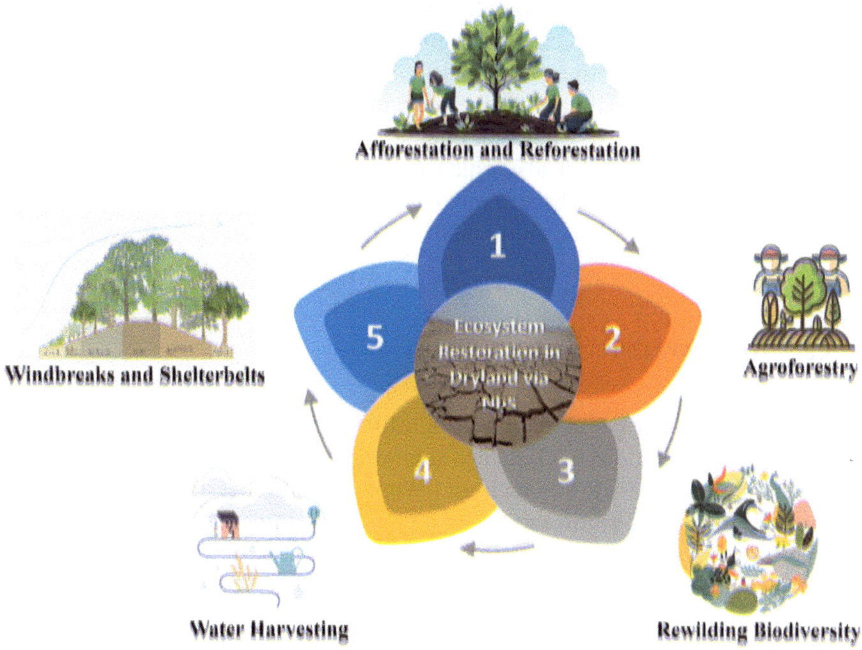

Fig. 3.1 Ecosystem restoration strategies in dryland via NbS

3.3.1 *Afforestation and Reforestation*

Afforestation and reforestation are important measures for increasing the resilience of dryland ecosystems. These approaches not only restore vegetation cover but also provide a variety of ecological and socioeconomic advantages that are critical for maintaining these eco-sensitive habitats. Afforestation is the process of planting new forests in places that did not have any before, whereas reforestation is the process of regenerating or replanting forest areas destroyed or damaged. Both of these processes signify NbS for the reclamation of degraded lands, which can be very beneficial in dryland settings. Afforestation and reforestation efforts in arid and semi-arid regions significantly enhance plant diversity by increasing vegetation cover and improving soil structure by reducing soil erosion and enhancing organic matter (OM) accumulation (Segura et al. 2021). They can neutralize soil pH, further enhancing soil fertility and promoting better water retention and nutrient cycling (Hong et al. 2018). These measures are necessary to stabilize the environment, which controls specific areas' temperature and humidity levels. In places prone to desertification and wind and soil erosion, it also helps prevent sand fixation—a vital concern. By reducing soil salinity and erosion, afforestation also helps to enhance the quality of the soil and promote healthier plant growth. Mixed-species afforestation can enhance biodiversity in dryland ecosystems through higher nitrogen acquisition and retention (Lang et al. 2014). Introducing diverse native species can help restore ecological balance and provide habitat for various wildlife, thereby increasing the resilience of these ecosystems to disturbances such as drought and land degradation (Reisman-Berman et al. 2019). Additionally, afforestation and reforestation are necessary for climate change adaptation in sensitive ecosystems like drylands (Seddon et al. 2020; Chausson et al. 2020). They can significantly enhance carbon sequestration in drylands, which is vital for maintaining ecosystem resilience and contributing to global efforts to reduce atmospheric CO_2 levels in the face of climate variability (Tal and Gordon 2010; Liu et al. 2022).

3.3.2 Agroforestry

The most comprehensive definition of agroforestry is provided by the Food and Agriculture Organization (FAO), which summarizes the definition from The Center for International Forestry Research and World Agroforestry (CIFOR-ICRAF) It defines agroforestry as a term encompassing land-use systems and technologies that intentionally incorporate woody perennials—such as trees, shrubs, palms, and bamboos—alongside crops and/or animals within the same land-management units, organized in a specific spatial arrangement or temporal sequence (Santoro et al. 2020). Agroforestry can be promoted as a sustainable NbS for land management in dryland ecosystems. Agroforestry systems exhibit interactions among their various components, resulting in unique landscapes, diversifying production, and enhancing both economic and environmental advantages for local communities (Paudel et al. 2022). Improving soil fertility and lowering soil erosion helps this combined approach to increase the resilience of the agricultural system in dry environments. Using a range of tree species—both vital for the long-term survival of ecosystems—agroforestry boosts biodiversity and helps regulate groundwater levels (Bari and Schofield 1991; Jose 2012). As a means of adapting to climate change, it provides natural defences against severe weather events, reducing sensitivity to heat and drought (Bayala et al. 2014). Moreover, including nitrogen-fixing trees increases soil fertility, lowering the demand for synthetic fertilizers and enhancing general system production (Sileshi et al. 2011). Agroforestry also significantly improves local standards of life by providing long-term solutions for desert communities and promoting a good cohabitation of environmental preservation and agriculture.

3.3.3 Rewilding

In drylands, rewilding is a method of ecological restoration whereby deteriorated habitats are restored by means of biological processes. The concept of rewilding has transformed and expanded since it first emerged. The approach typically involves the restoration of ecosystems by reintroducing species, serving as a sustainable method to mitigate the decline of biodiversity and its associated ecosystem services

(Nogués-Bravo et al. 2016). Initiatives like rewilding considerably improve the supply of ecosystem services like water retention and soil stability through natural plants and animals. Long understood as a leading cause of land deterioration in these regions is overgrazing. Rotational grazing and restricted cattle densities—two adaptive grazing management techniques NbS advocates—allow vegetation to recover and stop further degradation (Teague et al. 2011). Fernández et al. (2020) advocated that aridity is a crucial variable for determining cattle density, which is more restricted in drylands. The direct impact of water scarcity reduces pasture productivity in drylands. Therefore, rotational grazing and cattle density restriction can be proved effective NbS. Apart from helping many animals, these rebuilt ecosystems give local populations choices for sustainable food and fodder generation. Because they increase livelihoods by creating more employment in conservation and sustainable resource management (Sweeney et al. 2019), rewilding dryland projects are particularly important. Using NbS to establish protected zones in dryland restoration can be a very important tactic for maintaining biodiversity and improving ecosystem services in these locations. These sites are essential for landscape restoration, enhancing habitat connectivity, and safeguarding ecosystem resilience. They help store carbon and act as buffers against the consequences of climate change. Protected regions also help to control water supplies, which is crucial in drylands (Carroll and Ray 2021).

3.3.4 Water Harvesting

Managing limited water supplies in arid and semi-arid areas depends critically on water harvesting in dryland restoration. Collecting and storing runoff or precipitation for later use improves the water supply accessible for human needs and ecosystems. Among the techniques are simple contour bunds and check dams to more complex systems like rooftop collection or subterranean dams. Rainwater harvesting through these methods qualifies as the most significant NbS in drylands (Valdecantos et al. 2014). By holding water, a higher rate of evaporation or runoff in drylands can be checked to significant levels through these NbS measures, which help to recharge groundwater, reduce soil erosion, and encourage the growth of flora (Owuor et al. 2016).

3.3.5 Windbreaks and Shelterbelts

Incorporating windbreaks and shelterbelts is another useful NbS for restoring drylands. Although windbreaks and shelterbelts are used synonymously in agroforestry practices, a significant difference lies between them; as in dryland areas, windbreaks and shelterbelts are deliberate plantings of trees or plants serving as protective barriers to agricultural or even barren lands (Ramanan et al. 2023). These linear vegetation systems are designed to reduce wind erosion, a common issue in arid climates that can harm crops and degrade soil quality (Brandle et al. 2004). These barriers increase more suitable microclimates for agricultural output and the expansion of plants by intercepting and lowering wind speeds (Alemu 2016). By providing habitats and corridors for animals in otherwise divided environments, windbreaks and shelterbelts can also help preserve biodiversity (Harvey et al. 2005).

3.4 Economic Benefits and Challenges of Ecosystem Restoration in Drylands

Ecosystem restoration in drylands through NbS presents multifaceted economic benefits and challenges. NbS aim to restore ecosystems while enhancing livelihoods, yet various economic challenges can hinder their implementation. This section delves into the economic benefits of NbS in drylands and the challenges faced in their implementation, providing insights into practical strategies for sustainable development.

3.4.1 Economic Benefits of NbS in Drylands

3.4.1.1 Enhanced Agricultural Productivity

The integration of NbS, mainly through agroecology and sustainable land-management practices, has substantially improved agricultural productivity across various dryland regions globally (Table 3.1). Studies have indicated that agroforestry systems, which combine trees with crops, can increase crop yields by as much as 50% compared to conventional monoculture systems (Yadav et al. 2021). This boost in productivity is attributed mainly to enhanced soil fertility, improved water retention, and increased biodiversity, which contributes to pest and disease control. Furthermore, crop diversification through NbS not only stabilizes income streams for farmers but also bolsters food security in regions prone to climatic variability.

3.4.1.2 Job Creation and Green Economy

Investment in NbS can act as a catalyst for economic growth by creating numerous job opportunities, particularly in restoration projects, sustainable agriculture, and forest management. A comprehensive analysis suggests that for every $1 million invested in ecosystem restoration, approximately 30 jobs are generated (BenDor et al. 2015). This job creation is especially crucial in rural communities

Table 3.1 Assessment of potential economic benefits from nature-based solutions (NbS) in dryland ecosystem

Continents	Successful example of implementing NbS in drylands	Potential economic benefits availed	References
Africa	In the Sahel region, agroforestry and sustainable land management practices are employed as NbS under the name of "Great Green Wall Initiative" to improve agricultural productivity and resilience to climate variability	This initiative aims to restore 100 million hectares of degraded land, which is expected to generate up to 19 million green jobs and $90 billion in economic benefits through enhanced agricultural yields and ecosystem services	Goffner et al. (2019)
Asia	In tropical drylands of India, several NbS are followed, strongly adhering to the IUCN standards, such as sustainable forest management (SFM), agroforestry, integrated farming systems (IFS), holistic regenerative landscape management, etc.	The application of such NbS ensures varying economic benefits to the local population, ranging up to $4030 per hectare of land	Sharma et al. (2024)
Australia	In Australia, NbS, such as the restoration of wetlands and coastal ecosystems, has been shown to provide significant economic benefits by enhancing fisheries and tourism	Investing in wetland restoration can yield a return of $4 for every $1 spent, primarily through improved water quality and increased fish stocks	Yimer et al. (2024)
Europe	In the arid Mediterranean regions, NbS, such as sustainable land management and reforestation, have been recognized for their economic benefits in combating desertification	Implementation of NbS can reduce costs associated with land degradation, with potential savings of up to €1.5 billion annually through improved agricultural productivity and ecosystem services	Netti et al. (2024)
Latin America	In Brazil, to improve water security and reduce the economic cost of drought in dry and degraded lands, several NbS, such as reforestation and agroecological practices, have been implemented	Implementing NbS has the potential to reduce economic losses by 28% and increase net present value by BRL 144 million and BRL 632 million for water supply and carbon sequestration benefits	Ciasca et al. (2023)

with limited employment opportunities. For example, initiatives under the Bonn Challenge, which aims to restore 150 million hectares of degraded land by 2020, can potentially create over 2 million jobs globally (Verdone and Seidl 2017). Moreover, these jobs tend to be local, enhancing community resilience and fostering a transition toward a green economy that prioritizes sustainability and ecological health.

3.4.1.3 Climate Change Mitigation

NbS significantly contribute to climate change mitigation by enhancing ecosystems' carbon sequestration capabilities. Restoring natural habitats, such as forests, grasslands, and wetlands, can sequester vast amounts of carbon dioxide, offering substantial financial incentives through carbon credits and Payments for Ecosystem Services (PES). A meta-analysis found that NbS could potentially provide climate mitigation services valued at up to $4 trillion annually, with the costs of inaction being far greater (Vicarelli et al. 2024). Including such restoration efforts in national climate action plans can facilitate access to international funding and bolster local economies while addressing the global climate crisis.

3.4.1.4 Improved Water Resource Management

Implementing NbS, such as restoring natural hydrological processes and constructing water harvesting structures, can significantly enhance water availability in arid and semi-arid regions. Research indicates that check dams and contour bunding can improve groundwater recharge by as much as 20%, which is vital for sustaining agricultural activities and ensuring access to clean drinking water (Owuor et al. 2016). Enhanced water management not only supports food production but also reduces the economic risks associated with droughts and water scarcity, thus strengthening the resilience of local communities (Plaza et al. 2018).

3.4.1.5 Biodiversity and Ecosystem Services

The restoration of ecosystems through NbS fosters biodiversity, essential for maintaining ecosystem services that underpin economic activities. A thriving biodiversity can enhance services such as pollination, nutrient cycling, and pest control, which are critical for agricultural productivity. For instance, restored habitats can increase pollination services, resulting in yield increases of up to 20% in agricultural systems compared to degraded areas (Rey Benayas and Bullock 2012). This enhancement of ecosystem services not only boosts agricultural output but also contributes to the resilience of food systems in the face of climate variability, ensuring long-term food security for vulnerable populations (Bommarco et al. 2018).

3.4.2 Economic Challenges of NbS in Drylands

Despite the promising economic benefits, several challenges must be addressed to facilitate the successful implementation of NbS in drylands.

3.4.2.1 Initial Investment and Financial Barriers

One of the foremost challenges in implementing NbS is the initial financial investment required for restoration activities. The upfront costs for land preparation, planting, and maintenance can be prohibitive, particularly for smallholder farmers and local communities lacking access to credit (Berkes et al. 2000). In many cases, traditional financing mechanisms do not adequately support the long-term nature of restoration projects, leading to a gap in funding. Innovative financing solutions, such as blended finance, where public funds are used to attract private investments, and microcredit systems designed specifically for restoration initiatives, could provide the necessary capital and alleviate financial burdens on local stakeholders (Havemann et al. 2022).

3.4.2.2 Short-Term Versus Long-Term Gains

The disparity between short-term costs and long-term benefits presents a significant hurdle for stakeholders considering investment in NbS. Many restoration projects require a considerable period to yield observable economic returns, which can deter participation and support from local communities and investors focused on immediate results (Bestelmeyer et al. 2015). Policymakers and project leaders must effectively communicate the long-term advantages of restoration initiatives, highlighting successful case studies and providing evidence of the socioeconomic benefits derived from past projects. Ensuring community engagement and understanding throughout restoration can help align short-term expectations with long-term goals.

3.4.2.3 Market Access and Value Chains

For NbS to be economically viable, products derived from restored ecosystems must effectively reach markets. However, many rural areas suffer from inadequate infrastructure, poorly developed supply chains, and limited market access, which can restrict the financial benefits of restoration projects (Maestre et al. 2012). Developing robust value chains that connect producers with consumers is crucial for enhancing the economic viability of NbS. This may involve establishing cooperatives or partnerships that enable local communities to collectively market their products, enhancing bargaining power and ensuring fair compensation for ecosystem services.

3.4.2.4 Policy and Governance Challenges

Inconsistent policies and a lack of effective governance frameworks can hinder the scaling of NbS initiatives. The complexity of land tenure systems in dryland regions often complicates restoration efforts, as unclear property rights can discourage investment (Reed et al. 2015). To address these challenges, it is essential to establish clear policies that promote sustainable land-management practices and foster collaboration among diverse stakeholders, including government agencies, local communities, and nongovernmental organizations. Multi-level governance approaches, which involve coordination between local, regional, and national authorities, can facilitate effective implementation and support for NbS, ensuring alignment of efforts across various governance levels.

3.4.2.5 Socioeconomic Inequities

There is a risk that the benefits derived from NbS may not be equitably distributed among community members, particularly if marginalized groups, such as women and Indigenous populations, are excluded from decision-making processes (Djoudi et al. 2016). This inequity can undermine the sustainability of restoration efforts and hinder long-term success. To achieve equitable outcomes, it is crucial to ensure inclusive participation in the planning and implementation of NbS. Engaging local communities in the restoration process enhances social equity and fosters a sense of ownership and responsibility for the restoration outcomes. Incorporating traditional ecological knowledge into project design can also enhance the effectiveness and cultural relevance of NbS initiatives.

Ecosystem restoration in drylands through nature-based solutions offers substantial economic benefits, contributing to sustainable development and community resilience. Integrating NbS can lead to enhanced agricultural productivity, job creation, climate change mitigation, improved water management, and increased biodiversity. However, addressing the associated economic challenges, such as financial barriers, market access, and governance issues, is vital for successfully implementing these solutions. A holistic approach incorporating economic, social, and ecological dimensions will maximize the potential of NbS and enhance the livelihoods of dryland communities, ultimately contributing to global efforts to combat desertification and promote sustainable land use.

3.5 Conclusion

NbS present a promising approach to restoring drylands by enhancing ecosystem integrity, fostering resilience, and providing socioeconomic co-benefits. By employing techniques such as agroforestry, water harvesting, and sustainable grazing, NbS aims to increase plant cover, improve water availability, and restore soil health. These initiatives promote biodiversity conservation and enhance the ecosystems' capacity for carbon sequestration, erosion control, and groundwater replenishment. Crucially, NbS fosters resilience and sustainability, enabling communities and dryland ecosystems to better withstand natural stresses such as heatwaves and droughts. The involvement of local communities in these initiatives is essential, as it promotes social acceptance and cooperation by ensuring that restoration strategies align with indigenous knowledge and are culturally appropriate. This participatory approach not only empowers local stakeholders but also enhances the effectiveness of restoration efforts. However, implementing NbS faces significant challenges, including securing long-term funding, addressing land tenure disputes, and ensuring equitable participation in restoration initiatives. Despite these obstacles, the potential of NbS to restore degraded drylands remains substantial. By enhancing agricultural productivity, generating green investments, and creating job opportunities, NbS offer a cost-effective and sustainable alternative to traditional land-management practices. Moreover, as highlighted by the UN Decade on Ecosystem Restoration, the restoration of drylands through NbS contributes significantly to global efforts to achieve climate resilience and biodiversity conservation. In light of escalating environmental challenges worldwide, the advancement of NbS presents a critical opportunity to cultivate more resilient ecosystems and sustainable livelihoods in arid regions. By prioritizing NbS, we can promote ecological restoration that supports environmental health and enhances the socioeconomic fabric of dryland communities.

References

Alemu MM (2016) Ecological benefits of trees as windbreaks and shelterbelts. Int J Ecosys 6:10–13

Alikhanova S, Bull JW (2023) Review of nature-based solutions in dryland ecosystems: the Aral Sea case study. Environ Manag 72(3):457–472. https://doi.org/10.1007/s00267-023-01822-z

Aronson J, Blignaut JN, Milton SJ, Le Maitre D, Esler KJ, Limouzin A, Fontaine C, de Wit MP, Mugido W, Prinsloo P, van der Elst L, Lederer N (2010) Are socioeconomic benefits of restoration adequately quantified? A meta-analysis of recent papers (2000–2008) in restoration ecology and 12 other scientific journals. Restor Ecol 18(2):143–154. https://doi.org/10.1111/j.1526-100X.2009.00638.x

Bagwan WA (2023) Nature-based solutions (NbS) for dryland agriculture in semi-arid regions of Maharashtra, India: a short review with possible approaches for building climate resilience. In: Chatterjee U, Shaw R, Kumar S, Raj AD, Das S (eds) Climate crisis: adaptive approaches and sustainability. Springer Nature, Berlin, pp 277–291. https://doi.org/10.1007/978-3-031-44397-8_15

Bari MA, Schofield NJ (1991) Effects of agroforestry-pasture associations on groundwater level and salinity. Agrofor Syst 16(1):13–31. https://doi.org/10.1007/BF00053194

Barman A, Dutta S, Bera A, Saha P, Roy J, Roy Choudhury M, Bera M, Das S (2024) Synergizing sustainability: a critical review on harnessing agroforestry for biomass, carbon sequestration, and water-food-energy nexus. Energ Ecol Environ 9:1–35. https://doi.org/10.1007/s40974-024-00336-6

Bayala J, Sanou J, Teklehaimanot Z, Kalinganire A, Ouédraogo SJ (2014) Parklands for buffering climate risk and sustaining agricultural production in the Sahel of West Africa. Curr Opin Environ Sus 6:28–34. https://doi.org/10.1016/j.cosust.2013.10.004

BenDor T, Lester TW, Livengood A, Davis A, Yonavjak L (2015) Estimating the size and impact of the ecological restoration economy. PLoS One 10(6):e0128339. https://doi.org/10.1371/journal.pone.0128339

Berkes F, Colding J, Folke C (2000) Rediscovery of traditional ecological knowledge as adaptive management. Ecol Appl 10(5):1251–1262

Bestelmeyer BT, Okin GS, Duniway MC, Archer SR, Sayre NF, Williamson JC, Herrick JE (2015) Desertification, land use, and the transformation of global drylands. Front Ecol Environ 13(1):28–36. https://doi.org/10.1890/140162
Binyam AY, Desale KA (2015) Rainwater harvesting: an option for dry land agriculture in arid and semi-arid Ethiopia. Int J Water Res Environ Eng 7(2):17–28. https://doi.org/10.5897/IJWREE2014.0539
Bommarco R, Vico G, Hallin S (2018) Exploiting ecosystem services in agriculture for increased food security. Glob Food Sec 17:57–63. https://doi.org/10.1016/j.gfs.2018.04.001
Brandle JR, Hodges L, Zhou XH (2004) Windbreaks in North American agricultural systems. Agrofor Syst 61(1):65–78. https://doi.org/10.1023/B:AGFO.0000028990.31801.62
Carroll C, Ray JC (2021) Maximizing the effectiveness of national commitments to protected area expansion for conserving biodiversity and ecosystem carbon under climate change. Glob Chang Biol 27(15):3395–3414. https://doi.org/10.1111/gcb.15645
Chausson A, Turner B, Seddon D, Chabaneix N, Girardin CAJ, Kapos V, Key I, Roe D, Smith A, Woroniecki S, Seddon N (2020) Mapping the effectiveness of nature-based solutions for climate change adaptation. Glob Chang Biol 26(11):6134–6155. https://doi.org/10.1111/gcb.15310
Chazdon RL (2008) Beyond deforestation: restoring forests and ecosystem services on degraded lands. Science 320(5882):1458–1460. https://doi.org/10.1126/science.1155365
Chenoweth DA, Schlaepfer DR, Chambers JC, Brown JL, Urza AK, Hanberry BB, Board D, Crist M, Bradford JB (2023) Ecologically relevant moisture and temperature metrics for assessing dryland ecosystem dynamics. Ecohydrology 16(3):e2509. https://doi.org/10.1002/eco.2509
Ciasca BS, Klemz C, Raepple J, Kroeger T, Acosta EAP, Cho SJ, Barreto S, Bracale H, Cesário F (2023) Economic cost of drought and potential benefits of investing in nature-based solutions: a case study in São Paulo. Brazil Water 15(3):466. https://doi.org/10.3390/w15030466
Cohen-Shacham E, Walters G, Janzen C, Maginnis S (2016) Nature-based solutions to address global societal challenges. IUCN: Gland, Switzerland 97:2016–2036
Cohen-Shacham E, Andrade A, Dalton J, Dudley N, Jones M, Kumar C, Maginnis S, Maynard S, Nelson CR, Renaud FG, Welling R, Walters G (2019) Core principles for successfully implementing and upscaling nature-based solutions. Environ Sci Pol 98:20–29. https://doi.org/10.1016/j.envsci.2019.04.014
Debele SE, Kumar P, Sahani J, Martí-Cardona B, Mickovski SB, Leo LS, Porcù F, Bertini F, Montesi D, Vojinovic Z, Di Sabatino SD (2019) Nature-based solutions for hydro-meteorological hazards: revised concepts, classification schemes and databases. Environ Res 179(B):108799. https://doi.org/10.1016/j.envres.2019.108799
Djoudi H, Locatelli B, Vaast C, Asher K, Brockhaus M, Basnett Sijapati BB (2016) Beyond dichotomies: gender and intersecting inequalities in climate change studies. Ambio 45(Suppl 3):248–262. https://doi.org/10.1007/s13280-016-0825-2
Faivre N, Fritz M, Freitas T, de Boissezon B, Vandewoestijne S (2017) Nature-based solutions in the EU: innovating with nature to address social, economic and environmental challenges. Environ Res 159:509–518. https://doi.org/10.1016/j.envres.2017.08.032
Fernández PD, Kuemmerle T, Baumann M, Grau HR, Nasca JA, Radrizzani A, Gasparri NI (2020) Understanding the distribution of cattle production systems in the south American Chaco. J Land Use Sci 15(1):52–68. https://doi.org/10.1080/1747423X.2020.1720843
Glendenning CJ, van Ogtrop FF, Mishra AK, Vervoort RW (2012) Balancing watershed and local scale impacts of rain water harvesting in India – a review. Agric Water Manag 107:1–13. https://doi.org/10.1016/j.agwat.2012.01.011
Goffner D, Sinare H, Gordon LJ (2019) The Great Green Wall for the Sahara and the Sahel initiative as an opportunity to enhance resilience in Sahelian landscapes and livelihoods. Reg Environ Chang 19(5):1417–1428. https://doi.org/10.1007/s10113-019-01481-z
Gooden J, Pritzlaff R (2021) Dryland watershed restoration with rock detention structures: a nature-based solution to mitigate drought, erosion, flooding, and atmospheric carbon. Front Environ Sci 9:679189. https://doi.org/10.3389/fenvs.2021.679189

Guo Y, Chen Y (2024) A review of the impact of grazing on grassland ecosystems: research progress and prospects. Adv Res 4(3):455–473. https://doi.org/10.50908/arr.4.3_455

Harvey CA, Villanueva C, Villacís J, Chacón M, Muñoz D, López M, Ibrahim M, Gómez R, Taylor R, Martinez J, Navas A, Saenz J, Sánchez D, Medina A, Vilchez S, Hernández B, Perez A, Ruiz F, López F, Lang I, Sinclair FL (2005) Contribution of live fences to the ecological integrity of agricultural landscapes. Agric Ecosyst Environ 111(1–4):200–230. https://doi.org/10.1016/j.agee.2005.06.011

Havemann T, Negra C, Werneck F (2022) Blended finance for agriculture: exploring the constraints and possibilities of combining financial instruments for sustainable transitions. In: Desa G, Jia X (eds) Social innovation and sustainability transition. Springer Nature, pp 347–358. https://doi.org/10.1007/978-3-031-18560-1_23

Herrick JE, Lessard VC, Spaeth KE, Shaver PL, Dayton RS, Pyke DA, Jolley L, Goebel JJ (2010) National ecosystem assessments supported by scientific and local knowledge. Front Ecol Environ 8(8):403–408. https://doi.org/10.1890/100017

Hong S, Piao S, Chen A, Liu Y, Liu L, Peng S, Sardans J, Sun Y, Peñuelas J, Zeng H (2018) Afforestation neutralizes soil ph. Nat Commun 9(1):520. https://doi.org/10.1038/s41467-018-02970-1

Hooper DU, Chapin IIIFS, Ewel JJ, Hector A, Inchausti P, Lavorel S, Lawton JH, Lodge DM, Loreau M, Naeem S, Schmid B, Setälä H, Symstad AJ, Vandermeer J, Wardle DA (2005) Effects of biodiversity on ecosystem functioning: a consensus of current knowledge. Ecol Monogr 75(1):3–35. https://doi.org/10.1890/04-0922

IUCN (2016) IUCN resolutions, recommendations and other decisions. The World Conservation Congress. International Union for Conservation of Nature and Natural Resources Publications Unit, p 106

IUCN (2020) IUCN global standard for nature-based solutions: a user-friendly framework for the verification, design and scaling up of NbS (1st ed). International Union for Conservation of Nature and Natural Resources, International Union for Conservation of Nature. https://doi.org/10.2305/IUCN.CH.2020.08.en

Jose S (2012) Agroforestry for conserving and enhancing biodiversity. Agrofor Syst 85(1):1–8. https://doi.org/10.1007/s10457-012-9517-5

Kabisch N, Frantzeskaki N, Pauleit S, Naumann S, Davis M, Artmann M, Haase D, Knapp S, Korn H, Stadler J, Zaunberger K, Bonn A (2016) Nature-based solutions to climate change mitigation and adaptation in urban areas: perspectives on indicators, knowledge gaps, barriers, and opportunities for action. Ecol Soc 21(2):39. https://doi.org/10.5751/ES-08373-210239

Keesstra S, Nunes J, Novara A, Finger D, Avelar D, Kalantari Z, Cerdà A (2018) The superior effect of nature based solutions in land management for enhancing ecosystem services. Sci Total Environ 610–611:997–1009. https://doi.org/10.1016/j.scitotenv.2017.08.077

Kimmell LB, Fagan JM, Havrilla CA (2023) Soil restoration increases soil health across global drylands: a meta-analysis. J Appl Ecol 60(9):1939–1951. https://doi.org/10.1111/1365-2664.14459

Lal R (2004) Carbon sequestration in dryland ecosystems. Environ Manag 33(4):528–544. https://doi.org/10.1007/s00267-003-9110-9

Lang AC, von Oheimb GV, Scherer-Lorenzen M, Yang B, Trogisch S, Bruelheide H, Ma K, Härdtle W (2014) Mixed afforestation of young subtropical trees promotes nitrogen acquisition and retention. J Appl Ecol 51(1):224–233. https://doi.org/10.1111/1365-2664.12157

Li C, Fu B, Wang S, Stringer LC, Wang Y, Li Z, Liu Y, Zhou W (2021) Drivers and impacts of changes in China's drylands. Nat Rev Earth Environ 2(12):858–873. https://doi.org/10.1038/s43017-021-00226-z

Liu HY, Jay M, Chen X (2021) The role of nature-based solutions for improving environmental quality, health and well-being. Sustainability 13(19):10950. https://doi.org/10.3390/su131910950

Liu H, Xu C, Allen CD, Hartmann H, Wei X, Yakir D, Wu X, Yu P (2022) Nature-based framework for sustainable afforestation in global drylands under changing climate. Glob Chang Biol 28(7):2202–2220. https://doi.org/10.1111/gcb.16059

Lopez Porras GL, Stringer LC, Quinn CH (2020) Building dryland resilience: three principles to support adaptive water governance. Ecol Econ 177:106770. https://doi.org/10.1016/j.ecolecon.2020.106770

Maes J, Jacobs S (2017) Nature-based solutions for Europe's sustainable development. Conserv Lett 10(1):121–124. https://doi.org/10.1111/conl.12216

Maestre FT, Quero JL, Gotelli NJ, Escudero A, Ochoa V, Delgado-Baquerizo M, García-Gómez M, Bowker MA, Soliveres S, Escolar C, García-Palacios P, Berdugo M, Valencia E, Gozalo B, Gallardo A, Aguilera L, Arredondo T, Blones J, Boeken B et al (2012) Plant species richness and ecosystem multifunctionality in global drylands. Science 335(6065):214–218. https://doi.org/10.1126/science.1215442

Maestre FT, Solé RV, Singh BK (2017) Microbial biotechnology as a tool to restore degraded drylands. Microb Biotechnol 10(5):1250–1253. https://doi.org/10.1111/1751-7915.12832

Maginnis S, Jackson W, Dudley N (2004) Conservation landscapes: whose landscapes? Whose trade-offs? In: McShane TO, Wells MP (eds) Getting biodiversity projects to work: towards more effective conservation and development. Columbia University Press, pp 321–339

Mbow C, Van Noordwijk M, Luedeling E, Neufeldt H, Minang PA, Kowero G (2014) Agroforestry solutions to address food security and climate change challenges in Africa. Curr Opin Environ Sust 6:61–67. https://doi.org/10.1016/j.cosust.2013.10.014

Millennium Ecosystem Assessment (2005) Ecosystems and human well-being: desertification synthesis. World Resources Institute

Nelson DR, Bledsoe BP, Ferreira S, Nibbelink NP (2020) Challenges to realizing the potential of nature-based solutions. Curr Opin Environ Sust 45:49–55. https://doi.org/10.1016/j.cosust.2020.09.001

Nesshöver C, Assmuth T, Irvine KN, Rusch GM, Waylen KA, Delbaere B, Haase D, Jones-Walters L, Keune H, Kovacs E, Krauze K, Külvik M, Rey F, van Dijk J, Vistad OI, Wilkinson ME, Wittmer H (2017) The science, policy and practice of nature-based solutions: an interdisciplinary perspective. Sci Total Environ 579:1215–1227. https://doi.org/10.1016/j.scitotenv.2016.11.106

Netti AM, Abdelwahab OMM, Datola G, Ricci GF, Damiani P, Oppio A, Gentile F (2024) Assessment of nature-based solutions for water resource management in agricultural environments: a stakeholders' perspective in southern Italy. Sci Rep 14(1):24668. https://doi.org/10.1038/s41598-024-76346-5

Nkonya E, Mirzabaev A, Von Braun J (eds) (2016) Economics of land degradation and improvement–a global assessment for sustainable development. Springer Nature

Nogués-Bravo D, Simberloff D, Rahbek C, Sanders NJ (2016) Rewilding is the new Pandora's box in conservation. Curr Biol 26(3):R87–R91. https://doi.org/10.1016/j.cub.2015.12.044

Osaka S, Bellamy R, Castree N (2021) Framing "nature-based" solutions to climate change. WIREs Clim Change 12(5):e729. https://doi.org/10.1002/wcc.729

Owuor SO, Butterbach-Bahl K, Guzha AC, Rufino MC, Pelster DE, Díaz-Pinés E, Breuer L (2016) Groundwater recharge rates and surface runoff response to land use and land cover changes in semi-arid environments. Ecol Process 5(1):1–21. https://doi.org/10.1186/s13717-016-0060-6

Owusu S, Cofie O, Mul M, Barron J (2022) The significance of small reservoirs in sustaining agricultural landscapes in dry areas of West Africa: a review. Water 14(9):1440. https://doi.org/10.3390/w14091440

Paudel S, Baral H, Rojario A, Bhatta KP, Artati Y (2022) Agroforestry: opportunities and challenges in Timor-Leste. Forests 13(1):41. https://doi.org/10.3390/f13010041

Plaza C, Gascó G, Méndez AM, Zaccone C, Maestre FT (2018) Soil organic matter in dryland ecosystems. In: Garcia C, Nannipieri P, Hernandez T (eds) . Academic Press, The future of soil carbon, pp 39–70. https://doi.org/10.1016/B978-0-12-811687-6.00002-X

Powell B, Sandbrook C, Sunderland T, Tu TN (2015) The historical, environmental and socio-economic context of forests and tree-based systems for food security and nutrition. Forests Food 73

Raghavendra MP, Sharma MP, Ramesh A, Richa A, Billore SD, Verma RK (2020) Soil health indicators: methods and applications. In: Rakshit A, Ghosh S, Chakraborty S, Philip V, Datta

A (eds) Soil analysis: recent trends and applications. Springer Singapore, pp 221–253. https://doi.org/10.1007/978-981-15-2039-6_13
Ramanan S, Manojkumar S, Arunachalam A, Handa AK, Priyanka V (2023) Conceptual analysis of "windbreaks" and "shelterbelts" in agroforestry. Ind J Agrofor 24(2):39–45
Raymond CM, Frantzeskaki N, Kabisch N, Berry P, Breil M, Nita MR, Geneletti D, Calfapietra C (2017) A framework for assessing and implementing the co-benefits of nature-based solutions in urban areas. Environ Sci Pol 77:15–24. https://doi.org/10.1016/j.envsci.2017.07.008
Reed MS, Stringer LC, Dougill AJ, Perkins JS, Atlhopheng JR, Mulale K, Favretto N (2015) Reorienting land degradation towards sustainable land management: linking sustainable livelihoods with ecosystem services in rangeland systems. J Environ Manag 151:472–485. https://doi.org/10.1016/j.jenvman.2014.11.010
Reisman-Berman O, Keasar T, Tel-Zur N (2019) Native and non-native species for dryland afforestation: bridging ecosystem integrity and livelihood support. Ann For Sci 76(114). https://doi.org/10.1007/s13595-019-0903-2
Rey Benayas JM, Bullock JM (2012) Restoration of biodiversity and ecosystem services on agricultural land. Ecosystems 15(6):883–899. https://doi.org/10.1007/s10021-012-9552-0
Reynolds JF, Smith DMS, Lambin EF, Turner BL, Mortimore M, Batterbury SPJ, Downing TE, Dowlatabadi H, Fernández RJ, Herrick JE, Huber-Sannwald E, Jiang H, Leemans R, Lynam T, Maestre FT, Ayarza M, Walker B (2007) Global desertification: building a science for dryland development. Science 316(5826):847–851. https://doi.org/10.1126/science.1131634
Santoro A, Venturi M, Bertani R, Agnoletti M (2020) A review of the role of forests and agroforestry systems in the FAO globally important agricultural heritage systems (GIAHS) programme. Forests 11(8):860. https://doi.org/10.3390/f11080860
Seddon N (2022) Harnessing the potential of nature-based solutions for mitigating and adapting to climate change. Science 376(6600):1410–1416. https://doi.org/10.1126/science.abn9668
Seddon N, Chausson A, Berry P, Girardin CAJ, Smith A, Turner B (2020) Understanding the value and limits of nature-based solutions to climate change and other global challenges. Philos Trans R Soc London, Ser B 375(1794):20190120. https://doi.org/10.1098/rstb.2019.0120
Seddon N, Smith A, Smith P, Key I, Chausson A, Girardin C, House J, Srivastava S, Turner B (2021) Getting the message right on nature-based solutions to climate change. Glob Chang Biol 27(8):1518–1546. https://doi.org/10.1111/gcb.15513
Segura C, Jiménez MN, Fernández-Ondoño E, Navarro FB (2021) Effects of afforestation on plant diversity and soil quality in semiarid se Spain. Forests 12(12):1730. https://doi.org/10.3390/f12121730
Sharma M, Ashraf J, Mehta D, Pandey R (2024) Assessment of nature-based interventions adherence with IUCN global standards and an analysis of flow of associated ecosystem services in tropical drylands of India. Ecol Indic 159:111717. https://doi.org/10.1016/j.ecolind.2024.111717
Sileshi GW, Akinnifesi FK, Ajayi OC, Muys B (2011) Integration of legume trees in maize-based cropping systems improves rain use efficiency and yield stability under rain-fed agriculture. Agric Water Manag 98(9):1364–1372. https://doi.org/10.1016/j.agwat.2011.04.002
Sweeney OF, Turnbull J, Jones M, Letnic M, Newsome TM, Sharp A (2019) An Australian perspective on rewilding. Conserv Biol 33(4):812–820. https://doi.org/10.1111/cobi.13280
Tal A, Gordon J (2010) Carbon cautious: Israel's afforestation experience and approach to sequestration. Small-Scale For 9(4):409–428. https://doi.org/10.1007/s11842-010-9125-z
Teague WR, Dowhower SL, Baker SA, Haile N, DeLaune PB, Conover DM (2011) Grazing management impacts on vegetation, soil biota and soil chemical, physical and hydrological properties in tall grass prairie. Agric Ecosyst Environ 141(3–4):310–322. https://doi.org/10.1016/j.agee.2011.03.009
Tengö M, Brondizio ES, Elmqvist T, Malmer P, Spierenburg M (2014) Connecting diverse knowledge systems for enhanced ecosystem governance: the multiple evidence base approach. Ambio 43(5):579–591. https://doi.org/10.1007/s13280-014-0501-3

UNEP (2022) Draft resolution on nature-based solutions for supporting sustainable development [Technical report]. United Nations environment assembly of the United Nations Environment Programme

Valdecantos A, Fuentes D, Smanis A, Llovet J, Morcillo L, Bautista S (2014) Effectiveness of low-cost planting techniques for improving water availability to Olea europaea seedlings in degraded drylands. Restor Ecol 22(3):327–335. https://doi.org/10.1111/rec.12076

Verdone M, Seidl A (2017) Time, space, place, and the Bonn challenge global forest restoration target. Restor Ecol 25(6):903–911. https://doi.org/10.1111/rec.12512

Vicarelli M, Sudmeier-Rieux K, Alsadadi A, Shrestha A, Schütze S, Kang MM, Leue M, Wasielewski D, Mysiak J (2024) On the cost-effectiveness of nature-based solutions for reducing disaster risk. Sci Total Environ 947:174524. https://doi.org/10.1016/j.scitotenv.2024.174524

Vohland K, Barry B (2009) A review of in situ rainwater harvesting (RWH) practices modifying landscape functions in African drylands. Agric Ecosyst Environ 131(3–4):119–127. https://doi.org/10.1016/j.agee.2009.01.010

Wale HA, Dejenie T (2013) Dryland ecosystems: their features, constraints, potentials and managements. Res J Agric Environ Manag 2(10):277–288

Yadav SK, Banerjee A, Jhariya MK, Raj A, Khan N, Meena RS, Kumar S (2021) Agroecology towards environmental sustainability. In: Jhariya MK, Banerjee A, Meena RS, Kumar S, Raj A (eds) Sustainable intensification for agroecosystem services and management. Springer Singapore, pp 323–352. https://doi.org/10.1007/978-981-16-3207-5_10

Yan N, Marschner P, Cao W, Zuo C, Qin W (2015) Influence of salinity and water content on soil microorganisms. Int Soil Water Conserv Res 3(4):316–323. https://doi.org/10.1016/j.iswcr.2015.11.003

Yimer EA, De Trift L, Lobkowicz I, Villani L, Nossent J, van Griensven A (2024) The underexposed nature-based solutions: a critical state-of-art review on drought mitigation. J Environ Manag 352:119903. https://doi.org/10.1016/j.jenvman.2023.119903

Zhan W, Lian X, Liu J, Gentine P (2022) Inappropriateness of space-for-time and variability-for-time approaches to infer future dryland productivity changes. Front Environ Sci 10. https://doi.org/10.3389/fenvs.2022.1010269

Chapter 4
Why Less Visual and Undervalued Lands Need Urgent Restoration: Perspectives from India's Semi-arid Rangelands

Dada R. Dadas, Faraz Rupani, and Dipak Zade

Abstract Grasslands provide various ecosystem services for nature and humans to sustain livelihoods in the mountain, remote, and resource-poor regions. Grasslands are home to ecological diversity, pastoral groups, and small livestock holders. Despite their significant contributions, rangeland is a neglected landscape, and India is no exception. Due to the negligence of their contribution, rangelands are often termed wastelands in colonial documents and the same prejudices are percolated in the policy documents. With effect, the significance of grasslands has been poorly understood and acknowledged. With this prejudice, grasslands are converted into agricultural and industrial lands and massive plantations, affecting the generational access to grassland resources. This indicates that the valuation of grassland is done considering the mainstream perspective, wherein its real significance is neglected in land use planning. In this context, an attempt is made to explore the key question: why do we prioritise restoring landscapes that are visually striking or have direct monetary value while ignoring those that quietly sustain ecosystems and livelihoods? Rangelands, often dismissed as wastelands, hold immense ecological and economic significance. Yet, commercial interests increasingly threaten these commons, despite their vital role in supporting biodiversity and rural communities. This chapter examines the ongoing degradation of rangelands in India and highlights how an ecosystem-based adaptation (EbA) approach can drive meaningful restoration. By integrating ecological health with sustainable livelihoods, EbA offers a pathway to revitalise these landscapes and enhance community resilience in the face of climate change.

D. R. Dadas (✉) · F. Rupani · D. Zade
WOTR Centre for Resilience Studies (W-CReS), Watershed Organisation Trust (WOTR), Pune, Maharashtra, India
e-mail: dada.dadas@wotr.org.in; faraz.rupani@wotr.org.in; dipak.zade@wotr.org.in

Y. D. I. Khan et al. (eds.), *Land Restoration Through Ecosystem-Based Approach*, Disaster Resilience and Green Growth, https://doi.org/10.1007/978-981-95-0389-6_4

Keywords Grassland · Ecosystem-based adaptation · Pastoralist communities · Rangelands · Pastoralism

4.1 Introduction

Land is a crucial natural resource; rangelands cover 80 million square kilometers (UNCCD 2024). These rangelands include grasslands, savannahs, wetlands, and tundra , and are unsuitable for sustained cultivation and provide forage, water, and grazing cover for animals. Rangelands across Central, South, and Southwest Asia and Africa have been inhabited by traditional pastoral and agro-pastoral communities (Goldstein and Beall 1990). A vast majority, around 78%, of the rangelands typically lie in the drylands and are extremely important resources for the habitats of wild flora, fauna, and domestic livestock and have played an essential role in providing livelihoods and food security for millennia (Emanuelsson 2009) especially for resource-poor and smallholder people and pastoralists contributing towards income generation for the majority of 1.2 billion people who live under the poverty threshold in the developing countries (Briske 2017; FAO 2023).

The establishment of grasslands and grazing animals as major life forms extends well back into geological history, where fossil records of grasses have been known from the tertiary and true grasses that were well developed at least as long as 6–12 million years ago (Van Dyne et al. 1978). Grasslands were important to humans before they ever domesticated plants and animals, and a hunting-gathering culture existed and was sustained in the grassland ecosystem, an important segment of the traditional society. Soil biodiversity plays a key role in the rangeland; soil biodiversity roughly contributes between $1.5 and 13 trillion annually to ecosystem services (Van der Putten et al. 2004; European Soil Bureau Network 2005). Apart from their economic significance, these rangelands hold immense importance for the ecosystem by effectively mitigating the impact of greenhouse gases (GHGs) as they are more reliable in sinking carbon than forests (Dass et al. 2018), and also have a cultural significance for the indigenous communities, making them a key contributor to achieving Sustainable Development Goals: No Poverty (SDG1), Zero Hunger (SDG2), Climate Action (SDG13), and Life on Land (SDG15).

Internationally, there has been growing recognition of the importance of improving rangeland ecosystems, which have historically been under-prioritized. Projects like "*Healthy Ecosystems for Rangeland Development (HERD)*: Sustainable Rangeland Management Strategies and Practices" focus on revitalising the rangelands of Egypt and Jordan through sustainable management practices. HERD aims to restore these pastoral landscapes by enhancing ecosystem health and protecting the unique biodiversity of these regions, thereby supporting both environmental and community resilience. Similarly, in Kenya's arid and semi-arid lands (ASAL), the "*TWENDE: Towards Ending Drought Emergencies*" project is actively building resilience among livestock and other land use sectors. TWENDE promotes ecosystem-based adaptation (EbA) by restoring degraded rangelands and strengthening governance structures through capacity building and institutional support. This approach aims to enhance the effective management of rangeland ecosystems, ultimately improving the livelihoods of communities vulnerable to climate impacts and drought.

Rangelands in India occupy around 121 million hectares, from desert to alpine meadows in the Himalayas and often form a mosaic of diversified uses and ecological types. These rangelands play a key role in several livelihood options, one of which is primarily livestock rearing, and these rangelands are a key source for grazing. India's 40% of land surface area is used for grazing, which includes 17% from grasslands and 23% from forests. Out of the 121 million hectares, approximately 100 million hectares are considered to be underutilized including degraded forest lands, unsuitable lands for crop production, wastelands, and ravines (Malaviya et al. 2018).

Rangelands in the dryland areas are dispersed and are highly varied in nature. The livestock sector is an important sector in India with 20% of the world's livestock population contributing 4% to India's GDP and 26% to the agricultural GDP. The dependency of the majority of the semi-arid livestock population in India still depends upon grazing in the rangelands/pastures, desert areas, and rainfed farming systems (Heinrich Boll Foundation 2021), indicating the reliance of the sector on these undervalued lands. Besides livelihoods, rangeland also provides biodiversity values such as water run-off, groundwater recharge, soil moisture, vegetation cover, and other cultural services (Scanlon et al. 2005). Despite its enormous benefit, the global importance of the rangelands has often been favored in favor of forestry-based interventions by converting natural grasslands to plantation forest or any other use (Lahiri et al. 2023; Pandey 2019).

The overlook of policy has led to continuous land degradation resulting from unsustainable practice, resulting in depreciation of its value and quantum (Onyango et al. 2022). In order to highlight the issues related to rangelands, the United Nations declared 2026 as the International Year of Rangelands and Pastoralists (IYRP) with an objective to increase awareness and to promote investments in the restoration and sustainable management of the rangelands and to recognise and support the pastoralist communities and its contribution in the sustainable development process. For the pastoral communities, the governance system of land and natural resources is relatively loose, with collectivity, flexibility, adaptability, and multi-user uses being its core characteristics (Davies et al. 2016; Robinson and Flintan 2022; Flintan 2012). The increased pressures on the pastoralist lands due to competition for their use and access (Robinson and Flintan 2022; Flintan et al. 2011) are increasing their degradation at an alarming rate. UNCCD defines land degradation as *"the loss in arid, semi-arid and dry sub-humid areas of the biological or economic productivity and complexity of rainfed cropland, irrigated cropland or range, pasture, forest and woodlands resulting from land uses or from a process or combination of processes, including those arising from human activities and habitation patterns"*.

The objective of this chapter is to highlight the importance of rangeland for the semi-arid landscape of India, how sustainable measures like ecosystem-based adaptation (EbA) can help to reduce its degradation, and what more collective actions can be taken towards managing these rangelands to create sustainable ecosystems leading to income generation in the region.

4.2 Literature Review

4.2.1 Rangeland of India

The UNCCD defines rangelands as, *"natural or semi-natural ecosystems grazed by livestock and/or wild animals. Their vegetative cover comprises grasses, forbs, bushes, and shrubs, and may include open forests and agroforestry systems"*. Rangelands are made of seven biomes: 35% deserts and xeric shrublands, 26% tropical and subtropical grasslands, savannas, and shrublands, 15% tundra, 13% temperate grasslands, savannas, and shrublands, 6% montane grasslands and shrublands, 4% Mediterranean forests, and 1% flooded grasslands and savannas.

In India, rangelands are found in four ecoregions (Fig. 4.1) and have varied vegetation in each region spread across 121 million hectares of land with different vegetation types hosting wide varieties of flora and tribes who have been preserving these rangelands using their traditional knowledge (Table 4.1).

Grasslands in India occupy 24% of the land area (Rawat and Adhikari 2015), with its area declining by 31.5% from 18 million hectares to 12 million hectares from 2005 to 2015 (Pandey 2019) and are often mistaken as newly formed ecosystems which need no restoration efforts (Buisson et al. 2022). The total area available for grazing in the country is in the range of about 40% of the land area, and in states like Himachal Pradesh, Uttaranchal, Jammu & Kashmir, Meghalaya, Nagaland, and Arunachal Pradesh, over 70 per cent of the land area is utilized as grazing ground. Though the capacity of pasture lands is for grazing only a 31 million livestock population, on average, around 100 million grazes in the same area. This pressure has grown and is growing steadily—over 2% per annum—for the last two decades (Hazra and Saha 2015).

The Planning Commission on Grasslands and Deserts had set-up a task force which observed that the grasslands ecosystems were the most threatened in India and are still considered classified as "wastelands" and are aimed to make them use for other purposes rather than in its restoration (Lahiri et al. 2023) primarily through conversion to agriculture (Kher and Dutta 2021), tree-based plantation projects, and other mega projects (Arasumani et al. 2018).

India's past during colonial rule was in a state of decrepitude. The colonial rulers took advantage of land revenue (Guha1982), which led to the classification of lands into three parts: agricultural land which were taxed, forest lands used for revenue, and other lands were termed as waste as they did not give any direct benefit to the government (Balooni and Singh 2003). Grasslands fell into all the three categories (Rangarajan and Sivaramakrishnan 2014) and with the early plantation scheme by the colonial government led to conversion of treed savannas into closed canopy

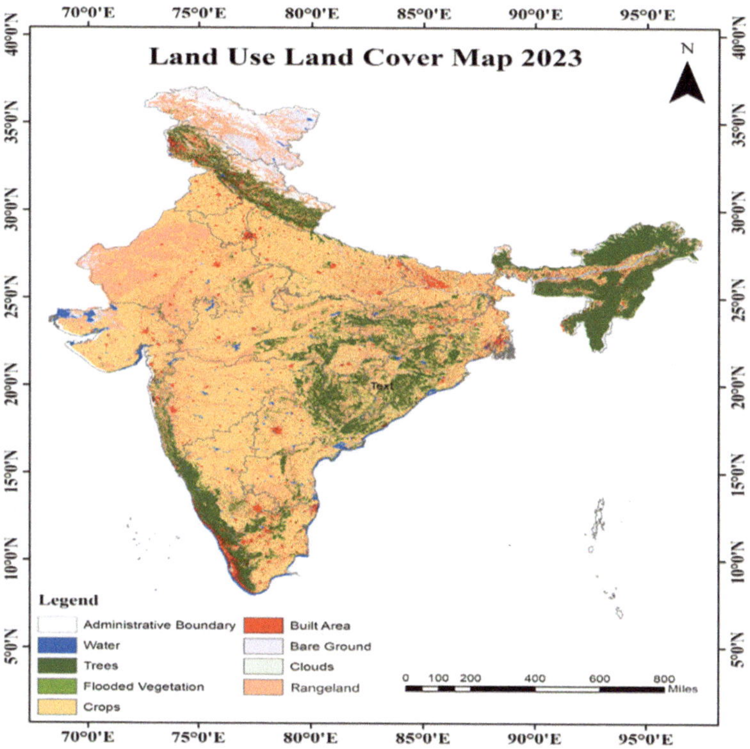

Fig. 4.1 Rangelands in India developed using data referenced from Karra et al. 2021 for the year 2023. (Map prepared by: Antara Gandi and Sowkhya Badatala). (For definitions of terms used in the map, refer to: ArcGIS Definitions)

forests (Agarwala et al. 2019) and those in the semi-arid or arid regions where growth and agricultural potential were low were considered as wastelands (Dove 2004). The deforestation rate was relatively greater under British rule (1880–1950) and early decades after independence, and then decreased after the 1980s due to government policies to protect the forests. In contrast to forests, cropland area increased from 92 million ha to 140.1 million ha during 1880–2010, experiencing a significant loss of grasslands/shrublands. A total of 26 million ha of forest areas (from 89 million ha in 1880 to 63 million ha in 2010) and 20 million ha of grasslands/shrublands (from 45 million ha to 25 million ha) decreased.

The colonial misconception regarding the grasslands as degraded forests (Kumar et al. 2020) continued postindependence. The National Forest Policy of 1952 was set up with a goal of making 33% of the country's land into forest cover (Joshi et al. 2011). The idea of using these lands to improve the condition of food insecurity gained pace immediately after India's independence (Watve et al. 2021). However, postcolonial policies aimed at reclaiming these lands to mitigate climate change, food insecurity, and ecological degradation led to the acceleration of the transformation of grasslands to annual crops and closed-canopy forests, but these targets

Table 4.1 Rangeland vegetation, tribes, and flora across Indian regions

Regions	Type of vegetation	Tribes	Flora
Semi Arid and Western Arid regions			
Rajasthan (Thar Desert) and surrounding areas	Grasses and shrubs	Bishnoi, Bhil, Meena, Garasia, Rabari	Grasses: *Cenchrus ciliaris, Dichanthium annulatum* Shrubs: *Capparis decidua* (ker), *Ziziphus mauritiana* (ber) Trees: *Prosopis cineraria* (khejri)
Gujarat (Banni Grasslands)	Grasses and shrubs	Maldhari, Ahir, Rabari	Grasses: *Dichanthium annulatum, Cenchrus biflorus* Shrubs: *Suaeda* spp., *Prosopis juliflora* Trees: *Salvadora persica*
Central and Southern Regions			
Maharashtra	Scrub and grasslands	Dangar, Maldhari, Mahadev Koli	Grasses: *Cymbopogon* spp., *Chloris montana*
Karnataka	Dwarf trees and grassland	Soliga, Hakki Pikki, Lambani	
Andhra Pradesh	Scrub and grasslands	Yanadi, Lambadi	
Madhya Pradesh	Savanna-type vegetation	Gond, Bhil, Baiga	Grasses: *Heteropogon contortus, Themeda triandra* Trees/shrubs: *Butea monosperma* (flame of the forest), *Acacia catechu*
Himalayan and Alpine Regions			
Jammu & Kashmir	Meadows	Gujjar, Bakarwal	Grasses: *Festuca* spp., *Poa alpina* Shrubs: *Juniperus* spp. Herbaceous plants: *Gentiana* spp.
Himachal Pradesh		Gaddi, Kinnaura	
Uttarakhand		Bhotiya, Jaunsari	
Eastern and North-Eastern India			
Assam	Open grasslands in the foothills	Bodo, Karbi, Mising	Grasses: Saccharum spontaneum (kans grass), Imperata cylindrica (lalang grass)
Arunachal Pradesh		Apatani, Nyishi, Adi	

were never realistic (Rana and Varshney 2022). Large-scale investments in afforestation programs were announced in the 1970s and 1980s (Saigal 2011), targeting grassland sites for tree planting can further aggravate the problems instead of it being solved (Dudley et al. 2020).

4.2.2 Degradation of Rangelands

Rangeland biomes contribute significantly towards the national economies as it covers 51% of Earth's land (Asner et al. 2004) and its degradation has a direct impact on the economy due to reduced biological and vegetation composition like reduced biomass, productivity of crops, less plant cover, species shift, biodiversity loss, and soil erosion (Zerga 2015). The degradation of these rangelands happens from natural and human factors, leading to a serious threat to the grazing lands (Abate and Angassa 2016). Human activities contribute more than 70% towards land, water, and soil resource degradation, whereas increasing climatic change leads to alteration of vegetation, soil, and water dynamics projected to make developing nation rangelands vulnerable to extreme weathers (IPCC 2019). The combined effect of human and climate on land degradation has led to lower pasture production (Daba and Mammo 2024) due to reduced grazing opportunities.

4.2.2.1 Conservation Politics and Perceived Rangeland Degradation

Rangelands are common property resources and the literature available on the rangeland, pasture, and pastoralism show that, in the recent two decades, many profound changes have occurred in the pastoral livelihood system (Miller 1997) and are often claimed to be overused by the pastoral communities (Dregne et al. 1991; Harris 2010; Mishra et al. 2001). It is assumed that grazing hurts the natural ecosystem, and its removal could result in better rangelands (Manier et al. 2006). In contrast to this assumption, grazing has had very minimal effects on plant species richness for a longer duration (Manier et al. 2006; Courtois et al. 2004; Loeser et. 2007) and pastoral communities have always sustainably managed their livestock and rangelands using their traditional ecological knowledge (TEK) and also through their local institutions (Bilal Butt 2011; Jun Li et al. 2007; Ostrom 1990; Adriansen 2008; Bhasin 2011; Mapinduzi et al. 2003; Fernandez-Gimenez 2000). Pastoralists are nowadays perceived as a burden rather than as a support for larger societies (Leloup 2006). The degradation of the rangelands has badly affected the pastoral production system and has negatively impacted the sustainability of the pastoral livelihood system (Miller 1997).

4.2.2.2 Anthropogenic and Climatic Pressures: Encroachment, Conflict, and Degradation of Rangelands

Rangelands in India and globally have been significantly impacted by the combined pressures of climate change, agricultural expansion, and tourism. The conversion of rangelands into cropland, particularly between 1880 and 2010, has reduced the area under grazing lands, creating competition over land resources and conflicts between pastoralists, crop farmers, and conservation institutions (Tamou 2017). Encroachment of arable farming has reduced grazing routes and watering points and disrupted the pastoral systems (Reid et al. 2000). Moreover, protected areas and wildlife sanctuaries have further restricted access to grazing resources and limited the movement of livestock.

Climate change further exacerbates these issues by altering monsoon patterns and increasing aridity, particularly in semi-arid and arid regions (Kassahun et al. 2008). This reduces soil moisture and nutrient availability, directly hindering vegetation growth and diversity (Hou et al. 2022; Zhang et al. 2023). Population growth and rising livestock numbers have added further strain, contributing to the degradation of ecosystems, including grasslands, forests, and water resources. These pressures disrupt ecological balance and threaten cultural richness, as rangelands host significant biodiversity, heritage sites, and indigenous languages (Karl 2011).

4.2.3 Traditional Knowledge and Adaptive Practices

Pastoralists have always spearheaded the management of rangelands with their wealth of knowledge obtained through years of experience on grazing lands and herding practices (Oba and Kotile 2001; Mapinduzi et al. 2003). Indigenous knowledge is an interrelationship of complex knowledge on human adaptive capacities, environmental changes, land use patterns, and decision-making systems of local capacities, which has been put to continuous use by these communities (Grice and Hodgkinson 2002). This knowledge can be used in synchrony with technical knowledge for management (Khwarae 2006) to allow communities to realise their capacity to monitor and respond to land degradation (Stringer and Reed 2007) and to create a deeper increase in understanding of environment and its changes (Dahlberg 2000). Any attempt to alter or improve the rangeland resource requires a deeper understanding of indigenous and local knowledge and perceptions of local communities (Solomon 2007). One such traditional practice of reserved land known as enclosures has been used by the pastoral communities in Tanzania to provide feed to the livestock in the dry season and promotes conservation of the environment (Mwilawa et al. 2008) through early recovery of vegetation (Angassa et al. 2010). Pastoral nomadism based on human and animal mobility is also an ecological response mechanism to the harsh conditions of an arid environment. Through seasonal and cyclical movement of grazing animals over great distances, these systems are capable of optimizing the use of limited and traditionally available food in the more arid areas (Koocheki 1994). When local resources are depleted, mobility enables herders to move on to use yet unexploited pastures and water sources of other localities (Koocheki and Gliessman 2005), increasing the pressures on those commons.

4.2.4 Rangeland Policies and Planning

The afforestation and protection of forests was initiated in the latter half of the twentieth century post-introduction of government policies for protecting forests across India, significantly impacting the grassland. Reconnaissance survey of the grasslands of India conducted during 1954 to 1962 by the Indian Council of Agricultural Research (ICAR) revealed that there are five major ecosystems across the country based on vegetation, which are governed by topography, climatic factors, elevation, latitude, and seasonal patterns of soil moisture, which are "*Sehima—Dichanthium grasslands, Dichanthium—Cenchrus—Lasiurus grasslands, Phragmites—Saccharum—Imperata grasslands, Themeda -Arundinella grasslands, and Temperate—Alpine grasslands*" (Roy et al. 2013).

In 1972, the Wildlife Protection Act was passed to establish legal protection for wildlife and their habitats, including grasslands and desert ecosystems. The National Commission on Agriculture (NCA) was later established in 1976 with further advocacy for stricter control and regulation of grazing to prevent overgrazing and degradation of grasslands through rotational grazing and improved management practices. In 1994, the Draft Grazing and Livestock Management Policy emphasized on guidelines for sustainable grazing management to balance livestock needs with the health of grassland ecosystems. The task force on Grasslands and Deserts in 2006–2007 formulated strategies to conserve grassland and desert ecosystems by balancing the needs of the pastoral communities. The task force emphasised restoration, protection of traditional grazing areas, and promoting sustainable livelihoods for communities who are dependent on these ecosystems. In 2014, the National Livestock Mission aimed at sustainable livestock management and fodder production measures to convert wasteland into productive pastureland.

India's commitment to Land Degradation Neutrality (LDN) targets restoring over 26 million hectares of degraded land to create a carbon sink of 2.5–3 billion tonnes. States like Gujarat aim to restore 20,000 hectares of land and remove Prosopis Juliflora, an invasive tree in the grassland of Kutch. The Maharashtra government plans to restore 33,200 ha of grassland under the "Forest Meadow and Fodder Area Development Program."

4.2.5 Ecosystem-Based Adaptation (EbA) in Rangeland Management

The United Nations Convention on Biological Diversity defines EbA as *"the use of biodiversity and ecosystem services as part of an overall adaptation strategy to help people adapt to the adverse effects of climate change"* (CBD 2009). It is part of the Nature-based solution (NbS) umbrella. The ecosystem approach primarily focuses on having an integrated approach towards managing key resources such as land, water, and living resources to promote sustainable use and conservation (Ariza et al. 2020), which are critical ways to culminate resilience and reduce the impact of climate change on the communities (Kupika et al. 2019; Nanfuka et al. 2020).

EbA approach's emphasis on sustainable management, as per scholars, can lead to improved social, economic, and environmental benefits for the entire ecosystem at varying scales (Munroe et al. 2012; Doswald et al. 2014; Woroniecki 2020). The management of resources is not exclusive to a gender; the role of gender participation of both men and women has been given increased attention in the EbA framework especially in the marginalized communities which was also noted by Woroniecki (2020). Although it has been observed that the participation of youth and children towards decision-making of natural resources will be limited (Tanyanyiwa and Chikwanha 2011). However, the process of the indigenous knowledge system is transferable as per study by Mawere (2013). It needs to engage the local interest of the people to achieve the objective of reducing degradation of natural resources (Boissie're et al. 2009; Conrad and Hilchey 2011).

It is quite evident that rangelands not only have direct economic benefits but also have a lot of indirect impact through ecosystem services, which are vital for creating a sustainable environment. However, various issues like recurrent climatic shocks, land-related conflicts, and infrastructure development badly affect rangeland and pastoralists. While the climate change impact is being released globally, the future trajectory of its severity will depend on the current action taken. Two such cases in the semi-arid region of Maharashtra light the need to take up sustainable measures that incorporate EbA principles.

4.3 Restoration of Rangelands Through the EbA Approach

There is evidence that restoring rangelands in the semi-arid regions of the world using NbS has led to higher economic returns, though it might show low or neutral short term returns but their long term are substantially high with studies showing social return of 5 dollars for every dollar invested in restoration of land within the span of 6 years (Nkonya et al. 2016; Reichhuber et al. 2019) (Boxs 4.1 and 4.2).

Box 4.1: Restoring Land's Resilience Through Ecosystem-Based Adaptation in Akole

Akole, a tribal-dominated block in the semi-arid region of Ahmednagar district, Maharashtra, is predominantly agricultural, with livestock as a critical secondary source of livelihood. Akole is a heavy rainfall zone, and the region faces challenges such as high-water runoff, which leads to soil erosion and diminished land productivity. Nature-based Solutions (NbS) were introduced to address these issues, offering sustainable and eco-friendly approaches to natural resource management. One key intervention involved constructing continuous contour trenches (CCT) across 11 hectares of land in Purushwadi, Khadki, Ghoti, and Pimpri villages. These trenches enhance water absorption, minimise runoff, and significantly reduce soil erosion, improving the land's health.

Recognising the importance of livestock to the community's economy, efforts were also made to expand grazing areas. Within the CCTs, two grass varieties, Adinath and Stylosanthes, which are native species, were planted to enrich pastures, ensuring better feed availability for livestock. This intervention has not only improved the health of the animals but also boosted dairy

(continued)

Box 4.1 **(continued)**

output, providing a positive ripple effect on the livelihoods of local communities. By integrating sustainable water and land management with livestock support, these initiatives have contributed to a more resilient agricultural system in Akole, creating long-term benefits for the environment and the people who depend on it.

Box 4.2: Community-Led Grassland Restoration and Livelihood Enhancement in a Village in Dhule

Lamkani village is located in the Dhule district of Maharashtra. The the first half of the twentieth century, the village was known for its grassland and good agricultural production. But due to a series of drought years especially, 1970s drought, amplified climate vulnerabilities of the village Lamkani. To combat the drought situation, WOTR donated around Rs. 1.5 million ($34,482.76 approx.) for watershed development to the Joint Forest Management Committee of the village and additionally, Rs. 65,000 ($1494.25 approx.) were contributed by the villagers of Lamkani to initiate CCT digging. To culminate adversities into opportunities, it took around 10 years. With long interventions, much of the ecosystem services, biodiversity, and livelihood opportunities improved for villagers.

With an improved grassland ecosystem, various animal and bird species were attracted, indicating the improved ecosystem of the village. The grassland restoration efforts were achieved using traditional approaches such as rotational grazing. This helped land to take a space to recover and rejuvenate. Using a rotational grazing approach, 50-hectare patches of the grassland were conserved at a time. Grazing was not allowed on the patch reserved for conservation. With effect, palatable species of grass were recovered and rainwater started percolating. This led to good quality of fodder for the livestock which further boosted dairy production of the region since 2011. Lamkani now supplies over 3000 liters of milk to different parts of the Dhule district. The economy improved as a result. The experiment of restoration in Lamkani is a significant symbol of community-led restoration which showcases how degraded grasslands can be converted into a healthy ecosystem and community wellbeing.

Benefit cost ratio (BCR) of EbA-based interventions can be useful in highlighting the effectiveness of cost and benefit for EbA-based interventions in comparison to hard engineering projects (Rizvi et al. 2015). Although the BCR has not been estimated for the above interventions, there is evidence from semi-arid countries across the globe on their BCR for rangeland interventions. Jordan's Hima project, which focused on the rangeland management of grazing land, had a BCR of 2.1

Table 4.2 Enhancing rangeland management interventions with ecosystem-based adaptation strategies

Sr. no	EbA pillar	Criteria	Scope for interventions
1	People adapt to climate change	Reduces social and environmental vulnerabilities to climate change (EbA criterion 1)	Rangeland restoration interventions should be based on the assessment of climate change and variability of the local region rather than using estimates at a large scale (Polley et al. 2013; Svejcar et al. 2016; Dierauer et al. 2019; Cartwright et al. 2020) to facilitate sustainable solutions. This would also need to build upon the capacity of stakeholders engaged in the restoration towards effective management of rangelands, leading to reduced environmental vulnerabilities and further sustaining rural livelihoods, resulting in reduced social vulnerability
		Generates societal benefits in the context of climate change adaptation (EbA criterion 2)	Communities need to divide their expected benefits from rangeland management into short-medium and long-term by primarily focusing on providing societal benefits in an equitable manner within the communities. These benefits should include livelihood provision, improved food security and market access, sustainable water provision, and reduced conflict over grazing and water resources. Furthermore, sustained efforts can lead to the generation of co-benefits in the longer term at the ecosystem scale, such as improved carbon sequestration, reduced water runoff, and diversification of livelihoods
2	Use of healthy ecosystem services and biodiversity	Restores, maintains, or improves ecosystem health (EbA criterion 3)	Rangeland management should focus on introducing local and indigenous species of livestock to help better adapt to climate change. Livestock keepers can play an important role in the management of the rangelands (Briske et al. 2015; Collins 2019) by keeping a look on issues like wildfire and invasive grasses. Perennial plant species can play an important role in climate adaptation practices (Halofsky and Peterson 2017; Halofsky et al. 2018; Ellsworth et al. 2024) in contrast to annual grasses, which could make rangelands vulnerable to climate change
3	Participatory governance	Is supported by policies at multiple levels (EbA criterion 4) Supports equitable governance and enhances capacities (EbA criterion 5)	Community-based governance of rangeland management needs to be embedded in the community that is being benefited from its restoration. Integrating EbA policy in existing policies can further make the intervention more sustainable and up-scalable by partnering with scientists to help analyse improved data analysis for improved decision-making (Allen et al. 2011; Littell et al. 2012; Joyce and Millar 2014; Marcot 2020; Rangwala et al. 2021). Furthermore, the local level governance strategy should primarily focus upon inclusivity by making it open across genders and by making it nondiscriminatory, transparent, accountable, and extendible from micro to macro level

Note: The information presented in this table is derived from the document "Making Ecosystem-based Adaptation Effective: A Framework for Defining Qualification Criteria and Quality Standards" (FEBA 2017)

(Myint and Westerberg 2015), whereas grazing reserve management in the Baligubalde & Bookh Rangelands of Somalia resulted in a BCR ranging between 6 and 10 (Hussein et al. 2021). Furthermore, indirect benefits generated from these interventions, like increased water infiltration and retention, environmental benefits such as nutrient cycling, and biomass standpoint (Weigelt 2021), can further augment the returns generated from EbA interventions.

The above examples of Akole and Lamkani from the semi-arid regions show the potential of EbA approaches in restoring rangelands, leading to improved local biodiversity, water, and soil health, eventually leading to the enhancement of rural livelihoods. However, the need for scope for broadening such initiatives lies in addressing the broader criteria and pillars of EbA (Table 4.2).

4.4 Discussion

The three elements of EbA participatory governance, adaptive capacities and ecosystem services and biodiversity play a crucial role in balancing the social, ecological, and economic needs of the communities. India, being the world's largest producer of milk and meat, contributes 50% and 70% of output, respectively, from the pastoral system (LIFE Network 2016), highlighting the significance of the economic value of rangelands for the Indian economy. Restoration of rangelands needs to adopt sustainable land management such as reseeding, natural regeneration, soil/water conservation, and water harvesting (Ouled Belgacem et al. 2019). Overgrazing and irrational grazing systems due to poor frameworks on the rights on common land, coupled with a changing climate, are leading to a loss in the carrying capacity of the land, resulting in a decrease of commons.

Rangeland restoration needs a longer duration for restoration, and it needs to be looked at from the local scale and then has to be tailored considering physical, climate, socio-cultural, and economic factors, varying regionally (Autonoma ´ De Yucat´ an et al. 2016) rather than macroscale to integrate the required management action. These actions are also usually hampered due to a lack of knowledge and experience (Flintan 2011), and an effective way to reduce rangeland degradation is to engage pastoralists in the implementation of any rangeland management projects (Fernandez-Gimenez 2000; Behmanesh et al. 2016). Involvement of local people, especially women who play a central role as natural resource collectors, users, and managers (Flintan 2008, 2011; Eneyew and Mengistu 2013; Johnson et al. 2016; Rota and Chakrabarti 2012; Kristjanson et al. 2014; Assé and Lassoie 2011), who have been affected in the restoration projects, can bring positive sociocultural effects (Blackstock and Richards 2007, Kunasekaran et al. 2022). This could improve their well-being (Fitri Faharuddin et al. 2020) by changing the social relations, leading to improvement in the social capital (Fu and Mao 2022).

4.5 Conclusion

From the Indian context, there is an urgent need to shift the viewpoint towards less visual land, as "wastelands" and start accepting them as resourceful land with economic potential through a sustainable lens. There is a need for investment in research towards restoring and maintaining rangeland ecosystems from an EbA perspective. There are financial, political, and technical constraints that are limiting the widespread adoption of EbA, and there is a need to develop a strong political will to adopt EbA approach to national and international governments through adaptation plans with proper designing and effective implementation of rangeland stewardship activities which will not only help in conserving and restoring rangeland but also lead to generating co-benefits. Underdevelopment of pastoralism due to a lack of unsustainable action plans deployed upon commons, due to the exclusion of pastoralist and other tribal communities using traditional and cost-effective interventions to conserve their commons. Engaging these communities can help restore rangelands by associating them in the decision-making process. It can lead to the protection of land rights and further create a link between government policies and their interventions towards restoration.

Commons are now under major threat posed by the race to become leaders in the "Green Economy", which has led to increased land grabbing of so-called wastelands to generate low-carbon footprint energy, leading to social and economic impact on the communities that have been dependent upon it from generation to generation. EbA practices can assist all the stakeholders in reaching the goal of a "Green Economy" by helping convert partial-sustainable measures to actual sustainable measures for the greater good. Furthermore, estimating the BCR of benefits generated to the whole society after restoring rangelands can bring in more interest in private sector investments to invest in sustainable management projects to improve resilience. The degradation of rangelands, if not controlled, can pose a serious threat in the region, leading to the forced migration of communities who have been dependent upon these lands, leading to the loss of their unique languages and cultural traditions, which are completely vanished, that are tied to their landscapes from generation to generation.

References

Abate T, Angassa A (2016) Conversion of savanna rangelands to bush-dominated landscape in Borana, Southern Ethiopia. Ecol Process 5(1):1–18. https://doi.org/10.1186/s13717-016-0049-1

Adriansen HK (2008) Understanding pastoral mobility: the case of Senegalese Fulani. Geographical Journal, 174(3):207–222

Agarwala M, DeFries R, Jhala YV, Qureshi Q (2019) Threats to coexistence of humans and forests in Central India

Allen CR, Fontaine JJ, Pope KL, Garmestani AS (2011) Adaptive management for a turbulent future. J Environ Manag 92:1339–1345. https://doi.org/10.1016/j.jenvman.2010.11.019

Angassa G, Oba G, Treydte AC, Waladji RB (2010) Role of traditional enclosure on diversity of herbaceous vegetation in semi-arid rangelands, Southern Ethiopia. Livestock Rural Res Dev 22(9):163

Arasumani M, Khan D, Das A, Lockwood I, Stewart R, Kiran RA, Muthukumar M, Bunyan M, Robin VV (2018) Not seeing the grass for the trees: timber plantations and agriculture shrink tropical montane grassland by two-thirds over four decades in the Palani hills, a Western Ghats sky Island. PLoS One 13(1):e0190003. https://doi.org/10.1371/journal.pone.0190003

Ariza-Montobbio P, & Cuvi N (2020) Ecosystem-based adaptation in Ecuador: Good practices for adaptive co-management. Ambiente & Sociedade, 23:e03152

Asner GP, Elmore AJ, Olander LP, Martin RE, Harris AT (2004) Grazing systems, ecosystem responses, and global change. Ann Rev Environ Resour 29:261–299

Assé R, & Lassoie JP (2011) Household decision-making in agroforestry parklands of Sudano-Sahelian Mali. Agroforestry Systems, 82(3):247–261

Balooni K, Singh K (2003) Financing of wasteland afforestation in India. Nat Res Forum 27(3):235–246. https://doi.org/10.1111/1477-8947.00058

Behmanesh B, Barani H, Sarvestani AA, Shahraki MR, Sharafatmandrad M (2016) Rangeland degradation assessment: a new strategy based on the ecological knowledge of indigenous pastoralists. Solid Earth 7(2):611–619

Bhasin V (2011) Pastoralists of Himalayas. Journal of Human Ecology, 33(3)

Blackstock KL, Richards C (2007) Evaluating stakeholder involvement in river basin planning: a Scottish case study. Water Policy 9(5):493–512

Briske DD (ed) (2017) Rangeland systems: processes, management, and challenges. Springer Open (Springer Series on Environmental Management), Cham

Briske DD, Joyce LA, Polley HW, Brown JR, Wolter K, Morgan JA, McCarl BA, Bailey DW (2015) Climate-change adaptation on rangelands: linking regional exposure with diverse adaptive capacity. Front Ecol Environ 13:249–256. https://doi.org/10.1890/140266

Buisson E, Archibald S, Fidelis A, Suding KN (2022) Ancient grasslands guide ambitious goals in grassland restoration. Science 377(6606):594–598

Boissière M, Sheil D, Basuki I, Wan M, & Le H (2009) Can engaging local people's interests reduce forest degradation in Central Vietnam?. Biodiversity and conservation, 18(10):2743–2757

Cartwright JM, Littlefield CE, Michalak JL, Dobrowski SZ (2020) Topographic, soil, and climate drivers of drought sensitivity in forests and shrublands of the Pacific northwest, USA. Sci Rep 10:18486. https://doi.org/10.1038/s41598-020-75273-5

Collins A (2019) Demographics, motivations, and experiences of producers participating in the environmental quality incentives program and the conservation stewardship program. University of Wyoming

Conrad CC, & Hilchey KG (2011) A review of citizen science and community-based environmental monitoring: issues and opportunities. Environmental monitoring and assessment, 176(1):273–291

Courtois DR, Perryman BL, Hussein HS (2004) Vegetation changes after 65 years of grazing and grazing exclusion. Rangeland Ecol Manag 57:574–582

Daba B, Mammo S (2024) Rangeland degradation and management practice in Ethiopia: a systematic review paper. Environ Sustain Indic 23:100413

Dahlberg AC (2000) Interpretations of environmental change and diversity: a critical approach to indications of degradation – the case of Kalakamate, Northeast Botswana. Land Degrad Dev 11:549–562

Dass P, Houlton BZ, Wang Y, Warlind D (2018) Grasslands may be more reliable carbon sinks than forests in California. Environ Res Lett 13:074027

Davies J, Herrera P, Ruiz-Mirazo J, Batello C, Hannam I, Mohamed-Katerere J, Nuesiri E (2016) Improving governance of pastoral lands. FAO

Dierauer JR, Allen DM, Whitfield PH (2019) Snow drought risk and susceptibility in the western United States and southwestern Canada. Water Resour Res 55:3076–3091. https://doi.org/10.1029/2018WR023229

Doswald N, Munroe R, Roe D, Giuliani A, Castelli I, Stephens J, ... & Reid H (2014) Effectiveness of ecosystem-based approaches for adaptation: review of the evidence-base. Climate and Development, 6(2):185–201
Dove MR (2004) Anthropogenic grasslands in Southeast Asia: sociology of knowledge and implications for agroforestry. Agrofor Syst 61(1):423–435
Dregne HE, Kassa M, Rosanov B (1991) A new assessment of the world status of desertification. Desertif Control Bull 20:6–18
Dudley N, Eufemia L, Fleckenstein M, Periago ME, Petersen I, Timmers JF (2020) Grasslands and savannahs in the UN decade on ecosystem restoration. Restor Ecol 28(6):1313–1317
Ellsworth LM, Gutierrez Yee L, Dittel JW, Sanchez DM, Antoninka A (2024) Sagebrush-associated bunchgrasses drive invasion resistance in a greenhouse experiment. Rangel Ecol Manag 92:24–33. https://doi.org/10.1016/j.rama.2023.09.004
Eneyew A, & Mengistu S (2013) Double marginalized livelihoods: invisible gender inequality in pastoral societies. Societies, 3(1):104–116
Emanuelsson U (2009) The rural landscapes of Europe: how man has shaped European nature. The Swedish Research Council
European Soil Bureau Network (2005) Soil Atlas of Europe. European Commission, Ispra, Italy
FAO (2023) Contribution of terrestrial animal source food to healthy diets for improved nutrition and health outcomes. FAO. https://doi.org/10.4060/cc3912en
Fernandez-Gimenez ME (2000) THE ROLE OF MONGOLIAN NOMADIC PASTORALISTS'ECOLOGICAL KNOWLEDGE IN RANGELAND MANAGEMENT. Ecological applications, 10(5):1318–1326
fitri Faharuddin A, Siti A, & Umi W (2020) Factors affecting the participation of tammekkawatang grazing pasture regency of Pinrang, South Sulawesi. Jurnal Riset Veteriner Indonesia (Journal of The Indonesian Veterinary Research)
Flintan F (2008) Women empowerment in pastoral societies. IUCN, Nairobi
Flintan F (2011) The causes, processes, and impacts of land fragmentation in pastoral areas. REGLAP, Oxfam, Nairobi
Flintan F, Tache B, Eid A (2011) Rangeland fragmentation in traditional grazing areas and its impact on drought resilience of pastoral communities: Lessons from Borana, Oromia and Harshin, Somali Regional States, Ethiopia. Regional Learning and Advocacy Programme
Flintan FE (2012) Making rangelands secure: past experience and future options
Fu T, & Mao S (2022) Individual social capital and community participation: an empirical analysis of Guangzhou, China. Sustainability, 14(12):6966
Grice AC, & Hodgkinson KC (2002) Challenges for Rangeland. Global rangelands: progress and prospects, 1
Goldstein MC, Beall CM, & Cincotta RP (1990) Traditional nomadic pastoralism and ecological conservation on Tibet's northern plateau. National Geographic Research, 6(2):139–156
Bertram M, Barrow E, Blackwood K, Rizvi AR, Reid H, and von Scheliha-Dawid S (2017) Making ecosystem-based adaptation effective: a framework for defining qualification criteria and quality standard. FEBA (Friends of Ecosystem-based Adaptation) technical paper developed for UNFCCC-SBSTA 46
Halofsky JE, Peterson DL (eds) (2017) Climate change vulnerability and adaptation in the Blue Mountains. Gen. Tech. Rep. PNW-GTR-939. USDA, Forest Service, Pacific Northwest Research Station, Portland, OR, USA, p. 331
Halofsky JE, Peterson DL, Ho JJ, Little NJ, Joyce LA (eds) (2018) Climate change vulnerability and adaptation in the Intermountain Region, part 1 and 2 (No. RMRS-GTR-375). U.S. Department of Agriculture, Forest Service, Rocky Mountain Research Station, Fort Collins, CO, USA
Harris RB (2010) Rangeland degradation on the Qinghai-Tibetan plateau: a review of the evidence of its magnitude and causes. J Arid Environ 74:1–12. https://doi.org/10.1016/j.jaridenv.2009.06.014
Hazra CR, Saha D (2015) Grazing concerns: overgrazing is leading to the degradation of pasturelands. Down to Earth

Holechek JL, Baker TT, Boren JC, & Galt D (2006) Grazing impacts on rangeland vegetation: what we have learned. Rangelands, 28(1):7–13

Hou Q, Ji Z, Yang H, Yu X (2022) Impacts of climate change and human activities on different degraded grasslands based on NDVI. Sci Rep 12(1):15918

Hussein M, Stringer L, Dallimer M, Aden A, Ali A (2021) Economics of land degradation initiative: an assessment of the economic impact of land degradation in Somaliland. GIZ: Bonn, Germany. Available from: www.eld-initiative.org

IPCC (2019) Climate change and land: an IPCC special report on climate change, desertification, land degradation, sustainable land management, food security, and greenhouse gas fluxes in terrestrial ecosystems. IPCC, Geneva

Joshi AK, Pant P, Kumar P, Giriraj A, & Joshi PK (2011) National forest policy in India: critique of targets and implementation. Small-Scale Forestry, 10(1):83–96

Joyce LA, Millar CI (2014) Improving the role of vulnerability assessments in decision support for effective climate adaptation. In: Sample VA, Bixler RP (eds) Forest conservation and management in the anthropocene: conference proceedings. Proceedings (no. RMRS-P-71). U.S. Department of Agriculture, Forest Service, Rocky Mountain Research Station, Fort Collins, pp 245–271

Karl, J.W., 2011. "Turning information into knowledge for rangeland management", Rangelands, 33(4), pp. 3–5, 3

Karra K, Kontgis C, Statman-Weil Z, Mazzariello JC, Mathis M, Brumby SP (2021) Global land use/land cover with sentinel 2 and deep learning. In: 2021 IEEE international geoscience and remote sensing symposium IGARSS. IEEE, pp 4704–4707

Kassahun A, Snyman HA, Smit GN (2008) Impact of rangeland degradation on the pastoral production systems, livelihoods, and perceptions of the Somali pastoralists in eastern Ethiopia. J Arid Environ 72(7):1265–1281

Kher V, & Dutta S (2021) Rangelands and crop fallows can supplement but not replace protected grasslands in sustaining Thar Desert's avifauna during the dry season. Journal of Arid Environments, 195:104623

Khwarae GM (2006) Community perceptions of rangeland degradation and management systems in Loologane and Shadishadi, Kweneng North, Botswana. Published Master thesis submitted to Norwegian University of Life Sciences, Norway

Kristjanson P, Waters-Bayer A, Johnson N, Tipilda A, Njuki J, Baltenweck I, MacMillan S (2014) Livestock and women's livelihoods. In Gender in agriculture: Closing the knowledge gap (pp. 209–233). Dordrecht: Springer Netherlands

Koocheki A, & Mohalati MN (1994) Feed value of some halophytic range plants of arid regions of Iran. In Halophytes as a resource for livestock and for rehabilitation of degraded lands (pp. 249–253). Dordrecht: Springer Netherlands

Kunasekaran P, Mostafa Rasoolimanesh S, Wang M, Ragavan NA, & Hamid ZA (2022) Enhancing local community participation towards heritage tourism in Taiping, Malaysia: Application of the Motivation-Opportunity-Ability (MOA) model. Journal of Heritage Tourism, 17(4):465–484

Kumar D, Pfeiffer M, Gaillard C, Langan L, Martens, C, & Scheiter S (2020) Misinterpretation of Asian savannas as degraded forest can mislead management and conservation policy under climate change. Biological Conservation, 241:108293

Kupika OL, Gandiwa E, Nhamo G, & Kativu S (2019) Local Ecological Knowledge on Climate Change and Ecosystem-Based Adaptation Strategies Promote Resilience in the Middle Zambezi Biosphere Reserve, Zimbabwe. Scientifica, 2019(1):3069254

Lahiri S, Roy A, & Fleischman F (2023) Grassland conservation and restoration in India: A governance crisis. Restoration Ecology, 31(4): e13858

Leloup S (2006) Investing in maintaining mobility in pastoral systems of the arid and semi-arid regions of sub-Saharan Africa. ALive: Partnership for Livestock Development, Poverty Alleviation and Sustainable Growth. accessed on 12th, July, 2021. http://www.fao.org/fileadmin/templates/lead/pdf/econf_06-10_mobility.pdf

LIFE Network (2016) Kullu call for the recognition of the importance of Common Pool Resources (CPRs) and Pastoralism for India's Livestock Sector. Kullu, Himachal Pradesh, India, LIFE Network

Littell JS, Peterson DL, Millar CI, O'Halloran KA (2012) U.S. National Forests adapt to climate change through science–management partnerships. Climate Change 110:269–296. https://doi.org/10.1007/s10584-011-0066-0

Malaviya, DR, Roy AK, & Kaushal P (2018) Rangelands/Grasslands of India: Current status and future prospects. Grasslands of the World, 235–252

Manier DJ, & Thompson Hobbs N (2006) Large herbivores influence the composition and diversity of shrub-steppe communities in the Rocky Mountains, USA. Oecologia, 146(4): 641–651

Mapinduzi AL, Oba G, Weladji RB, & Colman JE (2003) Use of indigenous ecological knowledge of the Maasai pastoralists for assessing rangeland biodiversity in Tanzania. African journal of Ecology, 41(4):329–336

Marcot BG (2020) The science and management of uncertainty: dealing with doubt in natural resource management. CRC Press, Boca Raton

Mawere M (2013) Environmental conservation through Ubuntu and other emerging perspectives. Langaa Rpcig

Miller D (1997) Range management and pastoralism: new perspective and their implications, ICIMOD, News Letter, no.27, Sprin.

Munroe R, Roe D, Doswald N, Spencer T, Möller I, Vira B, ... & Stephens J (2012) Review of the evidence base for ecosystem-based approaches for adaptation to climate change. Environmental Evidence, 1(1):13

Mwilawa AJ, Komwihangilo DM, & Kusekwa ML (2008) Conservation of forage resources for increasing livestock production in traditional forage reserves in Tanzania. African Journal of Ecology, 46:85–89

Myint MM, Westerberg V (2015) An economic valuation of a large-scale rangeland restoration project through the Hima system in Jordan. Report for the ELD initiative by. International Union for Conservation of Nature, Nairobi

Nanfuka S, Mfitumukiza D, & Egeru A (2020) Characterisation of ecosystem-based adaptations to drought in the central cattle corridor of Uganda. African Journal of Range & Forage Science, 37(4):257–267

Nkonya E, Anderson W, Kato E, Koo J, Mirzabaev A, von Braun J, Meyer S (2016) Global cost of land degradation. In: Economics of land degradation and improvement – a global assessment for sustainable development. Springer International Publishing, Cham, pp 117–165. https://doi.org/10.1007/978-3-319-19168-3_6

Oba G, Kotile DG (2001) Assessment of landscape-level degradation in Southern Ethiopia: pastoralists versus ecologists. Land Degrad Dev 12(6):461–475

Ouled Belgacem A, Ben Salem F, Gamoun M, Chibani R, & Louhaichi M (2019) Revival of traditional best practices for rangeland restoration under climate change in the dry areas: A case study from Southern Tunisia. International Journal of Climate Change Strategies and Management, 11(5):643–659

Ostrom E (1990) Governing the commons: The evolution of institutions for collective action. Cambridge university press

Pandey K (2019) India lost 31% of grasslands in a decade. Down to Earth. https://www.downtoearth.org.in/news/agriculture/india-lost-31-of-grasslands-in-a-decade-66643. Accessed on 12th Jan 2021

Polley HW, Briske DD, Morgan JA, Wolter K, Bailey DW, Brown JR (2013) Climate change and North American rangelands: trends, projections, and implications. Rangel Ecol Manag 66:493–511. https://doi.org/10.2111/REM-D-12-00068.1

Rana P, & Varshney LR (2021) Trustworthy predictive algorithms for complex forest system decision-making. Frontiers in Forests and Global Change, 3:587178

Rangarajan M, & Sivaramakrishnan K (2014) Shifting ground: people, animals, and mobility in India's environmental history. Oxford University Press

Rangwala I, Moss W, Wolken J, Rondeau R, Newlon K, Guinotte J, Travis WR (2021) Uncertainty, complexity and constraints: how do we robustly assess biological responses under a rapidly changing climate? Climate 9:177. https://doi.org/10.3390/cli9120177

Rawat GS, & Adhikari BS (2015) Ecology and management of grassland habitats in India. ENVIS bulletin: wildlife & protected areas, 17

Reichhuber A, Gerber N, Mirzabaev A, Svoboda M, López Santos A, Graw V, Stefanski R, Davies J, Vuković A, Fernández García MA, Fiati C, Jia X (2019) The land-drought nexus: enhancing the role of land-based interventions in drought mitigation and risk management. A report of the science-policy interface. United Nations Convention to Combat Desertification (UNCCD), Bonn, Germany

Reid R et al (2000) Land-use and land-cover dynamics in response to changes in climatic, biological and socio-political forces: the case of southwestern Ethiopia. Landsc Ecol 15:339–355

Rizvi AR, Baig S, Verdone M (2015) Ecosystems based adaptation: knowledge gaps in making an economic case for investing in nature based solutions for climate change. IUCN, Gland, p 48

Robinson L, & Flintan F (2022) Can formalisation of pastoral land tenure overcome its paradoxes? Reflections from East Africa. Pastoralism, 12(1):34

Rota A, Chakrabarti S, & Sperandini S (2012) Women and pastoralism. Livestock thematic papers, tools for project design. Rome, Italy: International Fund for Agricultural Development

Roy AK, & Singh JP (2013) Grasslands in India: Problems and perspectives for sustaining livestock and rural livelihoods. Tropical Grasslands-Forrajes Tropicales, 1(2):240–243

Saigal S (2011). Greening the 'wastelands': Evolving discourse on wastelands and its impact on community rights in India. In 13th Biennial conference of the international association for the study on commons

Scanlon BR et al (2005) Impact of land use and land cover change on groundwater recharge and quality in the southwestern US. Glob Chang Biol 11:1577. https://doi.org/10.1111/j.1365-2486.2005.01026.x

Solomon S (Ed.) (2007) Climate change 2007-the physical science basis: Working group I contribution to the fourth assessment report of the IPCC (Vol. 4). Cambridge university press

Stringer LC, & Reed MS (2007). Land degradation assessment in southern Africa: integrating local and scientific knowledge bases. Land Degradation & Development, 18(1):99–116

Svejcar T, Angell R, James J (2016) Spatial and temporal variability in minimum temperature trends in the western U.S. sagebrush steppe. J Arid Environ 133:125–133. https://doi.org/10.1016/j.jaridenv.2016.06.003

Tamou C (2017) Understanding relations between pastoralism and its changing natural environment 164 pages, accessed on 30th June, 2021-https://edepot.wur.nl/411051

Tanyanyiwa VI, & Chikwanha M (2011) The role of indigenous knowledge systems in the management of forest resources in Mugabe area, Masvingo, Zimbabwe. Journal of Sustainable Development in Africa, 13(3):132–149

UNCCD (2024) Global Land outlook thematic report on rangelands and pastoralism. United Nations Convention to Combat Desertification, Bonn

Onyango V, Masumbuko B, Somda J, Nianogo A, & Davies J (2022) Sustainable land management in rangeland and grasslands. Food & Agriculture Org.

Van Dyne GM, Smith FM, Czaplewski RL, Woodmansee RG (1978) Analysis and syntheses of grassland ecosystem dynamics. In: Singh JS, Gopal B (eds) Glimpses of ecology. International Scientific Publications, Jaipur, pp 1–79

van der Putten, WH, de Ruiter PC, Bezemer TM, Harvey JA, Wassen M, & Wolters V (2004) Trophic interactions in a changing world. Basic and Applied Ecology, 5(6):487–494

Watve A, Athreya V, & Majgaonkar I (2021) The need to overhaul wasteland classification systems in India

Weigelt J, Sinclair F, Mikulcak F, Lossack H (2021) Ecosystem-based adaptation in agriculture: how agroecology can contribute to tackling climate change. GLF White Paper

Woroniecki S (2020) Confronting the ecology of crisis: The interlinked roles of ecosystem-based adaptation and empowerment (Doctoral dissertation, Lund University)
Zerga B (2015) Rangeland degradation and restoration: A global perspective. Point Journal of Agriculture and Biotechnology Research, 1(2):37–54
Zhang M, Delgado-Baquerizo M, Li G, Isbell F, Wang Y, Hautier Y et al (2023) Experimental impacts of grazing on grassland biodiversity and function are explained by aridity. Nat Commun 14(1):5040

Chapter 5
Restoration of Tropical Drylands: Lessons from Tamil Nadu, India

R. J. Ranjit Daniels

Abstract Dryland ecosystems are components of semiarid biomes. These low-biomass ecosystems characterize landscapes or entire regions that receive 200–750 mm of rainfall in a year. Dryland ecosystems in tropical regions are more often secondary and result from human-induced desertification. Tamil Nadu is tropical with an average annual rainfall of less than 1000 mm. Half of its 38 districts receive annual rainfall less than the state's average, five of which receive less than 750 mm. The state is rich in biodiversity and has 40 wildlife-protected areas, including national parks, wildlife sanctuaries, and bird sanctuaries. Forty-five percent of the protected areas, including 3 of the state's 5 tiger reserves and 6 of 20 Ramsar sites, are located in the 19 drier districts. Large parts of the state fall in the semiarid agroecological zone. Due to its long agrarian history, different dryland farming systems have traditionally depended on the well-established network of wetlands. The palmyra palm is a keystone species in dryland agroecosystems. More than 50% of India's 102 million palmyra palms are in the state. Against this backdrop, the chapter discusses some of the challenges faced by the state in the restoration of its dryland ecosystems.

Keywords Ecosystem restoration · Biological invasions · Dryland ecosystem · Semiarid region

5.1 Introduction

Life is a continuum, and its fullness is the biosphere. Due to its vastness and the hierarchical nature of its components, comprehending the biosphere has never been easy. Commonly recognized components, such as genes, populations, species, communities, ecosystems, and biomes, vary in size and complexity. Ecologists have

R. J. R. Daniels (✉)
Trustee & Ecologist, Care Earth Trust, Chennai, Tamil Nadu, India

Y. D. I. Khan et al. (eds.), *Land Restoration Through Ecosystem-Based Approach*, Disaster Resilience and Green Growth, https://doi.org/10.1007/978-981-95-0389-6_5

traditionally resorted to a reductionist approach by studying one or more components at different functional and spatial scales.

A biome is a zone of life characterized by climate, soil, the principal vegetation type, associated animals, and the ecosystems that they sustain (Daniels and Krishnaswamy 2009). Using this broad definition, the biosphere has been divided into six or more biomes, of which arid and semiarid biomes are the most relevant to the discussion that follows. Arid biomes concern deserts where rainfall can be extremely low or even absent. Semiarid biomes, on the other hand, receive varying amounts of rainfall, which, according to some sources, is between 200 mm and 700 mm per year (Nadal-Romero et al. 2019). What characterizes semiarid biomes, however, is the relatively low aridity index. The aridity index is calculated as the ratio of annual precipitation to potential evapotranspiration. In semiarid biomes, the aridity index is between 0.2 and 0.5 (Thomas and Middleton 1997). The water imbalance leaves semiarid biomes in a perpetual state of water stress, resulting in low biodiversity, low productivity, highly seasonal ecosystem processes, and habitats with uniquely adapted species (Daniels and Krishnaswamy 2009).

5.2 Ecosystems Are Functional Units

An ecosystem is a dynamic system that includes both living organisms and the nonliving environment, each influencing the character of the other and both necessary for the maintenance of life (Odum 1971). The United Nations Convention on Biological Diversity has defined ecosystems as "dynamic complexes of plant, animal, and microorganism communities, and their nonliving environment interacting as a functional unit" (Daniels and Krishnaswamy 2009). From the two definitions, three important characteristics of ecosystems can be inferred: (1) ecosystems are dynamic, (2) there are interactions (or influences) that involve species and the nonliving environment, such as soil, water, wind, atmosphere, and sun, and (3) they function as organs that sustain the biosphere.

Five forms of interactions are common to all ecosystems. They are (1) association, (2) competition, (3) predation, (4) parasitism, and (5) disease. Without these interactions, ecosystems cannot exist. Commonly seen species associations are commensalism, mutualism, and symbiosis. Competition exists among species at various levels and can be intraspecific and interspecific. The role of competition in shaping biological communities and ecosystems is a well-researched subject (Whittaker 1975). Predators of various kinds are present in all ecosystems. They can be large mammalian predators such as the lion and tiger, or predatory birds, reptiles, frogs, and fishes, and even a wide range of invertebrates, including spiders, centipedes, scorpions, ants, wasps, beetles, and other insects. Parasites and diseases are as important as predators in keeping ecosystems healthy.

The difficulty in understanding and managing ecosystems stems from the fact that boundaries are not well-defined. Ecosystems are interconnected and form vast networks that together sustain the biosphere. Ecologists have, however, chosen to classify ecosystems as terrestrial and aquatic. More discrete divisions of terrestrial ecosystems are made using vegetation (for example, forests and grasslands). Aquatic ecosystems are classified as oceans, rivers, lakes, and wetlands. Further divisions are made in terrestrial and aquatic ecosystems using climate, altitude, salinity, etc. Sometimes, the boundaries are simply drawn to coincide with those of the biomes that serve as umbrellas to one or more local ecosystems (for example, tropical rainforests). The classification of ecosystems into discrete types is purely for tangible research (Daniels and Krishnaswamy 2009).

5.3 Vulnerability to Biological Invasions

Ecosystems pose several management challenges as both known and unknown factors influence them. Secondary ecosystems are porous and, among others, vulnerable to biological invasions. Management of invasive species is a major challenge (Williams 1996). Article 8(h) of the Convention on Biological Diversity categorically says, "contracting parties, as far as possible, prevent the introduction of, control or eradicate those alien species that threaten ecosystems, habitats or species." The fifth Conference of the Parties, held in Nairobi in May 2000, adopted specific guidelines for dealing with invasive species (Daniels and Krishnaswamy 2009).

Invasive species form mutualistic associations with native biodiversity. The seeds of the highly invasive alien species of plants, such as *Lantana camara* and *Prosopis juliflora*, two species that thrive in drier conditions, are dispersed by birds and mammals. How far and wide the highly mobile dispersal agents carry the seeds is poorly understood. Many restoration projects have suffered setbacks due to successive invasions by plants and animals, notwithstanding the huge investment of resources. It is, therefore, important to realize that secondary ecosystems, irrespective of the associated habitat or community, are influenced by ecological "externalities" that can be from faraway sources and even unknown.

5.4 Tropical Drylands

Tropical drylands are characterized by low biomass ecosystems due to the overall paucity of tall, woody vegetation. The ecosystems are commonly the products of sustained human impacts, fire, and desertification. The biomass of large herbivorous mammals can be high locally but seasonal, as witnessed in sub-Saharan East Africa. In India, large herbivores native to the tropical drylands are generally scarce except for a few species of mammals; chief among them is the blackbuck (*Antelope cervicapra*) that was found throughout the drier plains of the country till about 90 years ago (Krishnan 1972). The long history of decimating wild herbivorous mammals from tropical drylands has led to the ecosystem being dominated by domestic herbivores such as sheep and cattle. Over-grazing and the use of fire to induce fresh plant biomass have resulted in the local extinction of many native plants and animals, including the critically endangered Great Indian Bustard (*Ardeotis nigriceps*).

Tropical drylands in India are not as water-starved as typical semiarid biomes. Regions or landscapes categorized as drylands receive up to 750 mm of yearly rainfall (TNAU 2016). The structure and composition of the vegetation, when present, vary according to the history of human impacts and the resilience of the ecosystem. The most common vegetation type is secondary and varies from scrub to grasslands, dominated by species of thorny plants or xerophytes. Many of the species may be unpalatable or even toxic. The landscape can occasionally be barren and devoid of vegetation. They are wastelands according to the existing land classification systems in India.

5.5 Drylands of Tamil Nadu

Tamil Nadu is the southernmost Indian state. It has a geographical area of 130,058km^2. The average annual rainfall is a little less than 1000 mm. Half of its 38 districts receive annual rainfall lower than the state's average. The five driest districts receive less than 750 mm. Thoothukudi is a maritime district, while Erode, Tirupur, and Virudhunagar are districts that share the Western Ghats. Karur falls in the state's uplands, forming a cluster with Erode and Tirupur. Dry landscapes are locally found in moist districts, too. The Nilgiris district is the wettest in the state, with an annual rainfall of 1636.7 mm (Agrisnet 2019). The hilly district, however, includes dry areas with less than 800 mm and very wet areas that receive rainfall over 3500 mm (Daniels 2023).

Tamil Nadu has a coastline of 1076 km. The landmass has been placed in the semiarid agroecological zone, excluding the narrow coastal belt in the east and the eastern limits of the Western Ghats that extend into the western districts (Sehgal et al. 1992). However, the landmass is a mosaic of bioclimates with patches of vegetation in various stages of ecological degradation. The process of degradation may have started 2000 years ago. Early systems of classifying the landmass into distinct socioecological zones, such as that adopted by the Sangam period Tamils (300 BCE–300 CE), placed landscapes that have undergone extensive desertification under a distinct category called *paalai* (Daniels and Vinoth 2023). *Paalai* is derived from the Tamil word *paal* for milk. Species of plants that characterize the socioecological zone, such as *Manilkara hexandra, Wrightia tinctoria, Alstonia scholaris*, and *Calotropis gigantea*, produce milky-white latex. Consequently, some latex-bearing plants representing the zone have local Tamil names ending with the suffix *paala* or *paalai*.

Due to its long agrarian history, many traditional farming systems are entirely rainfed. Others depend on the intricate network of semi-natural and manmade wetlands classified as *eri, thaangal, kanmaai, oorani, thadaagam, kazhuveli, kulam, kuttai, keni,* etc., based on their size, depth, substrate, and utility. Farmers in the low-lying eastern parts of the state, including the coasts, have traditionally practised a rainfed system of dryland agriculture in which the Palmyra palm (*Borassus flabellifer*) serves as the "keystone" species. The centuries-old tradition of maintaining Palmyra palms and sustaining dryland agroecosystems and human livelihoods over generations has earned the species the status of "state tree."

5.6 Drylands and Biodiversity Conservation

Tamil Nadu has a well-balanced network of protected areas, including 5 national parks, 18 wildlife sanctuaries, and 17 bird sanctuaries. Some national parks and wildlife sanctuaries have been brought under biosphere reserves, of which two represent the Western Ghats and one the Gulf of Mannar in the Bay of Bengal. The state has also established 5 tiger reserves spread across 6 protected areas. The 17 bird sanctuaries are dedicated to the conservation of wetland birds. Tamil Nadu ranks first among the Indian states with 20 globally important wetlands designated as Ramsar sites. The Ramsar sites include most of the bird sanctuaries and one patch of high-elevation riparian evergreen forest in the Western Ghats.

What is of interest, however, is that 45% of the protected areas (including wildlife sanctuaries and bird sanctuaries) in the state fall in the 19 districts that receive less than 980 mm of annual rainfall (Table 5.1). They also include 2 biosphere reserves, 3 tiger reserves, and 6 Ramsar sites. Two southern districts, namely Tirunelveli and Ramanathapuram, have the maximum number of protected and other designated wildlife conservation areas. Erode, among the five driest districts, has a tiger reserve, a wildlife sanctuary, and a bird sanctuary, which is also a Ramsar site. Erode lies in the transition zone between the Western Ghats and Eastern Ghats. Two important tributaries of the river Cauvery, namely, Bhavani and Moyar, drain the Erode district.

5.7 Restoration of Drylands

The restoration of drylands in Tamil Nadu brings with it several socioecological challenges. One of the biggest challenges is controlling the invasion of *Prosopis juliflora*. This species of tropical American tree was introduced in the state during the 1960s as a source of fuelwood in rainfed and dry landscapes. Over the years, with the advent of LPG subsidies and rural outreach services, local people's dependence on the introduced species as fuel has sharply declined. The wood is currently used only for making charcoal locally on a small scale. *Prosopis juliflora*, being invasive, has overrun entire landscapes and regions in the state's dry and rainfed districts and has usurped grasslands inside protected areas. Efforts to eradicate the species have largely been unsuccessful despite huge investments.

Another socioecological challenge is the revival of the traditional use of the Palmyra palm and its role as a keystone species in the dryland agroecosystems. The

Table 5.1 Districts with rainfall less than 980 mm and their ecological significance

Name of district	Rainfall (mm)[a]	Biogeographic affinity	Protected areas[b]
Dindugal	977.5	Western Ghats	Kodaikanal WLS; Kadavur WLS
Salem	966.8	Eastern Ghats	–
Vellore	939.3	Eastern Ghats	–
Madurai	916.9	Western Ghats	Megamalai TR
Dharmapuri	898.5	Eastern Ghats	Cauvery North WLS; Cauvery South WLS
Perambalur	876.7	–	–
Sivagangai	874.7	–	Vettangudi BS
Pudukkottai	871.3	–	Viralimalai PS
Tirunelveli	836.9	Western Ghats	Kalakad-Mundanthurai TR; Agasthyamalai BR; Koonthankulam BS[c]; Gangaikondan WLS; Nellai WLS
Krishnagiri	827.3	Eastern Ghats	Cauvery North WLS; Cauvery South WLS
Ramanathapuram	824.9	East Coast	Gulf of Mannar Biosphere Reserve[c]; Kanjrirankulam BS[c]; Chitrangudi BS[c]; Melaselvanur-Keelaselvanur BS; Theerthangal BS; Sakkaraikotai BS
Tiruchy	803.6	–	–
Namakkal	792.1	Eastern Ghats	–
Theni	783.0	Western Ghats	Megamalai TR; Kodaikanal WLS
Erode	726.8	Western Ghats	Satyamangalam TR; Thanthai-Periyar WLS; Vellode BS[c]
Virudhunagar	713.7	Western Ghats	Grizzled Squirrel WLS; Megamalai TR
Thoothukudi	646.8	East Coast	Gulf of Mannar BR[c]; Vallanadu WLS
Karur	635.8	–	Kadavur WLS
Tirupur	618.8	Western Ghats	Nanjarayan BS[c]: Anamalai TR

BR biosphere reserve, *BS* bird sanctuary, *PS* peafowl sanctuary, *TR* tiger reserve, *WLS* wildlife sanctuary
[a]Agrisnet (2019)
[b]TN State Biodiversity Board & TN Biodiversity and Greening Project
[c]Ramsar site

Palmyra palm is found in the dry landscapes of the semiarid regions of peninsular India, extending as far north as Maharashtra and Orissa, where an estimated 102 million palms were known 25 years ago. More than 50% of the palms occurred in Tamil Nadu, mainly in the dry districts of Thoothukudi, Ramanathapuram, and Tirunelveli (Sankaralingam et al. 1999). Incidentally, the three southern districts contribute significantly to the state's biodiversity and conservation efforts (Table 5.1). Palmyra palms were traditionally used for tapping toddy. The tapping was done by a few backward communities who knew the art and had the skill of climbing the exceptionally tall palms. Since the sale of fermented palm toddy was banned in the state about 50 years ago, livelihoods have been lost, and the traditional jaggery made out of unfermented toddy has become a scarce commodity, leading to various forms of adulterated products and imitations reaching the markets. The natural regeneration of the palm is very slow. Since the primary revenue from the sale of fermented toddy and jaggery no longer provides for subsistence, less attention is given to the restoration of the palm-centric dryland agroecosystem. As a result, a unique tropical dryland ecosystem that thrived for centuries with the palm as a keystone species is slowly disintegrating.

Wetlands have played an important role in sustaining dryland ecosystems. The wetlands are seasonal and come to life only during the Indian monsoons. The majority of the wetlands depend on the northeast monsoon. This is also the season when migratory birds visit the state. The wetlands thus become seasonal habitats for thousands of resident and migratory birds. For this reason, the state has established 17 wetland bird sanctuaries and designated many as Ramsar sites. Resident wetland birds use some of these wetlands to breed. Wetland bird sanctuaries contribute to ecotourism in the state, as many visitors are delighted to see birds that gather in large numbers to breed or feed.

However, wetlands management is a multi-stakeholder business involving not only the forest department but also the public works department, fisheries, and irrigation departments. Sometimes, the Department of Agriculture is also involved. Multi-stakeholder management leads to conflicts of interest in issues that concern retaining the water until the migratory or breeding seasons are over, introducing and harvesting nonnative food fish, desilting, strengthening the bunds, encroachment of feeder channels, etc. Under-harvested nonnative food fish such as the giant African catfish (*Clarias gariepinnus*) and Mozambique tilapia (*Oreochromis mossambicus*) have become invasive locally. Desilting removes propagules of native submerged and emergent aquatic plants, many invertebrates, and fish (and eggs) that over-tide the summer by staying dormant in the silt. These plants and animals play important roles in the productivity of wetland ecosystems and in sustaining the aquatic-terrestrial food chain, which is crucial for migratory and breeding birds and the state's ecotourism.

The absence of natural predators is a significant challenge outside protected areas. Cheetah (*Acinonyx jubatus*) and gray wolf (*Canis lupus*), the largest mammalian predators of drylands, have disappeared from the state. The striped hyena (*Hyaena hyaena*), a predator-scavenger, is rare. Meso- and small mammalian predators such as the golden jackal (*Canis aureus*), Indian fox (*Vulpes bengalensis*), Asian palm civet (*Paradoxurus hermaphroditus*), and gray mongoose (*Uva edwardsii*) have been reduced to the levels of being ecologically extinct locally. Important scavengers such as the griffon vultures (*Gyps bengalensis* and *Gyps indicus*) are critically endangered and barely exist. Birds of prey, including raptors and owls, may play limited roles as predators. Wetland birds are predators of fish and other aquatic animals, but their roles are subject to the availability of aquatic habitats and, thus, seasonal.

Snakes have emerged as the key predators in the absence of larger terrestrial carnivores in dryland agroecosystems. However, as venomous snakes pose risks of unexpected encounters that can cause death and other secondary complications such as gangrene and amputation of limbs, people are wary and do not like to see snakes in their neighborhoods. The lack of awareness and access to timely medical help worsens fear and anxiety. More often than not, an accidental encounter results in the death of the snake, irrespective of whether it is venomous or not.

Human-wildlife conflicts in drylands that border wildlife-protected areas are a serious socioecological problem in the state. Social, because it affects life, livelihoods, and subsistence. Ecological because it is a form of direct competition for space and other limiting resources. Elephants (*Elephas maximus*) are fond of raiding dryland crops such as the finger millet or ragi (*Eleusine coracana*) (Balasubramanian et al. 1993). Some farmers have given up cultivation, unable to cope with the problems posed by large wildlife. In contrast, others have shifted to crops that are less palatable to elephants, some of which require irrigation. Irrigation in drylands is known to change local ecosystems permanently (Daniels et al. 1990; Daniels 2004).

5.8 Way Forward

Farmer-centric agricultural resource management was proposed by FAO around 30 years ago. The FAO-sponsored Farmer Centered Agriculture Resource Management (FARM) project focused on orienting farmers in the Asia-Pacific region to basic ecological concepts, such as the following: Sustainable agroecosystems need the services of biodiversity, including microorganisms. Biodiversity in the form of predators, parasites, and diseases has been successfully used to control invertebrate pests and rodents. Mutualistic associations, such as that of crops and pollinators, are also important. Some forms of competition can lead to socioeconomic losses. However, other forms of competition, involving native organisms and introduced pests, need to be identified and nurtured (Daniels 1995).

Anthropocentric restoration of drylands can be helpful. However, if restoration has to be sustainable, its scope has to be widened to include ecosystems and the major ecosystem processes. The most important ecosystem processes are the water cycle, energy cycle, nutrient cycle, and biotic succession. The water cycle is the most crucial process that maintains the integrity of drylands and the associated ecosystems. Supplementary water provisioning through irrigation can affect the other three processes, such as the energy cycle, nutrient cycle, and biotic succession, resulting in irreversible changes to the dryland ecosystems. The drylands of Tamil Nadu have been managed in a way that creates a delicate balance between traditional systems of farming and biodiversity conservation. The dry districts' many wetland bird sanctuaries and Ramsar sites bear testimony to this. Research in dryland agriculture and forestry should, therefore, broaden the scope to include more complex issues that concern dryland ecosystems, including human-wildlife conflicts and the restoration of associated traditional human livelihoods. Lower levels of productivity and biomass should not become obstacles to the restoration and sustainable management of drylands and the associated ecosystems.

References

Agrisnet (2019) II.6: Rainfall in 2019 – district-wise

Balasubramanian M, Baskaran N, Swaminathan S, Desai AA (1993) Crop raiding by Asian elephant (*Elephas maximus*) in the Nilgiri biosphere reserve, South India. In: Daniel JC, Datye HS (eds) A week with elephants. Bombay Natural History Society & Oxford University Press, Mumbai, pp 350–367

Daniels RJR (1995) Biodiversity indexing in agriculture: a training manual for the Asia-Pacific Farmer Centered Agriculture Resource Management. M. S. Swaminathan Research Foundation, Madras

Daniels RJR (2004) Interlinking of rivers: ecologists wake up! Curr Sci 87(8):1030–1031

Daniels RJR (2023) Biodiversity and conservation challenges. In: Hockings P (ed) The Nilgiri Hills: a kaleidoscope of people, culture and nature. Oriental BlackSwan, New Delhi, pp 15–22

Daniels RJR, Krishnaswamy J (2009) Environmental studies. Wiley, New Delhi

Daniels RJR, Vinoth B (2023) Birds of Tamil Nadu: a synopsis. Care Earth Trust, Chennai

Daniels RJR, Joshi NV, Gadgil M (1990) Changes in the bird fauna of Uttara Kannada, India, in relation to changes in land use over the past century. Biol Conserv 52:37–48

Krishnan M (1972) An ecological survey of the larger mammals of peninsular India. J Bombay Nat Hist Soc 69(3):469–473

Nadal-Romero E, Martinez-Murillo JF, Kuhn NJ (2019) Badlands dynamics in the context of global change

Odum EP (1971) Fundamentals of ecology, 3rd edn. Saunders, Philadelphia

Sankaralingam A, Hemalatha G, Ali AM (1999) A treatise on Palmyrah. ICAR-Central Plantation Crops Research Institute, Kasargod

Sehgal JL, Mandal DK, Mandal C, Vadivelu S (1992) Agroecological regions of India. National Bureau of soil survey and land use planning and. Oxford & IBH Publishing Co. Pvt Ltd, New Delhi

Thomas D, Middleton N (1997) World atlas of desertification, 2nd edn. United Nations Environment Program. https://wad.jrc.ec.europa.eu

TNAU (2016) Dryland agriculture: an introduction. TNAU AGRITECH Portal. https://agritech.tnau.ac.in

Whittaker RH (1975) Communities and ecosystems, 2nd edn. Collier-MacMillan, London

Williams M (1996) Biological invasions. Chapman and Hall, London

Chapter 6
Study of the Impact of Irrigation on Groundwater Quality in the Jam River Basin, Nashik and Ahilyanagar Districts, Maharashtra, India

U. G. Mhaske, Resham Bhalla, Chetan A. Patil, Pravin M. Nalawade, and G. D. Mhaske

Abstract The present study investigates the impact of irrigation on groundwater quality in the Jam River Basin, located in Nashik and Ahilyanagar districts, Maharashtra, India. 60 groundwater samples were collected and analyzed during the pre- and post-monsoon seasons of 2020 and 2021. Sampling sites were selected based on hydrogeological, geological, and land-use parameter variations. The study observed significant effects of irrigation on groundwater chemistry, including salinization, waterlogging, nitrate contamination, and boron toxicity. The findings reveal that groundwater salinization is a cumulative result of several interacting factors, including semi-arid climate, over-irrigation, chemical fertilizer usage, flood irrigation, and monoculture cropping, especially sugarcane. Nitrate contamination and boron toxicity were prominent in the northeastern part of the irrigated area and were linked to waterlogging and excessive fertilizer application. The study emphasizes the need for proper water and land resource management to mitigate the adverse effects of irrigation on groundwater quality.

Keywords Groundwater quality · Irrigation · Salinization · Waterlogging · Nitrate contamination · Boron toxicity · Jam River Basin

U. G. Mhaske
Department of Physics, HPT Arts and RYK Science College, Nashik, Maharashtra, India

R. Bhalla
Department of Zoology, LVH Arts, Science and Commerce College, Nashik, Maharashtra, India

C. A. Patil · P. M. Nalawade (✉) · G. D. Mhaske
Department of Environmental Science, KRT Arts, BH Commerce & AM Science (KTHM) College, Nashik, Maharashtra, India
e-mail: pmnalawade@kthmcollege.ac.in

Y. D. I. Khan et al. (eds.), *Land Restoration Through Ecosystem-Based Approach*, Disaster Resilience and Green Growth, https://doi.org/10.1007/978-981-95-0389-6_6

6.1 Introduction

The increasing demand for fresh water and frequent climate changes are significantly impacting the use of groundwater resources. To meet the growing need for food, water usage for agriculture has increased manifold, directly affecting both the quality and quantity of groundwater (Jackson et al. 2001; Wriedt et al. 2009; Huang et al. 2015). A significant portion of former East Germany has been converted to agricultural land. Due to land reunification and a boom in modern irrigation projects, there has been a large increase in irrigated land in recent decades (Simon 2009; Siebert et al. 2015). In areas lacking rivers and lakes but experiencing high water demand, groundwater resources face significant strain. The increased demand for irrigation and the challenge of maintaining a balance between groundwater levels, quality, and quantity currently pose a major concern (Germer et al. 2011; Gutzler et al. 2015). Similar challenges are expected to intensify with changing climatic patterns and irrigation demands (Kreins et al. 2015; Wolter & Kandian, 1997) Mainly in the northeastern part of Germany, groundwater is extensively used as an excellent source of drinking water, and at the same time, the conservation of ecological houses and wetlands is also undertaken (Merz et al. 2012). Groundwater temperatures and shallow aquifers are also influenced by climate change and urbanisation, potentially altering groundwater quality over time (Taylor & Stefan, 2009). Due to limited water availability during the agricultural season, the farmlands are significantly affected. To avoid agricultural losses caused by climate change, it is necessary to adopt appropriate management policies and plan measures to reduce potential damage. Therefore, in the water-scarce areas of northeastern Germany, crops grown in sandy soil are primarily sustained by the use of groundwater (Steidl et al. 2015). Therefore, a review of agricultural irrigation management policies is necessary, as it can reduce the adverse effects on groundwater and related ecosystems. Proper management of the water balance, water used for agricultural production, and groundwater levels is particularly crucial in water-scarce regions (Riediger et al. 2014). According to the study by Krämer et al. (2015), the demand for water for irrigation is expected to increase by 20 percent between 2051 and 2080 due to climate change. Ensuring the growing demand and the expected quality of groundwater will be difficult. Due to geochemical reactions occurring in groundwater reservoirs and the use of polluted water for agricultural irrigation, groundwater quality will deteriorate rapidly. Therefore, it is the need of the hour to check the quality of water used for agricultural irrigation and drinking purposes (Jacintha et al. 2016; Rawat et al. 2017).

Generally, water quality is indicated by the presence of major cations such as Na^+, Ca^{2+}, Mg^{2+}, K^+, and anions (Cl^-, SO_4^{2-}, HCO_3^-, CO_3^{2-}, NO_3^-) and soluble metals in the water. Similarly, water quality index models such as sodium adsorption ratio (SAR), percent sodium (Na%), residual sodium carbonate (RSC), residual alkalinity (RA), Kelly's ratio (KR), permeability index (PI), chloralkaline indices (CAI_1 and CAI_2), potential salinity (PS), magnesium hazard (MH), total dissolved solids (TDS), and total hardness (TH) are mainly used to check the quality of water used for irrigation purposes (Singh et al. 2015; Gautam et al. 2015). The impacts of excessive groundwater extraction, the effects of climate change on groundwater and related ecosystems, and the consequences of increased groundwater use on agricultural land, etc., do not appear to have been studied extensively (Taylor et al., 2012). Sustainable groundwater use is a value-driven process that preserves intra- and intergenerational equity and helps to maintain a balance between the economy, environment, and society (Gleeson et al. 2012). Water users need to view groundwater management from a positive perspective for economic and social development, the protection of ecosystems, and the services provided by ecosystems (Costanza et al. 1997). The functioning of environmental systems involves various processes that vary across different spatial and temporal scales and respond to environmental changes. In particular, decomposition, often measured in the context of leaf litter decomposition, is one of the most widely used functional variables for evaluating the impacts of environmental changes on the functioning of terrestrial and aquatic ecosystems (Makkonen et al. 2012). The groundwater recharge process in irrigated areas can sometimes have an adverse effect on groundwater quality. Groundwater is being significantly polluted due to many agriculture-related processes. The large-scale extraction of groundwater for agricultural irrigation is depleting groundwater reserves worldwide, and groundwater reserves in many areas have become contaminated (Ceron and Pulido-Bosch 1996). Globally, many problems related to agriculture are arising, including water quality, pesticide pollution, and the increasing nutritional load in water bodies (Mateo-Sagasta and Burke 2010). According to the available information, water is mainly used extensively for agriculture, industrialization, and human consumption (Gleick 1993). If a water expert is asked to determine the water demand for their country, they generally use the water extracted for economy-related uses such as agriculture and industrialization. Although this information is significant, it does not discuss the water required for human consumption and its consumption patterns. Many things needed in daily life are imported from other countries, meaning that the water required based on population can be several times greater than the total apparent need. The concept of water footprint has been newly introduced so that proper information regarding water usage in various sectors can be provided. It can be appropriately used to create a usage-based indicator system (Hoekstra and Hung 2002).

Primarily, the quality of groundwater used in agriculture depends on irrigation practices. While agricultural irrigation is essential for growing food, it significantly leads to saline land, nutrient leaching, and pesticide contamination, and this adversely affects other important factors. By properly studying the adverse effects on important factors, management policies can be developed. A robust irrigation system, pesticide-free farming, and continuous monitoring of groundwater quality can protect and conserve valuable resources like groundwater. Addressing this relationship between irrigation and groundwater quality is not merely an environmental concern but a crucial need for ensuring long-term water security and the viability of agriculture.

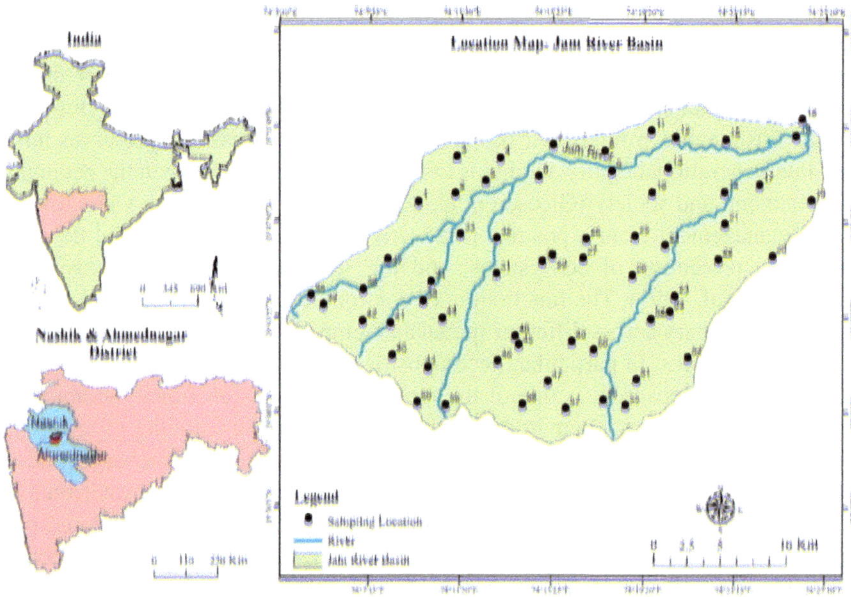

Fig. 6.1 Geospatial representation of groundwater monitoring locations in the study region

6.2 Materials and Methods

6.2.1 Study Area

The Jam River is a tributary of the Godavari River. Its basin lies in the Nashik and Ahilyanagar (formerly known as Ahmednagar) districts of Maharashtra, bounded by latitudes 19°44′27″N to 19°52′18″N and longitudes 74°6′28″E to 74°25′56″E in the Deccan Plateau. The area map and sampling sites are also studied in Fig. 6.1.

The Jam River originates on the high Mhasha Hill in Sinnar Tehsil, Nashik District, and meets the Godavari River close to Kopargaon Tehsil, Ahilyanagar District. The basin has 636.67 square kilometers in total. It travels 52.2 km in the Ahilyanagar district of Maharashtra, connecting the eastern part of Sinnar Tehsil with the southwest part of Kopargaon Tehsil. Numerous settlements in Nashik and Ahilyanagar districts benefit from this river for their drinking and irrigation needs.

6.2.2 Methodology

This study aimed to assess irrigation's effects on groundwater quality in the Jam River Basin. We use a random sampling method for sample collection. Groundwater sampling was conducted at 60 locations during the pre-monsoon (May) and post-monsoon (November) seasons in both 2020 and 2021. The sample size (60 samples) is determined based on geographical factors, including the area distance, resource availability, and statistical considerations. While it may not encompass all the dug and bore wells in the basin, the sampling strategy ensures that the collected data represents the study area. The samples were obtained from a combination of bore and dug wells, with 11 samples from bore wells and 49 from dug wells. The sampling locations were strategically selected based on geological, hydrogeological, and land-use variations across the study area. The coordinates of all groundwater sampling locations were recorded using a Global Positioning System (GPS) to ensure spatial accuracy. To visualize groundwater quality variations across the study area, hydrological and spatial distribution maps were generated using the Inverse Distance Weighting (IDW) interpolation method in ARC-GIS and ERDAS Imagine software. IDW is a deterministic spatial interpolation technique that estimates values at unsampled locations based on the proximity of known data points. It is particularly effective in mapping groundwater quality where the observed data is sparse and irregularly distributed. IDW can be used to monitor and manage hydrochemical parameters of groundwater resources. It can also strengthen river basin management systems for pollution monitoring and control. Similar predictive techniques like Artificial Neural Networks (ANNs) have been effectively applied for groundwater nitrate estimation in the Kadava River basin near Nashik (Wagh et al., 2017). IDW is one of several interpolation techniques that can be used to analyze groundwater flow and physicochemical parameter distribution. Another technique is kriging, which is a reliable method for interpolation when the data of samples has high spatial correlation.

6.2.3 Water Sample Collection and Analysis

The groundwater samples were systematically collected using clean, sterilized containers to ensure minimal contamination. Parameters like pH, electrical conductivity (EC), and total dissolved solids (TDS) were analyzed right at the water sample location (Hanna Instruments HI98194). A multitester instrument is a very useful device for testing the above parameters. These parameters provide a preliminary indication of groundwater salinity and ionic content.

Laboratory analyses were performed in the laboratory following the American Public Health Association (APHA 1995) as described in the Standard Methods for the Examination of Water and Wastewater, which provides comprehensive and standardized procedures for the accurate analysis of various water quality parameters. These protocols, specifically outlined in the Standard Methods for the Examination of Water and Wastewater, offer numerous benefits that make them a trusted resource in water analysis. The APHA protocol for titrimetric analysis was used to determine concentrations of chloride (Cl^-), total alkalinity (as $CaCO_3$), calcium (Ca^{2+}), magnesium (Mg^{2+}), and total hardness (TH). Flame photometry was employed to analyze sodium (Na^+) and potassium (K^+) ions. Spectrophotometric methods were utilized for the analysis of nitrate (NO_3^-), sulfate (SO_4^{2-}), phosphate (PO_4^{3-}), and fluoride (F^-).

6.2.4 *Data Analysis*

The obtained results were statistically analyzed using Statistical Package for the Social Sciences (SPSS) software for sound samples, and borewell samples were compared using a t-test. Also, cluster analysis has been performed to find the water samples' internal relationships and characteristics. Further, we have performed ANOVA to study spatial and temporal variations among hydrochemical parameters of groundwater to identify trends in groundwater quality changes before and after the monsoon seasons. The differences between nonirrigated and irrigated zones were highlighted by comparing average values for key parameters, including EC, TDS, Na^+, Cl^-, TH, Ca^{2+}, Mg^{2+}, and total alkalinity (TA). Previous reviews also emphasise the need for evaluating soil leaching, fertiliser runoff, and hydrochemical dynamics under various irrigation regimes (Kolaja et al., 1986). The findings were then interpreted to understand the impact of irrigation on groundwater quality, focusing on the following indicators:

- Salinization: Changes in the concentration of dissolved salts were studied based on TDS and significant cations and anions.
- Waterlogging: The depth of the water table in each region was measured and categorized using the Indian Ministry of Water Resources guidelines to determine the extent of waterlogged areas (Kumar et al. 2005).
- Nitrate Pollution: Nitrate levels were compared across seasons to assess the impact of fertilizer use on groundwater contamination.
- Boron Toxicity: The presence of boron in groundwater was evaluated, particularly in areas where salinization was observed.

6.3 Results and Discussion

The changes in land use and land cover (LULC) for the study area in 2021 are shown in Fig. 6.2. It reveals a landscape predominantly shaped by agriculture, which covers 403.92 sq. km, accounting for 63.44% of the total area. This highlights the region's heavy dependence on farming as the primary land use. Wastelands occupy 122.50 sq. km (19.24%), suggesting opportunities for reclamation or alternative utilization. Urbanization is evident, with built-up areas covering 28.13 sq. km (4.42%), primarily along transportation networks. Temporarily unused agricultural lands, classified as fallow, comprise 49.98 sq. km (7.85%). Vegetation spans a smaller portion, covering 13.24 sq. km (2.08%), while water bodies are limited to 18.91 sq. km (2.97%), reflecting the region's modest natural water resources. These findings, provided in Table 6.1, are a foundational understanding of the land-use patterns within the Jam River Basin, offering critical insights into the relationship between land use and groundwater quality, particularly in an irrigation-dominated landscape.

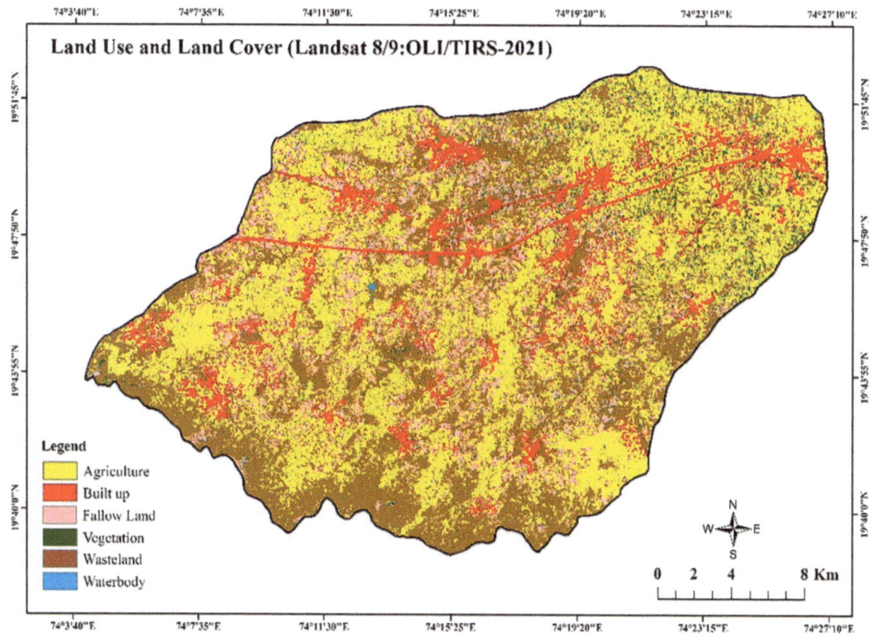

Fig. 6.2 Land use and land cover (LULC) classification of the study area

Table 6.1 Land use and land cover of study area

Sr. no.	LULC classes	Area (sq. km)	% Area
1.	Agriculture	403.92	63.44
2.	Built up	28.13	4.42
3.	Fallow land	49.98	7.85
4.	Vegetation	13.24	2.08
5.	Wasteland	122.50	19.24
6.	Waterbody	18.91	2.97
	Total	636.67	100.00

The results of the groundwater quality analysis are summarized in Tables 6.2 and 6.3 for the pre-monsoon and post-monsoon seasons of 2020 and 2021, respectively. The study revealed notable changes in groundwater chemistry between irrigated and nonirrigated zones.

6.3.1 Groundwater Salinization

Surface water, such as rivers and lakes combined, accounts for less than 1% of the freshwater reserves at the global level, while almost 12% is groundwater. Because of the reliance on groundwater for industrial, household, and agricultural purposes,

Table 6.2 Physicochemical characteristics of groundwater from study area (pre- and post-monsoon 2020)

	Pre-monsoon (May 2020)				Post-monsoon (November 2020)			
Parameters	Min	Max	Average	Standard deviation	Min	Max	Average	Standard deviation
pH	7.1	8.5	7.90	0.33	7.1	8.2	7.61	0.30
EC	108	6710	1763.82	1461.61	109	7170	1586.82	1403.7
TDS	97	3820	1054.81	870.58	76	4188	949.5	857.02
Na^+	20.18	1080.2	267.81	311.26	4.80	880.70	196.10	222.70
K^+	0.1	7.43	2.06	1.74	0.1	7.17	2.00	1.41
Ca^{2+}	60.52	561.12	236.24	112.38	48.09	370.1	131.12	65.37
Mg^{2+}	7.89	350.4	138.39	71.25	4.86	331.69	92.96	69.81
TH	320	2120	781.87	354.09	120	2110	492.87	351.48
Cl^-	98.6	1704	483.20	387.72	90.16	1682.32	419.80	332.35
F^-	0.17	1.21	0.66	0.20	0.41	1.37	0.66	0.22
TA	170	875	444.91	185.20	130	1365	472.17	245.21
CO_3^-	00	160	60.08	30.21	00	160	30.88	29.21
HCO_3^-	125	825	385.88	175.02	100	1325	442.74	238.44
SO_4^{2-}	33.1	1207.49	177.84	279.21	30.85	192.26	62.48	47.71
PO_4^{2-}	0.02	1.94	0.68	0.57	0.01	1.12	0.22	0.33
NO_3^-	5.17	66.14	26.27	15.09	3.1	79.69	29.86	19.28

Note: All significant ions and TDS are expressed in mg/L, pH on the scale, and EC in μS/cm

Table 6.3 Physicochemical characteristics of groundwater from study area (pre- and post-monsoon 2021)

Pre-monsoon (May 2021)					Post-monsoon (November 2021)			
Parameters	Min	Max	Average	Standard deviation	Min	Max	Average	Standard deviation
pH	7.2	8.4	7.9	0.32	7.1	8.7	7.6	0.31
EC	337	7780	1942.14	1499.07	111	6390	1627.62	1392.17
TDS	217	4590	1158.51	949.90	79	4109	987.70	882.18
Na^+	23.21	1120.99	276.89	329.10	8.24	890.4	193.60	225.64
K^+	0.00	17.16	2.02	2.61	0.00	6.26	1.60	1.37
Ca^{2+}	63.12	537.07	223.17	93.49	40.48	352.70	132.92	59.43
Mg^{2+}	21.87	410.8	137.6	73.81	1.21	417.96	79.25	71.96
TH	400	3245	1227.16	629.73	268	1760	716.06	381.30
Cl^-	17.04	1710.2	389.57	387.89	65.32	1668.5	469.19	355.08
F^-	0.13	1.17	0.66	0.20	0.46	1.35	0.67	0.21
TA	200	870	474.11	135.03	170	1295	487.17	211.63
CO_3^-	20	100	49.92	20.71	00	120	37.33	23.15
HCO_3^-	150	800	422.41	134.44	150	1225	450.32	205.13
SO_4^{2-}	13.77	1224.19	243.33	276.17	37.49	516.61	167.48	103.16
PO_4^{2-}	0.02	1.92	0.96	0.63	0.01	1.11	0.37	0.38
NO_3^-	0.82	66.71	25.86	15.84	3	75.67	29.18	19.91

Note: All significant ions and TDS are expressed in mg/L, pH on the scale, and EC in μS/cm groundwater supplies have been overexploited, and a lack of rainfall has resulted in water scarcity issues and salinity problems. The quality of groundwater is degraded by the discharge of effluent from industry and the leaching of fertilizers from agricultural activities (Pauw et al. 2012). Understanding groundwater's chemical composition and salinity is necessary before evaluating its suitability. In dissolved form, groundwater comprises seven primary chemical parameters. The major ions under investigation include calcium (Ca^{2+}), magnesium (Mg^{2+}), chloride (Cl^-), bicarbonate (HCO_3^-), sodium (Na^+), potassium (K^+), and sulfate (SO_4^{2-}). Once contaminated, groundwater quality is often difficult to restore due to slow subsurface flow, limited natural attenuation processes, and the persistence of certain pollutants. As a result, maintaining groundwater quality is critical for protecting and managing water resources. The hydrogeological situation in Tamil Nadu is not good because the majority of the state, 73%, is underlain by hard and soft rock or sedimentary formations in the eastern sections, and the remaining 27% is in neighboring areas (Abboud 2018).

Figure 6.3a, b illustrate the groundwater salinity classification based on TDS levels during the pre-monsoon and post-monsoon seasons, highlighting spatial variations in salinity across the study area. Tables 6.2 and 6.3 show that TDS values exceeded the threshold for freshwater (1000 mg/L) in several samples, particularly in the downstream portion of the basin where flood irrigation is common. The adverse effects of saline irrigation practices on crop productivity and soil health are

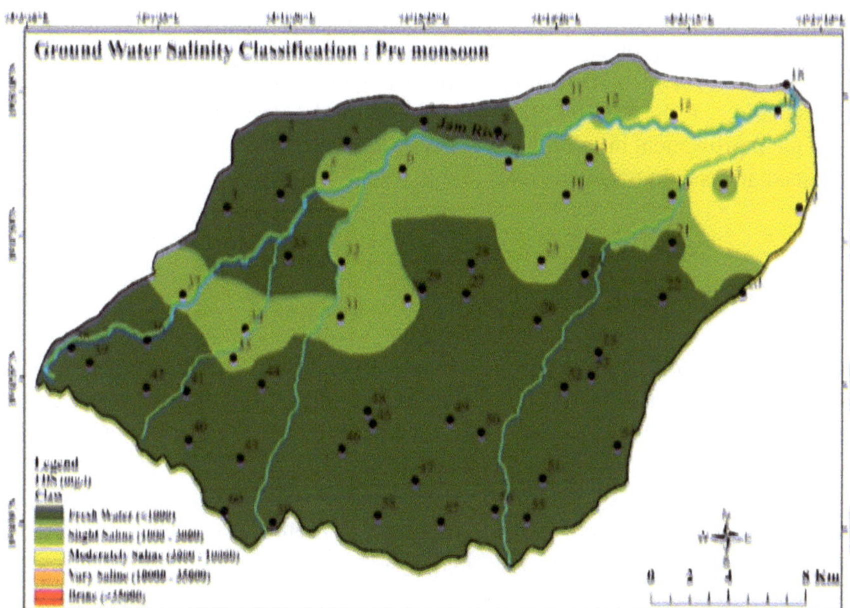

Fig. 6.3 (**a**) Groundwater salinity classification during pre-monsoon season based on TDS from the study area. (**b**) Groundwater salinity classification during the post-monsoon season based on TDS from the study area

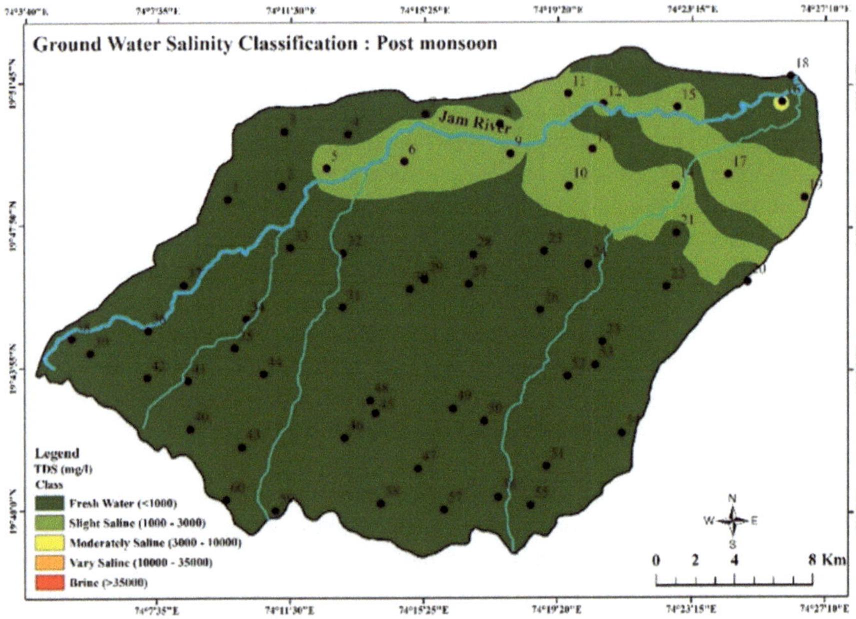

Fig. 6.3 (continued)

well established (Gupta & Gupta, 1997a; Astaraei & Chauhan, 1995). During the pre-monsoon season of 2020, 21.66% of the samples fell under the slightly saline category (TDS between 1000 and 3000 mg/L), while 6.66% were classified as moderately saline (TDS between 3000 and 10,000 mg/L). Post-monsoon values indicated an improvement in salinity levels due to dilution from rainfall, but salinization remained an issue in irrigated zones (Table 6.4).

6.3.2 *Waterlogging*

Waterlogging, another adverse consequence of irrigation, was observed in certain low-lying areas of the Jam River Basin, particularly in the downstream regions. Long-term waterlogging can lead to severe soil degradation, especially when sodic irrigation is involved (Sharma & Chauhan, 1994). The depth of the water table was used to categorize areas as waterlogged (<2 meters), potentially waterlogged (2–3 meters), or safe (>3 meters). In both pre- and post-monsoon seasons, 4.08% of the samples were found to be from waterlogged areas, while an additional 4.08% were from potentially waterlogged areas (Table 6.5).

Field observations suggest that seepage from irrigation canals and poor drainage in these areas contributed to the waterlogging, exacerbated by the cultivation of long-duration crops such as sugarcane. In arid and semi-arid regions, the expansion of irrigation systems often leads to an unintended rise in the groundwater table. This phenomenon can trigger waterlogging and soil salinization, ultimately causing land

Table 6.4 Groundwater salinity classification based on TDS

Nature of water	TDS (mg/l)	Total samples and %	Pre-monsoon sample numbers	Total samples and %	Post-monsoon sample numbers
Fresh water	<1000	43 (71.66%)	1–4, 7–8, 10, 13–14, 20–24, 26–29, 33, 36, 38–60	47 (78.33%)	1–4, 7–8, 10, 13–14, 20–30, 32–33, 35–36, 38–60
Slightly saline	1000–3000	13 (21.66%)	5–6, 9, 11–12, 17, 25, 30–32, 34–35, 37	12 (20%)	5–6, 9, 11–12, 15, 17–19, 31, 34, 37
Moderately saline	3000–10,000	4 (6.66%)	15–16, 18–19	1 (1.66%)	16
Very saline	10,000–35,000	Nil	Nil	Nil	Nil
Brine	>35,000	Nil	Nil	Nil	Nil

After Hem (1991)

Table 6.5 Categorization of waterlogged area

Category	Depth of WT(m)	Total samples and %	Sample numbers
Waterlogged	<2	2(4.08)	8, 16
Potentially waterlogged	2–3	2(4.08)	14, 56
Safe	>3	45(91.83)	1, 3–4, 7, 10–11, 13, 15, 17–26, 28–37, 39–42, 44–47, 49–50, 53–55, 57–60

degradation and, in severe cases, the abandonment of agricultural fields (Singh and Lal 2018). Such conditions demand improved drainage interventions to manage salinity and waterlogged soils (Singh, 2016; Kolaja et al., 1986). According to Singh et al. (2015), approximately one-third of irrigated farmland globally suffers from decreased productivity or complete crop loss due to these conditions.

6.3.3 *Nitrate Contamination*

Figure 6.4a, b represent the graphical representation of NO3 concentrations during the pre-monsoon and post-monsoon seasons of 2020 and 2021. The elevated nitrate levels were a concern in the irrigated areas, especially in the northeastern basin. Nitrate (NO_3^-) concentrations exceeded the safe drinking water limit of 45 mg/L in 11.66% of the samples during the pre-monsoon season of 2020, rising to 21.66% in the post-monsoon season. The increase in nitrate levels is attributed to the heavy use of nitrogen-based fertilizers, such as urea, in agricultural practices. Nitrate leaches quickly into the groundwater, particularly in over-irrigated fields where water percolates quickly through the soil, carrying dissolved nutrients (Li et al. 2015). Similar

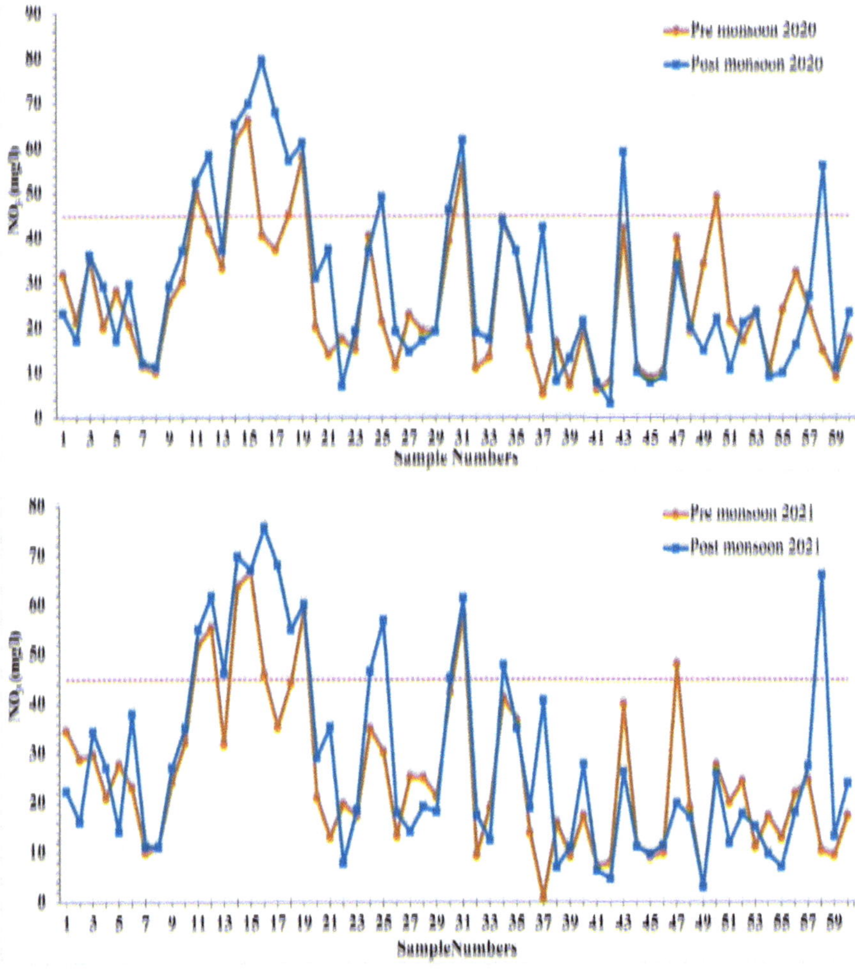

Fig. 6.4 (**a**) Graphical representation of NO_3 in pre-monsoon seasons of 2020 and 2021. (**b**) Graphical representation of NO_3 in pre-monsoon seasons of 2020 and 2021

conclusions were drawn by Hill (1982) and Sehmi et al. (2000) regarding nitrate transport in intensively farmed areas.

The post-monsoon data further indicate that nitrate contamination was more prevalent in irrigated areas than in nonirrigated zones. Villages like Madhi, Ghari, Chandekasare, and Pohegaon showed consistently high nitrate levels in the study area, indicating a direct correlation between fertilizer use, irrigation, and groundwater pollution. These findings are consistent with earlier research from the same basin indicating nitrate contamination trends (Mhaske et al., 2022). Similarly, where the groundwater level is optimum, it shows within the permissible limit.

6.3.4 Boron Toxicity

Boron is a naturally occurring element in groundwater, but elevated levels of boron can be toxic to plants. In this study, the samples from irrigated areas exhibited higher boron concentrations, particularly in the downstream regions affected by salinization. Salinization in the downstream areas is mainly caused by poor irrigation practices (over-irrigation and saline water use), high evaporation in arid climates, leaving salts behind, agricultural runoff and upstream land clearing, adding salts to the water, reduced river flow from dams or water diversion, concentrating salts, and seawater intrusion due to rising sea levels or over-extraction of groundwater. Downstream farmers use more boron because saline soils often cause boron deficiency, upstream leaching reduces boron availability, and downstream irrigation water may lack boron. Additionally, downstream crops may have higher boron demand, prompting increased application. Samples taken during both the pre- and post-monsoon seasons of 2020 and 2021 revealed boron concentrations exceeding the critical level of 1 mg/L in several locations, with values as high as 3 mg/L in some instances.

Figure 6.5a, b depicts the graphical representation of boron concentrations during the pre-monsoon and post-monsoon seasons of 2020 and 2021, highlighting temporal variation in its distribution. Boron toxicity was most pronounced in saline soils, where boron accumulates due to poor drainage and waterlogging, affecting soil physicochemical properties (Astaraei & Chauhan 1995) . Plants in these areas showed symptoms such as leaf burn and stunted growth (Naz et al. 2016). Boron is recognized as an essential micronutrient vital for optimal plant growth and development. It contributes to multiple physiological and biochemical processes, including

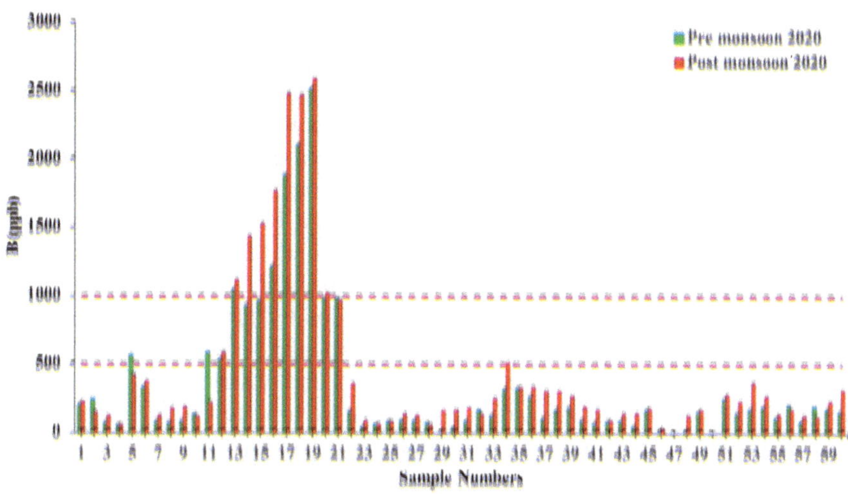

Fig. 6.5 (**a**) Graphical representation of boron in pre-monsoon seasons of 2020 and 2021. (**b**) Graphical representation of boron in post-monsoon seasons of 2020 and 2021

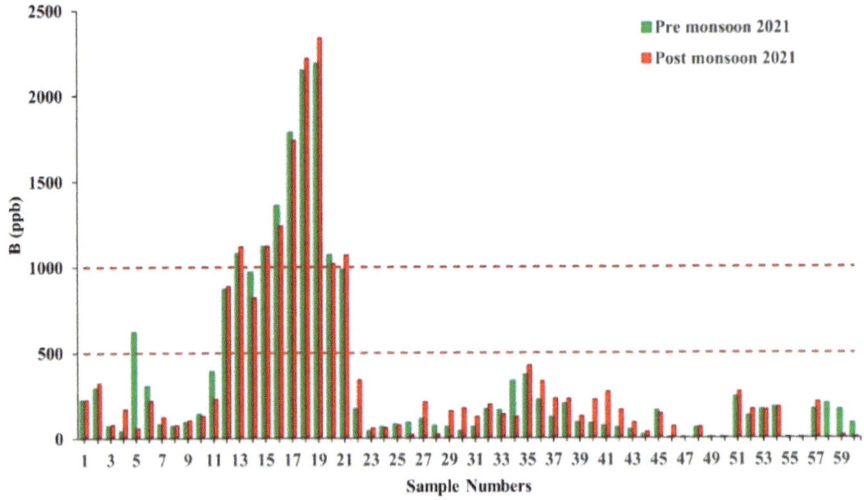

Fig. 6.5 (continued)

sugar transport, cell wall formation and lignification, carbohydrate metabolism, RNA and DNA synthesis, respiration, membrane integrity, and the regulation of metal-activated enzymes (Das et al. 2022). Boron toxicity in irrigated and saline soils is influenced by the Mg/Ca ratio, SAR values, and salinity levels of irrigation water (Astaraei & Chauhan, 1995; Gupta & Gupta, 1997b).

6.4 Conclusion

This research emphasizes the considerable influence of irrigation practices on the groundwater quality within the Jam River Basin, located in Maharashtra, India. Through the analysis of 60 groundwater samples collected over two consecutive years (2020 and 2021), several key issues related to groundwater quality were identified. Thus, the current study may assist regional water regulators in creating policies for managing water appropriateness in the Jam River Basin. The irrigation practices in the basin, particularly over-irrigation and flood irrigation, have contributed to groundwater salinization. A considerable percentage of groundwater samples exhibited elevated levels of total dissolved solids (TDS), indicating varying degrees of salinity. The downstream region, in particular, showed a high concentration of dissolved salts due to stagnant conditions and waterlogging. Poor natural drainage and seepage from irrigation canals have led to waterlogging in some low-lying regions. The cultivation of water-intensive crops such as sugarcane further exacerbates this issue. Waterlogging affects groundwater quality and results in soil degradation and reduced agricultural productivity. The heavy use of nitrogen-based fertilizers in irrigated fields has polluted the groundwater with nitrate. Elevated nitrate concentrations were observed, particularly in the northeastern basin, where irrigation return flows and fertilizer leaching contributed to groundwater contamination.

Excessive boron levels were detected in the downstream areas, where salinization and waterlogging were prominent. High boron concentrations and salinity significantly threatened plant health and agricultural productivity. The findings of this study underscore the need for sustainable water and land resource management practices in the Jam River Basin. The following recommendations are proposed to mitigate the adverse effects of implementing water-saving irrigation techniques, such as drip or sprinkler systems, which can help reduce waterlogging and salinization by minimizing water use and improving efficiency. Promoting the use of organic fertilizers and implementing controlled fertilizer application practices can help reduce nitrate contamination in groundwater. Constructing and maintaining efficient drainage systems in irrigated areas can alleviate waterlogging and prevent the accumulation of salts in the soil. Using gypsum and improved leaching practices in saline and alkaline soils can help reduce boron toxicity and enhance soil permeability. The study's findings serve as a basis for developing regional policies to improve groundwater management in the Jam River Basin, ensure sustainable agricultural practices, and protect the health of local ecosystems.

6.5 The Way Forward and Scalability of the Study

- Long-Term Monitoring: Implement continuous, real-time groundwater monitoring using IoT sensors to track TDS, nitrates, and boron levels, allowing timely interventions.
- Climate Change Impact: Assess how climate change and extreme weather events affect groundwater quality and availability, with potential modeling for future conditions.
- Hydrological Modeling: Develop integrated hydrological models to predict the impact of irrigation practices and land-use changes on water quality and scale findings to other regions.
- Alternative Cropping Systems: Explore water-efficient crops and crop rotation to reduce groundwater pressure and minimize nutrient contamination.
- Advanced Irrigation Techniques: Promote precision irrigation and automation using remote sensing to optimize water use and minimize salinity and waterlogging.
- Groundwater Recharge: Investigate artificial recharge methods, such as rainwater harvesting, and identify areas for groundwater replenishment.
- Policy Framework: Design water governance models, economic incentives, and policy guidelines for sustainable water management, adaptable to other basins.
- Public Awareness: Engage local communities through awareness campaigns and farmer training on sustainable practices and water conservation.

- Cross-Basin Studies: Conduct comparative studies between different river basins to identify common groundwater quality issues and scalable solutions.
- Sustainability Guidelines: Develop best practice guidelines for sustainable irrigation for regions with similar agricultural and hydrological conditions.

References

Abboud IA (2018) Geochemistry and quality of groundwater of the Yarmouk basin aquifer, North Jordan. Environ Geochem Health 40(4):1405–1435

APHA, AWWA, & WPCF (1995) Standard methods for the examination of water and wastewater, 19th edn. American Public Health Association

Astaraei AR, Chauhan RPS (1995) Effect of waters varying in Mg/Ca ratio, SAR and salinity on physico-chemical properties of soil. Indian J Soil Sci 40:19–24

Ceron JC, Pulido-Bosch A (1996) Groundwater problems resulting from CO2 pollution and overexploitation in the alto Guadalentin aquifer (Murcia, Spain). Environ Geol 28:223–228

Costanza R, d'Arge R, de Groot RS et al (1997) The total value of the world's ecosystem services and natural capital. Nature 387:253–260

Das RK, Mizan A, Zohra FT, Ahmed S, Ahmed KS, Hossain H (2022) Extraction of a novel tanning agent from indigenous plant bark and its application in leather processing. J Leather Sci Eng 4(1):18

Gautam SK, Maharana C, Sharma D, Singh AK, Tripathi JK, Singh SK (2015) Evaluation of groundwater quality in the Chotanagpur plateau region of the Subarnarekha River Basin, Jharkhand State, India. Sustain Water Qual Ecol 6:57–74. https://doi.org/10.1016/j.swaqe.2015.06.001

Germer S, Kaiser K, Oliver B, Huettl RF (2011) A review of water balance changes and responses of ecosystems and society in the Berlin-Brandenburg region. Erde 1–2:65–95

Gleeson T, Alley WM, Allen DM, Sophocleous MA, Zhou Y, Taniguchi M, VanderSteen J (2012) Towards sustainable groundwater use: setting long-term goals, backcasting, and managing adaptively. Ground Water 50(1):19–26

Gleick PH (ed) (1993) Water in crisis: a guide to the world's fresh water resources. Oxford University Press

Gupta SK, Gupta IC (1997a) Crop production in waterlogged saline soils. Scientific Publishers

Gupta SK, Gupta IC (1997b) Management of saline soils and waters. Scientific Publishers

Gutzler C, Helming K, Balla D, Dannowski R, Deumlich D, Glemnitz M, Knierim A, Mirschel W, Nendel C, Paul C (2015) Agricultural land use changes – a scenario-based sustainability impact assessment for Brandenburg, Germany. Ecol Indic 48:505–517

Hem JD (1991) Study and interpretation of chemical characteristics of natural waters (U.S. Geological Survey Water Supply Paper No. 2254)

Hill AR (1982) Nitrate distribution in the groundwater of the Alliston region of Ontario, Canada. Groundwater 20(6):696–702

Hoekstra AY, Hung PQ (2002) Virtual water trade: a quantification of virtual water flows between nations in relation to international crop trade. Value of Water Research Report Series No. 11, UNESCO-IHE Institute for Water Education, Delft, The Netherlands. http://www.waterfootprint.org/Reports/Report11.pdf

Huang T, Pang Z, Liu J (2015) Groundwater recharge mechanism in an integrated tableland system, North China Plain. Hydrogeol J 23:909–923. https://doi.org/10.1007/s10040-015-1243-8

Jacintha TGA, Rawat KS, Mishra A, Singh SK (2016) Hydrogeochemical characterization of groundwater of Penninsular Indian region using multivariate statistical techniques. Appl Water Sci. https://doi.org/10.1007/s13201-016-0400-9

Jackson RB, Carpenter SR, Dahm CN, McKnight DM, Naiman RJ, Postel SL, Running SW (2001) Water in a changing world. Ecol Appl 11(4):1027–1045

Kolaja V, Vrba J, Zurrnmann KH (1986) Control and management of agricultural impact on groundwater. In: Castany G, Groba E, Romjin E (eds) Impact of agricultural activities on groundwater, vol 5, p 197

Kreins P, Henseler M, Anter J, Herrmann F, Wendland F (2015) Quantification of climate change impact on regional agricultural irrigation and groundwater demand. Water Resour Manag 29:3585–3600

Kumar R, Singh RD, Sharma KD (2005) Water resources of India. Curr Sci:794–811

Li X, Li J, Xi B, Yuan Z, Zhu X, Zhang X (2015) Effects of groundwater level variations on the nitrate content of groundwater: a case study in Luoyang area, China. Environ Earth Sci 74:3969–3983

Makkonen M, Berg MP, Handa IT, Hättenschwiler S, van Ruijven J, van Bodegom PM, Aerts R (2012) Highly consistent effects of plant litter identity and functional traits on decomposition across a latitudinal gradient. Ecol Lett 15:1033–1041

Mateo-Sagasta J, Burke J (2010) Agriculture and water quality interactions: a global overview. SOLAW Background Thematic Report TR08. Food and Agriculture Organization of the United Nations, Rome. http://www.fao.org/nr/solaw/. Accessed 16 Sept 2016

Merz B, Kaiser K, Bens O, Emmermann R, Flühler H, Grünewald U, Negendank J (2012) Klimawandel und Wasserhaushalt. In: Hüttl RF et al (eds) Georessource Wasser – Herausforderung Globaler Wandel. Springer-Verlag

Mhaske GD, Nalawade PM, Bhalla R (2022) Groundwater quality evaluation in relation to nitrate pollution in the Jam River basin at Nashik and Ahilyanagar district, Maharashtra, India. Int J Ecol Environ Conserv 28:S491–S498

Naz T, Akhtar J, Iqbal MM et al (2016) Boron toxicity in salt-affected soils and effects on plants. In: Soil science: agricultural and environmental perspectives, pp 259–286

Pauw P, De Louw PG, Essink GO (2012) Groundwater salinisation in the Wadden Sea area of The Netherlands: quantifying the effects of climate change, sea-level rise, and anthropogenic interferences. Neth J Geosci 91(3):373–383

Rawat KS, Tripathi VK, Singh SK (2017) Groundwater quality evaluation using numerical indices: a case study (Delhi, India). Sustain Water Resour Manag 4:875. https://doi.org/10.1007/s40899-017-0181-9

Riediger J, Breckling B, Nuske RS, Schröder W (2014) Will climate change increase irrigation requirements in agriculture of Central Europe? A simulation study for northern Germany. Environ Sci Eur 26:1–13

Sehmi K, Suchet PA, Clauser N, Probst J (2000) Impact of nitrogen fertilizers on the natural weathering erosion process and fluvial transport in the Garonne basin. Appl Geochem 15:865–878

Sharma JP, Chauhan RPS (1994) Effect of sodic irrigation on nutrient and oil content of Palmarosa. J Indian Soc Soil Sci 42:335–337

Siebert S, Kummu M, Porkka M, Döll P, Ramankutty N, Scanlon BR (2015) A global data set of the extent of irrigated land from 1900 to 2005. Hydrol Earth Syst Sci 19:1521–1545

Simon M (2009) Die Landwirtschaftiche Bewässerung in Ostdeutschland seit 1949 – Eine historische Analyse vor dem Hintergrund des Klimawandels. Postdam Institute Climate Impact research (PIK), Potsdam

Singh A (2015) Soil salinization and waterlogging: a threat to environment and agricultural sustainability. Ecol Indic 57:128–130. https://doi.org/10.1016/j.ecolind.2015.04.027

Singh G, Lal K (2018) Review and case studies on drainage: an alternative drainage system to manage waterlogging and salinity. Irrig Drain 67:51–64. https://doi.org/10.1002/ird.2252

Singh SK, Srivastava PK, Singh D, Han D, Gautam SK, Pandey AC (2015) Modelling groundwater quality over a humid subtropical region using numerical indices, earth observation datasets, and X-ray diffraction technique: a case study of Allahabad district, India. Environ Geochem Health 37(1):157–180. https://doi.org/10.1007/s10653-014-9638-z

Steidl J, Schuler J, Schubert U, Dietrich O, Zander P (2015) Expansion of an existing water management model for the analysis of opportunities and impacts of agricultural irrigation under climate change conditions. Water 7(11):6351–6377

Taylor CA, Stefan HG (2009) Shallow groundwater temperature response to climate change and urbanisation. J Hydrol 375(3–4):601–612
Wagh VM, Panaskar DB, Muley AA (2017) Estimation of nitrate concentration in groundwater of Kadava river basin-Nashik district, Maharashtra, India by using artificial neural network model. Model Earth Syst Environ 3(1):36
Wolter HW, Kandian A (1997) Harnessing water to feed a hungry world. ICID J 46(1):1–20
Wriedt G, Van der Velde M, Aloe A, Bouraoui F (2009) Estimating irrigation water requirements in Europe. J Hydrol 373:527–544

Part II
Nature-Based Restoration Strategies for Climate Resilient Ecosystems

Chapter 7
Ecosystem-Based Approaches to Natural Resource Management: Adapting to Climate Change in Arid Regions

Archana Verma, Dheeraj Singh, M. B. Noor Mohamed, K. Shiran, A. Keerthika, N. V. Saresh, and Navjot Kaler

Abstract Dryland ecosystems are very fragile and vulnerable to significant climate change threats. Extreme and harsh weather conditions due to excessive heat and inadequate variable precipitation, strong winds, and low soil organic matter are some of the important contributors to the climate severity of the region. There is a need to explore the ecosystem-based approach for addressing and reducing the effects of climate change in dry regions. The combined efforts for integrated conservation, sustainable resource management, involvement of local community, and ecosystem-based solutions will enhance the resilience vis-à-vis preserving biodiversity of the region. The rehabilitation of damaged or deteriorated areas and sustainable land-use practices like integrated farming systems, agroforestry, soil conservation techniques, sustainable livestock management, etc., are essential for building adaptive capacity. These approaches will not only help increase the productive capacity but also provide several ecosystem services like carbon sequestration, increasing soil fertility, improving water table, air quality, etc. A demand exists for works that involve participants at multiple levels, ensuring that adaptation approaches are both ecologically sound and socially rightful. By implementing these ecosystem-based solutions, Indian arid regions can enhance their resilience in response to climate change, enhance livelihoods, and foster sustainable development while preserving vital ecosystem services. Eventually, a holistic ecosystem-based approach is a pressing need not to address the urgent demands of climate change but to offer long-term sustainability and adaptability in dryland ecosystems.

A. Verma · D. Singh (✉) · M. B. Noor Mohamed · K. Shiran · A. Keerthika
ICAR-Central Arid Zone Research Institute, Jodhpur, Rajasthan, India

N. V. Saresh
ARS, Keshwana, Agriculture University, Jodhpur, Rajasthan, India

N. Kaler
Dr Y S Parmar UHF, Solan, Himachal Pradesh, India

Y. D. I. Khan et al. (eds.), *Land Restoration Through Ecosystem-Based Approach*, Disaster Resilience and Green Growth, https://doi.org/10.1007/978-981-95-0389-6_7

Keywords Drylands · Nature-based solutions · Sustainability · Resilience · Land use

7.1 Introduction

Drylands cover 40% of the total land area of the earth (Pravalie 2016; Wang et al. 2022) and are characterized by high temperature, low and erratic precipitation, high wind velocity, water scarcity, and low organic matter (Naorem and Somasundaram 2023). Dryland areas are inhabited by over 2 billion people, accounting for approximately 35% of the world's population. More than half (55%) of people living in dryland regions reside in rural areas. Approximately half of the poorest people in the world live in dry areas (MEA 2005). Despite the environmental constraints, these regions support various biological diversity in both plants and animal species. The arid zones of India also support a rich diversity of plant, animal, and human communities, which have evolved to adapt to the severe climate of the region (Chaudhary et al. 2011). These regions are very important to support local livelihood in rural areas. These ecosystems support a significant proportion of the world population where agriculture, agroforestry, and livestock-based integrated farming systems are the primary bread-earning occupations. Furthermore, these farming practices provide various ecosystem services such as biodiversity preservation, soil and water conservation, and locking atmospheric carbon. However, in the era of climate change, drylands are more susceptible and prone to climate change; hence, sustainability and resilience are crucial to ensure food security (Baraj et al. 2024). These regions are more prone to rising temperatures, disturbed rainfall patterns, droughts, and other weather fluctuations (Venkateswarlu 2012; Yadav et al. 2020). Complex interactions between soil, vegetation, and water sources occur in these regions, which must be carefully addressed. Sustainability in the dried regions signals toward maintaining equilibrium between natural systems and their human usage. However, the resilience of these ecosystems will depend on their ability to withstand biotic and abiotic stresses due to changing climatic conditions. Agriculture and livestock are the stay of livelihood in these regions, which also fosters the sustainability and resilience of the regions; therefore, the increasing human pressure on natural resources needs to be worked carefully. Ecosystem-based approaches (EbA) hence assist in the restoration and functioning of dryland regions by amalgamating ecological principles into land-use practices and adaptation strategies and ensuring the conservation of natural resources promoting long-term sustainability and productivity (Sharma et al. 2024; Wang et al. 2024). This chapter provides insight into various ecosystem-based approaches for managing natural resources and preparing drylands for future climatic challenges. Understanding climatic aberrations, their long-term impacts, and developing nature-based solutions for adaptation and mitigation are essential. This includes promoting sustainable land use practices and effective community management by identifying key implementation factors to enhance climate adaptation in these regions.

7.2 Vulnerability of Dryland Ecosystems to Changing Climate

Globally, climatic shifts cause significant challenges and threats to the diverse ecosystems. Population in dryland regions depending on agriculture is at higher risk of climate variability. In Rajasthan, nearly 60% of the total cultivated land relies on rainfall for irrigation (Singh et al. 2019). By the year 2021–2050, temperature is going to rise by 2–2.5 °C in the state (Gopalakrishnan et al. 2011). Under such conditions, severe weather occurrence is defined as a pattern of extreme weather that persists for some time, such as a season. Changes in extreme weather events can be analyzed from two perspectives: changes in their frequency or changes in their intensity. Out of 504 extreme weather events and trends recorded worldwide from 400 peer-reviewed studies published in different countries of Asia, the following data was collected:

- Climate change contributed to an increased likelihood or severity in 71% of the 504 extreme weather events analyzed.
- About 9% of the events or trends became less likely or less severe due to climate change, indicating that 80% of all recorded events were influenced by human activity.
- Among 152 extreme heat events, climate change was found to have intensified or increased the likelihood of 93%.
- About 56% of the 126 rainfall or flooding events examined were made more frequent or severe due to human influence.
- Climate change played a role in 68% of the 81 drought events studied.

The temperature has exhibited significant variability, with an average increase of approximately 0.5 ° C during the last 30 years. The maximum rise was noticed from 1995 to 2000 (Singh and Kumar 2015). The annual temperature increased by 0.7 °C between 1901 and 2018; if the current pace of greenhouse gas emission continues, the average annual temperature is anticipated to increase by 50C by the end of the twenty-first century (Ravindra et al. 2024; Krishnan et al. 2020). The average rainfall ranges from 250 to 450 mm, and the rainfall pattern has shown considerable variability, indicating significant fluctuations over time. Water scarcity and stress conditions have been a key factor in droughts. Over the past 30 years, rainfall has decreased by 50 mm due to rising temperatures. However, in recent years, varied rainfall patterns and increases in the intensity of rainfall and floods have been noticed in fewer parts of Rajasthan. Additionally, certain areas, including western Bikaner, northern Jaisalmer, central Barmer, and western Jalore, have experienced a reduction in arid conditions (Singh et al. 2019). The arid climate belt has shifted eastward, accelerating land degradation and desertification. Scarce water resources further increase the region's vulnerability to climate change. Moreover, canal irrigation has exacerbated land degradation, leading to issues such as alkalization and salinization.

Recurrence of drought is a significant phenomenon that severely impacts food grain production and livestock, leading to reduced agricultural yields and increased vulnerability for farming communities. Rajasthan was affected by drought of varied intensity, with the highest frequency observed in the western and southern regions of the state between 1901 and 2002 (48 out of 102) (Rathore 2005). Moreover, in 2016, 19 out of 33 districts in Rajasthan experienced famine or drought conditions, highlighting the severe impact of climate variability on the region.

7.2.1 Impact of Climate Change on Agriculture

The correlation of foliage cover and primary productivity mainly depends on the timing and amount of rain (Zhao et al. 2021), temperature, soil productivity, atmospheric CO_2, solar radiation, fertilizer use, pesticides, and insecticides (Gholkar et al. 2014). The arid threshold with a value of 0.54 is crucial below which ecosystem the productivity changes (Berdugo et al. 2022). Any small change in aridity will have a considerable effect on the vegetation of drylands. The extension of dry areas will adversely affect carbon stock/sequestration (Huang et al. 2016). But how flora responds varies with different areas and is regulated by ecosystem functions, i.e., "drier gets drier and wetter gets wetter."

Overgrazing, land-use changes, and drought can also influence the direction of dryland evolution. Grazing contributes to soil erosion, vegetation loss, and nutrient cycling disruptions, ultimately accelerating erosion and land denudation. Agriculture also has a major impact on dryland ecosystems; i.e., improper agricultural practices, irrigation, and use of agrochemicals can negatively impact soil health and biodiversity (Cherlet et al. 2018). However, sustainable farming methods like agroforestry and conservation agriculture can aid in the restoration of dryland ecosystems by improving soil fertility, carbon stocking, and conservation of biodiversity (González-Sánchez et al. 2016).

In Rajasthan, from 1998 to 2016, the crop loss index score was lowest in Dungarpur (0.022) and highest in Ganganagar (0.643). Districts such as Sirohi, Banswara, Dausa, Udaipur, Rajsamand, Jalore, Karauli, Dholpur, and Pratapgarh reported the lowest levels of production losses. Hanumangarh, Ganganagar, and Jaisalmer had the highest sensitivity to the changing climate, with 11 of 33 districts being extremely sensitive overall (Singh et al. 2019). The details given in Table 7.1 highlight the threat of climate change on agriculture across various districts of Rajasthan.

Table 7.1 Effect of climate change impact and agricultural vulnerability in different districts of the West Indian arid zone

District	Vulnerability index	Degree of vulnerability
Sirohi	0.141	Less
Ajmer	0.394	Less
Banswara	0.463	Less
Dausa	0.498	Less
Kota	0.574	Average
Bhilwara	0.588	Average
Tonk	0.618	Average
Sikar	0.638	Average
Alwar	0.703	High
Ganganagar	0.913	High
Jaisalmer	0.946	High
Hanumangarh	0.973	High
Dungarpur	0.326	Low

Source: Singh et al. (2019)

7.3 Ecosystem-Based Approaches for Adaptation and Mitigation

Ecosystem-based approaches (EbA) have gained prominence as an approach that leverages natural ecosystems to mitigate climate-related risks and adapt to climate change impacts, emphasizing the sustainable use, conservation, and restoration of biodiversity and ecosystem services (Doswald and Osti 2011). EbA address both social and environmental vulnerabilities by enhancing ecosystem health and resilience, with particular utility in arid regions where ecosystems are susceptible to rising temperatures and erratic precipitation (weADAPT 2022). In these dryland regions, where agriculture, water resources, and biodiversity are at high risk, ecosystem-based strategies provide an adaptable, multifunctional solution to withstand climate variability, manage water scarcity, and promote biodiversity conservation.

EbA strategies are highly flexible and context-specific, offering adaptation benefits beyond conventional, engineered solutions. By harnessing ecosystems' inherent capabilities, EbA support livelihoods, biodiversity, and essential ecosystem services such as soil fertility, water retention, and carbon sequestration while remaining cost-effective and inclusive of community needs (Doswald and Osti 2011).

7.3.1 Definition and Principles of Ecosystem-Based Solutions

Ecosystem-based approaches, also referred to as "nature-based solutions," capitalize on natural processes to address both climate mitigation and adaptation (Colls et al. 2009). Defined by the Convention on Biological Diversity (CBD), EbA entail "the use of biodiversity and ecosystem services as part of an overall adaptation strategy to help people adapt to the adverse effects of climate change" (SCBD 2009). This approach fosters climate resilience while offering social, financial, and cultural advantages for local communities, making it highly relevant for arid landscapes vulnerable to human-induced and climate-related stressors.

The core principles guiding EbA interventions emphasize a landscape approach, community involvement, and enhancing ecosystems' adaptive capacity, often proving more sustainable and affordable than "gray" or engineered infrastructure solutions. These principles are outlined as follows:

- *Integrative Management*: EbA take a landscape-level view, integrating land, water, vegetation, and human activities to create resilient ecosystems. This principle underscores the interconnectedness of ecosystem components, which collectively provide vital services like soil stability, water retention, and biodiversity conservation.
- *Adaptive Capacity Building*: By enhancing ecosystem functions, EbA strengthen the capacity of natural systems and local communities to withstand climate variability and environmental shocks. For instance, restoring vegetation and improving soil health help ecosystems absorb and recover from severe weather events like droughts and floods.
- *Community Engagement and Inclusivity*: Effective EbA require active participation from local communities, who bring traditional knowledge and practical insights. Involving communities ensures that adaptation strategies align with local needs and priorities, make them more in tune with local cultures, and ensure long-term sustainability over time.
- *Sustainability and Cost-Effectiveness*: EbA often prove more cost-effective than engineered adaptation solutions, as it leverages natural processes for long-term climate resilience. Restoration projects, for instance, regenerate degraded lands and support biodiversity without the ongoing costs associated with hard infrastructure (SCBD 2009).

These principles highlight the potential of ecosystem-based adaptation to create multifunctional, resilient landscapes that support both human livelihoods and biodiversity.

7.3.1.1 Practical Applications in Arid Regions

In arid and semi-arid regions, ecosystem-based adaptation interventions involve various practices to strengthen the resilience of ecosystems and local communities. These measures are designed to address the specific challenges of dry climates, such as extreme temperatures, scarce water resources, and soil degradation.

Key ecosystem-based strategies in these areas include:

- *Natural Vegetation Cover Enhancement*: Increasing natural vegetation cover through reforestation, agroforestry, and grassland restoration helps stabilize soils, reduce erosion, and create microclimates supporting biodiversity and agriculture (WOTR 2022).

- *Soil Quality Improvement*: Soil conservation techniques such as contour plowing, mulching, and the use of organic amendments enhance soil structure and water retention, which are crucial for crop productivity in drylands. In Maharashtra, watershed interventions integrating soil conservation practices have successfully improved agricultural yields and bolstered ecosystem resilience to erratic rainfall (weADAPT 2022).
- *Water Resource Protection*: Protecting and managing water resources is crucial in arid areas wherever scarcity of water is a defining challenge. Techniques such as rainwater harvesting, check dams, and wetland restoration create sustainable water sources supporting human needs and ecological health. Wetland restoration in dryland regions has also proven effective for regulating water availability, reducing flood risks, and supporting groundwater recharge (Doswald and Osti 2011).
- *Buffer Zone Creation for Erosion and Flood Management*: Establishing vegetative buffer zones around water bodies and agricultural land can mitigate the impacts of flash floods and control erosion common in dry and semi-dry areas.

Ecosystem-based adaptation practices in arid regions vary across agricultural and nonagricultural lands while sharing several common strategies. In agricultural land, sustainable farming systems such as *Prosopis cineraria*-based agroforestry, drought-resistant crops, water harvesting techniques, and soil conservation practices are crucial for enhancing soil fertility, improving water retention, and increasing productivity while mitigating land degradation. In nonagricultural land, emphasis is placed on restoring degraded ecosystems, including grasslands and ravine lands, through sustainable livestock management, native vegetation conservation, and dune stabilization to maintain ecosystem services. Several EbA measures are common to both land types, including water resource management (e.g., rainwater harvesting and micro-catchments), biodiversity conservation (through the protection of native species), and carbon sequestration (via revegetation and sustainable land use), ensuring resilience against climate change and land degradation across the arid landscape.

7.3.2 Importance of Holistic Management

A holistic management approach integrates water, soil, vegetation, and human essentials to tackle the challenges of dryland ecosystems comprehensively. This approach is essential for EbA in arid and semi-arid regions, where managing resource scarcity and soil degradation requires adaptive strategies that consider the ecological and socioeconomic needs of the area (Doswald and Osti 2011).

- *Water and Soil Management*: Water scarcity and soil erosion are significant issues in dryland regions. Techniques like contour farming, mulching, and organic soil amendments are essential for conserving soil moisture, enhancing nutrient levels, and reducing erosion. In Maharashtra, watershed restoration projects have shown that these techniques improve water retention and protect agricultural productivity in areas prone to erratic rainfall (WOTR 2022).

- *Agroforestry and Mixed Farming Systems*: Agroforestry combines trees with crops or livestock in the same system, providing various ecosystem services like soil fertility. Enhancement, shade, and carbon sequestration: Implementing diverse agroforestry systems with native plant species as an adaptation strategy can also support biodiversity conservation. The integration of trees, like *Prosopis cineraria* in Rajasthan, stabilizes soils, creates microclimates, and offers economic benefits through products like fodder, fruit, and timber. This practice also diversifies income sources, contributing to economic resilience in regions with climate-sensitive agricultural systems.
- *Drought-Resilient Vegetation*: Including local drought-tolerant plants, such as *Prosopis cineraria*, can significantly enhance ecosystem resilience. Drought-tolerant species reduce soil temperature, improve nitrogen levels, and stabilize landscapes, making them vital for climate adaptation in arid regions. This strategy also supports biodiversity by providing habitats and resources for various species, thereby bolstering ecological health (SCBD 2009).

Holistic management in EbA recognizes the interconnected nature of ecosystems. It promotes adaptive measures that address environmental, economic, and social factors concurrently, thereby fostering resilience in vulnerable landscapes.

7.3.3 Case Studies of Successful Ecosystem-Based Interventions

Several case studies illustrate the effectiveness of EbA in India, showcasing how integrated conservation, sustainable resource management, and community engagement can build resilience to climate change.

- *Watershed Restoration*: The Watershed Organisation Trust (WOTR) in Maharashtra has implemented watershed restoration projects to address the increasing challenges of water scarcity and soil degradation. Through techniques like contour trenching and check dams, these projects improve soil and water retention, reduce erosion, and recharge groundwater, benefiting agricultural productivity and local biodiversity. This holistic intervention underscores the importance of water management in arid regions and highlights the role of EbA in stabilizing ecosystems and supporting rural livelihoods (weADAPT 2022). The government of Telangana launched Mission Kakatiya in the year 2015, which had the objective of restoring over 46,000 tanks and lakes across the state. The project has restored 27,665 water bodies by March 2018 by improving their water storage capacity and supporting irrigation for approximately 1.5 million acres of agricultural land. The yield of crops like cotton has gone up by 11.6%, maize by 6.7%, and paddy by 4.4% due to the desilting process providing nutrient-rich soil to farmers (Minampati and Sugandhar 2023).

- *Wetland Restoration in the Indian Himalayan Region (IHR)*: In the IHR, wetland restoration has proven critical for flood control, water storage, and biodiversity conservation. Rejuvenated wetlands function as natural reservoirs, capturing excess water during heavy rainfall and gradually releasing it, thereby reducing flood risks and maintaining stable groundwater levels. This ecosystem-based approach supports water availability and provides habitats for various species, demonstrating the multifunctional benefits of wetland conservation in areas experiencing climate stress (ICIMOD 2024).

These case studies demonstrate the wide-ranging benefits of ecosystem-based approaches in arid regions, including soil management, water conservation, carbon restoration, and economic resilience. The use of local species, community-driven projects, and sustainable land-use practices highlights the flexibility of EbA to adapt to different environmental and socioeconomic contexts, making it a versatile solution for climate adaptation in drylands. Ecosystem-based solutions like practicing sustainable land management, conserving forests and grassland, and restoring biodiversity provide actionable solutions to address climate change issues. In India's arid regions, millions of people depend on natural resources for their livelihoods. These solutions not only help protect communities from the harsh impacts of climate change but also work to preserve the country's incredible biodiversity (UNDP 2024). The possible ecosystem-based adaptation measures that can be followed in Indian arid conditions are given in Table 7.2.

7.4 Sustainable Land Management Strategies and Their Ecosystem Services

Sustainable land-use practices are "the adoption of land-use systems that through appropriate management practices enable land users to maximize the economic and social benefits from the land while maintaining or enhancing the ecological support functions of the land resources" (FAO). Sustainable land-use practices/systems (Fig. 7.1) in arid zones are essential for preserving ecological stability, promoting biodiversity, and supporting local communities. Adopting sustainable land-use practices in arid zones addresses environmental challenges and provides substantial ecosystem benefits. These practices enhance ecosystem health, community resilience, and sustainable livelihoods, and they can lead to long-term resilience in response to climate variability and resource scarcity for the future of arid landscapes.

Table 7.2 Possible ecosystem-based adaptation measures for arid regions (modified from SCBD, 2009 Technical Series No. 41)

Adaptation measure	Adaptive function	Co-benefits
Arid tree species-based agroforestry systems, e.g., *Prosopis cineraria*-based agroforestry system	Stabilization of sand dunes, improvement of soil fertility through nitrogen fixation	Social: improved fodder and fuel security for local communities; economic: income generation from pods, gum, and fuelwood; biodiversity: support for arid flora and fauna; mitigation: carbon sequestration
Water harvesting and micro-catchments, e.g., check dams, percolation ponds, Khadins	Conservation and efficient use of scarce water resources	Social: improved water availability for households; economic: enhanced agricultural productivity; biodiversity: support for aquatic and desert species; mitigation: reduction in land degradation and soil erosion
Drought-resistant cropping systems, e.g., Bajra (*Pennisetum glaucum*), cluster bean (*Cyamopsis tetragonoloba*)	Increased resilience to water scarcity	Social: improved food security; economic: stable agricultural income in dry years; biodiversity: preservation of local crop varieties; mitigation: soil carbon storage enhancement
Restoration of degraded grasslands, e.g., *Lasiurus sindicus*, *Cenchrus ciliaris*	Enhanced grazing capacity, improved soil structure	Social: support for pastoral livelihoods; economic: income from sustainable grazing practices; biodiversity: habitat for arid-region species; mitigation: soil carbon sequestration
Sustainable livestock management, e.g., Tharparkar cattle, Sirohi goat	Protection against overgrazing and habitat degradation, dung useful as organic manure, etc.	Social: livelihood enhancement for herders; economic: increased income from improved livestock productivity; biodiversity: maintenance of native vegetation; mitigation: reduced methane emissions
Conservation of indigenous plants	Protection of native species adapted to arid conditions	Social: use of traditional knowledge and medicinal plants; economic: market opportunities for native products; biodiversity: conservation of endemic species; mitigation: improved carbon storage/ sequestration potential
Soil and water conservation practices, e.g., Khadins (traditional water harvesting system), windbreaks and shelterbelts, application of organic manures, mulching, rainwater harvesting pits	Prevention of erosion, improvement of water infiltration and percolation	Social: increased agricultural potential; economic: reduced costs of land degradation; biodiversity: recovery of degraded ecosystems; mitigation: reduction in sedimentation and greenhouse gas emissions and loss of soil organic carbon

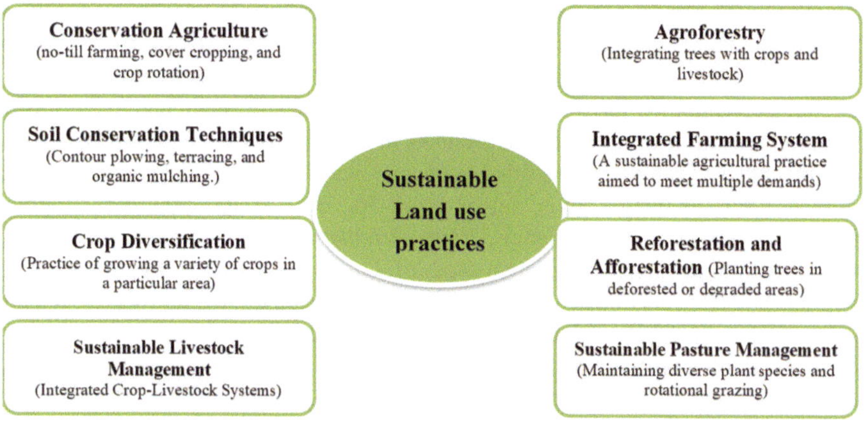

Fig. 7.1 Key sustainable land-use practices

7.4.1 Ecosystem Services from Sustainable Land-Use Practices

Sustainable land-use practices provide a range of ecosystem benefits that contribute to ecological balance and enhance the resilience of both natural and human systems. Here, ecosystem benefits of the major land-use practices are discussed.

7.4.1.1 Provisioning Services

Increased Crop Productivity In the drylands of Rajasthan, traditional agroforestry systems thrive with species like *Prosopis cineraria, Ziziphus mauritiana, Tecomella undulata, Acacia nilotica, Azadirachta indica,* and *Ailanthus excelsa* (Roy et al. 2011; Tanwar et al. 2019). Among these, *Prosopis cineraria* have shown remarkable benefits when integrated with annual crops, particularly millets, yielding an increase of 10–15% under its canopy. This species not only enhances crop productivity but also supports farmers by providing shade and improving soil moisture retention. Crops grow more effectively under *Prosopis cineraria*, with yield performance ranked as follows: pearl millet > cowpea>green gram>dew gram. In addition to enhancing the yields of dryland crops, these trees provide extra benefits through annual lopping, yielding 25–30 kg of dry leaves and 40–60 kg of fuelwood per tree (Bhati et al. 2008). In a similar way, forage production under *Tecomella undulata* was estimated to be 1.5 t/ha. Maximum grass biomass (13.9 t/ha) was found in *Prosopis cineraria* in peak growing season (Tewari et al. 1999). *Ziziphus mauritiana* cv Seb plantation produced 0.2 t/ha of grain and 0.8 t/ha quality fruits on the same unit of land when it was intercropped with mung bean. The net profit from the system was Rs.8000 per hectare, while sole crop gave Rs.4800 per hectare (Gupta 1997). *Ailanthus excelsa* was also intercropped with wheat, millet, barley, mustard, pulses, and guar crops as intercrop, and 25 quintal/ha of wheat and 12.5 quintal/ha of barley yield were recorded when sown in the intermittent space of 10 m × 10 m *A. excelsa* plantation (Jat et al. 2011).

Livelihood Improvement In arid regions, many farmers are considered marginal due to the challenging environmental conditions. However, certain trees provide valuable products such as gum, edible and nonedible oils, and resins, which contribute to increasing their income (Sharma 2000). Additionally, medicinal plants are an important resource of minor forest products in these areas, e.g., *Prosopis cineraria*, locally called as sangri, whose dried pods are sold to boost the local economy. The market value of sangri pods ranges from Rs. 500 to Rs. 700 per kilogram (Subbulakshmi et al. 2023). Sangri, along with the fruits of Gonda (*Cordia myxa*), Ker (*Capparis decidua*), and Kumat seeds of *Acacia senegal* locally called Kumath and Kachra (*Cucumis melo var. agrestis*), is a key element of the local vegetable delicacy known as "Panchkutta," which is mainly used among the people in arid ecosystems of Rajasthan. These ingredients are typically crushed, combined, and then fried after being seasoned to enhance their flavor (Panwar et al. 2014). In Salvadora-based agroforestry, an income of thirteen thousand rupees was earned by farmers in a year compared with a sole agricultural crop (Harsh and Tewari 2007; Subbulakshmi et al. 2019).

Soil Health Improvement The leaf litter, pruned branches, and root exudates from trees and shrubs decompose to form organic matter, enhancing the soil with nutrients and boosting its fertility. This continuous addition of organic matter improves soil structure, increases water retention, and boosts nutrient retention, thus promoting sustainable crop growth. The nutrient composition of sandy soils significantly improved after a decade of planting ber (Ziziphus) under the agri-horti system. The nutrient returns from the litterfall of ber followed this order: N > K > Ca > P (Awasthi and Singh 2010). Additionally, incorporating cowpea (*Dolichos biflorus*) as a cover crop in ber orchards enhanced the water-holding capacity of light soils by increasing the organic carbon content (Pareek and Vishal 1996). Reports indicate that a system centered around *Prosopis cineraria* can substantially improve soil microbial activity. Additionally, *P. cineraria* has been shown to play a crucial role in stabilizing sand dunes in Rajasthan and is capable of withstanding periodic burial (Gates and Brown 1988). Trees play a significant role in stabilizing the soil organic carbon by providing inputs of lignin-rich litter that decomposes slowly (Montagnini and Nair 2004). In comparison with a no-tree control in an arid location, fluxes of C, N, and P through microbial biomass have been shown to be higher in land-use systems based on *Prosopis cineraria*, followed by *Dalbergia sissoo*, *Acacia leucophloea*, and *Acacia nilotica* (Yadav et al. 2011; Keerthika et al. 2017). In *Prosopis cineraria-based* agroforestry system, nutrient input from litterfall of trees was 197.61 kg $ha^{-}1$ year $^{-1}$ nitrogen, 21.46 kg ha^{-1} year $^{-1}$ phosphorus, and 63.48 kg ha^{-1} year $^{-1}$ potassium. These nutrients upon decomposition help in fertility buildup of soils (Verma et al. 2021). In Gujarat's arid regions, agrisilvicultural systems based on *Ailanthus excelsa* and *Azadirachta indica* generated 26% and 59% more revenue than monocropping, respectively (Patel et al. 2008).

Fodder Production The nutritional quality of pasture areas can be improved by integrating legume crops and grass species, enhancing livestock productivity and ultimately increasing farmers' income. Additionally, trees in a silvi-pasture system contribute fodder, fruits, and gum providing further sources of income for farmers. This integration not only improves the overall health of the pasture but also supports a more sustainable agricultural system. The pasture species in the dry zones of India are highly drought-resistant, edible, and rich in nutrients. Common grasslands are essential for the farming systems of these regions, supplying fodder to livestock for a significant portion of the year, typically lasting 4 to 6 months. Fodder production can be significantly enhanced from 0.50 to 3.6 t/ha by growing a mixed grass species, legume crops, and trees (through suitable combinations and improved management practices) (Ramana 2013).

With 2.78 t/ha of dry forage from the pasture and 0.95 t/ha of green tree leaves, the three-tier silvi-pastoral system (*Cenchrus*, *Ailanthus excelsa*, and *Dichrostachys cinerea*) generated the highest average forage output (Tanwar et al. 2015). Table 7.3 depicts some important trees and shrubs with their potential fodder production.

Shelterbelts and Wind Breaks In arid regions, soil erosion by wind is a major challenge that can significantly impact agricultural productivity. Shelterbelts and windbreaks are effective strategies for mitigating these effects, improving farm conditions, and enhancing overall productivity. The net returns of 305.6% have increased by shifting from non-shelterbelt to shelterbelt systems on farms in the Jaisalmer district (Gajja et al. 2008). Evaporative demand studies at CAZRI demonstrated that three-row shelterbelts, consisting of *Acacia tortilis*, *Cassia siamea*, and *Prosopis juliflora* as side rows with *Albizia lebbeck* in the center, effectively reduced wind speed and wind erosion. Additionally, micro-shelterbelts of tall-growing plants provided significant protection, leading to a 20–40% increase in the yield of irrigated vegetable crops (Gupta and Ramakrishna 1988). The shelterbelts of *Acacia tortilis*, *Eucalyptus camaldulensis*, *Dalbergia sissoo*, and *Tecomella undulata* caused maximum reduction in wind speed from 21.5% to 36% on the downward side at a distance of 2H (H means average height of shelterbelt) (Prasad et al. 2013).

Table 7.3 Key fodder-producing species and their forage yield in the arid zone of Rajasthan

Species	Ideal spacing (meters)	Fresh fodder production and availability post-plantation
Prosopis cineraria	10 × 10	About 100 kg in 8–10 years
Colophospermum mopane	8 × 8	5–7 kg in 3–4 years
Ziziphus nummularia	5 × 5	3–5 kg in 3–4 years
Salvadora oleoides	12 × 12	80–100 kg in 8–10 years

Source: Sharma (2013)

7.4.1.2 Regulating Services

Carbon Sequestration In arid zones, deep-rooted climax tree species like *Prosopis cineraria*, which can reach depths of up to 70 meters, play a crucial role in ecosystem stability. Trees and grasses can dramatically boost soil carbon reserves from 24.3 Pg to 34.9 Pg on degraded soils (Subbulakshmi et al. 2019).

Among the silvicultural species, *Colophospermum mopane* had the maximum storage of carbon (45.84%), while *Ailanthus excelsa* had the lowest (43.61%). Agroforestry systems based on *Prosopis cineraria* demonstrated the highest average carbon stock compared to the other silvicultural species, highlighting their potential for enhancing carbon sequestration in arid regions. Agroforestry had a higher average carbon stock than horticulture plots, silvi species, and solo agriculture (Singh and Singh 2015). According to Verma et al. (2015), *Prosopis cineraria, Tecomella undulata*, and *Ziziphus rotundifolia* on agricultural land produced 0.70, 0.47, and 0.40 tons of carbon stock per hectare, respectively. Agroforestry systems' capacity to sequester carbon ranged from 61.97 Mg C ha^{-1} to 106.6 Mg C ha^{-1}. The carbon storage of solo cropping systems, on the other hand, which do not include tree components, was much lower, ranging from 33.34 to 35.52 Mg C ha^{-1} (Verma et al. 2024).

Biodiversity Conservation Agroforestry enhances biodiversity by offering habitat and food sources for various living beings. Multifaceted agroforestry systems establish diverse ecological niches, fostering the growth of native plants and supporting pollinators, birds, and small mammals. It has been discovered that trees in Rajasthan's agro-ecosystems supply habitat and food for numerous bird species, including the endangered cavity nesting birds. In dry western Rajasthan, traditional agroforestry systems have been shown to support a high level of species diversity and serve as a buffer for parks and protected areas (Tewari 2021). In order to reduce biotic and abiotic disturbance, the severely degraded silvi-pasture system in the Govardhanpura and Gokulpura cluster of villages in the Bundi district which has a somewhat poorer rainfall regime was replanted and fenced. Significant variations in plant species biodiversity were observed (Dixit et al. 2005).

Economic Resilience Diversification of income sources through multiple agricultural products reduces economic risk. The production of forest products other than timber from trees remains largely unaffected even when seasonal crops are not able to survive due to dry conditions, helping to maintain availability of food in such conditions. Moreover, certain trees provide additional non-timber products which contribute to the earning of farmers with small landholdings majorly reside in this hot, arid region (Sharma 2000; Tewari et al. 2014; and Subbulakshmi et al. 2019). An economic analysis of established agroforestry systems in western Rajasthan shows that these systems typically yield a positive net B:C ratio (Table 7.4). Besides, 25.4 million hectares of agroforestry land can generate employment of 943 million days annually with enhanced agroforestry systems (NRCAF 2007).

Table 7.4 Economic returns from major agroforestry systems of arid western Rajasthan

Agroforestry system	Returns (Rs ha^{-1})			Gross returns (Rs ha^{-1})	Net returns (Rs ha^{-1})	Net B:C ratio
	Crops	Fuel wood	Leaf fodder			
P. cineraria-A. nilotica-based	1230	4103	870	6203	4353	2.3
P. cineraria-based	3670	600	420	4690	3140	2.0
Ziziphus spp.-*P. cineraria*-based	1506	620	600	2726	1176	0.7
Ziziphus spp.-*P. cineraria-Salvadora* spp.-based	1400	500	500	2400	900	0.6

Source: Tewari et al. (2014)

7.5 Community Engagement and Empowerment for Adapting to Climate Change in Arid Regions

7.5.1 Role of Local Communities in Sustainability

Addressing the challenges posed by climate change is a pressing global concern, and local communities play a vital role in developing sustainable adaptation strategies, particularly in arid regions (Lasage et al. 2008; Khatri et al. 2013; Berman et al. 2020; Makate 2019). These communities often possess invaluable indigenous knowledge and practices that can inform effective climate change adaptation measures (Khatri et al. 2013; Kodirekkala 2018; Makate 2019; Metcalfe et al. 2020). Some reports recently underlined the significance of being aware of community viewpoints, adaptations, and the resilience of communities in arid zones. In Mexico, researchers found that population growth, higher temperatures, and reduced summer rainfall have increased the pressures on communities reliant on small-scale farming and fishing (Sarah et al. 2020).

7.5.2 Participatory Approaches and Stakeholder Engagement

Participatory approaches focus on involving local stakeholders, including communities, government departments, NGOs, and the private sector, to create more resilient and sustainable strategies for adaptation. Here are how these approaches and engagements can be structured and their benefits in the context of arid regions:

1. Stakeholder Mapping and Inclusion: Identifying key stakeholders is an important move for active involvement in the process of mapping. This involves understanding who is directly affected by climate change impacts in the region. In arid areas, stakeholders often include local communities (farmers, herders, indigenous groups), government bodies, research institutions, NGOs, and private companies.

Community-Led Vulnerability Assessments: Community members, especially indigenous groups, often possess invaluable knowledge about local climates, ecosystems, and traditional adaptation methods. Including this knowledge helps in accurately assessing vulnerabilities. CoDriVE-PD is a recombinant tool that can be used for vulnerability assessments which combines key elements from three well-established international research methodologies: the Participatory Tool on Climate and Disaster Risks, the Department for International Development (DFID)'s Sustainable Livelihoods Framework, and the Driver-Pressure-State-Impact-Response (DPSIR) model. It evaluates climate risks by examining a community's perception of its circumstances through the lens of the five livelihood capitals, drawing on resilience theory and a systems thinking approach to synthesize and interpret the results. This tool not only provides an accurate assessment of who is vulnerable, what the risks are, and why they exist, but also aids in the design of responses. These responses can either be incorporated proactively during the planning stage or as mid-course adjustments during implementation (Watershed Organisation Trust 2013, https://wotr.org/2017/09/26/1309/).

2. Collaborative Planning and Decision-Making: Joint Development of Adaptation Plans: Plans created with the input of all stakeholders, especially those most impacted, tend to be more practical and widely accepted. For example, water management approaches developed with the involvement of both agricultural and urban stakeholders in arid regions can help balance competing demands. In rural Maharashtra, a community-based vulnerability assessment tool was employed to examine climate risks and vulnerabilities across different social groups. The findings reveal that vulnerability is socially differentiated, varying across farmer categories and social groups. Factors such as caste and social status significantly influence access to resources, land ownership, livelihood choices, and strategies, which in turn affect vulnerability to climate change. The study confirmed that vulnerability assessments should be conducted at smaller scales, as climate risks can vary even within neighboring villages. This approach supports the development of targeted programs and policies to enhance the adaptive capacity of the rural poor and highlights the importance of integrating community engagement into academic research (Kuchimanchi et al. 2019).
3. Capacity Building and Skill Development: Training in adaptive practices that train community members in sustainable land and water management practices, such as soil and water conservation, and drought-tolerant crops, can increase local resilience. Engaging local institutions and enhancing their ability to support climate adaptation is key. Government bodies can benefit from training in risk assessment and resource management specific to arid regions.
4. Participatory Monitoring and Evaluation: Locals can participate in monitoring the effectiveness of adaptation measures, providing regular feedback on challenges and successes. Engaging communities in evaluating strategies ensures that adaptation measures remain relevant and effective over time.
5. Policy Advocacy and Inclusive Governance to support Policy Change: Stakeholder engagement can lead to more inclusive climate adaptation policies, and incorporating stakeholder feedback in Governance ensures that government actions align with local needs and environmental realities.

7.5.3 Benefits of Participatory Approaches in Arid Regions

- *Enhanced Resilience*: By involving stakeholders, adaptation plans are more tailored to local challenges, fostering resilience.
- *Ownership and Sustainability*: Stakeholders who contribute to creating adaptation strategies are more likely to support and sustain them.
- *Knowledge Exchange*: Participatory approaches facilitate knowledge sharing between experts and locals, blending traditional and scientific knowledge for innovative solutions.
- Examples of Successful Participatory Approaches in Arid Regions
- *Water User Associations*: These have been effective in managing scarce water resources through community-based systems in places like North Africa.
- *Participatory Drought Planning*: In India, stakeholder involvement in drought management strategies has led to community-driven water harvesting and soil management practices.

7.5.4 Integrating Customary Information with Current Practices for Climate Change Adoption in Arid Regions

Integrating local know-how with modern practices for climate change adaptation in arid regions can create powerful solutions that are both innovative and contextually appropriate. Traditional knowledge includes the long-held wisdom and practices of indigenous and local communities who have lived sustainably within arid environments. When combined with modern scientific approaches, these insights can enhance resilience and adaptability to climate stressors like extreme temperatures, droughts, and water scarcity. Figure 7.2 outlines the process of integrating traditional and modern practices, from identifying local knowledge to implementing and scaling up effective adaptation strategies. It emphasizes community involvement at each step, ensuring that solutions are practical, resilient, and sustainable.

7.5.5 Case Studies of Integrated Approaches

The Zai Pits of West Africa Farmers in Burkina Faso and Mali have used Zai Pits—shallow holes that concentrate water and nutrients around plants for centuries (Fig. 7.3). Researchers have introduced composting techniques to increase nutrient availability, significantly boosting crop yields in these drought-prone areas (Danjuma and Mohammed 2015).

The Hima System in Jordan An ancient grazing management system, Hima, has been revived and adapted using modern land management techniques. Combining rotational grazing with land restoration has improved pasture conditions and reduced desertification in Jordan (Myint and Westerberg 2014).

Fig. 7.2 Flowchart outlining the process of integrating traditional and modern practices

INTEGRATING TRADITIONAL KNOWLEDGE WITH MODREN PRECTICES
|
▼
IDENTIFY TRADITIONAL KNOWLEDGE AND PRACTICES
└─▶ Gather knowledge from local communities
└─▶ Identify traditional techniques in:
└─ Water management
└─ Agriculture
└─ Soil conservation
└─ Climate prediction
|
▼
ASSESS RELEVANCE OF TRADITIONAL PRACTICES FOR ADAPTATION
└─▶ Analyze practices based on:
└─ Local environmental challenges
└─ Climate change projections
|
▼
IDENTIFY MODERN TECHNOLOGIES & SCIENTIFIC KNOWLEDGE
└─▶ Select relevant modern practices such as:
└─ Advanced irrigation systems
└─ Drought-resistant crop varieties
└─ Climate monitoring technologies
|
▼
DESIGN INTEGRATED ADAPTATION STRATEGIES
└─▶ Blend traditional practices with modern technologies
└─▶ Ensure strategies are culturally acceptable and cost-effective
|
▼
IMPLEMENTATION & CAPACITY BUILDING
└─▶ Train local communities in integrated practices
└─▶ Provide resources and support for implementation
|
▼
PARTICIPATORY MONITORING AND EVALUATION
└─▶ Involve communities in monitoring effectiveness
└─▶ Adjust strategies based on community feedback and observed results
|
▼
SUSTAINABILITY AND UPSCALING
└─▶ Promote successful strategies across the region
└─▶ Foster policy support for integrated adaptation approaches

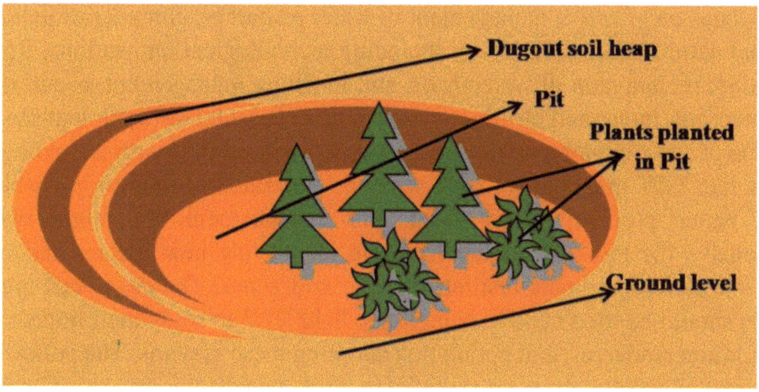

Fig. 7.3 Zai Pits method of planting

The Khadin System of Rajasthan The *Khadin* is an ancient system of land use established long time ago in the Jaisalmer district of Rajasthan, specifically suited for areas where rocky catchments are located near valley plains. In this system, embankments in lower elevations are created to channel and store rainwater, which then percolates into the ground, supporting crop cultivation. Excess water flows out through a spillway, while the main body of water remains in the Khadin throughout the monsoon season (Kolarkar et al. 1983).

7.6 Challenges and Future Perceptions

Ecosystem based approaches (EbA) pose numerous challenges for adaptation in arid zone in changing climate, but at the same time, it offers multiple opportunities also. The key constraints are social, economic, and policy issues, abrupt climatic situations, and need for continuous research and innovative solutions. The hurdles in implementing sustainable land-use management practices will be water scarcity, land degradation, and biodiversity loss. Sustainable water management is a big challenge due to limited water resources in these regions. Desertification and soil loss are the common constraints of the drylands. The tree-based land-use practices like agroforestry will help up to large extent in successful implementation of ecosystem-based solutions. Arid regions are characterized by unique and high stress tolerance species. However, there will be a challenge to conserve them due to problems of extreme weather aberrations, habitat fragmentation, and excess exploitation. Socioeconomic and policy issues can also limit the successful implementation of ecosystem based approaches. Need for short-term economic gain over long-term environment stability, conflicted interest of industries with local communities and stakeholders, and limited access of rural livelihoods to education technology and financial resources are some of key socioeconomic factors hindering the sustainable ecosystem restoration. Therefore, the future research and innovative solutions should focus on effective management of water resources, climate resilient agriculture, and nature-based solution by including technological innovations. Rainwater harvesting, reclamation of watersheds, soil moisture management, use of drought-resistant crops, integrated farming system approaches, and restoration of sand dunes are some of the common practices that need to be combined with technology innovations like GIS mapping, remote sensing, and satellite monitoring which will enable better productive and sustainable management of natural resources. Additionally, the research should focus on identifying new species and land-use systems to develop climate resilience in the arid zones. The developed innovative solution should be tried hand in hand at the field level to study their impact on ecosystem, water resources, and population adopting these systems. The policy support and successful policy setups are essential to implement ecological-based approaches and strategies.

7.7 Conclusion

In drylands referring mostly to arid and semi-arid regions, ecosystem-based adaptation or approachesEbA offer great scope and potential to deal with the challenges of changing climate. They provide effective and sustainable solutions to minimize the damage and overexploitation of natural resources and mitigate the risk factors. EbA will further emphasize the conservation and restoration of biodiversity of local flora and fauna. It is a multidimensional approach that not only enhances the resilience but also supports the livelihood over long term by promoting its sustainable use. Successful case studies in India and around the world show the effective implication of practices like watershed reclamation, agroforestry and mixed farming practices, and conservation of water and soils. However, for large-scale impact there is a need of strengthening community involvement and policy support. Focusing on these measures, EbA can help build climate resilience in vulnerable regions, offering a cost-effective, nature-based alternative to conventional solutions.

References

Awasthi OP, Singh IS (2010) Effect of ber and pomegranate plantation on soil nutrient status of typic torripsamments. Ind J Hort 67:138–142

Baraj B, Mishra M, Sudarsan D et al (2024) Climate change and resilience, adaptation, and sustainability of agriculture in India: a bibliometric review. Hel 10(8):e29586

Berdugo M, Gaitán J, Delgado-Baquerizo M et al (2022) Prevalence and drivers of abrupt vegetation shifts in global drylands. Proc Natl Acad Sci USA 119:e2123393119

Berman M, Baztan J, Kofinas G et al (2020) Adaptation to climate change in coastal communities: findings from seven sites on four continents. Clim Cha 159:1–16. https://doi.org/10.1007/s10584-019-02571-x

Bhati TK, Tewari JC, Rathore SS (2008) Productivity dynamics of integrated farming systems in western Rajasthan. In: Pratap N et al (eds) Diversification of arid farming systems. Arid Zone Research Association of India and Scientific Publishers, Jodhpur, pp 23–30

Chaudhary P, Bohra NK, Choudhary KR (2011) Conserving biodiversity of community forests and rangelands of a hot arid region of India. Land Use Poli 28:506–513

Cherlet M, Hutchinson C, Reynolds J et al (2018) World atlas of desertification. Publications Office of the European Union, Luxembourg

Colls A, Ash N, Ikkala N (2009) Ecosystem-based adaptation: a natural response to climate change. IUCN, Gland, p 16

Danjuma MN, Mohammed S (2015) Zai pits system: a catalyst for restoration in the dry lands IOSR. Jour of Agri Vet Sci 8(2/I):01–04

Dixit S, Tewari JC, Wani SP et al (2005) Participatory biodiversity assessment: enabling rural poor for better natural resource management. ICRISATP, India, p 16

Doswald N, Osti M (2011) Ecosystem-based approaches to adaptation and mitigation: good practice examples and lessons learned in Europe. Bundesamt für Naturschutz (BfN), Bonn

FAO (n.d.). https://www.fao.org/land-water/land/sustainable-land management/en/#:~:text=The%20United%20Nations%20defines%20sustainable,the%20maintenance%20of%20their%20environmental

Gajja BL, Prasad R, Mertia RS et al (2008) Impact of shelterbelt on net returns from agricultural production in arid Western Rajasthan. Agri Eco Res Rev 21(1):118–122

Gates PJ, Brown K (1988) *Acacia tortilis* and *Prosopis cineraria*: leguminous trees for arid areas. Outl on Agri 17:61–64

Gholkar MD, Goroshi S, Singh RP, Parihar JS (2014) Influence of agricultural developments on net primary productivity (NPP) in the semi-arid region of India: a study using GloPEM model. The International Archives of the Photogrammetry, Remote Sensing and Spatial Information Sciences, vol 40, pp 725–732

González-Sánchez EJ, Kassam A, Basch G et al (2016) Conservation agriculture and its contribution to the achievement of agri-environmental and economic challenges in Europe. AIMS Agric Food 1(4):387–408

Gopalakrishnan R, Jayaraman M, Ravindranath NH (2011) Regional Climate Modelling results for Rajasthan state. Presented at the workshop, Science-based policy options for climate change adaptation in Rajasthan. 24–25 February 2011. Organized by Rajasthan state pollution control Board, Jaipur, Rajasthan

Gupta JP (1997) Some alternative production systems and their management for sustainability. In: Gupta JP, Sharma BM (eds) Agroforestry for sustained productivity in arid regions. Scientific publishers, Jodhpur, pp 31–33

Gupta JP, Ramakrishna YS (1988) Role of shelterbelts in checking wind erosion and increasing crop production in sandy waste lands of western Rajasthan. Wastelands development and their utilization. Scientific Publishers, pp 271–280

Harsh LN, Tewari JC (2007) Agroforestry system in arid region (Rajasthan) of India. In: Puri S, Panwar P (eds) Agroforestry systems and practices. New India Publishing Agency, p 175

Huang J, Yu H, Guan X et al (2016) Accelerated dryland expansion under climate change. Nat Clim Chang 6:166–171

International Centre for Integrated Mountain Development (ICIMOD) (2024) Himalayan resilience enabling action programme. Available at: https://lib.icimod.org/record/36508

Jat RHS, Singh K, Mann JS (2011) Ardu (*Ailanthus sp*) in arid ecosystem: a compatible species for combating with drought and securing livelihood security of resource poor people. Indian J Tradit Knowl 10(1):102–113

Keerthika A, Gupta DK, Noor Mohamed MB, Jangid BL, Shukla AK, Choudhary KK (2017) Agroforestry for climate change mitigation and adaptation in hot arid region of Rajasthan. In: Parthiban KT, Jude Sudhagar R, Cinthia Fernandaz C, Suresh KK (eds) Agroforestry strategies for climate change: mitigation and adaptation. Jaya Publishing House, Delhi, pp 285–300

Khatri DB, Bista R, Gurung N (2013) Climate change adaptation and local institutions: how to connect community groups with local government for adaptation planning. J Liv 11(1):14–28. https://doi.org/10.3126/jfl.v11i1.8610

Kodirekkala KR (2018) Cultural adaptation to climate change among indigenous people of South India. Clim Cha 147:299–312. https://doi.org/10.1007/s10584-017-2116-8

Kolarkar AS, Murthy KNK, Singh N (1983) *Khadin* – a method of harvesting water for agriculture in the Thar Desert. J Arid Environ 6(1):59–66

Krishnan R, Sanjay J, Gnanaseelan C et al (2020) Assessment of climate change over the Indian region: a report of the ministry of earth sciences (MOES), Government of India. Springer Nature, p 22

Kuchimanchi BR, Nazareth D, Bendapudi R et al (2019) Assessing differential vulnerability of communities in the agrarian context in two districts of Maharashtra, India. Clim Dev 11(10):918–929. https://doi.org/10.1080/17565529.2019.1593815

Lasage R, Aerts J, Mutiso GCM et al (2008) Potential for community-based adaptation to droughts: sand dams in Kitui, Kenya. Phys Chem Earth Parts A/B/C 33(1–2):67–73. https://doi.org/10.1016/j.pce.2007.04.009

Makate C (2019) Local institutions and indigenous knowledge in adoption and scaling of climate-smart agricultural innovations among sub-Saharan smallholder farmers. Inter Jou Cli Cha Str Man 12(2):270–287

MEA. (2005) Millennium Ecosystem Assessment: Ecosystems and human well-being: Synthesis. www.millenniumassessment.org/documents/ document.356.aspx.pdf

Metcalfe SE, Birgit S, Doreen S et al (2020) Community perception, adaptation and resilience to extreme weather in the Yucatan Peninsula. Mexico Regi Env Cha 20:25. https://doi.org/10.1007/s10113-020-01586-w

Minampati VRR, Sugandhar P (2023) Rejuvenation of water resources Management for Sustainable Development: a case study on Mission Kakatiya in state of Telangana. Int J Environ Sci 9:19–29

Montagnini F, Nair PKR (2004) Carbon sequestration: an under exploited environmental benefit of agroforestry systems. Agrofor Syst 61:281–295

Myint MM Westerberg V (2014) An economic valuation of a large-scale rangeland restoration project through in Jordan. Report for the ELD Initiative by International Union for Conservation of Nature, Nairobi, Kenya. Available from: www.eld-initiative.org

Naorem A, Somasundaram J, Yash P et al (2023) Soil constraints in an arid environment – challenges, prospects, and implications. Agroforestry 13(1):220

NRCAF (2007) NRCAF perspective plan vision 2025. NRC for Agroforestry

Panwar D, Kailash P, Bharti CS (2014) Unripe pods of *Prosopis cineraria* used as a vegetable (sangri) in Shekhawati region. Int J Sci Eng Res 5(2):892–895

Pareek OP, Vishal N (1996) Coordinated fruit research in Indian arid zone: a two-decade profile. NRC for Arid Horticulture, Bikaner (India)

Patel JM, Jaimini SN, Patel SB (2008) Evaluation of neem and ardu based agri-silvi systems. In: Partap N, Singh MP, Kar A et al (eds) Diversification of arid farming systems. AZRAI and Scientific Publishers, Jodhpur, p 114

Prasad R, Mertia RS, Gajja L et al (2013) Morphological characteristics of tree shelterbelts and their impact on wind regimes, air temperature and soil quality. Indian J Agrofor 15(1):74–81

Pravalie R (2016) Drylands extent and environmental issues. A global approach. Eart Sci Rev 161:259–278

Ramana DBV (2013) Silvi-Pastoral and Horti-Pastoral models for small ruminant production. Central Research Institute for Dryland Agriculture, Hyderabad. Available at: http://www.crida.in/DRM1-Winter%20School/DBVR.pdf

Rathore MS (2005) State level analysis of drought policies and impacts in Rajasthan, India. IWMI Working Paper 93: Drought Series Paper 6. IWMI, Colombo, Sri Lanka

Ravindra K, Bhardwaj S, Ram C et al (2024) Temperature projections and heatwave attribution scenarios over India: a systematic review. Heliyon 10(4):e26431

Roy MM, Tewari JC, Ram M (2011) Agroforestry for climate change adaptations and livelihood improvement in Indian hot arid regions. Int J Agric Crop Sci 17(2):43–54

Sarah EM, Birgit S, Doreen SB et al (2020) Community perception, adaptation and resilience to extreme weather in the Yucatan peninsula, Mexico. Reg Env Cha 20:25

SCBD, Secretariat of the Convention on Biological Diversity (2009) Connecting biodiversity and climate change mitigation and adaptation: report of the second Ad Hoc technical expert group on biodiversity and climate change. Montreal, Technical Series No. 41, p 126

Sharma AK (2000) Addition in tradition: agroforestry in the arid zone of India. LEISA-INDIA 2(3):19–20

Sharma NK (2013) Fodder strategy for sustainable animal production in arid Rajasthan. Anna Ari Zon 52(2):101–108

Sharma M, Ashraf J, Mehta D et al (2024) Assessment of nature-based interventions adherence with IUCN global standards and an analysis of flow of associated ecosystem services in tropical drylands of India. Ecol Indic 159:111717

Singh RB, Kumar A (2015) Climate variability and water resource scarcity in drylands of Rajasthan, India. Geo Dis 2:1–10

Singh G, Singh B (2015) Rooting pattern and equations for estimating biomasses of *Hardwickia binata* and *Colophospermum mopane* trees in agroforestry system in Indian desert. Res Rev J Bot Sci 4(1):30–40

Singh NP, Singh S, Anand B et al (2019) Climate vulnerability assessment in semi-arid and arid region of Rajasthan, India: an enquiry into the disadvantaged districts. J Agrofor 21(2):197–202

Subbulakshmi V, Sheetal KR, Renjith PS et al (2019) NTFP based agroforestry to sustain income and employment generation activities of arid regions of Rajasthan. In: Parthiban KT, Ramah K, Sivakumar K et al (eds) Multifunctional agroforestry (ecosystem services). Part-I Jaya Publishing House, pp 291–324

Subbulakshmi V, Sheetal KR, Noor Mohamed MB et al (2023) Arid agroforestry for Thar Desert. In: Varghese N et al (eds) Natural resource management in the Thar Desert Region of Rajasthan. Springer, pp 155–192

Tanwar SPS, Kumar, S Roy MM (2015) Integrated farming system for higher income, security and optimum resource utilization in arid regions. Agri Care. Available at: http://agricare.kisanhelp.in/content/integrated-farming-system-higher-income-security-andoptimum-resource-utilization-arid

Tanwar SP, Kumar P, Verma A et al (2019) Carbon sequestration potential of agroforestry systems in the Indian arid zone. Curr Sci 117(12):2014–2022

Tewari JC (2021) Traditional and improved agroforestry systems in arid western Rajasthan with special reference to climate change adaptation and livelihood security. In: Shukla AK, Gupta DK, Jangid BL et al (eds) Horticulture based integrated farming systems. New India Publishing Agency, pp 159–178

Tewari JC, Bohra MD, Harsh LN (1999) Structure and production function of traditional extensive agroforestry system and scope of intensive agroforestry in Thar Desert. Indian J Agrofor 1(1):81–94

Tewari JC, Ram M, Roy MM et al (2014) Livelihood improvements and climate change adaptation through agroforestry in hot arid environment. In: Dagar J, Singh A, Arunachalam A (eds) Agroforestry systems in India: livelihood security and ecosystem services, vol 10. Advances in Agroforestry, pp 155–183

United Nations Development Programme (UNDP) (2024) India shows why ecosystem-based approaches are critical to global climate action. Available at: https://www.undp.org/

Venkateswarlu B (2012) Dryland agriculture: coping with climate change. Ann Ari Zon 51(3 & 4)

Verma A, Tewari JC, Partap R, Kumar S (2015) Carbon stock of Woody vegetation in hot arid region of Rajasthan. Int J Trop Agric 33(4):3019–3023

Verma A, Kumar P, Soni ML et al (2021) Litter production and litter dynamics in different agroforestry systems in the arid western region of India. Biol Agric Hortic 38(1):40–60. https://doi.org/10.1080/01448765.2021.1971110

Verma A, Naorem A, Keerthika A et al (2024) Biomass and carbon partitioning across different components of agroforestry systems in the arid zones of India. Agrofor Syst 98:3089–3108. https://doi.org/10.1007/s10457-024-01076-6

Wang L, Jiao W, MacBean N, at al. (2022) Dryland productivity under a changing climate. Natu Cli Cha 12:981–994

Wang Y et al (2024) Ecosystem management and sustainable livelihoods in drylands. In: Fu B, Stafford-Smith M (eds) Dryland social-ecological systems in changing environments. Springer, Singapore

Watershed Organisation Trust (2013) Community-driven vulnerability evaluation tool "CoDriVE-Programme designer" a handbook – incorporating vulnerability to climate change into project design and implementation. ISBN No: 978-81-86748-32-9v

Watershed Organisation Trust (WOTR) (2022) Towards a roadmap for ecosystem-based adaptation in Maharashtra, International Climate Initiative (IKI). Available at: https://www.international-climate-initiative.com/PUBLICATION1033

weADAPT (2022) Adapting to climate change in India: the value of an ecosystem-based response. (EbA Info Brief Series No. 1). 2020. International Climate Initiative (IKI). Available at: https://weadapt.org

Yadav RS, Yadav BL, Chhipa B et al (2011) Soil biological properties under different tree based traditional agroforestry systems in a semi-arid region of Rajasthan. India AgroforSyst 81(3):195–202

Yadav OP, Singh JP, Kakani RK et al (2020) Managing agrobiodiversity of Indian drylands for climate-change adaptation. Indian J Pla Gen Res 33(1):3–16

Zhao W, Yu X, Liu Y et al (2021) Distinct vegetation response to drying and wetting trends across an aridity threshold. Envi Res Com 3:025002

Chapter 8
Fostering the Future of Drylands: Balancing Productivity and Resilience Through Nature-Based Solutions

M. Saritha, Shrvan Kumar, Anand Kumar Naorem, Om Prakash Meena, Kamlesh Kumar Meena, and Dheeraj Singh

Abstract Climate change impacts are more conspicuous in natural systems like drylands, where the consequences translate mainly into changing rainfall and land degradation patterns, increased mean temperature, decreased agricultural productivity, and biome shifts. To cope with these challenges and the consequential food and water insecurity and poverty, there is a need for adaptive strategies that promote economic productivity, environmental health, and societal development through an ecosystem-based approach. This involves the integrated management of land, water, and bioresources to promote environmental resilience by restoring, maintaining, and improving ecosystems and to support sustainable livelihoods. As it considers all the goods and services of the ecosystem, such as recreational services, biodiversity conservation, water quality, and carbon storage, in addition to food, fuel, and fiber, this novel approach is more efficient than the traditional technical measures. As a result, this approach has gained importance in recent times in the research arena and in making policy decisions. Increasing the vegetation cover by planting drought-tolerant crops, resorting to water-efficient cropping practices, and facilitating groundwater management, afforestation and reforestation strategies, rehabilitation of mine areas, etc., are some of the dryland conservation and rehabilitation measures adopted for sustainable development. A proper assessment of the goods and services of drylands, along with capacity building and knowledge sharing, would enable the development of a cost-effective approach tailored specifically to the region. Healthy ecosystems with improved soil health, enhanced water security, and biodiversity conservation could foster a sustainable future for communities reliant on these fragile drylands.

Keywords Climate change · Drylands · Integrated management · Knowledge sharing · Sustainable livelihood

---

M. Saritha · S. Kumar · A. K. Naorem · O. P. Meena · K. K. Meena · D. Singh (✉)
ICAR-Central Arid Zone Research Institute, Jodhpur, Rajasthan, India

Y. D. I. Khan et al. (eds.), *Land Restoration Through Ecosystem-Based Approach*, Disaster Resilience and Green Growth, https://doi.org/10.1007/978-981-95-0389-6_8

8.1 Understanding Ecosystem-Based Approaches

Ecosystem-based approaches (EBAs) encompass comprehensive strategies that merge ecological, social, and economic dimensions to promote sustainable management of natural resources while enhancing human well-being. These approaches highlight the significance of ecosystem services (ES) in policy development, particularly within agriculture, where they strive to boost productivity and minimize environmental harm (Devi 2024). By incorporating ecological principles into agricultural techniques, EBAs encourage farmers to adopt methods that improve soil health and soil water retention and reduce reliance on chemical inputs. Furthermore, EBAs emphasize the need for managing human activities to maintain ecosystem health, ensuring that cumulative impacts remain below critical thresholds (Sander 2024). This holistic management approach not only aids in conserving biodiversity but also protects vital ecosystem functions that communities depend on for their livelihoods.

In addition to addressing agricultural challenges, EBAs leverage biodiversity and ecosystem services to strengthen community resilience against climate change. Practices such as reforestation and habitat protection serve as practical examples of how local communities can enhance their adaptive capacity. By restoring ecosystems, these strategies help mitigate the impacts of severe weather events like floods and droughts, which are becoming more frequent as a result of climate change. Nature-based solutions (NbS) play an integral role in this framework by promoting ecosystem connectivity and conservation while also tackling broader societal issues like climate change and biodiversity loss (Yazdandoost 2023). Overall, EBAs represent a robust framework for fostering sustainable development and ecological integrity across various sectors, ensuring that both natural systems and human communities thrive together.

The development of ecosystem-based approaches (EBA) in environmental governance has undergone significant transformation, particularly through international legal frameworks. The concept originally arose from the necessity to conserve natural ecosystems and gained momentum through various international agreements leading up to the 1992 Convention on Biological Diversity (CBD), which established the ecosystem approach as a comprehensive strategy for integrated resource management (Suietnov 2022). A crucial turning point occurred in the 1980s, when the ecosystem approach began to gain recognition within marine environmental law, addressing the shortcomings of fragmented management practices (Enright and Boteler 2020). This shift reflects a growing awareness of the interconnections between ecological systems and human activities, highlighting the importance of adaptive management and collaboration among stakeholders (Enright and Boteler 2020).

Moreover, the incorporation of ecosystem services into agricultural policy demonstrates the approach's applicability across multiple sectors, aiming to improve sustainability and resource efficiency while tackling challenges related to valuation and implementation (Devi 2024). This historical evolution of EBA indicates a significant paradigm shift toward more proactive and integrative environmental governance, as it aligns with the recognition that effective management demands an all-encompassing view that considers both ecological health and human needs (Liu et al. 2024).

8.2 Effects of Climate Change on Ecosystems

8.2.1 Alterations in Biodiversity

Climate change impacts biodiversity mainly through mechanisms such as habitat loss, shifts in species distributions, and increased susceptibility to diseases and invasive species (Fig. 8.1). Rising temperatures and extreme weather events disrupt ecosystems, particularly in vulnerable areas like coastal regions, resulting in changes to community structures and the potential for species extinctions (Khatsuriya et al. 2024; Santojanni et al. 2023). Furthermore, climate change intensifies existing pressures, such as variations in salinity and decreased rainfall, which pose additional threats to both plant and animal populations (Muhsin et al. 2023). These effects weaken ecosystem resilience, as biodiversity plays a vital role in sustaining ecosystem functions and services. A diverse ecosystem is better equipped to adapt to environmental changes; however, many current conservation initiatives tend to focus narrowly on species richness, often overlooking broader biodiversity aspects that enhance resilience (Vasiliev 2022). Consequently, a comprehensive approach that incorporates various levels of biodiversity is crucial for developing effective conservation and adaptation strategies (Wang et al. 2024).

It highlights key aspects of biodiversity affected by climate change, including habitat loss and fragmentation, altered species distributions, changes in species

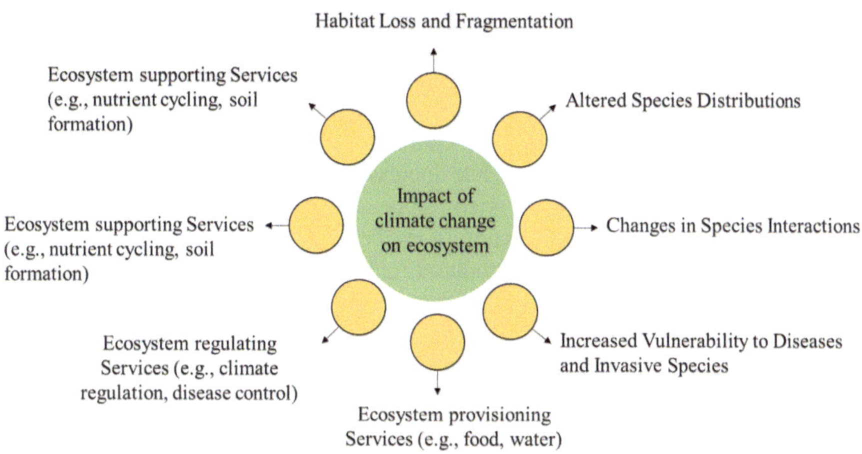

Fig. 8.1 Climate change impacts on ecosystem services (Figure created by authors)

interactions, and increased vulnerability to diseases and invasive species. Additionally, the diagram depicts the consequential effects on ecosystem services, categorizing them into provisioning services (like food and water supply), regulating services (like climate control and disease management), supporting services (comprising nutrient cycling and soil development), and cultural services (covering recreation and spiritual advantages).

Increasing temperatures and shifts in precipitation patterns are key factors in habitat loss and fragmentation, significantly affecting biodiversity across various ecosystems. Research indicates that climate change diminishes habitat connectivity, exemplified by the substantial reduction of core habitats for species such as the brown-eared pheasant, which may face a potential habitat loss of up to 97.63% under projected climate scenarios (Cao and Shi 2022). In a similar vein, the endangered *Aconitum coreanum* is expected to experience habitat decrease of at least 80%, resulting in heightened fragmentation and weakened patch cohesion (Park et al. 2023). The case of the American pika, which has been extirpated from its core habitat in California, illustrates how even seemingly suitable environments can become uninhabitable due to increasing temperatures and diminishing snowpack (Stewart et al. 2017). Additionally, climate-induced changes in hydrology have significantly reduced available habitats for Chinook salmon, adversely affecting both their spawning and juvenile habitats (Crozier et al. 2021). A meta-analysis highlights that the combined effects of habitat loss and climate change are especially severe in regions experiencing high maximum temperatures, emphasizing the urgent necessity for integrated conservation strategies (Mantyka-Pringle et al. 2012).

Rising temperatures and changing precipitation patterns play a significant role in shifting species distributions through intricate interactions among various environmental factors. As global temperatures rise, species are generally expected to move toward cooler areas, such as higher latitudes and elevations. However, empirical data reveals considerable variability in these movements, with only 46.6% of species exhibiting the anticipated migrations (Rubenstein et al. 2023). Changes in precipitation can either promote or impede these shifts; for instance, studies have shown that increased summer temperatures and altered snowfall patterns impact the elevational movements of plant species in Western North America (Harsch and HilleRisLambers 2016). Additionally, within avian populations, rising temperatures typically drive species upslope, while increased precipitation may lead them to migrate downslope, resulting in varied responses across different regions (Tingley et al. 2012). This complex interaction of temperature and precipitation, combined with species-specific tolerances, produces multi-directional shifts that complicate the assumption of straightforward poleward movements (VanDerWal et al. 2013). Therefore, understanding these dynamics is essential for accurately predicting future biodiversity patterns in the context of climate change scenarios (Lauchlan and Nagelkerken 2020).

8.2.2 *Changes in Ecosystem Services*

Climate change profoundly affects provisioning services, particularly food and water, through mechanisms that disrupt ecosystems and agricultural productivity. Weather extremes, such as droughts and floods, result in decreased agricultural yields and pose significant threats to food security by altering optimal planting seasons and increasing pest populations (Delgado-Torres et al. 2024; Semeraro et al. 2023). In rural areas, dependence on forest resources for livelihood is jeopardized as climate-driven tree mortality reduces the availability of these essential resources (Arenas-Wong et al. 2023). Moreover, tropical regions encounter serious challenges, including crop failures, loss of forage, and reduced water supply due to rising temperatures and ocean acidification. These changes highlight the urgent need for adaptation strategies, such as sustainable water management and innovative agricultural practices, to bolster resilience and ensure the continued flow of ecosystem services (Semeraro et al. 2023; Weil et al. 2023).

Climate change substantially affects ecosystem regulating services, mainly by altering temperature, changing precipitation patterns, and increasing occurrences of extreme weather events. These alterations disrupt crucial functions such as carbon sequestration, water regulation, and disease control. For example, rising temperatures and shifting rainfall patterns can intensify droughts and floods, adversely affecting biodiversity and the resilience of ecosystems, which diminishes their capacity to regulate climate and manage disease vectors (Shivanna 2022). Additionally, climate change induces shifts in species distributions and alters the timing of biological events, disrupting ecological interactions and affecting essential services like pollination and seed dispersal (Pandey and Ghosh 2023). The degradation of urban ecosystems further complicates these dynamics, as urban areas contend with unique challenges, such as the "urban heat island effect," which can negatively impact air quality and public health (Pandey and Ghosh 2023).

Climate change profoundly affects ecosystem supporting services, especially nutrient cycling and soil formation, via several mechanisms. Changes in climatic conditions, like increasing temperatures and shifting precipitation patterns, disrupt microbial communities that are vital for nutrient transformation and cycling, resulting in decreased organic carbon and nitrogen levels in soils (Das et al. 2024). Such alterations can cause soils to transition from being carbon sinks to carbon sources, thereby increasing greenhouse gas emissions. Furthermore, climate change influences soil processes, including erosion, structural degradation, and hydrological dynamics, which in turn affect nutrient availability and cycling efficiency (Chatterjee et al. 2024). The interaction between abiotic factors and microbial activity is essential, as microbial communities facilitate decomposition and mineralization, thereby regulating nutrient dynamics (Singh et al. 2021b). Overall, the intricate relationships between climate change and soil processes highlight the need for adaptive management strategies to mitigate the negative impacts on ecosystem services.

8.2.3 Vulnerable Ecosystems (Drylands)

Climate change exerts a profound influence on dryland ecosystems and biodiversity through various mechanisms, including altered precipitation patterns, increased aridity, and shifts in species composition. Variability in precipitation—rather than merely average annual levels—emerges as a critical driver that affects soil multifunctionality and microbial diversity. This variability can lead to a decline in beneficial species and essential ecosystem functions, ultimately compromising soil health and resilience (Guo et al. 2023). For example, fluctuations in rainfall can disturb the fragile equilibrium of nutrient cycling and soil structure, making it increasingly difficult for plants and microorganisms to thrive in these environments.

As climatic conditions continue to grow drier and warmer, ecosystems may experience unprecedented transitions toward characteristics typically associated with drylands. This shift impacts crucial ecological processes, such as carbon cycling and plant productivity, which are vital for maintaining ecosystem stability and function (Grünzweig et al. 2022). For instance, the reduction in moisture availability may inhibit plant growth, thereby limiting the ability of these ecosystems to sequester carbon and regulate atmospheric conditions. This could lead to a feedback loop where diminished plant productivity further exacerbates climate change by reducing carbon storage capacity.

Long-term studies have shown that extreme weather events—such as prolonged droughts and wildfires—intensify these changes by restructuring plant communities and diminishing primary production (Hudson et al. 2022). Such events can cause immediate damage to vegetation, but their long-term effects are often more insidious, leading to shifts in species dominance and even permanent changes in the ecosystem structure. The loss of resilient species and the introduction of more drought-sensitive species can fundamentally alter the dynamics of these ecosystems, making recovery increasingly difficult.

Furthermore, the risk of decline among tree species in warm drylands is alarmingly high. Projections indicate that between 44% and 88% of tree species may face significant threats due to climate-induced aridification (Cartereau 2023). These declines not only endanger biodiversity but also undermine the ecosystem services provided by these trees, such as carbon sequestration, habitat for wildlife, and soil stabilization. The loss of tree cover can exacerbate soil erosion and desertification, further diminishing the ecological integrity of these landscapes.

These interconnected factors highlight the vulnerability of dryland ecosystems to the multifaceted impacts of climate change. As these ecosystems face increasing stressors, the urgent need for targeted conservation efforts becomes evident. Strategies that enhance resilience, such as promoting native species restoration, implementing sustainable land management practices, and improving water conservation, are essential for mitigating the adverse effects of climate change and safeguarding biodiversity in these critical regions.

8.3 Ecosystem Resilience and Adaptation

8.3.1 Overview of Ecosystem Goods and Services Relevant to Drylands

While offering diverse goods and services, dryland ecosystems are often not as fully acknowledged as other terrestrial ecosystems. These regions play a crucial and sometimes distinctive role in supporting plant and animal life and human populations. "Drylands" often conjures up pictures of large, seemingly lifeless regions. Numerous misconceptions often surround the concept of drylands. Table 8.1 describes three common misunderstandings.

Drylands support human livelihoods by providing forage for livestock, which produce meat, dairy, wool, and leather. These regions are also crucial for food production, as many major crops including wheat, barley, sorghum, and millets, originated in drylands. Wild varieties from these regions now serve as essential genetic resources for breeding drought-resistant crops. Although freshwater resources in drylands are often scarce and fluctuate in availability, they remain essential for drinking, irrigation, and sustaining wetland ecosystems.

Drylands offer habitat to species that are well-adapted to harsh conditions. These ecosystems support a diverse range of life, from microorganisms and insects like ants and grasshoppers to reptiles and large carnivores like cheetahs and leopards. Due to their vast expanse, drylands are capable of storing significant amounts of

Table 8.1 Drylands: myth and reality

Sr. no.	Misunderstanding for drylands	Reality
1.	Desolate and barren areas where human survival is not feasible	Sustained people's livelihoods for millennia Home to nearly 2 billion people globally and supports numerous modern cities
2.	Incapable of sustaining plant and animal life	Drylands pose challenges to the survival of plants and animals, yet many species have developed unique adaptations to thrive in these environments, coping with the harsh climate and fluctuating water availability A typical plant adaptation is the growth of deep, widespread root systems Animals may adapt by going dormant, seeking shade, or retreating underground to shelter during the hottest times of the day
3.	Deteriorated as a result of improper use and excessive human activity	These environments are both sensitive and resilient. At the same time, overuse can result in significant degradation, making it difficult to revive the region as it is already resource-poor and gets trapped in the vicious cycle of resource depletion. Some drylands naturally feature low productivity, limited vegetation and wildlife, and infertile soil, even in the absence of human influence

Source: White and Nackoney (2003)

carbon, primarily in the soil rather than in the vegetation. Improving the carbon sequestration potential of drylands could be a potential strategy for mitigating global warming by reducing atmospheric CO_2 levels.

Drylands provide a vital source of wood fuel for cooking and hold significant potential for wind and solar energy generation. In certain regions, they are rich in mineral fuels like natural gas and petroleum. Aesthetically, drylands are often characterized by expansive, scenic landscapes. As tourism destinations, they offer opportunities for activities such as wildlife photography and hunting.

8.3.1.1 Drylands, Forage, and Livestock

Today, drylands are essential for providing forage to support domestic livestock production. They sustain large herds of cattle, sheep, goats, and camels, which supply humans with meat, milk, wool, and leather products. Very high densities of livestock in drylands have always been found in India (greater than 40 TLUs [tropical livestock units] per square kilometer) (Thornton et al. 2002). The total livestock population in the country (536.76 million) has increased 10.5% over 2003, as per the 20th livestock census (2019). Livestock is the second most significant contributor of farm income in arid regions, and the livestock enterprise has gradually shifted from subsistence to commercial mode (Manjunatha et al. 2019; Hajong et al. 2023). Livestock can contribute to maintaining soil fertility by improving nutrient retention and water-holding capacity and fostering a favorable environment for microflora and fauna. However, if drylands are overgrazed, it can lead to soil compaction and erosion, resulting in a decline in soil fertility, organic matter, and water retention.

8.3.1.2 Soils in Drylands

The capacity of drylands to generate forage for livestock is partially determined by soil conditions. The Global Assessment of Human-Induced Soil Degradation (GLASOD), along with the more recent Assessment of the Status of Human-Induced Soil Degradation in South and Southeast Asia (ASSOD), are initiatives aimed at qualitatively evaluating soil degradation. An estimated 25–35% of the world's drylands are degraded, with around 20 million hectares of fertile land losing productivity annually and with over 250 million people directly affected (IUCN Issues Brief 2017). According to the information on the extent and severity of global land degradation by ISRIC—World Soil Information (International Soil Reference and Information Centre), most of the areas were moderately degraded (47%), followed by lightly degraded (27.4%) and strongly degraded (2.6%) (Edrisi et al. 2022). The impact of land degradation was observed to be more pronounced in the Asia–Pacific and African regions compared to the Americas and Europe.

Human activities are significantly impacting drylands through practices such as agriculture and livestock grazing, urbanization and settlement, tourism, global warming, desertification, mining, biomass burning and fire, introduction of non-native species, and wildlife exploitation.

Drylands, usually defined as "the regions with an Aridity Index (i.e., the ratio of annual precipitation to annual potential evapotranspiration) below 0.65, currently cover 41% of global land surface and are home to 38% of the global population" (Lu et al. 2018). Soils of drylands are highly vulnerable to rapid erosion, leading to fertile topsoil being lost. Productivity in rainfed drylands has stayed low, primarily due to the limited utilization of key inputs such as fertilizers, supplementary irrigation, high-quality seeds, and plant protection chemicals—particularly for dryland crops, in contrast to those cultivated in irrigated regions.

Rainfed drylands are especially vulnerable to the risks associated with climate change, leading to significant losses (Table 8.2). Land use changes and degradation are major contributors to global greenhouse gas emissions. Degradation results in higher carbon emissions in two key ways: the destruction of vegetation reduces biomass, and increased soil erosion accelerates carbon release. Erosion lowers primary productivity, which diminishes the soil's ability to store carbon while also directly releasing stored organic matter. While not all carbon from eroded soil is instantly released into the atmosphere, the overall effect of erosion is likely to result in higher carbon emissions (Millenium Ecosystem Assessment 2005). Several estimates have been made regarding the carbon emission levels from the degradation of soils in drylands at various scales. Globally, it is estimated that dryland ecosystems release between 0.23 and 0.29 gigatonnes (Gt) of carbon annually into the atmosphere, accounting for approximately 4% of global emissions from all sources combined (Lal 2001; Millenium Ecosystem Assessment 2005).

Table 8.2 Climate-change-associated risks in drylands

Climate-change-associated risks	Adverse effect	References
Land use change: Pasture/rangeland to agricultural land/barren	Greenhouse gas emissions	Srinivasarao et al. (2019)
Land degradation: Loss of biomass due to destruction of vegetation	Increasing carbon emissions	Srinivasarao et al. (2019)
Erosion: Soil erosion	*Emission in two ways*: Reducing primary productivity (the ability of the soil to produce, expressed in terms of carbon stored above ground and below ground; it is estimated with various methods), thus diminishing the carbon storage potential of the soil Direct losses of stored organic matter	Srinivasarao et al. (2019)
Eroded soil: While all the carbon in eroded soil is not released right away	The overall impact of erosion is expected to result in higher carbon emissions	Millenium Ecosystem Assessment (2005)

8.3.1.3 Productivity Constraints in Rainfed Drylands

Rainfed drylands face several challenges that hinder the productivity of these ecosystems. These ecosystems are particularly fragile due to the recurring droughts and the intensified overexploitation of scarce resources.

Rainfed drylands are particularly vulnerable to climatic fluctuations, leading to issues such as:

- Erosion of fertile soil
- Deficient nutrient levels
- Depleted organic carbon content
- Low productivity level

The increased frequency of droughts is a significant threat with severe impacts on rainfed drylands. Drought refers to a condition where rainfall is significantly lower than the typical amount for a given area (Pandey and Bhandari 2007), which varies from one region to another depending upon the climatic situations, water resources, farming methods, and other activities of the area (Prasad 1998).

Drought is nearly an inevitable occurrence in arid regions with less than 500 mm average yearly rainfall. In semiarid regions (with 500–750 mm mean annual rainfall), droughts happen in 40–60% of years, primarily due to inadequate rainfall or insufficient moisture in soil between consecutive rainfall events. Many areas in India are vulnerable to moderate to severe droughts, presenting a significant threat to food security.

In drylands, high oxidation rates and faster erosion result in low soil organic carbon (SOC) concentrations and decreased health of soil (Srinivasarao et al. 2015). Additionally, drylands are susceptible to issues such as acidity, alkalinity, and salinity, which aggravate the deterioration of soil health.

The key restrictions which form a serious challenge to the livelihood of dryland farmers include limited farm holding size, inadequate mechanization, low socioeconomic status, minimal risk-bearing capacity, restricted access to financing, and inadequate infrastructure. Moreover, limitations in moisture accessibility in the field result in poor nutrient use efficiency. Disparity between the nutrient inputs, their removal by crops, and losses contributes to the decline in the dryland soil fertility. Fertilizer application, along with soil moisture content, is critical in enhancing crop productivity and increasing nutrient and water use efficiency (Srinivasarao et al. 2013b).

8.3.2 Vital Role of Nutrient Management in Agriculture Resilient to Climate Change

Drylands are among the most vulnerable locations to climate change and human activity. Desertification and land degradation will become more likely in emerging nations like India due to temperature rise, intensified warming, aridity increase, and population growth.

Table 8.3 Nutrient management strategies in climate-resilient rainfed agriculture systems in India's drylands

Sr. no.	Nutrient management strategy	Contribution to		
		Adaptation	Mitigation	Reduction in input cost/ profitability
1.	Site-specific nutrient management (SSNM)	L	H	VH
2.	Integrated nutrient management (INM)	H	H	VH
3.	Neem-coated urea	VL	VH	H
4.	Placement of fertilizers	M	H	H
5.	Foliar spray	H	L	H
6.	Crop residue recycling	VH	H	H
7.	Vermi-compost addition/ composting	H	VH	H
8.	Green manuring/green leaf manuring	H	H	H
9.	Mulch-cum-manuring	VH	VH	H
10.	Bio-fertilizers	L	M	H
11.	Tank silt application	M	M	M
12.	Balanced nutrition	L	M	M
13.	Conservation systems	H	M	M
14.	Manure management	H	H	H
15.	Fertigation	L	M	M
16.	Nanofertilizers	L	H	M

VH very high, *H* high, *M* medium, *L* low, *VL* very low
Source: Srinivasarao et al. (2003)

In drylands, nutrient management is a vibrant approach to enhance the soil health and its sustainability as well as crop productivity. A few strategies that would assist dryland farmers in mitigating the risks of changing climate, increasing net profit per rupee of investment, and ultimately improving livelihoods are increasing soil organic matter, applying fertilizers judiciously, and improving agronomic management (Table 8.3). Techno-economic assessments of nutrient management practices have revealed that integrated and organic management demonstrate a higher BC ratio (1.01–2.26), comparatively a higher yield, and lower GHG emissions (Gholkar et al. 2022). Proper plant nutrition is crucial for enhancing productivity amidst the effects of climate change, including droughts (early, mid-season, and terminal droughts), tolerance to heat and cold waves, minimizing crop lodging caused by cyclones and floods, and managing coastal salinity due to seawater intrusion (Table 8.4).

Table 8.4 Critical role of proper plant nutrition in mitigating impacts of climate change on agriculture

Optimum plant nutrition	Climate change adaptation
Nitrogen	Rejuvenation after stress
Phosphorus	Better root growth and drought tolerance
Potassium	Stress tolerance and water relations
Ca, Mg, and S	Enzyme activation reduces photo-oxidation damage and improves drought tolerance
Micronutrients	Enhance CHO and nitrogen metabolism, reduce ROS and reduce photo-oxidation
Beneficial nutrients	Alleviate oxidative stress by regulating the activity of antioxidant enzymes

Source: Srinivasarao et al. (2003)

8.3.3 Strategies to Tackle the Climate Change Impacts

Various technological approaches to nutrient management support climate-resilient agriculture in rainfed dryland areas of India. Properly installed, these nutrient management technologies can definitely reduce the risk of climate change, increase profitability in vulnerable areas, and reduce the cost of cultivation.

Climate adaptation led mitigation strategies in nutrient management:

- Improved soil organic carbon
- Optimum/balanced nutrition
- Slow release of nutrients
- Enhanced nutrient use efficiency
- Nutrient availability at critical crop/plant growth
- Minimizing crop residue burning
- Harnessing N_2 fixation
- Soil-water-nutrient synergy

8.3.3.1 Enhancing Soil Organic Matter

Numerous factors affect the fertility and quality of dryland soils as well as their capacity to provide ecosystem services. Among these, soil organic matter (SOM) is of significant relevance. This is because SOM in drylands is made up of a vast global carbon reservoir and affects almost all of the soil's characteristics and ecosystem activities.

Additionally, SOM encourages the development of aggregates with mineral particles, which helps to improve soil structure, porosity, and water-holding capacity. This reduces the severity of water shortages and protects soils from erosion and compaction. Moreover, organic matter is a major source of plant nutrients and contributes significantly to the soil's ability to buffer acid-base. Furthermore, soil organic matter, a significant source of food for a wide range of species, is closely linked to soil biological activity and diversity. In drylands, carbon stock estimation is a very tedious task owing to their spatial variability, heterogeneous soil characteristics, and fluctuations in moisture availability, which, together with the human impacts and methodological challenges, make it highly resource-intensive (Lal 2004; Okin and Gillette 2001). Inappropriate land use, inadequate soil management, or farming methods can degrade soil quality and potentially emit carbon dioxide (CO_2) into the atmosphere (Lal 2004). As a result, adequate SOM management is critical for maintaining soil productivity and ensuring food security, increasing carbon sequestration, and limiting atmospheric CO_2 rise. Nearly all soil and crop management strategies impact SOM. Crop yield improvement potential per unit ton of soil organic carbon has been calculated to be 170 kg ha^{-1} among dryland crops in India (Srinivasarao et al. 2019).

8.3.3.2 Reducing Crop Residue Burning and Reusing Organic Wastes

Following crop harvest, a large biomass of crop residue is generated, which provides a variety of benefits such as animal feed, soil mulching, manure preparation, rooftop materials for rural huts, and fuel for household and industrial purposes. Every year, around 90–140 Mt. of crop leftovers are burnt on-farm to prepare the ground for the next crop (Bhuvaneshwari et al. 2019). While crop leftovers are the primary source of organic matter to enhance soil health and boost crop productivity. However, farmers believe that burning crop leftovers aids in the management of weeds, pests, and diseases. In the short term, the availability of nutrients like P and K increases, and soil acidity decreases, leading to the depletion of nutrients such as N and S, and also organic matter, along with a decrease in soil microbial activity. Crop residue burning is a significant contributor to the emission of hydrocarbons, including greenhouse gases (GHGs), which ultimately contribute to pollution of the environment.

Converting organic waste into biochar through pyrolysis is a viable alternative that can accelerate the soil carbon sequestration rate and improve soil quality and minimize farm waste (Srinivasarao et al. 2012, 2013a). Transforming waste biomass into biochar would shift substantial quantities of carbon from the active pools to the dormant pools, providing a valuable scope to influence the carbon cycle. Using biochar as a soil conditioner is seen as a promising strategy to address human-driven climate change.

8.3.3.3 Leveraging Soil-Water-Nutrient Interplay

In rainfed dryland conditions, conservation of soil moisture is closely linked to the efficient use of fertilizers. Water harvesting systems, like tanks with facilities to capture seepage water for irrigation, are promising methods in rainfed areas to improve yield and nutrient use efficiency (Singh et al. 2019). Optimizing soil water-nutrient interactions is crucial for improving nutrient efficiency and minimizing losses.

8.3.3.4 Green Leaf Manuring

Green leaf manuring species adapt effectively to various soil types and play a crucial role in enhancing soil fertility and productivity. Several trees, like neem (*Azadirachta indica*), gulmohar (*Delonix regia*), copper pod (*Peltophorum ferrugenum*), and karanj (*Pongamia glabra*), are easily grown in diverse parts of the country; their leaves can serve as both mulch and a source of organic matter in rainfed agro-ecological systems. Furthermore, as these materials decompose, they can readily deliver sufficient amounts of plant nutrients like nitrogen, phosphorus, and potassium to the crop. According to research, tree leaf manuring is a low-cost, environmentally sound technology. For instance, in a rainfed setting, Gliricidia could be used as a pesticide, rodenticide, and repellent in addition to supplying green biomass. Gliricidia is highly adaptable in marginally degraded fields and reduces N_2O and CO_2 emissions into the environment and contributes to the reduction of greenhouse gases (Srinivasarao et al. 2011, 2017).

8.3.3.5 Nutrient Application Based on Soil Testing

Using soil-test-based recommendations of fertilizers for crops helps achieve higher yields and benefit-cost ratios by applying nutrients according to the specific nutrient deficiencies. Depending on the crop-based assessments of native availability of plant nutrients integrated with empirical models developed by various researchers to predict crop responses based on nutrient interactions (Witt et al. 1999), field-specific recommendations for balanced quantities of nutrients could be provided. A long-term study by Girija et al. (2008) investigated the effectiveness of soil-test-based optimal fertilizer applications in achieving higher finger millet yield in South Indian semiarid Alfisols. These recommendations help maintain and improve soil fertility while also boosting crop production and nutrient use efficiency. Enhanced fertilizer efficiency, especially for nitrogen (N), can reduce N_2O emissions from agricultural fields. Phosphorus (P) recommendations based on soil tests promote optimal root development, while potassium (K) nutrition plays a key role in improving water management within plants.

8.3.3.6 Conservation Agriculture (CA)

Conservation agriculture is now widely accepted in the agricultural sector around the world as an alternative to conventional agriculture due to several benefits to soil health and the environment. Adopting Conservation Agriculture (CA) enhances water infiltration, reduces soil erosion, and moderates soil temperature. It also helps suppress weeds, improve soil aggregation, alleviate soil compaction, and increase organic matter content in the surface soil. Additionally, CA reduces greenhouse gas emissions and lowers production costs (Srinivasarao et al. 2015). Intense land degradation caused by runoff-induced soil erosion in rainfed, arid, and semiarid regions necessitates conservation agriculture urgently to maintain sustainability. Improvements in soil carbon are linked to multiple benefits that contribute to increased agricultural productivity. Findings from long-term experiments revealed that every ton of soil organic carbon accumulation increases crop productivity by 100–350 kg per hectare across different dryland crops (Srinivasarao et al. 2013a).

8.3.3.7 Precise Nutrient Application and Variable Rate Technology (VRT)

Precision agriculture (PA), as the name suggests, is an effective technology for cultivating crops with greater precision or efficiency while preserving moisture and nutrients within the root zone. The three main methods through which PA can help achieve this goal are:

- Gathering spatial data on existing field conditions (e.g., remote sensing, canopy size, or yield measurement)
- Applying precise amounts of fertilizer to crops at the right time and in the right locations
- Keeping detailed records of all fertilizer applications for mapping spatial and temporal patterns

Thus, precision agriculture (PA) enables us to identify the optimal locations for nutrient placement, determine the appropriate amounts to apply, and track nutrient applications using accumulation logs and GIS maps.

Variable-rate technology (VRT) enables the application of fertilizers, irrigation water, and other soil amendments at varying rates throughout the field, eliminating the need for manual equipment adjustments. Sensor-based precision tools, such as sensor-driven variable-rate fertilizer applicators combined with spectral reflectance sensors, further enhance this process (GreenSeeker). Using an NDVI-based variable-rate fertilizer applicator in crops like wheat and rice resulted in an estimated 8–15% reduction in urea usage (Singh et al. 2017). In addition to reducing input costs, this technology would also decrease N_2O emissions, ultimately lowering greenhouse gas (GHG) emissions.

8.3.3.8 Adaptation-Led Mitigation to Lower GHG Emissions

In the context of climate change, adaptation and mitigation are two key concepts. According to the IPCC definition, “adaptation refers to adjustments made in natural or human systems to cope with actual or anticipated climate impacts, aiming to minimize harm or take advantage of beneficial opportunities.” On the other hand, climate change mitigation refers to efforts aimed at reducing the rate of climate change, primarily by reducing greenhouse gas emissions and encouraging practices that extract these gases from the air. Adapting to weather fluctuations and extreme occurrences is essential for minimizing vulnerability to the long-term effects of climate change. According to Srinivasarao (2018), “key technologies for mitigation of greenhouse gases are, opting resilient crops and cropping systems including contingency planning, soil health management by increasing carbon sequestration, balanced and timely use of fertilizers, soil and water conservation measures, and mitigating the GHG from livestock by improving the feeding practices."

Climate change is an urgent global challenge which, comprising multiple impacts, poses a risk to our environment, economy, and society.

Conventionally, modifications being observed in the climate of Rajasthan are beyond the natural climate variability. Rajasthan is among the regions with the highest climate susceptibility, greatest vulnerability, and minimum adaptive capacity due to the highest probability of drought occurrence in the nation, scarcity in water resources, and a highly uneven rainfall distribution both temporarily and spatially.

8.3.3.9 Agroforestry

Agroforestry has emerged as a valuable strategy for both mitigating and adapting to climate change (Mosquera-Losada et al. 2018). It directly mitigates climate change by enhancing carbon sinks through capture of atmospheric carbon via photosynthesis and its storage in biomass and soil (Albrecht and Kandji 2003) (Table 8.5). The carbon sequestration capacity of agroforestry systems varies significantly (0.29 to 15.21 Mg ha^{-1} annually) and is influenced by factors like environmental conditions, tree age, species diversity, and management techniques (Ramachandran Nair et al. 2009). Integrating trees into agricultural land also eases the strain on natural forests by decreasing the demand for harvesting forest trees for timber and fodder. This helps reduce tree felling and forest clearance, thereby indirectly combating climate change (Meijer et al. 2016).

Table 8.5 Role of agroforestry in adapting to climate change

Agroforestry system impact	References
The ability to reduce nitrogen (N) loss through soil erosion, runoff, and leaching, as well as to minimize gaseous nitrogen losses, such as emissions of nitric oxide (NO) and nitrous oxide (N_2O)	Muchane et al. (2020), Kim and Isaac (2022)
By storing nitrogen (N) in the soil, this approach reduces the need for additional nitrogen inputs on farmlands. This not only decreases farmers' reliance on external nitrogen sources but also helps mitigate climate change by lowering nitrogen-related emissions	Kim and Isaac (2022)
Improving fodder quality enhances its nutritional content and digestibility, which can help reduce methane (CH_4) emissions in livestock farming from enteric fermentation	Thornton and Herrero (2010), Torres et al. (2017)
Increasing farm profitability through strategies like agroforestry helps mitigate the effects of damage caused by wind and water by shielding farms. It also diversifies the monetary resources, providing greater resilience against climate-related challenges	Molua (2005), Mbow et al. (2014)
Effectively managed agroforestry practices can enhance crop productivity by fostering a conducive microclimate, enhancing soil fertility, increasing soil water retention, boosting crop pollination, and serving as valuable habitats for beneficial organisms	Muthuri et al. (2005), Bentrup et al. (2019), Kim and Isaac (2022)

India, as one of the most vulnerable countries to climate change (Panda 2009; Datta et al. 2022) and the third-largest contributor of greenhouse gas emissions globally (with agriculture contributing 18% of the country's total emissions), has significant potential to address the continuing climate crisis through the widespread adoption of agroforestry practices (Sapkota et al. 2019).

Nath et al. (2021) observed that the rates of carbon sequestration in different agroforestry systems in India were significantly higher (0.03-0.06 Pg per year) than the annual growth rate of total agricultural emissions in India (0.015 Pg per year CO_2 eq.). The study concluded that expanding agroforestry areas by 30% could substantially lower the emissions from the nation's agricultural and industrial sectors due to its significant carbon sequestration potential.

Moreover, empirical studies have shown that farmers can strengthen their climate resilience through agroforestry by improving their ability to adapt to climate change and benefiting from a range of ecosystem services (Jana et al. 2015; Pandey et al. 2017; Datta and Behera 2023).

8.3.3.10 Integrated Farming Systems and Climate Preparedness

The Integrated Farming Systems (IFS) rely less on external inputs due to the greater recycling capacity of these systems. This implies that small and mixed farms may be less susceptible to climate change-induced crop failure because of the better variety of farm products. The IFS systems from various locations reported a net reduction in GHG emissions due to increased carbon sequestration, as found in a study given in the Annual Report of AICRP on Integrated Farming System (2019). In Palampur, the system consisting of horticulture, field and fodder crops, dairy, vermicompost, boundary plantations, and kitchen gardening resulted in a GHG emission of −1787 kg CO_2 eq. ha^{-1}. Jorhat reported a GHG emission of −3175 kg CO_2 eq. ha^{-1} from a system that included field and horticulture crops, livestock, poultry, fisheries, apiary, vermicompost, biogas, liquid manure, and FYM production. In Kayani, a system involving crops, dairy, horticulture, vermicompost, biogas, and fishery resulted in −4517 kg CO_2 eq. ha^{-1} of GHG emission. The Raipur system, which combined field and horticulture crops, livestock, poultry, fisheries, mushrooms, vermicompost, boundary plantations, and kitchen gardening, emitted −7713 kg CO_2 eq. ha^{-1}. Finally, the Telangana system, based on crops, livestock, and a hortipasture IFS model, resulted in the highest GHG emission at −27,036 kg CO_2 eq. ha^{-1} (Annual report of AICRP on Integrated Farming System 2019).

Further, these systems demonstrated that the enhanced residue recycling and presence of trees help mitigate the negative impacts of climate change by enhancing the sequestering of carbon in the soil and tree biomass. Similarly, Salton et al. (2014) found positive net GHG emissions in conventional systems and negative emissions in IFS, which was largely attributed to higher carbon sequestration in IFS models that counterbalanced N_2O emissions.

The IFS method has been shown to lower CH_4 absorption, as documented by Dong et al. (2000), Liu et al. (2007), and Schönbach et al. (2012). CH_4 absorption in the IFS systems of Brazil was 68% less compared to conventional single cropping systems, while in temperate plains, Chen et al. (2011) observed a 30% reduction in CH_4 absorption with IFS. The negative effect of IFS on CH_4 absorption could be attributed to enhanced nutrient recycling within the system through organic farming methods. This, in turn, may have enhanced the prevalence and activity of methanotrophs, which play a role in methane oxidation, and potentially reduced air diffusion, which may have hindered CH_4 movement (Chen et al. 2011). IFS is encouraged as an effective strategy to enhance productivity, rehabilitate degraded pastures, and lower greenhouse gas emissions (Gil et al. 2015; Paramesh et al. 2022). Better agronomic practices and biodiversity conservation play a key role in creating resilient, efficient, and sustainable agricultural systems while also minimizing environmental pollution. Bell et al. (2014) highlighted that integrating perennial forages with cropping systems, such as agroforestry, provides various strategies for mitigating climate change by boosting carbon sequestration and nutrient cycling.

Barbosa et al. (2015) identified IFS as an effective method to minimize greenhouse gas emissions and loss of nutrients by facilitating nutrient cycles and using crop residues to feed animals. Sunderland (2011) suggested that incorporating multipurpose trees into farming systems not only offers food and income to small and marginal farmers, serving as a livelihood source, but also helps sequester atmospheric carbon.

Ecosystem services provided by IFS:

- Nutritional security
- Food security
- Better nutrient recycling
- Soil quality enrichment
- Biodiversity conservation
- Enhanced profitability
- Increased employment generation
- Climate resilience

8.3.3.11 Afforestation

Drylands in Rajasthan face considerable environmental challenges, with a critical need for efficient strategies to mitigate climate change. Gupta et al. (2024) developed a model to estimate rates of carbon sequestration based on the growth patterns of specific tree species. The study reveals that about 40% of the land in Rajasthan is ideal for afforestation measures, and their potential carbon sequestration rates could vary from 2 to 8 tons ha^{-1} annually. Tree species such as *Azadirachta indica* and *Prosopis cineraria* are effective short-term carbon stores, while *Phyllanthus emblica* and *Ziziphus mauritiana* are appropriate for long-term sequestration. This emphasizes the need for "targeted afforestation" with appropriate tree species as a nature-based solution (NbS). The findings offer a framework to strengthen resilience and guide climate adaptation strategies in arid regions, underlining the immense potential for carbon sequestration through afforestation in Rajasthan.

However, important considerations are warranted regarding the survival of newly planted trees. As per the National Afforestation Programme (NAP) of the Ministry of Environment and Forest (MoEF), the survival percentage in afforestation programs was estimated as 87.23% with 5.02% standard error (https://fsi.nic.in/uploads/documents/doc_3661_monitoring_of_area_coverage_and_survival.pdf). Analysis of tree survival and growth data from 176 restoration sites in Asian tropics and subtropics revealed that, on average, nearly 18% of the planted saplings perished within the first year, and this figure increased to 44% after 5 years. The rates of survival, however, varied significantly across sites and species, with some locations having more than 80% of tree survival after 5 years, while in others, a similar percentage had perished (Banin et al. 2022). Maximizing the recovery of species may require silvicultural interventions and assistance in natural regeneration. Enrichment planting of locally adapted tree species should be given priority (Bentos et al. 2020). Moreover, the heterogeneous edaphic conditions and environmental gradients in the areas should be considered for successful afforestation.

8.3.4 Climate Change Mitigation and Adaptation Through Indian Policy Approach

Agroforestry has the prospect of becoming a central component of India's policy strategy for climate change mitigation as well as adaptation. As part of its mitigation efforts, India's updated Nationally Determined Contributions (NDCs), by 2030, seek to lower the emission intensity of its GDP by 45% compared to 2005 levels, where the AFOLU (agriculture, forestry, and other land use) sector contributes to 8% of the total national GHG emissions (Mohan Kumar and Aravindakshan 2022). Also, by 2030, India targets reducing land degradation and desertification by reclaiming 26 million ha of land (NAAS Newsletter 2021). To reach this target, the focus is on generating "additional carbon sinks of 2.5 to 3 billion tonnes of CO_2 equivalent through expanded forest and tree cover by 2030" (PIB 2022), although agroforestry and more tree cover on agricultural lands play an important role in achieving this goal (Nath et al. 2021).

Additionally, agroforestry can support the achievement of NDC targets and complement India's National Action Plan on Climate Change (NAPCC), which offers a policy framework for national climate action, as well as State Action Plans on Climate Change (SAPCC) (Singh et al. 2021a). Agroforestry, as a sustainable and eco-friendly land use system that combines trees, crops, and livestock, is acknowledged as an efficient approach to facilitate the efforts undertaken to combat climate change.

8.4 Community Involvement and Stakeholder Engagement

The world's poorest populations are particularly vulnerable, facing the brunt of climate-related changes with little to help them cope. Climate change jeopardizes their livelihoods and rights, especially for those who are dependent on biological diversity and ecosystem services for essential needs like food, water, and shelter. Moreover, climate change undermines traditional coping strategies and threatens food security, heightening the risk of famine and exacerbating vulnerabilities to extreme weather events like droughts, floods, and disease outbreaks. The anticipated impacts on natural resources, species, and ecosystems will limit development options at both local and national levels while putting additional pressure on already-strained terrestrial, freshwater, and marine habitats. Declining land fertility and productivity due to erosion and salinization will adversely affect rural communities, further diminishing their prospects for sustainable development and deepening poverty through reduced income opportunities. As a result, many vulnerable communities are being pushed further into poverty and insecurity, threatening to reverse decades of progress in development (Arenas-Wong et al. 2023). Community involvement and stakeholder engagement, hence, are crucial components of ecosystem-based approaches, as they foster collaboration and build trust among diverse groups. Tools like "CoDriVE (Community-Driven Vulnerability Evaluation)—Programme Designer" enable communities to express their experiences of being affected by climatic and non-climatic forces, identify and evaluate their vulnerabilities or "development gaps," and inspire them to plan and implement adaptive measures to enhance resilience and reduce vulnerability (CoDriVE – Programme Designer 2013). By actively involving local communities, government bodies, NGOs, and businesses, these approaches ensure that the knowledge, values, and needs of all stakeholders are integrated into decision-making processes (Eikelenboom and Long 2023). This inclusive dialogue enhances the effectiveness of environmental management by drawing on local expertise and promoting sustainable practices that resonate with community priorities.

Two emerging approaches to climate change adaptation that have gained prominence in recent years are community-based adaptation (CBA) and ecosystem-based adaptation (EBA) (Girot et al. 2012). Enhancing the resilience of both social and ecological systems is essential in the context of a changing climate. Recently, interest in adaptation strategies has grown, but these approaches have often been fragmented and overly narrow in focus (Pandey et al. 2015). Researchers worldwide are currently advocating for evaluating ecosystem services using local residents' knowledge of ecological characteristics to inform conservation policies (Lee and Ahn 2023). This local knowledge, often expressed qualitatively, can complement data-driven assessments like indicators and modeling approaches (Pita et al. 2016). Engaging stakeholders in ecosystem-based strategies not only enhances resilience but also empowers communities to take ownership of their natural resources, ultimately fostering more sustainable outcomes for both people and the environment (Conde et al. 2004; Wells et al. 2023).

8.4.1 Participatory Approaches in Ecosystem Management

Residents are directly impacted by environmental changes in their vicinity, and their local knowledge can play a vital role in identifying ecological assets, and by employing participatory techniques, communities can verify issues like climate change and loss of biodiversity, generate valuable data for environmental management, and strengthen their capacity to respond to environmental challenges (Lee and Ahn 2023). This active involvement not only enhances understanding but also cultivates a sense of ownership and responsibility toward local ecosystems. Evaluating ecosystem services through resident participation effectively uncovers essential information in areas lacking environmental data, and it is often more cost-effective than traditional methods of collecting ecological data. Lee (2021) has suggested the application of "spatial text mining" as a method to quantify verbal data and map participants' knowledge. Through these evaluations, the author has identified hotspot areas for four categories of ecosystem services: provisioning services related to crop production, regulating services for air and water quality, supporting services for animal and plant habitats, and cultural services focused on leisure and recreation (Lee 2021). This identification of hotspots is crucial for targeted conservation efforts and sustainable resource management.

Benefits of stakeholder engagement, as laid out by Twigg (1999) and Conde et al. (2004), include the following. Collaborative efforts can significantly strengthen communities and enhance adaptive capacity by raising awareness of local issues and developing strategies to address them. This process reinforces local organizations and builds confidence, skills, and the ability to work together, ultimately increasing individuals' potential to reduce their vulnerability. Empowered individuals are better equipped to face various challenges, both individually and collectively. Involving stakeholders in planning—through priority-setting and expressing preferences—aligns with their right to influence decisions that affect their lives. While stakeholder engagement may take more time than conventional, externally driven approaches, it tends to be more sustainable, as it allows for ideas to be tested and refined before implementation.

8.4.2 Integrating Information from Diverse Sources, Such as Traditional, Local, and Scientific Knowledge

Under ecosystem-based adaptation (EBA), local communities and indigenous people can provide valuable traditional knowledge and adaptation practices, drawing from their experiences in coping with climate variability. When integrated with full participation and free, prior, and informed consent, this knowledge can significantly enhance current adaptation efforts. For instance, communities may utilize local crop varieties and cultivation techniques that are resilient to extreme weather, as well as methods for storing seed stocks that are suitable for their specific contexts. However, some of the changes currently affecting local communities, as well as those anticipated in the future, may extend beyond their historical experiences. Adapting to these changes might necessitate technological solutions from outside the communities. By combining local observations of weather patterns, hazards, and impacts with scientific forecasts and scenario analysis, communities can better assess their current and future vulnerabilities and plan effective responses (Girot et al. 2012). Kuchimanchi et al. (2019) assessed the vulnerability of communities to climatic fluctuations using a community-engagement tool and found that risks associated with climate differ even within small groups of villages and, hence, suggested the need for vulnerability assessments at smaller scales.

8.4.3 Nature-Based Solutions in the Dryland Ecosystems: Case Studies

Alikhanova and Bull (2023) carried out a comprehensive analysis of the nature-based solutions in dryland ecosystems and found that afforestation with native trees and agroforestry were the cost-effective strategies for combating desertification, as well as combating climate change by increasing carbon sequestration potential. Agroforestry is often recommended as a solution for the rehabilitation of marginal and salt-affected irrigated lands. It plays a crucial role in ensuring food security in areas facing a scarcity of water. Furthermore, agroforestry helps diversify income sources, restore biodiversity, improve carbon sequestration, and alleviate pressure on existing natural forests. Rangeland restoration offers another climate mitigation strategy, leveraging the carbon storage capacity of grasslands while also improving the livelihood of local pastoralists in degraded areas (Yao et al. 2021).

The Ecosystem-based Adaptation Plans of the UN Environment Programme suggest the following good practices for encouraging community engagement:

- Ownership of the process at the local level
- Utilizing existing coordination structures at the regional and local level
- Continuous adjustment according to local development planning processes
- Participatory process in the identification of adaptation measures

Through these measures, the UNEP is assisting the government of The Gambia in adapting to climate change through a major project titled "*Large-scale Ecosystem-based Adaptation in The Gambia: Developing a Climate-Resilient, Natural Resource-Based Economy.*" Supported by the Green Climate Fund, the project seeks to "restore 12,788 hectares of degraded forests, savannas, and mangroves, as well as 3,000 hectares of farmland to enhance the resilience of rural villages against storm surges and floods"—a strategy referred to as "'ecosystem-based adaptation." Similarly, UNEP is also carrying out ecosystem-based adaptation in El Salvador, Mexico, and Jamaica to build climate-resilient systems.

8.4.3.1 Case Study 1: Application of Stakeholder Engagement and Ecosystem Approach in the Kano-Maradi Transboundary Region of Nigeria and Niger (Mortimore et al. 2008)

The drylands, often extensive and crossing national borders, exhibit environmental heterogeneity and require multifaceted approaches for understanding and management. These areas, characterized by significant year-to-year variability, particularly in rainfall, demand flexibility and adaptability from their inhabitants, shaping their livelihoods more than the natural resource carrying capacities. Economic interdependence with more humid regions is vital for sustainable living. Research on drylands has often fallen short in influencing policy, highlighting a need for decentralized environmental planning and effective local institutions. This case study, focusing on the Niger-Nigeria border, examines lessons learned from the ecosystem approach (EA) in sustainably managing dryland ecosystems and supporting local livelihoods. The region, historically seen as degraded and economically marginalized, is the site of various development projects and a recent long-term program funded by the Global Environment Facility. The study area encompasses a range of villages, reflecting a gradient of rainfall and agricultural practices. Key objectives include equitable access to resources, biodiversity conservation, and institutional sustainability. Methods involved structured discussions, biodiversity inventories, and participatory mapping, culminating in a micro-project aimed at enhancing community engagement in ecosystem management.

Effective ecosystem management requires a deep understanding of the interactions among various stakeholder groups, fostering consensus on managing ecosystem services, and facilitating negotiations among competing interests. This approach emphasizes local empowerment and the role of external agencies in providing support rather than imposing solutions. The study examines the regional ecology of five villages along the Niger-Nigeria border: Tumbau and Dagaceri in Nigeria, and Magami, Serken Hausa, and Dan Kulu in Niger. Stakeholders include resident communities and transhumant pastoralists, all sharing interests in vegetation, land, and water sustainability. There is growing awareness among stakeholders of the economic value of ecosystem resources, with previously non-monetized resources, like grasses, becoming scarce and entering the market. Farmers recognize that the presence of trees enhances soil fertility and reduces erosion, prompting discussions on the need for more trees in agricultural areas. In Dan Kulu, strategies for livelihood diversification include heavy investment in livestock and protecting valuable trees. Population growth has intensified the demand for arable land, leading to permanent cultivation and a decline in traditional fallowing practices. Residents emphasize the necessity of enhancing tree cover on all land types to safeguard the environment.

Key findings from this project revealed that local knowledge and experience, which have supported rural communities until now, are critically important for ecosystem management in the future. Risk insurance through diversified livelihood portfolios is a condition of survival in Sahelian environments, and agriculture, livestock keeping, and ecosystem management should always be understood in this context. Short- and long-term migration to urban centers will continue to be essential in many households. Biodiversity is not disregarded, as sometimes assumed by outsiders, but valued and deliberately conserved by the community within its capabilities. This is true for agricultural biodiversity (which is still rich alongside the uptake of introduced crop varieties) as well as "wild" biodiversity (which is preserved, for example, in trees planted and protected on farmlands). There is a transition underway in many communities from more extensive and destructive to more intensive and conservative ecosystem management. The support of this evolution should be an objective of development interventions. Technological inventories—including innovations brought by development projects in the area—are severely constrained by a poverty of financial capital. Human and social capital development, which are increasingly the focus of development interventions, offer much scope through institutional initiatives coupled with incentives provided in an enabling economic and policy environment. Knowledge management systems should safeguard and mobilize local knowledge alongside research-based knowledge, ensuring that it is fully accessible to local communities and supportive, rather than directive, of professional departments of government.

8.4.3.2 Case Study 2: Community-Driven Climate Change Adaptation Through Afforestation in Bangladesh (Ecosystem-based Adaptation (EbA) Good Practice Case Studies 2021)

The Community-Based Adaptation to Climate Change through Coastal Afforestation (CBACC-CF) project in Bangladesh is a vital example of how vulnerable communities can be empowered to tackle the impacts of climate change. Bangladesh, despite not being a major contributor to climate change, is highly susceptible to its effects, particularly in the coastal regions, where rising sea levels, cyclones, and flooding pose serious threats. The project focused on reducing this vulnerability by enhancing both environmental and community resilience. The project was implemented in March 2009 and spanned a period of 6 years to realize the accomplishments.

The project aimed to strengthen the resilience of coastal communities by restoring and replanting degraded mangroves and wetland ecosystems, which act as natural barriers against storms, tidal surges, and flooding. These ecosystems also provide a range of livelihood opportunities, making them crucial for the socioeconomic well-being of local populations. By implementing this project, the Government of Bangladesh (GoB) sought to:

- Enhance the protective capacity of coastal ecosystems
- Create sustainable livelihoods for vulnerable communities
- Promote climate change adaptation through community-based, nature-driven solutions

The major outcomes of the project were as follows:

1. Livelihood Diversification: More than 85,000 individuals received training in planting techniques and land management, helping them diversify their income

sources. The project also included cash-for-work programs, which provided immediate financial benefits to local people while involving them directly in the restoration efforts.

2. Coastal Protection: The restoration of mangrove forests led to increased siltation and better coastal surge protection, particularly along 14 kilometers of coastline. The mangroves helped mitigate the impacts of cyclones and tidal surges, reducing damage to both infrastructure and livelihoods.
3. Land Use Model: The project introduced a new model for sustainable land use, which aimed to rehabilitate barren coastal lands, transforming them into productive, ecologically and economically valuable areas.
4. Lessons and Best Practices: The project generated lessons on how to integrate ecosystem-based adaptation (EBA) with community-based adaptation (CBA), which can be replicated and scaled up in other parts of Bangladesh, the region, and internationally.

The main factors that led to the success of the project were:

1. *Cooperation Across Stakeholders*: Successful implementation required cooperation between a range of stakeholders, including local governments, national ministries, and community organizations. Collaborative governance was crucial to the project's effectiveness.
2. *Community Involvement*: The active participation of local communities was key to the project's success. Ensuring that local people were involved in the design and execution of the project, particularly through the integration of EbA and CBA, made sure that the adaptation measures were both environmentally effective and also culturally suitable and beneficial to local livelihoods.
3. *Education and Training*: Educating local communities on climate adaptation strategies and training them in sustainable practices increased their capacity to cope with climate impacts. These educational efforts strengthened the project's long-term success by enhancing local knowledge and adaptive capacity.
4. *Integrating Climate Adaptation with Restoration*: The project demonstrated how climate adaptation strategies can be incorporated into existing environmental restoration efforts, such as mangrove replanting, thereby creating multiple co-benefits for both ecosystems and local communities.

This project demonstrates a successful model for integrating ecosystem-based adaptation with community-driven approaches to climate resilience and highlights the importance of stakeholder collaboration, community engagement, and capacity building in tackling climate change in vulnerable regions. The lessons learned from this project could be applied to similar coastal regions worldwide, especially in Least Developed Countries (LDCs) that confront unequal risks from changes in climate, despite contributing little to the problem. The project is not only an example of how natural ecosystems can be restored to protect vulnerable coastal communities but also a testament to the power of community-led climate change adaptation in building resilience.

8.5 Challenges and Barriers

Ecosystem-based approaches to understanding natural ecosystems are often complex and difficult to implement due to their similar and widespread nature (Cowan Jr et al. 2012). A number of key indicators stand in the way of reducing urgent climate risks beyond climate change (IPCC Sixth Assessment Report 2022). Despite the potential of ecosystem-based approaches, they are frequently neglected in the National Adaptation Action Plans (NAPAs) of 44 developing countries, reinforcing the case that only 22% of the plans include activities to improve ecosystem projects. Key challenges of the ecosystem-based approach include governance, stakeholder participation, long-term management, and the uncertainty of climate change (Ojea 2015).

8.5.1 Empirical Challenges

These areas are often desperately affected by extreme weather conditions, drought, and resource scarcity (Huang et al. 2017). In terms of productivity impacts, lower productivity and higher subsistence levels lead to greater degradation risks and create more challenges for dryland ecosystem management (Wang et al. 2024). Adapting ecosystem-based approaches in drylands also suffers from a lack of reliable data and a poor understanding of these complex systems. Many dryland habitats are understudied, leading to inconsistent information on species interactions, soil health, and water availability, which may well influence management strategies (Wang et al. 2024). Furthermore, differences in ecological conditions across different drylands (D'Odorico and Bhattachan 2012; Ukkola et al. 2021) make it difficult to improve practices and draw commonly used conclusions. Monitoring and evaluating the effects of ecosystem interventions is further complicated by the uncertainty of the interconnectedness of human processes (Cowan Jr et al. 2012). These experiences can lead to a lack of sufficient evidence to support decision-making, hinder the dissemination of effective practices, and reduce overall performance in these complex environments. Solving these issues will require further research, knowledge sharing, and management adaptations tailored to the unique conditions of drylands.

8.5.2 Economic Challenges

Communities may lack the financial resources to invest in good practices or infrastructure, which can discourage investment in conservation and development (Rode et al. 2016). For local residents, the initial costs associated with transitioning to ecosystem management (e.g., infrastructure, resource development, and technology) can be prohibitive (Ekins and Zenghelis 2021). According to the UNEP Adaptation Gap Report (2020), "annual adaptation costs in developing countries are estimated at US$70 billion, expected to reach US$140–300 billion in 2030 and US$280–500 billion in 2050, with the worldwide investments that presently degrade ecosystems exceeding conservation efforts by 600-852 billion USD annually" (UNEP 2022). In addition, economic instability and dependence on agricultural production lead to greater financial instability, making it difficult for communities to prioritize long-term ecological investments over immediate economic needs. The costs of data collection and assessment, resource management, and recovery efforts are also significant. For example, large-scale afforestation in deserts/drylands without adequate planning will result in significant economic losses and poor plant survival (Alikhanova and Bull 2023). To overcome these financial challenges, new financial mechanisms, such as payments for ecosystem services and public-private partnerships, are vital to increase the economic value of drought-affected communities while promoting sustainable development.

8.5.3 Legal Challenges

Ecosystem-based approaches in drylands often stem from complex land use and management systems. In many drylands, land ownership is unclear or contested, which can lead to conflicts among stakeholders and hinder conservation efforts (Turner et al. 2022). Furthermore, existing laws do not recognize or support indigenous rights and voices, as well as the ecological knowledge and practices of local communities. Administrative policies can also limit access to resources and make it difficult for communities to participate in cultural activities. Using an ecosystem-based approach requires cooperation in the use of legal instruments to address fundamental changes in the border governance system and the development of cross-border collaboration (Platjouw 2019; O'Hagan 2020). Furthermore, the lack of policies that address the interaction between health and social well-being can lead to fragmented management strategies. To overcome these legal challenges, reforms are needed to strengthen governance, clarify land rights, and create legal frameworks that support the integration of ecosystem-based approaches into local and national policies.

8.5.4 Societal Challenges

These challenges greatly affect the effectiveness of ecosystem-based approaches in drylands, which are often based on social, cultural, and community participation. Often, local people are not aware of the benefits of this process, and there needs to be a way to ensure multi-stakeholder and community participation (ADB Core Environment Program 2015). Traditional culture and lifestyles can also interfere with the new system, making it difficult to support community identity. Capacity and expertise must be expanded to realize the potential of ecosystem-based approaches (ADB Core Environment Program 2015). In addition, socioeconomic inequality can lead to greater tensions between different groups, as weaker communities may feel excluded from decision-making or lack access to the resources needed to participate. Population pressures and migration patterns can also degrade local ecosystems, making it difficult to maintain sustainable resources. When planning and implementing ecosystem strategies to address these social problems, it is important to encourage dialogue, strengthen education and capacity building, and ensure that the voices of all people in the community are heard.

Addressing these challenges requires a comprehensive understanding of the ecological and socioeconomic context to improve the resilience and sustainability of dryland environments. Trade-offs that provide short-term economic benefits and access to natural resources are often necessary (Cowan Jr et al. 2012). Ecosystem-based approaches, therefore, require greater interaction between policymakers and managers and a greater emphasis on human ecosystems (Milner-Gulland 2012).

8.6 Future Direction and Innovation

With 90% of the dryland population living in developing countries, ensuring the sustainability of drylands and improving their livelihoods are critical to global development (Lu et al. 2018). Therefore, technological and cultural advancements are needed to transform dryland science and strengthen the link between science, management, and knowledge in this ecosystem. New methods of data collection, management, and analysis processes that include large and diverse datasets are needed to improve decision-making for sustainable management (Browning et al. 2015). Also, cultural changes that support a collaborative environment for sharing and using resources are important for sustained innovation (Peters et al. 2015). Furthermore, how well policymakers integrate research findings into policy is a key determinant of dryland ecosystem dynamics and biodiversity standards.

8.6.1 Technological Advancements

There is a critical need for improved understanding of the ecological dynamics unique to these environments, including water availability, soil health, and biodiversity patterns. Given the increasing financial constraints, the need for diversified data streams with high sampling frequencies is required in drylands (Browning et al. 2015). Cost-effective strategies are required to facilitate local-scale translation of plans. For instance, placing inexpensive digital cameras at fixed locations can monitor soil changes, forest greenness, photosynthesis, and carbon fluxes, aiding in the prediction of environmental changes (Richardson 2023). Also, collection of environmental information using sensor networks can supplement, enhance, or replace traditional data collection methods (Collins et al. 2006). The fact that soil attributes affect soil color has been employed in the use of digital cameras to determine soil salt content, soil surface roughness, soil organic carbon, and soil moisture contents (Xu et al. 2019; Heil et al. 2022; Diaz et al. 2024). In this process, the raw photograph is analyzed to extract the color components: red (R), green (G), blue (B), and gray (Gr), and the digital number value of each pixel within these components is then calculated that exhibits distinct correlation with soil characteristics (Heil et al. 2022). Advanced remote sensing technologies and models will be useful in collecting and analyzing large datasets for more accurate assessments of ecosystem status. These innovations will improve the ability to collect and analyze data in drylands.

8.6.2 Cultural Advancements

A collaborative framework that includes local communities and research stakeholders is essential to ensure strategies are culturally meaningful and applicable and also not disturbing the emotional, social, and religious beliefs unless they are harming the ecosystems. "Social capital" has implications for building community resilience and collective management (Carmen et al. 2022). Greater knowledge and education about the ecological benefits of drylands can enable local people to participate in conservation. Open, collaborative, and cooperative research will be particularly useful for drylands in remote and large, spatially diverse landscapes that also face the challenges of the needs of a growing population (Browning et al. 2015). Creating networks and platforms for discussion and exchange will also create social cohesion and enable different voices to contribute to decision-making processes related to their environment. Ultimately, this cultural development will help create a sense of shared responsibility for dryland management.

8.6.3 Economic Advancements

Understanding the impact chain (how many factors affect ecosystems) and the role of factors in this chain is important for identifying effective investment areas to reduce impact (UNEP 2022). By focusing on these elements, we can better allocate resources and strategies to improve ecosystem health and protect against climate change. UNEP (2022) identified solutions to support investment in ecosystem-based adaptation, such as knowledge and awareness of the differential impacts and risks of climate adaptation, policy, coordination of activities, and participation by government agencies based on gender and local standards. The policy development process can be more effective and gain community support by using funds, including public and private partnerships, financial bonds, utility insurance, and other risks (UNEP 2022). New financial mechanisms to support these initiatives, such as payments for ecosystem services, can support sustainable practices and enable local communities to benefit from the protection of their ecosystems. Investing in green technologies can be beneficial for reducing environmental impact. Furthermore, creating market opportunities for sustainable products can provide financial support for conservation so that communities can thrive while keeping ecosystems healthy. By combining economic interests with environmental stewardship, we can support a sustainable future for drylands.

8.7 Conclusion

The future direction of the ecosystem-based approach is to support advances in technology, data analytics, and community engagement to improve environmental conservation. Innovations such as remote sensing, artificial intelligence, and machine learning will lead to increased monitoring of the health and functioning of ecosystems, allowing for time-sensitive monitoring. Integrating public research and local knowledge into decision-making will foster greater collaboration among stakeholders and ensure that diverse perspectives on conservation are considered. Moving forward, investing in natural capital and supporting new financial institutions will be critical to ensure that ecosystems thrive in the face of climate change. Ultimately, embracing these approaches not only safeguards our planet's ecological integrity but also secures the well-being of communities that depend on these vital resources for their livelihoods.

References

20th Livestock Census (2019) All India Report: Ministry of Agriculture, Department of Animal Husbandry, Dairying & Fisheries, Krishi Bhawan, New Delhi, India

ADB Core Environment Program (2015) Ecosystem-based approaches to address climate change challenges in the Greater Mekong subregion. ADB Publication Stock No. ARM157168-2. https://www.adb.org/sites/default/files/publication/158165/ecosystem-based-approaches-gms.pdf Accessed 31 Jan 2025

Albrecht A, Kandji ST (2003) Carbon sequestration in tropical agroforestry systems. Agric Ecosyst Environ 99:15–27

Alikhanova S, Bull JW (2023) Review of nature-based solutions in dryland ecosystems: the Aral Sea case study. Environ Manag 72:457–472. https://doi.org/10.1007/s00267-023-01822-z

Annual report of AICRP on Integrated farming system (2019) ICAR-Indian Institute of Farming Systems Research, Modipuram, Uttar Pradesh, India

Arenas-Wong R, Robles-Morua A, Bojórquez A, Martínez-Yrízar A, Yepez EA, Álvarez-Yépiz JC (2023) Climate-induced changes to provisioning ecosystem services in rural socioecosystems in Mexico. Weather Clim Extrem 41:100583. https://doi.org/10.1016/j.wace.2023.100583

Banin LF, Raine EH, Rowland LM et al (2022) The road to recovery: a synthesis of outcomes from ecosystem restoration in tropical and sub-tropical Asian forests. Philos Trans R Soc Lond Ser B Biol Sci 378:20210090. https://doi.org/10.1098/rstb.2021.0090

Barbosa FA, Soares Filho BS, Merry FD et al (2015) Cenários para a pecuária de corte amazônica. IGC/UFMG, 1–154

Bell LW, Moore AD, Kirkegaard JA (2014) Evolution in crop–livestock integration systems that improve farm productivity and environmental performance in Australia. Eur J Agron 57:10–20. https://doi.org/10.1016/j.eja.2013.04.007

Bentos TV, Nascimento HE, Vizcarra M, Williamson GB (2020) Effects of experimental lightgaps and topography on enrichment plantings in a central Amazonian secondary forest. Acta Amaz 50:192–198

Bentrup G, Hopwood J, Adamson NL, Vaughan M (2019) Temperate agroforestry systems and insect pollinators: a review. Forests 10:981

Bhuvaneshwari S, Hettiarachchi H, Meegoda JN (2019) Crop residue burning in India: policy challenges and potential solutions. Int J Environ Res Public Health 16:832. https://doi.org/10.3390/ijerph16050832

Browning DM, Rango A, Karl JW et al (2015) Emerging technological and cultural shifts advancing drylands research and management. Front Ecol Environ 13:52–60. https://doi.org/10.1890/140161

Cao A, Shi X (2022) The effects of climate change on habitat connectivity: a case study of the Brown-eared pheasant in China. Land 11:806. https://doi.org/10.3390/land11060806

Carmen E, Fazey I, Ross H et al (2022) Building community resilience in a context of climate change: the role of social capital. Ambio 51:1371–1387. https://doi.org/10.1007/s13280-021-01678-9

Cartereau M (2023) Tree biodiversity of warm drylands is likely to decline in a drier world. Glob Change Biol 29:3707–3722. https://doi.org/10.1111/gcb.16722

Chatterjee D, Das SR, Saha S, Sarkar A, Pathak H (2024) Impacts of climate change on soil processes. In: Pathak H, Chatterjee D, Saha S, Das B (eds) Climate change impacts on soil-plant-atmosphere continuum. Advances in global change research, vol 78. Springer, Singapore, pp 3–36

Chen W, Wolf B, Zheng X et al (2011) Annual methane uptake by temperate semiarid steppes as regulated by stocking rates, aboveground plant biomass and topsoil air permeability. Glob Change Biol 17:2803–2816. https://doi.org/10.1111/j.1365-2486.2011.02444.x

CoDriVE –Programme Designer (2013) A Handbook, Watershed Organisation Trust (WOTR), Pune, Maharashtra. https://weadapt.org/knowledge-base/vulnerability/community-driven-vulnerability-evaluation/

Collins SL, Bettencourt LMA, Hagberg A et al (2006) New opportunities in ecological sensing using wireless sensor networks. Front Ecol Environ 4:402–407

Conde C, Lonsdale KG, Nyong A et al (2004) Engaging stakeholders in the adaptation process: Technical Paper 2. Accessed online https://www4.unfccc.int/sites/NAPC/Country%20Documents/General/apf%20technical%20paper02.pdf. Accessed 31 Jan 2025

Cowan JH Jr, Rice JC, Walters CJ et al (2012) Challenges for implementing an ecosystem approach to fisheries management. Mar Coast Fish 4:496–510. https://doi.org/10.1080/19425120.2012.690825

Crozier LG, Burke JB, Chasco BE et al (2021) Climate change threatens Chinook salmon throughout their lifecycle. Commun Biol 4:222

Das D, Ray P, Datta SP (2024) Nutrient cycling and climate change. In: Pathak H, Chatterjee D, Saha S, Das B (eds) Climate change impacts on soil-plant-atmosphere continuum. Advances in global change research, vol 78. Springer, Singapore, pp 37–62

Datta P, Behera B (2023) Climate change adaptation through agroforestry: empirical evidence from Indian Eastern Himalayan foothills. In: Bandh SA (ed) Strategizing agricultural management for climate change mitigation and adaptation. Springer International Publishing, Cham, pp 167–181

Datta P, Behera B, Rahut DB (2022) Climate change and Indian agriculture: a systematic review of farmers' perception, adaptation, and transformation. Environ Chall 8:100543. https://doi.org/10.1016/j.envc.2022.100543

Delgado-Torres C, Octenjak S, Marcos R, et al (2024) Co-production of multi-annual climate services to support food and wine production resilience. EMS Annual Meeting Abstracts. https://doi.org/10.5194/ems2024-62

Devi PI (2024) Ecosystem service-based approaches in agricultural policy making. Indian J Agric Econ 79:94

Diaz FJ, Ahmad A, Parra L, Sendra S, Lloret J (2024) Low-cost optical sensors for soil composition monitoring. Sensors 24:1140. https://doi.org/10.3390/s24041140

D'Odorico P, Bhattachan A (2012) Hydrologic variability in dryland regions: impacts on ecosystem dynamics and food security. Philos Trans R Soc Lond Ser B Biol Sci 367:3145–3157. https://doi.org/10.1098/rstb.2012.0016

Dong Y, Zhang S, Qi Y, Chen Z, Geng Y (2000) Fluxes of CO_2, N_2O and CH_4 from a typical temperate grassland in Inner Mongolia and its daily variation. Chin Sci Bull 45:1590–1594. https://doi.org/10.1007/BF02886219

Ecosystem-Based Adaptation (EbA) Good Practice Case Studies (2021) Ecosystem-based adaptation through South-South co-operation. UNEP-IEMP, Chinese Academy of Sciences, Beijing, pp 3–9

Edrisi SA, Sarkar P, Son J, Prakash NT, Baral H (2022) Assessing the realization of global land restoration: a meta-analysis. Anthr Sci 1:179–194. https://doi.org/10.1007/s44177-022-00018-0

Eikelenboom M, Long TB (2023) Breaking the cycle of marginalization: how to involve local communities in multi-stakeholder initiatives? J Bus Ethics 186:31–62. https://doi.org/10.1007/s10551-022-05252-5

Ekins P, Zenghelis D (2021) The costs and benefits of environmental sustainability. Sustain Sci 16:949–965. https://doi.org/10.1007/s11625-021-00910-5

Enright SR, Boteler B (2020) The ecosystem approach in international marine environmental law and governance. In: O'Higgins TG, Lago M, DeWitt TH (eds) Ecosystem-based management, ecosystem services and aquatic biodiversity, 1st edn. Springer Nature, Berlin, pp 333–352

Gholkar M, Thombare P, Koli U, Kumbhar N (2022) Techno-economic assessment of agricultural land remediation measures through nutrient management practices to achieve sustainable agricultural production. Environ Chall 7:100492. https://doi.org/10.1016/j.envc.2022.100492

Gil J, Siebold M, Berger T (2015) Adoption and development of integrated crop–livestock–forestry systems in Mato Grosso, Brazil. Agric Ecosyst 199:394–406. https://doi.org/10.1016/j.agee.2014.10.008

Girija A, Maruthi Sankar GR, Krishna Gandhi B (2008) Optimization of fertilizers at varying soil test values and crop seasonal rainfall for finger millet under semi-arid Alfisols. Indian J Dryland Agric Res Dev 23:48–55

Girot P, Ehrhart C, Oglethorpe J (2012) Integrating community and ecosystem-based approaches in climate change adaptation responses. Ecosystems & Livelihoods Adaptation Network (ELAN). http://www.careclimatechange.org/files/adaptation/ELAN_IntegratedApproach_150412.pdf

Grünzweig JM, Boeck HJD et al (2022) Dryland mechanisms could widely control ecosystem functioning in a drier and warmer world. Nat Ecol Evol 6:1064–1076. https://doi.org/10.1038/s41559-022-01779-y

Guo X, Li H, Wang Z, Yang H, Hu C, Song L, Wang W (2023) Climate change impacts dryland multifunctionality: a cascade of precipitation variability to species replacement in multitrophic microbiota. Preprint at https://www.biorxiv.org/content/10.1101/2023.12.19.572331v1

Gupta SK, Kanga S, Meraj G et al (2024) Optimizing land use for climate mitigation using nature based solution (NBS) strategy: a study on afforestation potential and carbon sequestration in Rajasthan, India. Discov Geosci 2:36. https://doi.org/10.1007/s44288-024-00046-w

Hajong D, Patel AK et al (2023) Adoption of artificial insemination among large ruminants in arid Western Rajasthan, India. Indian J Ext Educ 59:65–69

Harsch MA, HilleRisLambers J (2016) Climate warming and seasonal precipitation change interact to limit species distribution shifts across western North America. PLoS One 11:e0159184

Heil J, Jörges C, Stumpe B (2022) Evaluation of using digital photography as a cost-effective tool for the rapid assessment of soil organic carbon at a regional scale. Soil Security 6:100023. https://doi.org/10.1016/j.soisec.2021.100023

Huang J, Li Y, Fu C et al (2017) Dryland climate change: recent progress and challenges. Rev Geophys 55:719–778. https://doi.org/10.1002/2016RG000550

Hudson A, Peters DPC, Blair JM et al (2022) Cross-site comparisons of dryland ecosystem response to climate change in the US long-term ecological research network. Bioscience 72:889–907. https://doi.org/10.1093/biosci/biab134

IPCC Sixth Assessment Report (2022) Working Group-II – impacts, adaptation and vulnerability. https://www.ipcc.ch/report/ar6/wg2/downloads/report/IPCC_AR6_WGII_HeadlineStatements.pdf. Accessed online on 22 Jan 2025

IUCN Issues Brief (2017) Drylands and land degradation. https://iucn.org/sites/default/files/2022-07/drylands_and_land_degradation_issues_brief_0.pdf. Accessed 31 Jan 2025

Jana S, Roy J, Marambe B et al (2015) Home gardens in the Paschim Medinipur District of West Bengal in India: a land use system with multiple benefits. Int J Environ Sustain Dev 14:191–206

Khatsuriya PA, Makwana VM, Dodia PP (2024) Biodiversity resilience amidst climate shifts: a global perspective. In: Idris S (ed) Global challenges for the environment and climate ChangeAdvances in environmental engineering and green technologies book series. IGI Global, pp 75–99. https://doi.org/10.4018/979-8-3693-2845-3.ch005

Kim DG, Isaac ME (2022) Nitrogen dynamics in agroforestry systems: a review. Agron Sustain Dev 42:1–18

Kuchimanchi BR, Nazareth D, Bendapudi R, Awasthi S, D'Souza M (2019) Assessing differential vulnerability of communities in the agrarian context in two districts of Maharashtra, India. Clim Dev 11:918–929. https://doi.org/10.1080/17565529.2019.1593815

Lal R (2001) World cropland soils as a source or sink for atmospheric C. Adv Agron 71:145–174

Lal R (2004) Soil carbon sequestration impacts on global climate change and food security. Science 304:1623–1627

Lauchlan SS, Nagelkerken I (2020) Species range shifts along multistressor mosaics in estuarine environments under future climate. Fish Fish 21:32–46. https://doi.org/10.1111/FAF.12412

Lee JH (2021) Setting the governance of a participatory ecosystem service assessment based on text mining the language of stakeholders' opinions. J Environ Manag 284:112003

Lee JH, Ahn S (2023) Ecosystem service evaluation based on local knowledge of residents using spatial text-mining. Sci Rep 13:22747. https://doi.org/10.1038/s41598-023-49612-1

Liu C, Holst J, Brüggemann N et al (2007) Winter-grazing reduces methane uptake by soils of a typical semi-arid steppe in Inner Mongolia, China. Atmos Environ 41:5948–5958

Liu J, Zhao Y, Chen X, Akarsha S (2024) Ecosystem-based management approaches for watershed conservation and geosustainability. Heliyon 10:e29800. https://doi.org/10.1016/j.heliyon.2024.e29800

Lu N, Wang M, Ning B, Yu D, Fu B (2018) Research advances in ecosystem services in drylands under global environmental changes. Curr Opin Environ Sustain 33:92–98. https://doi.org/10.1016/j.cosust.2018.05.004

Manjunatha BL, Shamsudheen M, Sureshkumar M et al (2019) Occupational structure and determinants of household income of pastoralists in Banni grasslands in Gujarat. Indian J Anim Sci 89:453–458

Mantyka-Pringle C, Martin TG, Rhodes JR (2012) Interactions between climate and habitat loss effects on biodiversity: a systematic review and meta-analysis. Glob Change Biol 18:1239–1252. https://doi.org/10.1111/J.1365-2486.2011.02593.X

Mbow C, Smith P, Skole D, Duguma L, Bustamante M (2014) Achieving mitigation and adaptation to climate change through sustainable agroforestry practices in Africa. Curr Opin Environ Sustain 6:8–14

Meijer SS, Sileshi GW, Catacutan D, Nieuwenhuis M (2016) Agroforestry and deforestation in Malawi: inter-linkages between attitudes, beliefs and behaviours. Agrofor Syst 90:645–658

Millenium Ecosystem Assessment (2005) Ecosystems and human well-being desertification synthesis. World Resources Institute, Washington. https://www.millenniumassessment.org/documents/document.355.aspx.pdf. Accessed 31 Jan 2025

Milner-Gulland EJ (2012) Interactions between human behaviour and ecological systems. Philos Trans R Soc Lond Ser B Biol Sci 367:270–278. https://doi.org/10.1098/rstb.2011.0175

Mohan Kumar B, Aravindakshan S (2022) Carbon footprints of the Indian AFOLU (Agriculture, forestry, and other land use) sector: a review. Carbon Footprints 2:1. https://doi.org/10.20517/cf.2022.04

Molua EL (2005) The economics of tropical agroforestry systems: the case of agroforestry farms in Cameroon. Forest Policy Econ 7:199–211

Mortimore M, Ariyo J, Bouzou IM, Mohammed S, Yamba B (2008) Niger and Nigeria: the Maradi-Kano region. A dryland case study of local natural resource management. In: Shepherd G (ed) The ecosystem approach: learning from experience. IUCN, Gland, Berlin, pp 23–58

Mosquera-Losada MR, Santiago-Freijanes JJ, Rois-DíAz M et al (2018) Agroforestry in Europe: a land management policy tool to combat climate change. Land Use Policy 78:603–613

Muchane MN, Sileshi GW, Gripenberg S et al (2020) Agroforestry boosts soil health in the humid and sub-humid tropics: a meta-analysis. Agric Ecosyst Environ 295:106899. https://doi.org/10.1016/j.agee.2020.106899

Muhsin EA, Hussain BAHA, Hammadi A, Khalaf HA (2023) Climate change and biodiversity: a review on understanding the global and local impacts of warming on the ecosystems. Al-Kufa Univ J Biol 15:10–18. https://doi.org/10.36320/ajb/v15.i2.11986

Muthuri CW, Ong CK, Black CR, Ngumi VW, Mati BM (2005) Tree and crop productivity in grevillea, Alnus and paulownia-based agroforestry systems in semiarid Kenya. For Ecol Manag 212:23–39

NAAS Newsletter (2021) National Academy of Agricultural Sciences, New Delhi

Nath AJ, Sileshi GW, Laskar SY et al (2021) Quantifying carbon stocks and sequestration potential in agroforestry systems under divergent management scenarios relevant to India's nationally determined contribution. J Clean Prod 281:124831

O'Hagan AM (2020) Ecosystem-based management (EbM) and ecosystem services in EU law, policy and governance. In: O'Higgins TG, Lago M, DeWitt TH (eds) Ecosystem-based management, ecosystem services and aquatic biodiversity: theory, tools and applications. Springer, Cham, pp 353–372

Ojea E (2015) Challenges for mainstreaming ecosystem-based adaptation into the international climate agenda. Curr Opin Env Sust 14:41–48

Okin GS, Gillette DA (2001) Distribution of vegetation in wind-dominated landscapes: implications for wind erosion modeling and landscape processes. J Geophys Res 106:9673–9683

Panda A (2009) Assessing vulnerability to climate change in India. Econ Polit Wkly 44:105–107

Pandey S, Bhandari H (2007) Drought: an overview. In: Pandey S, Bhandari H, Hardy B (eds) Economic costs of drought and rice farmers coping mechanisms: cross-country comparatives analysis. International Rice Research Institute, Los Banos, pp 31–42

Pandey B, Ghosh A (2023) Urban ecosystem services and climate change: a dynamic interplay. Front Sustain Cities 5:1281430. https://doi.org/10.3389/frsc.2023.1281430

Pandey R, Meena D, Aretano R et al (2015) Socio-ecological vulnerability of smallholders due to climate change in mountains: agroforestry as an adaptation measure. Chang Adapt Soc Ecol Syst 2:26–41. https://doi.org/10.1515/cass-2015-0003

Pandey R, Aretano R, Gupta AK, Meena D, Kumar B, Alatalo JM (2017) Agroecology as a climate change adaptation strategy for smallholders of Tehri-Garhwal in the Indian Himalayan region. Small-Scale For 16:53–63

Paramesh V, Ravisankar N, Behera U et al (2022) Integrated farming system approaches to achieve food and nutritional security for enhancing profitability, employment, and climate resilience in India. Food Energy Secur 11:e321. https://doi.org/10.1002/fes3.321

Park SU, Koo KA, Tho J-H, Jung C (2023) Projection of potential habitat change and fragmentation of the endangered species *Aconitum coreanum* under climate change. J Clim Res. https://doi.org/10.15531/ksccr.2023.14.1.067

Peters DPC, Havstad KM, Archer SR, Sala OE (2015) Beyond desertification: new paradigms for dryland landscapes. Front Ecol Environ 13:4–12

PIB (2022) Creation of additional carbon sink. Ministry of Environment, Forest and Climate Change, Government of India

Pita P, Fernández-Vidal D, García-Galdo J, Muíño R (2016) The use of the traditional ecological perception of fishermen, cost-effective tools and participatory models in artisanal fisheries: towards the co-management of common octopus in Galicia (NW Spain). Fish Res 178:4–12
Platjouw FM (2019) Dimensions of transboundary legal coherence needed to foster ecosystem-based governance in the Arctic. Mar Policy 110:103666. https://doi.org/10.1016/j.marpol.2019.103666
Prasad P (1998) Famines and droughts: survival strategies. Rawat Publication, Jaipur
Ramachandran Nair PK, Nair VD, Mohan Kumar B, Haile SG (2009) Soil carbon sequestration in tropical agroforestry systems: a feasibility appraisal. Environ Sci Pol 12:1099–1111
Richardson AD (2023) PhenoCam: an evolving, open-source tool to study the temporal and spatial variability of ecosystem-scale phenology. Agric For Meteorol 342:109751. https://doi.org/10.1016/j.agrformet.2023.109751
Rode J, Wittmer H, Emerton L, Schröter-Schlaack C (2016) 'Ecosystem service opportunities': aa practice-oriented framework for identifying economic instruments to enhance biodiversity and human livelihoods. J Nat Conserv 33:35–47. https://doi.org/10.1016/j.jnc.2016.07.001
Rubenstein MA, Weiskopf SR, Bertrand R et al (2023) Climate change and the global redistribution of biodiversity: substantial variation in empirical support for expected range shifts. Environ Evid 12:7. https://doi.org/10.1186/s13750-023-00296-0
Salton JC, Mercante FM, Tomazi M et al (2014) Integrated crop-livestock system in tropical Brazil: toward a sustainable production system. Agric Ecosyst Environ 190:70–79. https://doi.org/10.1016/j.agee.2013.09.023
Sander G (2024) European approaches support an essential definition of ecosystem-based management and demonstrate its implementation for the oceans. Ocean Dev Int Law 54:421–447. https://doi.org/10.1080/00908320.2023.2301105
Santojanni FB, Miner H, Hain H, Sutton G (2023) The impact of climate change on biodiversity in coastal ecosystems. Jurnal Ilmu Pendidikan dan Humaniora. https://doi.org/10.35335/jiph.v12i3.9
Sapkota TB, Vetter SH, Jat ML (2019) Cost-effective opportunities for climate change mitigation in Indian agriculture. Sci Total Environ 655:1342–1354
Schönbach P, Wolf B, Dickhöfer U et al (2012) Grazing effects on the greenhouse gas balance of a temperate steppe ecosystem. Nutr Cycl Agroecosyst 93:357–371. https://doi.org/10.1007/s10705-012-9521-1
Semeraro T, Scarano A, Leggieri A, Calisi A, Caroli MD (2023) Impact of climate change on agroecosystems and potential adaptation strategies. Land 12:1117. https://doi.org/10.3390/land12061117
Shivanna KR (2022) Climate change and its impact on biodiversity and human welfare. Proc Indian Natl Sci Acad 88:160–171
Singh KK, Dubey AK, Singh RC (2017) Energy in conservation agriculture. In: Belavadi VV, Agriculture KNN, Gangadharappa NR (eds) Agriculture under climate change – threats, strategies and policies. Allied Publishers Pvt Ltd, India, pp 280–288
Singh D, Choudhary MK, Meena M, Kumar C (2019) Rain water harvesting for food and livelihood security: a case study from Pali, India. Open Agric 4:767–777
Singh C, Solomon D, Rao N (2021a) How does climate change adaptation policy in India consider gender? An analysis of 28 state action plans. Clim Pol 21:958–975
Singh R, Patel K, Chaurasia M (2021b) Climate change and nutrients dynamics of soil. In: Choudhary DK, Mishra A, Varma A (eds) Climate change and the microbiome. Soil biology, vol 63. Springer, Cham, pp 521–549. https://doi.org/10.1007/978-3-030-76863-8_27
Srinivasarao C (2018) Climate change and extreme climatic events: impacts and adaptation strategies for agriculture. The Andhra Agric J 65:253–259
Srinivasarao C, Prasad JVNS, Vittal KPR, Venkateshwarlu B, Sharma KL (2003) Role of optimum nutrition in drought management. Indian J Fertil 48:105–112
Srinivasarao C, Venkateswarlu B, Dinesh Babu M et al (2011) Soil health improvement with Gliricida green leaf manuring in rainfed agriculture. Central Research Institute for Dryland Agriculture, Hyderabad

Srinivasarao C, Venkateswarlu B, Lal R et al (2012) Soil carbon sequestration and agronomic productivity of an Alfisol for a groundnut based system in a semiarid environment in southern India. Eur J Agron 43:40–48

Srinivasarao C, Venkateswarlu B, Lal R, Singh AK, Kundu S (2013a) Sustainable management of soils of dryland ecosystems for enhancing agronomy productivity and sequestering carbon. Adv Agron 121:253–329

Srinivasarao C, Venkateswarlu B, Lal R, Singh AK, Kundu S, Jakkula VS (2013b) Carbon sink capacity and agronomic productivity of soils of semiarid regions of India. In: Lal R, Stewart BA (eds) Principles of sustainable soil management in agroecosystems. CRC Press, pp 423–476

Srinivasarao C, Lal R, Prasad JVNS et al (2015) Potential and challenges of rainfed farming in India. Adv Agron 133:113–181

Srinivasarao C, Indoria AK, Sharma KL (2017) Effective management practices for improving soil organic matter for increasing crop productivity in rainfed agroecology of India. Curr Sci 112:7

Srinivasarao C, Lakshmi CS, Kundu S (2019) Critical role of nutrient management in climate adaptive dryland farming systems in India. Indian J Fertil 15(11):1220–1233

Stewart JAE, Stewart JAE, Wright DH, Heckman K (2017) Apparent climate-mediated loss and fragmentation of core habitat of the American pika in the Northern Sierra Nevada, California, USA. PLoS One 12:e0181834. https://doi.org/10.1371/JOURNAL.PONE.0181834

Suietnov Y (2022) Ecosystem approach in international environmental law: from the idea of conservation of natural ecosystems to the emergence of a holistic concept. Problems Legality:116. https://doi.org/10.21564/2414-990x.157.256982

Sunderland TCH (2011) Food security: why is biodiversity important? Int For Rev 13:265–274

Thornton PK, Herrero M (2010) Potential for reduced methane and carbon dioxide emissions from livestock and pasture management in the tropics. Proc Natl Acad Sci 107:19667–19672

Thornton PK, Kruska RL, Henninger N et al (2002) Mapping poverty and livestock in the developing world. ILRI, Nairobi

Tingley MW, Koo MS, Moritz C, Rush AC, Beissinger SR (2012) The push and pull of climate change causes heterogeneous shifts in avian elevational ranges. Glob Change Biol 18:3279–3290. https://doi.org/10.1111/J.1365-2486.2012.02784.X

Torres CMME, Jacovine LAG et al (2017) Greenhouse gas emissions and carbon sequestration by agroforestry systems in southeastern Brazil. Sci Rep 7:1–7

Turner WC, Périquet S, Goelst CE et al (2022) Africa's drylands in a changing world: challenges for wildlife conservation under climate and land-use changes in the Greater Etosha landscape. Glob Ecolo Conserv 38:e02221. https://doi.org/10.1016/j.gecco.2022.e02221

Twigg J (1999) The age of accountability: community involvement in disaster reduction. Paper presented at the sixth annual conference of the international emergency management society (TIEMS '99), on the theme of "Contingencies, emergency, crisis and disaster management: defining the agenda for the third millennium", Delft, The Netherlands, 8–11 June 1999

Ukkola AM, De Kauwe MG, Roderick ML, Burrell A, Lehmann P, Pitman AJ (2021) Annual precipitation explains variability in dryland vegetation greenness globally but not locally. Glob Change Biol 27:4367–4380. https://doi.org/10.1111/gcb.15729

UNEP (2022) Beyond Ecosystem-based Adaptation gaps and barriers. Overcoming barriers in different regions, no-regret and low-regret adaptation options and avoiding maladaptation. 2nd Workshop under the Glasgow–Sharm el-Sheikh Work Programme on the Global Goal on Adaptation, 31 August 2022, pp 3

UNEP Adaptation Gap Report (2020) United Nations Environment Programme. https://www.unep.org/resources/adaptation-gap-report-2020. Accessed 31 Jan 2025

VanDerWal J, Murphy HT, Kutt AS et al (2013) Focus on poleward shifts in species' distribution underestimates the fingerprint of climate change. Nat Clim Chang 3:239–243. https://doi.org/10.1038/NCLIMATE1688

Vasiliev DV (2022) The role of biodiversity in ecosystem resilience. IOP Conf Ser Earth Environ Sci 1072(1):012012. https://doi.org/10.1088/1755-1315/1072/1/012012

Wang Y, Liu Y, Shan L et al (2024) Ecosystem management and sustainable livelihoods in drylands. In: Fu B, Stafford-Smith M (eds) Dryland social-ecological systems in changing environments. Springer Nature, pp 139–157

Weil C, Johnson JA, Chaplin-Kramer R (2023) Modeling climate change impact on agro-ecosystem services. In: Nendel C (ed) Modelling climate change impacts on agricultural systems. Science Publishing, Cambridge, pp 3–38

Wells EC, Lehigh GR, Vidmar AM (2023) Stakeholder engagement for sustainable communities. In: Brinkmann R (ed) The Palgrave handbook of global sustainability. Palgrave Macmillan, Cham, pp 1143–1155. https://doi.org/10.1007/978-3-031-01949-4_10

White RP, Nackoney J (2003) Drylands, people, and ecosystem goods and services: a web-based geospatial analysis. World Resources Institute. Available via http://pdf.wri.org/drylands.pdf Accessed 21 Jan 2025

Witt C, Dobermann A, Abdulrachman S et al (1999) Internal nutrient efficiencies of irrigated lowland rice in tropical and subtropical Asia. Field Crop Res 63:113–138

Xu L, Zheng CL, Wang ZC, Nyongesah MJ (2019) A digital camera as an alternative tool for estimating soil salinity and soil surface roughness. Geoderma 341:68–75

Yao Y, Fu B, Liu Y, Wang Y, Song S (2021) The contribution of ecosystem restoration to sustainable development goals in Asian drylands: a literature review. Land Degrad Dev 32:4472–4483. https://doi.org/10.1002/ldr.4065

Yazdandoost MY (2023) Innovative nature-based solutions for ecosystems resilience, connectivity and conservation, Iran. J Global Resour 9:31. https://doi.org/10.46587/jgr.2023.v09i02.004

Chapter 9
Adapting to Climate Change in Dryland Ecosystems; Mitigation and Resilience Through Ecosystem-Based Adaptation

Chandan Kumar, Dheeraj Singh, Om Prakash Meena, and Anand Kumar Naorem

Abstract Rapid climate change has driven nations and organizations to develop strategies for adapting to shifting environmental conditions, prompting the need for evidence-based adaptations that guide focus and maximize effectiveness. Ecosystem-based adaptation (EbA), which integrates ecosystem resilience into climate adaptation policies, is increasingly adopted worldwide. These adaptations are especially relevant in arid regions where communities rely heavily on natural resources for daily needs (Seddon et al. 2016). EbA strategies encompass biodiversity and ecosystem services as essential elements for managing climate-related challenges within broader adaptation frameworks. These adaptations have evolved to incorporate sustainable management, conservation, and restoration practices, offering socioeconomic co-benefits for local communities. Notable examples include the conservation and restoration of forests and vegetation to stabilize slopes and prevent landslides, the regulation of water flows to reduce flash flooding risk, and the restoration of dryland ecosystems to moderate temperature shifts and protect against soil erosion. Additionally, dryland EbA strategies aid in adapting to changing rainfall patterns, providing critical resilience in the face of climate variability.

Keywords Ecosystem-based adaptation (EbA) · Climate resilience · Dryland ecosystem · Biodiversity conservation · Natural resource management

C. Kumar
ICAR-CAZRI, KVK, Pali-Marwar, Rajasthan, India

D. Singh (✉) · O. P. Meena · A. K. Naorem
ICAR-CAZRI, Jodhpur, Rajasthan, India

Y. D. I. Khan et al. (eds.), *Land Restoration Through Ecosystem-Based Approach*, Disaster Resilience and Green Growth, https://doi.org/10.1007/978-981-95-0389-6_9

9.1 Introduction

Dryland ecosystems face considerable threats due to inadequate management practices and extreme climate conditions. Developing adaptation strategies for these regions requires a thorough understanding of the impacts of climate change, as these ecosystems will be crucial in ensuring food security and alleviating poverty, hunger, and malnutrition in the coming decades. Effective management of agricultural resources in drylands is thus urgently needed. These ecosystems, consisting of a diverse array of terrestrial biomass, display wide heterogeneity in topography, climate, geological features, and biological conditions. Found on every continent in both the northern and southern hemispheres, drylands support approximately 35–40% of the global population (over 2 billion people) in dryland areas, with varying population distributions across different aridity zones. Of these, only 1–5% live in hyper-arid regions, 20–25% in arid zones, and 70–75% in semi-arid regions (Jellason et al. 2022). Countries with extensive dryland areas include, but are not limited to, India, Afghanistan, Pakistan, Iran, Syria, China, Argentina, Brazil, Mexico, Peru, Chile, Australia, the United States, Canada, Uzbekistan, Kazakhstan, and significant portions of Sub-Saharan Africa (Rathore et al. 2017). This chapter primarily evaluates various technologies for climate change modification and resilience within dryland ecosystems. One major constraint in these regions is soil salinity, particularly in enclosed basins where surface water lacks adequate drainage. In drylands, agriculture is both a driver and a victim of extreme climate variability, characterized by high temperatures, water scarcity, low humidity, sparse vegetation, and vulnerability to soil erosion. These conditions contribute to frequent droughts and a reduction in overall crop yields. Droughts are anticipated to continue with similar frequency and geographic distribution shortly. Climate variability has significantly impacted farmers' adaptive capacities, increasing their susceptibility to future weather-related adversities. Human-induced emissions of greenhouse gases, especially carbon dioxide (CO_2) from the combustion of fossil fuels like coal, oil, and gas, have contributed to a significant rise in Earth's surface temperature. Global surface temperature was 1.09 [0.95 to 1.20]°C higher in 2011–2020 than in 1850–1906, with larger increases over land (1.59 [1.34 to 1.83]°C) than over the ocean (0.88 [0.68 to 1.01]°C). Global surface temperature in the first two decades of the twenty-first century (2001–2020) was 0.99 [0.84 to 1.10]°C higher than 1850–1900. Global surface temperature has increased faster since 1970 than in any other 50-year period over at least the last 2000 years (IPCC 2023). This warming trend poses extensive risks to both human societies and ecosystems, including accelerated species extinction rates, adverse health impacts, shifts in biomass, reductions in agricultural productivity, and extensive flooding due to sea-level rise (Scarano 2017). Climate change adaptation and mitigation/resilience remain among the most pressing challenges for humanity, particularly for highly vulnerable ecosystems. Drylands face severe water scarcity and increasingly erratic rainfall, mountains experience rapid temperature increases that contribute to glacial melt and subsequent water shortages, and coastal regions are exposed to frequent, intense storms and rising sea levels that result in coastal erosion and saltwater intrusion. Implementing adaptation strategies is essential to enhance the resilience of populations in these sensitive regions, particularly as they are the most exposed to climate change's impacts (Scarano 2017).

An ecosystem-based adaptation (EbA), which utilizes ecosystems and their services to support climate adaptation efforts, presents a promising means to bolster the resilience of vulnerable communities (Reid et al. 2019). Governments, funding bodies, and research institutions increasingly recognize EbA as a nature-based response to climate challenges. These adaptations serve as an effective strategy, offering additional benefits related to biodiversity preservation and climate mitigation (CBD 2009). However, there are gaps in understanding and applying EbA. First, as a relatively recent approach, EbA lacks long-term data and lessons learned. Second, it has diverse applications that vary significantly by ecosystem. Third, EbA often requires decades to yield substantial results, slowing the development of best practices for effective implementation (Munroe et al. 2012). Finally, measuring EbA's socioeconomic benefits can be challenging, making it difficult to assess its real impact on communities. Therefore, detailed documentation of EbA outcomes, both successes and failures, is essential for advancing knowledge in this area (Fig. 9.1).

People in dryland regions have considerable potential to contribute to climate change mitigation, as they possess adaptive skills that can enhance the sustainable

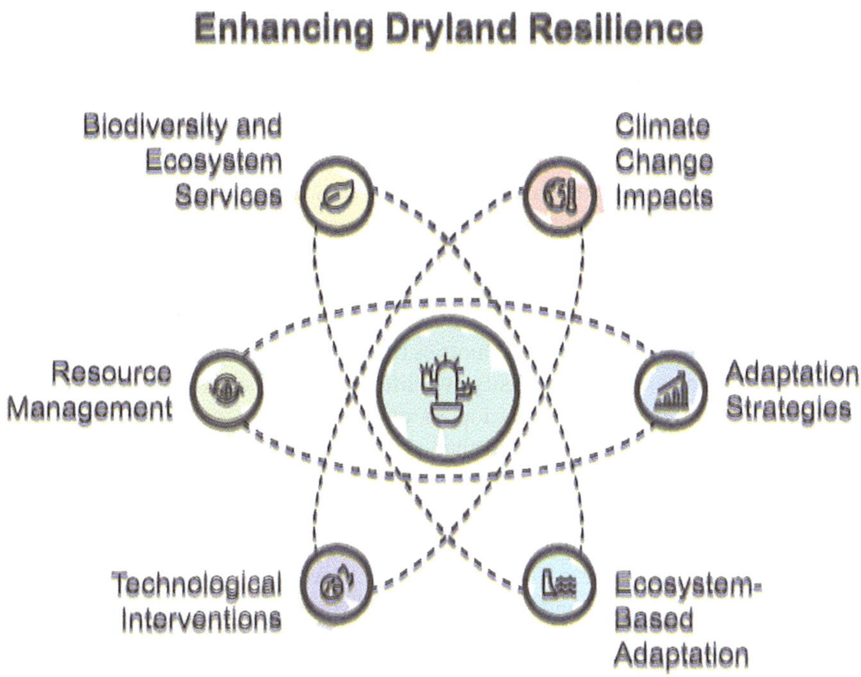

Fig. 9.1 Multifaceted strategies to enhance the resilience of drylands. (Scarano 2017)

use of natural resources, develop diverse livelihoods, and share valuable expertise, such as in extensive grazing resource management. Drylands contain significant yet underutilized resources that could bolster climate resilience, including solar radiation for crop growth, solar and wind energy, minerals, oil and gas reserves, and, in certain areas, unexplored groundwater or surface water sources. Furthermore, these regions offer notable non-agricultural land use options, such as agro-tourism, wildlife habitats, and non-timber forest products, while certain areas also support high-value, intensive agricultural production (Jellason et al. 2022). Drylands also host diverse biodiversity and ecosystem services, which play a crucial role in climate adaptation and mitigation while sustaining local livelihoods. Resilient ecosystems, with their capacity to withstand and recover from extreme events, contribute significantly to adaptation efforts. To preserve ecosystem health and resilience in the face of climate change, it is essential to identify existing gaps and employ ecosystem-based adaptation (EbA) strategies alongside technology-driven interventions to address these challenges (Seddon et al. 2016).

9.2 Features of Dryland Ecosystem

Drylands are characterized by limited water availability, where precipitation is balanced by surface evaporation and plants' transpiration (evapotranspiration). According to the United Nations Environment Program (UNEP), drylands are defined by an aridity index (AI), which is the ratio of precipitation (P) to potential evapotranspiration (PET). Areas with an aridity index (AI) less than 0.65 are classified as drylands, and they globally share an area with a population, as shown in Fig. 9.2 (Safriel et al. 2005). Africa and Asia together account for two-thirds of the world's drylands, with Africa comprising 41.3% (13.1%) and Asia (13.2%) following closely. Other regions include Australia (4.4%), Europe (2.1%), North America (4.9%), and South America (3.6%). Developing countries are home to a significantly higher proportion of drylands, making up 72% of global dryland areas, and this percentage increases with aridity, with nearly all hyper-arid lands located in developing nations. Substantial dryland areas are also found in the United States, Australia, and Southern European countries (Mortimore et al. 2009). Due to the vast diversity in climate, soils, biodiversity, wildlife, and human activity across drylands, providing a comprehensive definition or characterization is challenging. These global areas and populations are further classified into four types based on their specific aridity index (AI):

However, aridity and rainfed farming are common features across dryland regions globally (Swiderska et al. 2018). The primary limitation to productivity in these areas is moisture stress, making rainwater harvesting structures, mulching to reduce evaporation losses, and using drought-tolerant crops essential for effective dryland ecosystem management.

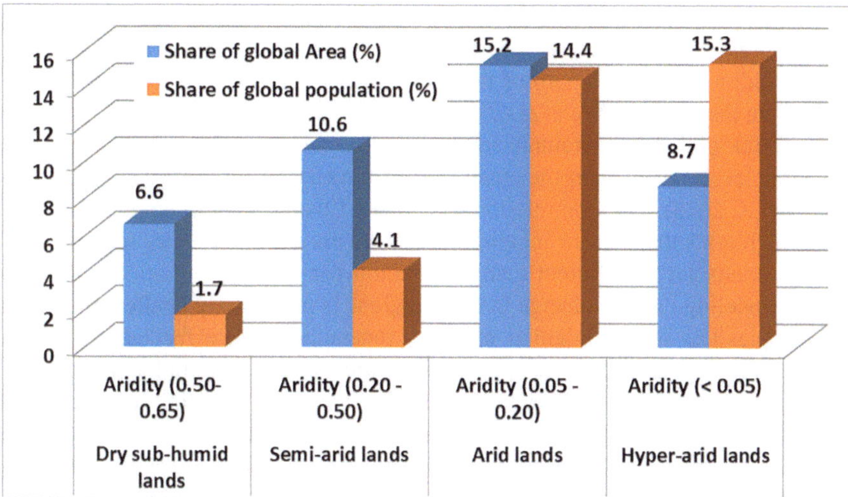

Fig. 9.2 The dryland ecosystem types and global share of area and population. (Safriel et al. 2005)

9.2.1 Climatic Characteristics of Drylands

In India, nearly 57% of the net cultivated area consists of drylands. The Food and Agriculture Organization (FAO) classifies these drylands into three categories based on annual precipitation: arid (less than 50 cm), semi-arid (50–75 cm), and dry sub-humid (75–150 cm). These categories also depend on the length of time required for annual crops to grow, which is determined by the number of days when precipitation exceeds half of the potential evapotranspiration (PET) plus the period needed to use 100 mm of water from excess precipitation stored in the soil profile. According to this criterion, regions with growing periods of 1–59 days, 60–119 days, and 120–179 days are classified as semi-arid, arid, and sub-humid, respectively. Collectively, these regions account for nearly 45% of India's drylands, with arid regions covering 7%, semi-arid regions 20%, and dry sub-humid regions 18% (Ghosn 2006).

Dryland ecosystems are characterized by inadequate and inconsistent precipitation, low rainfall, low productivity, short growing seasons, high evapotranspiration rates, high wind velocity, and intense solar radiation (Ogbaga et al. 2020). While some of these regions are dry and cold, others are excessively hot. Despite these challenging biophysical conditions, drylands support a substantial population that depends primarily on agriculture and related activities, leading to the overuse of the region's finite natural resources. The sustainability of agricultural production in these areas is increasingly threatened by several factors, including environmental pollution (soil and water), climate change, declining farm profitability, low input-use efficiency (fertilizer, water, energy, and labor), degradation of natural resources (land, water, and biodiversity), and a lack of sufficient manpower. Arid and semi-arid regions experience extended dry periods during the crop-growing season, resulting in higher crop failure rates. While sub-humid regions may also face dry spells during the growing season, the likelihood of crop failure is significantly lower. Additionally, the length of the rainy season decreases with increasing distance from the equator (Wu et al. 2019).

By 2100, global temperatures are projected to rise by at least 2 °C above pre-industrial levels, even under the most optimistic scenarios for greenhouse gas emissions (IPCC 2018). The Commission on Climate Change and Development (CCCD 2008) highlights that significant climate changes are expected in dry regions; however, forecasts across different scenarios exhibit considerable variation and uncertainty. Modeling climate change in these regions is particularly challenging due to the complexities of precipitation patterns, vegetation-climate feedbacks, and the direct physiological impacts of CO_2 on plant life. As a result, predicting how dryland ecosystems will respond to increased CO_2 and global warming remains highly uncertain. The historical record of drylands has been examined as a potential guide for future events. Since water is the most limited natural resource in these regions, prolonged rainfall interruptions can lead to severe crises, including catastrophic famine, as experienced in Africa during the 1970s and 1980s. Approximately 41% of the global population resides in drier areas where drought is common, and extended periods of abnormally low rainfall are considered "extreme" climatic events. Climate modeling studies predict that the average rainfall in many dry regions will decrease further, exacerbating the challenges faced by these areas (Yan et al. 2019).

9.2.2 *Soils of Drylands*

The ability of a soil profile to store water is largely determined by its depth. However, in drylands, the presence of a "hardpan" layer often limits soil depth, restricting its water retention capacity. This hardpan, which typically lies between 5 and 60 cm below the surface, can vary in composition depending on the effects of vegetation and climate on the underlying parent rock. Drylands tend to generate coarse-textured soils characterized by low organic matter, which further limits water retention and high permeability. Excessive soil strength and rigid pore structures in these soils hinder root growth, reducing the plant's ability to access water and nutrients effectively (Ndheleve et al. 2017). The productivity of dryland soils is generally low due to intense weathering of minerals, poor water retention, limited root development, and high nutrient leaching. To improve productivity, soil management strategies must focus on enhancing root growth conditions and ensuring a consistent and adequate supply of nutrients and water. These soils are typically low in nitrogen, low to medium in phosphorus, and medium to high in potash, making effective nutrient management critical for sustainable farming practices in dryland areas. Nitrogen, in particular, plays a key role in improving productivity, with its availability closely tied to the region's rainfall patterns. Studies by Kipkemoi et al. (2021) and others have shown that adequate rainfall significantly increases productivity with nitrogen application. The organic matter content of dryland soils is typically low, as organic material does not accumulate significantly in these areas. As soils are intensively used for crop production, this limited organic matter is quickly depleted (Kassas 2002).

Dryland soils exhibit significant variation between dry sub-humid and hyper-arid regions, yet they share several common characteristics. Swiderska et al. (2018) noted that these soils typically display alkalinity, calcium carbonate in the subsoil, clayey textures in depressions, and sandy or gravelly compositions in upland areas. They are generally low in fertility, primarily due to low organic matter content. In dryland ecosystems, where surface water does not flow out of closed basins, salts tend to accumulate, leading to soil salinization. It is estimated that 25–30% of irrigated lands are affected by salinization, a process often exacerbated by human activities. Water used for irrigation in drylands commonly contains salts in varying concentrations. Over the course of a year, up to 10 tons of salts per hectare can accumulate through irrigation, with most of these salts remaining in the soil after evapotranspiration removes the water (Jellason et al. 2022). If these salts are not removed through internal drainage or leached into the subsoil, they will build up in the surface soil, potentially hindering plant growth.

Excess irrigation water must be applied when leaching is insufficient to manage salinity in dryland climates. However, if natural or artificial drainage systems are unable to remove excess water, the leached salts are likely to return and re-salinate the surface soil. In Fig. 9.3, soil of the dryland ecosystem faces many challenges, including stoniness, shallowness, compaction from tillage equipment, surface soil hardening, susceptibility to wind and water erosion, limited drainage capacity, etc. (Rao and Ryan 2004).

9.2.3 *Biodiversity in Drylands*

Dryland biodiversity plays a crucial role in the global fight against poverty, climate change, and desertification. Despite the perception that drylands are barren and unappealing, they are home to a surprising range of biodiversity and unique ecosystems. In fact, dryland soils may harbor more biodiversity than tropical rainforests. The biodiversity of drylands is not solely based on species diversity, as seen in tropical forests, but focuses more on ecosystem diversity (CBD 2009). Native plants (such as cacti, succulents, acacia, baobab, and various grasses) and animals (which make up 64% of bird species, 55% of mammals, and 25% of amphibians) in arid regions have developed a variety of adaptations to survive harsh conditions, including frequent droughts (Kassas 2002). Although dryland ecosystems tend to be stable over long periods, seasonal and annual fluctuations in plant and animal populations can occur due to variations in local climate. Human activities, such as deforestation and other forms of vegetation disturbance, can degrade the land's productive capacity and cause permanent damage to ecosystems (Swiderska et al. 2018). One of the most significant human impacts is cultivation, which disrupts the natural ecosystem

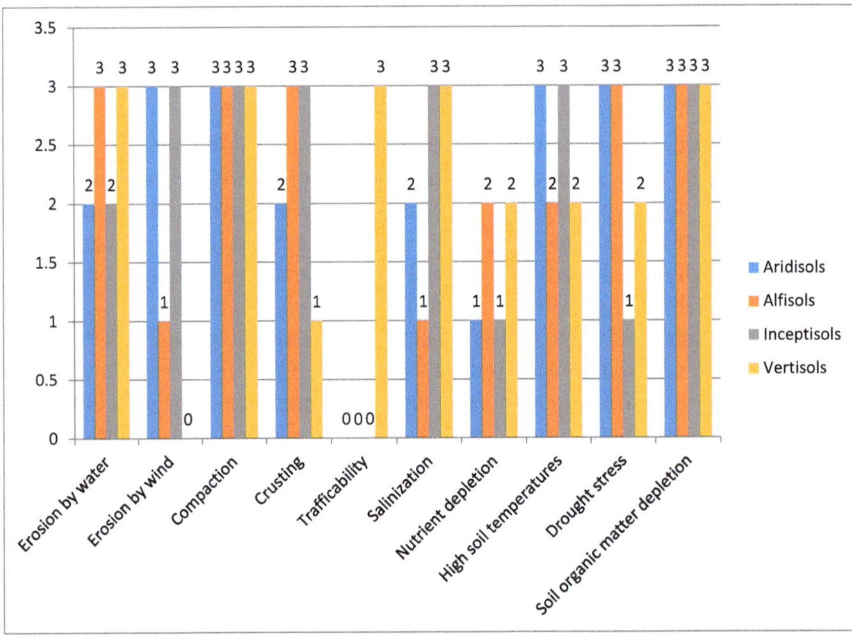

Fig. 9.3 Soil-related constraints in dryland ecosystem. (Rao and Ryan 2004) (0 = none, 1 = low, 2 = moderate, 3 = severe)

it replaces but does not necessarily endanger long-term productivity. Water is essential for all life, and dryland ecosystems have adapted to cope with water scarcity. As aridity increases, the potential biomass decreases, with fewer species able to adapt to the changing conditions. Some plants, such as those that store water in their tissues or those with waxy coatings to reduce water loss, are well-suited to these environments. Other adaptations include plants with narrow leaves or tissues that can tolerate desiccation without harm (Zeng et al. 2021).

However, dryland biodiversity is under threat from several human activities, including overexploitation of resources, poor water management, inappropriate agricultural practices, soil erosion, invasive species, habitat conversion, poverty, and climate change (Ghosn 2006). These threats undermine the ecosystem services that dryland biodiversity provides, which are vital to local communities for food, fuel, and water, as well as global benefits such as carbon sequestration. Yet, ecosystem services in drylands are often undervalued, and these services' cultural and spiritual significance is frequently overlooked. It is essential to raise awareness of the value of these services and implement appropriate protections to prevent their loss. Drylands in regions such as the United States, southern Europe, and Central Asia are also recognized for their rich cultural heritage.

9.2.4 Demographic Characteristics of Drylands

Drylands are regions where primary production is generally limited by water availability. Population density is typically low in hyper-arid zones, often less than 1 person per square kilometer, while it increases to around 5 people per square kilometer in arid areas and roughly 10 per square kilometer in semi-arid regions. Agriculture is the primary occupation for 65–70% of people in dryland areas, followed by 5–10% working in animal-based industries and 15–20% residing in urban areas (Mortimore et al. 2009). Dryland regions, particularly in West Asia, Sub-Saharan Africa, and South Africa, host over 75% of the global population in these areas, with countries like Mexico, Peru, Bolivia, and Australia also having a significant percentage of their populations living in drylands. As of 2015, over half of the people living in dry regions were located in African countries, with nations such as Ethiopia, South Africa, Burkina Faso, and Morocco, along with India, Iran, China, Turkey, Afghanistan, and Mexico, having some of the highest population densities in dryland areas. By 2050, the populations in drylands of China, Burkina Faso, Iran, and Turkey are expected to double (Mortimore et al. 2009). One notable demographic trend in drylands is the growing movement of people to urban slums, particularly in areas facing economic hardship and health challenges. These regions often have the highest infant mortality rates, the lowest per capita GDP, and the lowest human well-being indices. Despite these hardships, the 1990s saw the fastest population growth in drylands, contributing to the growing impoverishment in these areas. Over half of the world's impoverished population now resides in dryland regions (Kummu et al. 2018).

Economic development in dryland regions is typically lower than the global average, although dryland nations contribute significantly to global GDP. Between 1990 and 2015, dryland nations accounted for more than 25% of the world's GDP, with substantial variations across continents (Kummu et al. 2018). Asia, Europe, and North America have seen the most significant economic growth in dryland areas, with Asia showing a more uniform growth distribution between dryland and non-dryland regions. However, dryland regions in Africa have generally experienced slower growth compared to humid countries. In many Asian and African nations, dryland areas contribute more than half of the national GDP, with regions such as Western Asia, Eastern Europe, East China, and the United States standing out for their economic development in drylands. Rapid demographic changes, whether increases or decreases, complicate resource management and planning in these regions (CCCD 2008).

9.3 Climate Change in Dryland Ecosystem

The dryland ecosystem is confronted with significant challenges due to climate change, which affects both ecosystem health and human well-being. Altered rainfall patterns will disrupt water distribution and supply, while vegetation loss may lead to

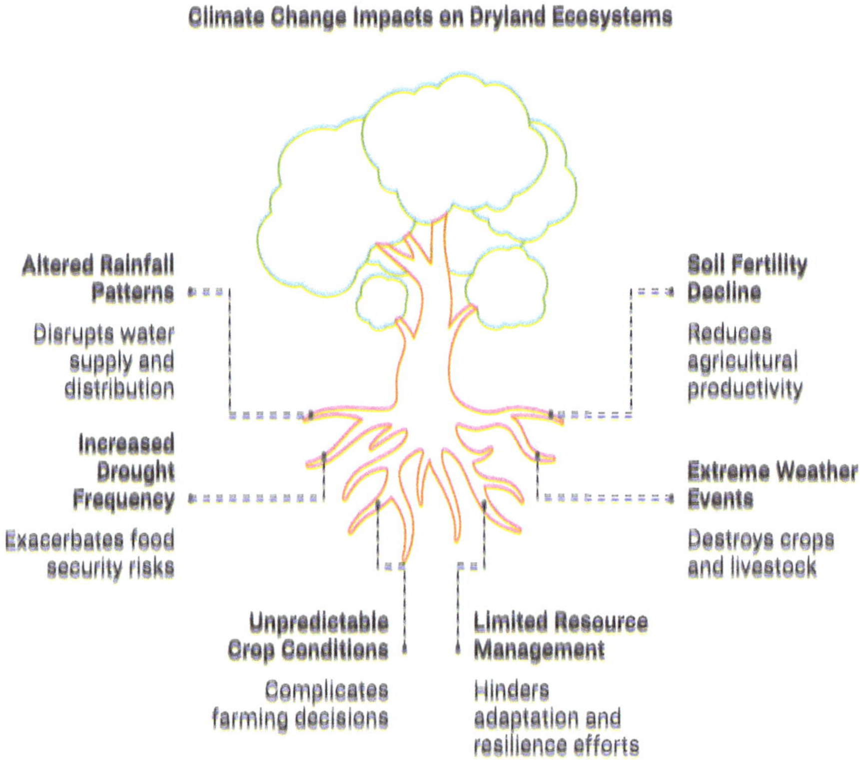

Fig. 9.4 Possible impacts of climate change on dryland ecosystems. (Ahmed et al. 2022)

soil hardening, increased water runoff, and a reduction in groundwater recharge. Climate change is expected to intensify drought frequency and severity in dry regions, exacerbating both biotic and abiotic stresses for populations and species already grappling with extreme conditions. Rather than just rising temperatures, the effects of climate change will manifest in changes to hydrological cycles, characterized by more frequent droughts and, unfortunately, an elevated risk of flooding (Fig. 9.4). The most vulnerable populations, often residing in arid regions, are expected to experience the most severe impacts of these climate-related challenges. The primary effects of climate change on dryland ecosystems include:

- Declines in agricultural productivity and crop yields, which put dryland nations' food security at risk
- Crops that are completely destroyed by extreme weather conditions, like protracted droughts and heavy floods
- A gradual, widespread decline in soil fertility brought on by erosion-induced soil carbon loss, increased organic matter breakdown due to high temperatures, decreased soil moisture content, and decreased storage capacity
- Lower animal productivity due to heat dissipation and a shortage of quality feed and fodder

- More unpredictable rainfall patterns, challenging to choose appropriate crops with different crop durations, and challenges in determining when to sow and harvest
- Changes in the threat of pests and diseases for crops and animals when temperatures rise
- Availability of less water in areas that are already water-scarce, along with heavy rainfall events that cause more water to be lost through runoff, etc.
- Agro-ecological changes and the risk posed by newly attacking plant and animal species
- Decline and transformation of important crop species' habitats, which lowers their biodiversity
- The loss of vegetative land cover and changes in rainfall patterns that result in unpredictable grassland production make farmers less secure

These impacts will further limit the livelihoods of rural communities in arid regions, exacerbating poverty, reducing opportunities for income generation, and driving higher rates of migration. As highlighted, there is an urgent need to improve the management of natural resources such as land, water, and biodiversity and strengthen people's ability to anticipate and adapt to future climate conditions.

9.4 Mitigation and Resilience to Climate Change in Dryland Ecosystem

Dryland ecosystems need innovative, cutting-edge technologies as well as ecosystem-based adaptation (EbA) strategies to reduce climate risks and foster sustainable development, thereby enhancing resilience. The following are adaptaions for mitigating and building resilience to climate change in dryland ecosystems, aimed at improving productivity and optimizing resources.

9.4.1 Precision Farming

India's severe water shortage is made worse by droughts brought on by climate change and unpredictable rainfall patterns. Water efficiency can be improved by alternative technologies like drip irrigation and precision ecosystem farming, which lower water use and increase yields. The precision ecosystem farming method makes use of technology and data to use resources in a focused manner. Initiatives like adjusting irrigation schedules using sensors and real-time data are assisting growers in making well-informed decisions and cutting down on water waste. The main obstacle to farmers using adaptation measures to lessen climate change was the inability to promptly access weather forecast data, specifically about temperature, relative humidity, and rainfall. There was uncertainty among farmers over the selection of suitable crops because weather information was unavailable. Furthermore, they are unable to obtain farm inputs on time, such as fertilizers, high-quality seeds, and timely credit availability. As temperatures rose, rainfall decreased, monsoon failures increased, and water stress conditions worsened, farmers faced shifting climatic circumstances. Reduced yields, more crop failures, dropping water levels in open and bore wells, and a rise in insect and disease outbreaks were some ways farmers saw the effects of climate change in dryland ecosystems (Ahmed et al. 2022). Nonetheless, farmers demonstrated climate change resistance by using adaptation techniques such as crop diversification, inter/mixed cropping, adjusting the date of planting, choosing drought-tolerant and alternate cropping types, and implementing soil and rainwater conservation structures. Farmers would be more resilient to climate change if the government improved access to and timely distribution of weather forecasts, essential agricultural inputs, and the crop insurance system (Dupdal et al. 2020).

9.4.2 Crop Diversification

In dryland ecosystems, traditional crops like cotton and sorghum were replaced by horticulture crops like pomegranate, ber, anola, date palm, drumstick, and citrus fruits, as well as mixed/intercropping and crop diversification techniques with short-duration crops like green gram, chickpea, groundnut, pearl millet, and cowpea in an effort to combat climate change. Additionally, in order to increase earnings, farmers used accessible bore water for protective irrigation and rainfall-saving techniques to grow alternative crops like tomato, brinjal, and chilli. Additionally, farmers produced crop varieties that were resistant to heat and drought in order to resist climate change and maintain crop harvests. Furthermore, farmers adopted rainfall and soil conservation techniques such as bunding and farm ponds to retain rainwater on-site, which reduced soil loss and boosted profile soil moisture for improved crop development. According to earlier studies by Dupdal et al. (2020), farmers who adopted better crop management techniques, such as conserving soil and rainwater, in accordance with weather-based agromet advisory services under changing climatic conditions observed higher yields and more financial benefits. The mixed/intercropping systems of sorghum + pigeon pea, groundnut + sorghum, and chickpea + sorghum, together with the prompt sowing of crops as soon as the monsoon began, were the farm management techniques used by dryland farmers.

9.4.3 Moisture Conservation

The primary goal of moisture conservation is to minimize the loss of water from soils through transpiration (water loss via plants) and evaporation (water loss directly from the soil), collectively known as evapotranspiration. Maintaining soil moisture is essential for reducing the amount of water required for crop irrigation, making it a key strategy for enhancing productivity. This is especially important in dryland regions where groundwater and/or rainfall resources for irrigation are limited or declining due to climate change or other factors. Sandy soils in drylands are particularly susceptible to moisture loss through deep percolation because of their low moisture retention capacity and high infiltration rates. The ICAR-CAZRI has investigated the potential of subsurface moisture barriers to address this issue. Materials like vermiculite, asphalt, pond sediments, and bentonite clay were tested for their ability to reduce deep percolation of soil water. Results showed that 50–70% of the total rainfall was retained in the root zone when a 5 mm thick layer of bentonite clay or pond sediments was placed at 60 cm depth. According to Ado et al. (2020), using asphalt as a subsurface barrier (2 mm thick at 60 cm depth) increased pearl millet yields by 40–50% and reduced deep percolation by one-sixth. However, these techniques have not been widely adopted in farmers' fields. In addition to moisture stress, high soil temperatures pose a significant challenge to agricultural development in arid regions. Mulches have been found effective in reducing weed growth and positively influencing the soil's hydrothermal cycles. In hot, arid regions, evaporation from the soil surface accounts for 20–43% of evapotranspiration in various crops (Olazabal et al. 2018).

Rathore et al. (2017) demonstrated that applying grass mulch at a rate of 6 t/ha significantly increased the yields of green grams, moth beans, and cluster beans. Mulching has also been shown to effectively conserve moisture in winter crops grown during the wet season. In addition to reducing evaporation losses, mulching promotes water infiltration, inhibits weed growth, reduces surface runoff, and regulates soil temperature. These ecosystem-based adaptation (EbA) strategies contribute to mitigating the effects of climate change by increasing agricultural yields and income. A combination of crop variety, rainwater harvesting, and soil conservation techniques has been shown to result in higher yields. According to Ravi Shankar et al. 2013), to sustain crop yields and improve income in dryland ecosystems, farmers should consider purchasing insurance, adjusting sowing dates, intercropping with drought-resistant varieties such as chickpeas, constructing rainwater harvesting structures, and cultivating early-maturing and drought-resistant varieties.

9.4.4 Energy Conservation

The key to addressing the climate crisis lies in transitioning to renewable energy sources. The gradual shift toward renewable energy has already contributed to reducing greenhouse gas emissions and enhancing carbon sequestration. Fossil fuel combustion for heat and electricity generation is a major source of greenhouse gases, which trap solar heat and contribute to global warming. Over 75% of global greenhouse gas emissions and nearly 90% of carbon dioxide emissions come from burning fossil fuels such as coal, oil, and gas, making them the largest contributors to climate change. To mitigate this, we must end our reliance on fossil fuels and invest in clean, affordable, sustainable, and reliable alternatives. Renewable energy sources, such as solar, wind, hydropower, waste, and geothermal energy, are abundant, naturally replenishing, and contribute little to no greenhouse gas emissions (IRENA 2019). Solar energy, in particular, is widely available and offers significant potential for reducing our carbon footprint. Hybrid systems, combining small wind turbines and solar-powered devices, could further optimize energy generation in many regions. In India, around 12–13% of the total energy comes from renewable sources like wind, solar, and biomass. However, coal remains the dominant energy source, accounting for over 60% of total energy production.

Over the past four decades, the ICAR-Central Arid Zone Research Institute (CAZRI) has developed numerous innovative solar technologies tailored for dryland ecosystems. These include agri-voltaic systems for solar farming, solar dryers, non-tracking solar cookers, animal feed solar cookers, three-in-one solar devices, solar PV pumping systems, solar PV dusters, PV winnowers and driers, solar PV sprayers, and solar distillation units. Similarly, the ICAR-Central Research Institute for Dryland Agriculture (CRIDA) has been working on developing basic renewable energy-based tools and systems aimed at supporting smallholder farmers. Additionally, biogas has been explored as a supplementary energy source in experiments involving biomass-based herbal dryers. These innovations contribute significantly to improving energy access, enhancing productivity, and fostering sustainability in dryland agriculture and ecosystems (Rathore et al. 2017).

9.4.5 Dryland Ecotourism

In dryland regions globally, key challenges such as hunger, poverty, water scarcity, and soil degradation persist. Factors like excessive cultivation, overgrazing, deforestation, inadequate irrigation, and rising temperatures contribute to the deterioration of these ecosystems. Desertification, one of the most severe consequences of land degradation in drylands, harms crops, livestock, and the millions of people who rely on these ecosystems for survival. Ecotourism in drylands presents a promising adaptation to address these challenges while supporting the achievement of sustainable development goals, such as zero hunger, zero poverty, zero land degradation, and climate change mitigation. By promoting both farm and off-farm income, ecotourism has proven to be an effective tool in improving livelihoods and conserving ecosystems. In developed countries, ecotourism has been successfully integrated into rangelands, balancing conservation efforts with the economic needs of local communities. Even in developing nations, where most farmers are smallholders with limited access to markets and state services, community-based ecotourism initiatives have achieved significant ecological benefits. By fostering sustainable tourism, these projects contribute to ecosystem preservation and provide vital economic opportunities for local populations (Mugari et al. 2020).

9.5 Adaptation to Climate Change in Dryland Ecosystem

The world's most vulnerable communities, particularly those living in dryland ecosystems, are increasingly suffering due to climate change. Ecosystem-based adaptations (EbA), which leverage biodiversity and ecosystem services, play a crucial role in helping local communities adapt to the adverse effects of climate change. EbA includes strategies such as soil restoration, climate management, hydrological regulation, and the provision of essential resources like food, construction materials, and medicinal plants. This strategy is gaining traction among governments and international funders and can be applied across various ecosystem types and scales. Climate change adaptation is typically designed to mitigate the harmful impacts of climate change at the local level, often through adjustments in crop choices, cultivation techniques, or management practices. Some adaptation strategies also aim to counterbalance output losses by capitalizing on the positive impacts of climate change, such as longer growing seasons or increased rainfall. Resilience is an important component of these adaptation efforts, though it often focuses primarily on resilience to temperature and precipitation variations. The Intergovernmental Panel on Climate Change (IPCC 2014) defines adaptation as the process of adjusting systems to reduce the impacts of climate change while maintaining or enhancing ecosystem services.

In dryland regions, adaptation is essential for populations who rely on ecosystem services for their livelihoods. However, rapid climate change is outpacing traditional adaptation strategies, making it harder for communities to adjust. Ecosystem-based adaptation (EbA) with strong

biodiversity and effective protection and management systems can mitigate climate change's adverse effects on both ecosystems and humans (Mills et al. 2020). Experts predict that by the end of the twenty-first century, the world's drylands could expand by up to 10%, with 80% of this growth occurring in developing countries. Drylands are home to people and animals that have developed unique adaptations to the challenging climate. However, climate change, manifested through altered precipitation patterns and soil degradation, is eroding these coping mechanisms. As a result, food security, access to water, and poverty will likely worsen in these regions. To combat these issues, nations need to restore dryland ecosystems, adopt sustainable land management practices, and implement significant emission reduction strategies.

Dryland ecosystems, including agricultural lands, pastures, and forests, are characterized by seasonal and geographical variability in rainfall. Pastures, which dominate in drylands, cover much of the world's dryland area. Over millennia, both microbes and human populations have evolved to survive the harsh climate conditions in these regions. For example, animals like the Bactrian camel in Central Asia and the African elephant have developed physiological and behavioral adaptations. Similarly, human populations have adapted through a variety of resource management strategies. In developing countries such as India, these coping strategies need to be enhanced to address climate change. Adaptation in drylands often involves modifying agricultural practices, restoring the ecosystem, and implementing traditional practices that enhance resilience to extreme events. These practices include biodiversity restoration, diversifying livestock herds (e.g., mixing cattle, camels, sheep, or goats), adopting water-efficient cropping techniques, planting drought-resistant crops, and supporting livestock mobility across borders and communal lands. Communities should be actively engaged in policy processes that direct dryland development, ensuring that strategies align with local needs and regional leadership systems. In drylands, mobility is an essential yet complex adaptation strategy. It relies on the careful management of water resources and periodic grazing. Livestock mobility should be recognized and supported by dryland development strategies, as it offers an important means of adapting to climate change and ensuring sustainable land management.

References

Ado AM, Savadogo P et al (2020) Farmers' perceptions and adaptation strategies to climate risks and their determinants: insights from a farming community of Aguie district in Niger. GeoJournal 85:1075–1095

Ahmed M, Hayat R, Ahmad M (2022) Impact of climate change on dryland agricultural systems: a review of current status, potentials, and further work need. Inter J Plant Prod 16:341–363. https://doi.org/10.1007/s42106-022-00197-1

Commission on Climate Change and Development (CCCD) (2008) Climate change and drylands. Stockholm, Sweden, p 1–10

Convention on Biological Diversity (CBD) (2009) Connecting biodiversity and climate change mitigation and adaptation: report of the second ad hoc technical expert group on biodiversity and climate change. Technical Series No. 41. Secretariat of the Convention on Biological Diversity (CBD). Montreal, Canada, p 126

Dupdal R, Dhakar R, Rama RCA et al (2020) Farmer's perception and economic impact assessment of agromet advisory services in rainfed regions of Karnataka and Andhra Pradesh. J Agromet 22(3):258–265

Ghosn AA (2006) Dryland biodiversity: international developments and directions. Workshop on biodiversity and the oil and gas industry in arid environments: North Africa and the Middle East. Abu Dhabi, 5 April 2006, pp 1–20

IPCC (2014) Climate change, 2014: synthesis report. In: Pachauri RK (ed) Contribution of working groups I, II and III to the fifth assessment report of the intergovernmental panel on climate change. IPCC, Geneva, p 151

IPCC (2018) Summary for policymakers. In: Masson V (ed) Global warming of 1.5°C. An IPCC special report on the impacts of global warming of 1.5°C above pre industrial levels and related global greenhouse gas emission pathways, in the context of strengthening the global response to the threat of climate change. World Meteorological Organization, Geneva, pp 1–32

IPCC (2023) Summary for policymakers. In: Climate change 2023: synthesis report. Contribution of Working Groups I, II and III to the Sixth assessment report of the intergovernmental panel on climate change [Core Writing Team, H. Lee and J. Romero (eds.)]. IPCC, Geneva, Switzerland, pp 1–34. https://doi.org/10.59327/IPCC/AR6-9789291691647.001

IRENA (2019) Climate change and renewable energy: national policies and the role of communities, cities and regions (Report to the G20 Climate Sustainability Working Group (CSWG)), International Renewable Energy Agency, Abu Dhabi

Jellason NP, Salite D, Conway JS et al (2022) A systematic review of smallholder farmers' climate change adaptation and enabling conditions for knowledge integration in sub-Saharan African (SSA) drylands. Environ Dev 43:100733. https://doi.org/10.1016/j.envdev.2022.100733

Kassas M (2002) Aridity, drought, and desertification: roles of science. In: Ryan J (ed) Desert and dryland development: challenges and potential in the new millennium. Proceedings of the sixth international conference on the development of dry lands 22–27 August 1999, Cairo, Egypt. International Center for Agricultural Research in the Dry Areas (ICARDA), Aleppo, Syria, pp 45

Kipkemoi I, Michaelides K, Rosolem R et al (2021) Climatic expression of rainfall on soil moisture dynamics in drylands. Hydrol Earth Syst Sci Discuss:1–24

Kummu M, Taka M et al (2018) Gridded global datasets for gross domestic product and human development index over 1990–2015. Sci Data 5:180004. https://doi.org/10.1038/sdata.2018.4

Mills AJ, Tan D et al (2020) Ecosystem-based adaptation to climate change: lessons learned from a pioneering project spanning Mauritania, Nepal, the Seychelles, and China. Plants, People, Planet 00:1–11. https://doi.org/10.1002/ppp3.10126

Mortimore M, Anderson S, Cotula L et al (2009) Dryland opportunities: a new paradigm for people, ecosystems and development, IUCN, Gland, Switzerland; IIED, London, UK and UNDP/DDC, Nairobi, Kenya, p 86

Mugari E, Masundire H et al (2020) Adapting to climate change in semi-arid rural areas: a case of the Limpopo basin part of Botswana. Sustainability 12

Munroe R, Roe D, Doswald N et al (2012) Review of the evidence base for ecosystem-based approaches for adaptation to climate change. Environ Evid 1(1):1–11. https://doi.org/10.1186/2047-2382-1-13

Ndheleve S, Nakin M et al (2017) Impacts of supplemental irrigation as a climate change adaptation strategy for maize production: a case of the eastern Cape Province of South Africa. Afr J Online (AJOL) 43:222–228

Ogbaga CC, Athar HUR, Amir M et al (2020) Clarity on frequently asked questions about drought measurements in plant physiology. Scientific African 8:e00405

Olazabal M, Chiabai A, Foudi S et al (2018) Emergence of new knowledge for climate change adaptation. Environ Sci Pol 83:46–53

Rao SC, Ryan J (2004) Challenges and strategies for dryland agriculture, crop science society of America, American Society of Agronomy, 677 South Segoe Road, Madison, Wisconsin 53711 USA

Rathore VS, Kumar M, Yadava ND et al (2017) Dryland agriculture and secondary salinization in canal commands of arid Rajasthan. J Soil Salinity Water Qual 9(1):30–46

Ravishankar K, Nagasree K, Maruthi S et al (2013) Farmers' perceptions and adaptation measures towards changing climate in South India and role of extension in adaptation and mitigation to Changing Climate. Extension Bulletin No. 03/2013. Central Research Institute for Dryland Agriculture, Hyderabad, p 28

Reid H, Jones XH, Porras I et al (2019) is ecosystem-based adaptation effective? Perceptions and lessons learned from 13 project sites, Iied

Safriel U, Adeel Z, Niemeijer D et al (2005) Chapter 22: dryland systems. In: Hassan R (ed) Millennium ecosystem assessment. Vol. 1. Ecosystems and human well-being: current state and trends. World Resources Institute, Washington, pp 623–662

Scarano FR (2017) Ecosystem-based adaptation to climate change: concept, scalability and a role for conservation science. Perspect Ecol Conserv 15(2):6573. https://doi.org/10.1016/j.pecon.2017.05.003

Seddon N, Reid H et al (2016) Ecosystem-based adaptation: a win-win formula for sustainability in a warming world? IIED Briefing paper, http://pubs.iied.org/17364IIED

Swiderska K, King-Okumu C et al (2018) Ecosystem-based adaptation: A handbook for EbA in mountain, dryland and coastal ecosystems. Retrieved from http://www.unep-iemp.org/file/upload/2019/03/26/1554354083.pdf

Wu Y, Wang Q, Wang H et al (2019) Shelterbelt poplar forests induced soil changes in deep soil profiles and climates contributed their inter-site variations in dryland regions, north eastern China. Front Plant Sci 10

Yan Y, Liu Q, Zhang Q et al (2019) Adaptation of dominant species to drought in the inner Mongolia grassland – species level and functional type level analysis. Front Plant Sci 10

Zeng H, Wu B, Zhang M et al (2021) Dryland ecosystem dynamic change and its drivers in Mediterranean region. Curr Opin Environ Sustain 48:59–67

Chapter 10
Soil-Based Dryland Restoration Practices in Southern Africa: A Systematic Review

Lazarus Chapungu and Kelvin Zhanda

Abstract Drylands in Southern Africa are increasingly grappling with issues like land degradation, loss of biodiversity, and reduced agricultural productivity, all worsened by climate change and human activities. Soil-based restoration practices have emerged as vital strategies to address these challenges, providing opportunities for ecosystem recovery and sustainable land management. This study offers a systematic review of soil-based restoration practices in the region. Using the PRISMA framework, we examine 41 peer-reviewed studies focusing on agroforestry, conservation agriculture, soil amendments, water harvesting, revegetation, and afforestation. The results indicate that these practices improve soil health, stabilize ecosystems, and enhance water and nutrient cycles, with agroforestry and conservation agriculture being particularly effective in reducing soil erosion and boosting fertility. Nonetheless, resource limitations, land tenure challenges, and low adoption rates impede broader implementation. This chapter underscores the importance of integrated strategies, including community-based management, policy incentives, and targeted investments in drought-resistant species and soil restoration technologies. By tackling these issues, Southern Africa can utilize soil-based restoration to promote biodiversity, support livelihoods, and strengthen resilience against the impacts of climate change.

Keywords Ecosystem-based approach · Drylands · Soil health · Soil-based restoration practices · Southern Africa

L. Chapungu
Institute for Corporate Citizenship, University of South Africa, Pretoria, South Africa

K. Zhanda (✉)
Department of Architecture and Real Estate, University of Zimbabwe, Harare, Zimbabwe

Y. D. I. Khan et al. (eds.), *Land Restoration Through Ecosystem-Based Approach*, Disaster Resilience and Green Growth, https://doi.org/10.1007/978-981-95-0389-6_10

10.1 Introduction

There are ongoing policy discussions and efforts on how to restore degraded dryland landscapes and to enhance their resilience and productivity globally. The United Nations Decade on Ecosystem Restoration (2021–2030) marks a period to identify and share best practices for "preventing, stopping, and reversing" ecosystem degradation on each continent and in all oceans (United Nations Environment Programme 2021). This goes a long way in achieving global goals by combating climate change, ending poverty, and preventing mass extinction. The United Nations Convention to Combat Desertification (UNCCD) also upholds efforts on restoring degraded land and soil, combating desertification, and attempting to attain land degradation neutrality by 2030. In addition, the Food and Agriculture Organization of the United Nations (FAO) organized a stakeholder meeting in Malawi from the 29th of July to the 2nd of August 2024 to discuss the challenges and measures for drylands in the region (FAO 2024). The main objective of the initiative was to reduce and reverse land and ecosystem degradation, desertification, and deforestation in drylands while supporting livelihoods. The program included 11 African countries, of which seven were Southern African countries.

Africa's drylands, comprising nearly a third of global drylands, span 19.6 million square kilometers (Turner et al. 2022) and support over 525 million people, representing over 40% of Africa's population (Global Centre on Adaptation 2021). Southern African drylands, covering 2.2 million square kilometers or 84% of Africa's dryland landscapes, constitute 3.6% of global drylands (FAO 2024; David et al. 2022). Countries like Zimbabwe and Zambia are dominated by semiarid and dry sub-humid zones, while Botswana and Angola have significant arid zones. South Africa and Namibia also feature extensive arid and hyper-arid areas, with Namibia having 8% in hyper-arid zones (David et al. 2022; Liu and Zhou 2021). Malawi, Lesotho, and Mozambique are nearly evenly split between dry sub-humid and semiarid zones (David et al. 2022; FAO 2024).

Climate change exacerbates challenges in these landscapes, leading to rising temperatures, precipitation variability, and associated risks such as soil moisture depletion, drought, floods, and increased water stress (David et al. 2022; Prăvălie 2016; Wu et al. 2024). Anthropogenic factors, including rapid urbanization and poor agricultural conservation practices (for example, conventional tillage, toxic chemicals, overgrazing, and monocropping), drive land degradation, biodiversity loss, and diminished ecosystem services, despite urbanization aiding poverty alleviation (Liu et al. 2019). Natural factors like veld fires and flash floods further degrade soils (David et al. 2022; Li et al. 2024). Inefficient reforestation and the collapse of indigenous conservation systems worsen the situation.

Rapid population growth in the region compounds these pressures, threatening biodiversity and ecosystem services (Turner et al. 2022; FAO 2024). Despite these challenges, Africa's drylands retain relatively intact ecosystems compared to others globally (Turner et al. 2022). Notable regional variations exist; desertification is most pronounced in Namibia (Liu and Zhou 2021) and Botswana (Omari et al. 2023; Akinyemi et al. 2021).

While the restoration of degraded drylands is a complex process (Kimmell et al. 2023), there are various ways to undertake it with a direct or indirect consideration of soil characteristics. Soil plays a dominant role in sustaining Earth's life (Fatichi et al. 2020) by linking the hydrologic and atmospheric processes to the biogeochemical and biosphere cycles. The bulk of scholarships investigating soil management effects on biological and chemical properties of dryland soils (Kimmell et al. 2023; Farrell et al. 2020; Costantini et al. 2016) build a strong basis for building illustrative models of the land degradation and restoration practices (González et al. 2023; Kimmell et al. 2023). Soil health and quality are the foundation for designing restoration strategies (Costantini et al. 2016), using soil properties to monitor and evaluate the outcomes of restoration practices on the functioning of the dryland ecosystem. In this study, soil-based restoration practices, modern or traditional, are quite critical. Soil health underpins the competence of the soil in dryland ecosystems to function in a way that promotes plant and animal health, conserves environmental quality, and nourishes biological productivity (Bünemann et al. 2018; González et al. 2023).

Dryland soil-based restoration requires the quantitative evaluation of biological, physical, and chemical soil health parameters (Baer et al. 2012). Many historical dryland restoration practices are known to enhance soil organic carbon (SOC) and soil biodiversity (Laban et al. 2018; Martin 2017), generally focusing on maintaining ground cover, management of soil nutrients and water, and methods for erosion control. SOC and soil biodiversity are significant for dryland ecosystem function. SOC levels in dryland soils are usually low and probably at an exorbitant state (Kimmell et al. 2023) for land restoration processes. However, Bünemann et al. (2018) and Arije et al. (2024) demonstrate a huge SOC sequestration potential in arid and semiarid drylands with ecosystem restoration. With soils in global water-starved areas contributing an estimated one-third to global stocks of SOC and soil biodiversity (Baer et al. 2012), it holds high potential for SOC recovery (Ghimire et al. 2022). Therefore, SOC and soil biodiversity are the best ways for fragile ecosystems to recover productivity through improved nutrient and water use efficiency. Notwithstanding this importance, practices to improve soil health and functions are often disregarded in public land policy programs (Costantini et al. 2016). Dryland socio-ecological studies in Africa show that presently, dryland management is one of the most pressing issues that needs to be addressed to restore and protect the ecosystems in the semiarid drylands of all regions of Africa (Li et al. 2024).

Sustainable revegetation practices with native agroforestry, pastoralism, and conservation agriculture have been implemented widely in the drylands. Yet, these practices are ineffective (Kramer and Sheil 2024), and their adoption is low in the global low-income drylands, wherein the human population exceeds the capacity of drylands' water security and agricultural productivity. Yang et al. (2021) argue that despite the success of agroforestry practices in promoting plant establishment, they may not improve soil health and quality conditions without other alternatives, such as active soil-based restoration methods. While scientific evidence shows various dynamics of passive restoration's reactivation of soil biogeochemical cycles (Costantini et al. 2016), active restoration is critical for enhancing dryland soil parameters. Active soil-based dryland restoration is a viable alternative (Kimmell et al. 2023; Farrell et al. 2020), meaning systematic labor-intensive soil restoration methods aimed specifically at improving soil health, functions, and structure without the direct involvement of plants. This includes water retention agents, erosion control structures, and the use of microbial inoculations (González et al. 2023). Most importantly, active restoration practices are essential to realize positive soil-based outcomes, especially in agricultural drylands, whereas passive restoration practices result in adverse soil restoration outcomes (Miguel et al., 2020; Laban et al. 2018). However, collectively, active and passive soil-based restoration practices deal with several abiotic (low moisture and nutrient availability, as well as soil erodibility) and biotic (diminished beneficial soil microbial communities) obstacles to dryland ecosystems and biodiversity recovery, while vegetation-based practices alone only address plant propagation.

This chapter explores the dynamics of soil-based restoration practices in the drylands of Southern Africa. The key research questions include (i) what are the prevalent soil-based restoration methods, (ii) what are the factors that contribute to the success or failure of soil-based restoration practices, and (iii) what are the sustainable strategies for maintaining the health of dryland soil ecosystems. The chapter provides valuable insights for policymakers, researchers, and practitioners, aiding in the formulation of a comprehensive restoration policy for Southern Africa and other global drylands. The findings emphasize the potential for enhanced biodiversity, improved soil health, better ecosystem services, and increased collaboration among multiple stakeholders.

10.2 Methodology

This study is based on a methodology that uses the Preferred Reporting Items for Systematic Literature Review and Meta-Analysis (PRISMA) as its framework. This approach promotes transparency and reduces reviewer bias by outlining a detailed methodology that guides the literature review process. Additionally, it allows for clear reporting of the review's objectives, procedures, and conclusions. However, this methodology does have limitations, particularly concerning the precision of systematic search algorithms (Hiebl 2023).

10.2.1 Inclusion/Exclusion Criteria

The eligibility criteria used in this study were premised on the study context, thematic focus, publication type, and accessibility of the selected articles. Table 10.1 shows further details regarding the eligibility criteria applied in this study.

Table 10.1 Eligibility criteria adopted for document screening

Category	Included	Excluded	Rationale for inclusion/ exclusion
Context of interest	Articles focused on the effectiveness, challenges, and socio-environmental impacts associated with soil-based restoration practices in the drylands of Southern Africa	Articles focused on issues other than soil-based restoration practices	Articles focusing on issues other than soil-based restoration practices did not meet the criteria to provide answers to the study's research questions
Type of study	Peer-reviewed journal articles, systematic reviews, and meta-analyses, with a scientific or evidence-based approach	Commentaries, editorials, opinion pieces, and non-peer-reviewed publications	Peer-reviewed research and literature review articles have scientific rigor with trusted and expert-checked results and conclusions
Data quality	Studies with robust methodologies, clear data sources, and transparent reporting	Studies with unclear methodologies, poor data transparency, or lacking scientific rigor	Studies with unclear methodologies and poor data transparency may misrepresent facts, and wrong conclusions may be drawn from the study
Language	Publications available in English	Publications available in a non-English language	The authors are not familiar with other languages
Accessibility	Open access	Non-open access	Open access guarantees access to articles for analysis
Research methodologies	Empirical studies, modeling studies, case studies, and qualitative or quantitative analyses	Studies with unclear methodologies, poor data transparency, or lacking scientific rigor	Articles with clear methodologies provide reliable results and conclusions

10.2.2 Database Selection, Document Identification, and Screening

The identification of documents was carried out through a systematic process for selecting databases following the approach adopted by Chapungu et al. (2024) (Fig. 10.1). The criteria for choosing these databases included both the quality and quantity of available records, as well as their historical use in previous systematic literature reviews. Based on these factors, Scopus was identified as the top database, providing a strong collection of peer-reviewed publications focused on climate-induced displacement. Importantly, Scopus included over 23,774 documents on this subject. Research conducted by Homrich et al. (2018) also supports the idea that

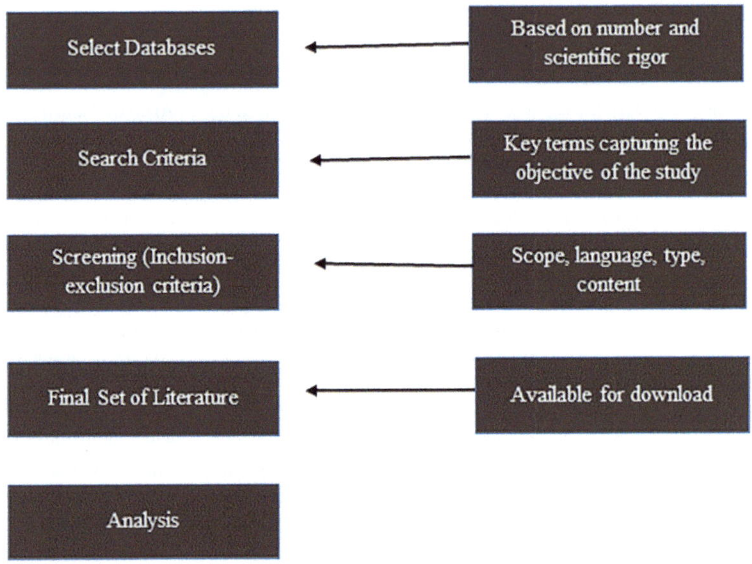

Fig. 10.1 Database selection and article screening process. (Source: Adapted from Chapungu et al. 2024)

Scopus is one of the leading databases for high-quality, peer-reviewed articles. For this study, Scopus was considered both adequate and suitable, as several earlier studies have utilized it to conduct thorough and reliable systematic literature analyses (Yang et al. 2017; Figueroa-Domecq et al. 2015).

The search terms in Scopus's Advanced Search portal were developed using the block-building method, which involved breaking down the topic into equally important thematic blocks centered around key terms. This method produced a document-term matrix that effectively identified alternative search terms for each main block term. Once the key terms and their alternatives were finalized in the matrix, they were input into the Scopus Advanced Search portal using the appropriate Boolean operators and word structures. The search terms were applied to document titles, abstracts, and keywords. The primary search string in Scopus was as follows:

> TITLE-ABS-KEY ("dryland restoration" OR "soil-based restoration practices" OR management OR regeneration OR rehabilitation) AND ("soil fertility" OR "soil moisture conservation" OR agroforestry OR "conservation agriculture" OR "soil amendments")

To refine the results and ensure that only relevant scientific articles are selected, Scopus-based filters were used according to predefined inclusion and exclusion criteria. These filters encompassed parameters such as subject area, publication year, keywords, document type, language, and open access status. The detailed search string, presented in Box 10.1, was subsequently utilized to extract relevant articles for abstract screening.

Box 10.1: The Search String Used in Scopus Database

TITLE-ABS-KEY (("dryland restoration" OR "soil-based restoration practices" OR management OR regeneration OR rehabilitation) AND ("soil fertility" OR "soil moisture conservation" OR agroforestry OR "conservation agriculture" OR "soil amendments")) AND (LIMIT-TO (SUBJAREA, "AGRI") OR LIMIT-TO (SUBJAREA, "ENVI") OR LIMIT-TO (SUBJAREA, "EART") OR LIMIT-TO (SUBJAREA, "SOCI")) AND (LIMIT-TO (DOCTYPE, "ar")) AND (LIMIT-TO (LANGUAGE, "English")) AND (LIMIT-TO (AFFILCOUNTRY, "Zimbabwe") OR LIMIT-TO (AFFILCOUNTRY, "South Africa") OR LIMIT-TO (AFFILCOUNTRY, "Malawi") OR LIMIT-TO (AFFILCOUNTRY, "Zambia") OR LIMIT-TO (AFFILCOUNTRY, "Madagascar") OR LIMIT-TO (AFFILCOUNTRY, "Mozambique") OR LIMIT-TO (AFFILCOUNTRY, "Namibia") OR LIMIT-TO (AFFILCOUNTRY, "Swaziland") OR LIMIT-TO (AFFILCOUNTRY, "Lesotho") OR LIMIT-TO (AFFILCOUNTRY, "Botswana")) =703.

The scientific articles found using the search string in Box 10.1 were exported as a CSV file and then loaded into Excel for a more in-depth analysis based on their abstracts. A typical double-screening procedure was used, with both co-authors independently evaluating the abstracts. This strategy enabled an objective selection of the most appropriate documents. Cohen's Kappa statistic was used to test the process's dependability. Cohen's Kappa is a statistical measure that assesses the degree of agreement between two raters when categorizing items into various groups (Chicco et al. 2021). Accounting for the potential of chance agreement makes it a more reliable measure than simply calculating percentage agreement. Cohen's Kappa is calculated using the following formula:

$$K = \frac{Po ? Pe}{1 ? Pe}$$

where

K is Cohen's Kappa.

Po is the observed agreement between raters (the percentage of times both raters agree).

Pe is the expected agreement by chance alone (calculated based on the marginal probabilities of each rater).

The computed coefficient of agreement was 0.96, which indicates an almost perfect level of agreement. This high Kappa value highlighted the reliability of the screening process, enabling the subsequent analysis of the eligible documents to move forward with strong confidence in the accuracy of the selection. Articles identified as eligible through the abstract double-screening method were then subjected to full-text screening, focusing on quality assessment and contextual relevance.

After screening the full text, the documents that met the search criteria were analyzed to extract both qualitative and quantitative data that explain soil-based restoration practices in Africa. The analysis adhered to the structured six-step thematic analysis process outlined by Naeem et al. (2022), which includes (1). reading the manuscripts and selecting quotations, (2). annotating, (3). conceptualizing the data, (4). segmenting the data, (5). analyzing the segments, and (6). writing. Initially, the 47 screened documents were reviewed to gain preliminary insights and identify recurring themes related to soil-based restoration practices in dryland regions of Africa, resulting in a summary of the document content. During the annotation phase, relevant words, phrases, and sections were labeled with codes to highlight trends, key opinions, processes, policy developments, notable events, and other important factors for organizing the data.

Following the annotation, the data were organized by aligning with key themes identified throughout the documents. The codes generated during this process were categorized into main groups and subgroups, while any insignificant codes were discarded. During the data segmentation phase, these categories were arranged in a spreadsheet, where the relationships among them were labeled and described. The segmented data were then analyzed to pinpoint key issues for the final write-up. Insights gained from reading the documents were utilized to outline the soil-based restoration practices in the dryland regions of Southern Africa. Various metrics were compiled and analyzed using comparative methods, including the frequency calculation of specific variables and the grouping of articles based on their research designs and methodologies. This thorough process contributed to the results section, offering a detailed perspective on the African dryland landscape.

10.3 Results and Discussions

From the systematic literature review process, a total of 23,774 were initially identified using a predefined search criterion. The articles were initially screened through Scopus-based filters to remain with articles that directly address the proposed research questions. A total of 23,071 were screened out, and 703 documents remained. The 703 articles were subjected to full abstract screening, and consequently, 578 were deemed ineligible for full-text analysis. Consequently, 125 articles were sought for full-text retrieval and analysis. However, 6 articles could not be retrieved. This resulted in 119 articles being subjected to full-text analysis. Due to reasons stated in Fig. 10.2, including wrong settings, methodological problems, language, and others, 78 documents were deemed ineligible for inclusion in the study. This resulted in 41 articles being eligible for inclusion in the study. Figure 10.2 presents a flowchart illustrating the article screening process.

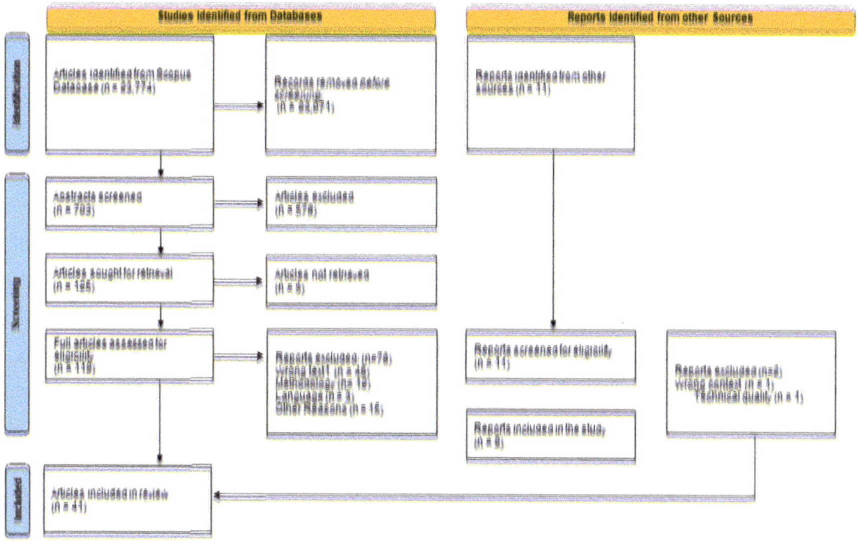

Fig. 10.2 PRISMA flow diagram showing the screening process of documents to come up with studies for inclusion. (Adapted from Page et al. 2021)

10.3.1 Soil-based Restoration Practices

The systematic review of literature highlighted a variety of soil-based restoration practices being used in the dryland areas of Southern Africa. These practices tackle significant issues related to land degradation and are frequently customized to fit the specific ecological and socioeconomic conditions of the region. Important strategies include agroforestry systems, conservation agriculture, soil amendments, erosion control measures, water harvesting techniques, and efforts for revegetation. Table 10.2 shows some of the soil-based practices, their description, the countries where they have been practiced, and some source documents where the case studies were presented.

As shown in Table 10.2, there are five soil-based restoration practices that are prevalent in Southern Africa, practiced mainly in Zimbabwe, Zambia, Botswana, South Africa, and Namibia. These practices are discussed in the next sections, focusing on their advantages, factors, effectiveness, and various other dynamics.

10.3.1.1 Agroforestry Systems

Agroforestry systems play a crucial role in restoring soils in Southern African drylands, where agricultural productivity and ecosystem health have been severely affected by degraded soils and unpredictable climate conditions. Castle et al. (2022) highlight that agroforestry, which involves integrating trees with crops or livestock, provides numerous ecological and socioeconomic advantages while tackling the

Table 10.2 Soil-based practices in Southern Africa and their descriptions

Practice	Description	Countries practised	Sources
Agroforestry system	Integrating trees with crops or livestock to improve soil fertility, prevent erosion, and enhance productivity	Zambia, Malawi, Zimbabwe	Makumba et al. (2007), Ajayi et al. (2011), Nyoka et al. (2015a, b)
Conservation agriculture	This includes practices such as minimum tillage, crop rotation, and mulching to improve soil structure, reduce erosion, and enhance water retention	South Africa, Zimbabwe	Mupangwa et al. (2013), Garrity et al. (2010)
Erosion control measures	Techniques such as stone bunds, terraces, and check dams to reduce soil loss and stabilize degraded areas	Namibia, Botswana, Zambia	Nyoka et al. (2015a, b), Garrity et al. (2010)
Water harvesting	Practices like zai pits and contour bunding to capture and retain water, improving soil moisture and crop yields in arid regions	Zimbabwe, Namibia	Mupangwa et al. (2013), Makumba et al. (2007)
Revegetation and forestry	Planting native vegetation and trees to restore soil stability, increase organic carbon levels, and enhance biodiversity	Botswana, South Africa, Zimbabwe	Ajayi et al. (2011), Garrity et al. (2010)

Source: Authors' compilation from literature

root causes of land degradation. A major benefit of agroforestry systems is their capacity to improve soil fertility and increase organic matter. For instance, *Faidherbia albida*, a nitrogen-fixing tree often utilized in agroforestry, significantly enhances soil nutrients. Various studies indicate that *F. albida* boosts nitrogen levels in the soil, benefiting both the trees and nearby crops, which results in greater yields without the need for chemical fertilizers (Makumba et al. 2007). Additionally, the natural shedding of leaves during the growing season creates a mulch layer that helps retain moisture and reduces soil erosion.

Agroforestry systems are essential for combating soil erosion, a significant issue in the semiarid and arid areas of Southern Africa. The root systems of trees help to stabilize the soil, making it less vulnerable to both wind and water erosion. For example, research conducted in Zambia shows that agroforestry practices, like planting contour tree rows, have led to a decrease in topsoil loss and an increase in water infiltration, particularly in areas with steep slopes (Nyoka et al. 2015a, b). These methods not only protect the soil but also foster microclimatic conditions that are beneficial for crop growth. Additionally, water use efficiency, which is vital for dryland farming, is improved through agroforestry. The tree canopy minimizes surface evaporation while facilitating better rainwater absorption into the soil. In Zimbabwe, farmers who engage in agroforestry with native tree species have reported enhanced drought resilience in their crops, primarily due to improved soil moisture and lower rates of evapotranspiration (Mupangwa et al. 2013). This highlights the significance of agroforestry as a strategy for adapting to climate challenges in water-scarce areas.

In addition to its biophysical benefits, agroforestry offers economic and social advantages, making it a sustainable choice for smallholder farmers. By integrating tree crops with food crops, farmers can diversify their income sources, which helps reduce their vulnerability to market fluctuations or climate-related shocks. For instance, incorporating fruit-bearing trees like mangoes and oranges into agroforestry systems not only enhances food security but also generates new income opportunities through the sale of excess produce (Ajayi et al. 2011). Furthermore, the firewood and timber sourced from agroforestry systems help ease the pressure on natural forests, contributing to broader landscape restoration efforts. However, despite its proven benefits, the adoption of agroforestry systems encounters several challenges. Socioeconomic factors, such as the lengthy period before trees yield produce, can discourage some farmers from embracing these systems. Additionally, limited access to seedlings and technical expertise further hampers implementation, especially in rural and resource-limited areas (Garrity et al. 2010). Institutional support, including policy incentives and training programs, has been recognized as a vital element in addressing these obstacles.

Agroforestry systems present a significant opportunity for restoring soils in Southern African drylands, providing a comprehensive approach to combat land degradation, water shortages, and economic instability. The existing literature robustly advocates for its implementation, highlighting its ability to boost soil health, reduce erosion, and enhance community livelihoods. Nevertheless, expanding agroforestry practices necessitates tackling the socioeconomic and institutional barriers that hinder broader adoption. By incorporating local knowledge and offering focused policy support, Southern Africa can better harness the potential of agroforestry to revitalize the drylands of Southern Africa.

10.3.1.2 Conservation Agriculture

Conservation agriculture (CA) is widely acknowledged in the literature as an effective method for restoring soil health in the drylands of Southern Africa. It is characterized by key practices such as minimum soil disturbance (no-till farming), maintaining permanent soil cover (through mulching), and implementing crop rotation. These principles provide a sustainable way to enhance soil health, reduce erosion, and increase resilience against climate challenges. A significant focus in the literature is on how conservation agriculture improves soil structure and organic matter. By avoiding conventional tillage, CA reduces soil disruption, which allows natural processes to maintain a stable soil structure. This leads to better water infiltration and retention, making soils more resilient to drought conditions. Mupangwa et al. (2013) found that CA systems in semiarid areas of Zimbabwe notably improve soil moisture levels, resulting in higher crop yields during droughts. Furthermore, using organic residues for mulching helps in decreasing water evaporation and aids in building organic carbon in the soil, which enhances soil fertility over time (Twomlow et al. 2008).

Erosion control is a key advantage noted in the literature. Using crop residues as ground cover and minimizing soil disturbance helps prevent topsoil loss due to wind or water. Research from South Africa shows that farmers who implement conservation agriculture (CA) practices see lower soil erosion rates, especially in areas with heavy seasonal rainfall (Grabowski et al. 2018). These results highlight CA's potential to address a major issue in dryland farming—the erosion of fertile topsoil. The literature also emphasizes the role of crop rotation in improving soil fertility and disrupting pest and disease cycles. For instance, rotating crops with legumes enhances nitrogen levels in the soil, which decreases reliance on synthetic fertilizers (Giller et al. 2009). In Zambia, CA systems that include legume rotations have demonstrated significant improvements in soil nutrient levels and crop yields, providing a dual advantage for smallholder farmers aiming to rehabilitate degraded soils while ensuring food security (Ngoma et al. 2016).

In addition to its biophysical advantages, conservation agriculture (CA) plays a significant role in enhancing the socioeconomic resilience of farming communities. It lessens the labor requirements, particularly for plowing, which is often a labor-intensive task in traditional farming systems. This is particularly beneficial for resource-limited farmers and households led by women, who frequently encounter labor shortages. Moreover, the economic feasibility of CA is evident through reduced input costs, stemming from a decreased dependence on fertilizers and the utilization of crop residues for mulching (Mazvimavi and Twomlow 2009). However, despite these benefits, the literature points out several obstacles to the broader adoption of conservation agriculture. Socioeconomic factors, such as limited access to suitable tools and knowledge, impede implementation, particularly among smallholder farmers. For instance, in Malawi, adoption rates remain low due to the high upfront costs associated with specialized equipment like no-till planters, along with a lack of financial incentives for farmers (Grabowski and Kerr 2014). Additionally, there is a persistent resistance to moving away from traditional farming practices in many communities, highlighting the necessity for focused education and awareness initiatives. Another issue is the competition for crop residues, which are vital for mulching in CA systems but are often repurposed as livestock feed. This competing demand for residues creates trade-offs that make it challenging to consistently apply CA principles (Ngwira et al. 2012). Overcoming these challenges will require policy support, including subsidies for CA equipment, training programs for farmers, and incentives for sustainable management of residues.

10.3.1.3 Soil Amendments

Soil amendments are increasingly acknowledged in the literature as essential tools for restoring degraded land in the drylands of Southern Africa. These amendments, which encompass biochar, compost, and manure, are used to revitalize depleted soils and boost fertility, and improve both the physical and biological characteristics of soils in arid and semiarid areas. Among the most extensively researched soil amendments in this region is biochar, a carbon-rich substance produced through the pyrolysis of organic biomass. Its capacity to retain nutrients and moisture makes biochar especially advantageous in dryland environments, where soils frequently lack nutrients and are susceptible to water loss. Studies conducted in Zambia and Malawi demonstrate that incorporating biochar into sandy soils significantly enhances soil organic carbon (SOC) levels, cation exchange capacity, and water retention (Cornelissen et al. 2013). Additionally, the long-term stability of biochar in the soil helps maintain fertility over time, decreasing the necessity for frequent applications compared to other organic materials.

Compost and manure are widely utilized as organic amendments to improve soil health. Compost adds organic matter to the soil, which boosts microbial activity and nutrient cycling. Research conducted in Zimbabwe shows that using compost can enhance crop yields and soil fertility in degraded areas, with 15–145% (Kugedera et al. 2023) significant productivity gains observed in maize and sorghum (Mupangwa et al. 2008). Likewise, manure serves as a cost-effective resource for smallholder farmers, supplying vital nutrients like nitrogen and phosphorus while also enhancing soil structure and its ability to retain water (Chivenge et al. 2011). In Botswana, combining livestock farming with manure application has effectively revitalized overgrazed and nutrient-poor rangelands. Mineral fertilizers are frequently applied alongside organic amendments to tackle serious nutrient deficiencies, particularly nitrogen and phosphorus. In Malawi, integrated soil fertility management (ISFM) practices that combine organic inputs such as compost with small amounts of mineral fertilizers have been found to improve nutrient availability and increase crop yields (Vanlauwe et al. 2010). This method is especially advantageous for farmers with limited resources, as it lessens reliance on expensive synthetic fertilizers.

The use of soil amendments not only helps to correct nutrient deficiencies but also plays a crucial role in preventing erosion and desertification. Organic amendments like compost and biochar enhance soil structure, making it more resistant to both wind and water erosion. Research conducted in South Africa shows that applying manure and compost to degraded soils can significantly lower erosion rates, especially in regions that experience heavy rainfall (Lal 2015). Improved soil stability also encourages the growth of vegetation, which further reduces erosion and helps restore ecosystem functions. However, despite these advantages, the use of soil amendments in Southern Africa faces several obstacles. A significant challenge is the limited availability and high cost of quality amendments, particularly for smallholder farmers. For instance, producing and transporting biochar requires initial investments that may be out of reach for resource-limited households (Cornelissen et al. 2013). Additionally, while manure is often easily accessible in rural areas, its quality and nutrient content can vary widely, which can diminish its effectiveness (Chivenge et al. 2011).

Cultural practices and gaps in knowledge are significant barriers to the widespread use of soil amendments. Many farmers are unaware of the long-term advantages of organic amendments and tend to stick with traditional farming methods that can worsen soil degradation. Initiatives that focus on educating farmers and establishing demonstration plots have been recognized as essential strategies to tackle this challenge (Vanlauwe et al. 2010). In summary, the literature highlights the promise of soil amendments as a viable restoration approach for degraded soils in Southern African drylands. Amendments such as biochar, compost, and manure not only enhance soil fertility but also improve water retention and reduce erosion, providing both ecological and economic advantages. However, overcoming obstacles like resource availability, costs, and the spread of knowledge is vital for increasing the adoption and effectiveness of these practices. With the right policy support and community involvement, soil amendments could significantly contribute to the restoration of Southern Africa's drylands and promote sustainable agriculture.

10.3.1.4 Water Harvesting

Water harvesting is an essential soil-based restoration method that has gained significant attention in discussions about addressing water scarcity and enhancing soil productivity in the drylands of Southern Africa. It involves the collection, storage, and efficient use of surface runoff or rainfall, tackling the intertwined issues of limited water supply and land degradation, which makes it especially relevant for the region's arid and semiarid environments. One of the most well-known methods in Southern Africa is the use of zai pits, which are small planting basins designed to capture and hold rainfall. These structures improve water infiltration and decrease runoff, creating better conditions for crop growth even in degraded soils. Research conducted in Zimbabwe has demonstrated the effectiveness of zai pits in restoring unproductive lands, with findings indicating notable increases in maize and sorghum yields when this method is applied (Motsi et al. 2004). Zai pits are particularly beneficial in regions with unpredictable rainfall, as they optimize the use of limited water resources while also enhancing soil structure through the addition of organic matter.

Contour bunding and terracing are widely recognized water harvesting techniques utilized throughout the region. Contour bunds involve creating ridges that follow the natural contours of slopes, which helps to minimize soil erosion and capture water, allowing it to seep into the ground. In Botswana and Namibia, contour bunding has been used to address soil erosion and improve moisture retention in grazing areas, resulting in better vegetation cover (Banda et al. 2016). Likewise, terracing, although it requires significant labor, has been effective in areas with steep slopes by preventing water runoff and promoting groundwater recharge. Another method highlighted in the literature is the building of small check dams and sand dams in riverbeds to collect and store seasonal water flows. These structures help to slow down the movement of water, allowing sediments and nutrients to settle while forming small reservoirs that can be accessed during dry spells. Research from semiarid regions of South Africa shows that check dams can greatly enhance soil moisture and support the growth of vegetation downstream (Moyo et al. 2017). Sand dams, in particular, are valued for their capacity to minimize evaporation by storing water beneath the sand surface, thus providing a dependable water source for both agricultural and domestic needs in arid areas.

The integration of in-field rainwater harvesting (IRWH) systems is becoming increasingly popular in Southern Africa. IRWH involves adjusting traditional tillage practices to focus on capturing and holding rainwater in cultivated areas. Research from South Africa indicates that IRWH systems can boost crop yields by as much as 50%, especially for crops that are sensitive to water shortages, like maize and wheat (Botha et al. 2007). This approach is cost-effective and can be carried out using materials that are readily available, making it a viable option for smallholder farmers. However, despite their benefits, the uptake of water harvesting techniques encounters several obstacles. A major challenge is the labor-intensive nature of methods such as zai pits and terracing, which can hinder widespread adoption, particularly in communities with limited labor availability (Motsi et al. 2004). Furthermore, building check dams and sand dams requires technical skills and initial financial investment, which may not be easily accessible to rural farmers (Moyo et al. 2017).

10.3.1.5 Revegetation and Afforestation

Revegetation and afforestation are crucial strategies for restoring soil health and combating land degradation and desertification in the drylands of Southern Africa. These methods involve planting native grasses, shrubs, and trees to stabilize the soil, boost biodiversity, and enhance ecosystem services. Revegetation aims to reestablish native plants, such as grasses like Eragrostis tef and shrubs like Salvadora persica, to fight soil erosion and rehabilitate degraded rangelands, particularly in Namibia and Botswana. In contrast, afforestation focuses on introducing drought-resistant tree species such as Acacia senegal and Faidherbia albida, which help fix nitrogen, improve soil organic matter, and reduce runoff. These practices also contribute to better microclimates, increased soil moisture, and greater biodiversity, which in turn benefit local agricultural systems. However, there are challenges to widespread implementation, including high seedling mortality rates due to drought, limited access to water, land tenure complications, and grazing pressures. To address these issues, strategies such as using drought-resistant species, promoting community-based management, and integrating restoration efforts with income-generating activities like agroforestry and beekeeping are essential. Additionally, policy support for planting materials and technical assistance is vital for scaling these initiatives. In conclusion, revegetation and afforestation serve as effective methods for restoring degraded landscapes in Southern Africa, providing solutions to issues like soil erosion, biodiversity decline, and reduced agricultural productivity. By tackling existing barriers and ensuring sustainable management practices, we can enhance their effectiveness in fighting desertification and improving ecosystem resilience.

10.3.2 Policies and Programs Guiding Soil-Based Restoration in Southern Africa

Soil-based restoration in Southern Africa is shaped by various policies and programs at national, regional, and international levels, all aimed at addressing land degradation and desertification while boosting biodiversity and ecosystem services. Table 10.3 presents some of the policies and programs in Southern Africa.

The African Union's Agenda 2063 and initiatives like the Great Green Wall offer a continental framework for sustainable land management, which has inspired regional projects in Namibia and Botswana. Additionally, the Southern African Development Community (SADC) protocols encourage collaboration on afforestation, agroforestry, and sustainable rangeland management as part of wider climate change mitigation efforts. National Action Programs under the UN Convention to Combat Desertification (UNCCD) serve as the foundation for country-level initiatives, incorporating conservation agriculture, afforestation, and community-led land management. Countries like Zimbabwe and Botswana focus on restoring degraded landscapes through methods such as planting native vegetation and managing communal grazing areas. Conservation agriculture initiatives, such as Zambia's Conservation Agriculture Scaling Up (CASU) project, advocate for minimum tillage, crop rotation, and cover cropping to enhance soil fertility and minimize erosion.

Community-Based Natural Resource Management (CBNRM) programs in Namibia, Botswana, and Zimbabwe enable local communities to take charge of managing and restoring degraded lands by blending traditional knowledge with contemporary methods. Initiatives like public-private partnerships and corporate social responsibility efforts, including afforestation projects by companies in South Africa, offer additional resources and expertise to bolster these initiatives. Policies that tackle the water-energy-food nexus, such as South Africa's National Water Act, further weave soil restoration into a wider framework of environmental management. However, challenges remain, including insufficient funding, weak enforcement, and conflicts over land tenure. To address these issues, it is essential to enhance regional coordination, promote inclusive community participation, and invest in research focused on drought-resistant species and restoration technologies. With focused support and adaptive management, these policies and programs can play a vital role in achieving sustainable land restoration and resilience throughout Southern Africa.

10.3.3 Practices, Limitations, and Solutions

The need to understand the limitations and solutions for specific soil-based restoration practices in Southern Africa cannot be overemphasized. Table 10.4 provides a comprehensive overview of various activities aimed at restoring dryland ecosystems in the region. It identifies specific restoration practices such as agroforestry, soil

Table 10.3 Policies and programs guiding soil-based restoration in Southern African drylands

Policy/program	Description	Examples	Source document
Southern African Development Community (SADC) Regional Action Programme (RAP)	A regional framework under the UNCCD addressing desertification and promoting sustainable land management	Support for the Great Green Wall Initiative in Southern Africa	SADC RAP for Combating Desertification (UNCCD)
South Africa: LandCare Program	Government initiative promoting soil conservation, sustainable farming, and land rehabilitation	Subsidized restoration practices like conservation agriculture in rural South Africa	South Africa LandCare Program Official Reports
Namibia: National Drought Policy and Strategy	Policy addressing the impacts of recurring droughts through water harvesting and soil restoration	Implementation of water harvesting techniques in arid regions of Namibia	Namibia National Drought Policy Documents
Zimbabwe: Conservation Agriculture Policy	Encourages sustainable farming practices such as minimum tillage, crop rotation, and mulching to combat soil erosion	Promoting conservation agriculture among smallholder farmers in Zimbabwe	Zimbabwe Conservation Agriculture Policy Framework
Comprehensive Africa Agriculture Development Programme (CAADP)	A framework for improving agricultural productivity and soil health across Africa, supporting policy and funding mobilization	Investments in conservation agriculture projects in Zambia and Malawi	African Union CAADP Documents
Namibia's Community-Based Natural Resource Management (CBNRM) Program	Empowers local communities to manage and restore degraded lands, integrating traditional and modern restoration method	Community-driven revegetation and soil management efforts in Namibia	Namibia CBNRM Program Reports
South Africa: Working for Water and Working for Land Programs	Public works programs create jobs while implementing revegetation and soil erosion control projects	Revegetation and soil erosion control projects implemented in South Africa	South Africa Government Working for Water and Land Program Reports
African Union's Agenda 2063	A pan-African policy framework for sustainable land management and combating desertification	National commitments to sustainable land management in Southern Africa	African Union Agenda 2063 Policy Framework
UNCCD Country Action Plans	Country-specific action plans under the UNCCD framework targeting soil restoration and combating desertification	Zimbabwe and Botswana are developing soil restoration action plans	UNCCD Country Action Plan Reports

mulching, terracing, and water harvesting, along with the countries where these practices are implemented, including South Africa, Zambia, Zimbabwe, and Malawi, among others. Additionally, the table highlights the key limitations hindering the expansion of these practices, such as limited access to resources, lack of technical expertise, and inadequate policy support. To address these hurdles, Table 10.4 suggests practical solutions like establishing community nurseries, promoting rotational grazing, offering subsidies, and improving infrastructure. This resource serves as a valuable tool for policymakers, practitioners, and researchers seeking to enhance dryland restoration efforts in Southern Africa.

10.4 Discussion

This study emphasizes the crucial role of soil-based restoration practices in tackling issues like land degradation, biodiversity loss, and decreasing agricultural productivity in the drylands of Southern Africa. Practices such as agroforestry, conservation agriculture, soil amendments, water harvesting, revegetation, and afforestation are essential for improving soil health, restoring ecosystem services, and enhancing resilience to climate change. However, their effectiveness is affected by various ecological, social, and economic factors. Agroforestry systems have shown considerable promise in boosting soil fertility, minimizing erosion, and increasing water use efficiency in the area. The incorporation of nitrogen-fixing trees, like Faidherbia albida, enhances soil organic matter and nutrient availability, resulting in improved crop yields (Makumba et al. 2007; Nyoka et al. 2015a, b). Furthermore, agroforestry improves microclimatic conditions, which helps reduce evapotranspiration and enhances soil moisture retention (Mupangwa et al. 2013). Despite these advantages, the adoption of these practices is still low due to the lengthy time required to see economic returns, a lack of technical knowledge, and limited access to quality seedlings (Ajayi et al. 2011). These challenges underscore the necessity for institutional support, including policy incentives and training programs for farmers.

Conservation agriculture (CA) practices, including minimum tillage, crop rotation, and mulching, are effective in improving soil structure, reducing erosion, and enhancing water retention. For instance, research conducted in Zimbabwe and South Africa shows that CA practices significantly boost soil moisture levels and decrease erosion in semiarid regions (Mupangwa et al. 2013; Grabowski et al. 2018). Additionally, crop rotation, particularly with legumes, promotes nitrogen fixation, which lessens the reliance on synthetic fertilizers (Giller et al. 2009). Nonetheless, CA encounters challenges such as competition for crop residues, which are frequently used as livestock feed, and resistance to adopting new farming methods (Ngwira et al. 2012). Tackling these challenges necessitates focused educational initiatives and subsidies for CA tools and materials.

Soil amendments like biochar, compost, and manure play a crucial role in rejuvenating nutrient-depleted soils. Biochar, in particular, has demonstrated significant potential for boosting soil organic carbon, enhancing cation exchange capacity, and

Table 10.4 Soil-based dryland restoration activities in Southern Africa: practices, limitations, and solutions

Restoration activity	Countries practiced	Limitations/hurdles	Suggestions to overcome limitations
Agroforestry	Zambia, Zimbabwe, Malawi	Limited access to tree seedlings and agroforestry training; land tenure conflicts	Establish community-based nurseries; strengthen land tenure policies; provide training programs
Soil mulching and cover cropping	South Africa, Namibia, Botswana	Lack of knowledge on benefits; limited access to mulching materials	Conduct awareness campaigns; develop local supply chains for mulching materials
Terracing and Contour farming	Lesotho, Eswatini, Tanzania	High labor intensity; limited government support for implementation	Promote mechanized solutions; offer subsidies or incentives for labor costs
Manure and compost application	Zimbabwe, Mozambique, Angola	Inadequate availability of organic materials; transportation challenges	Promote community composting initiatives; improve infrastructure for transport
Water harvesting techniques	Namibia, South Africa, Botswana	Limited investment in water storage infrastructure; variability in rainfall	Build communal water reservoirs; develop policies for water resource management
Revegetation and grass planting	Namibia, Angola, Zambia	Overgrazing by livestock; poor enforcement of grazing management	Implement rotational grazing practices; educate communities on grazing management
Biochar application	South Africa, Zimbabwe	Limited access to biochar production technology; high initial investment costs	Support local biochar production initiatives; provide subsidies for equipmen
Soil erosion control measures	Lesotho, Tanzania, Zambia	Lack of technical expertise; insufficient funding for projects	Train local technicians; seek donor funding and public-private partnerships
Crop diversification and rotation	Malawi, Mozambique, South Africa	Farmer reluctance to change traditional practices; lack of access to diverse seeds	Conduct demonstration farms; establish seed banks for diverse crops
Integrated nutrient management	Zimbabwe, Zambia, South Africa	High cost of fertilizers; poor awareness of integrated methods	Provide fertilizer subsidies; conduct training on integrated nutrient techniques

improving water retention in arid and semiarid regions (Cornelissen et al. 2013). Organic amendments such as compost and manure are not only cost-effective but also widely available in many rural areas, although their variable quality can sometimes limit their effectiveness (Chivenge et al. 2011). The high costs associated with biochar production and transportation present a major hurdle, highlighting the necessity for investments in local production facilities and programs to raise farmer awareness. Water harvesting techniques, including zai pits, contour bunding, and sand dams, are particularly effective in tackling water scarcity and boosting soil productivity in drylands. However, the labor-intensive nature of these methods and the requirement for technical expertise can make adoption challenging. Integrating these techniques into larger community-based projects could improve their scalability and overall impact.

Revegetation and afforestation play a crucial role in restoring degraded landscapes, enhancing biodiversity, and stabilizing soils. In countries like Botswana, Namibia, and Zimbabwe, native species such as Eragrostis tef, Salvadora persica, Acacia senegal, and Faidherbia albida have been successfully utilized to rehabilitate degraded rangelands and agricultural areas (Nyoka et al. 2015a, b; Mongwaketse et al. 2018). These efforts also improve microclimates, which in turn benefit nearby agricultural practices. However, challenges such as high seedling mortality from drought, limited water access, and land tenure conflicts often hinder implementation. To overcome these obstacles, it is essential to combine drought-resistant species with community-based management and integrate these efforts with income-generating activities like agroforestry and beekeeping. Although these soil restoration practices show great potential, their effectiveness hinges on tackling systemic issues such as resource limitations, socioeconomic challenges, and insufficient institutional support. Successful implementation necessitates collaboration among multiple stakeholders, supportive policies, and targeted investments in research and innovation. For example, promoting drought-resistant plant varieties and creating affordable technologies for soil restoration could greatly improve adoption rates and outcomes. Additionally, incorporating restoration practices into broader frameworks, such as the water-energy-food nexus and climate adaptation strategies, can enhance their socio-ecological benefits.

10.5 Conclusion

This study highlights the essential role of soil-based restoration practices in tackling the widespread issues of land degradation, biodiversity loss, and decreasing agricultural productivity in the drylands of Southern Africa. Techniques such as agroforestry, conservation agriculture, soil amendments, water harvesting, revegetation, and afforestation have been shown to improve soil health, restore ecosystem services, and enhance resilience against climate change impacts. These methods play a crucial role in stabilizing soils, enhancing water and nutrient cycles, and promoting the recovery of biodiversity. However, despite their potential, the broad implementation of these practices faces challenges such as limited resources, socioeconomic barriers to adoption, and difficulties in incorporating them into current land management systems. Overcoming these challenges necessitates collaboration among multiple stakeholders, focused policy measures, and investments in research and innovation. For example, encouraging the use of drought-resistant plant species, offering technical and financial assistance to farmers, and promoting community-based management systems can greatly improve the scalability and effectiveness of these practices. By embracing integrated and sustainable strategies, Southern Africa can utilize soil-based restoration practices to revitalize degraded landscapes, support local livelihoods, and advance broader goals of sustainable development and climate resilience. These initiatives are crucial not only for restoring ecological balance but also for ensuring the long-term productivity and sustainability of dryland ecosystems in the region.

References

Ajayi OC, Place F, Akinnifesi FK, Sileshi G (2011) Agricultural success from Africa: the case of fertilizer tree systems in Southern Africa. Int J Agric Sustain 9(1):129–136

Akinyemi FO, Ghazaryan G, Dubovyk O (2021) Assessing UN indicators of land degradation neutrality and proportion of degraded land for Botswana using remote sensing based national level metrics. Land Degrad Dev 32(1):158–172. https://doi.org/10.1002/ldr.3695

Arije D, Ghimire R, Bista P, Angadi SV, Gard CC (2024) Soil organic carbon recovery and soil health in semi-arid drylands with years of transition to perennial grasses. J Arid Environ 225:105263. https://doi.org/10.1016/j.jaridenv.2024.105263

Baer SG, Heneghan L, Eviner VT (2012) Applying soil ecological knowledge to restore ecosystem services. In: Wall DH, Bardgett RD, Behan-Pelletier V, Herrick JE, Jones TH, Ritz K, Six J, Strong DR, van der Putten WH (eds) Soil ecology and ecosystem services. Oxford University Press. https://doi.org/10.1093/acprof:oso/9780199575923.003.0032

Banda T, Mphangwe N et al (2016) Contour bunding as a soil and water conservation strategy in semi-arid Botswana. Land Degrad Dev 27(1):154–163

Botha JJ, van Rensburg LD, Anderson JJ et al (2007) Infield rainwater harvesting: a case study in improving water productivity in smallholder dryland farming systems. Water SA 33(3):273–278

Bünemann EK, Bongiorno G, Bai Z, Creamer RE, De Deyn G, de Goede R, Fleskens L, Geissen V, Kuyper TW, Mäder P, Pulleman M, Sukkel W, van Groenigen JW, Brussaard L (2018) Soil quality – a critical review. Soil Biol Biochem 120:105–125. https://doi.org/10.1016/j.soilbio.2018.01.030

Castle SE, Miller DC, Merten N, Ordonez PJ, Baylis K (2022) Evidence for the impacts of agroforestry on ecosystem services and human well-being in high-income countries: a systematic map. Environ Evid 11(1):10. https://doi.org/10.1186/s13750-022-00260-4

Chapungu L, Togo M, Raimundo I (2024) Prospects and challenges for adopting artificial intelligence and digital tools in climate resilience in rural Southern Africa: a systematic review. In: Matsa M, Chapungu L, Nhamo G (eds) Climate change resilience in rural Southern Africa. sustainable development goals series. Springer, Cham. https://doi.org/10.1007/978-3-031-73600-1_15

Chicco D, Warrens MJ, & Jurman G (2021) The coefficient of determination R-squared is more informative than SMAPE, MAE, MAPE, MSE and RMSE in regression analysis evaluation. PeerJ. Computer science, 7:e623. https://doi.org/10.7717/peerj-cs.623

Chivenge P, Murwira H, Giller KE, Mapfumo P, Six J (2011) Long-term impact of organic and mineral inputs on soil organic matter fractions and maize productivity under smallholder farming in Zimbabwe. Agric Ecosyst Environ 146(1):23–33

Cornelissen G, Pandit NR, Taylor P et al (2013) Biochar as a soil amendment: a review of its agronomic and environmental effects in Southern Africa. Soil Tillage Res 127:69–78

Costantini EAC, Branquinho C, Nunes A, Schwilch G, Stavi I, Valdecantos A, Zucca C (2016) Soil indicators to assess the effectiveness of restoration strategies in dryland ecosystems. Solid Earth 7(2):397–414. https://doi.org/10.5194/se-7-397-2016

David RM, Rosser NJ, Donoghue DNM (2022) Remote sensing for monitoring tropical dryland forests: a review of current research, knowledge gaps and future directions for Southern Africa. Environ Res Commun 4(4):042001. https://doi.org/10.1088/2515-7620/ac5b84

FAO (2024) Enhancing efforts for productive and resilient dryland landscapes in Southern Africa. FAORAF. Retrieved October 22, 2024, from https://www.fao.org/africa/news-stories/news-detail/enhancing-efforts-for-productive-and-resilient-dryland-landscapes-in-southern-africa/en

Farrell HL, Léger A, Breed MF, Gornish ES (2020) Restoration, soil organisms, and soil processes: emerging approaches. Restor Ecol 28(S4):S307–S310. https://doi.org/10.1111/rec.13237

Fatichi S, Or D, Walko R, Vereecken H, Young MH, Ghezzehei TA, Hengl T, Kollet S, Agam N, Avissar R (2020) Soil structure is an important omission in Earth System Models. Nat Commun 11(1):522. https://doi.org/10.1038/s41467-020-14411-z

Figueroa-Domecq, C, de Jong A, & Williams AM (2020) Gender, tourism & entrepreneurship: A critical review. Annals of Tourism Research, 84, 102980. https://doi.org/10.1016/j.annals.2020.102980

Garrity DP, Akinnifesi FK, Ajayi OC et al (2010) Evergreen agriculture: a robust approach to sustainable food security in Africa. Food Secur 2(3):197–214

Ghimire R, Clay DE, Thapa S, Hurd B (2022) More carbon per drop to enhance soil carbon sequestration in water-limited environments. Carbon Manag 13(1):450–462. https://doi.org/10.1080/17583004.2022.2117082

Giller KE, Witter E, Corbeels M, Tittonell P (2009) Conservation agriculture and smallholder farming in Africa: the heretics' view. Field Crop Res 114(1):23–34

González JMF, Febles Díaz JM, Amaral-Sobrinho NMB, Zonta E, Maura Santiago AV (2023) An approach to determine the neutrality of soil degradation in karst regions. Geoderma Reg 34:e00685. https://doi.org/10.1016/j.geodrs.2023.e00685

Grabowski P, Kerr JM (2014) Resource constraints and partial adoption of conservation agriculture by hand-hoe farmers in Malawi. Int J Agric Sustain 12(1):37–53

Grabowski P, Twomlow S, Haggblade S (2018) Adoption patterns and productivity impacts of conservation agriculture in Southern Africa. J Soil Water Conserv 73(2):114–123

Hiebl MRW (2023) Sample selection in systematic literature reviews of management research. Organizational Research Methods, 26(2):229–261. https://doi.org/10.1177/1094428120986851

Homrich AS, Galvão G, Abadia LG, & Carvalho MM (2018) The circular economy umbrella: Trends and gaps on integrating pathways. Journal of Cleaner Production, 175:525–543. https://doi.org/10.1016/j.jclepro.2017.11.064

Kimmell LB, Fagan JM, Havrilla CA (2023) Soil restoration increases soil health across global drylands: a meta-analysis. J Appl Ecol 60(9):1939–1951. https://doi.org/10.1111/1365-2664.14459

Kramer K, Sheil D (2024) Restoring deforested drylands for a wetter future – harnessing trees for credits, climate and water. Front Forests Global Change 7. https://doi.org/10.3389/ffgc.2024.1371117

Kugedera AT, Badza T, Odindo AO (2023) Integrated soil fertility management practices for improved crop production in smallholder farming systems of semi-arid areas: a synthetic review. Cogent Food Agric 9(1):2256137. https://doi.org/10.1080/23311932.2023.2256137

Laban P, Metternicht G, Davies J (2018) Soil biodiversity and soil organic carbon: keeping drylands alive. IUCN, Gland, p 24

Lal R (2015) Restoring soil quality to mitigate soil degradation. Sustainability 7(5):5875–5895

Li F et al (2024) Dryland social-ecological systems in Africa. In: Fu B, Stafford-Smith M (eds) Dryland social-ecological systems in changing environments. Springer, Singapore. https://doi.org/10.1007/978-981-99-9375-8_9

Liu X, Zhou J (2021) Assessment of the continuous extreme drought events in Namibia during the last decade. Water 13(20). https://doi.org/10.3390/w13202942

Liu ZF, Ding MH, He CY et al (2019) The impairment of environmental sustainability due to rapid urbanization in the dryland region of northern China. LandUrban Plan 187:165–180. https://doi.org/10.1016/j.landurbplan.2018.10.020

Makumba W, Akinnifesi FK et al (2007) Long-term impact of a gliricidia-maize intercropping system on carbon sequestration in Southern Malawi. Agric Ecosyst Environ 118(1–4):99–108

Martin DM (2017) Ecological restoration should be redefined for the twenty-first century. Restor Ecol 25(5):668–673. https://doi.org/10.1111/rec.12554

Mazvimavi K, Twomlow S (2009) Socioeconomic and institutional factors influencing the adoption of conservation farming by vulnerable households in Zimbabwe. Agric Syst 101(1):20–29

Miguel MF, Butterfield HS, Lortie CJ (2020) A meta-analysis contrasting active versus passive restoration practices in dryland agricultural ecosystems. PeerJ 23(8):e10428. https://doi.org/10.7717/peerj.10428

Mongwaketse T, Moeletsi ME et al (2018) Revegetation as a strategy for land rehabilitation in Botswana: opportunities and challenges. Land Degrad Dev 29(2):125–135

Motsi KE, Chuma E, Mukamuri BB (2004) Rainwater harvesting for sustainable agriculture in communal lands of Zimbabwe. Phys Chem Earth 29(15–18):1069–1073

Moyo T, van Tol JJ et al (2017) The role of sand dams in water harvesting and soil conservation in arid environments. J Soil Water Conserv 72(3):210–219

Mupangwa W, Twomlow S, Walker S (2008) The influence of conservation agriculture practices on the productivity of maize and sorghum under semi-arid conditions in Zimbabwe. Field Crop Res 105(3):248–258

Mupangwa W, Twomlow S, Walker S (2013) The influence of conservation agriculture practices on soil moisture dynamics and crop productivity in semi-arid Zimbabwe. Agric Water Manag 118:1–9

Naeem M, Ozuem W, Howell K, & Ranfagni S (2023) A Step-by-Step Process of Thematic Analysis to Develop a Conceptual Model in Qualitative Research. International Journal of Qualitative Methods, 22. https://doi.org/10.1177/16094069231205789 (Original work published 2023)

Ngoma H, Angelsen A, Yesuf M (2016) Productivity and welfare effects of conservation agriculture. Food Policy 62:152–161

Ngwira AR, Aune JB, Thierfelder C (2012) On-farm evaluation of conservation agriculture in Malawi: soil health, yield, and economic benefits. Soil Tillage Res 124:244–250

Nyoka I, Chikowo R, Mapfumo P (2015a) Afforestation and its role in soil rehabilitation in Southern Africa: lessons from Zimbabwe. Land Degrad Dev 26(6):529–540

Nyoka I, Chikowo R, Mapfumo P (2015b) Erosion control and soil health improvement through agroforestry: evidence from Zambia. Land Degrad Dev 26(6):529–540

Omari MK, Ziervogel G, New M (2023) Drought governance: a cross-level governance analysis in Botswana. Clim Risk Manag 42:100557. https://doi.org/10.1016/j.crm.2023.100557

Page MJ, McKenzie JE, Bossuyt PM, Boutron I, Hoffmann TC, Mulrow CD, Shamseer L, Tetzlaff, JM, Akl EA, Brennan SE, Chou R, Glanville J, Grimshaw JM, Hróbjartsson A, Lalu MM, Li T, Loder EW, Mayo-Wilson E, McDonald S, McGuinness LA, … Moher D (2021) The PRISMA 2020 statement: An updated guideline for reporting systematic reviews. BMJ, 372, Article n71. https://doi.org/10.1136/bmj.n71prismastatement.org+15

Prăvălie R (2016) Drylands extent and environmental issues. A global approach. Earth Sci Rev 161:259–278. https://doi.org/10.1016/j.earscirev.2016.08.003

Turner WC, Périquet S, Goelst CE, Vera KB, Cameron EZ, Alexander KA, Belant JL, Cloete CC, du Preez P, Getz WM, Hetem RS, Kamath PL, Kasaona MK, Mackenzie M, Mendelsohn J, Mfune JKE, Muntifering JR, Portas R, Scott HA, Strauss WM (2022) Africa's drylands in a changing world: challenges for wildlife conservation under climate and land-use changes in the Greater Etosha Landscape. Global Ecol Conservation 38:e02221. https://doi.org/10.1016/j.gecco.2022.e02221

Twomlow S, Urolov JC et al (2008) Lessons from the field – conservation agriculture in Southern Africa. J SAT Agric Res 6:1–11

United Nations Environment Programme (2021) Ecosystem restoration for people, nature and climate: becoming #GenerationRestoration. United Nations

Vanlauwe B, Descheemaeker K et al (2010) Integrated soil fertility management in sub-Saharan Africa: unravelling local adaptations. Soil Biol Biochem 42(4):10–23

Wu B, Smith WK, Zeng H (2024) Dryland dynamics and driving forces. In: Fu B, Stafford-Smith M (eds) Dryland social-ecological systems in changing environments. Springer, Singapore. https://doi.org/10.1007/978-981-99-9375-8_2

Yang Q, Li M, Zu Z, Ma Z (2021) Has the stilling of the surface wind speed ended in China? Sci. China Earth Sci. 64:1036–1049

Yang ECL, Khoo-Lattimore C, & Arcodia C (2017) A systematic literature review of risk and gender research in tourism. Tourism Management, 58:89–100. https://doi.org/10.1016/j.tourman.2016.10.011

Chapter 11
Nature-Based Disaster Risk Management Approach in the Sudano-Sahelian Zone of Cameroon

Fréderic Saha and Thierry Gaitan Seutchueng Tchuenga

Abstract In the context of increasing risks and disasters in the world in general, and Cameroon in particular, it is necessary to explore all management options. Considering the general ecological crisis and the global vision of development, as summarized in the Sustainable Development Goals, the orientation toward nature-based solutions is the most relevant. This is a somewhat new concept and approach for Cameroon. However, the all-out search for risk reduction and adaptation solutions reveals initiatives that deserve to be encouraged, strengthened, and replicated. Likewise, Cameroon presents advantages for the implementation of other strategies presented by the IUCN, and they have already been put into practice in other areas of the world. This study identifies and describes potential nature-based solutions for disaster risk prevention in the Sudano-Sahelian zone of Cameroon. The investigation approach was exploratory, descriptive, and analytical in nature. It appears that to reduce risks (prevention), the restoration of wetlands is necessary; in fact, hydraulic developments in these plains have increased the risk. In the current context of humidity, it is necessary to increase the submergence space to reduce the pressure on the dikes. Reforestation or betterment of the creation of plant dikes is also desired. Protecting dikes by revegetating them also has advantages. In terms of adaptation, the creation of reservoirs or overflow zones to contain water during periods of precipitation and flooding is important. This helps contain excess water and prevent flooding. At the agricultural level, changes in practices are necessary.

Keywords Nature-based solutions · Disaster risk reduction · Sudano-Sahelian zone · Cameroon · Flood · Drought

---

F. Saha (✉)
Faculty of Arts, Letters and Human Sciences, The University of Yaoundé 1, Yaoundé, Cameroon

T. G. Seutchueng Tchuenga
CIFOR-ICRAF, ReSI-NoC Project, Garoua, Cameroon

Agricultural Research Institute for Development, Yaoundé, Cameroon

Y. D. I. Khan et al. (eds.), *Land Restoration Through Ecosystem-Based Approach*, Disaster Resilience and Green Growth, https://doi.org/10.1007/978-981-95-0389-6_11

11.1 Introduction

Nature-based solutions (NBS) are actions that rely on ecosystems to address societal challenges (IUCN 2019). For ecosystems in good ecological condition, it is a question of protecting them to continue to benefit from their services. For those experiencing changes due to human activities, sustainable management avoids degradation and helps maintain functionality. For ecosystems that have already been degraded, restoration is recommended to resume their productivity in terms of ecosystem services (Cohen-Shacham et al. 2016). These behaviors and operations (protection, sustainable management and restoration of nature) must help humanity to better cope with climate change, risks and disasters, food insecurity, health, etc. (IUCN 2019). NBS has long been marginalized in favor of traditional engineering solutions. In 2009, when REDD+ was proposed as a relevant option for mitigating climate change, the concept of NBS gained renewed interest (UNEP 2020).

Climate change results in increased natural risks across the Earth (IPPC 2023). This is how the number of natural disasters increased in recent years. Material damage and loss of life have also increased exponentially. Notably, hydroclimatic risks experienced the strongest growth. Efforts made in terms of risk reduction, preparation, and management of disaster situations remain insufficient. The United Nations Office for Disaster Risk Reduction (UNDRR) presents disaster risk reduction as a tool for achieving Sustainable Development Goals. Indeed, it will be impossible to eradicate poverty (SDG1) and hunger (SDG2) and even to achieve good health and well-being (SDG3) on earth if we continue to suffer so many disasters every year. Good risk and disaster management leads to results on all other sustainable development objectives. If risk management is done through nature-based solutions, the results are even more relevant with the gains on SDG 15, which urges countries to protect and restore terrestrial ecosystems while promoting their sustainable use.

Cameroon is implementing a set of strategies to address natural risks (Bang 2022). These strategies include grey engineering and ecological infrastructure. With the increase in disasters, the commitment to achieving the SDG targets by 2030, and the renewed attention to DFS, it is necessary to revisit the potential of each option. This chapter draws on literature review and field observations to identify and analyze the potential of NBS for disaster risk management in the northern lowlands. The hypothesis formulated from the outset is that this area of Cameroon presents enormous potential for disaster risk reduction and adaptation through nature-based solutions. Unfortunately, this potential is underexploited; hence, the poor performance and the continued increase in material and human damage are caused in particular by flooding and drought, which are the two risks concerned by this research.

11.1.1 Context of the Study and Objectives

The lowlands of northern Cameroon are shared between the administrative regions of the North and the Far North (Fig. 11.1). They are characterized by a tropical climate with a Sudanian nuance in the south and Sahelian in the north. In agro-ecological terms, this is the Sudano-Sahelian zone. The dry season lasts between eight and 10 months, and the watercourses are intermittent, hence the drought problem. In addition, this area experiences many years of rainfall shortage, which exacerbates the problem of drought. Maximum temperatures are sometimes above 40 °C, particularly during the dry season, hence the heat wave problem (MINEPDED 2015). July and August receive almost 90% of the annual rainfall. Flooding is inevitable because the soil is mainly clayey, loamy, and sandy. After nearly two decades of deficits (since 1970), this area has experienced a return to humidity in recent years (Ekolu et al. 2022). Thus, the start of the rainy season

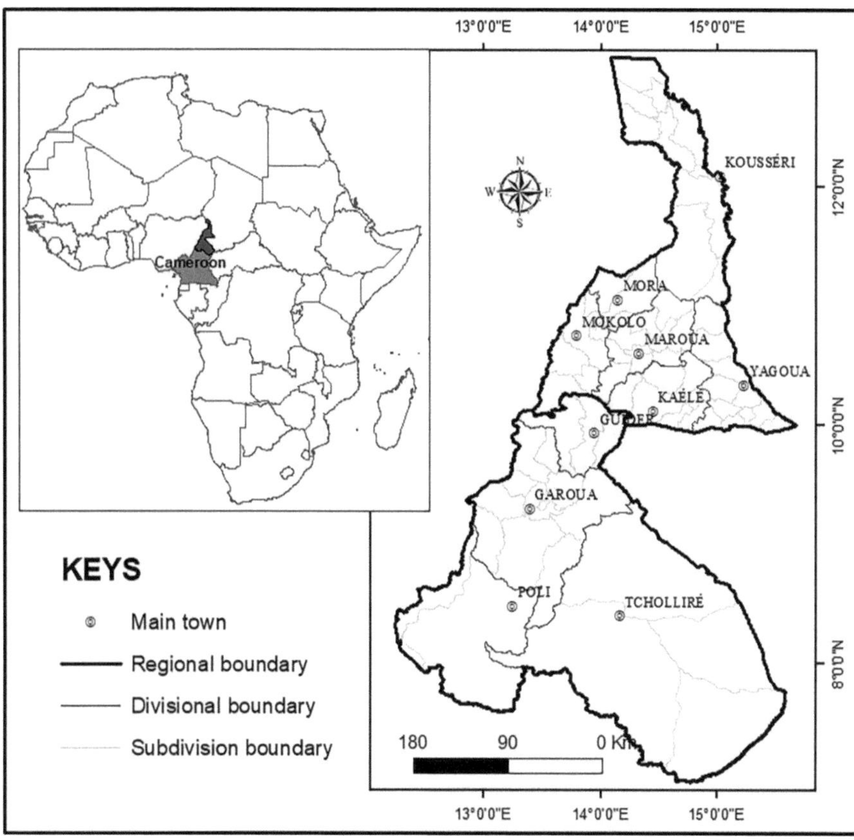

Fig. 11.1 Location of North and Far North region in Central Africa. (The author has prepared the map based on GIS layers from the interactive forest atlas of Cameroon, 2014)

was earlier, the quantities were clearly above the 1970–1990 average, and episodes of extreme rain (> 50 mm in 24 h) were becoming more frequent (Saha 2020). These factors explain the occurrence and resurgence of flooding in the area. In the development process of this zone, some hydraulic infrastructures have been put in place. These are the Maga and Lagdo lakes, whose dikes represent dangers for local populations. The excessive recharge of these lakes is also dangerous for populations. Given the arid or semiarid nature of the area, most towns were created along rivers. We can note the cases of Garoua on the banks of the Bénoué, Maroua with the Tsanaga and Kaliao mayo, Yagoua, Pouss, Kousseri on the left bank of the Logone, etc. These towns experience river flooding annually (see Table 11.1).

Cameroon has not yet implemented a real risk and disaster management strategy. For example, Tchindjang et al. (2022) noted that an early warning system does not yet exist. Thus, most actions visible on the ground are part of the response in times of disaster.

Table 11.1 Some cases of disasters due to floods and drought in the lowlands of northern Cameroon

Year	Event	Area affected	Damages
1977	Flood	Maroua, Gazawa	7 dead
1988	Flood	Maroua, Kousseri	10 dead, 165 people homeless
1991	Flood	Pouss et Kousseri	2 dead, 200 people homeless
1994	Flood	Maroua, Mokolo, Kousseri	41 dead, 1004 people homeless
1997	Flood	Maroua	7 dead, 248 people homeless
2000	Flood	Maroua	17 dead, 206 people homeless
2005	Flood	Maroua	14 dead, 453 people homeless
2007	Flood	Maroua and Mayo Danay	8 dead, 4421 people homeless
2010	Flood	Pouss	12 dead, 3000 people homeless
2012	Flood	Logone and Chari	20 dead, 20,000 people homeless
	Flood	Garoua	26,000 people affected
2015	Flood	Maroua, Gazawa, Ndoukoula	5 dead
2016	Flood	Garoua	6 dead
2017	Flood	Zina, Mora, Makary	8 dead, 750 people homeless
		Garoua	2 dead
2008	Flood	Garoua	9 dead
2003	Flood	Poli	5 dead, 8720 people homeless
2021	Flood	Garoua	3 dead
2023	Flood	Garoua	1 dead
2024	Flood	Far North	30 dead and 400,000 peoples affected

Source: Saha et al. (2020), completed

11.2 Methodological Approach

11.2.1 The Study Area

The northern lowlands have an estimated population of nearly 7 million inhabitants (based on data from the third population and housing census in Cameroon in 2005). This is almost a quarter of the total population of Cameroon (the World Bank places the population of Cameroon at 28 million in 2022). Like Cameroon, it is a young population with an average age of approximately 15 years (BUCREP 2010). Economic activities are primarily agriculture, livestock, and fishing, which comprise nearly 85% of the population. The equipment was rudimentary. Agriculture is essentially rain-fed and off-season, depending directly on precipitation behavior. The cattle herd is the largest in the country. The livestock system is more nomadic, following grazing routes throughout the year. In 2014, the poverty rate in Cameroon was 38.5%, and the North and Far North regions presented 74% and 68%, respectively, making the northern lowlands the poorest geographical unit in Cameroon (NIS 2015). Nearly 65% of housing in these two regions is precarious. The walls are made of sheet metal, beaten earth or straw, the roofs are made of straw and/or thatch, and the floors are made of uncoated earth (DSCN 2002). These constructions are very fragile and collapse when flooded.

11.2.2 Study Approach

The lowlands of northern Cameroon were chosen for this study because of the high occurrence of catastrophic floods and their semiarid character, exposing it to different forms of drought. Socioeconomically, this area has poor scores for most indicators (NIS 2015). The strong pressure on resources and unsustainable practices are at the origin of environmental degradation, exacerbating risks. The absence of an operational strategy for risk and disaster management in this area leads to enormous human and material losses almost every year. Through an exploratory, descriptive, and analytical approach, this study assesses the potential of the lowlands of Northern Cameroon in terms of risk management options through nature-based solutions.

11.2.2.1 Literature Review

The Sudano-Sahelian zone of Cameroon very early benefited from the attention of both the authorities and scientists in the environmental field; this is because of its rich biodiversity, particularly wildlife, and also the fragility of the environment. This is how the Bénoué national park had already been created as a reserve by the French during the colonial period in 1932. The Waza national park was created in 1968 and recognized as a biosphere reserve by UNESCO in 1979. Benefiting from

Table 11.2 Communities consulted with questionnaire

Region	Subdivision	Neighborhood/village	Number of participants
Far North	Kousseri	Madagascar 2	13
	Logone-Birni	Zimado	11
	Zina	Magouya	12
	Maga	Pouss	10
	Kai-Kai	Keléo	20
North	Lagdo	Lopéré	14
	Garoua	Ouro Labo	23
Total			103

this environment where the objective of conservation or sustainable management is displayed, numerous projects have been carried out. Their relationships are used in this study. The recent creation of the University of Maroua increases the scientific attention paid to the area. Thus, we can note the documents (Wesseling et al. 1994; Arcadis-Euroconsult and Ere développement 1997; de Noray 2002; IUCN 2003) of the Waza Logone project that we consulted at the Center for the Study of the Environment and of Development (CEDC) in Maroua. Recent work as part of the emergency flood control project has also left numerous reports which can be consulted online on the website of the World Bank which was the main financial partner. On social and economic aspects, the reports of the national statistics institute were consulted.

11.2.2.2 Data Collection from Communities

Flooding is one of the major concerns in the Sudano-Sahelian zone of Cameroon. Familiarization with risk contributes to the good perception of this problem by populations. Prevention and adaptation efforts are visible in floodplains annually affected by devastating floods. Field data was collected using questionnaires in seven subdivisions. In total, 103 households were interviewed (see Table 11.2). The questionnaire addressed the issue of flood risk management in the area. Questions on risk perception, proposals for options for preventing disaster situations and adaptation were collected for this study.

11.3 Findings and Analysis

Overall, risk management takes place in three phases: before, during, and after. Pre-disaster intervention aims to reduce risk. Actions are oriented toward prevention. Considering the fact that, regardless of the efforts made, zero risk does not exist, we must prepare for disaster situations in order to intervene effectively to avoid loss of human life and heavy material losses. After a disaster, natural and human systems must be adjusted in the form of adaptation to build resilience. Nature-based solutions can not only reduce risks (ECO-DRR) but enrich adaptation options. Without this being in the context of nature-based solutions, these strategies for the most part have been treated in different studies. Thus, a significant amount of documentation, including gray literature, was used.

11.3.1 Ecosystem-Based Disaster Risk Reduction in Low Dry Land of Cameroon

The lowlands of northern Cameroon present enormous potential in terms of options for reducing the risks of flooding and drought through nature-based solutions. The restoration of wetlands, reforestation, and the creation of plant dikes are among strategies for reducing the risks of flooding and drought with enormous benefits for ecosystems. Table 11.3 summarizes each NBS advantage for risk management and the positive impacts on the ecosystem.

Table 11.3 Advantage for risk management and positive impact on the ecosystem for NBS presented

Strategy	Advantage for risk management (flood and drought)	Positive impact on the ecosystem
Wetland restoration	Absorbs excess water and reduces flooding Stores and improves water quality and reduces the risk of drought	Eliminates invasive species harmful to the ecosystem
Reforestation	Reduces soil erosion, improves water infiltration, slows runoff, and reduces flooding Improves air quality, promotes the establishment of a humid microclimate during drought periods	Improves air quality Habitat for wildlife Carbon sequestration and fight against climate change Fight against desertification and prevent the advance of the desert Temperature reductions Protection against winds
Creation of plant dikes	Filters rainwater and protects against flooding Stabilizes the banks of the watercourse and reduces flooding Reduces water evaporation and prevents drought	Habitat for wildlife Improves biodiversity
Vegetation of banks and dikes	Reduces erosion Slows flow and retains water Reduces the cost of maintaining infrastructure and increases its sustainability	Habitat for wildlife Ecological corridor

11.3.1.1 Restoration of Wetlands

Apart from the Mandara Mountains, the Sudanese zone of Cameroon is mainly made up of lowlands. Figure 11.2 shows an altimeter configuration of the area. The lowest lands are between 56 and 200 m, described in the literature as the Bénoué trench (Bessong 2012). Then, there is the plain of Waza Logone, also called the great Yaéré (large swamp), and the plain of Diamaré at the foot of the Mandara mountains.

Today, the average population density in the floodplain is very high (more than 100 inhabitants/km^2 in the town of Garoua). Between 1977 and 1982, Cameroon installed a hydroelectric dam. It is a water reservoir with a submerged surface area of nearly 700 km ^{2} and capacity of 6 billion m3 (Tchana 2002). The objective of this development was to produce electrical energy, rice production, and fish farming. A similar development was made in the Waza Logone plain with installation of the Maga hydro-agricultural dam between 1978 and 1980. The water surface retained by this dike varies between 360 km^2 at maximum coast and 120 km^2 at the end of the dry season (Seignobos and Raugel 2000). Apart from its main dike maintaining the body of water over a distance of 27 km between Pouss and Girvidig, another dike had been set up between Yagoua and Tékélé over a distance of 70 km.

These developments have profoundly modified the hydrological functioning of the areas concerned. In the Bénoué Plain, Ngounou Ngatcha et al. (2002) noticed the suppression of floods, the reduction of flood-prone areas, and the drying up of numerous natural submersible basins. It should be noted that these developments were made during a period of worsening rainfall that began in 1970 (Olivry 1997). In addition, these developments have drawn migrants from other regions of Cameroon and neighboring countries (Arditi 1985). The administration supported this economic development process in the plains by creating new administrative units to ensure the supervision of populations and activities. The most serious flooding in the area is currently caused by the following:

- Breaking of the dike (flooding downstream of the lake)
- Excessive recharge of reservoir lakes (upstream flooding)

Therefore, the ambition of containing floodwaters in these reservoirs does not always work. With climate change and the increase in extreme rain, infrastructure is increasingly at risk. Rivers claim entire floodplains. Following the 2012 flood, Cameroon invested with the support of the World Bank nearly 54 billion FCFA in the repair of these structures and their strengthening; however, this did not end catastrophic flooding. The case of 2024 is illustrative (see Table 11.1).

In his thesis on the assessment of flood risk and the resilience of populations in the Far North region, Saha et al. (2020) relayed the desires of populations who wanted to live with water. In fact, local communities decry their exclusion in the decision-making process in plain management. For the latter, the state has failed in its policy aimed at combating water because it is the cause of catastrophic floods and reduces the economic productivity of the area.

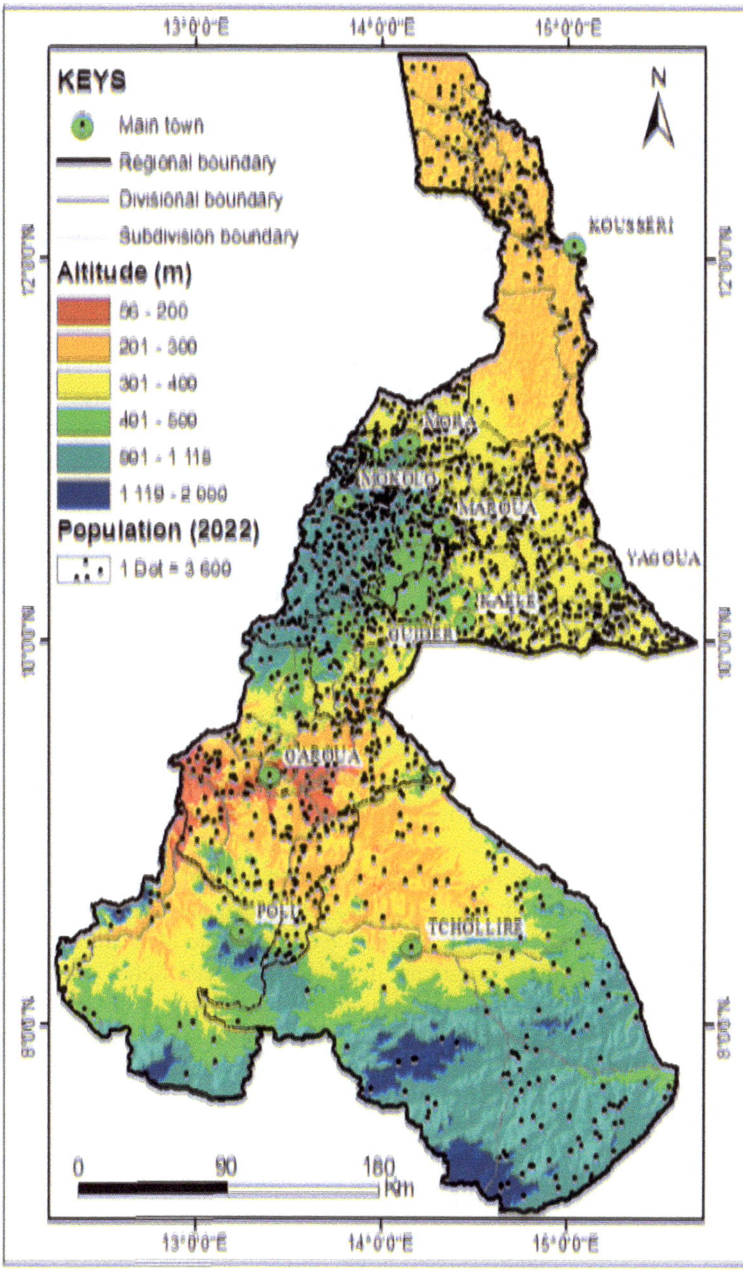

Fig. 11.2 Altimetry and density in the Northern and Far North regions of Cameroon. (The authors have prepared the map by using open-source data from ASTER GDEM and central bureau of census and population studies)

The International Union for Conservation of Nature (IUCN) noted serious consequences following the installation of the Maga hydro-agricultural dam. In fact, this development reduced flooding in the plain, leading to a decline in the productivity of pastures and fishing, a water deficit for livestock in the dry season and, above all, a general degradation of the environment (Arcadis-Euroconsult and Ere development 1997). In the face of these consequences, the IUCN set up the Waza Logone project, one of the objectives of which was hydrological rehabilitation of the plain. Thus, in 1994, through openings on the side dike of the Maga Dam, the re-flooding of the plain started to the great joy of the populations (de Noray 2002).

The restoration of wetlands makes it possible to put an end to flash floods, which cause significant annual damage in the Sudano-Sahelian zone of Cameroon. This allows floods to gradually submerge the plains, providing populations the opportunity to protect themselves by taking shelter. The use of gray infrastructure (dykes) to contain the flood of the Logone in its minor bed and also to store the flows coming from the Mandara Mountains through the Mayo Tsanaga in the Maga Lake creates an overflow, especially with the return of humidity in the area. It should be noted that the rehabilitation of wetlands does not mean the destruction of hydro-agricultural and/or electricity reserves. The idea is to direct the water or better the surplus water into the plain to avoid too much pressure on the dikes and sudden floods.

11.3.1.2 Environmental Preservation and Reforestation

Vegetation constitutes an important protection against natural disasters. It protects against mass movements and erosion, contributes to infiltration, and prevents flooding. In the Sahelian zone, vegetation acts as a barrier to desertification and the progression of violent winds. Very early on, Cameroon opted to create protected areas to preserve the environment. Trees also produce wood and other resources. The expected ecosystem services guide the choice of species to plant. In the Far North Region, several government programs, notably the "Green Sahel" operation, are responsible for planting thousands of trees each year (Wandji et al. 2013). In addition, Civil Society Organizations and development partners contribute to reforestation. Between 2008 and 2012, more than 2.5 million trees were planted in the Far North Region (Table 11.4).

11.3.1.3 The Creation of Plant Dikes

Riverbank erosion is a major problem in the Sudano-Sahelian zone of Cameroon. This is even more worrying in areas where the soils are sandy clay and therefore not very coherent. In the rainy season, riverbank landslides are observed. The diagnostic report of the Urban Development Plan for the city of Maroua notes that bank erosion is one of the biggest problems in this city (IDEA Consult 2016). This results in flooding of riparian areas, including roads, houses, water wells, and agricultural

Table 11.4 Reforestation by the Sahel Green project in the Far North Region 2008–2012

Year	Sites	Division	Area (ha)	Planted plants	Total plants/ year	Success rate
2008	Léra	Mayo-Kani	1500	240,000	480,000	62
	Mada	Logone-et-Chari	1500	240,000		40
2009	Goussor	Mayo-Kani	1000	160,000	480,000	67
	Maltam	Logone-et-Chari	1000	160,000		40
	Daiba-Kalfou	Mayo-Danay	1000	160,000		65
2010	Djidel-Bogo	Diamaré	500	80,000	560,000	55
	Bipaing	Mayo-Kani	500	80,000		63
	Berkédé	Mayo-Kani	500	80,000		70
	Ouro-Daban	Mayo-Danay	500	80,000		72
	Adjiri-Mora	Mayo-Sava	500	80,000		50
	Katikimé-Darak	Logone-et-Chari	700	112,000		78
	Djénéné-Goulfey	Logone-et-Chari	300	48,000		75
2011	Lamtari-Kaélé	Mayo-Kani	500	80,000	560,000	72
	Beungel-Boboyo	Mayo-Kani	500	80,000		60
	Matchoualta-Lara	Mayo-Kani	500	80,000		60
	Karena-Darak	Logone-et-Chari	500	80,000		75
	Tcherfeke-Yagoua	Mayo-Danay	500	80,000		65
	Endala-Kerawa	Mayo-Sava	500	80,000		55
	Boulah-Mokong	Mayo-Tsanaga	500	80,000		70
2012	Gabane-Lara	Mayo-Kani	500	80,000	480,000	70
	Boboyo II-Kaélé	Mayo-Kani	200	32,000		50
	Matchoualta II	Mayo-Kani	500	80,000		80
	Boula-Mokong II	Mayo-Tsanaga	500	80,000		80
	Bogo-Ouro Yaya	Diamaré	300	48,000		72
	Doukoula	Mayo-Danay	500	80,000		67
	Kossa-Mora	Mayo-Sava	500	80,000		80
			16,000	2 560,000	2 560,000	65

Source: Saha et al. (2020)

plots. Infrastructure such as bridges is also weakened. In the Bénoué basin, Mbele Abbo (2021) noted an acceleration of morphogenic processes following the installation of the Lagdo dam. The phenomenon is present is fluvial erosion based on recursive sliding of the banks. The most prominent solution to protect these banks is the installation of rubble walls to slow the erosive action of the current at the base of the banks. As a strategy to combat flooding in the town of Maroua, the PDU provides for the protection of 15,570 linear meters of banks on the Tsanaga and Kaliao

A) Reforestation of the banks of the Logone river | B) Vetiver on the embankment

Fig. 11.3 Few examples of vegetation use in flood risk mitigation. (a) Reforestation of the banks of the Logone river. (b) Vetiver on the embankment. (Authors' own photographs)

mayos (IDEA Consult 2016). Similar initiatives are being observed on several other rivers across the area.

It should be noted that several national and international organizations working in the field of nature protection and sustainable resource management carry out different activities in this area of Cameroon. Here and there we observe projects fighting against the factors of environmental degradation. Along watercourses whose banks are subject to erosion, Ipomea asarifolia known for its ability to quickly stabilize the soil is planted. Eucalyptus trees (*Eucalyptus globulus*) are planted in places when, in addition to stabilizing the banks, we also want to drain a wetland (see Fig. 11.3a). As part of the Emergency Flood Control Project in the Far North region (PULCI), in 2017, the NGO ABIOGeT (Actions for Biodiversity and Land Management) used vegetation technics to stabilize the dike and the banks of Lake Maga. Thus, over a distance of 17 km corresponding to 23.8 ha of embankment area was vegetated. The plants used are Vetiver, Jatropha, and Ipomea (see Fig. 11.3b).

It is difficult to identify real vegetated dikes on the ground set up to deal with the flooding problem. The B&C photos show a sort of hybridization between gray infrastructure (dyke) and vegetation. Vetiver and Ipomoea were used to protect the dike and prevent erosion. The lack of success of this strategy is explained by the fact that it is preventive, but Bang (2022) notes that Cameroon invests very few resources in risk prevention. Actions are rather reactive and post-disaster, particularly with reconstruction (Aka et al. 2017). Thus, actions are envisaged in emergencies, and the dikes and the banks' protection by rubble walls adapt better. Setting up a vegetated dike takes years; time for plants to grow. By directing disaster risk management policy toward prevention, it would be interesting today to:

- Demarcate areas that are too degraded on the banks to set up plant dikes.
- Set up plant dikes to reinforce existing engineering infrastructure.
- Also reinforce this hybridization by revegetating the dikes to avoid their rapid deterioration.

11.3.2 Ecosystem-Based Risk Adaptation in Low Dry Land of Cameroon

Whatever the efforts made in risk prevention, achieving zero risk is impossible. Thus, it is complementary to reduce vulnerability and increase resilience to risks. This involves adjusting natural, human, and even socioeconomic systems to mitigate the harmful effects of risks. Usually, to cope with floods, the populations of the Sudano-Sahelian region of Cameroon develop a set of adaptation strategies: raising construction sites, setting up water evacuation channels, constructing on stilts, and adjusting drainage systems, agropastoral production (pasture cultivation and storage, adoption of improved short-cycle seeds), etc. When it comes to drought, water reserves are often put in place. Some of these techniques must be promoted given their benefits for ecosystems and other social and economic advantages, especially as this area faces high poverty.

11.3.2.1 Create Reservoirs and Overflow Areas

In the plains of the Sudano-Sahelian zone of Cameroon, floods are pluvial, fluvial but also by overflowing of reservoir lakes. If today with extreme phenomena the quantities of water are increasing, it is necessary to set up reservoirs. In the NBS approach, a reservoir appears like a pond. In certain areas the absence of slope explains the stagnation of surface water after rains. When the rains continue, flooding is inevitable. In this context, we should choose the lowest area of each neighborhood to make it the place of accumulation of rainwater. A system of drainage canals was then placed in place to direct water to this pond. It should be noted that in many neighborhoods of towns in these areas the first inhabitants settled on the highest land. The densification of the habitat gradually led to the occupation of lowlands which were once the outlets of the waters. In other cases, due to the phenomenon of silting up, these pits were gradually filled in. In the absence of the possibility of draining the water to watercourses, the rehabilitation of these ponds is a solution. In certain cases where ponds do not exist, an altimeter study can make it possible to choose a space that can be developed. These ponds will at the same time be water reserves to cope with drought in the dry season. There are numerous benefits to the ecosystem: recharge of the water table, habitat for wildlife, and creation of a microclimate by increasing humidity. On a socioeconomic level, a water reserve supports agriculture and many other activities that can be implemented.

Overflow zones are considered for flood and drought resilience. It should be noted that the floodplain is a natural overflow zone for a watercourse. Failing to restore these plains as proposed in the risk reduction options, overflow zones can be planned. It may be a question of demarcating part of the flood plain and rehabilitating it or of setting up pits into which excess water can be dumped during times of flooding. It should be noted that the districts located in the lower plains of Logone and Bénoué are the most affected by the flooding of these rivers. Capturing overflow water into reservoirs would save lives and also minimize material damage. Lake Maga has a recharge system via the Logone. It was planned that during periods of flooding, water from the Logone would pass through an overflow to be stored in this lake. But as the lake has another source of supply (Mayo Tsanaga), it happens that the lake experiences overflows at the same time as the Logone. For lakes that frequently overflow and cause flood losses, the same approach is to find a reservoir for excess water. We can either deepen the lake, as envisaged for Lake Maga, which has already lost more than half of its depth due to the phenomenon of silting (author), or widen or install additional pits for managing surpluses of water during flood periods. It should be noted that increasing the quantities of water available on surfaces has numerous advantages for the ecosystem and the fight against drought, as already mentioned above.

11.3.2.2 Changes in Agro-pastoral Practices to Adapt to Risks

Faced with the resurgence of more frequent and more violent risks, but also environmental degradation, it is vital to make changes in production practices (Grigorieva et al. 2023; Shaffril et al. 2024). It should be noted that agricultural drought and floods are the cause of the loss of thousands of hectares of agricultural operations in the Sudano-Sahelian zone of Cameroon. However, risks also represent opportunities that can be exploited. In the face of drought (shortening of the rainy season), it is often recommended to adopt short-cycle varieties (Ntali et al. 2023). Regarding flooding, it is recommended to avoid practicing speculations with intolerance to water in usually flooded plains. Following the floods of 2015 in Zina, nearly 80% of the population of this district did not grow rain-fed crops in 2016 (Laborde et al. 2018). Likewise, several varieties of short-term cereals are increasingly promoted in the area by the Agricultural Research Institute for Development (IRAD) (Tchuenga et al. 2024). Drought-resistant plants consume less water throughout their cycle. They also make it possible to maintain agricultural activity on dry plots, which slows down desertification. This is also an advantage for maintaining the biodiversity of the arid zone. This practice also limits the dependence on irrigation, which sometimes has enormous negative environmental consequences (Loyer 1989).

Among the strategies proposed by the populations for disaster risk reduction in the Sudano-Sahelian zone, we find nature-based solutions in good positions (Fig. 11.4). Reforestation to improve infiltration and stabilize the banks of watercourses is proposed by 17% of those questioned. They also propose releasing the most flood-prone areas to increase the tidal space of water surfaces; this means the restoration of wetlands. During periods of flooding, people do not forget that their area is, above all, characterized by a long period of drought. Thus, most strategies revolve around storing surplus water from the rainy season. By recommending the release of drains and even the eviction of populations living too close to rivers or lakes, the populations surveyed recognize the responsibility of communities in creating the risk. There is a desire to live with water, hence the need to combine rapid-result engineering strategies and NBS whose impacts are generally medium and long term.

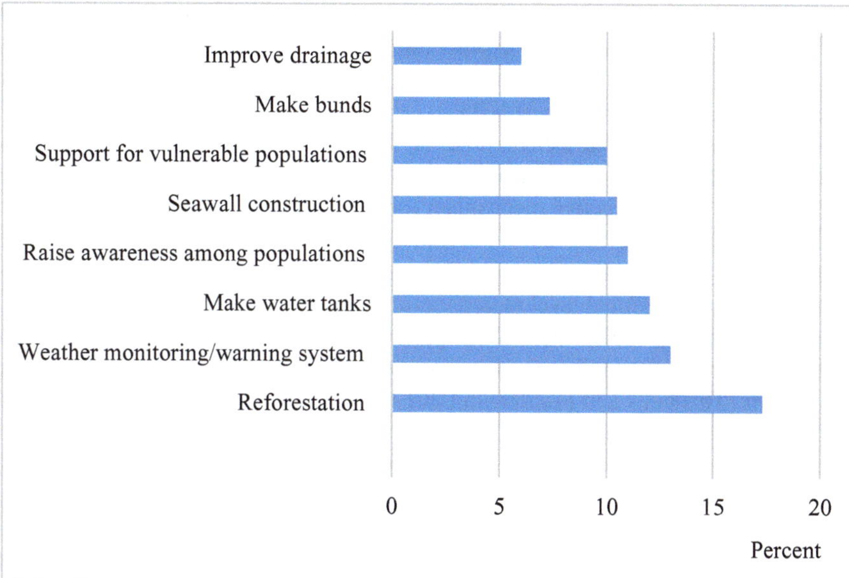

Fig. 11.4 Proposals from the population for better management of flood risk in the area. (the author has prepared the graph)

11.4 Discussion of Results

Nature-based solutions for disaster risk management are mostly daring. They do not always align with the socioeconomic development guidelines as defined by the authorities. This brings up to date the difficulty of reconciling the economic, social, and ecological pillars within the framework of sustainable development, particularly with regard to strong sustainability (Tsayem Demaze 2009). For example, this was the case for wetland restoration. These wetlands have been modified to boost their economic production. Their restoration presents itself as a back pedal, a step back in time. In addition, the developments made in wetlands have enabled the establishment of infrastructure and even settlements which would be lost in the event of total restoration. Thus, taking into account the social, economic, and even cultural context is essential for the support of local communities.

This situation of confusion is encouraged by the lack of coherence in sectoral policies. The lack of interministerial coordination and a clear ambitious national strategy to integrate nature-based solutions is also to be deplored. Han and Kuhlicke (2021) noted that in South Korea, nature-based solutions do not benefit from an adequate framework for their implementation. However, concepts such as green infrastructure, low impact development (LID), or ecosystem-based approach are used. In urban areas, Ershad Sarabi et al. (2019) spoke of institutional fragmentation ("sectoral silos") as one of the four barriers to the uptake and implementation of nature-based solutions in urban settings. In fact, each sectorial administration works in accordance with its own vision using its own means and even a language that is not shared by the others. Thus the NBS initiative and even their monitoring being intersectorial are neglected. In some cases, failures in law enforcement also act as barriers to the implementation of NBS.

On the economic level, we can note the difficulties of access to public, private, and even international financing for the implementation of NBS. The same is true of the absence of an effective mechanism for valorizing ecosystem services (water, pollination, carbon storage, etc.). African countries in general and Cameroon in particular are experiencing enormous difficulties in mobilizing financial resources implemented at the global level as economic instruments for environmental protection and restoration (Saha 2019). NBS projects are long term (Cowan et al. 2010). The public nature of benefits whose value is difficult to capture does not encourage investment (Toxopeus and Friedemann 2017). In addition, the experimental nature of NBS creates uncertainty regarding the results, making it a significant risk for investors who prefer safer investments. Cameroon, like other developing countries, devotes only a few resources to investment each year. The infrastructure needs are enormous, hence the preference for projects with quick results and high impact for the populations. The case of the Sudano-Sahelian zone of Cameroon is all the more worrying given the delay in all economic indicators.

Thus, the NBS that we have proposed are faced with a set of obstacles. Their implementation requires a paradigm shift. It is necessary to involve universities, research centers, municipalities, and other grassroots organizations to raise awareness among populations to act to preserve and enhance ecosystem services. The implementation of NBS generally involves several sectorial administrations. If these projects are integrated into regional and local development plans, this could help avoid these sectorial silos. Cameroon plans its development essentially at three levels: national, regional, and municipal. If at the political level, the NBS are adopted, this could emerge in the orientations for updating the regional plans for sustainable development of the territory. Likewise, municipal development plans ensured effective implementation at local levels. A good administrative system is then necessary to monitor these projects in the long term beyond the mandates of local elected officials.

11.5 Conclusion

The objective of this contribution was to highlight the potential of nature-based solutions for disaster risk management in the Sudano-Sahelian environment in Cameroon. It appears that NBS can be used for risk reduction as well as for adaptation. In terms of risk reduction, wetland restoration has many benefits, both in terms of reducing floods and droughts and enormous gains for ecosystems. It should be noted that this area is dominated topographically by lowlands which are the flood plains of the two main rivers in the area, the Logone and the Bénoué. Technological developments (dams and dikes) are the causes of flash floods responsible for the greatest losses. The restoration of wetlands avoids this excessive "imprisonment" of water in lakes and courses by allowing progressive flooding of the plains.

Vegetation and reforestation are other risk-reduction techniques. Its advantage is that it increases infiltration, reduces runoff, and humidifies the atmosphere. Vegetating dikes and dams help limit erosion. Likewise, more ambitious projects such as plant dikes would be advantageous. In terms of adaptation, the creation of reservoirs and overflow zones seems to be an approach that must be promoted. In the absence of lowlands as an outlet for runoff, it is necessary to set up reservoirs towards which the precipitation water should be directed. For rivers and lakes that overflow during floods, overflow zones make it possible to collect excess water to limit pressure on dams and dikes, as well as uncontrolled flooding.

References

Abioget (2017) ABIOGeT experience in stabilizing the embankment of the dike and the banks of Lake Maga for flood prevention in the Mayo-Tsanaga sub-watershed (Far North Region of Cameroon) https://www.abioget.org/img_documents/document20/m-thodologie-v-g-talisation-berges-trives-05ao-t-2018.pdf. Accessed 16 Oct 2024

Aka FT, Buh GW, Fantong WY et al (2017) Disaster prevention, disaster preparedness and local community resilience within the context of disaster risk management in Cameroon. Nat Hazards 86:57–88

Arcadis-Euroconsult et Ere développement (1997) Projet-Waza, Logone Cameroun, external evaluation mission https://iucn.org/sites/default/files/2022-05/projet_wasa_logone_cameroun_fr_1997.pdf p92 Accessed 26 Oct 2024

Arditi C (1985) Quelques réflexion socioéconomiques sur la riziculture irrigué dans le nord du Cameroun (SEMRY I et SEMRY II). Bulletin de l'Association française des anthropologues n°20. Recherche et/ou développement. Pp59–82. Accessed 12 Nov 2024

Bang HN (2022) Policy and institutional frameworks for disaster risk management in Cameroon: challenges and prospects. In: Azibo Balgah R, Ndzifon Kimengsi J (eds) Disaster management in Sub-Saharan Africa: policies, institutions and processes. Emerald Publishing Limited, Leeds, pp 97–122. https://doi.org/10.1108/978-1-80262-817-320221004

Bessong M (2012) Paléoenvironnements et diagénèse dans un réservoir gréseux d'âge crétacé du Fossé de la Bénoué au Nord Cameroun: les grès de Garoua (Doctoral dissertation, Poitiers). Accessed 5 Nov 2024

BUCREP (2010) Third general census of population and housing. Presentation report of the final results, (Central Bureau of Censuses and Population Studies, Cameroon), Yaoundé, Cameroon. p 65. Accessed 21 Oct 2024

Cohen-Shacham E, Walters G, Janzen C, Maginnis S (eds) (2016) Nature-based solutions to address global societal challenges. IUCN, Gland, p 97. https://doi.org/10.2305/IUCN.CH.2016.13.en

Cowan C, Epple C, Korn H, Schliep R, Stadler J (2010) Working with nature to tackle climate change. In Report of the Enca/BfN workshop on "Developing ecosystem-based approaches to climate change – why, what and how"

de Noray M-L (2002) Waza Logone UICN, Gland, Switzerland and Cambridge, United Kingdom, p 128. https://portals.iucn.org/library/node/8034. Accessed 12 Oct 2023

DSCN (2002) Conditions de vie des ménages et profil de pauvreté à l'Extrême-Nord Cameroun en 2001. Study carried out within the framework of PREPAFEN, p 133. Accessed 12 Sept 2024

Ekolu J, Dieppois B, Sidibe M, Eden JM, Tramblay Y, Villarini G et al (2022) Long-term variability in hydrological droughts and floods in sub-Saharan Africa: new perspectives from a 65-year daily streamflow dataset. J Hydrol 613:128359

Grigorieva E, Livenets A, Stelmakh E (2023) Adaptation of agriculture to climate change: a scoping review. Climate 11:202. https://doi.org/10.3390/cli11100202

Han S, Kuhlicke C (2021) Barriers and drivers for mainstreaming nature-based solutions for flood risks: the case of South Korea. Int J Disaster Risk Sci 12:661–672. https://doi.org/10.1007/s13753-021-00372-4

IDEA Consult (2016) Urban Planning Master Plan (PDU) for the city of Maroua. Diagnostic assessment. p 240

International Union for Conservation of Nature (IUCN) (2019) NBS: concept, practices and benefits

IPPC (2023) Synthesis report of the ipcc sixth assessment report (AR6) Longer Report, p 85. https://report.ipcc.ch/ar6syr/pdf/IPCC_AR6_SYR_LongerReport.pdf

IUCN. (2003). Waza Logone floodplain, Cameroon: economic benefits of wetland restoration. Case Studies In Wetland Valuation, p 6

Laborde S, Mahamat A, Moritz M (2018) The interplay of top-down planning and adaptive self-organization in an African floodplain. Hum Ecol 46(2):171–182

Loyer J-Y (1989) Dégradation saline des sols, induite par l'irrigation en domaine sahélien. In M Eldin & P Milleville (Eds.), Le risque en agriculture (1). IRD Editions. https://doi.org/10.4000/books.irdeditions.16266

Mbele Abbo F (2021) Erosion fluviale et mutations morphométriques du drainage dans le secteur aval du barrage de Lagdo (Bassin de la Bénoué, Nord-Cameroun). Proc IAHS 384:113–119

MINEPDED (2015) National Plan for Adaptation to Climate Change. Document prepared with the support of Japan, GIZ, le GWP et le PNUD. p 154. Accessed 10 Nov 2024

National Institute of Statistic (NIS) (2015) Multiple Indicator Cluster Survey (MICS5), 2014, Final report. Yaoundé, Cameroon. Accessed 13 Nov 2024

Ngounou Ngatcha B, Nijtchoua R, Naah E (2002) Le barrage de Lagdo (Nord-Cameroun): impact sur les plaines d'inondation de la Bénoué. In: Orange D, Arfi R, Kuper M, Morand P, Poncet Y (eds) Gestion intégrée des ressources naturelles en zones inondables tropicales. IRD; CNRST, Paris; Bamako, pp 455–474

Ntali YM, Lyimo JG, Dakyaga F (2023) Trends, impacts, and local responses to drought stress in diamare division, northern Cameroon. World Dev Sustain 2:100040

Olivry JC (1997) Long term effects of rain shortage: the ill rivers of Western and Central Africa. In: FRIEND General Report/UNESCO IHP V, Édit. UNESCO, Paris, p. 158162. Accessed 13 Nov 2024

Saha F (2019) Economic instruments for mitigating climate change in Cameroon. Geo-Eco-Trop 43(1):89–102

Saha F (2020) Les inondations dans la région de l'extrême-nord du Cameroun en contexte de changements climatiques. PhD thesis, the University of Yaounde 1, p 409

Saha F, Tchindjang M, Dzana J-G, Nguemadjita D (2020) Natural hazards in the far-north region of Cameroon and the hydrological extremes dynamics of the Chari-Logone system. Physio-Géo. https://doi.org/10.4000/physio-geo.10719

Sarabi SE, Han Q, Romme AGL, de Vries B, Wendling L (2019) Key enablers of and barriers to the uptake and implementation of nature-based solutions in urban settings: a review. Resources 8:121. https://doi.org/10.3390/resources8030121

Seignobos C, Raugel B (2000) La pêche dans le lac de Maga: localisation, effectifs et techniques. In: Seignobos C et al (eds) Atlas de la province de l'extrême nord Cameroun, p 124–127

Shaffril HAM, Samah AA, Samsuddin SF, Ahmad N, Tangang F, Sidique SFA et al (2024) Diversification of agriculture practices as a response to climate change impacts among farmers in low-income countries: a systematic literature review. Clim Serv 35:100508

Tchana M (2002) Présentation sommaire des aménagements hydrauliques au Cameroun, Yaoundé, DGRST-CRH Report

Tchindjang M, Saha F, Mbevo Fendoung P (2022) Evaluation des systèmes d'alerte précoce en république du Cameroun. Expert report from the French Red Cross, p 75

Tchuenga GTS, Tchindjang M, Ngome PIT, Degrande A, Basga SD, Saha F (2024) Agricultural innovations and adaptations to climate change in the northern Cameroon region. Sustainability 16(22):10096. https://doi.org/10.3390/su162210096

Toxopeus H, Friedemann P (2017) Characterizing nature-based solutions from a business model and financing perspective. (May). Retrieved from https://naturvation.eu/sites/default/files/news/files/naturvation_characterizing_naturebased_solutions_from_a_business_model_and_financing_perspective.pdf

United Nations Environment Programme (UNEP) (2020) The economics of nature-based solutions: current status and future priorities. United Nations Environment Programme Nairobi. Accessed 3 Nov 2024

Wandji A, Moukitek OG, Takoucheu SUB (2013) La lutte contre la sècheresse dans la Région de l'Extrême-Nord du Cameroun 1975–2012. DIPES II dissertation of the ENS of Maroua, p 140

Wesseling J, Drijver C, Naah E, Namba ZA (1994) Waza Logone flood restoration study: identification of options for Reflooding. Delft Hydraulics and Centre of Environmental Science, Leyden

Chapter 12
Avian Diversity and Feeding Guild of Inland Wetland in a Semiarid Region: Implications for Wetland Conservation and Restoration

Saurabh Purohit, Krishanmurti, Manendra Singh, Omkar M. Hande, Ashwini M. Wadhu, and Y. D. Imran Khan

Abstract Wetlands provide numerous ecosystem services and are critical foraging, roosting, and breeding habitats for resident and migratory birds. Climate change, wetland fragmentation and degradation, and anthropogenic pressure continuously impose threats on avian diversity. This chapter covers a case study on avifaunal diversity and feeding guilds in the inland wetland of the semiarid region of the Deccan Plateau, India. A total of 81 bird species belonging to 15 orders and 37 families, from which 68 resident and 13 migratory species were recorded and identified. Among the avian species, two species—Black-headed Ibis (*Threskiornis melanocephalus)* and Asian Woolly-necked Stork (*Ciconia episcopus*)—are near threatened (NT), and one species (River tern *Sterna aurantia*) is vulnerable (VU), and four species—Black-winged Kite (*Elanus caeruleus)*, Brahminy Kite (*Haliastur indus)*, Indian Peafowl (*Pavo cristatus),* and Eurasian Spoonbill (*Platalea leucorodia*)—are categorized as Schedule I of the Wildlife Protection Act (1972). An analysis of the feeding guilds of avifauna revealed that 33% (27 species) were carnivorous, 27% (22 species) were insectivorous, and 17% (14 species) were omnivorous. The present study is the first assessment of the avifaunal diversity in Koulasnala wetland. It provides a foundation for long-term studies on the conservation of the region's VU and NT bird species through ecosystem-based adaptation (EbA). Then, conservation plans can be implemented in other wetlands located in semiarid regions.

Keywords Wetlands · Avifaunal diversity · Feeding guilds · Semiarid region

---

S. Purohit · Krishanmurti · M. Singh · O. M. Hande · A. M. Wadhu · Y. D. I. Khan (✉)
WOTR Centre for Resilience Studies (W-CReS), Watershed Organisation Trust (WOTR), Pune, Maharashtra, India
e-mail: saurabhpurohit10@gmail.com; ikeducationkhan08@gmail.com

Y. D. I. Khan et al. (eds.), *Land Restoration Through Ecosystem-Based Approach*, Disaster Resilience and Green Growth, https://doi.org/10.1007/978-981-95-0389-6_12

12.1 Introduction

Wetlands are an intricate web of complex interactions between aquatic and terrestrial ecosystems. These inland water bodies are biologically diverse and the most productive ecosystem, sustaining under diverse climatic and geological conditions (Wu and Chen 2020; Mandishona and Knight 2022; Singh et al. 2022; Wasserman and Dalu 2022). The Ramsar Convention on Wetlands (RCW) estimated that wetlands cover 12.1 million square kilometers (9% of the global geographical area), of which 54% are permanent and 46% are seasonally inundated (Davidson et al. 2023; RCW 2018). A total of 1308 wetlands (Ramsar wetland–89, significant wetland–111, and other wetlands–1108) are present in India (www.indianwetlands.in), covering 3–4% of the Indian geographical area (Sandilyan 2022). Wetlands provide provisional (food, fresh water, fiber, fuel, medicinal, genetic) (Pal and Debanshi 2023), regulating (climate regulation, water regulation (hydrological flows), erosion control, natural hazard protection) (Zhang et al. 2021), cultural (recreational, aesthetic, educational, spiritual, and inspirational) (Margaryan et al. 2022), and supporting services (nutrient cycling, soil formation) (Mushet and Roth 2020). Wetlands have also been dubbed "nature's supermarkets" due to the extensive food chain and profuse biodiversity that they support (Mitsch et al. 2015).

Inland wetlands are classified into natural inland wetlands and artificial or man-made inland wetlands. Man-made inland wetlands are crucial for biodiversity conservation and provide stopover and breeding habitat for migratory birds. Moreover, these critically important ecosystems face challenges and threats from climate change, anthropogenic activities, and the alteration of ecosystem processes (Gajera et al. 2012; Pal and Singha 2023). About ~81% of inland wetland species populations and 36% of coastal or marine species have declined globally since 1970 (RCW 2018). Thus, over 10% of inland and coastal wetland species are highly threatened.

The wetland avifauna occupy multiple trophic levels and play a crucial role in the mass and energy fluxes between aquatic and terrestrial food chains/webs of wetland nutrient cycles, making them an integral part of the wetland ecosystem (Rajashekara and Venkatesha 2018; Panda et al. 2021; Pattnaik et al. 2022). Avifauna are dependent on wetlands for feeding, breeding, nesting, and raising their young ones (Woldemariam et al. 2018; Vishwakarma et al. 2021). Avian species are sensitive to environmental vagaries and indicators for ecological research, and they are considered useful markers of biodiversity in a particular area (Alexandrino et al. 2016; Barik et al. 2022). Birds play an important part in disseminating ecosystem services (Kronenberg 2014), and they depend on ecological processes for survival. In India, 1332 species belonging to 485 genera, 113 families, and 26 orders have been reported (Praveen et al. 2020), of which 78 species of birds are endemic to India, constituting about 6% of the country's bird diversity. About 380 species under 221 genera, 82 families, and 22 orders are reported from the Indian state of Telangana. Earlier studies quantified avifaunal diversity in Indian wetlands (Datta 2011; Kumar and Gupta 2013; Balkhande et al. 2014; Lakshmi et al. 2015; Jalander et al. 2015; Mazumdar 2019; Gayen et al. 2020; Neelgund and Kadadevaru 2020; Singh et al. 2020; Panda et al. 2021; Vishwakarma et al. 2021; Mandal et al. 2022; Pattnaik et al. 2022).

Dryland waterbodies, known for their resilience to extreme conditions, are increasingly affected by human activities and climate change (Parra et al. 2021). Consequently, the avian habitat for nesting and shelter is also disrupted; thus, the migratory species population is continuously decreasing (Nautiyal et al. 2020). Therefore, seed dispersal, plant-bird mutualism, significantly declined (Young et al. 2012). A significant challenge lies in the delayed societal and political responses to scientific warnings, hindering timely mitigation efforts. Thus, quantifying avian diversity in dryland wetlands is crucial for assessing ecosystem health and conservation needs, as birds serve as bioindicators of environmental changes and habitat quality (Maclean et al. 2014). Restoring dryland wetlands and avian diversity through an ecosystem-based adaptation (EbA) approach focused on rebuilding ecological balance by restoring water flows, recharging groundwater, and reconnecting floodplains (Mitsch and Gosselink 2015). Replanting native vegetation stabilizes soil and enhances bird habitat quality (Davidson 2016) while managing invasive species and diversifying wetland zones to support resident and migratory avifauna (Sekercioglu 2006). Community involvement and sustainable land-use practices help to protect wetlands as avian habitats while also supporting livelihoods (Haque et al. 2021). Regular bird monitoring guides conservation strategies, making EbA a practical, long-term solution for wetland restoration.

12.2 Materials and Methods

12.2.1 Study Area

Koulasnala wetland is an existing medium irrigation project across the Koulasnala River near Sawargaon village and Jukkal mandal of Kamareddy district, Telangana (Fig. 12.1). The project was commissioned in the year 1999 to benefit 19 villages in Jukkal and Bichkunda mandals in the Godavari basin. It has a gross storage of 35.04 Mcum with a maximum flood discharge of 1869 m^3/s (Cumec: a cubic meter per second) and an ayacut of 9000 acres. This study area experiences a tropical climate, is geographically located in a semiarid region, and has mostly hot and dry climates (Dayanandam et al. 2021). Temperature ranges from 4–45 °C in summer to 15–29 °C in the winter season (Edukondal et al. 2019). The wetland is dominated by Bush Morning Glory *(Ipomoea carnea)* on all sides. Other species are pale knotweed (*Persicaria lapathifolia)*, elephant grass (*Typha domingensis)*, and purple nutsedge (*Cyperus rotundus)*. Several climbers are also observed, e.g., velvet bean (*Mucuna pruriens)*, ceylon caper (*Capparis zeylanica),* etc., with scattered tree species, e.g., gum arabic (*Acacia nilotica)*, tamarind (*Tamarindus indica)*, Flame of Forest (*Butea monosperma)*, Sacred Fig (*Ficus religiosa)*, Tanner's Cassia (*Senna auriculata)*, etc. The marshy area and water provide a suitable habitat for the avifaunal diversity.

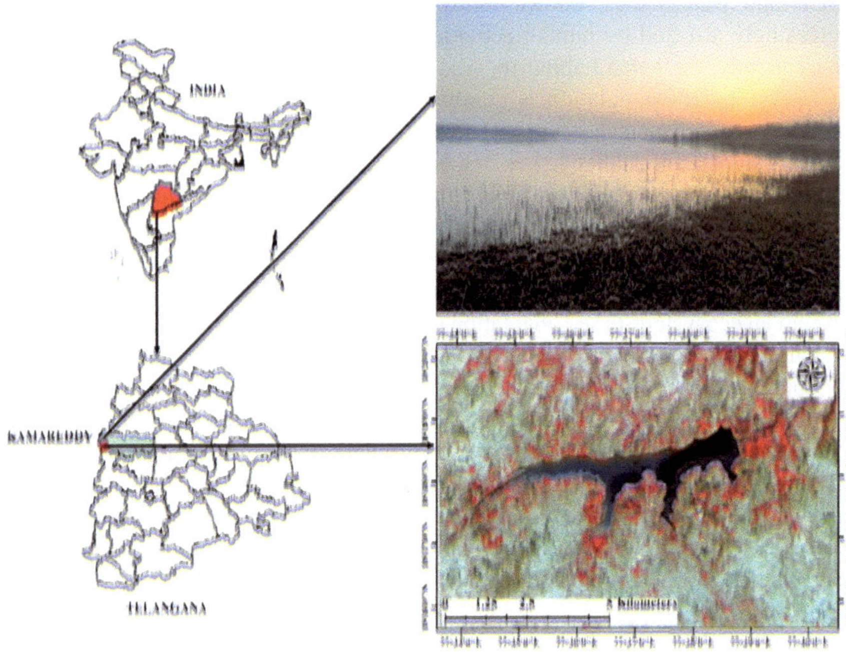

Fig. 12.1 Map showing the location of Koulasnala wetland on Landsat 9 FCC (19/02/23)

12.2.2 Methodology

Observations were made in February 2023, in two stages: the first stage took place during trips to the wetland from 06:30 to 09:30 in the morning, and the second stage took place in the evening from 16:00 to 18:00. February has the highest species richness due to the arrival of the migratory birds (Ulman and Singh 2022). The direct observation method was used to assess the overall number of bird species in and around the wetland (Nautiyal et al. 2015). Observations were made using a Croma binocular 10 × 42, and the Nikon P900 bridge camera took photos. The Merlin bird identification application (Horn et al. 2015), which is based on a deep learning neural network (Weinstein 2018), was used in combination with a bird identification guide for the Indian subcontinent (Grimmett et al. 2013). Proper distance between the birds and the observer was maintained during the observation because birds are extremely sensitive to human activities (Vishwakarma et al. 2021). The relative diversity index (RDI) of families was calculated adopting the following formula (La Torre-Cuadros et al. 2007):

$$\text{RDI} = \frac{\text{Number of bird species in a family}}{\text{Total number of species}} \times 100$$

12.3 Results and Discussion

In the present survey, 81 avifauna species belonging to 15 orders and 37 families were recorded (Table 12.1). In this, 68 resident and 13 migratory species were recorded, including two near threatened (NT) species, which are the Black-headed Ibis (*Threskiornis melanocephalus*), the Asian Woolly-necked Stork (*Ciconia episcopus*), and one vulnerable (VU) River tern (*Sterna aurantia*) species. According to the Wildlife Protection Act (1972), four species fall in the Schedule I—Black-winged Kite (*Elanus caeruleus*), Brahminy Kite (*Haliastur indus*), Indian Peafowl (*Pavo cristatus*), and Eurasian Spoonbill (*Platalea leucorodia*). All three species of Ibis—Black-headed Ibis (*Threskiornis melanocephalus*), Red-naped Ibis (*Pseudibis papillosa*), and Glossy Ibis (*Plegadis falcinellus*)—can be seen in this wetland. In our analysis, the dominant avifauna species were found in the order Passeriformes, 30.9% ($n = 25$), followed by Pelecaniformes, 14.8% ($n = 12$), Anseriformes, 9.9% ($n = 8$), and Charadriiformes, 9.9% ($n = 8$) (Table 12.2).

The present study found that out of 37 families, Anatidae and Ardeidae have maximum number of species and RDI value, i.e., eight species (RDI = 9.88%), followed by Motacillidae with five species (RDI = 6.17%), Columbidae and Threskiornithidae with four species each (RDI = 4.94%), Alcedinidae, Hirundinidae Muscicapidae, Phasianidae and Rallidae with three species each (RDI = 3.7%), Accipitridae, Alaudidae, Charadriidae, Ciconiidae, Cuculidae, Laridae, Leiothrichidae with two species each (RDI = 2.47%), Aegithinidae, Bucerotidae, Campephagidae, Cisticolidae, Coraciidae, Dicruridae, Estrildidae, Laniidae, Megalaimidae, Nectariniidae, Ploceidae, Podicipedidae, Pycnonotidae, Recurvirostridae, Rostratulidae, Sturnidae and Upupidae with one species each (RDI = 1.23%), respectively (Table 12.3). Ulman and Singh (2022) also found that Anatidae is the dominant family in the unprotected wetlands of the Ayodhya district of Uttar Pradesh, India.

Previous studies have also estimated the avifaunal diversity in different wetlands in India. Jalander et al. (2015) reported 65 bird species belonging to 38 families in the Nizam Sagar Project, at Nizamabad district of Telangana, around 40 km from the present study area. Mohammad and Krishna (2016) reported 40 bird species in the Wyra and Paler reservoirs of Khammam district, Telangana, India. Sailu et al. (2017) reported 166 bird species from Ameenpur Lake, Telangana state. Rao et al. (2014) reported 232 species of birds in Kolleru Wetland, Andhra Pradesh. Prasad et al. (2014) reported 164 bird species belonging to 53 families in the Manjeera Wildlife Sanctuary, Telangana. Harisha and Hosetti (2018) reported the occurrence of a total of 51 wetland bird species belonging to 15 families and eight orders in Komaranahalli Lake, Davanagere, Karnataka.

Birds' feeding guilds reveal crucial information about species' ecological relationships and are thus valuable for assessing the ecological drivers of community change in scientific studies. An analysis of the feeding guilds of these birds revealed that 33% (27 species) were carnivorous, 27% (22 species) were insectivorous, 17%

Table 12.1 Status of avifaunal diversity in Koulasnala Project, Telangana

S. no	Common name	Scientific name	Feeding guilds	IUCN status	WPA status
	Order: Accipitriformes; family: Accipitridae				
1	Black-winged Kite	*Elanus caeruleus*	C	LC	Schedule I
2	Brahminy Kite	*Haliastur indus*	C	LC	Schedule I
	Order: Anseriformes; family: Anatidae				
3	Green-winged Teal	*Anas crecca*	H	LC	Schedule IV
4	Garganey	*Spatula querquedula*	O	LC	Schedule IV
5	Northern Shoveler	*Spatula clypeata*	O	LC	Schedule IV
6	Eurasian Wigeon	*Mareca penelope*	H	LC	Schedule IV
7	Bar-headed Goose	*Anser indicus*	H	LC	Schedule IV
8	Northern Pintail	*Anas acuta*	H	LC	Schedule IV
9	Indian Spot-billed Duck	*Anas poecilorhyncha*	H	LC	Schedule IV
10	Cotton Pygmy Goose	*Nettapus coromandelianus*	O	LC	Schedule IV
	Order: Bucerotiformes; family: Anatidae				
11	Indian Grey Hornbill	*Ocyceros birostris*	O	LC	Schedule IV
	Order: Bucerotiformes; family: Upupidae				
12	Eurasian Hoopoe	*Upupa epops*	O	LC	Schedule IV
	Order: Charadriiformes; family: Charadriidae				
13	Little Ringed Plover	*Charadrius dubius*	I	LC	Schedule IV
14	Red-wattled Lapwing	*Vanellus indicus*	MG	LC	Schedule IV
	Order: Charadriiformes; family: Laridae				
15	River Tern	*Sterna aurantia*	C	VU	Schedule IV
16	Whiskered Tern	*Chlidonias hybrida*	C	LC	Schedule IV
	Order: Charadriiformes; family: Recurvirostridae				
17	Black-winged Stilt	*Himantopus himantopus*	I	LC	Schedule IV
	Order: Charadriiformes; family: Rostratulidae				
18	Greater Painted-snipe	*Rostratula benghalensis*	C	LC	Schedule IV
	Order: Charadriiformes; family: Scolopacidae				
19	Wood Sandpiper	*Tringa glareola*	C	LC	Schedule IV
20	Marsh Sandpiper	*Tringa stagnatilis*	C	LC	Schedule IV
	Order: Ciconiiformes; family: Ciconiidae				
21	Asian Openbill	*Anastomus oscitans*	C	LC	Schedule IV
22	Asian Woolly-necked Stork	*Ciconia episcopus*	C	NT	Schedule IV
	Order: Columbiformes; family: Columbidae				
23	Laughing Dove	*Spilopelia senegalensis*	G	LC	Schedule IV
24	Eurasian Collared- Dove	*Streptopelia decaocto*	G	LC	Schedule IV
25	Yellow-footed Green-Pigeon	*Treron phoenicopterus*	G	LC	Schedule IV

(continued)

Table 12.1 (continued)

S. no	Common name	Scientific name	Feeding guilds	IUCN status	WPA status
26	Rock Pigeon	*Columba livia*	G	LC	Schedule IV
	Order: Coraciiformes; family: Alcedinidae				
27	White-throated Kingfisher	*Halcyon smyrnensis*	C	LC	Schedule IV
28	Common Kingfisher	*Alcedo atthis*	C	LC	Schedule IV
29	Pied Kingfisher	*Ceryle rudis*	C	LC	Schedule IV
	Order: Coraciiformes; family: Meropidae				
30	Asian Green Bee-eater	*Merops orientalis*	I	LC	Schedule IV
31	Blue-tailed Bee-eater	*Merops philippinus*	I	LC	Schedule IV
	Order: Coraciiformes; family: Coraciidae				
32	Indian Roller	*Coracias benghalensis*	C	LC	Schedule IV
	Order: Cuculiformes; family: Cuculidae				
33	Greater Coucal	*Centropus sinensis*	I	LC	Schedule IV
34	Common Hawk-Cuckoo	*Hierococcyx varius*	NR	LC	Schedule IV
	Order: Galliformes; family: Phasianidae				
35	Jungle Bush-Quail	*Perdicula asiatica*	NR	LC	Schedule IV
36	Indian Peafowl	*Pavo cristatus*	O	LC	Schedule I
37	Grey Francolin	*Ortygornis pondicerianus*	NR	LC	Schedule IV
	Order: Gruiformes; family: Rallidae				
38	Eurasian Coot	*Fulica atra*	H	LC	Schedule IV
39	White-breasted Waterhen	*Amaurornis phoenicurus*	O	LC	Schedule IV
40	Purple Swamphen	*Porphyrio porphyrio*	O	LC	Schedule IV
	Order: Passeriformes; family: Aegithinidae				
41	Common Iora	*Aegithina tiphia*	O	LC	Schedule IV
	Order: Passeriformes; family: Alaudidae				
42	Rufous-tailed Lark	*Ammomanes phoenicura*	O	LC	Schedule IV
43	Ashy-crowned Sparrow-Lark	*Eremopterix griseus*	I	LC	Schedule IV
	Order: Passeriformes; family: Campephagidae				
44	Small Minivet	*Pericrocotus cinnamomeus*	I	LC	Schedule IV
	Order: Passeriformes; family: Cisticolidae				
45	Ashy Prinia	*Prinia socialis*	I	LC	Schedule IV
	Order: Passeriformes; family: Dicruridae				
46	Black Drongo	*Dicrurus macrocercus*	I	LC	Schedule IV
	Order: Passeriformes; family: Estrildidae				
47	White-throated Munia	*Euodice malabarica*	G	LC	Schedule IV
	Order: Passeriformes; family: Hirundinidae				
48	Barn Swallow	*Hirundo rustica*	I	LC	Schedule IV

(continued)

Table 12.1 (continued)

S. no	Common name	Scientific name	Feeding guilds	IUCN status	WPA status
49	Red-rumped Swallow	*Cecropis daurica*	I	LC	Schedule IV
50	Wire-tailed Swallow	*Hirundo smithii*	I	LC	Schedule IV
	Order: Passeriformes; family: Laniidae				
51	Long-tailed Shrike	*Lanius schach*	I	LC	Schedule IV
	Order: Passeriformes; family: Leiothrichidae				
52	Common Babbler	*Argya caudata*	O	LC	Schedule IV
53	Jungle Babbler	*Argya striata*	O	LC	Schedule IV
	Order: Passeriformes; family: Motacillidae				
54	Western Yellow Wagtail	*Motacilla flava*	I	LC	Schedule IV
55	Grey Wagtail	*Motacilla cinerea*	I	LC	Schedule IV
56	White Wagtail	*Motacilla alba*	I	LC	Schedule IV
57	Paddyfield Pipit	*Anthus rufulus*	I	LC	Schedule IV
58	White-browed Wagtail	*Motacilla maderaspatensis*	I	LC	Schedule IV
	Order: Passeriformes; family: Muscicapidae				
59	Oriental Magpie-Robin	*Copsychus saularis*	I	LC	Schedule IV
60	Indian Robin	*Copsychus fulicatus*	I	LC	Schedule IV
61	Pied Bushchat	*Saxicola caprata*	I	LC	Schedule IV
	Order: Passeriformes; family: Nectariniidae				
62	Purple Sunbird	*Cinnyris asiaticus*	N	LC	Schedule IV
	Order: Passeriformes; family: Ploceidae				
63	Baya Weaver	*Ploceus philippinus*	O	LC	Schedule IV
	Order: Passeriformes; family: Pycnonotidae				
64	Red-vented Bulbul	*Pycnonotus cafer*	I	LC	Schedule IV
	Order: Passeriformes; family: Sturnidae				
65	Brahminy Starling	*Sturnia pagodarum*	O	LC	Schedule IV
	Order: Pelecaniformes; family: Ardeidae				
66	Grey Heron	*Ardea cinerea*	C	LC	Schedule IV
67	Purple Heron	*Ardea purpurea*	C	LC	Schedule IV
68	Indian Pond-Heron	*Ardeola grayii*	C	LC	Schedule IV
69	Cattle Egret	*Bubulcus ibis*	MG	LC	Schedule IV
70	Intermediate Egret	*Ardea intermedia*	C	LC	Schedule IV
71	Great Egret	*Ardea alba*	C	LC	Schedule IV
72	Little Egret	*Egretta garzetta*	C	LC	Schedule IV
73	Black-crowned Night-Heron	*Nycticorax nycticorax*	C	LC	Schedule IV
	Order: Pelecaniformes; family: Threskiornithidae				
74	Black-headed Ibis	*Threskiornis melanocephalus*	C	NT	Schedule IV
75	Glossy Ibis	*Plegadis falcinellus*	C	LC	Schedule IV
76	Red-naped Ibis	*Pseudibis papillosa*	C	LC	Schedule IV

(continued)

Table 12.1 (continued)

S. no	Common name	Scientific name	Feeding guilds	IUCN status	WPA status
77	Eurasian Spoonbill	*Platalea leucorodia*	C	LC	Schedule I
	Order: Piciformes; family: Megalaimidae				
78	Coppersmith Barbet	*Psilopogon haemacephalus*	F	LC	Schedule IV
	Order: Podicipediformes; family: Podicipedidae				
79	Little Grebe	*Tachybaptus ruficollis*	C	LC	Schedule IV
	Order: Suliformes; family: Phalacrocoracidae				
80	Little Cormorant	*Microcarbo Niger*	C	LC	Schedule IV
81	Great Cormorant	*Phalacrocorax carbo*	C	LC	Schedule IV

Table 12.2 Order wise total number of avian species and their status

Order	Resident	Migratory	Species no.	Percent
Passeriformes	21	4	25	30.9
Pelecaniformes	12	0	12	14.8
Anseriformes	2	6	8	9.9
Charadriiformes	6	2	8	9.9
Coraciiformes	6	0	6	7.4
Columbiformes	4	0	4	4.9
Galliformes	3	0	3	3.7
Gruiformes	2	1	3	3.7
Accipitriformes	2	0	2	2.5
Bucerotiformes	2	0	2	2.5
Ciconiiformes	2	0	2	2.5
Cuculiformes	2	0	2	2.5
Suliformes	2	0	2	2.5
Piciformes	1	0	1	1.2
Podicipediformes	1	0	1	1.2

(14 species) were omnivorous, 9% (seven species) were herbivorous, 6% (five species) were granivorous, 2% (two species) were multiple guilds and not recorded, and 1% (one species) were frugivorous and nectarivorous, respectively (Fig. 12.2). Harisha et al. (2021) reported that insectivorous, carnivorous, and omnivorous were the dominant feeding guilds of birds in Daroji Sloth Bear Sanctuary, Karnataka. Jangral and Vashishat (2022) reported that carnivorous and insectivorous feeding guilds comprise approximately half of the feeding guilds in Keshopur Chhamb Community Reserve, Punjab.

The presence of carnivorous bird species indicates the presence of fish, invertebrates, amphibians, and reptiles in the area. The Black-winged kite was seen as attacking the Indian Spot-billed Duck (*Anas poecilorhyncha*). Kingfishers and River terns were also seen in search of their prey. The larger birds, such as the Grey Heron (*Ardea cinerea*), Purple Heron (*Ardea purpurea*), and Glossy ibis, were

Table 12.3 Relative diversity index (RDI) of various avian families at Koulasnala wetland, Telangana

Family	No. of species	Relative diversity index (RDI)
Anatidae	8	9.88
Ardeidae	8	9.88
Motacillidae	5	6.17
Columbidae	4	4.94
Threskiornithidae	4	4.94
Alcedinidae	3	3.7
Hirundinidae	3	3.7
Muscicapidae	3	3.7
Phasianidae	3	3.7
Rallidae	3	3.7
Accipitridae	2	2.47
Alaudidae	2	2.47
Charadriidae	2	2.47
Ciconiidae	2	2.47
Cuculidae	2	2.47
Laridae	2	2.47
Leiothrichidae	2	2.47
Meropidae	2	2.47
Phalacrocoracidae	2	2.47
Scolopacidae	2	2.47
Aegithinidae	1	1.23
Bucerotidae	1	1.23
Campephagidae	1	1.23
Cisticolidae	1	1.23
Coraciidae	1	1.23
Dicruridae	1	1.23
Estrildidae	1	1.23
Laniidae	1	1.23
Megalaimidae	1	1.23
Nectariniidae	1	1.23
Ploceidae	1	1.23
Podicipedidae	1	1.23
Pycnonotidae	1	1.23
Recurvirostridae	1	1.23
Rostratulidae	1	1.23
Sturnidae	1	1.23
Upupidae	1	1.23

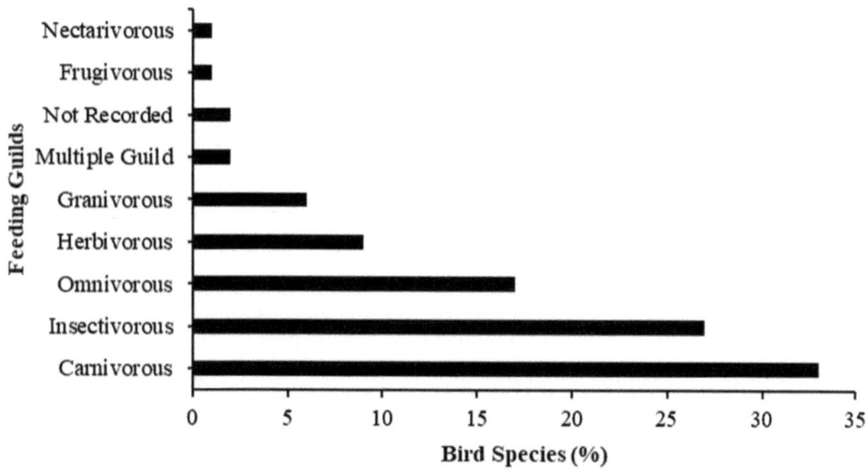

Fig. 12.2 Feeding guilds (%) of birds at Koulasnala wetland

observed standing and waiting for the prey. Grey Herons prefer larger fish over smaller ones (Jangral and Vashishat 2022). It was also observed that the Asian openbills were mostly feeding upon molluscs. Little wading birds, such as Cattle egrets (*Bubulcus ibis)* and Red-wattled lapwings (*Vanellus indicus)*, were observed foraging in shallow water. This is likely due to the ease with which prey may be captured in such an environment and the greater concentration of prey along the shoreline. Insectivorous bird communities act as natural insect pest control in agriculture, horticulture, and forest ecosystems. The primary source of nutrition for herbivorous bird groups was vegetation, which included herbs, shrubs, shoots, roots, and leaves of plants. The Eurasian coots (*Fulica atra)* were observed foraging on land and plunging into the water for food. Eurasian wigeon (*Mareca penelope)* and Bar-headed goose (*Anser indicus)* were mostly seen as roosting. The vegetation surrounding the reservoir provides the ideal habitat for the avifauna. Even though there are villages all around the wetland, we did not see any locals hunting there.

Avifauna is crucial to maintaining ecosystem health and balance by contributing to seed dispersal, habitat regeneration, pest control, nutrient cycling, and ecological stability. About 25% of ecosystem services are provided by common birds. However, the levels of provision of different services highly depend on the relationship, which is generally linear and strongly asymptotic (Gaston 2022). Frugivorous birds aid in seed dispersal by consuming fruits and excreting seeds, promoting regeneration, plant and genetic diversity, and habitat connectivity, which is crucial in dryland wetlands (Tella et al. 2020). Insectivorous birds, such as swallows, flycatchers, and woodpeckers, act as natural pest regulators (Sekercioglu 2006). Avian species contribute to nutrient cycling through guano supply that enriches the soil with nitrogen and phosphorus, and significantly increases soil conductivity by adding NO_3^-, NH_4^+, K^+, and PO_4^{3-} ions concentrations (Zwolicki et al. 2013). Waterbirds also enhance wetland productivity by redistributing nutrients (Manny et al. 1994), while scavengers such as vultures accelerate decomposition and control diseases. Nectar-feeding birds, like hummingbirds, support pollination and biodiversity, especially in arid regions where pollinators are very less (Fenster et al. 2004). The relationship between waterbirds and wetland habitats is affirmed for the value of wetland ecological restoration and conservation (Qiu et al. 2024). Bird populations and species are influenced by the water level of a wetland (McNicholl 2018). Inland wetland losses primarily occurred due to water abstraction and drainage for agriculture, causing hydrological droughts (Hu et al. 2017). This effect on avifauna and aquatic microbial communities is more pronounced in arid and semiarid wetlands (Santofimia et al. 2024). Therefore, native biodiversity protection plans like REDD+ are crucial to reducing land degradation (Siril et al. 2022). In arid or semiarid regions, precipitation seasonally occurs and is highly variable and substantially less than the evaporation rate; thus, groundwater discharge is a major factor for balancing water salinity and wetland ecology (Jolly et al. 2008). Wetland degradation is an undeniable reality usually caused by anthropogenic activities, which are considered the main factor (Hu et al. 2017). Globally, wetland loss was about 64–71% from the nineteenth century to the early twenty-first century (Davidson 2014). Therefore, the International Union for Conservation of Nature (IUCN) highlighted that EbA for wetlands includes restoration, management, and conservation plans to help different communities adapt to climate change (IUCN 2017). The EbA for wetlands can be achieved by flood protection, water and food security, habitat conservation, and strategies for climate change mitigation (IUCN 2017; Nyathi et al. 2024; Sack and Thorn 2024). The EbA approach can help to increase or restore avifaunal diversity in dryland by habitat restoration, ecosystem management, and reforestation (Colls et al. 2009; Gauthier et al. 2021). Moreover, birds' habitat tracking and information on micro-evolution help to create effective EbA plans (Jacques-Hamilton 2025).

The present results showed that the avifaunal biodiversity is rich in the Koulasnala wetland. Thus, wetland is suitable for ecotourism through birdwatching as a cultural service with lower investment costs, more dependency, and fewer negative side effects on the target environment, with ecotourists' connection to birdwatching activities and the development of natural empathy (Liu et al. 2021). Moreover, the wetland area can become an important tool for the nature-based solution (NbS) in conserving natural resources. Birdwatching can improve the economic status of the surrounding villages and encourage them to conserve the wetland. Since this wetland provides water for irrigation and household needs, there must be a balance between different livelihood options (agriculture, fishery, and ecotourism) with the common objective of the development of a sustainable ecosystem.

12.4 Conclusion

The present study provides preliminary data on the avifaunal diversity of the wetland. This will attract more birdwatchers and diversify the livelihood options for the surrounding villages. The complementary relationship between different livelihood options will help provide a conducive environment for biodiversity to thrive. More extensive long-term scientific investigations and monitoring are required to further understand the ecological state, seasonal abundance, diversity, and conservation of birds in this specific location.

Acknowledgments Authors express heartfelt gratitude to HSBC Software Development (India) Private Limited for their generous help in funding the project. Special thanks to Dr. Marcella D'Souza, Director, W-CReS, and Mr. Dipak Zade, Thematic Lead (Social Science), for their unwavering support and encouragement. The authors would also want to acknowledge the support of the management of WOTR and its implementation team in Telangana (Jyothirmayee Kandula, M. Raviprasad, Vadithya Sudhakar, and Jagdish Patil). Most importantly, this study would not have been completed without the active support from the community of the villages (Babulgaon and adjacent villages) around the wetland. Finally, the authors would like to thank W-CReS colleagues, without whom this research work would not have been possible.

Annexure: Birds of Koulasnala

Black-headed Ibis

Red-naped Ibis

Glossy Ibis

Asian Openbill

Asian Woolly-necked Stork

Grey Heron

River Tern

Greater Painted-Snipe

Ashy-crowned Sparrow-Lark

Green-winged Teal

Bar-headed Goose

Northern Pintail

References

Alexandrino ER, Buechley ER, Piratelli AJ, de Barros KMPM, de Andrade MR, Şekercioğlu ÇH, Silva WR et al (2016) Bird sensitivity to disturbance as an indicator of forest patch conditions: an issue in environmental assessments. Ecol Indic 66:369–381. https://doi.org/10.1016/j.ecolind.2016.02.006

Balkhande JV, Jalander V, Sainath MG (2014) Check list of birds from Satapur water body, Renjal Mandal Dist. Nizamabad, Telangana. Int J Sci Res 3:549–551

Barik S, Saha GK, Mazumdar S (2022) Influence of land cover features on avian community and potential conservation priority areas for biodiversity at a Ramsar site in India. Ecol Process 11:25. https://doi.org/10.1186/s13717-022-00369-x

Colls A, Ash N, Ikkala N (2009) Ecosystem-based adaptation: a natural response to climate change. IUCN, Gland, p 16

Datta T (2011) Human interference and avifaunal diversity of two wetlands of Jalpaiguri, West Bengal, India. J Threat Taxa 3:2253–2262

Davidson NC (2014) How much wetland has the world lost? Long-term and recent trends in global wetland area. Mar Freshw Res 65(10):934–941

Davidson NC (2016) Wetland losses and the status of wetland-dependent species. Mar Freshw Res 67(5):699–709

Davidson NC, Finlayson CM, Fluet-Chouinard E (2023) The extent and distribution of the world's wetlands. In: Ramsar Wetlands, Elsevier, pp 91–114

Dayanandam G, Rao RYV, Raju ARN, Geetha R, Sudha K, Sudhakar G et al. (2021) Weather and climatology of Telangana. Telangana State Development Planning Society (TSDPS)/ Directorate of Economics & Statistics (DES), Hyderabad, pp 47–52

Edukondal A, Ramu M, Krishna GH, Muralidhar M (2019) Geophysical investigation for groundwater in Jukkal-Bichukonda area in Nizamabad district, Telangana state, India: a case study. Int J Innov Res Technol 6:261–268

Fenster CB, Armbruster WS, Wilson P, Dudash MR, & Thomson JD (2004) Pollination syndromes and floral specialization. Annu. Rev. Ecol. Evol. Syst., 35(1):375-403

Gajera NB, Mahato AKR, Kumar VV (2012) Wetland birds of arid region: a study on their diversity and distribution pattern in Kachchh. Columban J Life Sci 13(1–2):47

Gaston KJ (2022) Birds and ecosystem services. Curr Biol 32(20):R1163–R1166

Gauthier S, May B, Vasseur L (2021) Ecosystem-based adaptation to protect avian species in coastal communities in the Greater Niagara region, Canada. Climate 9(6):91

Gayen J, Ghosh PK, Datta D (2020) Avifauna habitability of wetlands in an intensive cultural landscape: insights from the Ichhamati floodplains, India. Environ Monit Assess 192:329. https://doi.org/10.1007/s10661-020-08306-1

Grimmett R, Inskipp C, Inskipp T (2013) Birds of the Indian Subcontinent: India, Pakistan, Sri Lanka, Nepal, Bhutan, Bangladesh and the Maldives. Bloomsbury Publisher, pp 1–527

Haque MM, Rahman MA, Das SK (2021) Avian diversity and habitat use in a tropical wetland: implications for conservation. Wetl Ecol Manag 29(1):89–102

Harisha MN, Hosetti BB (2018) Status and conservation issues of wetland birds in Komaranahalli Lake, Davanagere District, Karnataka, India. J Threat Taxa 10:11290–11294

Harisha MN, Samad KA, Hosetti BB (2021) Conservation status, feeding guilds, and diversity of birds in Daroji sloth bear sanctuary, Karnataka, India. J Threat Taxa 13:18738–18751

Horn VG, Branson S, Farrell R, Haber S, Barry J, Ipeirotis P, Perona P, Belongie S (2015) Building a bird recognition app and large scale dataset with citizen scientists: the fine print in fine-grained dataset collection. In 2015 conference on computer vision and pattern recognition (CVPR), pp 595–604. https://doi.org/10.1109/CVPR.2015.7298658

Hu S, Niu Z, Chen Y, Li L, Zhang H (2017) Global wetlands: potential distribution, wetland loss, and status. Sci Total Environ 586:319–327

IUCN (2017) Ecosystem-based Adaptation, 28 rue Mauverney, CH-1196 Gland, Switzerland, https://iucn.org/resources/issues-brief/ecosystem-based-adaptation

Jacques-Hamilton R, Valcu M, Kwon E, Versluijs TS, & Kempenaers B (2025) Measurement error in remotely sensed fractional snow cover datasets: implications for ecological research. Environmental Research: Ecology, 4(1):015005

Jalander V, Balkhande JV, Sainath G (2015) Avifaunal diversity of Nizam Sagar project, Dist. Nizamabad, Telangana state. Int J Pure Appl Biosci 3:487–491

Jangral S, Vashishat N (2022) Feeding guild structure of birds at Keshopur Chhamb wetland, Gurdaspur. Indian J Entomol:1–6

Jolly ID, McEwan KL, & Holland KL (2008) A review of groundwater–surface water interactions in arid/semi-arid wetlands and the consequences of salinity for wetland ecology. Ecohydrology: Ecosystems, Land and Water Process Interactions, Ecohydrogeomorphology, 1(1):43–58

Kronenberg J (2014) Environmental impacts of the use of ecosystem services: case study of birdwatching. Environ Manag 54:617–630. https://doi.org/10.1007/s00267-014-0317-8

Kumar P, Gupta SK (2013) Status of wetland birds of Chhilchhila Wildlife Sanctuary, Haryana, India. J Threat Taxa 5:3969–3976

La Torre-Cuadros MDLA, Herrando-Pérez S, Young KR (2007) Diversity and structural patterns for tropical montane and premontane forests of Central Peru, with an assessment of the use of higher-taxon surrogacy. Biodivers Conserv 16:2965–2988. https://doi.org/10.1007/s10531-007-9155-9

Lakshmi BB, Rao BT, Rao KR, Naidu YP (2015) Diversity, distribution and status of birds of Kolleru Lake – a Ramsar site in Andhra Pradesh. Int J Innov Res Sci Eng Technol 4:5759–5784

Liu T, Ma L, Cheng L, Hou Y, Wen Y (2021) Is ecological birdwatching tourism a more effective way to transform the value of ecosystem services? A case study of birdwatching destinations in Mingxi County, China. Int J Environ Res Public Health 18:12424. https://doi.org/10.3390/ijerph182312424

Maclean IMD, Holt RD, Donald PF (2014) Wetlands, birds, and climate change: importance of temperature and precipitation variability. Biol Conserv 177:64–75

Mandal MH, Yasmin B, Roy A, Ghosh S, Siddique G (2022) Investigating present status of floodplain wetlands as habitat of water birds and its determinants: an experience from lower part of deltaic West Bengal, India. Wetlands 42:70. https://doi.org/10.1007/s13157-022-01591-1

Mandishona E, Knight J (2022) Inland wetlands in Africa: a review of their typologies and ecosystem services. Prog Phys Geogr Earth Environ 46:547–565. https://doi.org/10.1177/03091333221075328

Margaryan L, Prince S, Ioannides D, Röslmaier M (2022) Dancing with cranes: a humanist perspective on cultural ecosystem services of wetlands. Tour Geogr 24:501–522. https://doi.org/10.1080/14616688.2018.1522512

Manny BA, Johnson WC & Wetzel RG Nutrient additions by waterfowl to lakes and reservoirs: predicting their effects on productivity and water quality. Hydrobiologia 279:121–132 (1994). https://doi.org/10.1007/BF00027847

Mazumdar S (2019) Composition of avian communities in a human-modified wetland Okhla Bird Sanctuary, India: with notes on conservation initiatives. Proc Zool Soc 72:319–333. https://doi.org/10.1007/s12595-017-0239-6

McNicholl MK (2018) Avian wetland habitat functions affected by water level fluctuations. In: Coastal wetlands. CRC Press, pp 87–98

Mitsch WJ, Gosselink JG (2015) Wetlands, 5th edn. Wiley

Mitsch WJ, Bernal B, Hernandez ME (2015) Ecosystem services of wetlands. Int J Biodivers Sci Ecosyst Serv Manag 11:1–4. https://doi.org/10.1080/21513732.2015.1006250

Mohammad JM, Krishna PV (2016) Avifaunal diversity of Wyra and Paler Reservoir of Khammam District, Telangana, India. Species 17:160–174

Mushet DM, Roth CL (2020) Modeling the supporting ecosystem services of depressional wetlands in agricultural landscapes. Wetlands 40:1061–1069. https://doi.org/10.1007/s13157-020-01297-2

Nautiyal S, Bhaskar K, Khan YDI (2015) Biodiversity of semiarid landscape: baseline study for understanding the impact of human development on ecosystems, 1st edn. Springer Int Publ, pp 1–398

Nautiyal S, Kaechele H, Babu MSU, Khan YDI, Dutta H (2020) Avian conservation under two management regimes: a case study from the viewpoint of governance and social dimension. Environ Syst Decis 40:148–169. https://doi.org/10.1007/s10669-019-09745-z

Neelgund HD, Kadadevaru G (2020) Avifaunal diversity of some selected water bodies of Khanapur Taluka, Belagavi District, Karnataka, India. J Threat Taxa 12:15572–15586

Nyathi D, Ndlovu J, Hadebe S (2024) Challenges of implementing urban ecosystem-based adaptation to climate change in developing countries: a systematic review. In: Cities of tomorrow: urban resilience and climate change preparedness, pp 249–266

Pal S, Debanshi S (2023) Exploring the effect of wastewater pollution susceptibility towards wetland provisioning services. Ecohydrol Hydrobiol 23:162–176. https://doi.org/10.1016/j.ecohyd.2022.12.003

Pal S, Singha P (2023) Linking river flow modification with wetland hydrological instability, habitat condition, and ecological responses. Environ Sci Pollut Res 30(5):11634–11660

Panda BP, Das AK, Jena SK, Mahapatra B, Dash AK, Pradhan A, Parida SP (2021) Habitat heterogeneity and seasonal variations influencing avian community structure in wetlands. J Asia-Pac Biodivers 14:23–32. https://doi.org/10.1016/j.japb.2020.10.001

Parra G, Guerrero F, Armengol J et al (2021) The future of temporary wetlands in drylands under global change. Inland Waters 11(4):445–456. https://doi.org/10.1080/20442041.2021.1936865

Pattnaik D, Singh C, Lenka R (2022) Avifaunal diversity and species composition in the largest unexplored wetland (Kirtipur) of Ganjam District, Odisha. Proc Zool Soc 75:437–448. https://doi.org/10.1007/s12595-022-00456-z

Prasad KK, Ramakrishna B, Srinivasulu C, Srinivasulu B (2014) Taxonomic updates to the checklists of birds of India, and the south Asian region – 2020. Indian Birds 16:12–19

Praveen J, Jayapal R, Pittie A (2020) Taxonomic updates to the checklists of birds of India, and the South Asian region – 2020. Indian Birds 16(1):12–19

Qiu J, Zhang Y, Ma J (2024) Wetland habitats supporting waterbird diversity: conservation perspective on biodiversity-ecosystem functioning relationship. J Environ Manag 357:120663

Rajashekara S, Venkatesha MG (2018) Impact of urban threats and disturbance on the survival of waterbird communities in wetlands of Bengaluru City, India. Proc Zool Soc 71:336–351. https://doi.org/10.1007/s12595-017-0217-z

Ramsar Convention on Wetlands (2018) Global wetland outlook: state of the world's wetlands and their services to people. Ramsar Convention Secretariat, Gland

Rao VV, Kumar RS, Surender G, Narayana BL (2014) Diversity, abundance and variation of water birds at Kolleru Wetland in Andhra Pradesh. Biol Forum Int J 6:62–76

Sack H, Thorn J (2024) Confronting climate at the water's edge: marine and coastal nature-based solutions in Southern Africa. In: Handbook of nature-based solutions to mitigation and adaptation to climate change. Springer Int Publ, Cham, pp 1–30

Sailu G, Narayana BL, Ramaiyan D, Naresh B, Rao VV, Adepu H, Khandelwal R, Devulapalli P, Krishna P (2017) Faunal diversity of Ameenpur Lake, Telangana state, India: a biodiversity heritage site. J Entomol Zool Stud 5:512–526

Sandilyan S (2022) Alien fish species in Indian inland wetlands: current status and future challenges. Wetl Ecol Manag 30:423–437. https://doi.org/10.1007/s11273-022-09870-8

Santofimia E, González-Toril E, de Diego G, Rincón-Tomás B, Aguilera Á (2024) Ecological degradation of a fragile semi-arid wetland and the implications in its microbial community: the case study of las Tablas de Daimiel National Park (Spain). Sci Total Environ 924:171626. https://doi.org/10.1016/j.scitotenv.2024.171626

Sekercioglu CH (2006) Increasing awareness of avian ecological functions. Trends Ecol Evol 21(8):464–471

Singh J, Antil S, Goyal V, Malik V (2020) Avifaunal diversity of Tilyar Lake, Rohtak, Haryana, India. J Threat Taxa 12:15909–15915

Singh Y, Singh G, Khattar JS, Barinova S, Kaur J, Kumar S, Singh DP (2022) Assessment of water quality condition and spatiotemporal patterns in selected wetlands of Punjab, India. Environ Sci Pollut Res 29:2493–2509. https://doi.org/10.1007/s11356-021-15590-y

Siril S et al (2022) Role of REDD+ in reducing land degradation and achieving SDGs. In: Panwar P, Shukla G, Bhat JA, Chakravarty S (eds) Land degradation neutrality: achieving SDG 15 by Forest management. Springer, Singapore. https://doi.org/10.1007/978-981-19-5478-8_16

Tella JL, Hiraldo F, Pacífico E, Díaz-Luque JA, Dénes FV, Fontoura FM, Guedes N, Blanco G (2020) Conserving the diversity of ecological interactions: the role of two threatened macaw species as legitimate dispersers of "megafaunal" fruits. Diversity 12(2):45

Ulman Y, Singh M (2022) Avifaunal diversity in unprotected wetlands of Ayodhya district, Uttar Pradesh, India. J Threat Taxa 14:21561–21578

Vishwakarma A, Anthony FM, Tiwari S, Choubey S (2021) Avifaunal diversity of winter season in Kopra Reservoir of Bilaspur, Chhattisgarh, India. Proc Zool Soc 74:118–126. https://doi.org/10.1007/s12595-020-00349-z

Wasserman RJ, Dalu T (2022) Tropical freshwater wetlands: an introduction. In: Dalu T, Wasserman RJ (eds) Fundamentals of tropical freshwater wetlands. Elsevier, pp 1–22. https://doi.org/10.1016/B978-0-12-822362-8.00024-4

Weinstein BG (2018) A computer vision for animal ecology. J Anim Ecol 87:533–545. https://doi.org/10.1111/1365-2656.12780

Woldemariam W, Mekonnen T, Morrison K, Aticho A (2018) Assessment of wetland flora and avifauna species diversity in Kafa zone, southwestern Ethiopia. J Asia-Pac Biodivers 11:494–502. https://doi.org/10.1016/j.japb.2018.08.003

Wu C, Chen W (2020) Indicator system construction and health assessment of wetland ecosystem – taking Hongze Lake wetland, China as an example. Ecol Indic 112:106164. https://doi.org/10.1016/j.ecolind.2020.106164

Young LM, Kelly D, Nelson XJ (2012) Alpine flora may depend on declining frugivorous parrot for seed dispersal. Biol Conserv 147(1):133–142

Zhang X, He S, Yang Y (2021) Evaluation of wetland ecosystem services value of the Yellow River Delta. Environ Monit Assess 3(6):353. https://doi.org/10.1007/s10661-021-09130-x

Zwolicki A, Zmudczyńska-Skarbek KM, Iliszko L et al (2013) Guano deposition and nutrient enrichment in the vicinity of planktivorous and piscivorous seabird colonies in Spitsbergen. Polar Biol 36:363–372. https://doi.org/10.1007/s00300-012-1265-5

Chapter 13
Reviving Groundwater Ecosystem Health: Industrial Effluent Impacts on Water Quality and Agricultural Sustainability in Solapur's Chincholi Industrial Zone

Shrikant Mukate, Dipak Panaskar, Vasant Wagh, Yogesh Deshpande, and Uday Sahu

Abstract Water is an essential component of life, which should be available in sufficient amounts both qualitatively and quantitatively to meet the demands of the end user. Every purpose needs different amounts and quality of water. The quality standards need defined by the agencies need to be strictly followed for their respective purposes. Anthropogenic and natural sources affect the water quality. Industrial outputs such as effluents play a crucial role in affecting the quality of soil and water. The water suitability for irrigation is defined by the indices such as SAR, MH, RSC, %Na, PI, and KR. Following this, it is essential to know the corrosion potential of water before using the groundwater for industrial purpose. Several indices like CR, RSI, PSI, and L-S Index were calculated through the AquaChem software to identify corrosion potential. To identify the possible sources of contaminants, statistical methods were adopted. Correlation matrix, cluster analysis (CA), and principal component analysis (PCA) were performed to determine the water quality harming parameters. Water quality is mostly affected in a region where industries are functional. Groundwater is good for agriculture use except for a few samples. Very few groundwater samples showcase the high corrosion potential, whereas the rest show their fitness for industrial applications.

S. Mukate (✉)
Sustainable Sapiens, Pune, Maharashtra, India

D. Panaskar · V. Wagh
School of Earth Sciences, SRTM University, Nanded, Maharashtra, India

Y. Deshpande
School of Open Learning, SPPU University, Pune, Maharashtra, India

U. Sahu
Department of Geology, Toshniwal College of Arts, Commerce and Science, Hingoli, Maharashtra, India

Y. D. I. Khan et al. (eds.), *Land Restoration Through Ecosystem-Based Approach*, Disaster Resilience and Green Growth, https://doi.org/10.1007/978-981-95-0389-6_13

Keywords Industrial discharge · Groundwater analysis · Drinking · Irrigation · Corrosion · Statistical analysis

13.1 Introduction

Water is a very crucial component for the sustenance of life on Earth. Water quantity and quality are very important. The qualitative analysis of the water is essential to define its usability for different purposes. The concentration of different chemical components present in water defines the use of water for a particular activity. Water is used for drinking and for domestic, industrial, and agricultural purposes. For drinking purposes, the qualitative presence or absence of chemical ions, biological entities, and suspended solids is strictly defined worldwide by the World Health Organization (WHO) and the Bureau of Indian Standards (BIS) in India. For domestic purposes, quality standards are flexible. Industrial activities need lots of water, which needs to be managed efficiently for specific purposes. In the case of agriculture, the increased salinity of water affects crop production and soil productivity. So, it is imperative to identify the salinity potential of water for salinity. The saline water creates a chemical scarcity where water is available, but plants cannot absorb it due to high salinity. These scenarios define the importance of the quality of water for different activities.

Water Quality Indices (WQI) are one of the most promising tools used to date to define the suitability of water potability. It is a mathematical tool that converts large data sets into single-dimensionally informative numbers. It integrates a number of variables to produce a single value for effectively communicating the information (Horton 1965). Generally, water quality indices are developed to summarize and report the quality of water based on scientific criteria in an understandable way to decision-makers, policy planners, and local inhabitants for the management of water resources (Mukate et al. 2019). Horton (1965) proposed the concept of WQI, which was further modified by the National Sanitation Foundation (NSF) in 1970. The NSF developed general WQI (NSFWQI) to provide a standardized method to evaluate the water quality based on physicochemical parameters (Brown et al. 1970). Mukate et al. (2019) developed the Integrated Water Quality Index (IWQI), using desirable and permissible limits to generate results. IWQI is very easy to calculate and generate unbiased results as there is no need to define weights. Egbueri (2020) and Islam et al. (2020) have used the IWQI to check the consistency of the index with other indices. The results concluded that IWQI is more consistent and reliable.

Water suitability for agriculture depends on its availability and sodium, calcium, magnesium, potassium, and bicarbonate concentration. Many researchers have tried to assess groundwater's potential for irrigation (Wagh et al. 2016; Panaskar et al. 2016). Soil quality and crop productivity depend on the irrigation water (Singh and Singh 2008). Industries need plentiful water for production; incrustation and corrosion property are tested before use (Raghunath 1982; Karanth 1987; Das et al.

2016). Water is considered suitable for industrial use only if it doesn't form any scales and is non-corrosive, as they use a lot of steel and iron material in their infrastructure. The computation of saturation indices of water is significant in assessing the tendency of water to precipitate out or dissolve calcium carbonate (Singh et al. 2006). The quality evaluation of groundwater for corrosion and adhesion potential is important because farmers and industrialists use iron pipes to pump groundwater for irrigation and distribution.

Many authors have used statistical methods to identify the water quality degrading factors (Boateng et al. 2016; Mukate et al. 2019; Panaskar et al. 2016; Hamzah et al. 2017; Sethi et al. 2012). The correlation matrix in hydro-geochemistry is used to identify the degree of correlation among the different physicochemical parameters affecting groundwater quality (Aryafar and Ardejani 2013). The K-means clustering algorithm helps to partition the groups of similar composition amongst a multivariate data set (MacQueen 1967). The cluster analysis (CA) identifies groups of similar traits based on like characteristics within a class and dissimilarities between different classes (Lattin et al. 2003). A series of multivariate methods were used to identify the similar data sets (Danielsson et al. 1999). The groups are divided into similarity levels, and the dendrogram is developed where observations are merged.

This chapter discusses the suitability of water for different purposes and statistical analyses of chemical parameters present in water. Data has been presented spatially and temporally, along with appropriate graphic formats to help the readers on water quality trends and industrial impacts on groundwater quality. The research objectives of this chapter are to identify the groundwater quality impacts on potability, irrigation, and corrosion potential through well-developed indices and to recognize the hotspot to avoid the deterioration of other resources.

13.2 Material and Methodology

In 2014, 49 groundwater samples were collected from in and around the Solapur city industrial area named Chincholi Maharashtra Industrial Development Corporation (onwards MIDC). Eleven physicochemical parameters were analysed to assess the water quality (APHA 2005). Drinking water suitability, i.e. potability, has been evaluated through the mostly used conventional WQI (Horton 1965) and newly developed Integrated Water Quality Index (IWQI) (Mukate et al. 2019). Irrigation water quality is based on irrigation water suitability indices such as sodium adsorption ratio (SAR), magnesium hazard (MH), residual sodium carbonate (RSC), per cent sodium (% Na), Permeability Index (PI), and Kelley's ratio (KR). The corrosion or incrustation is mainly influenced by the imbalance of pH, EC, HCO_3, Cl, and SO_4 (Anon 1983). To identify the corrosive and aggressive nature of groundwater, indices like the corrosion ratio (CR), saturation pH (pHs), Langelier Saturation Index (LSI), Ryznar Stability Index (RSI), Puckorius Scaling Index (PSI), and Larson-Skold Index (L-S Index) were calculated. MS Excel, SPSS, and R programming were used to perform statistical analysis. This helps to organize the data and understand the controlling factors. The correlation matrix, principal component analysis, and cluster analysis were performed to ascertain the role of individual parameters in influencing water quality.

13.2.1 Study Area

Geographically, the study area is bounded by 750 45′ to 750 52′ E longitude and 170 40′ to 170 50′ N latitude. It is located only 15 km northwest of the Solapur District headquarter. Solapur is the southernmost district of Maharashtra state, sharing a boundary with the Telangana and Karnataka states. The textile products and jowar are very famous throughout India. Hydrologically, the study area is a part of the Upper Krishna River Basin: Sina River, which is a major tributary of Bhima River (Fig. 13.1). Kondi, Pakni, and Chincholi are the major settlements having around 15,000 population. It mostly uses groundwater for drinking, domestic, and irrigation purposes. The major crops that are grown are sugarcane, Jowar, sugarcane, groundnut, soybean, wheat, etc. The Chincholi industrial area was established in 1994 by the Maharashtra Industrial Development Corporation (MIDC). Since then, industrial units in pharmaceutical, engineering, and food processing sectors have been functioning smoothly. This industrial area is equipped with two common effluent treatment plants; only one is functional. The field visits showed that industrial wastewater in natural drainages and small pits was not treated in many places. Due to such activities, local farmers and residents face many challenges that scientific investigations need to address.

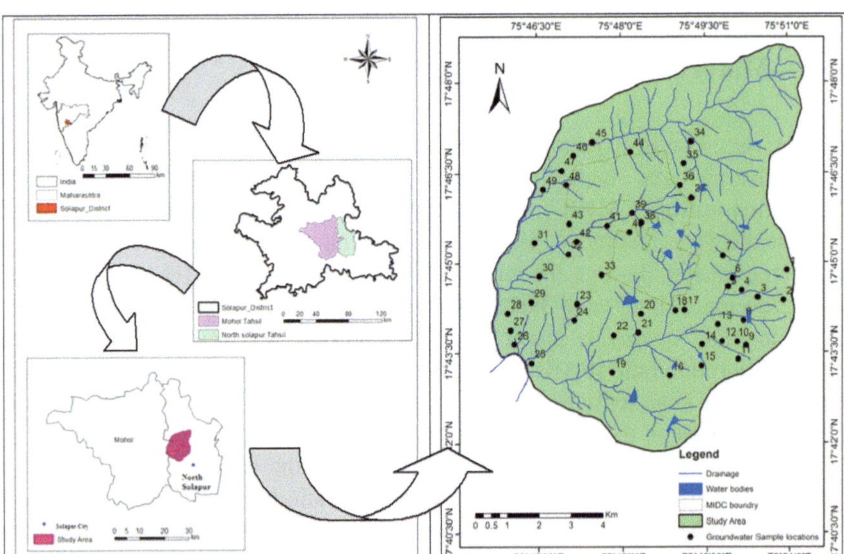

Fig. 13.1 Geographical location of the study area and water sample locations

13.3 Results and Discussion

13.3.1 Water Quality Index (WQI)

The lowest value indicates excellent quality, and the highest value suggests the unsuitability of water for drinking purposes. The lowest WQI value is observed at sample numbers 11 and 1 in both the seasons. In the pre-monsoon season, the highest value was observed at sample number 41. It is observed that samples collected from Kondi settlement were found suitable for drinking in all the seasons (sample numbers 1–17), while samples (sample numbers 18–33) from Pakni village show the impact of agricultural activities due to enrichment of minerals in the groundwater samples, and because of that, water quality is deteriorated. Some higher peaks are observed at sample numbers 22, 25, 26, and 27. Samples from Chincholi industrial area (sample numbers 34–49) show great variation in water quality among them, samples from upper (sample numbers 33–37) and lower reaches (sample numbers 45–49) are not affected, but samples (sample numbers 38, 39, 41, and 44) in the vicinity of industrial area are heavily affected.

The WQI results have been classified into five categories, from excellent to unsuitable (Table 13.1). It reveals that 67.35% samples fall in the good category, 20.41% of groundwater samples belong to the marginal category, and 6.12% and 4.08% of groundwater samples are poor and unsuitable for drinking in the pre-monsoon season. In the post-monsoon season, 69.39% of groundwater samples are of good quality, 20.41% of samples are marginal, 8.16% of samples are poor, and 2.04% of samples are unsuitable for drinking purposes. Throughout the study area, the majority of the samples fall in the good category; a considerable number of samples belong to the marginal category, and very few samples are poor and unsuitable for drinking. Stormwater run-off from agricultural and industrial areas and domestic run-off are deteriorating the groundwater quality (Mukate et al. 2015, 2018). The leaching of industrial effluents mainly influences samples 39, 41, and 44. Samples 26 and 27 are affected by the leaching of agrochemicals, mainly potassium, nitrate, and chloride.

Table 13.1 Classification of groundwater samples based on WQI values for 2014

		Pre-monsoon 2014		Post-monsoon 2014	
Category	Class	% of samples	Sample numbers	% of samples	Sample numbers
Excellent	0–50	2.04	11		
Good	50–100	67.35	1–10, 12–21, 23, 28–31, 34, 36, 42, 43, 46–49	69.39	1–20, 23, 24, 28, 29, 33, 34, 36, 37, 42, 43, 46–49
Marginal	100–200	20.41	22, 24–26, 32, 33, 35, 37, 40, 45	20.41	21, 22, 25, 30–32, 35, 38, 40, 45
Poor	200–300	6.12	27, 38, 39	8.16	26, 27, 39, 44
Unsuitable	>300	4.08	41, 44	2.04	41

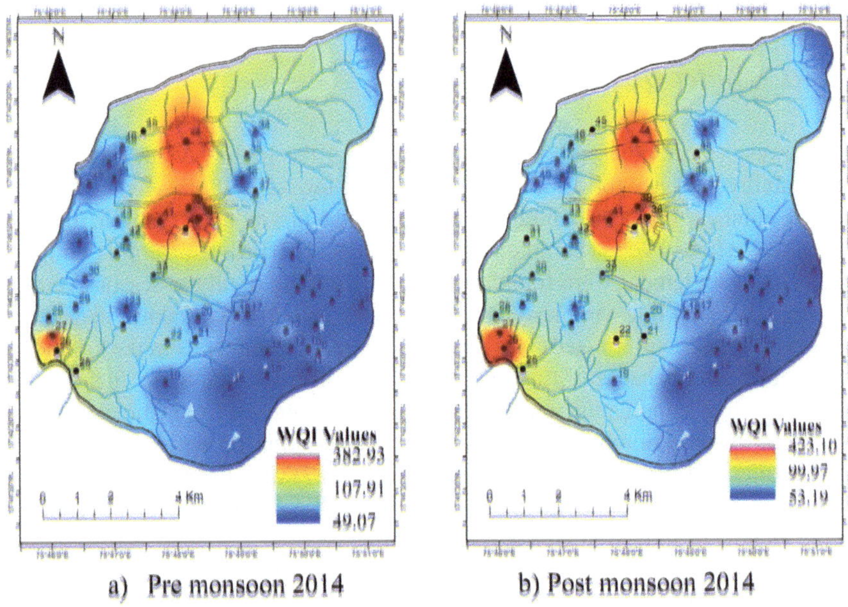

Fig. 13.2 Spatial distribution of WQI of groundwater samples. (**a**) Pre-monsoon 2014. (**b**) Post-monsoon 2014

Samples located in the neighbourhood of industrial establishments and a few samples on the bank of Sina River deteriorated in all the seasons, owing to industrial activities (Fig. 13.2). The drainage flow of the basin is in the direction of Northeast to Southwest. The samples in the upper reaches (sample numbers 35, 36, 37, and 38) are suitable for drinking; hence, they are not affected by industrial effluents, and samples from lower reaches are still in good quality (sample numbers 28–32, 42, 43, and 46–49). On the other hand, the samples located in the Central region and on the periphery of the MIDC area (sample numbers 38, 39, 40, 41, 44, and 45) are severely contaminated due to the release of industrial wastewater and represent their unsuitability for drinking Shankar et al. 2008). In addition, a few groundwater samples in the South and Southwest directions also show their unsuitability for drinking due to agriculture and domestic activities. The eastern side of the study area has good quality of water, while the southern part has moderate water quality in all the seasons.

13.3.2 Integrated Water Quality Index (IWQI)

This Water Quality Index uses the desirable and permissible limits of BIS for Ca, Mg, Na, K, Cl, SO_4, NO_3, and TDS (BIS 2012). The IWQI value ranges from 1.07 to 13.69, with an average value of 2.97 in pre-monsoon and 0.92–17.65 with an average of 3.17 in the post-monsoon season (Fig. 13.3). Majority of the samples

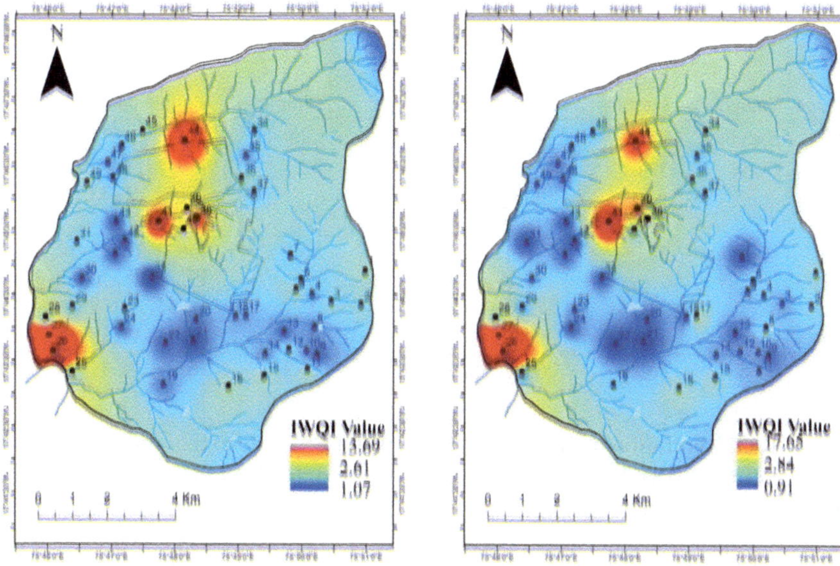

a) Pre monsoon 2014

b) Post monsoon 2014

Fig. 13.3 Spatial distribution of IWQI of groundwater samples. (**a**) Pre-monsoon 2014. (**b**) Post-monsoon 2014

have value of more than 1 (except sample number 21 in post-monsoon), which shows that there is not even a single sample belonging to an excellent category at the 20% deficit. The values of the few samples are very close to IWQI value 2, which increases the vulnerability of the samples. It is imperative to identify the enriched components and their possible sources to improve the water quality in future. It was also observed that a few samples (1, 15, 34) were contaminated by domestic waste, (48, 49) by agricultural waste, and (39, 41, 44) by industrial activities.

In the post-monsoon season, only sample no. 21 was excellent, and 32.65% samples were found suitable for drinking in both seasons of 2014 (Table 13.2). The marginal category (IWQI value 2–3) contains most groundwater samples, i.e. 53.06% and 48.98% in pre- and post-monsoon seasons. The marginal category samples suggest that the pollution level is low and can be utilized for restricted activities. In the study area, 4.08% of samples during pre-monsoon and 8.16% of groundwater samples in the post-monsoon season belong to the poor category, i.e. IWQI value 3–5. Domestic activities mainly affect these samples (Mukate et al. 2017). Sample number 36 is affected by Ca, Mg, Na, SO_4, Cl, and NO_3. In contrast, sample numbers 40 and 45 are impacted by Ca, Mg, Na, SO_4, NO_3, and TDS located on the periphery of the industrial area, representing the influence of industrial activities. The sample numbers 26, 27, 39, 41, and 44 (10.20%) have IWQI values more than 5 in all the seasons, backed from excessive concentration of cations like Ca, Mg, and K, and anions like Cl and NO_3. It focuses on ion exchange processes like Ca and Mg substituted by Na and Cl; and nitrates were replaced by sulfates. The

Table 13.2 Classification of groundwater samples based on IWQI values for 2014

Range	Category	Pre-monsoon 2014		Post-monsoon 2014	
		% of the samples	Sample number	% of the samples	Sample number
0–1	Excellent	–		2.0	21
1–2	Good	32.65	8–10, 13, 14, 19–22, 24, 30, 32, 33, 35, 42, 43	32.6	7, 9, 13, 18, 20, 22, 24, 25, 29, 31–33, 42, 43, 47, 48
2–3	Marginal	53.06	1–7, 11, 12, 15–18,23, 25, 28, 29, 31, 34, 36, 37, 45–49	46.9	1–6, 8, 10–12, 14, 15, 19, 23, 28, 30, 35–37, 40, 45, 46, 49
3–5	Poor	4.08	39, 40	8.2	16, 17, 34, 38
>5	Unsuitable	10.20	26, 27, 38, 41, 44	10.2	26, 27, 39, 41, 44

highest K concentration in sample numbers 26 and 27 is due to the irrigation run-off (Mukate et al. 2018). Samples numbers 39, 41, and 44 show the uppermost IWQI value, and these samples are found in industrial areas, which suggests that the discharge of unprocessed or partly processed industrial effluents is percolating into the groundwater. In this study, a few surface water tanks are filled with industrial effluents, and their impacts are found on the nearby groundwater extracting wells such as sample numbers 20, 21, and 22.

Impact of irrigation run-off is clearly observed in the Southwest region. This section is under intense agricultural practices and watered with Ujani canal water system, Sina river water, and borewells for many years. The use of fertilizers and inappropriate sealing of dug and borewells cause the mixing of run-off into the water resources. The 26th number sample from Pakni village is very close to the septic chamber of the domestic hygiene system, which results in increased concentrations of Cl and NO_3. Excretion in open space is observed near sample 27, located in the outskirts of the Pakni village. Sample numbers 26 and 27 also have high potassium and chloride, which are possibly added by geogenic foundations and fertilizers (Pawar et al. 2008). The pollution spread is limited to sample number 46, but considering the present data, sample numbers 32, 38, 42, 43, 47, 48, and 49 will be affected in future. Hence, there is a need to mitigate the industrial discharges affecting the water quality and to manage the aquatic resources with proper sealing.

13.3.3 Groundwater Quality Indices for Irrigation Purpose

13.3.3.1 Sodium Adsorption Ratio (SAR)

The sodium adsorption measures the sodium or alkali hazard to the produce. The increased SAR value reduces the soil permeability and affects the soil configuration (Todd 1980). Generally, water with low SAR value is supposed to be appropriate for watering because high value improves the spreading of colloids or clays when it

Table 13.3 Classification of groundwater samples based on irrigation indices

Index	Range	Category	Percentage of groundwater samples	
			Pre-monsoon 2014	Post-monsoon 2014
SAR	<10	Excellent	100	100
	10–18	Good	0	0
	18–26	Doubtful	0	0
	>26	Unsuitable	0	0
MH	<50	Suitable	40.82	87.76
	>50	Unsuitable	59.18	12.24
RSC	<1.25	Good	100	89.80
	1.25–2.5	Medium	0	4.08
	>2.5	Bad	0	6.12
%Na	0–20	Excellent	22.45	14.29
	20–40	Good	42.86	46.94
	40–60	Permissible for drinking	24.49	26.53
	60–80	Doubtful	8.16	12.24
	>80	Unsuitable	2.04	0
PI	>75	Excellent	4.08	4.08
	25–75	Good	71.43	73.47
	<25	Unsuitable	24.49	22.45
KR	<1	Suitable	81.63	81.63
	>1	Unsuitable	18.37	18.37

interacts with the soil and may swap calcium and magnesium ions in the soil resulting in damage to the soil structure and drop in its ability to conduct water and air (Richards 1954). In the case of the study area, the SAR value ranges from 0.48 to 8.52, with an average of 2.40, and 0.76–7.21, with an average of 2.67 in pre- and post-monsoon seasons, respectively. It also suggests that the average SAR value increases in post-monsoon seasons compared to pre-monsoon seasons, possibly due to the mixing of surface run-off enriched with agricultural and industrial waste. Water with SAR values <10 are considered to be excellent, 10–18 as good, 18–26 as fair, and > 26 as unsuitable (Table 13.3) (Richards 1954). In the present study, all the groundwater samples fall in excellent class except one sample. In the MIDC area during the study periods, the SAR value of all the groundwater samples was less than 10, suggesting an excellent source for irrigation.

13.3.3.2 Magnesium Hazard (MH)

The magnesium hazard is also called as the magnesium adsorption ratio, which expresses the excessive amount of magnesium over the sum of calcium and magnesium (Raghunath 1982). Calcium and magnesium are related to soil combination and friability, which maintain a state of balance in most waters and are indispensable plant nutrients. The increased quantity of Ca and Mg in groundwater changes the soil pH, affecting phosphorous availability (Shammiri et al. 2005). MH values deviate from 9.63 to 81.31 (avg. 54.78) in pre-monsoon and 15.51–64.92 (avg. 40.57) in the post-monsoon season. The MH >50 suggests the increased soil alkalinity, which affects the overall crop yield, while MH < 50 depicts the water is appropriate for irrigation. 40.82% and 87.76% of groundwater samples are suitable in pre- and post-monsoon seasons (Table 13.3).

13.3.3.3 Residual Sodium Carbonate (RSC)

The residual sodium carbonate Index measures the quantity of bicarbonate and carbonate ions in excess of Ca + Mg (alkaline earth). Water with high contents of carbonate and bicarbonate inclines to precipitate the calcium and magnesium and leads to an augmented comparative proportion of sodium, which results in the formation of sodium bicarbonate (Sadashivaiah et al. 2008). The low RSC suggests that the build-up of sodium is not possible in the soil, which is good for irrigation. The RSC value varies from −40.23 to 0.90 (avg. −5.56) and −58.65 to 3.64 (avg. −4.39) in the pre- and post-monsoon season. A negative value suggests a low risk of sodium accumulation due to a balanced level of calcium and magnesium, while a positive value indicates that the bicarbonate and carbonate will remove free calcium and magnesium in the soil, thereby making sodium accumulation possible (Attibu 2014). If the RSC value of water samples exceeds 2.5, then the water becomes unsuitable for irrigation, while a value between 1.25 and 2.5 is slightly suitable, and a value <1.25 suggests good water quality (USDA 1954). It is detected that all the groundwater samples from pre-monsoon seasons are good for irrigation. In the post-monsoon season, 89.80% of groundwater samples are good, 4.08% are marginally suitable, and 6.12% are unsuitable for irrigation (Table 13.3).

13.3.3.4 Per Cent Sodium (%Na)

Sodium content plays a significant role in the cataloguing of water for irrigation because its interaction with soil results in the blockage of particles, thereby plummeting the permeability of water and respirational gases (Todd and Mays 2005; Sundaray et al. 2009). Sodium is a key parameter that defines the suitability of water for plants; excessive sodium reacts with soil through the process of base exchange, thus reducing permeability (Obiefuna and Sheriff 2011). The %Na value varies from 9.78 to 80.26 and 12.71 to 78.91, with averages of 35.71 and 38.28 in pre- and post-monsoon seasons, respectively. Around 22.5% and 14.3% of groundwater samples were found excellent in pre- and post-monsoon seasons (Table 13.3). Most of the groundwater samples fall in the good group, 42.86% and 46.94% in pre- and post-monsoon seasons, respectively. However, 24.49% of the samples were found permissible for drinking in pre-monsoon seasons and 26.53% in post-monsoon seasons. The samples having % Na values in between 60 and 80 are doubtful for irrigation, and 8.16% and 12.24% groundwater samples fall in this category in pre- and post-monsoon seasons. A few samples belong to doubtful categories, such as 2.04% groundwater samples in the pre-monsoon season.

13.3.3.5 Permeability Index (PI)

Soil permeability is affected by sodium, calcium, magnesium, and bicarbonate contents of the soil. Hence, the PI is measured as an essential index for the purpose of groundwater appropriateness for crop irrigation (Doneen 1964). The permeability index value of groundwater samples ranges from 15.50 to 95.99 with an average of 50.80, and 18.49 to 100.62 with an average of 55.38, respectively, in the pre- and post-monsoon season. Rendering to permeability indices, the groundwater samples are classified into three different classes, viz. class I (> 75), class II (25–75), and class III (<25). Class I and II water are categorized as good for irrigation with 75% or more of maximum permeability (Table 13.3). Class III water is unsuitable with 25% of maximum permeability. In the present study, 4.08% of the samples are excellent, 71.43 and 73.47% of the samples are good, and 24.49 and 22.45% of the groundwater samples are unsuitable in pre- and post-monsoon seasons of the year 2014, respectively.

13.3.3.6 Kelley's Ratio (KR)

This ratio measures the quantity of sodium against the amount of concentrations of calcium and magnesium (Kelley 1951). The ratio > 1 suggests the unnecessary concentration of sodium, which reduces soil permeability and alters other soil properties, whereas a ratio < 1 proves water suitability for irrigation. In the case of the study area, the values of Kelly's ratio show a very narrow range, from 0.1 to 4.05 (avg. 0.72) and 0.14 to 3.72 (avg. 0.78) in pre- and post-monsoon seasons, respectively. It suggests that the average value is close to one in all the seasons. Hence, the alkali hazard will damage the soil properties of the study area. The value of KR > 1 indicates an excess level of sodium in groundwater, which is unsuitable for irrigation. 81.63% of groundwater samples were found suitable, and 18.37% of the samples were unsuitable for irrigation in both seasons (Table 13.3).

13.3.4 Groundwater Quality for Industrial Purpose

13.3.4.1 Industrial Suitability of Water Based on EC, Cl, SO_4, and CR

The mechanism of corrosion is electrochemical in nature. Iron forms ferrous ions in anodic areas and transfers electrons to cathodic areas. The process neutralizes hydrogen ions, leading to atomic hydrogen deposition on the metal surface (Clarke and Barnes 1969). Reducing sulfide is a pertinent feature of groundwater corrosion and incrustation (Clarke and Barnes 1964).

The reduction of sulfate (McKinney 1962) is given by the following equation:

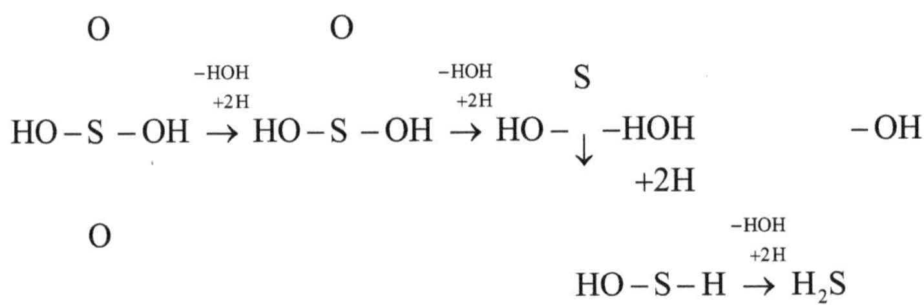

Sulfate-reducing organisms accelerate the reaction between atomic hydrogen and sulfate, stripping away the hydrogen layer from the cathode which promotes corrosion. (Clarke and Barnes 1969).

Electrical conductivity value is high (>1500 µS/cm) in 77.55% and 75.51% of the groundwater samples, which represents the corrosive nature of water in pre- and post-monsoon seasons, respectively (Table 13.4). The chloride is mainly prejudiced by industrial wastewater discharge. Water samples with high chloride (>500 mg/l) are located in the vicinity of industrial establishments. The corrosive nature of groundwater is higher (10.20%) in pre-monsoon and decreases (8.16%) in the post-monsoon season due to the addition of fresh water.

The incrustation is a stiff cement-like deposition, which is not detrimental but it guards the metals from corrosion. It is caused due to the precipitation of carbonates of calcium and magnesium or their sulfates and silicates (Agarwal 2012). The heavy

Table 13.4 Classification of groundwater samples based on industrial suitability during 2014

		Pre-monsoon		Post-monsoon		
Parameters	Range	Number of samples	% of samples	Number of samples	% of samples	Category
EC	<1500	11	22.45	12	24.49	Non-corrosive
	>1500	38	77.55	37	75.51	Corrosive
Cl	<500	44	89.80	45	91.84	Non-corrosive
	>500	5	10.20	4	8.16	Corrosive
SO_4	<100	26	53.06	26	53.06	Non-incrusting
	>100	23	46.94	23	46.94	Hard-incrusting
CR	<1	0	0	0	0	Non-corrosive
	>1	49	100	49	100	Corrosive

The EC is expressed in µS/cm, Cl and SO_4 in mg/l

incrustation leads to reducing water carrying capacity of pipes. The sulfate concentration in groundwater depicts that 46.94% samples in both the seasons are hard-incrusting. The hard-incrusting aquifers found along the downstream of industrial area may be due to the penetration of untreated industrial effluents.

The corrosive nature is also calculated in terms of corrosivity ratio (CR) based upon the comparative profusion of strong acid to weak acid (Ryznar 1944; Raman 1985). The CR of the groundwater samples is computed by using the following equation:

$$\mathrm{CR} = \left[\left(\mathrm{Cl} - /35.5 \right) + 2\left(\mathrm{SO42} - /96 \right) / 2\left(\mathrm{HCO3} - + \mathrm{CO32} - /100 \right) \right]$$

All the values are expressed in mg/l.

If the corrosivity ratio (CR) exceeds 1, it is not appropriate for industrial functions due to its corrosive nature. All the samples exceed the threshold limit of 1, so they are not well suited for industrial operations in all seasons.

13.3.4.2 Langelier Saturation Index (LSI)

It is defined as the variance between actual pH (measured) and calculated pHs (Langlier 1936). It is related to the saturation of a calcium carbonate layer or formation of scale over metal pipes, boilers, and other components. Corrosive water results in the system failures and health problems due to the dissolution of different heavy metals. It also affects the water conduction capacity of pipes which affects the performance of pipes and other systems. Langelier developed an equation for pH at which water is saturated with calcium carbonate; it expresses the calcium carbonate solubility and bicarbonate dissociation.

$$\mathrm{LSI} = \mathrm{pH} - \mathrm{pHs}$$

where

pH is the measured water pH.
pH_s is the pH at saturation in calcite or calcium carbonate.

A positive LSI suggests water is saturated with $CaCO_3$ and has potential for scale formation which is non-corrosive in nature. A protecting layer of $CaCO_3$ is predictable to form on the pipe wall (Benson 2009; Zia et al. 2008). However, water having LSI values < -0.5 is considered as aggressive. Aggressive water will not develop a shielding layer of calcium carbonate, which is corrosive in nature (Leitz and Guerra 2013). The LSI value varies from −0.84 to 1.68 (avg. 0.55) in pre-monsoon and −0.38 to 1.08 (avg. 0.61) in the post-monsoon season. The result suggests that all the groundwater samples are scale-forming and less corrosive in nature.

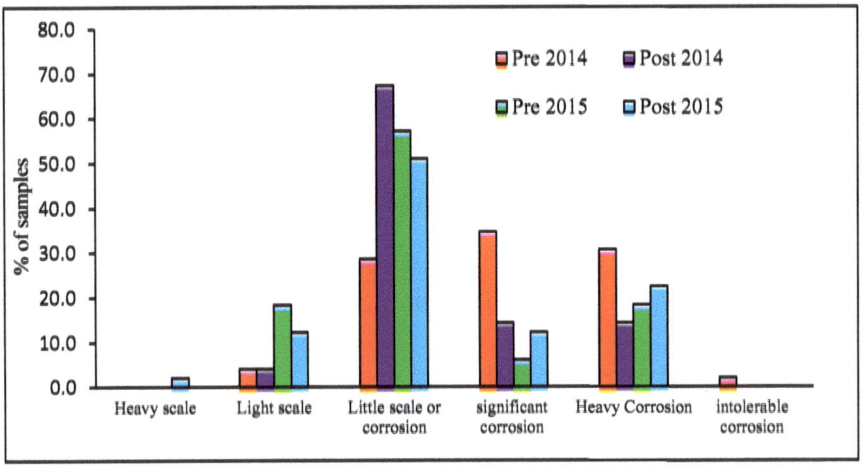

Fig. 13.4 Classification of RSI of groundwater based on Carrier (1965)

13.3.4.3 Ryznar Stability Index (RSI)

It is mainly used to measure the scale thickness in water systems to predict the consequences of water chemistry. The RSI values have been classified by Ryznar (1944) and Carrier (1965) to assess the impacts of water chemistry on metals.

$$\text{RSI} = 2.\text{pHs} - \text{pH}(\text{measured})$$

RSI indicates the aggressiveness of the water, but Carrier suggests about the scale and corrosion potential of the water (Figs. 13.4 and 13.5). The Carrier classification depicts that groundwater samples have little scale or corrosion property in majority of the samples, i.e. 28.57%; 67.34% in pre- and post-monsoon seasons, respectively. Heavy corrosion has been observed in 30.61% and 14.28% of groundwater samples in pre- and post-monsoon seasons, respectively, and sample number 17 in the pre-monsoon season shows intolerable corrosion (Fig. 13.4). The Ryznar classification (Fig. 13.5) depicts that 69.39% and 48.98% samples from pre/post-monsoon seasons fall in the aggressive group, which means there is no risk of corrosion in all the seasons. But sample numbers 17 and 47 represent the water is very aggressive and thus unsuitable for transportation through pipes.

13.3.4.4 Puckorius Scaling Index (PSI)

The Puckorius Scaling Index (PSI) is a variation of the RSI (Puckorius and Brooke 1991). This index also states an empirical relationship by using a calculated pH, referred to as the equilibrium pH. PSI depicts the possibility of establishing the relationship between supersaturated water and sedimentation by considering the

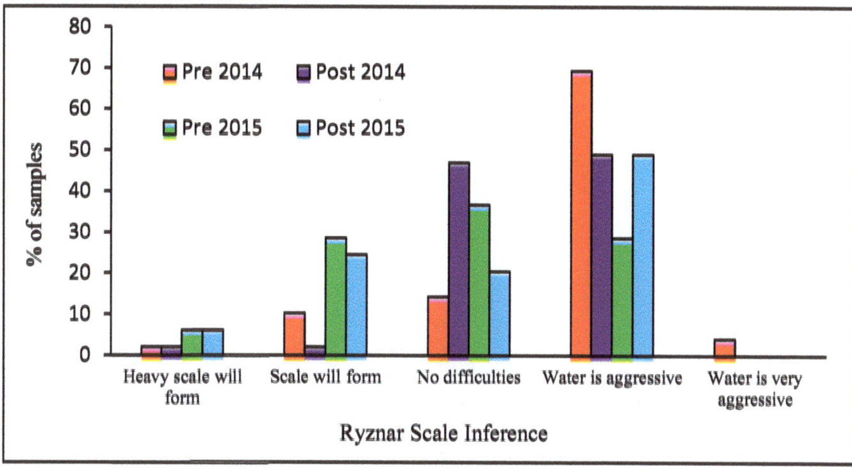

Fig. 13.5 Classification of RSI of groundwater based on Ryznar (1944)

calculated pH and equilibrium pH. The PSI values in the Chincholi industrial area varies from 5.65 to 9.15, with an average of 7.69 in the pre-monsoon season, and 4.61 to 8.27, with an average of 6.78 in the post-monsoon season. If the PSI value is fewer than 6, water inclines to have sedimentation, while the value more than 6 represents a corrosive nature (Zia et al. 2008; Silbert 2006). The PSI considers scaling as likely to occur if its value is <6 and as likely to dissolve scale if its value is >7. Figure 13.5 shows that 95.95% of samples in pre- and 95.92% in post-monsoon seasons are scale-forming and very few samples are corrosive in nature. The spatial spreading map illustrates that sample number 44, located in the industrial area, shows the corrosive nature in the post-monsoon season (Fig. 13.6).

13.3.4.5 Larson-Skold Index (L-S Index)

The Larson-Skold Index is mainly used to assess the water corrosiveness (Larson and Skold 1958). It is found that excessive amounts of chloride and sulfate upsurge the corrosivity, and alkalinity tends to reduce. The index suggests the ratio of sulfate and chloride equivalents per million (epm) to the combined alkalinity of bicarbonate and carbonate.

$$\text{Larson} - \text{Skold Index} = \left(\text{Cl} + \text{SO}_4\right) / \left(\text{HCO}_3 + \text{CO}_3\right)$$

LSI varies from 0.31 to 9.45 in the pre-monsoon and 0.24 to 9.5 in the post-monsoon season of the year 2014. The following guidelines have interpreted the Larson-Skold Index (Table 13.5). It is observed that the majority of the samples show high rates of localized corrosion that may be anticipated in all the seasons, followed by corrosion rates that may be advanced than expected. Figure 13.7 shows that groundwater samples with L-S Index value >1.2 are higher and L-S Index value <0.8 are lowest in all the seasons except the post-monsoon season.

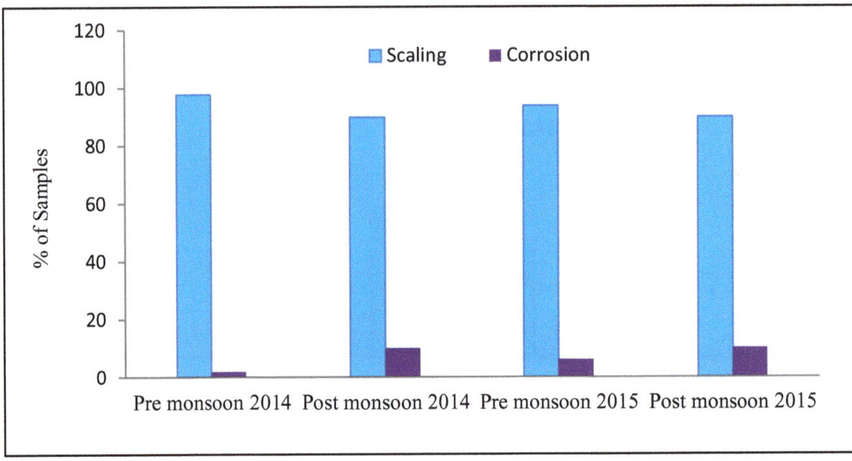

Fig. 13.6 Classification of groundwater samples based on PSI

Table 13.5 Groundwater classification criterion based on Larson-Skold Index values (1958)

Index value	Inference
Index <0.8	Chlorides and sulfate possibly will not interfere with natural film formation
0.8 < index <1.2	Chlorides and sulfates may interfere with natural film formation. Higher than preferred corrosion rates might be anticipated
Index >1.2	The leaning towards high corrosion rates of a local type should be estimated as the index increases

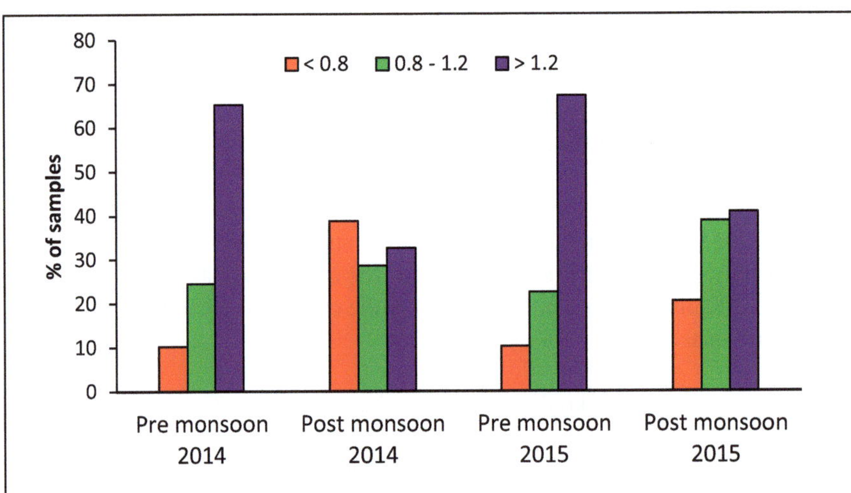

Fig. 13.7 Classification of groundwater samples based on the Larson-Skold Index

13.3.5 *Multivariate Statistical Analysis*

13.3.5.1 Correlation Matrix (CM)

The correlation value >0.7 is considered strong; r = 0.7 to 0.5 as moderate; and $r < 0.3$ as weakly correlated. The high positive correlation of EC and TDS with Na, Ca, Mg, Cl, NO_3, TH, and SO_4 finds the salinity affecting factors and the dissolution of salts in both seasons. These environments contribute to the increased value of electrical conductivity. Dissolved salts of calcium and magnesium contribute to total hardness (TH). A strong correlation was detected with Ca ($r = 0.99$) and Mg ($r = 0.98$) in the pre-monsoon season. The moderate correlation of SO_4 with EC and TDS ($r = 0.65$) and NO_3 ($r = 0.67$) indicates that nitrogen-rich fertilizers affect water quality. Chloride and nitrate are strongly correlated with Ca and Mg and moderately correlated with Na. K positively correlates with CO_3 ($r = 0.58$) (Table 13.6).

Table 13.7 presents the correlation matrix for the post-monsoon season, suggesting a strong correlation between electrical conductivity (EC) and Ca, Mg, Cl, NO_3, TDS, and TH, with a moderate correlation for Na and SO_4. Ca strongly correlates with Mg, Cl, TDS, and TH, while Mg shows high correlation with Cl ($r = 0.89$), TDS ($r = 0.88$), and TH ($r = 0.88$). The Ca-Mg correlation depicts the oxidation-reduction and ion exchange processes in groundwater. Agricultural inputs, domestic waste, and industrial discharge contribute to Na and Cl variations, elevating TDS, which is also contributed by weathering and erosion processes.

13.3.5.2 Cluster Analysis (CA)

The cluster analysis demonstrates that unpolluted samples are high (63.08%) in post-monsoon in comparison with the pre-monsoon (51.03%) season, while slightly polluted samples increase in pre-monsoon (32.65%) than the post-monsoon season (20.40%) (Table 13.8). The maximum numbers of samples from these two categories are situated in residential or barren land use, where anthropogenic activities are meagre. However, the moderately polluted samples (10.20%) and highly polluted samples (6.12%) remain the same in both seasons of 2014. The samples belonging to the moderately and highly polluted categories are situated in industrial and agricultural areas: percolation of industrial effluent and agricultural waste in recent times.

All the samples are differentiated into four different clusters (Fig. 13.8). These clusters are further classified based on the chemical composition and spatial location of these samples. The cataloguing of groundwater samples is done based on the degree of pollution in that particular sample, such as unpolluted, slightly polluted, moderately polluted, and highly polluted samples (Table 13.8). The principal component analysis is used to reduce the dimensionality of data and discover the causative factor impacting quality.

Table 13.6 Correlation analysis of physicochemical parameters of groundwater for the pre-monsoon season of 2014

	pH	EC	TDS	TH	Ca	Mg	Na	K	CO_3	HCO_3	Cl	NO_3	SO_4
pH	1												
EC	−0.21	1											
TDS	−0.21	1	1										
TH	−0.23	**0.93**	**0.93**	1									
Ca	−0.20	**0.82**	**0.82**	**0.88**	1								
Mg	−0.23	**0.90**	**0.90**	**0.96**	**0.72**	1							
Na	0.00	**0.66**	**0.66**	0.45	0.33	0.48	1						
K	−0.05	0.22	0.22	0.09	0.03	0.11	0.31	1					
CO_3	0.08	−0.04	−0.04	−0.07	0.01	−0.10	0.06	**0.58**	1				
HCO_3	0.17	0.34	0.34	0.39	0.32	0.39	0.37	−0.06	−0.10	1			
Cl	−0.18	**0.96**	**0.96**	**0.95**	**0.81**	**0.94**	**0.66**	0.14	−0.08	0.37	1		
NO_3	−0.21	**0.68**	**0.68**	**0.57**	**0.51**	**0.54**	**0.60**	0.36	0.05	0.37	**0.62**	1	
SO_4	−0.39	**0.78**	**0.78**	**0.68**	**0.65**	**0.63**	0.49	0.38	0.07	0.15	**0.65**	**0.65**	1

Highlighted values show substantial correlations

Table 13.7 Correlation analysis of physicochemical parameters of groundwater for the post-monsoon season of 2014

	pH	EC	TDS	TH	Ca	Mg	Na	K	CO_3	HCO_3	Cl	NO_3	SO_4
pH	1												
EC	−0.55	1											
TDS	−0.55	1	1										
TH	−0.47	**0.86**	**0.86**	1									
Ca	−0.47	**0.85**	**0.85**	**0.99**	1								
Mg	−0.46	**0.84**	**0.84**	**0.98**	**0.95**	1							
Na	−0.51	**0.59**	**0.59**	0.40	0.38	0.42	1						
K	−0.22	0.27	0.27	0.03	0.04	0.03	0.46	1					
CO_3	0.18	−0.17	−0.17	−0.34	−0.36	−0.31	0.29	0.28	1				
HCO_3	−0.40	**0.51**	**0.51**	0.45	0.42	0.47	**0.53**	0.39	0.06	1			
Cl	−0.51	**0.88**	**0.88**	**0.98**	**0.97**	**0.97**	**0.52**	0.10	−0.29	0.44	1		
NO_3	−0.37	**0.69**	**0.69**	**0.56**	**0.57**	0.54	**0.58**	0.45	0.03	0.45	**0.60**	1	
SO_4	−0.50	**0.65**	**0.65**	0.42	0.42	0.40	**0.70**	0.44	−0.07	0.33	**0.47**	**0.67**	1

Highlighted values show substantial correlations

Table 13.8 Groundwater sample categories based on the degree of pollution in 2014

Degree of pollution	Pre-monsoon	% of the samples	Post-monsoon	% of the samples
Unpolluted	1–17, 19, 20, 23, 36, 37, 43, 48, 49,	51.03	1–20, 23, 24, 33, 34, 36, 37, 43, 46–49	63.28
Slightly polluted	18, 21, 22, 24, 26, 28, 29, 30–35, 42, 46, 47,	32.65	21, 22, 28, 29–32, 35, 40, 42,	20.40
Moderately polluted	25, 27, 39, 40, 45,	10.20	25, 26, 27, 39, 45,	10.20
Highly polluted	38, 41, 44	6.12	38, 41, 44	6.12

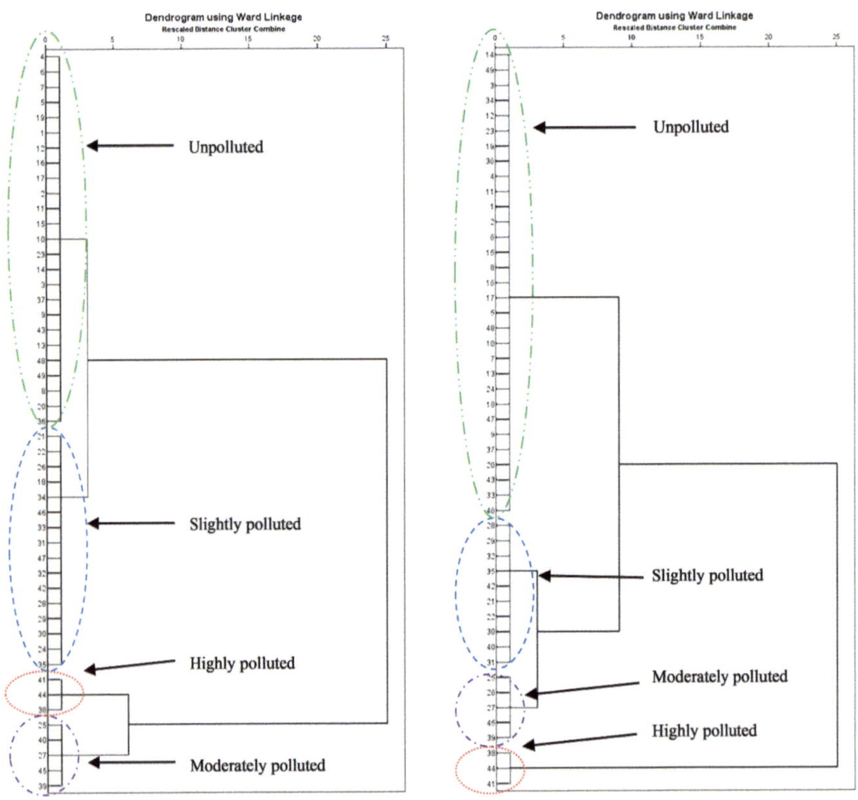

Pre monsoon 2014 Post monsoon 2014

Fig. 13.8 Sample-wise clusters in 2014

13.3.6 Principal Component Analysis (PCA)

The Varimax with Kaiser Normalization is implemented with five repetitions in pre-monsoon and six repetitions in post-monsoon. Table 13.9 and Fig. 13.9 represent the rotated component matrix including loading factors, eigenvalue, % variance, and % cumulative variance. Three factors in pre-monsoon seasons and two factors in post-monsoon seasons were extracted. The first factor contributes cumulative % of 51.65 and 44.13, while the second factor contributes 66.63 and 72.98, respectively.

The principal component analysis (PCA) depicts high positive loadings of EC, Ca, Mg, Cl, NO_3, and TH, while pH is insignificant across all seasons. Cl and NO_3 owing to industrial effluents and domestic waste inputs depict the strong anthropogenic contribution. Na and Cl show high loadings because of their solubility (Rao et al. 2014), while TH is limited by Mg, Cl, and SO_4, revealing permanent water hardness. Ca, Na, Cl, and SO_4 suggest mineral weathering, wastewater discharge, and cation exchange are important processes impacting water quality (Raju et al. 2016).

The Na and Ca ion exchange exhibits groundwater interactions, stemming from rock-water interactions (Drever 1988). Na and Cl quantities are contributed by agricultural run-off, industrial discharge, and sewage. Potassium, appearing in PC 2 across seasons, is mainly because of the use of potassium-rich fertilizers. High HCO_3 suggests monsoonal aquifer recharge from running water and weathering, while CO_3 loading depicts organic matter degradation (Raju et al. 2016). High Mg, Na, Cl, and SO_4 suggest water is influenced by anthropogenic inputs, while Na, Cl, SO_4, and HCO_3 increase the salinity and alkalinity.

Table 13.9 Rotated principle components analysis

	Components				
	Pre-monsoon			Post-monsoon	
Parameters	1	2	3	1	2
pH	−0.297	0.040	**0.795**	−0.470	−0.445
EC	**0.975**	0.075	0.026	**0.799**	0.494
TH	**0.960**	−0.088	−0.006	**0.963**	0.198
Ca	**0.850**	−0.080	−0.054	**0.956**	0.187
Mg	**0.922**	−0.084	0.022	**0.943**	0.206
Na	**0.622**	0.321	0.356	0.257	**0.837**
K	0.160	**0.897**	−0.060	−0.148	**0.783**
CO_3	−0.097	**0.813**	0.012	−0.545	0.460
HCO_3	0.421	−0.126	**0.673**	0.340	**0.581**
Cl	**0.965**	−0.020	0.089	**0.935**	0.289
NO_3	**0.717**	0.319	0.089	0.457	**0.678**
SO_4	**0.788**	0.286	−0.271	0.352	**0.707**
Total	6.20	1.80	1.31	5.30	3.46
% of variance	51.65	14.97	10.91	44.13	28.86
Cumulative %	51.65	66.63	77.54	44.13	72.98

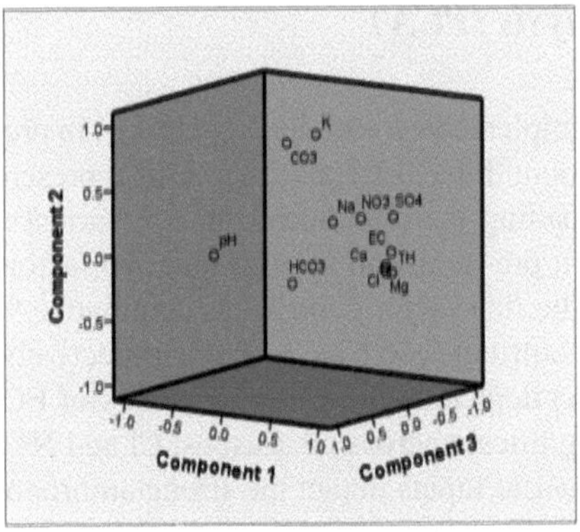

a). Pre monsoon 2014

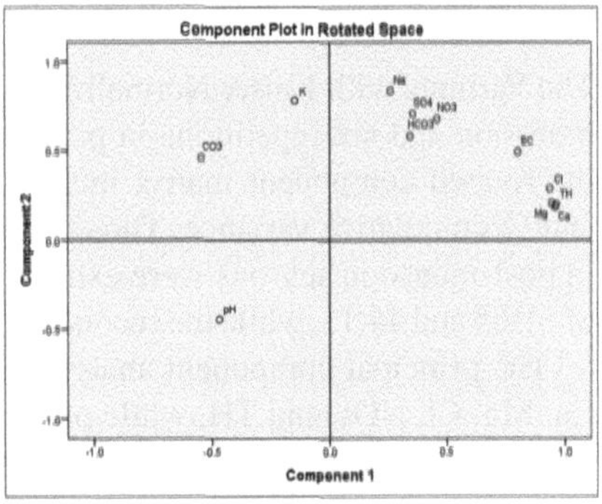

b). Post monsoon 2014

Fig. 13.9 Component plot in rotated space. (**a**) Pre-monsoon 2014. (**b**) Post-monsoon 2014

13.3.7 Delineation of Pollution Zones

The groundwater pollution zones and polluted surface water tanks have been identified on the basis of a number of parameters exceeding the permissible limits in all the seasons. Three pollution zones and four surface water tanks have been identified throughout the study area (Fig. 13.10). Out of three different pollution zones, two zones have been identified on the western part of the Chincholi MIDC area. These two groundwater pollution zones are separated by the water divide, which is inferred on the basis of the drainage pattern of the study area. These two groundwater pollution zones reveal the impact of the discharge of untreated industrial effluent on the aquifer. The third groundwater pollution zone is located in the Southwest part on the bank of Sina River. This zone is surrounded by agricultural fields, which suggests that agricultural activities influence the groundwater quality in that area. For the present research work, four different surface water tanks were selected to identify the influence of the Chincholi MIDC area on the surface water quality. All the surface water tanks are dejectedly polluted and are not suitable for any use.

In summary, industrial effluent discharge has impacted the groundwater at a few locations and the surface water quality in the vicinity of the industrial establishments. Samples located far from the industrial establishments were found in their natural state, except for a few samples from the southwestern region, which are influenced by agricultural activities. These results suggest how the landscapes are being altered with the change in land use for different purposes in the neighbourhood. Three different land-use types showcase the very unique behaviour in terms of their impact on groundwater. So, it is very imperative to decode such interactions while implementing or changing the land use from the existing one to restore the ecosystems healthy and providing the much needed ecosystem services.

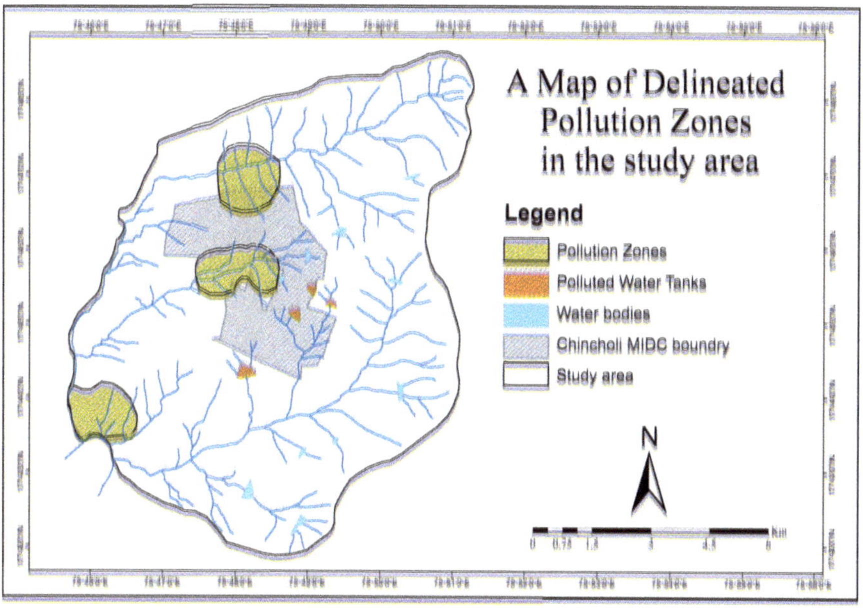

Fig. 13.10 Delineation of groundwater pollution zones and surface water tank

13.4 Conclusions

The findings from the Chincholi Industrial Zone underscore the urgent need for ecosystem-based approaches to restore groundwater health and ensure agricultural sustainability in dryland regions. The Integrated water quality Index (IWQI) proved to be a more precise tool than conventional WQI, as it accounts for the specific concentrations and permissible limits of contaminants, providing better insights into ecosystem changes. Since agriculture is the backbone of this region's livelihood, safeguarding water and soil quality is essential for sustaining economic and ecological balance. The study highlights irrigation water quality assessments' preliminary but critical role in identifying potential threats to soil and plant health. Rapid industrial expansion increases the demand for cost-effective, accessible, and scalable tools to monitor soil, water, and ecosystem health. Furthermore, industrial wastewater treatment's ethical and technical challenges raise pertinent questions for sustainable water management. Reusing treated industrial water in agriculture presents an opportunity to enhance water-use efficiency and reduce dependency on groundwater, but this must be pursued without compromising groundwater ecosystem integrity. Unregulated industrial effluent discharge poses a significant risk to groundwater systems, leading to contamination, corrosion, and economic losses due to frequent infrastructure failures. The problem is exacerbated by the intentional mixing of effluent with rainwater run-off, making it imperative to establish stringent monitoring mechanisms. A robust, community-driven approach is necessary to safeguard natural ecosystems, restore degraded land, and support sustainable development in dryland regions. This requires widespread awareness among industrialists, farmers, youth, and local stakeholders about industrial pollution's long-term environmental and human health consequences. Implementing an ecosystem-based land and water management approach will be key to mitigating degradation and fostering resilience in dryland landscapes.

Acknowledgments The work was funded by Department of Science and Technology (DST), Govt. of India, New Delhi for sanctioned grants under INSPIRE fellowship to Shrikant Mukate. The authors also wish to thanks to Director, School of Earth Sciences for provided necessary facilities during research.

References

Al-Shammiri M, Al-Saffar A, Bohamad S, & Ahmed M (2005) Waste water quality and reuse in irrigation in Kuwait using microfiltration technology in treatment. Desalination, 185(1–3):213–225

Agarwal VC (2012) Groundwater hydrology. PHI Learning Pvt. Ltd, p 259

APHA American Public Health Asso (2005) Standard methods for the examination of water and wastewater, 21st edn, New York

Aryafar A, Ardejani FD (2013) R-mod factor analysis, a popular multivariate statistical technique to evaluate water quality in Khaf-Sangan basin, Mashhad, northeast of Iran. Arab J Geosci 6(3):893–900

Attibu PK (2014) Assessment of salinity and sodicity of water used for irrigation in three sub-catchments in the Upper Athi River Basin, Kenya (Doctoral dissertation)

Balaji E, Nagaraju A, Sreedhar Y, Thejaswi A, Sharifi Z (2017) Hydrochemical characterization of groundwater in around Tirupati area, Chittoor District, Andhra Pradesh, South India. Appl Water Sci 7(3):1203–1212

Benson AS (2009) Evaluation of iron corrosion release models for water distribution systems. Doctoral dissertation, Virginia Tech

BIS (2012) Indian standards specification for drinking water, BIS: 10500: 2012. http://www.cgwb.gov.in/Documents/WQ-standards.pdf

Boateng TK, Opoku F, Acquaah SO, Akoto O (2016) Groundwater quality assessment using statistical approach and water quality index in Ejisu-Juaben municipality, Ghana. Environ Earth Sci 75(6):489

Brown RM, McClelland NI, Deininger RA, Tozer RG (1970) A water quality index: do we dare. Water Sewage Works 117(10):339–343

Carrier Air Conditioning Company (1965) Handbook of air conditioning system design. McGraw-Hill Books, New York

Clarke FE, Barnes I (1964) Preliminary evaluation of corrosion and encrustation mechanisms in tube wells of the Indus Plains, West Pakistan. U.S. Geol. Survey Open-File Report, p 81

Clarke FE, Barnes I (1969) Evaluation and control of corrosion and encrustation in tube wells of the Indus Plains, West Pakistan. Geol. Surv. Water-Supply Pap, p 1608

Collins R, Jenkins A (1996a) The impact of agricultural land use on stream chemistry in the Middle Hills of the Himalayas, Nepal. J Hydrol 185:71–86

Collins R, & Jenkins A (1996b) The impact of agricultural land use on stream chemistry in the middle hills of the Himalayas, Nepal. Journal of Hydrology, 185(1–4):71–86

Danielsson Å, Cato I, Carman R, Rahm L (1999) Spatial clustering of metals in the sediments of the Skagerrak/Kattegat. Appl Geochem 14(6):689–706

Das PP, Sahoo HK, Mohapatra PP (2016) An integrative geospatial and hydrogeochemical analysis for the assessment of groundwater quality in Mahakalapara block, Odisha, India. Environ Earth Sci 75(2):1–18

Doneen LD (1964) Water quality for agriculture. Department of Irrigation, University of California, Davis

Drever JI (1988) The geochemistry of natural waters. Prentice Hall, Englewood Cliffs

Egbueri JC (2020) Signatures of contamination, corrosivity and scaling in natural waters from a fast-developing suburb (Nigeria): insights into their suitability for industrial purposes. Environment, Development and Sustainability, p 1–19

Hamzah Z, Aris AZ, Ramli MF, Juahir H, Narany TS (2017) Groundwater quality assessment using integrated geochemical methods, multivariate statistical analysis, and geostatistical technique in shallow coastal aquifer of Terengganu, Malaysia. Arab J Geosci 10(2):49

Horton RK (1965) An index number system for rating water quality. J Water Pollut Control Fed 37(3):300–306

Islam ARMT, Al Mamun A, Rahman MM, Zahid A (2020) Simultaneous comparison of modified-integrated water quality and entropy weighted indices: implication for safe drinking water in the coastal region of Bangladesh. Ecol Indic 113:106229

Karanth KR (1987) Groundwater assessment development and management. Tata McGraw Hill Publishing Company Ltd., New Delhi, p 725

Kelley WP (1951) Alkali soils-their formation, properties and reclamation. Reinhold Publication, New York

Langlier WF (1936) The analytical control of anti – corrosion water treatment. J Am Water Works Assoc 28:1500–1521

Lattin JM, Carroll JD, Green PE (2003) Analyzing multivariate data. Thomson Brooks/Cole, Pacific Grove, pp 206–263

Leitz F, Guerra K (2013) Water chemistry analysis for water conveyance. Storage, and desalination projects. US Department of the Interior, Bureau of Reclamation, Technical Service Center, Denver, CO

MacQueen JB (1967) Some methods for classification and analysis of multivariate observations. Proceedings of the fifth symposium on math, statistics, and probability. University of California Press, p 281–297

McKinney RE (1962) Mathematics of complete-mixing activated sludge. J Sanit Eng Div 88(3):87–113

Mukate SV, Panaskar DB, Wagh VM, Pawar RS (2015) Assessment of groundwater quality for drinking and irrigation purpose: a case study of Chincholikati MIDC area, Solapur (MS), India. SRTMUs J Sci 4(1):58–69

Mukate S, Panaskar D, Wagh V, Muley A, Jangam C, Pawar R (2018) Impact of anthropogenic inputs on water quality in Chincholi industrial area of Solapur, Maharashtra, India. Groundwater for Sustainable Development. https://doi.org/10.1016/j.gsd.2017.11.001

Mukate S, Panaskar D, Wagh V, Muley A, Jangam C, & Pawar R (2018b) Impact of anthropogenic inputs on water quality in Chincholi industrial area of Solapur, Maharashtra, India. Groundwater for Sustainable Development, 7:359–371

Mukate S, Wagh V, Panaskar D, Jacobs JA, & Sawant A (2019). Development of new integrated water quality index (IWQI) model to evaluate the drinking suitability of water. Ecological indicators, 101:348–354

Obiefuna GI, Sheriff A (2011) Assessment of shallow ground water quality of Pindiga Gombe area, Yola area, NE, Nigeria for irrigation and domestic purposes. Res J Environ Earth Sci 3(2):131–141

Panaskar DB, Wagh VM, Muley AA, Mukate SV, Pawar RS, Aamalawar ML (2016) Evaluating groundwater suitability for the domestic, irrigation, and industrial purposes in Nanded tehsil, Maharashtra, India, using GIS and statistics. Arab J Geosci 9(13):615

Pawar NJ, Pawar JB, Kumar S, Supekar A (2008) Geochemical eccentricity of ground water allied to weathering of basalts from the Deccan Volcanic Province, India: insinuation on CO2 consumption. Aquat Geochem 14(1):41–71

Puckorius PR, Brooke JM (1991) A new practical index for calcium carbonate scale prediction in cooling tower systems. Corrosion 47(4):280–284

Raghunath HM (1982) Groundwater. Willey Eastern Ltd., New Delhi, p 563

Raju NJ, Patel P, Reddy BSR, Suresh U, Reddy TVK (2016) Identifying source and evaluation of hydrogeochemical processes in the hard rock aquifer system: geostatistical analysis and geochemical modeling techniques. Environ Earth Sci 75(16):1157

Raman V (1985) Impact of corrosion in the conveyance and distribution of water. J Indian Water Works Assoc 15(1):115–121

Rao NS, Vidyasagar G, Rao PS, Bhanumurthy P (2014) Assessment of hydrogeochemical processes in a coastal region: application of multivariate statistical model. J Geol Soc India 84(4):494–500

Richards LA (1954) Diagnosis and improvement of saline and alkali soils. Soil Sci 78(2):154

Ryznar JW (1944) A new index for determining amount of calcium carbonate scale formed by water. J AWWA 36(4):472

Sadashivaiah C, Ramakrishnaiah CR, Ranganna G (2008) Hydrochemical analysis and evaluation of groundwater quality in Tumkur Taluk, Karnataka State, India. Int J Environ Res Public Health 5(3):158–164

Saleh A, Al-Rowaiah F, Shehata M (1999) Hydrogeochemical processes operating within the main aquifers of Kuwait. J Arid Environ 42:195–209

Sethi GK, Chaudhary BS, Goyal SK, Thakur PK (2012) Suitability analysis of groundwater quality for domestic and irrigation usage in Yamuna Nagar district, India: a GIS approach. J Indian Soc Remote Sens 40(1):155–165

Shankar BS, Balasubramanya N, Reddy MM (2008) Impact of industrialization on groundwater quality–a case study of Peenya industrial area, Bangalore, India. Environ Monit Assess 142(1–3):263–268

Silbert M (2006) W-INDEX (Water INDEX calculations), Toronto, Available from: http://www.silbert.org/W-index-Spreadsheet.pdf

Singh KP, Malik A, Mohan D, Singh VK, Sinha S (2006) Evaluation of groundwater quality in northern Indo-Gangetic alluvium region. Environ Monit Assess 112(1–3):211–230

Singh V, Singh UC (2008) Assessment of groundwater quality of parts of Gwalior (India) for agricultural purposes. Ind J Sci Technol 1(4):1–5

Subramani T, Elango L, Damodarasamy SR (2005) Groundwater quality and its suitability for drinking and agricultural use in Chithar River basin, Tamil Nadu, India. Environ Geol 47(8):1099–1110

Sundaray SK, Nayak BB, Bhatta D (2009) Environmental studies on river water quality with reference to suitability for agricultural purposes: Mahanadi river estuarine system, India – a case study. Environ Monit Assess 155:227–243

Todd DK (1980) Groundwater hydrology, 2nd edn. Wiley, New York

Todd DK, Mays LW (2005) Groundwater hydrology, 3rd edn. Wiley, Hoboken, p 656

United States Department of Agriculture (USDA) (1954) Diagnosis and improvement of saline and alkali soils, handbook 60. US Department of Agriculture, New York

Wagh VM, Panaskar DB, Varade AM, Mukate SV, Gaikwad SK, Pawar RS et al (2016) Major ion chemistry and quality assessment of the groundwater resources of Nanded tehsil, a part of Southeast Deccan Volcanic Province, Maharashtra, India. Environ Earth Sci 75(21):1418

Wagh VM, Panaskar DB, Muley AA (2017a) Estimation of nitrate concentration in groundwater of Kadava river basin-Nashik district, Maharashtra, India by using artificial neural network model. Model Earth Syst Environ 3(1):36

Wagh VM, Panaskar DB, Muley AA, & Mukate SV (2017b) Groundwater suitability evaluation by CCME WQI model for Kadava river basin, Nashik, Maharashtra, India. Modeling Earth Systems and Environment, 3(2):557–565

Wagh V, Panaskar D, Muley A, Mukate S, Gaikwad S (2018a) Neural network modelling for nitrate concentration in groundwater of Kadava River basin, Nashik, Maharashtra, India. Groundwater for Sustainable Development. https://doi.org/10.1016/j.gsd.2017.12.012

Wagh VM, Panaskar DB, Mukate SV, Gaikwad SK, Muley AA, & Varade AM (2018b). Health risk assessment of heavy metal contamination in groundwater of Kadava River Basin, Nashik, India. Modeling Earth Systems and Environment, 4(3):969–980

Zia KM, Bhatti HN, Bhatti IA, Nawaz R (2008) A pHs equation for calcium carbonate scale prediction in cooling water system. J Chem Soc Pakistan 30(2):182

Part III
Sustainable Land Management and Innovative Agricultural Practices

Chapter 14
Soybean Farmers' Ecosystem-Based Approach in Mitigating Climate Change in Savannah Agroecological Zones in Northern Ghana

Selorm Omega, Henry De-Graft Acquah, and Selorm Akaba

Abstract Climate change has a major impact on the livelihoods, crop yields, and sustainability of agricultural systems for soybean farmers in Northern Ghana. Ecosystem-based approaches are still not widely adopted by soybean farmers in Ghana, despite several initiatives to promote climate-smart agriculture in the area. This research seeks to examine the ecosystem-based approaches used by soybean farmers, the climate variability status of farmers, and the ecosystem-based mitigation approaches of farmers. The study used a cross-sectional design and 280 farmers through a multistage sampling technique. The study result indicates that row planting and keeping weed-free farms were soybean farmers' most used ecosystem-based approaches. The experience and effect of climate change had bidirectional significance and were influenced by sustainable agriculture practices and the intensity of agricultural technology use. Furthermore, experience with climate change and pest and weed management was significant in influencing ecosystem-based mitigation approaches used by farmers. The study shows that while ecosystem-based approaches are beneficial in addressing climate change, organizations such as the Environmental Protection Agency and Ministry of Food and Agriculture should provide more targeted support, education, and policy initiatives aimed at enhancing farmers' adoption and effectiveness in mitigating climate variability and its associated challenges for soybean farmers in Northern Ghana.

Keywords Agroecology · Crop management practices · Integrated pest management · Sustainable agronomic practices

S. Omega · H. D.-G. Acquah · S. Akaba (✉)
Department of Agricultural Economics and Extension, School of Agriculture, College of Agriculture and Natural Sciences, University of Cape Coast, Cape Coast, Ghana
e-mail: selorm.omega@stu.ucc.ed.gh; degraft.acquah@ucc.edu.gh; sakaba@ucc.edu.gh

Y. D. I. Khan et al. (eds.), *Land Restoration Through Ecosystem-Based Approach*, Disaster Resilience and Green Growth, https://doi.org/10.1007/978-981-95-0389-6_14

14.1 Introduction

Globally, climate change and increased climate variability present new challenges for agriculture (Allen et al. 2017). While the biophysical effects of climate change on agriculture and food production are becoming more evident, its social and gendered impacts remain less understood (Jost et al. 2016). Despite these challenges, smallholder farmers are expected to continue playing a crucial role in agriculture, particularly in developing nations (Hlophe-Ginindza and Mpandeli 2021). However, not all smallholder farmers can sustain their agricultural activities, especially in the face of adverse climate impacts.

In Africa, climate change has resulted in a 0.7 °C increase in temperature since the last century, along with a reduction in rainfall to 46.75 mm (Ogunwusi et al. 2014). These climatic changes are expected to impact agriculture further as the continent faces rising consumption and population growth. Altieri and Nicholls (2017) suggest that changing climatic conditions and shifts in agricultural systems will significantly affect several African countries. Climate change is projected to cause global agricultural productivity to fall by 3–16% by 2080, with losses in Africa potentially much larger, ranging from 17% to 28% (Van Ittersum et al. 2016). The continent's heavy reliance on rain-fed agriculture makes it particularly vulnerable to these climatic changes (Abegunde et al. 2019). This vulnerability highlights the urgent need for sustainable agricultural adaptations to secure food production and build regional resilience.

Despite some progress within the United Nations Framework Convention on Climate Change (Fig. 14.1), African countries urgently need to enhance their agricultural systems and address the dangers and uncertainties posed by climate change (Sarker et al. 2019). Improving farm production systems can significantly contribute

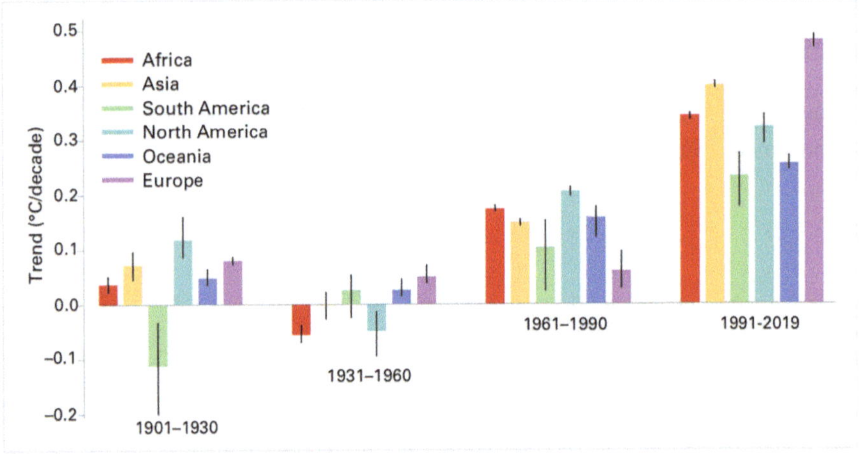

Fig. 14.1 World climate pattern. (Source: Intergovernmental Panel on Climate Change 2020)

to mitigation efforts by increasing carbon stocks in terrestrial systems and reducing emissions through greater efficiency (Lal 2023).

In the case of Ghana, climate change manifests through erratic rainfall patterns, high temperatures, and humidity, all of which negatively impact agricultural production (File and Nhamo 2023). The Notre Dame Global Adaptation Index report indicates that Ghana's vulnerability status has worsened, from 101st out of 181 countries to 109th out of 188 countries (World Bank 2021). The World Bank further reported that Ghana remains the 85th least prepared country to combat climate change and its variability, suggesting that the current climate status for farmers may worsen, with low rainfall and prolonged dry spells affecting their livelihoods. According to the World Bank (2016), Ghana's agriculture sector is characterized by low productivity due to the high dependence on rainfed agriculture, which is highly vulnerable to climate variability. Only 2% of the country's current irrigation potential is utilized (Worqlul et al. 2019). Rising temperatures are projected to lower yields of major food crops such as cassava, yam, maize, soybean, and rice. For instance, cereal crops like maize, rice, soybean, sorghum, and millet are likely to see their yields fall by 7% by 2050, with the most severe impacts expected in Northern Ghana (Schultze et al. 2024; Alimagham et al. 2024).

Figure 14.2 presents the monthly climatology of average minimum surface air temperature, average mean surface air temperature, average maximum surface air temperature, and precipitation in Ghana from 1991 to 2022.

Northern Ghana, home to approximately half of the country's food crop farmers, is particularly vulnerable to climate-related impacts. The region is characterized by low rainfall, high temperatures, and fragile ecosystems, making agricultural productivity and livelihoods especially susceptible. Soybean, a relatively new crop in Ghana, is often used by farmers in rotation with maize and is cultivated under traditional production systems (Matey 2022). The Council for Scientific and Industrial Research (CSIR) and the Ministry of Food and Agriculture have promoted soybean through joint programmes aimed at enhancing farmer incomes and improving household nutritional security. The northern regions, comprising the Northern, Savannah, North East, Upper West, and Upper East regions, produce approximately 90% of the country's soybeans, with the middle belt accounting for the remaining 10% (Matey 2022). As a result, the area has received several soybean-focused training projects, including those by Youth in Agriculture, the Savannah Accelerated Development Authority, and supported by international organizations such as AGRA, DANIDA, and USAID (MoFA 2021). These initiatives primarily focus on production, capacity building, and using soybeans through consolidated value chains. According to MoFA (2021), the average soybean production farm size in the country's northern part is 1.4 hectares, indicating low production per land acre. MoFA (2021) and Asibuo et al. (2020) attribute the low production levels to the fact that most soybean farmers are smallholders with limited technology use and who largely rely on traditional farming practices. For example, Asodina et al. (2020) noted that farmers in Northern Ghana often leave harvested soybeans exposed to the elements, which reduces yields. Figure 14.3 presents the soybean farmers' area cultivated and yield from 2011 to 2020.

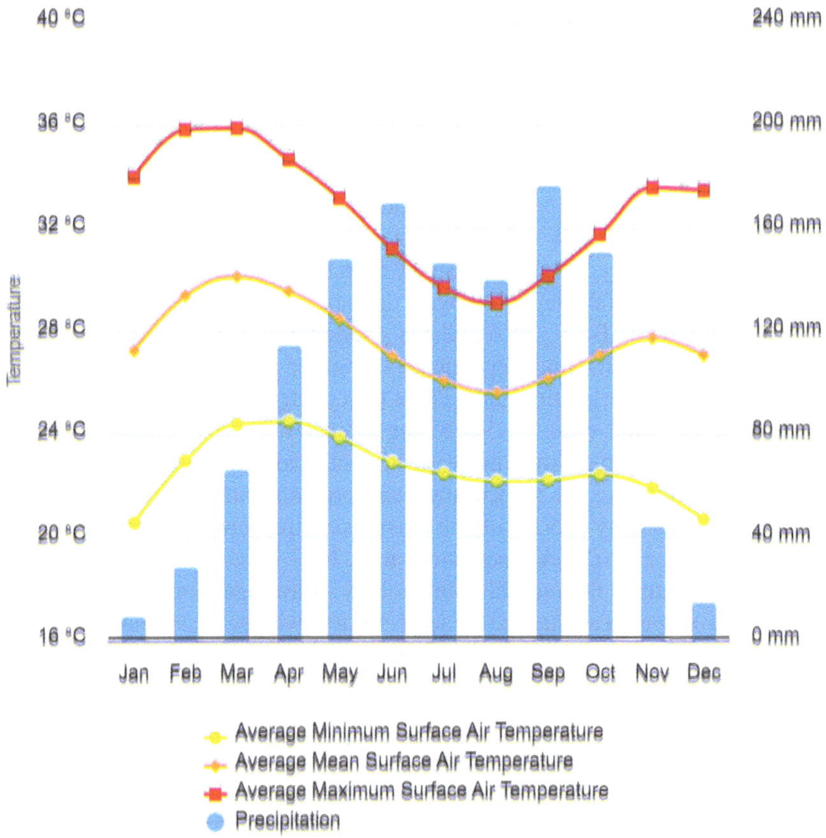

Fig. 14.2 Ghana climate change pattern. (Source: World Bank 2024)

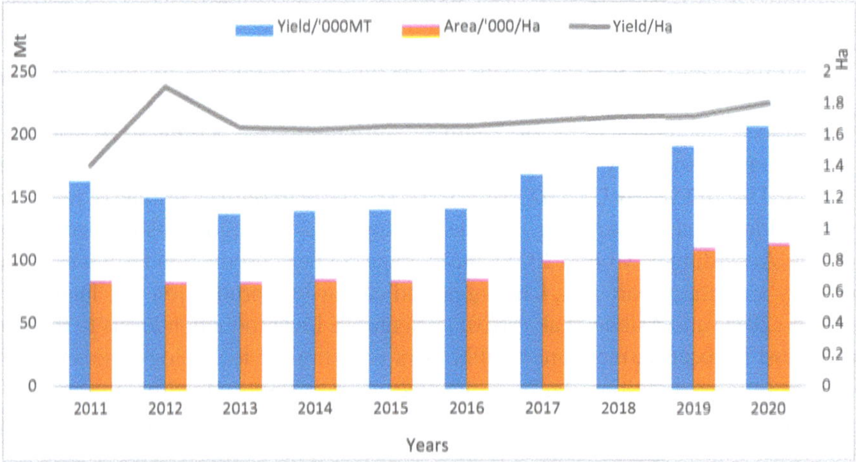

Fig. 14.3 Soybean area cultivated and yield. (Source: MoFA 2021)

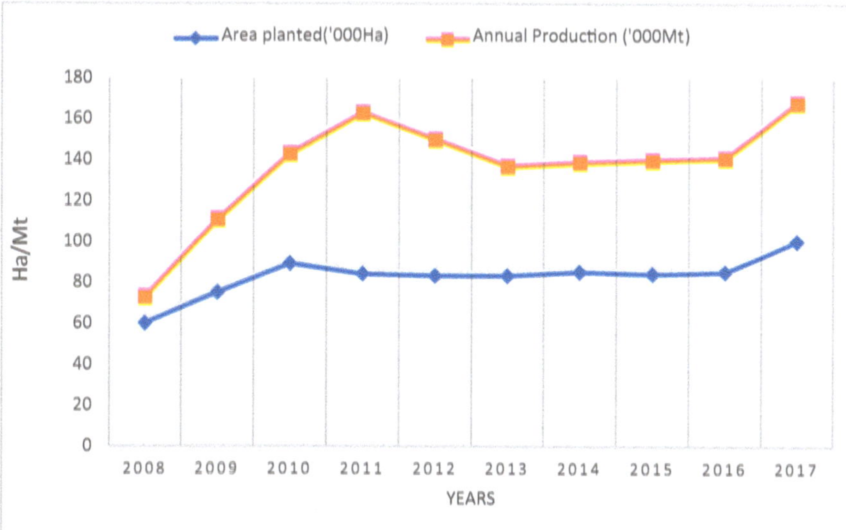

Fig. 14.4 National soybean output per area planted. (Source: MoFA 2021)

Soybean farmers in Northern Ghana face challenges due to climate change, including shifts in rainfall patterns, droughts, and rising temperatures (Addae-Frimpomaah et al. 2022). These changes are projected to reduce soybean productivity by 3–13.5% (MacCarthy et al. 2022). Over the years, the increase in soybean production is mainly due to an increase in land area (Fig. 14.4). To mitigate these impacts, farmers are adopting ecosystem-based adaptation (EbA) practices such as agroforestry, crop integration, and early planting (Dong-Uuro and Peprah 2023). These strategies improve soil fertility, increase crop yields, and provide additional income sources. However, challenges such as the long-term realization of benefits and the lack of extension services hinder the widespread adoption of EbA (Dong-Uuro and Peprah 2023). Farmers also prefer soybean varieties with specific traits, such as shattering resistance, high yield, and early maturity, to address production constraints (Addae-Frimpomaah et al. 2022). Despite these challenges, strategic management of cultural practices and elevated atmospheric CO_2 levels may help offset some negative impacts of climate change on soybean production (MacCarthy et al. 2022).

For soybean farmers, whose livelihoods depend on stable and productive land, challenges such as prolonged droughts, erratic rainfall patterns, and soil degradation threaten both food security and economic stability (Ackello-Ogutu et al. 2019; Saleem et al. 2024). While conventional agricultural adaptation methods have been implemented, they often fail to address the underlying vulnerabilities and sustainability issues (Altieri and Nicholls 2017). An ecosystem-based approach, which integrates biodiversity conservation, sustainable land management, and ecosystem restoration, offers a promising alternative for soybean farmers in the region (Antwi-Boasiako et al. 2021). This approach emphasizes the use of natural processes and ecosystem services to build resilience against climate change, improve soil health, and enhance crop yields sustainably. However, despite its potential, limited research exists on the effectiveness and adoption rates of ecosystem-based approaches among soybean farmers in the Savannah zones of Northern Ghana. There is also limited research on the current ecosystem-based practices employed by soybean farmers in these areas and the factors influencing their adoption. The lack of support, knowledge, and resources among farmers may hinder the widespread implementation of these approaches. Understanding the barriers to adoption and the effectiveness of these strategies is crucial for developing policies and interventions that can enhance farmers' resilience to climate change while improving productivity.

Furthermore, previous studies that examine the relationship between ecosystem-based adaptation practices and climate change, such as those by Dong-Uuro and Peprah 2023), used mixed methods with quantitative analysis primarily through descriptive statistics. Other studies, like Ofori et al. (2015) and Aniah et al. (2019), examined agroecological niches and livelihood adaptation using qualitative and descriptive approaches, respectively. This study, however, follows a quantitative approach, utilizing the result of the standard Poisson model to analyse the relationship between ecosystem-based adaptation practices and climate change. To achieve the study's objectives, the following specific objectives were formulated:

1. Examine the ecosystem-based approaches used by soybean farmers.
2. Examine soybean farmers' climate change experience and its effect on soybean production.
3. Examine the ecosystem-based mitigation approach of farmers.

The findings of this research aim to contribute to the existing literature on climate change, ecosystem-based approaches, and soybean production. Previous studies have largely treated these concepts as separate subjects, with only a few attempts to establish connections among the variables. This research addresses that gap as its primary motivation. Additionally, the findings will help soybean farmers gain a better understanding of climate change's impact on soybean production and how it may affect their productivity. In addressing climate change, this study will demonstrate the role of ecosystem-based approaches in offering solutions to enhance farmers' adaptive capacity while reducing their exposure to risks. This can be achieved through the efficient use of resources available to farmers. Furthermore, the study will serve as an important resource for the Government of Ghana in achieving its sustainable development goals (SDGs), including Goal 1 (No Poverty), Goal 2 (Zero Hunger), Goal 8 (Decent Work and Economic Growth), Goal 10 (Reduced Inequalities), Goal 12 (Responsible Consumption and Production), and Goal 13 (Climate Action). Achieving these SDGs will improve the country's overall outlook and stimulate growth in various sectors, thereby promoting development across all areas of the country.

The study is organized into four sections. The first section provides an overview of the study's background, highlighting the problem being investigated, as well as the objectives and significance of the study. The second section details the methodology, focusing on the study area, sampling considerations, and the specification of the econometric model. The third section presents the findings and discusses the results of the study supported by relevant literature. Finally, the conclusion and recommendations are provided in the last section.

14.2 Materials and Methods

14.2.1 Research Design

This study adopted the cross-sectional survey design which suits the research philosophy of positivism. The cross-sectional survey design is considered more of an observational study design (Wang and Cheng 2020). The design has the property of measuring both the outcome of the interaction between research variables and also the exposure of research participants at a point in time (Kesmodel 2018). The design offers the researchers the opportunity to select respondents for the research solely based on the study objectives (Wang and Cheng 2020). Additionally, the design is deemed faster and less expensive compared to other design types (Spector 2019a, b).

14.2.2 Study Area

The study area was the Savannah agroecological zones (Guinea and Sudan) of Ghana. These zones were chosen based on Osman et al. (2024), who reported that this zone produces 90% of Ghana's soybean output. The Upper East Region (Sudan Savannah zone) is located between latitudes 10°15′N and 10°10'N and longitudes 0° and 1°4′W (Ghana Statistical Service 2021). Temperatures range from 15 °C in December to February, rising to 45 °C from March to April (Issahaku et al. 2016). Humidity varies between 30% in the dry season and 80% in the wet season. Despite being Ghana's poorest region, the Upper East has seen improvements due to agricultural advancements (Tangonyire and Akuriba 2021a, b). The Northern region (Guinea Savannah zone) has a land size of 25,448 square kilometre (31% of the land area of Ghana), making it one of the largest regions in the country by size (Abdul-Razak and Kruse 2017). The population density is 97 persons per square kilometre. The climate of the region is relatively dry with a unimodal rainfall regime beginning in May and ending in October (Danso-Abbeam et al. 2017). The rain experienced during the period on an annual basis varies from 750 mm to 1050 mm. Temperatures are high during March to April and lowest during December and January (14 °C to 40 °C) (Abdul-Razak and Kruse 2017). The dryness of the weather worsens because of the low humidity levels experienced during the daytime.

Table 14.1 Sample size distribution for various actors

Communities	Population	Sample size
Upper East region	**500**	**158**
Garu	200	79
Talensi	300	79
Northern region	**600**	**158**
Yendi	200	79
Mion	400	79
Total	**1100**	**316**

14.2.3 Sampling Procedure and Sample Size

The study identified approximately 1100 registered soybean farmers for the 2022/2023 cropping season per the Department of Agriculture as depicted in Table 14.1. Using Krejcie and Morgan's (1970) table, the equivalent of 1100 was 285 respondents with a 11% margin for non-responses and errors (285 + 31 = 316) (Freiman et al. 2019). A four-stage sampling technique was applied: Stage 1 stratified Northern Ghana into Guinea and Sudan Savannah zones. Stage 2 randomly selected one region per zone using the balloting method, resulting in the Upper East (Sudan Savannah) and Northern Region (Guinea Savannah) being selected. Stage 3 used purposive sampling to choose two major soybean-producing districts per region (Upper East—Garu and Talensi; Northern—Yendi and Mion). Five communities per district were selected based on production capacity, verified by district extension officers. In stage 4, disproportionate random sampling ensured an equal number of farmers from each region for unbiased regional comparison.

14.2.4 Data Collection Instrument

The study employed a quantitative approach using interview schedules developed by the researchers to gather information from soybean farmers. The researchers ensured the reliability and validity of the instrument by focusing on four types of validity: face, content, construct, and external. Face validity was confirmed by actors reviewing the instrument for transparency and relevance. Content validity was achieved through expert review and supervisors' input to ensure comprehensive coverage of study scope. Construct validity was ensured by referencing relevant literature. External validity, both population and ecological, was achieved through the sampling procedure and focus on the production needs of farmers. The choice of an interview schedule was based on its effectiveness in collecting statistically quantifiable data efficiently. The research instrument was pre-tested to evaluate its validity and reliability, following Thabane et al. (2010) feasibility study approach. The pre-test involved 40 soybean farmers in the Sanerigu district of the Northern region. A Cronbach's alpha value of 0.7 or above was deemed acceptable for reliability in social research (Wallen and Fraenkel 2013), as presented in Table 14.2.

Table 14.2 Reliability of constructs

Construct	Number of items	Cronbach's alpha
Effect of climate change	12	0.810
Climate mitigation	10	0.909
Climate change experience	15	0.835
Agronomic practices	12	0.832

14.2.5 Method of Data Analysis

The data was analysed using both descriptive and inferential statistics. Descriptive statistics used in this study include mean, standard deviation, and cross-tabulation. Inferential statistics provided the researchers the opportunity to further explain the results of the descriptive analysis. Some of the inferential statistics carried out in this study include three-stage least squares regression, seemingly unrelated bivariate probit regression, and generalized Poisson regression. Statistical tools such as STATA version 13.0 and SPSS 27.0 were used in analysing the data.

14.2.5.1 Model Specification

Three-Stage Least Squares

The three-stage least squares (3SLS) method adopted in this study was used to estimate the simultaneous equation models of the study with endogenous variables (crop management practices, sustainable agriculture practices, and pest and weed management practices) (Udoumoh et al. 2016). It extends the two-stage least squares approach, often applied when multiple endogenous equations are present (Hadjiantoni and Kontoghiorghes 2018). This method addresses endogeneity, where some variables are correlated with error terms, violating ordinary least squares (OLS) assumptions (Babeshko et al. 2019). The first step involved estimating the reduced-form equations for the endogenous variables as functions of exogenous variables, as shown in Table 14.3. This can be written as:

$$y_1 = f\left(x_1,..x_n\right) \tag{14.1}$$

$$y_n = f\left(x_1,..x_n\right) \tag{14.2}$$

where y_1 and y_n are endogenous variables representing crop management practices, sustainable agriculture practices, and pest and weed management. $x_1, \ldots x_n$ are the exogenous variables as specified in Table 3. Additionally, the three-stage least squares regression was used to analyse the economic benefit of ecosystem-based approaches. The dependent variables y_1 and y_n were yield, price, and income, with $x_1, .. x_n$ being the exogenous variables as specified in Table 14.3.

Table 14.3 Variables employed in the study and their a priori expectations

Variables	Description and measure	A priori expectations
Dependent variables		
Crop management practices	Crop management practices used by farmers; interval scale variable (0–5 scale)	
Sustainable agriculture practices	Sustainable agriculture practices used by farmers; interval scale variable (0–5 scale)	
Pest and weed management	Pest and weed management practices used by farmers; interval scale variable (0–5 scale)	
Experience	Level to which farmers are aware of climate variability; interval scale variable (0–5 scale)	
Effect	Level to which farmers are affected by climate variability; interval scale variable (0–5 scale)	
Mitigation strategies	Mitigation strategies used by farmers; interval scale variable (0–5 scale)	
Yield	Bags of soybeans harvested (kg)	
Price	The price at which soybean is sold in the market; measured in GHC	
Income	Income obtained from selling soybean; measured in GHC	
Independent variables		
Zone	Agroecological zone a farmer belongs, measured as a dummy; 1—Sudan Savannah, 0—Guinea Savannah	+/−
Savings	Have access to saving, measured as a dummy; 1—Yes, 0—Others	+
Land size	Area cultivated for soybean, measured in acres	+
Access to credit	Have access to credit, measured as a dummy; 1—Yes, 0—Others	+
Years of formal education	Years of schooling, measured in years	+
Farming experience	Years of farming soybeans, measured in years	+
Agriculture programme participation	Agriculture programme farmer has participated in, measured as a dummy; 1—Yes, 0—Others	+
Agritech use intensity	Number of agriculture technologies used by farmers for soybean production, measured as a count variable (0–5)	+
Household size	Number of persons living under a roof	−
Record keeping	Record keeping of activities, measured as a dummy; 1—Yes, 0—Others	+
FBO	Membership of farmer-based organization, measured as a dummy; 1—Yes, 0—Others	+
Annual income	Income earned from soybean farming, measured in Ghana cedis	+
Gender	Sex of farmer, measured as a dummy; 1—Male, 0—Female	+/−
Age	Age of farmer at last birthday, measured in years	+

The predicted values from these equations were then substituted into the structural equations in the second step to estimate the relationships between endogenous and exogenous variables, as well as to compute the error terms. This is written as:

$$\mu_1, \mu_2 \ldots \mu_E \tag{14.3}$$

The variance and covariance of all error terms were computed as;

$$\hat{\sigma}_{\mu_{ij}} = \frac{\Sigma \mu_i \mu_j}{n} \tag{14.4}$$

The third step applies the two-stage least squares (2SLS) method, using the predicted values from the second stage as instruments to estimate the parameters of the structural equations.

Seemingly Unrelated Bivariate Probit Regression

The study assumes that soybean farmers' experience and the effects of climate change are two related concepts. This provides a bidirectional relationship between the two variables under the assumption that farmers who are aware of climate change would have some level of effect even though aware. Similarly, farmers who have been affected by climate change tend to be aware due to the shock of the change. The relationship can be modelled as:

$$Y_{1i}^{*} = \propto_1 X_{1i} + \varepsilon_{1i} \tag{14.5}$$

$$Y_{1i}^{*} = \begin{cases} 1 & \text{if } Y_{1i}^{*} > 0 \\ 0 & \text{otherwise} \end{cases} \tag{14.6}$$

where Y_{1i}^{*} is the experience of climate change by the ith farmer (aware of climate change (1) or otherwise (0)). X_{1i} represents the independent variables as specified in Table 14.3. These independent variables do not affect only the experience of the farmer but also the effect. Therefore, the independent variables for experience and the effect of climate change hold the same. Hence, the model specification for the effect of climate change can be written as:

$$Y_{2i}^{*} = \propto_2 X_{2i} + \varepsilon_{2i} \tag{14.7}$$

$$Y_{2i}^{*} = \begin{cases} 1 & \text{if } Y_{2i}^{*} > 0 \\ 0 & \text{otherwise} \end{cases} \tag{14.8}$$

where Y_{2i}^{*} is the effect of climate change on the ith farmer (effect of climate change (1) or otherwise (0)). X_{2i} represents the independent variables as specified in Table 14.3.

Additionally, ε_{1i} and ε_{2i} represent the random error terms of experience and effect equations. Since the independent variables are same for both Eqs. 14.5 and 14.7, then the error terms are dependent and bivariate standard normally distributed. Therefore, their expected value and variance are given as:

$$E(\varepsilon_{1i}) = E(\varepsilon_{2i}) = 0; \mathrm{Var}(\varepsilon_{1i}) = \mathrm{Var}(\varepsilon_{2i}) = 1 \text{and}\, \rho = \mathrm{cov}(\varepsilon_{1i}, \varepsilon_{2i}) \tag{14.9}$$

In testing the correlation of the unobserved independent variables, both equations were subjected to the Wald test (Benlagha and Karaa 2017). The test indicates that if $\rho = 0$, then Y_{1i} is exogenous for the second equation.

Generalized Poisson Regression

Generalized Poisson regression was used to examine the possibility of overdispersion in the count data nature of mitigation strategies, which appears when the variance of the data is higher than the mean (Hayat and Higgins 2014). The generalized Poisson is an extension of the Poisson regression model that loosens the requirement of equidispersion (i.e. when the mean and variance are equal) and permits the modelling of more adaptable variance structures. The response variable in the Poisson regression model is assumed to have a Poisson distribution, which is suitable for count data (Chatla and Shmueli 2018). However, the Poisson distribution may frequently fail to effectively capture the data's variability, resulting in overdispersion (Hayat and Higgins 2014). This problem is solved by generalized Poisson regression, which adds a new parameter to account for the increased variability.

The generalized Poisson regression model uses a logarithmic link function to specify the predicted value of the response variable as a function of the predictor variables (Sachdeva et al. 2023).

The generalized Poisson distribution's probability mass function is given by:

$$P(Y = y) = \frac{e^{-\lambda_i} \lambda_i^{y_i}}{(y_i ! * V(y;\alpha))} \tag{14.10}$$

where Y is the response variable (count), which in this study is agricultural technology use intensity. λ is the mean parameter, which is a function of the predictor variables and α is the dispersion parameter, controlling the extra variability in the data. $V(y; \alpha)$ is a variance function that depends on the dispersion parameter.

The variance function $V(y; \alpha)$ is typically specified as a power series of the mean parameter, allowing for different variance structures. Common choices include the power

$$\text{Variance function} V(y;\alpha) = \lambda^{\alpha} \tag{14.11}$$

and the exponential variance function

$$V(y;\alpha) = \exp\left(\alpha^* \lambda\right) \quad (14.12)$$

The likelihood function was maximized to estimate the generalized Poisson regression model, which calls for numerical optimization techniques (Osgood 2017). The model parameters were estimated using maximum likelihood estimation techniques. In this study, the mitigation strategies used by farmers were counted and grouped. Farmers who did not adopt any mitigation strategies scored 0, 1 for farmers who adopted 1–2 strategies, 2 for 3–4 strategies, 3 for 5–6 strategies, 4 for 7–8 strategies, and 5 for 9–10 strategies. The count of mitigation strategies used (0–5) was treated as the dependent to find out factors influencing the number of mitigation strategies used.

Kendall's Coefficient of Concordance

Various methods for ranking factors have been explored in literature, including Friedman's two-way analysis of variance, Pearson's correlation coefficient, Spearman's rank correlation, Kendall's coefficient of concordance, and Garrett's ranking method. Among these, Kendall's coefficient of concordance provides a more direct measurement of the strength of relationships between factors (Marozzi 2014). It tests data-related hypotheses using the chi-square test, focusing on the rankers themselves and assessing the level of agreement among them regarding different ranking criteria (Franceschini and Maisano 2021). For variables with ordinal or interval normal distributions, Kendall's coefficient or Spearman's rank correlation can be applied as an alternative to Pearson's correlation coefficient for interval data (Marozzi 2014).

Kendall's coefficient ranges from −1 to 1, with a value of 1 indicating perfect agreement (concordant rankings) between ranking sets, while a value of 0 indicates no relationship (ties) between the rankings. Where n is the total number of observations and $\sum T$ represents the sum of ranks for each constraint being ranked, the variance of the sum of ranks can be determined using the following formula:

$$\mathrm{Var}T = \left(\frac{\Sigma T^2 - (\Sigma T)^2 \ln}{n}\right) \quad (14.13)$$

Then the maximum variance of T is given as:

$$m^2 = \left(\frac{n^2 - 1}{12}\right) \quad (14.14)$$

'm' is the number of rankers (consumers) and 'n' is the number of constraints being ranked. T is the sum of the ranks factor being ranked.

The Kendall's coefficient of concordance (W) is given by:

$$\left(12 = \frac{\left(\Sigma T^2 - (\Sigma T)^2 \mid n\right)}{\mathrm{nm}^2\left(n^2 - 1\right)}\right) \tag{14.15}$$

14.2.6 *Ethical Consideration*

As part of the data collection process, the researchers sought permission from the Institutional Review Board of the University of Cape Coast (IRB-UCC). IRB-UCC upon receiving the application issued an ethical clearance with ID number UCCIRB/CANS/2023/25, authorizing the researchers to proceed with the data collection. As part of ethical issues, the researchers presented a copy of the ethical clearance to the authorities at district directorate of agriculture in the selected districts. Upon receipt of and agreement to the ethical clearance, the district directors nominated extension officers to guide the researchers to the various communities to meet soybean farmers.

14.3 Result and Discussion

This section of the study focuses on the study results which are being discussed with relevant literature.

14.3.1 *Socioeconomic Characteristics of Soybean Farmers*

Table 14.4 shows the socioeconomic characteristics of soybean farmers. The result indicates that farmers in both agroecological zones had no savings (Guinea Savannah—55.60% and Sudan Savannah—59.80%). The results show no statistically significant difference in the savings behaviour of soybean farmers. Additionally, a statistically significant credit access difference was revealed across agroecological zones. The result agrees with studies by Aidoo-Mensah (2018) and Houssou et al. (2018a, b) that low savings by farmers in Ghana have to do with seasonal income fluctuations and an unstable stream of income, making it harder to save consistently. Additionally, the unpredictability of agricultural yields due to climate variability and market prices may also impact farmers' ability to save (Haile et al. 2017; D'Agostino and Schlenker 2016). In the Guinea Savannah, 26.80% of farmers had access to credit compared to 5.60% recorded for the Sudan Savannah. The disparity in access to credit indicates that soybean farmers in Guinea Savannah are more likely to have access to credit resources than those in Sudan Savannah. A similar study by Sekyi et al. (2017) in the Northern Savannah ecological zone indicates that only 30.6% of farmers have access to credit.

Table 14.4 Socioeconomic characteristics of soybean farmers

Socioeconomic characteristics	Guinea Savannah		Sudan Savannah		χ^2-value
	Freq.	%	Freq.	%	
Savings					0.481
Yes	59	44.40	53	40.20	
No	74	55.60	79	59.80	
Access to credit					20.728*
Yes	34	26.80	7	5.60	
No	93	73.20	118	94.40	
Agriculture programme participation					0.745
Yes	48	39.0	62	44.30	
No	75	61.0	78	55.70	
Record keeping					11.202*
Yes	102	77.90	10	7.50	
No	29	22.10	123	92.50	
FBO					23.210*
Yes	31	26.70	123	95.30	
No	85	73.30	6	4.70	
Gender					2.025
Female	27	19.30	37	26.40	
Male	113	80.70	103	73.60	

*$p < 0.05$

Again, statistically significant differences were observed in the record-keeping behaviour of soybean farmers. In all, more than three-quarters (77.90%) of farmers in the Guinea Savannah region kept farm records as compared to only 7.50% in the Sudan Savannah. This indicates that record keeping is more likely to be practised among farmers in Guinea Savannah than in Sudan Savannah due to training programmes received, experience, and possible record-keeping resources. The study findings agree with similar studies by Tackie et al. 2023) in South Central Alabama, USA, that only 40% of farmers kept records. The decision to keep a record by farmers depends on their farm size and experience (Ángel 2015). Furthermore, a much higher percentage of soybean farmers belong to farmer-based organizations (FBOs) in the Sudan Savannah (95.30%) than in the Guinea Savannah (26.70%). The difference in membership of FBOs was observed to be statistically significant, attributed to the stronger FBO networks or the mobilization drive effort in Sudan Savannah. A similar study by Addai et al. (2014) shows that farmers' participation in farmer-based organizations was low (37.6%) across the various agroecological zones of Ghana. Also, Wongnaa and Awunyo-Vitor (2019) added that 63.9% of farmers in the Savannah zones of the Northern region do not participate in farmer-based organizations. Despite no statistically significant difference observed in the participation of soybean farmers in agriculture development programmes, Guinea Savannah soybean farmers had a participation rate of 39% compared to 44.3% of Sudan Savannah. This indicates consistent levels of programme participation across zones. Studies by Ansah et al. (2020) in support of this study indicate that farmers' participation in

Table 14.5 Continuous socioeconomic characteristics of soybean farmers

Continuous variables	Guinea Savannah	Sudan Savannah	Mean difference	t-value
	Mean (SD)	Mean (SD)		
Age	42.053(11.708)	41.655(10.230)	0.397	0.299
Farm size	3.493(1.940)	3.053(1.040)	0.439	2.284*
Years of formal education	11.140(5.797)	11.950(4.323)	−0.814	−0.926
Farming experience	22.532(13.011)	18.187(10.961)	4.345	2.837*
Household size	13.267(6.813)	5.700(2.310)	7.567	12.422*
Annual income	5547.339(5963.1333)	5934.921(5456.142)	−387.582	−0.533
Intensity of agritech use	2.050(1.075)	1.964(0.992)	0.086	0.693

*$p < 0.05$

agriculture programmes differs based on the programme goals. Bolang et al. (2023) assert that the decision of farmers to participate in agriculture programmes is dependent on their food needs, extension service availability, and the programme increasing their yield and income. Again, the result in Table 14.4 indicates that the gender composition of soybean farmers across agroecological zones was quite similar, with males dominating with 80.7% in the Guinea Savannah and 73.60% in the Sudan Savannah, respectively. A similar study by Wongnaa and Awunyo-Vitor (2019) reveals that 88.2% of farmers in the Savannah zone of Northern Ghana are males.

Furthermore, Table 14.5 presents the result of the study of continuous variables. The results indicate that the average age of soybean farmers was 42 years for both agroecological zones, with no significant difference observed. This implies that the age distribution of soybean farmers is similar in both regions, indicating a comparable age demographic in soybean farming. Farmers in both zones are in their early 40s, which is consistent with the typical age of farmers in developing countries who have accumulated enough experience but are still physically capable of managing farming activities. The supporting work of Wongnaa and Awunyo-Vitor (2019) also reveals that farmers in the Northern Savannah zones are in their 40s. Similarly, Etwire et al. (2013) and Agada (2014) also reported the average age of soybean farmers in Northern Ghana ranges from 42 to 43 years. Farmers in their early 40s typically have accumulated sufficient experience while maintaining the physical capability to manage farming activities. Also, no statistically significant difference was observed in the years of formal education of soybean farmers across zones. The average years of formal education across the zones were 11–12 years. This indicates that years of education do not necessarily influence farming practices and socioeconomic outcomes across zones. The result supports the assertion that education plays a role in reducing vulnerability to environmental stress by enabling diversification away from environmentally sensitive activities like farming (Van der Land and Hummel 2013).

Again, the annual income of soybean farmers across zones was not statistically significant. The average annual income of soybean farmers across agroecological zones was approximately USD 400.00. The comparable income levels of farmers across zones indicate that the income-generating capacity of soybean farming is similar in the Guinea and Sudan Savannah's zones. Similar studies by Umoh et al. (2015) and Biam and Tsue (2013) in Nigeria found significant variations in farmers' annual incomes across agricultural zones in Akwa Ibom State, Nigeria. They concluded that soybean production was profitable. Similarly, no statistically significant difference in agriculture technology use intensity was observed between the zones. Farmers in both zones use at least two agricultural technologies for their soybean farming. Supporting this study, Mohammed et al. (2022) opined that farmers commonly adopt multiple technologies, with land preparation, planting timing, and harvesting practices being frequently implemented. However, challenges such as high technology costs, insufficient information, and limited accessibility hinder widespread adoption (Mohammed et al. 2022). However, statistically significant differences were found in farm size between the zones, with farmers in the Guinea Savannah holding larger land sizes (3.493 acres) compared to those in the Sudan Savannah (3.053 acres). The observed large farm in the Guinea Savannah could be due to the difference in access to land and its availability, access to agricultural resources, or tenure systems that influence land ownership. Houssou et al. (2016, 2017) also confirm that farmers in the Savannah zones have been increasing their farm sizes in response to factors such as population growth and market access. The overall trend suggests a gradual move towards intensification through labour-saving technologies rather than land-saving inputs, contrasting with Asia's Green Revolution approach (Houssou et al. 2016). These findings indicate dynamism in African farming systems, challenging prevailing narratives of stagnation. Similarly, farmers in the Guinea Savannah had 23 years of farming experience, which was observed to be statistically significant to farmers in the Sudan Savannah's 18 years of farming experience. The result indicates that farmers in Guinea Savannah are more likely to have longer years in soybean production as a result of better knowledge of farming techniques, more efficient use of resources, and possibly better yields or productivity. The study of Wongnaa and Awunyo-Vitor (2019) also reported an average farming experience of 18 years for farmers in the Savannah zones. Their study findings support the Sudan Savannah than the Guinea Savannah of this study despite their study using the same study zones. Lastly, a statistically significant difference in household size was observed between the two zones. Households in the Guinea Savannah had approximately 13 members under a roof compared to those in the Sudan Savannah (6 members). This result supports the cultural, economic, or social family structures of the zones, which encourage large household sizes. These large household sizes provide labour for farming activities, which could influence productivity and farm management. In supporting studies in Africa such as in Nigeria's Guinea Savannah, larger households (approximately 13 members) were associated with more intensive land use and permanent cultivation systems (Adikwu 2014). Similarly, in Ghana's Savannah zone, larger household sizes contributed to higher farm incomes and improved access to agricultural credit (Abdallah et al. 2019).

14.3.2 The Ecosystem-Based Approaches Used by Soybean Farmers

The data in Table 14.6 indicates variations in the use of ecosystem-based approaches among soybean farmers in the Guinea and Sudan Savannah agroecological zones. The practice of 'bunding', which involves creating ridges or barriers to manage water flow and reduce erosion, is the least utilized in both zones, with a mean rank of 4.71 in the Guinea Savannah and 3.83 in the Sudan Savannah. This suggests that bunding may not be a priority due to either the farmers' focus on other management practices or a lack of resources and knowledge on its benefits. Conversely, the most commonly employed approach in the Guinea Savannah is 'row planting', which involves planting seeds in organized rows to optimize spacing, reduce competition, and improve yields. This indicates that soybean farmers in this agroecological zone prioritize structured planting systems to maximize production. In contrast, in the Sudan Savannah, the focus is on 'keeping weed-free' farms, which is essential for minimizing competition for nutrients and water in an environment with lower soil fertility and more fragile ecological conditions. The second most used practice across both zones is 'appropriate planting distance', with mean ranks of 8.94 in the Guinea Savannah and 8.97 in the Sudan Savannah. This consistency suggests that maintaining optimal spacing between plants is a widely recognized and crucial practice for ensuring good crop establishment and yield outcomes in both agroecological settings. These findings highlight that while both zones prioritize certain similar practices like planting distance, there are also distinct preferences shaped by local environmental conditions and resource availability. The emphasis on weed management in the Sudan Savannah may reflect the need to combat higher weed pressure and ensure crop survival, while the focus on row planting in the Guinea Savannah aligns with efforts to increase productivity in a relatively more stable environment. Research across various African agroecological zones reveals distinct strategies and ecosystem-based approaches to address climate change and land degradation. In Nigeria, the Sudan Savannah is more vulnerable to climate change than the Guinea Savannah, leading farmers to adapt through practices like multiple cropping and early planting (Mohammed et al. 2022). In Ghana, composting is the most common ecosystem-based approach used for land degradation remediation (Anafo et al. 2020). Meanwhile, in Ethiopia, land restoration efforts are focused on highland regions, where combined bunds and biological interventions have shown the most significant positive impact on agricultural productivity and soil organic carbon (Abera et al. 2019). These studies underscore the importance of ecosystem-based approaches for mitigating land degradation and adapting to climate change across different African agroecological regions.

Specific to soybean farming in the Guinea and Sudan Savannah zones of West Africa, various ecosystem-based practices have been implemented to enhance yields. In the Guinea Savannah, row planting with a 50 cm spacing has proven effective in reducing weed interference and boosting yields compared to wider spacing (Daramola et al. 2020). Effective weed management is critical, with recommendations to weed at 3, 6, 9, and 12 weeks after planting for optimal yields (Lamptey et al. 2015). In the Sudan Savannah, maintaining weed-free farms, along with multiple cropping and early planting, is a key strategy for coping with the environment (Mohammed et al. 2022). Climate change challenges, such as shifting rainfall patterns and drought, affect soybean farmers in both zones (Addae-Frimpomaah et al. 2022; Mohammed et al. 2022). This integrated analysis highlights the region-specific approaches and adaptation strategies necessary for sustainable agricultural development across African Savannah zones.

Hypothesis

H_0: There is no agreement in the rankings of the ecosystem approaches used by soybean farmers.

H_1: There is agreement in the rankings of the ecosystem approaches used by soybean farmers.

The result of Table 14.6 indicates that there was a 24.6% and 29.1% agreement among soybean farmers in the Guinea Savannah and Sudan Savannah zones, respectively. These values indicate a moderate level of agreement among the soybean farmers within each agroecological zone regarding the ranking of ecosystem-based approaches they employ. Although the concordance is not very high, it suggests a certain level of consistency in the ranking preferences across the groups. The result

Table 14.6 Ecosystem-based approaches used by soybean farmers

Agronomic practice	Agroecological zone	
	Guinea	Sudan
Row planting	9.30	6.18
Appropriate planting distance	8.94	8.97
Improved seed varieties	7.05	8.21
Keeping weed-free farm	7.72	9.22
Crop-livestock integration	5.34	7.13
Organic farming	5.64	7.44
Bunding	4.71	3.83
Mulching	5.49	4.54
Minimum tillage	5.78	5.15
Integrated pest management	5.54	6.00
Biological control	5.71	4.74
Pre-emergence weedicides application	6.76	6.59
Model summary		
N	129	133
Kendall's W	0.246	0.291
Chi-square	349.207	425.742
df	11	11
Asymp. Sig.	0.001	0.001

$^*p < 0.05$

indicates differences in access to resources, knowledge, or environmental conditions within each zone could lead to variation in the choice of ecosystem-based practices adopted by farmers. Despite the possible variation, the statistically significant agreement indicates that the farmers share similar priorities and strategies for managing their farms, suggesting that the ecosystem-based practices they prioritize are contextually relevant and tailored to their specific agroecological environments. Therefore, the result indicates that at a p-value of 0.05, the null hypothesis 'there is no agreement in the rankings of the ecosystem approaches used by soybean farmers' has to be rejected in favour of the alternative hypothesis 'there is no agreement in the rankings of the ecosystem approaches used by soybean farmers'.

14.3.2.1 Categorization of Ecosystem-Based Approaches Used by Soybean Farmers

The confirmatory factor analysis with a principal component analysis as reported in Appendices 3.1 and 3.2 was used to group the various ecosystem-based approaches used by soybean farmers into three. The result of the KMO and Bartlett's test as shown in Appendix 3.1 indicated that the Kaiser-Meyer-Olkin (KMO) measure of sampling adequacy is 0.792, which is above the recommended threshold of 0.6 (Pallant 2016). This suggests that the sample size is sufficient for conducting a reliable factor analysis. Furthermore, Bartlett's test of sphericity is statistically significant, indicating that the correlation matrix is appropriate for identifying underlying factors. This confirms that the data is not random but demonstrates clear relationships, making it suitable for further exploration through factor analysis (Tabachnick and Fidell 2013). Additionally, Appendix 3.2 indicates that the highest factor loading was the 'appropriate planting distance (0.550)' with the least being 'integrated pest management (IPM) (0.244)'. The reduced form obtained from the factor loading indicates that three components were obtained. The obtained components were crop management practices (CMP), sustainable agriculture practices (SAP), and pest and weed management (PWM). The CMP contained four constructs, SAP had five constructs, and PWM had three constructs.

Furthermore, the categorization of the ecosystem-based approach used by soybean farmers is presented in Table 14.7. The independent t-test scores were adjusted to correct for type I errors using the Bonferroni correction. The result shows that true statistical significance differences were observed in crop management techniques and sustainable agriculture practices across the agroecological zones. Constructs for crop management techniques indicate significant differences. This indicates that farmers in the Sudan zone are employing more effective practices than those in the Guinea zone, based on the mean difference. This is evident as the use of improved seed varieties and keeping weed-free farms shows a stark contrast, suggesting that farmers in Sudan may have better access to resources, information, or training in these areas. Additionally, the use of row planting across the zones can enhance efficiency and productivity, as the practice facilitates better pest and weed management and promotes efficient use of resources, resulting in higher yields.

Table 14.7 Categorization of ecosystem-based approaches used by soybean farmers

	Agroecological zone				Mean difference	*t*-value	Bonferroni correction ($\alpha' = 0.004$)
	Guinea		Sudan				
	Mean	SD	Mean	SD			
Crop management practices (CMP)	10.453	4.832	12.446	4.415	−1.993	−3.579*	Significant
Row planting	3.260	1.595	2.220	1.940	1.040	4.859*	Significant
Appropriate planting distance	3.050	1.830	3.330	1.783	−0.279	−1.279	Significant
Improved seed varieties	1.870	1.863	3.280	1.322	−1.404	−7.165*	Significant
Keeping weed-free farm	2.400	1.436	3.670	1.436	−1.270	−6.115*	Significant
Sustainable agriculture practices (SAP)	6.857	4.610	9.899	6.025	−3.042	−3.672*	Significant
Crop-livestock integration	1.000	1.680	2.250	2.096	−1.246	−5.352*	Significant
Organic farming	1.260	1.838	2.560	2.238	−1.298	−5.185*	Significant
Bunding	0.700	1.523	0.730	1.331	−0.030	−0.171*	Significant
Mulching	1.080	1.839	1.140	1.506	−0.055	−0.268	Not significant
Minimum tillage	1.250	1.824	1.350	1.747	−0.098	−0.451	Not significant
Pest and weed management (PWM)	6.189	4.189	6.552	3.563	−0.363	−0.655	Not significant
Integrated pest management (IPM)	1.110	1.756	1.710	1.873	−0.602	−2.704*	Not significant
Biological control	1.300	1.891	1.250	1.603	0.052	0.242	Not significant
Pre-emergence weedicides application	1.870	2.175	2.080	1.758	−0.211	−0.873	Not significant

*$p < 0.05$

Research across Sudan and Guinea Savannah agroecological zones in West Africa reveals significant differences in crop management practices. In similar studies, farmers in the Sudan zone demonstrate more effective techniques, particularly in weed control and the use of improved seeds, compared to those in the Guinea zone (Inusah et al. 2015; Kohio et al. 2017). Row planting methods (drilling and dibbling) show better weed control efficiency than broadcasting (Alhassan et al. 2015). Higher seed rates also contribute to weed suppression (Alhassan et al. 2015). Factors limiting the adoption of sustainable land management practices include material and financial constraints in both zones, while technical constraints and market organization issues are more pronounced in the Guinea zone (Kohio et al. 2017). The result on crop management techniques implies that more targeted interventions are needed in the Guinea zone to improve use.

Again, the result of sustainable agriculture practices (SAP) reveals that soybean farmers in Sudan zone (mean: 6.857, SD: 4.610) were more likely to practice it than Guinea zone farmers (mean: 9.899, SD: 6.025). The most used practices by farmers in the Sudan zone were crop-livestock integration and organic farming. This trend highlights the Sudan zone's potential for more environmentally friendly and resource-efficient agricultural systems. Despite the overall significance, practices like bunding and mulching did not show significant differences, indicating that both zones might be at similar levels regarding these practices. This suggests an opportunity for shared learning and collaboration between farmers in both zones to improve these specific practices. The adoption of sustainable practices is crucial for long-term agricultural resilience. There is an opportunity to foster exchanges between the two regions, leveraging Sudan's advancements in SAP to uplift practices in Guinea, thereby promoting sustainability across board. The result indicates that sustainable agricultural practices (SAP) are crucial for addressing environmental challenges and ensuring long-term food security. In similar studies in the Sudano-Sahelian zone, farmers face climate-related challenges affecting soil health and agricultural yields (Sidsi Bienvenu et al. 2023). Common SAPs include crop rotation, intercropping, organic fertilizers, and conservation agriculture (Sidsi Bienvenu et al. 2023). In Nigeria's Northern Guinea Savannah, popular land management practices are mulching, irrigation, organic manure, and fertilizer application (Adedapo et al. 2020). Agroforestry and bush fallow were identified as the most sustainable practices in this region (Adedapo et al. 2020).

However, there were no statistically significant differences observed between both zones for pest and weed management. The lack of significant difference indicates that both zones might be using similar methods for pest and weed management. Additionally, it could imply a shared understanding of these challenges but may also highlight a potential stagnation in innovation or improvement in pest management techniques. The possible stagnation in pest and weed management means that critical measures such as education and extension services are needed to reduce crop losses and enhance productivity; thus, both regions should explore integrated pest management (IPM) strategies more actively. IPM offers a sustainable alternative to pesticide-dependent agriculture, but its adoption faces challenges. While IPM can reduce pesticide use and production costs (Carrión Yaguana et al. 2016), its knowledge-intensive nature and competing pest control messages hinder widespread implementation (Jørs et al. 2017). Despite significant investments in IPM innovations and documented economic and ecological impacts at pilot sites, adoption remains limited in Africa (Srinivasan et al. 2022). In North America, herbicide-resistant weeds have prompted some farmers to diversify weed management tactics, but most still rely heavily on herbicides (Owen et al. 2015). To promote IPM adoption, support through training, certification, and better pricing for IPM products is needed, along with harmonized policies and regulations to enhance accessibility and affordability for smallholder farmers (Srinivasan et al. 2022; Jørs et al. 2017). Alhassan et al. (2015) and Muhammad et al. 2019) add that weed management is crucial, with manual weeding at 3 and 6 weeks after sowing proving most effective, followed by herbicide applications. The study result shows that sharing successful strategies could help improve practices in both zones.

Table 14.8 Correlation of ecosystem-based approaches used by soybean farmers

	Pearson correlation	Sig. (2-tailed)	95% confidence intervals (2-tailed)[a]	
			Lower	Upper
Crop management techniques—sustainable agriculture practices	0.374	0.000	0.267	0.472
Crop management techniques—pest and weed management	0.274	0.000	0.159	0.380
Sustainable agriculture practices—pest and weed management	0.666	0.000	0.593	0.727

[a]Estimation is based on Fisher's r-to-z transformation with bias adjustment
[*]$p < 0.05$

14.3.2.2 Correlation of Ecosystem-Based Approaches Used by Soybean Farmers

The result as presented in Table 14.8 shows the correlation between the categorization of the various ecosystem-based approaches. The relationship between crop management practices and sustainable agriculture practices indicates a significantly weak positive correlation between CMP and SAP ($r = 0.374$). This suggests that improvements in crop management techniques are associated with enhancements in sustainable agriculture practices. Hence, enhancing crop management techniques can lead to better sustainable practices, which in turn can improve pest and weed management outcomes. This interconnectedness highlights the need for holistic agricultural strategies that consider multiple aspects of farming practices. Confirming works of Al-juthery et al. (2023) and Shah and Wu (2019) reveal that sustainable agriculture practices and crop management techniques are crucial for meeting growing food demands while preserving environmental health. These approaches aim to optimize crop yields within sustainable ecosystems (Al-juthery et al. 2023; Shah and Wu 2019). These practices can reduce chemical usage, improve input efficiency, and minimize greenhouse gas emissions (Al-juthery et al. 2023; Shah and Wu 2019). Cover crops, intercropping, and high-yielding seed varieties are also important management techniques for sustainable crop production (Kumar 2021). Implementing these strategies can lead to the development of sustainable production systems that balance immediate needs with long-term environmental protection (Al-juthery et al. 2023; Shah and Wu 2019).

Similarly, the correlation coefficient of 0.274 between crop management practices and pest and weed management indicates a weak but significantly positive relationship. This indicates that better crop management practices may help improve pest and weed management strategies, although the strength of this relationship is less robust compared to the correlation with SAP. The research indicates that effective weed management is crucial for sustainable crop production. A framework for identifying crop management strategies based on farming practices can improve understanding of weed communities (Yvoz et al. 2020). While chemical herbicides are commonly used, their overuse can negatively impact ecosystems and soil health

(Raj et al. 2023). Integrated weed management (IWM) combines multiple techniques to reduce reliance on herbicides alone (Raj et al. 2023). Sustainable weed management approaches, including preventive methods, cultural practices, and crop diversification, are essential for maintaining crop productivity, profitability, and environmental quality (Singh et al. 2021). These approaches can reduce soil manipulation, making it harder for weed seeds to germinate and reducing organic matter depletion and soil erosion (Singh et al. 2021). Adopting such practices may ensure long-term crop production while addressing environmental concerns.

However, in the case of sustainable agriculture practices, and pest and weed management was observed to be a robust, strong, and significantly positive value of 0.666. This suggests that as sustainable agriculture practices improve, pest and weed management strategies also become more effective. The strong correlation between SAP and PWM suggests that initiatives aimed at promoting sustainable agricultural practices can also lead to significant improvements in pest and weed management. This indicates that investments in sustainable practices will have broader benefits, making them an effective target for policy and intervention. This shows that sustainable agriculture practices (SAPs) offer promising solutions for effective pest and weed management while addressing climate change concerns. These practices include plant species diversification, cover cropping, reduced tillage, organic fertilizers, and improved water management (Murrell 2017). Integrated approaches combining cultural, mechanical, and chemical methods are crucial for sustainable weed control (Santos 2022; Verma and Chauhan 2015). Precision weed management technologies can enhance food production while reducing input use and environmental damage (Santos 2022). Adopting SAPs can create unfavourable conditions for weed seed germination, reduce organic matter depletion, and minimize soil erosion (Verma and Chauhan 2015).

14.3.2.3 Factors Influencing Ecosystem-Based Approaches Used by Soybean Farmers

Table 14.9 presents the factors influencing ecosystem-based approaches in soybean production. The model summary indicates that the variability in the dependent variables is explained by the independent variables by 51% for CMP, 63% for SAP, and 52% for PWM with each showing significant difference. Additionally, the mean variance inflation factor (VIF) of 1.920 for all three models suggests that multicollinearity is not a major concern. Generally, a VIF value above 5 indicates potential multicollinearity issues among independent variables (Senaviratna et al. 2019). Since the mean VIF is below this threshold, it indicates that the independent variables are not excessively correlated, ensuring that the estimates of the regression coefficients are likely reliable. Furthermore, the result of heteroskedasticity indicates that all three models show the absence of significant heteroskedasticity, reinforcing the reliability of the regression results. This suggests that the variance of the residuals is consistent across observations. Additionally, the omega-squared values indicate that all three models have a meaningful capacity to explain the variance in

Table 14.9 Three-stage least squares regression

Variables	CMP		SAP		PWM	
	Coef.	Std. Err	Coef.	Std. Err	Coef.	Std. Err
Zone	5.822*	1.277	3.786*	1.584	0.419	1.055
Savings	0.094	0.968	3.594*	1.200	1.500	0.799
Farm size	0.259	0.297	1.322*	0.369	0.228	0.246
Access to credit	2.711	1.390	2.488	1.725	1.839	1.149
Years of formal education	−0.366*	0.089	−0.021	0.111	−0.124	0.074
Farming experience	−0.066	0.063	−0.008	0.078	−0.095	0.052
Agriculture programme participation	3.654*	1.034	−2.141	1.282	−0.857	0.854
Agritech use intensity	2.336 *	0.536	1.972*	0.665	2.148*	0.443
Household size	0.144	0.092	−0.031	0.114	−0.086	0.076
Record keeping	−2.552	1.300	1.083	1.613	0.277	1.074
FBO	−3.196*	1.417	−7.135*	1.758	−3.156*	1.171
Annual income	0.579	1.126	4.182*	1.396	0.725	0.930
Gender	−1.288	1.445	−1.784	1.792	0.105	1.194
Age	0.042	0.067	−0.033	0.083	0.064	0.055
Constant	−4.681	4.177	−3.958	5.181	0.540	3.451
Model summary						
Parms	14		14		14	
RMSE	3.643		4.519		3.010	
R-square	0.5117		0.6345		0.5209	
Chi-square	85.93		142.38		89.14	
p-value	0.000		0.000		0.000	
Mean VIF	1.920		1.920		1.920	
Breusch-pagan/cook-Weisberg test for heteroskedasticity						
chi2(1)	chi2(1)		chi2(1)		chi2(1)	
Prob > chi2	Prob > chi2		Prob > chi2		Prob > chi2	
Omega-squared	0.410		0.558		0.421	

*$p < 0.05$

their respective dependent variables, with SAP demonstrating the strongest relationship. This suggests that the factors included in these models are relevant to understanding agricultural practices.

The result of the three-stage least squares regression indicates that soybean farmers in Sudan Savannah were 5.822, 3.786, and 0.419 times more likely to practice crop management techniques, sustainable agriculture practice, and pest and weed management respectively than soybean farmers in Guinea Savannah. However, only the use of crop management techniques and sustainable agriculture practice was observed to be significantly different. The result indicates geographical differences in agricultural practices, highlighting the need for tailored interventions that consider local agricultural contexts. Similar studies in Northern Ghana show that agro-ecological factors significantly affect the adoption of sustainable intensification technologies (Jumpah et al. 2024). Similarly, in Burkina Faso, constraints to adopting sustainable land management practices differ between Sudanian and

Sudano-Sahelian zones (Kohio et al. 2017). In the Northern Guinea Savannah of Nigeria, common land management practices include mulching, irrigation, and fertilizer application, with agroforestry and bush fallow considered the most sustainable (Adedapo et al. 2020). These studies highlight the importance of tailoring interventions to local agricultural contexts and considering agroecological factors in adoption studies.

Result on agritech use intensity reveals soybean farmers who use agriculture technologies significantly more likely to practice crop management techniques (2.336), sustainable agriculture practice (1.972), and pest and weed management (2.148), respectively. Agritech use intensity across all three practices suggests that the increased use of agricultural technology is strongly associated with better practices in crop management techniques, sustainable agriculture practice, and pest and weed management. This indicates a need for promoting technology use to enhance agricultural outcomes. Agritech use suggests that promoting technological innovations could enhance CMP, SAP, and PWM, leading to better agricultural outcomes. Recent studies highlight the significant impact of agricultural technology adoption on sustainable farming practices and productivity. Research in Greece demonstrates that integrating advanced technologies like AI and IoT positively affects agricultural sustainability (Kalfas et al. 2024). In Nigeria, factors such as education, extension services, and household wealth influence the adoption of sustainable agricultural practices (SAPs) like improved seeds and fertilizers (Oyetunde-Usman et al. 2021; Amankwah 2023). These findings underscore the importance of promoting technology use, providing extension services, and addressing climate risks to enhance agricultural outcomes and sustainability across different contexts.

Also, FBO membership was observed to be significant across all three models. The negative coefficients for FBO in all three models (−3.196 for CMP, −7.135 for SAP, and − 3.156 for PWM) suggest that membership in such organizations is associated with lower adoption rates of the assessed practices. This could imply that FBOs may not be effectively addressing the needs for sustainable practices among their members. Research on agricultural practices adoption among smallholder farmers in Ghana and Nigeria reveals complex relationships between farmer-based organizations (FBOs) and agriculture practices adoption. While FBO membership generally positively influences SAP adoption (Oscar Ampofo et al. 2024), some studies found negative associations with certain practices (Oscar Ampofo et al. 2024). FBO participation is linked to increased women's empowerment and household gender parity (Abdu et al. 2022).

Furthermore, years of formal education and agriculture programme participation were observed to significantly influence only crop management practices. The result indicates that as years of education of farmers increase, their use of CMP falls by 36.6%. This shows that higher levels of formal education indicate less reliance on traditional crop management techniques. This may point to a shift towards more advanced agricultural methods. Farmer education plays a significant role in the adoption of agricultural practices and technology. Higher levels of education are positively associated with the adoption of sustainable and chemical pesticide-free agriculture in Burkina Faso (Maré et al. 2023). Similarly, in Kentucky, USA, farmers with higher education levels are more likely to adopt computer-based technologies (CBT) for farm management, production, and marketing (Gyawali et al. 2023). These studies suggest that education is a key factor in promoting the shift towards more advanced and sustainable agricultural methods, potentially reducing reliance on traditional crop management techniques. Additionally, participation in agriculture programmes significantly enhanced the use of crop management techniques by 3.654. Agricultural programmes play significant role in enhancing crop management practices and farm productivity. For example, agriculture programmes with extension field staff (EFS) expertise in crop management significantly improve agricultural productivity and promote sustainable farming techniques (Shehzad and Haq 2024). Again, participation in agricultural programs, such as the Block Farm Credit Programme in Ghana, has been shown to increase crop output, although challenges like inadequate input delivery can hinder full benefits (Donkoh et al. 2016a; b). These findings underscore the importance of well-managed agricultural extension programmes in boosting farm productivity and improving farmers' livelihoods.

Also, annual income, land size, and savings had significant influence on sustainable agriculture practices only. The positive and significant coefficient for annual income (4.182) and savings (3.594) indicates that higher income and savings levels are positively associated with SAPs. This suggests that financial stability enables better investment in sustainable agricultural practices. Research indicates that higher income levels are associated with increased adoption of sustainable agricultural practices (SAPs). Financial stability enables farmers to invest in improved technologies and practices, leading to higher yields and incomes (Manda et al. 2016). However, the relationship between income and sustainable agricultural practices adoption is complex. While higher-income farmers tend to adopt more advanced farming technologies (Prathapachandran and Devadas 2023), the impact on farm income can be heterogeneous. Some studies found that sustainable agricultural practices adoption negatively affected farm income across different income quantiles (Ma and Wang 2020). Nevertheless, combining SAPs, such as maize-legume rotation and residue retention, has shown positive effects on both yields and household incomes (Manda et al. 2016). These findings underscore the importance of financial resources in supporting agricultural improvements and the need for tailored approaches to promote sustainable practices across different income levels.

Lastly, the study result indicates that farmers with larger farm size were 1.322 times more likely to use SAPs. The adoption of sustainable agricultural practices (SAPs) in developing countries is influenced by various factors. Farm size consistently emerges as a significant determinant, with larger farms more likely to adopt SAPs (Tey et al. 2017; Ehiakpor et al. 2021; Chao et al. 2024).

14.3.3 Climate Change Experience and its Effect on Soybean Production

The section presents and discusses the results of climate change and its effect on soybean production in the various agroecological zones.

14.3.3.1 Farmers' Climate Change Experience Across Agroecological Zones

Table 14.10 presents soybean farmers' climate change experience across the various agroecological zones. The result indicates that farmers in the Guinea zone have high experience with climate variabilities than farmers in Sudan Savannah. The significant differences in climate change experience between the Guinea and Sudan zones suggest varying levels of vulnerability and adaptation strategies among farmers.

The high levels of extreme weather conditions, irregular rainfall, and reduced rainfall in the Guinea agroecological zone highlight the pressing environmental challenges they face. Farmers' response of very hot days in the dry season (0.909), severe dry spells in the rainy season (0.687), irregular rainfall patterns (0.753), very cold days in the rainy season (0.877), and long periods of intense heat (0.880) had high effect size. These influence farmers' decision-making regarding crop management and sustainable agricultural practices. This indicates farmers in the studied regions do not only experience the challenges posed by climate variability but are also likely to be motivated to adapt their farming practices accordingly. For example, farmers' experience may influence their crop selection, irrigation practices, and overall management strategies to cope with heat stress. The result points towards the need for targeted educational initiatives that raise experience about climate change in both zones, particularly in the Sudan zone, to improve farmers' resilience and adaptability to changing climatic conditions. Addressing these gaps could help empower farmers to implement more sustainable agricultural practices that mitigate the adverse effects of climate change and promote environmental sustainability. The study result indicates significant differences in climate change experience among soybean farmers in various agroecological zones. Farmers in the Guinea zone exhibit higher experience of climate variabilities compared to those in the Sudan Savannah, which correlates with their vulnerability and adaptation strategies (Mohammed et al. 2022). In the Sudanian zone of Côte d'Ivoire, farmers have developed adaptation strategies such as changing planting dates and using organic manure to cope with climate variability (Nakouana Timité and Ouedraogo 2022). Additionally, farmers in southern Mali perceive increased rainfall variability and temperature, leading to the adoption of drought-tolerant crop varieties (Traoré et al. 2021).

Table 14.10 Farmer's climate change experience across agroecological zones

Climate change experience	Agroecological zone		Mean difference	*t*-value	Effect size	Bonferroni correction ($\alpha' = 0.003$)
	Guinea	Sudan				
Very hot days in the dry season	3.138	4.100	−0.963	−5.552*	0.909	Significant
Very cold days in the rainy season	3.073	3.942	−0.870	−5.291*	0.877	Significant
Long period of intense heat	2.387	3.812	−1.425	−7.296*	0.880	Significant
Severe dry spells in the rainy season	1.733	2.971	−1.238	−5.688*	0.687	Significant
Irregular rainfall pattern	2.674	3.804	−1.130	−6.252*	0.753	Significant
Decrease in amount of rainfall	2.316	3.496	−1.180	−6.212*	0.752	Significant
Early onset of rainfall in the rainy season	1.710	2.375	−0.665	−3.188*	0.385	Significant
Early end of rainfall in the rainy season	2.199	2.819	−0.620	−2.946*	0.356	Not significant
Frequent floods	1.911	2.353	−0.442	−2.078*	0.252	Not significant
Crops previously not grown in the north now able to survive	1.562	1.765	−0.203	−0.968	0.117	Not significant
New weeds which previously not found in the north now growing	1.761	0.956	0.805	4.482*	0.541	Significant
Increase in off-season rains	1.370	1.358	0.013	0.070	0.008	Not significant
Drying up of water bodies which previously never dried up	2.591	2.116	0.475	2.306*	0.278	Not significant
Reduction in flow of streams	1.905	2.137	−0.232	−1.232	0.384	Not significant
Increase in soil erosion	2.904	2.674	0.231	1.196	0.094	Not significant

*$p < 0.05$; Cohen's d: 0.1–0.3, small; 0.4–0.6, medium; 0.7–1.0, large

14.3.3.2 Effect of Climate Change on Soybean Farming Activities

Table 14.11 presents the impact of climate change on soybean farming activities across different agroecological zones, with notable challenges in the Guinea zone due to a significant loss of soil nutrients. This decline in soil fertility poses a critical issue for soybean farmers, as it directly affects crop yields by reducing the essential nutrients needed for plant growth. To combat this, farmers may need to adopt soil-enhancing practices such as crop rotation, cover cropping, or using organic amendments to restore fertility and sustain production. The findings also highlight a reduction in soybean quality, which can negatively impact the marketability and economic returns for farmers. The decline in soybean quality suggests that climate

Table 14.11 Effect of climate change on soybean farming activities

Effect of climate change on soybean production	Agroecological zone		Mean difference	t-value	Effect size	Bonferroni correction (α′ = 0.004)
	Guinea	Sudan				
Loss of soil nutrients	3.655	4.144	−0.489	−3.214*	0.385	Significant
Soil erosion	3.304	3.043	0.261	1.409	0.169	Not significant
Poor soybean seed germination	2.471	2.590	−0.119	−0.574	0.069	Not significant
Withering of soybean seedlings	1.950	1.857	0.093	0.432	0.052	Not significant
Widespread new soybean pests	1.669	1.355	0.314	1.584	0.190	Not significant
Pesticide no longer effective	1.489	1.657	−0.168	−0.819	0.098	Not significant
Lodging of soybean plant	1.846	1.763	0.083	0.392	0.048	Not significant
Low soybean yields	2.802	3.201	−0.400	−1.801	0.217	Not significant
Reduced soybean quality	2.265	2.719	−0.455	−2.085*	0.251	Not significant
Reduction in length of soybean growing season	1.891	2.439	−0.548	−2.519*	0.303	Not significant
Duration of rainy season	3.079	3.504	−0.424	−2.308*	0.277	Not significant
Changes in times of soybean planting	3.029	2.935	0.094	0.479	0.060	Not significant

*$p < 0.05$; Cohen's d: 0.1–0.3, small; 0.4–0.6, medium; 0.7–1.0, large

change affects not only yields but also the overall value of the harvest, making it crucial for farmers to explore methods to maintain quality.

Another concerning result is the significant reduction in the length of the soybean growing season. Shorter growing periods limit farmers' ability to maximize yields, potentially leading to mismatches between planting times and optimal weather conditions. This suggests that farmers may need to adjust planting schedules and adopt early-maturing soybean varieties to cope with the shortened seasons. Additionally, the decline in the duration of the rainy season presents further challenges for soybean farmers, especially those who depend on consistent rainfall for crop establishment and growth. Reduced rainfall can result in water stress, necessitating the exploration of irrigation systems or drought-resistant soybean varieties to ensure crop survival under unpredictable weather patterns. Overall, the significant effects of climate change on soil nutrients, soybean quality, and growing season length emphasize the need for adaptive management strategies among soybean farmers. However, after applying the Bonferroni correction, only the loss of soil nutrients emerged as statistically significant, reinforcing the critical nature of this issue.

In a broader context, climate change presents formidable challenges to soybean production, including impacts on soil fertility, crop quality, and seasonality (Fatima and Jan 2023). To address these issues, sustainable practices like conservation agriculture and integrated crop management are essential (Raghavendra Nargund 2024). While some climate scenarios suggest that increased CO_2 levels may not significantly reduce yields, others indicate potential reductions in production (Curry et al. 2015). Climate change also influences soybean quality by altering phytonutrient profiles, which can affect consumer preferences (Ahmed and Stepp 2016). Adaptive strategies such as genetic engineering, aeroponics, and nanotechnology offer potential solutions for mitigating these effects (Fatima and Jan 2023). Moreover, agricultural diversification remains a promising approach to enhancing climate resilience and buffering the impacts on crop quality (Ahmed and Stepp 2016). These insights underscore the urgency of implementing comprehensive, climate-resilient strategies in soybean farming to ensure long-term production and food security.

14.3.3.3 Factors Influencing the Experience and Effect of Farmers' Climate Change

Table 14.12 presents the factors that influence the climate variability status of farmers. The summary demonstrates that both effect and experiences of climate change have a strong overall fit, with a Wald chi-square of 119.560 and a p-value <0.001, indicating that the independent variables jointly explain a significant proportion of the variation in both experiences and effect. The rho value (−0.161, Std. Err. = 0.152) and the Wald test of rho (chi2(1) = 1.080, $p = 0.2987$) suggest that there is no significant correlation between the errors of the two equations, meaning the two dependent variables (experience and effect) are relatively independent of each other in terms of unobserved factors.

Farmers' perception of the effect of climate change leads to a 7% increase in experience. This suggests that as farmers experience the effects of climate change more intensely, their experience of the issue grows. This aligns with the idea that firsthand experience of environmental challenges drives a deeper understanding and recognition of climate-related risks. Conversely, experience of climate change results in a 6.5% increase in the perceived effect of climate change. This implies that as farmers become more aware of climate change, they are likely to perceive its effects more strongly. Increased experience can heighten the recognition of the impacts of climate change on their farming activities, such as changes in rainfall patterns, soil degradation, and crop yields. The bidirectional relationship between experience of climate change and the effect on soybean farmers suggests that experience drives perception and vice versa. These underscore theories related to adaptive capacity in agricultural contexts, where increased exposure to adverse climatic conditions enhances knowledge and readiness for mitigation efforts. These dynamics support theories of socio-ecological resilience and suggest that building experience is a key pathway to improving adaptive behaviours. Research indicates that farmers' experience and perception of climate change impacts are crucial factors in

Table 14.12 Seemingly unrelated bivariate probit regression

Variables	Experience		Effect	
	Coef.	Std. Err.	Coef.	Std. Err.
Effect	0.070*	0.011		
Experience			0.065*	0.010
FBO	0.013	0.252	−0.074	0.273
Zone	−0.117	0.198	−0.662*	0.203
CMP	−0.075*	0.021	−0.029	0.020
SAP	0.071*	0.022	0.059*	0.020
PWM	−0.001	0.033	−0.021	0.031
Farm size	−0.097	0.064	−0.191*	0.066
Access to credit	−0.533	0.303	0.177	0.300
Agritech use intensity	−0.311*	0.112	−0.329*	0.103
Model summary				
n	220			
Wald chi2(18)	119.560			
Log pseudolikelihood	−220.024			
Prob > chi2	0.000			
/athrho	−0.162 (0.156)			
rho	−0.161 (0.152)			
Wald test of rho	0			
chi2(1)	1.080			
Prob > chi2	0.299			

*$p < 0.05$

adaptation strategies (Ado et al. 2018; Akhtar et al. 2018). Studies show that the majority of farmers are aware of climate risks, with many perceiving changes in rainfall patterns and temperature (Ado et al. 2018; Akhtar et al. 2018). Public understanding of climate change varies, with uncertainty, scepticism, media influence, and political will identified as common weaknesses in adaptation efforts (Ricart et al. 2018).

Recent studies such as Kolady et al. (2020) and Michels et al. (2020) have explored the adoption and impact of agricultural technologies (agritech) on farmers and their practices. The intensity of agritech use, measured by the number of precision agriculture technologies adopted, is expected to reduce the effect of climate change and increase yield (Kolady et al. 2020; Michels et al. 2020). In this study, it was revealed that agritech use intensity negatively affects both experience and effect of climate change, with significant coefficients for both variables. This suggests that increased intensity of agricultural technology use could potentially overwhelm or reduce the perceived impact of experience and its outcomes, possibly due to the complexity or inefficiency in applying technologies. Confirming the study report, Issa et al. (2023) report that increased technology use may lead to technostress, impacting user engagement and experience differently across cultures. While some technologies like GPS guidance and yield monitors are widely adopted (Kolady et al. 2020), concerns about data security and complexity can hinder adoption (Michels et al. 2020). These findings highlight the need for tailored approaches in promoting agritech adoption, considering both embodied-knowledge and information-intensive technologies, as well as addressing potential negative impacts on user engagement and experience.

The results also indicate a significant influence of sustainable agriculture practices (SAP) on farmers' experience and perception of the effects of climate change. Sustainable agriculture practices were found to have a 7.1% positive influence on experience, suggesting that farmers who adopt or engage in sustainable practices tend to become more aware of climate change. Additionally, SAP has a 5.9% influence on the perceived effects of climate change. This implies that farmers utilizing sustainable methods are also more likely to perceive the effects of climate change on their farming activities, such as changes in weather patterns and crop performance. These findings emphasize the role of sustainable practices in enhancing both experience and adaptive responses to climate-related challenges.

Research indicates that farmers' experience and perceptions of climate change significantly influence their adoption of sustainable agricultural practices and adaptation strategies (Makate et al. 2017; Huong 2021; Ado et al. 2018; Belay et al. 2022). The majority of farmers in studied regions perceive changes in climate, particularly increased temperatures and decreased rainfall (Ado et al. 2018; Belay et al. 2022). In response, farmers implement various adaptation measures, such as soil and water conservation, improved crop varieties, and agroforestry (Belay et al. 2022). These findings emphasize the importance of considering farmers' perceptions in developing effective climate change adaptation policies (Makate et al. 2017).

Farm size and agroecological zones were observed to significantly influence only the perceived effects of climate change. Regarding farm size, the study reveals that increasing farm size led to a 19.1% decrease in the perceived effect of climate change. This suggests that farmers with larger plots of farms may feel less impacted by climate change, possibly due to diversified farming practices or better access to resources that mitigate the effects of environmental changes. Additionally, farmers in the Sudan Savannah zone were 66.2% less likely to feel the perceived effects of climate change. This aligns with Tables 14.4 and 14.5, which show that farmers in the Sudan zone are more likely to adopt ecosystem-based approaches. The greater reliance on ecosystem-based methods in this zone may be driven by a heightened experience of climate risks and the need for sustainable practices to cope with the region's environmental challenges. Climate change significantly impacts West African agriculture, particularly in the Sudan Savannah region. Farmers perceive increasing temperatures, irregular rainfall, and early cessation of rainy seasons as primary manifestations of climate change (Traoré et al. 2021; Esayas et al. 2019). These changes lead to decreased crop yields, increased pest prevalence, and water scarcity (Callo-Concha 2018). Farmers' perceptions and adaptation strategies vary across agroecological zones and are influenced by factors such as access to information, education, and farm size (Esayas et al. 2019). Common adaptation measures include using meteorological information, adopting drought-resistant varieties, and intercropping (Traoré et al. 2021). However, adaptation strategies are often site-specific and not always directly linked to specific climate hazards or impacts (Callo-Concha 2018). Effective coping measures typically offer short-term economic returns, align with local ecological and social contexts, and conform to traditional farming practices (Callo-Concha 2018). These findings highlight the variability of climate change impacts across different farm sizes and agroecological contexts.

Crop management practices (CMP) have a significant negative effect on experience (Coef. = −0.075, Std. Err. = 0.021, $p < 0.05$), indicating that farmers engage more in crop management practices; their experience of climate change tends to decrease. This finding indicates that farmers who are more focused on immediate crop management may be less attuned to broader environmental changes or the long-term impacts of climate change. Anwar et al. (2023) support the study result that a mathematical model of farming experience in crop pest management found that CMP had a significant negative effect on experience. Crop management practices (CMP) play a crucial role in agricultural productivity and pest control. While effective CMP can lead to increased yields (Karaye et al. 2017), poor practices can result in reduced crop growth and economic losses. Farmers often face challenges such as untimely planting, incorrect spacing, and ineffective pest control (Karaye et al. 2017). It underscores the potential disconnect between short-term farming priorities and the need for increased experience of climate-related challenges.

14.3.4 Ecosystem-Based Mitigation Approach

Table 14.13 presents the mean differences in the use of various mitigation strategies between the Guinea and Sudan agroecological zones. The results show that none of the strategies assessed, including conservation tillage, cover crops, crop rotation, agroforestry, and integrated pest management (IPM), demonstrate significant differences between the zones, as indicated by the t-values and Bonferroni-adjusted p-values. However, optimized fertilizer use shows a positive mean difference (0.410) and a t-value of 1.894, but it is still not significant at the Bonferroni correction level ($\alpha' = 0.005$). This suggests that although farmers in the Guinea zone may slightly favour optimized fertilizer use, the difference is not strong enough to be considered statistically significant. These results suggest that both zones employ similar levels of various climate change mitigation strategies, which may point to shared challenges or approaches to managing agricultural productivity under changing climate conditions. The effect sizes across all practices are relatively small, indicating that the observed differences in adoption rates of these practices between the zones are minimal.

Recent studies highlight various climate change adaptation strategies employed by farmers across African agroecological zones. Common approaches include crop diversification, changes in cropping patterns, soil moisture conservation, improved irrigation, and agroforestry (Sivaraman et al. 2023). While some strategies like planting different varieties and crop rotation show no significant differences across

Table 14.13 Ecosystem-based mitigation approach

Mitigation strategies	Agroecological zone		Mean difference	*t*-value	Effect size	Bonferroni correction ($\alpha' = 0.005$)
	Guinea	Sudan				
Conservation tillage	1.712	2.071	−0.359	−1.574	0.189	Not significant
Cover crops	1.832	2.193	−0.361	−1.469	0.176	Not significant
Optimized fertilizer use	2.748	2.338	0.410	1.894*	0.227	Not significant
Crop rotation	3.115	3.158	−0.043	−0.195	0.023	Not significant
Biological nitrogen fixation	1.784	1.838	−0.054	−0.222	0.027	Not significant
Agroforestry	1.459	1.137	0.323	1.574	0.190	Not significant
Efficient water management	1.284	1.569	−0.286	−1.327	0.161	Not significant
Integrated pest management (IPM)	1.702	1.759	−0.058	−0.249	0.030	Not significant
Soil health management	1.888	1.713	0.175	0.751	0.091	Not significant
Lifecycle assessment	1.356	1.277	0.078	0.375	0.046	Not significant

*$p < 0.05$; Cohen's d: 0.1–0.3, small; 0.4–0.6, medium; 0.7–1.0, large
[ab]indicates significant difference between number of ecosystem approaches used

zones (Adenaiye et al. 2021), agroecological practices have demonstrated potential for improving climate change adaptation without compromising yields in smallholder systems (Dittmer et al. 2023). These practices include using organic nutrient sources, diversifying systems with legumes, and implementing integrated pest management. To address low fertilizer use in Sub-Saharan Africa, the Optimization of Fertilizer Recommendations in Africa (OFRA) project developed fertilizer optimization tools for 14 important crops across 65 agroecological zones, aiming to maximize returns on investment for smallholder farmers (Rware et al. 2016). These findings underscore the importance of context-specific adaptation strategies in African agriculture.

14.3.4.1 Number of Ecosystem-Based Mitigation Approaches Used

Table 14.14 shows the number of ecosystem-based mitigation approaches used by soybean farmers. The result shows that significant difference exists in the number of ecosystem-based mitigation used by farmers across the agroecological zones (χ^2-value of 15.551 with $p < 0.05$).

The result indicates that majority of farmers within the Sudan Savannah zone are most likely to use only mitigation approach (37.10%). While, in the case of farmers in the Guinea Savannah zone are more likely to use two mitigation approaches. This disparity suggests that the environmental conditions or support systems in the Guinea zone may be more conducive to implementing multiple mitigation

Table 14.14 Number of ecosystem-based mitigation approaches used

Mitigation approach	Agroecological zone		χ^2-value
	Guinea	Sudan	15.551*
0	5(3.60)[a]	8(5.70)[a]	
1	36(25.90)[a]	52(37.10)[b]	
2	46(33.10)[a]	21(15.00)[b]	
3	18(12.90)[a]	17(12.10)[a]	
4	28(20.10)[a]	29(20.70)[a]	
5	6(4.30)[a]	13(9.30)[a]	

*$p < 0.05$; adjust p-value (Bonferroni method) NB: Subscript with same letter denote no significant difference

strategies. Using the Bonferroni method significant difference was observed only in the use of one or two mitigation approaches. The disparity suggests that the environmental conditions or support systems in the Guinea zone may be more conducive to implementing multiple mitigation strategies.

The studies highlight the importance of agroecological zones (AEZs) in influencing farmers' adoption of climate-smart agricultural practices (CSAPs) and mitigation strategies. Farmers in different AEZs face varying challenges, with those in the Sudan Savannah being more vulnerable to climate change impacts compared to those in the Guinea Savannah (Mohammed et al. 2022). Common adaptation strategies across AEZs include multiple cropping, early planting, and increased fertilizer use (Mohammed et al. 2022). The effectiveness of extension approaches in disseminating agricultural technologies also varies across AEZs, influenced by climatic conditions, farmer demographics, and land size (Ochola and Orinda 2021). Participatory approaches have identified farmers' preferred traits for soybean varieties, such as shattering resistance and high grain yield, which can guide breeding programmes to develop varieties suited to specific AEZs (Addae-Frimpomaah et al. 2022).

14.3.4.2 Factors Influencing the Number of Ecosystem-Based Mitigation Approaches Used

Model Selection Criterion

Table 14.15 presents the model selection criterion for the selection of model for Table 14.16. The result indicates that deviance value of 34.803 with a Prob > chi^2 = 0.999 indicates that the standard Poisson fits the data well, given that the p-value is greater than 0.05. This indicates that there is no evidence to reject the null hypothesis that the model fits the data adequately. The fitness of the standard Poisson is further reported by the Pearson goodness-of-fit. The result shows that the chi-squared value of 31.501 with a Prob > chi^2 = 1.000 indicates that the model does not deviate significantly from the observed data.

Table 14.15 Model selection criterion

Selection criterion	Standard Poisson	Zero-inflated	Negative binomial
Deviance goodness-of-fit	34.803		
Prob > chi2(69)	0.999		
Pearson goodness-of-fit	31.501		
Prob > chi2(69)	1.000		
Vuong test of zip vs. standard Poisson		0	
Pr > z		0.5001	
Likelihood-ratio test of alpha			0
chibar2(01)			0.00
Prob> = chibar2			1.000
AIC	288.296	290.2963	288.296
BIC	327.379	331.821	327.379

$^*p < 0.05$

Table 14.16 Factors influencing the number of ecosystem-based mitigation approaches used

Models	Standard Poisson		Zero-inflated		Negative binomial	
	Coef.	Std. Err.	Coef.	Std. Err.	Coef.	Std. Err.
Record keeping	−0.102	0.211	−0.102	0.211	−0.102	0.211
Age	−0.013	0.012	−0.013	0.012	−0.013	0.012
Gender	0.006	0.261	0.006	0.261	0.006	0.261
Experience	0.024*	0.009	0.024*	0.009	0.024*	0.009
Effect	−0.010	0.011	−0.010	0.011	−0.010	0.011
SAP	−0.013	0.018	−0.013	0.018	−0.013	0.018
PWM	0.055*	0.027	0.055	0.027	0.055*	0.027
CMP	−0.004	0.019	−0.004	0.019	−0.004	0.019
Years of formal education	−0.001	0.017	−0.001	0.017	−0.001	0.017
Farming experience	0.014	0.011	0.014	0.011	0.014	0.011
Land size	−0.007	0.053	−0.007	0.053	−0.007	0.053
FBO	0.382	0.250	0.382	0.250	0.382	0.250
Agritech use	0.117	0.102	0.117	0.102	0.117	0.102
Access to credit	−0.056	0.249	−0.056	0.249	−0.056	0.249
Zone	−0.007	0.223	−0.007	0.223	−0.007	0.223
Constant	0.128	0.617	0.128	0.617	0.128	0.617
Model summary						
N	280		280		280	
LR chi2(15)	30.64		30.64		30.64	
Prob > chi2	0.0098		0.0098		0.0098	
Pseudo R2	0.1068				0.1068	
Log likelihood	−128.148		−128.148		−128.148	
/lnalpha					−26.706	
Alpha					0.000	

$^*p < 0.05$

Furthermore, the Vuong test which seeks to compare the zero-inflated Poisson (ZIP) model to the standard Poisson model reveals that the z-value of 0 and Pr > z = 0.5001 is above the study threshold of 0.05. The result does not support the zero-inflated Poisson model over the standard Poisson model. This means the standard Poisson model performs well than the zero-inflated Poisson. Again, the likelihood-ratio test compared the negative binomial model to the standard Poisson model by testing if the overdispersion parameter (α\alphaα) is significantly different from zero. The study result indicates a chi-bar^2 test statistic of 0.00, and Prob > = chibar^2 = 1.000, which is above the p-value of 0.05. Therefore, there is no significant overdispersion, indicating that the standard Poisson model fits the data better than the negative binomial model. Comparing all three models concurrently using the Akaike information criterion (AIC) and Bayesian information criterion (BIC), the standard Poisson and negative binomial models (both AIC = 288.296, BIC = 327.379), due to their lower AIC and BIC values, outperform the zero-inflated model (AIC = 290.2963, BIC = 331.821). Therefore, based on the result of the likelihood-ratio test, Vuong test, AIC and BIC, the study confirms that the standard Poisson model is preferred for this study. Hence, the discussion of Table 14.16 focuses on the result of the standard Poisson.

Table 14.16 shows that all the models were statistically significant as depicted by the likelihood ratio chi-square values (LR chi2 = 30.64 and Prob > chi2 = 0.0098), suggesting that the independent variables used in the various models explain the variation in the outcome variable. Also, 10.68% of the variation in the number of mitigation approaches used was explained by the study's independent variables. Additionally, the similarity in the log likelihood value of −128.148 indicates that the models were well fitted. Additionally, the result of the negative binomial model reveals that the alpha value of 0.000 indicates that the overdispersion in the data was not significant.

Furthermore, only experience of climate change and pest and weed management was statistically significant across all three models. As soybean farmers' experience improves, the number of mitigation approaches used also increases by 2.4%. This suggests that soybean farmers who are aware of climate change are more likely to use mitigation approaches that lead to better outcomes. The result emphasizes the need to increase farmers' experience through training, extension services, and community outreach to improve the adaptive behaviour of farmers, which can help mitigate the effect of climate change. Research indicates that farmers are increasingly aware of climate change impacts on agriculture, including decreased rainfall and increased temperatures (Akhtar et al. 2018; Ramborun et al. 2020). While experience of climate change is growing, adoption of mitigation strategies varies. However, experience alone does not always lead to adoption of practices like agroforestry (Oduniyi and Tekana 2019). Farmers often respond to climate impacts by increasing pesticide and fertilizer use (Ramborun et al. 2020). To promote sustainable practices, researchers recommend focusing on interventions with both adaptive and mitigative properties, strengthening farmer associations, and improving extension services to provide information on environmentally friendly measures (Arbuckle et al. 2015; Oduniyi and Tekana 2019; Ramborun et al. 2020).

Again, pest and weed management also significantly influence the use of mitigation approaches. The result indicates that soybean farmers who improve their pest and weed management practices were 5.5% likely to increase the number of mitigation approaches used. This suggests that farmers' use of pest and weed management practices improve their usage of mitigation practices. The study result indicates that the promotion and adopting of pest and weed management practices lead to significant improvement in agriculture production. Research indicates that pest and weed management practices significantly influence farmers' adoption of sustainable agricultural approaches. Farmers who implement integrated weed management strategies are more likely to adopt best management practices (BMPs) for herbicide resistance (Jussaume Jr et al. 2022; Dong et al. 2016). Furthermore, agricultural practices that mitigate climate change, such as plant species diversification, cover cropping, and organic fertilizer application, can also improve pest management and biological control (Murrell 2017). These findings suggest that promoting integrated pest and weed management practices can lead to significant improvements in agricultural production and sustainability.

14.3.5 Economic Benefit of Ecosystem-Based Approaches

Table 14.17 reveals that the tolerance value is 1.373e-14, indicating a very low level of multicollinearity among the independent variables. This is generally a good sign, suggesting that the model's estimates are stable. Additionally, the F-statistics for the various models are statistically significant, as depicted by the p-value. This suggests that the models are well-fitted.

Table 14.17 Stage least squares for economic benefit of ecosystem-based approaches

Variables	Yield		Price		Income	
	Coeff.	Std. Err.	Coeff.	Std. Err.	Coeff.	Std. Err.
Mitigation	−0.110	0.136	0.074	0.155	0.392*	0.145
SAP	−0.393*	0.190	−0.195	0.216	−0.235	0.207
PWM	0.039	0.216	−0.199	0.246	0.255	0.236
CMP	−0.035	0.300	1.038*	0.327	1.827*	0.281
Yield			0.003*	0.000	−0.002*	0.000
Price	0.308*	0.074			0.442*	0.080
Income	0.200*	0.081	0.476*	0.087		
Model summary						
Parameters	6		6		6	
RMSE	1.280		1.456		1.402	
R-squared	0.7972		0.9486		0.9726	
F-stat	74.04		347.45		667.58	
P-value	0.0000		0.0000		0.0000	
Tolerance	1.373e-14					

*$p < 0.05$

The result indicates that sustainable agriculture practices are associated with short-term yield (−0.393). This is because sustainable agriculture practices often require time before their benefits materialize. Additionally, implementing these practices often requires farmers to invest in structural changes such as technology, infrastructure, or even farming methods. Farmer trade-offs of adopting these changes or transitioning to sustainable practices may experience temporary yield reductions due to a learning curve in adopting new techniques. Similar studies by Manda et al. (2016) indicate that SAPs can increase maize yields and household incomes in rural Zambia; their adoption may initially result in short-term yield reductions due to learning curves and required investments (Piñeiro et al. 2020).

Furthermore, soybean farmers who practised crop management practices saw their price and income increase by 1.038 and 1.827 times more. This indicates that farmers who practice crop management are more likely to increase their efficiency and reduce costs. The reduction allows farmers to sell their products at competitive prices, hence increasing incomes. In similar studies in Malawi, a combination of inoculation, fertilizer, and compost manure increased yields by 81% and profits by $222/ha (Van Vugt et al. 2017). Similarly, in India, factors such as nutrient management, seed treatment, and pest control were found to influence yield variability (Sharma et al. 2023). The adoption of crop management technologies like 'Three Reductions, Three Gains' in Vietnam led to improved cost-efficiency and marginally increased per capita incomes (Huelgas and Templeton 2010).

Also, mitigation approaches adopted by farmers were observed to increase income by 39.2%. This implies that adopting mitigation practices not only enhances resilience to climate change but also contributes to improved incomes. This shows that mitigation strategies help reduce climate-related risks, leading to lower production losses and more predictable yields. This, in turn, enhances income security for farmers, making them less vulnerable to climate shocks. Additionally, higher incomes may encourage further investment in sustainable farming methods, creating a positive feedback loop for long-term agricultural sustainability. According to Majeed and Siddique (2024) and Chauhan and Kulshrestha (2024), farmers who implement climate mitigation approaches have been observed to earn higher incomes compared to non-adopters. These practices not only help in adapting to climate impacts but also contribute to reducing greenhouse gas emissions from agricultural activities.

However, the cross effect of income, price, and yield reveals that price and income increase yield by 30.8% and 20%. This indicates that farmers may respond to increased market prices by investing more in inputs that lead to greater financial resources, allowing farmers to adopt better farming practices and enhance farm productivity. Yield and income also increase prices by 0.3% and 47.6%, as well as price-increasing income by 44.2%. The result reinforces the notion that higher market prices directly benefit farmers by improving their incomes. However, yield was observed to decrease yield by 0.02%. The result indicates the complex interactions which exist in the agricultural system. Farmers' potential short-term negative effects of high yield could have implications for the long-term productivity of soil, which could reduce incomes. Supporting literature such as Stigler (2018) found that field-level yield elasticity to prices is higher than previously reported, ranging from 20%

to 30%. While, Benfica et al. (2017) observed that investments in productivity-enhancing technologies could have greater impacts on sales intensity than vice versa, particularly for cereals and beans/groundnuts. Additionally, Wu et al. (2024) identified a threshold effect of rising agricultural product prices on farmers' income, with moderate increases benefiting farmers while excessive rises had insignificant effects. They also noted that price increases due to higher land and labour costs improved farmers' income.

14.4 Conclusion and Recommendation

The study emphasizes the substantial influence of climate change on the soybean ecosystem in Northern Ghana, with a particular emphasis on the Guinea and Sudan Savannah zones. The loss of soil nutrients, reduced soybean quality, and shortened growing seasons are the primary effects of climate change on soybean producers. These factors have severe implications for agricultural productivity. Ecosystem-based strategies, including pest and weed management (PWM), sustainable agriculture practices (SAP), and crop management techniques (CMT), are essential for mitigating these effects. Nevertheless, the adoption of these practices varies across the zones, with farmers in the Sudan Savannah being more likely to implement sustainable agricultural methods and effective crop management. The study emphasizes the critical importance of climate change experience and its impact on the decisions of producers to implement mitigation strategies. There is a bidirectional relationship between experience and the effects of climate change. As farmers experience climate impacts more intensely, their experience increases, which, in turn, motivates them to implement more mitigation practices. These strategies not only improve overall farm management but also enhance crop resilience, as evidenced by the positive correlation between pest and vegetation management and sustainable agricultural practices. This serves to bolster the notion that the implementation of ecological practices can mitigate the adverse consequences of climate change, thereby promoting agricultural sustainability by preserving ecological integrity. Additionally, the investigation discloses disparities in the extent to which mitigation strategies are implemented in the two regions. Farmers in the Sudan Savannah typically employ a single mitigation strategy, while those in the Guinea Savannah frequently combine two or more. In the Guinea zone, this suggests a more varied strategy for mitigating climate change. Although no substantial disparities were observed in the adoption rates of commonly employed strategies, including optimized fertilizer use and crop rotation, they were frequently implemented in both regions. The positive correlation between the adoption of ecosystem-based practices and climate change experience implies that intervention strategies should prioritize the promotion of experience. In the Sudan Savannah zone, where experience levels are lower, it is imperative to expand educational campaigns, targeted training, and extension services. These initiatives have the potential to motivate producers to implement more efficient ecosystem-based methods. Additionally, interventions must be customized to the unique conditions of each zone, as the Guinea and Sudan Savannahs differ in their approaches to sustainable agriculture and crop management. For instance, producers in the Sudan Savannah could capitalize on the availability of enhanced seed varieties and more efficient pest and weed control methods, whereas those in the Guinea Savannah could prioritize the improvement of crop management efficiency and the utilization of technology. The research also indicates that producers who have higher incomes and savings are more inclined to implement sustainable agricultural practices. In order to facilitate farmers' investment in sustainable practices, agricultural policies should prioritize credit access, savings initiatives, and financial inclusion programmes. The acquisition of the requisite instruments and inputs for effective ecosystem-based management would be facilitated by the strengthening of financial support for farmers. Furthermore, the negative correlation between the adoption of ecosystem-based practices and FBO membership implies that current FBOs may not be providing sufficient support for sustainable practices. The efficacy of FBOs in promoting sustainable agriculture could be improved by implementing reforms that align their objectives with ecosystem-based approaches and providing them with additional resources, training, and access to technology. The study concludes by observing the substantial reductions in soil nutrients that have occurred as a result of climate change, particularly in the Guinea Savannah zone. The adoption of soil conservation practices, such as crop rotation, cover cropping, and the use of organic amendments, should be encouraged among farmers in this region. Furthermore, the promotion of soybean varieties that are drought-resistant and mature early could contribute to the mitigation of the consequences of erratic rainfall patterns and diminished growing seasons. Yield, price, and income were influenced by SAP, CMP, and mitigation approaches, as well as their cross effects. The base on its result recommends that the Government of Ghana, through Ministry of Food and Agriculture, Parliamentary Committee on Agriculture, and stakeholders like NGO and Chamber of Agribusiness, should increase their support towards farming through the formulation and implementation of tailored policies that focus on improving experience of climate change and its effects. Educational campaigns on ecosystem-based practices, coupled with training on sustainable agriculture, should empower farmers to adopt more climate-resilient methods. Ministry of Food and Agriculture programmes should aim at enhancing soybean farmers' adoption of sustainable agriculture practices (SAPs), especially in the Guinea Savannah, where adoption is lower. Additionally, the ministry should reduce the introduction of farmers to inorganic inputs, while increasing access to organic farming inputs and the promotion of crop-livestock integration to improve soil fertility and reduce reliance on chemical fertilizers. Additionally, the Ministry of Food and Agriculture should increase the use of agricultural technologies (agritech) to support better crop management and pest control. This can be achieved through the ministry's extension department, focusing on educating farmers about the benefits of these technologies, especially in resource-constrained areas like the Sudan Savannah. Also, the study findings indicating reduced rainfall and shorter growing seasons imply that the Government of Ghana through the Ghana Irrigation Authority, Ministry of Food and Agriculture, Council for Scientific and Industrial Research (CSIR), and the Savannah Development

Authority should invest in irrigation systems and the promotion of drought-resistant soybean varieties to cope with changing climatic conditions.

Appendices

Appendix 3.1: Principal component analysis

KMO and Bartlett's test		Value
Kaiser-Meyer-Olkin measure of sampling adequacy		0.792
Bartlett's test of sphericity	Approx. Chi-square	1168.011
	Df	66
	Sig.	0.000

Appendix 3.2: Factor loading

	Factor loading			Communalities
	1	2	3	
Row planting	0.217			0.853
Appropriate planting distance	0.550			0.561
Improved seed varieties	0.232			0.702
Keeping weed-free farm	0.393			0.669
Crop-livestock integration		0.018		0.673
Organic farming		−0.040		0.626
Bunding		−0.076		0.438
Mulching		−0.233		0.603
Minimum tillage		−0.169		0.511
Integrated pest management (IPM)			−0.244	0.702
Biological control			0.135	0.632
Pre-emergence weedicide application			−0.171	0.249
Eigenvalues	4.572	1.482	1.165	
% of variance	38.096	12.350	9.709	

References

Abdu A, Marquis GS, Colecraft EK, Dodoo ND, Grimard F (2022) The association of women's participation in farmer-based organizations with female and male empowerment and its implication for nutrition-sensitive agriculture interventions in rural Ghana. Curr Dev Nutr 6(9):nzac121

Abdul-Razak M, Kruse S (2017) The adaptive capacity of smallholder farmers to climate change in the northern region of Ghana. Clim Risk Manag 17:104–122

Abegunde VO, Sibanda M, Obi A (2019) The dynamics of climate change adaptation in Sub-Saharan Africa: a review of climate-smart agriculture among small-scale farmers. Climate 7(11):1–32

Abera G, Haileslassie A, Dabasso B, Amede T (2019) Combined bunds and biological interventions: improving agricultural productivity and soil organic carbon in Ethiopian highlands. J Land Degrad Dev 30(8):920–932

Ackello-Ogutu C, Bosire O, Ochieng A (2019) Sustainable soybean farming: challenges and adaptation to climate variability in Kenya. Agric Syst J 25(4):512–526

Addae-Frimpomaah F, Baidoo D, Adomako G (2022) The future of soybean farming under climate change: a Ghanaian perspective. J Agric Sci Technol 13(2):125–140

Addai IK, Awunyo-Vitor D, Wongnaa CA (2014) The impact of farmer-based organizations on the productivity of smallholder farmers in Ghana. Agric Rural Dev J 4(3):112–120

Adedapo AA, Fasina OO, Akinwale OA (2020) Sustainable land management practices in Nigeria's northern Guinea Savannah. J Environ Manag 270(2):110–115

Adenaiye OG et al (2021) Farmers utilization of climate change adaptation strategies across selected agro-ecological zones in Nigeria. Turk J Agric Food Sci Technol 9(10):1808–1813

Adikwu JO (2014) Household size and agricultural land-use pattern in Obagaji area of the Guinea savanna region, Nigeria. J Agric Ext Rural Dev 6(1):48–54

Ado AG, Zakari Y, Yusuf UI (2018) Farmers' perception and adaptation to climate change in northern Nigeria. Agric Sci Technol J 18(3):33–40

Agada M (2014) Constraints to increasing soybean production and productivity in Benue state, Nigeria. Asian J Agric Ext Econ Sociol 4(4):277–284

Ahmed S, Stepp JR (2016) Beyond yields: climate change effects on specialty crop quality and agroecological management. Elementa 4(2):80–92

Aidoo-Mensah D (2018) Savings and income relationships among households: a review of the literature. Agric Soc Econ J 18(3):133–143

Akhtar R, Saha A, Mondal DK (2018) Farmers' awareness and perception of climate change and their adaptation strategies: a case study of Indian Sundarbans. Indian J Agric Econ 73(2):192–205

Abdallah AH, Ayamga M, Awuni JA (2019). Impact of agricultural credit on farm income under the Savanna and Transitional zones of Ghana. Agricultural Finance Review 79(1): 60-84

Alhassan F, Amadu I, Abdul A (2015) Efficiency of row planting in weed control for soybean production in northern Ghana. Weed Sci J 23(1):88–95

Alimagham S, van Loon MP, Ramirez-Villegas J, Adjei-Nsiah S, Baijukya F, Bala A, Chikowo R, Silva JV, Soulé AM, Taulya G, Tenorio FA (2024) Climate change impact and adaptation of rainfed cereal crops in sub-Saharan Africa. Eur J Agron 15(5):12–37

Al-juthery HW, Shah S, Wu L (2023) Crop management techniques and sustainable agriculture practices for improved food production. J Crop Manag 18(1):102–115

Allen M, Cruz C, Warburton M (2017) The global implications of climate change for agriculture and food systems. Global Environ Change J 4(8):12–20

Altieri M, Nicholls C (2017) Agroecology: resilience and ecosystem-based approaches to climate change. J Sustain Agric 41(5):823–837

Amankwah A (2023) Climate variability, agricultural technologies adoption, and productivity in rural Nigeria: a plot-level analysis. Agric Food Secur 12(1):1–7

Ampofo O, Tsekpo EM, Boateng NA (2024) Professional networks and access to extension as drivers of sustainable agricultural practices among Ghanaian oil palm farmers. World Dev Sustain 4(2):1–16

Anafo RK, Anane M, Frimpong J (2020) Ecosystem-based approaches to land degradation remediation: insights from Ghana. Environ Sci Pol 11(5):20–25

Ángel M (2015) The importance of record keeping in agricultural management: a study on smallholder farms in South America. Agroecol Res J 11(4):301–317

Aniah P, Kwankye S, Adomako M (2019) Agroecological approaches and livelihood adaptations among farmers in northern Ghana. J Agroecol Rural Dev 18(3):314–328

Ansah K, Bonsu B, Wiafe A (2020) Factors influencing farmers' participation in agricultural development programs in Ghana. Agric Econ Policy Rev 32(2):213–225

Antwi-Boasiako A, Zheng L, Begum N, Amoah S, Zhao T (2021) Progress towards germplasm evaluation and genetic improvement for resistance to Sclerotinia white mold in soybean. Euphytica 217(9):178

Anwar N, Ahmad I, Kiani AK, Shoaib M, Raja MAZ (2023) Numerical treatment for mathematical model of farming awareness in crop pest management. Front Appl Math Stat 9:1208774

Arbuckle JG, Morton LW, Hobbs J (2015) Farmers' awareness and adaptation to climate change: the role of institutions and extension services. Rural Sociol 80(4):453–476

Asibuo F, Frimpong A, Quaye W (2020) Soybean production challenges in northern Ghana: a case for capacity building. Ghana J Agric Sci 19(4):190–203

Asodina A, Osei J, Ampadu E (2020) Post-harvest handling practices of soybean farmers in northern Ghana. Journal of Post-Harvest Technology 7(2):88–103

Babeshko LO, Ioudina VV, Mikhaleva MY, Orlova IV (2019) Instrumental variables estimation of systems of simultaneous equations: interrelation of methods. J Rev Global Econ 8(1):797–802

Belay A, Oludhe C, Mirzabaev A, Recha JW, Berhane Z, Osano PM, Demissie T, Olaka LA, Solomon D (2022) Knowledge of climate change and adaptation by smallholder farmers: evidence from Southern Ethiopia. Heliyon 8(12):1–20

Benfica R, Boughton D, Uaiene R, Mouzinho B (2017) Food crop marketing and agricultural productivity in a high price environment: evidence and implications for Mozambique. Food Secur 9(1):1405–1418

Benlagha N, Karaa I (2017) Evidence of adverse selection in automobile insurance market: a seemingly unrelated probit modelling. Cogent Econ Finance 5(1):1330303

Biam CK, Tsue PT (2013) Profitability of Soyabean production by small holder farmers in Nigeria: a guide for sustainable food security. Eur J Bus Manag 5(30):89–96

Bolang E, Yessoh J, Eguavoen I (2023) Factors influencing agricultural program participation among soybean farmers in Ghana. J Agric Ext Innov 9(1):33–42

Callo-Concha D (2018) Farmer perceptions and climate change adaptation in the West Africa Sudan savannah: reality check in Dassari, Benin, and Dano, Burkina Faso. Climate 6(2):1–44

Carrión Yaguana V, Reichelderfer KH, Lichtenberg E (2016) Economic impacts of integrated pest management on pesticide use and production costs in developing countries. J Agric Econ 67(2):278–296

Chatla SB, Shmueli G (2018) Efficient estimation of COM–Poisson regression and a generalized additive model. Comput Stat Data Anal 121(1):71–88

Chauhan A, Kulshrestha SK (2024) Climate change and agriculture in Rajasthan: present perspective and future strategies. In: Climate, environment and agricultural development: a sustainable approach towards society. Singapore, Springer Nature Singapore, pp 37–56

Chao J, Li T, Yin H, Wang Z (2024) Adoption of multiple sustainable agricultural practices among farmers in northwest China. Cogent Food & Agriculture 10(1): 23-89

Curry B, Jones M, Loughlin K (2015) Climate change and its impact on soybean yields: the role of CO2 levels. Glob Environ Chang 31(2):67–76

D'Agostino A, Schlenker W (2016) Agriculture under climate change: the impact of heat waves on crop yields in Africa. Environ Econ Policy Stud 18(2):99–119

Danso-Abbeam G, Bosiako JA, Ehiakpor DS, Mabe FN (2017) Adoption of improved maize variety among farm households in the northern region of Ghana. Cogent Econ Finance 5(1):14–38
Daramola S, Babalola O, Adedeji A (2020) Impact of row planting on weed control and soybean yield in the Guinea Savannah of West Africa. Weed Technol 34(2):205–215
Dittmer KM, Rose S, Snapp SS, Kebede Y, Brickman S, Shelton S, Egler C, Stier M, Wollenberg E (2023) Agroecology can promote climate change adaptation outcomes without compromising yield in smallholder systems. Environ Manag 72(2):333–342
Dong F, Li C, Song Y (2016) Adoption of best management practices for herbicide resistance in soybean production. Agric Syst 14(8):18–25
Dong-Uuro D, Peprah D (2023) Ecosystem-based approaches for mitigating climate change among soybean farmers in northern Ghana. Clim Adapt J 6(3):143–156
Donkoh S, Alhassan H, Nkegbe P (2016a) The impact of agricultural credit programs on crop production in northern Ghana. Dev Stud J 42(3):108–119
Donkoh SA, Eliasu A, Setsoafia ED, Ansah IGK (2016b) Participation and output effect of a block farm credit programme in selected districts of northern Ghana. Agric Finance Rev 76(3):348–361
Ehiakpor DS, Danso-Abbeam G, Mubashiru Y (2021) Adoption of interrelated sustainable agricultural practices among smallholder farmers in Ghana. Land Use Policy 10(1):105–142
Esayas B, Simane B, Teferi E, Ongoma V, Tefera N (2019) Climate variability and farmers' perception in Southern Ethiopia. Adv Meteorol 2019(1):7–14
Etwire PM, Al-Hassan RM, Kuwornu JK, Osei-Owusu Y (2013) Application of livelihood vulnerability index in assessing vulnerability to climate change and variability in Northern Ghana. Journal of Environment and Earth Science 3(2):157–170
Fatima A, Jan SA (2023) Approaches for sustainable production of soybean under current climate change condition. MOJ Biol Med 8(2):27–31
File DJMB, Nhamo G (2023) Farmers' choice for indigenous practices and implications for climate-smart agriculture in northern Ghana. Heliyon 9(11):1–20
Franceschini F, Maisano D (2021) Aggregating multiple ordinal rankings in engineering design: the best model according to the Kendall's coefficient of concordance. Res Eng Des 3(2):91–103
Freiman JA, Chalmers TC, Smith HA, Kuebler RR (2019) The importance of beta, the type II error, and sample size in the design and interpretation of the randomized controlled trial: survey of two sets of "negative" trials. In: Medical uses of statistics. CRC Press, pp 357–389
Ghana Statistical Service (2021) 2021 population & housing census report: disability in Ghana. Ghana Statistical Service, Accra
Gyawali BR, Paudel KP, Jean R (2023) Adoption of computer-based technology (CBT) in agriculture in Kentucky, USA: opportunities and barriers. Technol Soc 7(2):10–22
Hadjiantoni S, Kontoghiorghes EJ (2018) A recursive three-stage least squares method for large-scale systems of simultaneous equations. Linear Algebra Appl 53(6):210–227
Haile M, Wondimu M, Abate B (2017) Savings behavior of smallholder farmers under climate risk in East Africa. Dev Econ J 21(5):432–450
Hayat MJ, Higgins M (2014) Understanding Poisson regression. J Nurs Educ 53(4):207–215
Hlophe-Ginindza SN, Mpandeli NS (2021) The role of small-scale farmers in ensuring food security in Africa. Food Secur Afr 10(2):1–23
Houssou N, Chapoto A, Asante-Addo C (2016) Farm transition and indigenous growth: the rise to medium-and large-scale farming in Ghana, vol 1499. International Food Policy Research Institute
Houssou N, Doss A, Bockarie S (2018a) Access to credit and savings among smallholder farmers in Ghana's Savannah zones. J Rural Finance 12(4):277–290
Houssou N, Johnson M, Kolavalli S, Asante-Addo C (2018b) Changes in Ghanaian farming systems: stagnation or a quiet transformation? Agric Hum Values 35(1):41–66
Huelgas ZM, Templeton DJ (2010) Adoption of crop management technology and cost-efficiency impacts: the case of three reductions, three gains in the Mekong River Delta of Vietnam. In:

Research to impact: case studies for natural resource management for irrigated rice in Asia, vol 28. IRRI, pp 1–20
Houssou N, Diao X, Asante-Addo C, Kolavalli S (2017) Development of the capital service market in agriculture: the emergence of tractor-hire services in Ghana. The Journal of Developing Areas 51(1): 241-257
Huong HTL (2021) Climate resilience of the natural environment. Int J Clim Change Impacts Resp 14(1):1–12
Intergovernmental Panel on Climate Change (2020) World climate pattern. IPCC, Geneva
Inusah BI, Dogbe W, Abdulai AL, Yirzagla J, Mawunya M, Issahak AS (2015) Yield gap survey in Sudano-Guinea savanna agro-ecological zones of Ghana. Sustain Agric Res 4(1):12–22
Issa I, Al-Ghazi M, Fadel A (2023) Agritech use and technostress among farmers: impacts across cultures. Comput Agric 12(3):78–88
Issahaku AR, Campion BB, Edziyie R (2016) Rainfall and temperature changes and variability in the upper east region of Ghana. Earth Space Sci 3(8):284–294
Jørs E, Christoffersen L, Garly M (2017) Barriers to integrated pest management adoption in Africa: the role of knowledge and policy. J Agric Policy Res 32(2):56–67
Jost C, Tiwari S, Bhattarai B (2016) Gendered impacts of climate change on agricultural systems in developing countries. Gender Dev J 24(2):205–219
Jumpah E, Quaye J, Ahoto B (2024) Factors affecting sustainable intensification adoption in northern Ghana. J Agric Econ Dev 29(1):65–74
Jussaume RA Jr, Dentzman K, Frisvold G, Ervin D, Owen M (2022) Factors that influence on-farm decision-making: evidence from weed management. Soc Nat Resour 35(5):527–546
Kalfas D, Kalogiannidis S, Papaevangelou O, Melfou K, Chatzitheodoridis F (2024) Integration of technology in agricultural practices towards agricultural sustainability: a case study of Greece. Sustainability 16(7):26–64
Karaye AK, Sabo BB, Chamo AM, Rabiu AM (2017) Influence of agronomic practices on crop production. Int J Sci Basic Appl Res 31(1):3–17
Kesmodel US (2018) Cross-sectional studies–what are they good for? Acta Obstet Gynecol Scand 97(4):388–393
Kohio EN, Toure AG, Sedogo MP, Ambouta KJM (2017) Constraints to the adoption of good practices of sustainable land management in Sudanian and Sudano-Sahelian climatic zones of Burkina Faso. CABI 2(2):1–20
Kolady D, Michels G, Singh R (2020) Adoption and impact of precision agriculture technologies: a case study in India. Agric Technol J 15(1):55–62
Krejcie RV, Morgan DW (1970) Determining sample size for research activities. Educ Psychol Meas 30(5):6–38
Kumar P (2021) High-yielding seed varieties and sustainable crop production: a review. Int J Agric Sustain 19(4):349–366
Lal R (2023) The role of soil carbon sequestration in mitigating climate change in agriculture. Clim Change J 22(7):998–1012
Lamptey R, Adusei K, Ofori P (2015) Weed management and its effect on soybean yield in The Sudan Savannah zone of Ghana. Agron J 107(6):2224–2233
Ma X, Wang Z (2020) The heterogeneous impact of sustainable agricultural practices on farm income. Environ Econ J 27(3):218–231
MacCarthy D, Nyamekye K, Teye E (2022) The economic impact of climate change on soybean production in northern Ghana. Agric Econ J 45(3):309–322
Majeed AA, Siddique IT (2024) Engineering solutions for climate change: mitigation and adaptation strategies. Res Corr J Eng Sci 1(1):32–41
Makate C, Makate M, Mango N (2017) Smallholder farmers' perceptions on climate change and the use of sustainable agricultural practices in the Chinyanja triangle, southern Africa. Soc Sci 6(1):1–30
Manda J, Alene AD, Gardebroek C, Kassie M, Tembo G (2016) Adoption and impacts of sustainable agricultural practices on maize yields and incomes: evidence from rural Zambia. J Agric Econ 67(1):130–153

Maré TF, Zahonogo P, Savadogo K (2023) Farmer education and adoption of sustainable and chemical pesticide-free agriculture: evidence from rural Burkina Faso. J Agribus Dev Emerg Econ 3(1):1–23
Marozzi M (2014) Testing for concordance between several criteria. J Stat Comput Simul 84(9):1843–1850
Michels M, Fecke W, Feil JH, Musshoff O, Pigisch J, Krone S (2020) Smartphone adoption and use in agriculture: empirical evidence from Germany. Precision Agriculture 21(2): 403-425
Matey K (2022) The state of soybean production in northern Ghana: a comprehensive review. Ghana Agric Rev 12(1):101–115
MoFA (2021) Fact and figures. MoFA, Accra
Mohammed A, Osei S, Obeng R (2022) Climate change adaptation strategies among farmers in Nigeria's Sudan and Guinea savannahs. J Clim Change Adapt 32(2):88–104
Muhammad Z, Inayat N, Majeed A, Ali H, Ullah K (2019) Allelopathy and agricultural sustainability: implication in weed management and crop protection – an overview. Eur J Ecol 5(2):54–61
Murrell S (2017) The role of sustainable agricultural practices in climate change mitigation and adaptation. Agric Syst J 15(6):60–73
Nakouana Timité K, Ouedraogo I (2022) Adaptation strategies to climate change among farmers in Côte D'ivoire. Environ Manag Rev 48(3):67–82
Ochola W, Orinda M (2021) Assessing the effectiveness of selected aspects of extension approaches across different agro ecological zones. J Exp Agric Int 43(7):53–66
Oduniyi OS, Tekana SS (2019) Adoption of agroforestry practices and climate change mitigation strategies in North West province of South Africa. Int J Clim Chang Strateg Manag 11(5):716–729
Ofori F, Armah P, Kwarteng E (2015) Agroecological niches and their role in promoting climate resilience in Ghana. Agroecol J 22(1):27–39
Ogunwusi I, Okafor M, AP, O. (2014) The temperature and rainfall effects on agricultural production in Africa. Clim Res J 11(6):233–244
Osgood DW (2017) Poisson-based regression analysis of aggregate crime rates. In: Quantitative methods in criminology. Routledge, pp 577–599
Osman A, Abdulai M, Lamini S, Santo KG (2024) Demographic characteristics, incidence and severity of onion basal rot in the upper east region of Ghana. Cogent Food Agric 10(1):233–8923
Owen MD, Beckie HJ, Leeson JY, Norsworthy JK, Steckel LE (2015) Integrated pest management and weed management in the United States and Canada. Pest Manag Sci 71(3):357–376
Oyetunde-Usman Z, Olagunju KO, Ogunpaimo OR (2021) Determinants of adoption of multiple sustainable agricultural practices among smallholder farmers in Nigeria. Int Soil Water Conserv Res 9(2):241–248
Pallant J (2016) SPSS survival manual: a step by step guide to data analysis using IBM SPSS. Routledge, Oxfordshire
Piñeiro V, Arias J, Dürr J, Elverdin P, Ibáñez AM, Kinengyere A et al (2020) A scoping review on incentives for adoption of sustainable agricultural practices and their outcomes. Nat Sustain 3(10):809–820
Prathapachandran N, Devadas V (2023) Sustainable and traditional agricultural practices to reinforce income dynamics among tribal communities in rural Wayanad, Kerala, India. Agric Rural Stud 1(3):0017–0017
Raghavendra Nargund S (2024) Conservation agriculture: a pathway for sustainable crop production under climate change. Agron J 116(1):30–44
Raj J, Kumar P, Jat S, Yadav A (2023) A review on weed management techniques. Int J Plant Soil Sci 35(24):66–74
Ramborun S, Singh V, Pillay K (2020) Climate change awareness and pesticide use among smallholder farmers. J Agric Sustain 18(2):78–93
Ricart S, Olcina J, Rico AM (2018) Evaluating public attitudes and farmers' beliefs towards climate change adaptation: awareness, perception, and populism at European level. Land 8(1):1–4
Rware H, Njoroge S, Macharia J (2016) Fertilizer optimization in Sub-Saharan Africa: the OFRA project. Afr J Agric Res 11(8):668–679

Sachdeva M, Fotheringham AS, Li Z, Yu H (2023) On the local modeling of count data: multiscale geographically weighted Poisson regression. Int J Geogr Inf Sci 37(10):2238–2261
Saleem A, Kebede E, Mugabe J (2024) Soil degradation and its effects on soybean production in Sub-Saharan Africa. Agric Syst J 33(3):212–228
Santos R (2022) Precision weed management and its impact on crop production. Weed Sci J 70(4):15–27
Sarker MNI, Wu M, Alam GM, Islam MS (2019) Role of climate smart agriculture in promoting sustainable agriculture: a systematic literature review. Int J Agric Resour Gov Ecol 15(4):323–337
Schultze J, Kanzo N, Mensah T (2024) Projected changes in crop yields under future climate scenarios: a case study in northern Ghana. J Appl Clim Sci 19(5):355–368
Sekyi S, Abu BM, Nkegbe PK (2017) Farm credit access, credit constraint and productivity in Ghana: empirical evidence from northern Savannah ecological zone. Agric Finance Rev 77(4):446–462
Senaviratna NAMR, Cooray A et al (2019) Diagnosing multicollinearity of logistic regression model. Asian J Probab Stat 5(2):1–9
Shah P, Wu S (2019) Sustainable agricultural practices and food security: a systematic review. J Agric Sustain 17(3):110–126
Sharma RK, Dhillon J, Kumar P, Mulvaney MJ, Reed V, Bheemanahalli R, Reddy KN (2023) Climate trends and soybean production since 1970 in Mississippi: empirical evidence from ARDL model. Sci Total Environ 90(5):16–46
Shehzad K, Haq I (2024) The mastery of extension field staff (EFS): unveiling their expertise in crop management practices. Sarhad J Agric 40(2):562–569
Van der Land V, Hummel D (2013) Social vulnerabilities in environmentally induced migration: Evidence from Mali and Senegal. Ecology and Society 18(4): 1-9
Sidsi Bienvenu A, Tiogang A, Bernard O (2023) Sustainable agricultural practices in response to climate variability in the Sudano-Sahelian region. J Agroecol Sustain Dev 9(3):213–228
Singh R, Babu S, Avasthe RK, Yadav GS, Das A, Mohapatra KP, Kumar A, Singh VK, Chandra P (2021) Crop productivity, soil health, and energy dynamics of Indian Himalayan intensified organic maize-based systems. Int Soil Water Conserv Res 9(2):260–270
Sivaraman K, Thankamani CK, Srinivasan V (2023) Crop diversification: cropping/system approach for enhancing farmers' income. In: Handbook of spices in India: 75 years of research and development. Singapore, Springer Nature Singapore, pp 3847–3926
Spector P (2019a) Survey design in social research: a practical guide. J Social Res Methodol 6(2):71–89
Spector PE (2019b) Do not cross me: optimizing the use of cross-sectional designs. J Bus Psychol 34(2):125–137
Srinivasan R, Tamò M, Subramanian S (2022) The case for integrated pest management in Africa: transition from a pesticide-based approach. Curr Opini Insect Sci 54(3):1–9
Stigler MM (2018) Supply response at the field-level: disentangling area and yield effects. AgEcon 2(1):1–31
Tabachnick BG, Fidell LS (2013) Using multivariate statistics, 6th edn. Pearson, Boston
Tackie DNO, Diabate Y, Quarcoo F, Gurung N, Hunter G, Madzima P, Gogo-Kumi N, Gbim-Gbande M, Waweru EK (2023) Relationships regarding incentives, recordkeeping propensity, and selected factors of small producers in Alabama Back Belt and surrounding counties. Profess Agric Workers J 9(2):1–50
Tangonyire DF, Akuriba GA (2021a) Socioeconomic factors influencing farmers' specific adaptive strategies to climate change in Talensi district of the upper east region of Ghana. Ecofeminism Clim Change 2(2):50–68
Tangonyire P, Akuriba M (2021b) Improving livelihoods through soybean production in Ghana's poorest regions. Dev Econ J 8(2):138–151
Tey YS, Li E, Bruwer J, Abdullah AM, Brindal M, Radam A, Ismail MM, Darham S (2017) Factors influencing the adoption of sustainable agricultural practices in developing countries: a review. Environ Eng Manag J 16(2):1–20

Thabane L, Ma J, Chu R, Cheng J, Ismaila A, Rios LP, Robson R, Thabane M, Giangregorio L, Goldsmith CH (2010) A tutorial on pilot studies: the what, why and how. BMC Med Res Methodol 10(2):1–10
Traoré L, Bello OD, Togola A, Balogoun I, Chabi F, Yabi I, Ahoton EL, Saïdou A (2021) Farmers' perceptions of climate change and adaptation strategies on sorghum productivity in the Sudanian and Sahelian zones of Mali. Afr Crop Sci J 29(1):157–175
Udoumoh EF, Ohioma BEOO, Subeno TO (2016) Comparative analysis of three-stage least squares and multivariate regression method of estimating parameters of simultaneous equation models. J Basic Appl Sci Res 6(7):7–14
Umoh NE, Offiong RA, Ekpe IA, Yta EM (2015) Assessment of income poverty among farmers in Akwa Ibom state, Nigeria. J Emerg Trends Econ Manag Sci 6(8):345–349
Van Ittersum M, Bouma J, Stoorvogel J (2016) Global food security under climate change: the role of sustainable intensification. Global Food Secur J 7(1):23–30
Van Vugt D, Franke AC, Giller KE (2017) Participatory research to close the soybean yield gap on smallholder farms in Malawi. Exp Agric 53(3):396–415
Verma S, Chauhan B (2015) Integrated approaches to sustainable weed management. Crop Prot J 30(2):34–44
Wallen NE, Fraenkel JR (2013) Educational research: a guide to the process. Routledge
Wang X, Cheng Z (2020) Cross-sectional studies: strengths, weaknesses, and recommendations. Chest 158(1):S65–S71
Wongnaa CA, Awunyo-Vitor D (2019) Scale efficiency of maize farmers in four agro ecological zones of Ghana: a parametric approach. J Saudi Soc Agric Sci 18(3):275–287
World Bank (2016) Mapping donor support for low carbon development in low-income countries. World Bank, Washington
World Bank (2021) Action plan on climate change adaptation and resilience: managing risks for a more resilient future. World Bank, Washington
World Bank (2024) Climate change knowledge portal: monthly climatology of average minimum surface air temperature, average mean surface air temperature, average maximum surface air temperature & precipitation 1991–2022; Ghana. Washington, World Bank
Worqlul A, Collick A, Steenhuis T (2019) Maximizing irrigation potential in Ghana: a case study of the White Volta Basin. J Water Resour Dev 35(3):418–435
Wu J, Zhang M, Yang X, Wu B (2024) Effects of land and labor costs growth on agricultural product prices and farmers' income. Land 13(11):1–23
Yvoz S, Petit S, Biju-Duval L, Cordeau S (2020) A framework to type crop management strategies within a production situation to improve the comprehension of weed communities. Eur J Agron 11(5):12–69

Chapter 15
Mitigating Climate Change Through Sustainable Farming: An Organic Approach for Dryland Resilience in Maharashtra

Sandhya Sreekumar, Prachi Nautiyal, Sheetal Walvekar, Ganesh Deshmukh, K. Gireesan, Enosha Arshad, and Swati Nirala

Abstract Climate change poses significant challenges to dryland ecosystems, exacerbating water scarcity, soil degradation, and agricultural vulnerability. Sustainable farming practices, particularly organic farming, have emerged as viable solutions to mitigate these impacts while ensuring long-term agricultural resilience. This study explores the role of organic farming in enhancing soil fertility, conserving water, and reducing reliance on chemical inputs. By integrating ecosystem-based approaches, such as agroforestry, crop diversification, and soil conservation techniques, organic farming fosters biodiversity and strengthens climate adaptation in arid regions. The study adopts a mixed-methods approach, incorporating field interviews, and participatory rural appraisals to assess the effectiveness of these practices and their operational challenges in Maharashtra's dryland regions. Findings highlight the socioeconomic benefits of organic farming, including improved farmer incomes, enhanced food security, and strengthened community-led initiatives. However, challenges such as limited market access, inadequate policy support, and labor shortages hinder widespread adoption. The study concludes with policy recommendations to bridge the gap between research and implementation, advocating for farmer-led innovations, training modules designed by expert farmers, and institutional support to scale up sustainable agricultural models in dryland ecosystems.

S. Sreekumar (✉) · P. Nautiyal · S. Walvekar · G. Deshmukh · E. Arshad · S. Nirala
Department of Economics and Public Policy, Dr. Vishwanath Karad MIT World Peace University (MIT WPU), Pune, Maharashtra, India
e-mail: sandhya.sreekumar@mitwpu.edu.in

K. Gireesan
Department of Public Administration, Rajiv Gandhi National Institute of Youth Development (Institute of National Importance), Ministry of Youth Affairs and Sports, Government of India, Regional Centre, Chandigarh, Punjab, India

Y. D. I. Khan et al. (eds.), *Land Restoration Through Ecosystem-Based Approach*, Disaster Resilience and Green Growth, https://doi.org/10.1007/978-981-95-0389-6_15

Keywords Organic farming · FPOs · Dryland · Climate resilience

15.1 Introduction

Agricultural sustainability in dryland regions remains a pressing concern, particularly in Maharashtra, where semi-arid conditions and erratic rainfall pose significant challenges to farming communities (Todmal 2023). Securing and improving rural livelihoods requires risk management and productivity enhancement through diversification and sustainable intensification. Limitations and degradation of natural resources, especially water scarcity and increasing desertification, are the primary biophysical restrictions. This study examines the impact of sustainable farming practices within these dryland ecosystems, highlighting the socioeconomic and environmental implications. The transition from traditional, ecologically balanced agricultural systems to intensive, profit-driven models following the Green Revolution has led to over-exploitation of groundwater, soil degradation, and increased vulnerability to climate change (Ahamad et al. 2023; Singh et al. 2022). This revolution over-exploitation of groundwater for irrigation, worsening the region's agricultural water budget ($P = I + ET + R$, where P = precipitation, I = infiltration, ET = evapotranspiration, R runoff) (Jothiprakash et al. 2003). Addressing these challenges necessitates a shift toward a climate-resilient and ecosystem-based approach (EBA) that prioritizes resource conservation, biodiversity, and farmer empowerment (Bagwan 2023).

Organic farming has emerged as a promising EBA for dryland regions, offering a sustainable alternative to conventional methods. Enhancing soil fertility, reducing water usage, and minimizing chemical inputs support ecological balance and long-term productivity. It also strengthens farmer resilience against climate variability, ensuring economic and environmental sustainability in arid landscapes. The novelty of this study lies in its integrated approach to understanding dryland agriculture through the lens of climate resilience, economic sustainability, and social capital. Unlike previous research, such as Nautiyal et al. (2010) on soil physicochemical properties and Kumar et al. (2023) on irrigation efficiency, which primarily focuses on individual components such as soil health or irrigation efficiency, this study holistically evaluates the interplay between environmental conservation, farmer-led initiatives, and policy support mechanisms. By examining traditional knowledge systems and modern agroecological practices, the research presents an innovative framework for achieving long-term sustainability in Maharashtra's dryland farming sector.

Methodologically, the study employs a mixed-methods approach, incorporating qualitative field interviews with farmers, participatory rural appraisals, and thematic analysis of policy interventions. Data collection spans multiple districts within Maharashtra's semi-arid zones, ensuring a diverse representation of farming experiences and challenges. Integrating empirical field data with secondary literature allows for a comprehensive analysis of both micro-level farming practices and macro-level governance structures.

The key findings of this research indicate that sustainable farming practices, such as organic farming, offer viable solutions to the ecological and economic crises faced by dryland farmers (El-Beltagi et al. 2022; Paul and Nangare 2024). However, barriers such as limited market access, labor shortages, and inadequate institutional support hinder widespread adoption (Wollni et al. 2010). To address these gaps, the study proposes four key objectives: (1) to assess the socioeconomic viability of organic and climate-resilient farming models, (2) to analyze the role of FPOs and cooperative networks in promoting sustainable practices, (3) to examine challenges and benefits of adapting and practising organic farming in drylands, and (4) to provide policy recommendations for strengthening institutional frameworks supporting dryland agriculture. This study aims to contribute to the broader discourse on sustainable agricultural development in semi-arid regions by bridging the gap between research, policy, and grassroots farming innovations.

15.1.1 Drylands

Considering the Aridity Index (AI), which is calculated as the ratio of precipitation (P) to potential evapotranspiration (PET) during a multiannual period, drylands are defined by an overall climatic water deficit below the 0.65 mm/mm threshold (Middleton and Thomas 1997). Comprising 41% of the earth's surface, they are significant terrestrial habitats home to 38% of the world's population, 44% of its agriculture, 50% of its cattle, and 35% and 20% of its plant and variety hotspots, respectively. Drylands comprise 6.6%, 10.6%, 15.2%, and 8.7% of the world's land area, respectively, and are commonly categorized as semi-arid ($0.20 < AI < 0.50$), arid ($0.05 < AI < 0.20$), hyper-arid ($AI < 0.05$), and dry subhumid ($0.50 < AI < 0.65$). (Prăvălie 2016).

15.1.2 Drylands of Maharashtra

Four divisions of the state of Maharashtra are technically classified as drylands, which are semi-arid or dry subhumid regions. The spatial distribution of drylands in Maharashtra is depicted on the map below (Fig. 15.1). The dryland zones and the corresponding areas they cover (square kilometers) are shown in Table 15.1.

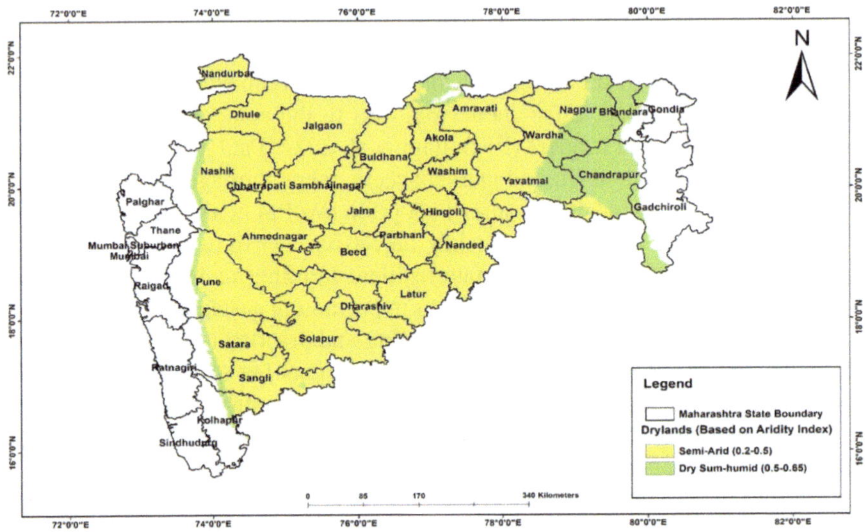

Fig. 15.1 Spatial distribution of drylands in Maharashtra (Based on Global Aridity Index v3). (Source: Anand et al. 2009)

Table 15.1 Division-wise area statistics for drylands of Maharashtra

S. no.	District name	Semi-arid (km^2)	S. no	District name	Semi-arid (km^2)
Marathwada division			**Western division**		
1.	Chatrapati Sambhajinagar	10,189	1.	Pune	10,561
2.	Jalna	7691	2.	Satara	7261.8
3.	Latur	7197.9	3.	Sangli	7912.8
4.	Nanded	10,536	4.	Solapur	14,879
5.	Dharashiv	7585.2	5.	Kolhapur	1302.8
6.	Parbhani	6154.6	**Vidarbha division**		
7.	Hingoli	4912.3	1.	Amravati	9066.1
8.	Beed	10,629	2.	Akola	5443.2
Khandesh division			3.	Bhandara	
1.	Nashik	10,626	4.	Wardha	4152.1
2.	Dhule	7197.1	5.	Chandrapur	1363.3
3.	Jalgaon	11,765	6.	Yeotmal	10,512
4.	Ahilya Nagar	16,360	7.	Gondia	–
5.	Nandurbar	5618.7	8.	Gadchiroli	–
			9.	Nagpur	4252.9
			10.	Washim	5162.6
			11.	Buldhana	9736.4

15.1.3 Reasons for Drylands in Maharashtra

The arid and semi-arid regions of Maharashtra and other parts of India face several challenges that contribute to dry land expansion and limited agricultural productivity. The primary causes of an arid climate include high temperatures exceeding 40 °C during summer, delayed and erratic rainfall, high wind speeds, frequent droughts, soil erosion, and deep-water tables. The soils in these regions are often shallow, with low water-holding capacity and fertility. Key factors contributing to the increase in dry land areas include deforestation, inappropriate agricultural practices, overgrazing, over-exploitation of groundwater, and unequal distribution of surface irrigation (Sarkar and Baswas 1980).

Natural and man-made factors cause drylands in Maharashtra, especially in Vidarbha and Marathwada. Low and unpredictable rainfall, intensified by climate change, is a significant factor. Shifting monsoon patterns, rising temperatures, and location in the Western Ghats' rain shadow make rainfall patterns more irregular. Further, increased evapotranspiration rates have exacerbated drought conditions (Todmal 2023). Changes in the Aridity Index (AI) due to global warming indicate a progressive decline in soil moisture retention, reducing the agricultural viability of these regions. Here, soil degradation, including salinization and desertification, has accelerated due to climate variability, making it more unsuitable for cultivation without irrigation. Traditional water management methods like ponds and tanks have degraded, making water distribution and storage inefficient.

Another critical factor is the disruption of natural groundwater recharge systems due to deforestation, over-extraction of groundwater for agriculture, and climate-induced variations in precipitation. Erratic rainfall leads to flash floods and prolonged dry spells, preventing effective groundwater recharge. In many places, wells run dry or provide low-quality water (Ahamad et al. 2023). The extensive production of water-intensive crops like sugarcane worsens this issue, leading to ecosystem imbalances and biodiversity loss (Singh et al. 2022).

Deforestation and loss of vegetative cover due to climate-driven land-use changes, including increasing infrastructure, urbanization, and agriculture, have weakened the region's natural ability to retain water and combat desertification (Srivastava and Chinnasamy 2021). The role of natural carbon sinks, such as grasslands and agroforestry systems, has declined, making the ecosystem more vulnerable to extreme weather events.

The delicate ecosystem of Maharashtra's drylands has been further strained by rising temperatures, erratic monsoons, and an increase in the frequency of extreme weather events like heatwaves and droughts (Bagwan 2023). Its topography and geology also influence the region's arid conditions. Hard, rocky terrain and plateaus unsuitable for retaining water make up a large portion of the state, particularly the interior portions. Agricultural activities and unsustainable farming methods have expanded onto marginal and less fertile soils in rural areas. Due to these reasons, drylands have become a recurring problem in Maharashtra, endangering livelihoods, agriculture, and the sustainability of the environment over the long term (Upwanshi et al. 2023).

15.1.4 Agricultural Practices in the Drylands of Maharashtra

In Maharashtra's arid regions, particularly in areas like Marathwada and Vidarbha, climate-resilient farming techniques have emerged as a response to prolonged droughts, rising temperatures, and declining soil fertility. Farmers are increasingly adopting EBA, integrating traditional knowledge with modern sustainable practices. Drought-resistant crops such as oilseeds (groundnut, sesame), pulses (chickpea, pigeon pea, etc.), and millets (sorghum, pearl millet, finger millet) have gained prominence, ensuring food security despite erratic rainfall (Paul and Nangare 2024; Satyavathi et al. 2021).

Further, mixed cropping and intercropping systems reduce risk and enhance ecosystem resilience (Kumar et al. 2023). Conservation tillage techniques, such as zero-tillage and biochar applications, are practised to maintain soil moisture and reduce carbon loss. Mulching with organic materials prevents evaporation and enhances carbon sequestration in soil, improving long-term fertility (Paradkar et al. 2023). Water conservation strategies, such as rainwater harvesting, agricultural ponds, drip irrigation, contour bunding, and check dams, are being adopted to improve groundwater recharge (Kale et al. 2024). These practices contribute to landscape restoration and ecosystem-based climate resilience, reducing farmers' vulnerability to extreme weather events.

Agroforestry has emerged as a nature-based climate adaptation strategy, where multi-tier cropping systems combine trees, shrubs, and crops to enhance biodiversity, restore degraded lands, and stabilize local microclimates. Species like neem, acacia, and moringa act as natural windbreaks and help regulate soil moisture levels (Anand et al. 2022; Pande et al. 2023; Shelar et al. 2023). Dryland agricultural systems incorporate livestock. Sheep, goats, and cattle are raised with crops to provide manure for fertilization and as a backup source of income, particularly in drought years when crops do not produce (Raina et al. 2023). Thus, due to these methods and traditional knowledge, Maharashtra's dryland farming is gradually becoming adaptable to unpredictable rainfall and drought circumstances.

15.1.5 Organic Farming as a Solution to Drylands' Issues

As an EBA strategy, organic farming enhances soil health, conserves water, and builds resilience against climate shocks. By promoting closed-loop nutrient cycles, organic farming minimizes dependence on external inputs and enhances natural soil fertility (Gnanaprakasam and Vanisree 2022). Beltagi et al. (2022) found that eco-friendly techniques such as bio-mulching and vermicomposting can reduce agricultural water demand by up to 50%, which is crucial for climate-stressed regions. The integration of cover crops, agroforestry, and natural fertilizers enhances carbon sequestration and biodiversity, strengthening soil structure and water retention capacity (Nautiyal et al. 2010; Thawale et al. 2021; Venugopalan et al. 2010). These practices improve environmental health and contribute to economic resilience by reducing input costs and enhancing crop yields. However, the adoption of such methods remains limited due to knowledge gaps and resistance to change among farmers, as highlighted by Zhang et al. (2024), who found that farmer education and training are crucial for the widespread adoption of sustainable practices.

Adaptive organic farming systems focus on biodiversity conservation and climate mitigation. They reduce the need for irrigation by improving the soil's water-holding capacity, a perfect strategy for Maharashtra's water-scarce areas. Polyculture, adoption of native plant species, and crop diversification strategies offer protection against climate-induced crop failures, ensuring farmers' livelihoods remain stable in unpredictable conditions. Furthermore, organic farming significantly reduces reliance on expensive synthetic chemicals that contribute to soil degradation and water contamination, fostering long-term sustainability using locally accessible organic inputs (Dhale et al. 2024).

However, the transition to climate-smart organic farming faces economic and knowledge barriers. While organic methods enhance long-term resilience, initial investment costs and certification challenges hinder adoption. Financial incentives for EBA, such as carbon credits for sustainable land management, could improve the viability of organic farming. Encouraging community-driven conservation programs, farm-led innovations, bridging the gap between government, consumers, and farmers, and agroecological innovations will further support organic farming in dryland agriculture (Wollni et al. 2010).

15.2 Methodology

15.2.1 Study Area

In western and central India, Maharashtra features a diverse agro-climatic landscape with a 720 km coastline along the Arabian Sea. The state experiences a tropical monsoon climate, marked by intense summers from March to May, followed by the monsoon season in June, with rainfall exceeding 400 mm in the Sahyadri ranges but dropping to 70 mm in the western plateau districts (Government of Maharashtra 2017). The Marathwada and Vidarbha regions receive slightly higher rainfall but remain prone to drought and water scarcity due to climate variability and inconsistent monsoons. Agriculture dominates the rural economy, yet disparities in rural infrastructure and access to irrigation remain critical challenges (Dash et al. 2023; Dound et al. 2018).

The study focuses on the Pune, Satara, and Solapur districts, as well as the arid and semi-arid regions of Maharashtra. Agriculture in the Pune, Satara, and Solapur districts is shaped by their semi-arid and arid conditions, with over 3000 villages dependent on a mix of rainfed and irrigated farming. The region's topography varies from 400 to 1500 meters above sea level, influencing soil moisture retention and productivity. While sugarcane dominates irrigated areas across these districts, non-irrigated lands primarily support cereals, pulses, and vegetables. In Pune, the undulating terrain and semi-arid climate make agriculture highly dependent on monsoon rainfall, leading to fluctuations in cropping patterns (CRIDA 2011a). Similarly, in Satara, the proximity to the Western Ghats creates distinct rainfall variations, with the western region benefiting from higher precipitation. At the same time, the eastern part remains drier, necessitating irrigation for sugarcane and dryland farming for cereals and pulses (CRIDA 2011b). Solapur experiences harsher arid conditions, with a greater reliance on irrigation for sustaining sugarcane cultivation, while non-irrigated areas primarily support drought-resistant crops such as sorghum and millets. The district's agricultural output is susceptible to monsoon variability, requiring adaptive strategies to mitigate water scarcity and ensure sustainability (CRIDA 2011c).

15.2.2 Data Collection and Analysis

This study employed a qualitative research approach, utilizing semi-structured interviews and field observations to explore themes related to dryland agriculture, agricultural market access, and organic farming practices. The research was conducted between May 2024 and August 2024 across selected villages in Maharashtra's arid and semi-arid regions. Site selection followed a two-stage process: first, villages were shortlisted based on their reliance on rural water systems and geographic location. Second, the lead researcher conducted field visits to assess market access, water system efficiency, adoption of improved agricultural practices, and farmers' willingness to participate in the study. Based on these assessments, 12 villages were selected to ensure diversity in location and farming practices.

Data collection involved 50 semi-structured interviews with organic farmers from both arid and semi-arid regions and field observations to complement interview findings. Participants included individual organic farmers, members, and Farmer Producer Organizations (FPOs) founders who had worked closely with Krishi Vikas Kendras (KVKs). The interviews covered key topics such as challenges in organic farming, government support, and FPO governance. The semi-structured format allowed for open-ended discussions, ensuring thematic consistency while giving participants sufficient time to reflect and respond. Field observations provided additional context by examining farming techniques, financial needs, and institutional support mechanisms. Triangulation was employed to enhance the credibility of findings by cross-referencing interview data with field observations and secondary sources, ensuring consistency in reported themes. Additionally, member checking was conducted by sharing preliminary findings with select participants to validate the accuracy of interpretations.

Thematic analysis was used to process and interpret the qualitative data. Following Creswell's (2015) "hand coding" technique, interview responses were categorized through iterative grouping and regrouping. The analysis involved familiarization with transcripts, initial coding to identify key patterns, clustering related ideas into broader themes, and refining these themes using Saldaña's (2021) approach. This structured coding helped establish conceptual links between different themes, providing a nuanced understanding of the challenges and opportunities in organic farming. The final themes were synthesized to highlight governance structures, policy implications, and farmer experiences, offering a comprehensive insight into the role of FPOs and government interventions in organic agriculture.

15.3 Findings and Discussion

(a) *Sustainable Farming Practices and Environmental Health*: This theme combines discussions about natural farming methods (e.g., homemade fertilizers, use of native seeds, and crop diversification), soil health, water conservation techniques, and sustainable farming innovations. It emphasizes practices that protect the environment, improve soil fertility, reduce water usage, and reduce dependence on chemical inputs. It also highlights the importance of long-term ecological sustainability in farming (Knowler and Bradshaw 2007).

During an interview, a farmer emphasized the use of natural farming methods, particularly making homemade potash using Bael fruit, jaggery, and water. This promotes an eco-friendly, cost-effective alternative to chemical fertilizers. Traditional farming knowledge is prioritized over modern techniques, with the speaker learning from previous experts (Interview, farmer, June 2024). The conversation delves into the environmental impact of current farming practices, particularly using genetically modified (GM) seeds, like soybeans, which deplete soil health. The speaker advocates for the use of native seeds, which not only reduce the need for frequent seed purchases but also improve soil fertility and carbon content, ensuring long-term sustainability (Interview, farmer, June 2024).

(b) *Economic Viability and Challenges in Agriculture*

More financially robust farms will hesitate to embrace organic farming to evade the adverse effects of this technology, like crop unpredictability. Studies indicated that closeness to metropolitan regions enhances the likelihood of adopting organic farming (Chatzimichael et al. 2014; Koesling et al. 2008). Intermittent investigations indicated that aridity (Genius et al. 2006) or cattle density (Läpple and Van Rensburg 2011; Läpple and Kelley 2015) may influence adoption. The study highlights that farmers even today hesitate to take up organic farming since its economic benefits are not projected and are not widely known. Solving this issue can help farmers depend less on additional financial support from governments or banks (Läpple and Van Rensburg 2011).

Challenges include the financial struggles, such as the high costs of chemicals and inputs, labor shortages, and the economic risks of transitioning to organic farming. Interviews highlight profitability matters, with examples of profitable crops (e.g., moong over soybeans), value addition, and product diversification (e.g., transforming crops into multiple products). Despite these opportunities, challenges like market access and the financial burden of certification for organic farming are also discussed. *"We used to get a good amount in growing roses in 1987, but the prices have dropped significantly with a steep increase in input costs. Only through organic farming and using natural inputs can we bring down the input cost. While facing losses in the beginning, I had sold 1 acre of land and also took the loan."* (Interview, farmer, August 2024)

(c) *Labor Issues and Agricultural Innovation*: The shortage of farm labor, the preference for alternative careers, and the shift toward low-maintenance single crops or monoculture are central to this theme. Innovation in farming practices to reduce labor dependency (e.g., growing bamboo or mahogany) and the challenges in finding and retaining labor are key points. There is also a need for efficient labor management and gender equality in wages within agriculture. *"Finding labour here is not easy at all. I'm considering shifting away from growing pomegranates and figs. I've planted mangoes on 15 acres and replaced figs with mangoes. I just started this year. The labour problem is especially challenging with figs and pomegranates, as they are labour-intensive crops. Sometimes we also have labour comes all the way from Hingoli (a city more than 570kms away from the farm) and we have to bear their living expenses."* (Interview, farmer, June 2024).

(d) *Government Policies, Institutional Support, and Farmer Engagement*: Research indicates that the demographic features of farmers, including age and education, and household variables like family size, may significantly influence the transition to organic farming. Furthermore, the structural attributes of a farm, such as its size, proximity to metropolitan areas, or specialization, may either promote or hinder adoption. It is essential to acknowledge that the elements characterizing the financial condition of farms were insufficiently addressed in surveys about organic farming. Similarly, farmers' perspectives on the environment and financial outcomes significantly influence their adoption of practices, as do the diverse sources of knowledge regarding organic farming.

This theme covers the critical relationship between government policies and local farming communities, including current agricultural support systems' inefficiencies and the disconnect between policymakers and on-the-ground realities. It includes the importance of locally driven initiatives, collaboration with government departments, and the role of farmer organizations in supporting farmers. A study discovered that joining farmer-based groups makes people more likely to use environmentally friendly methods. Further, demand-driven policy initiatives can help bridge the gap among government, consumers, and farmers (Wollni et al. 2010; Zhang et al. 2024).

Farmers also expressed frustration with subsidies for chemical inputs and the desire for policies supporting organic farming and rural infrastructure. Collaboration with government departments, like the forest department, and efforts to create programs to support farmers' innovation are key parts of a speaker's initiatives. The speaker suggests that these programs often fail to account for local conditions or alternative farming methods, which can lead to a disconnect between government initiatives and the real needs of farmers on the ground (Interview, farmer, August 2024).

(e) *Resistance to Change, Knowledge Gaps, and Education*: Evidence about the influence of farmers' information sources on the adoption of conservation practices in the region is sparse. Giovanopoulou and others (Giovanopoulou et al. 2011) determined that affiliation with professional cooperatives significantly diminishes the likelihood of adoption. In contrast, Gachango and others (Gachango et al. 2015) demonstrated that farmers' awareness of ecological conditions and strategies to mitigate nitrogen and phosphorus runoff does not influence adoption rates. Likewise, there is a lack of knowledge regarding the impact of conservation technology characteristics on adoption.

This theme highlights the reluctance of farmers to adopt new, sustainable practices despite their potential benefits. It focuses on the knowledge gaps in farming, particularly in organic methods, and the lack of professional support or training for farmers. This is highlighted by the speaker's struggle to convince farmers to shift from high-yield crops like soybeans to more sustainable and profitable alternatives. The need for greater access to education, innovation, and training in agricultural practices is emphasized, especially to engage younger generations and encourage them to pursue careers in farming (Interview, farmer, August 2024).

(f) *Community, Trust, and Local Empowerment*: This theme focuses on community, trust, and local empowerment in fostering sustainable farming practices. Many farmers complained about KVK officers lacking in-depth knowledge about organic farming or being biased toward conventional methods. A farmer went ahead and founded a leading farmers' group (Sangh, an FPO) aimed at overcoming systemic challenges like market access and price inequality. Compared to the multitudes of chemical farming-promoting FPOs, there were hardly any FPOs for organic farming. Neither could they sustain beyond a few years, nor was their focus on solving core operational issues with organic farming (Interview, farmer, July 2024).

Farmers emphasized the importance of collective action, farmer organizations, and initiatives like self-help groups (SHGs) to support local farming communities. The theme also touches on the role of women in rural development, the need for farmer-led innovations, and the desire for greater self-sufficiency at the local level. Trust in both government institutions and farming groups plays a critical role in adopting new farming practices.

(g) *Water Conservation and Efficiency*: Natural farming techniques can reduce water consumption by 50%. The use of earthworms and natural methods for water retention in soil is presented as a solution to water-scarce regions, especially during summers (Beltagi et al. 2022). There is emphasis on the reliance on alternative sources like tankers during drought periods. Water storage tanks (e.g., 1 crore-liter tanks) and gravity-fed irrigation systems are emphasized as a solution to water scarcity. The practice of mulching to conserve water (Demo and Asefa Bogale 2024) is another point of focus, reflecting the importance of water conservation techniques in farming, particularly in areas facing water shortages. *"I became interested in organic farming because there is no water here. If we supply chemicals and if the water content is inadequate, the plants will die quickly. I opted for organic farming as it requires minimal/no usage of chemicals. We are doing mulching in the fields. We save 50 percent of water from the mulching."* (Interview, farmer, July 2024).

15.4 Discussion

This research explored the potential of organic farming as a sustainable solution for dryland farming in Maharashtra, India. The findings reveal a complex interplay of economic, social, and environmental factors influencing farmers' decisions and the overall agricultural landscape in the region. While organic farming presents numerous opportunities for dryland farmers, several challenges must be addressed to facilitate a successful transition.

(a) Primary data collected highlight natural farming methods and practices like homemade fertilizers, crop diversification, and using native seeds as effective eco-friendly alternatives to chemical inputs, providing a holistic approach to agriculture to enhance soil fertility, reduce environmental harm, and promote resilience. By integrating traditional knowledge with modern sustainable practices, agriculture can achieve ecological balance, ensuring the viability of farming for future generations. This knowledge is fragmented and must be documented and disseminated to all.
(b) There exists the complex interplay of various economic (farm income and uncertainty in economic benefits), geographic (proximity to metropolitan region), and environmental (aridity, cattle density) factors in shaping sustainable agriculture farming practices. Farmers' interviews highlighted a potential increase in benefits with input cost reduction through organic farming adoptions. Farmers' interviews reflect that demographic features of farmers, household characteristics, structural attributes of a farm, financial condition, and farmers' perspectives and knowledge on the environment and financial outcomes significantly influence the transition to organic farming but are insufficiently addressed in surveys about organic farming. Further, product diversification and value-added services generating multiple sources of income can smooth and hasten the transition.
(c) Farmers expressed uneasiness in getting affordable and timely labor, which highlights the need to revisit the existing labor management and gender inequality in wages within agriculture from the perspective of the organic farming ecosystem.
(d) This disconnect between government policies and the on-the-ground realities of farmers results in frustrations with subsidies and rural infrastructure for organic farming in the state. A demand-driven approach that supports local innovations with on-the-ground interdepartmental collaborations can help bridge this policy

gap and misalignment. KVKs and other government agencies should take input from successful organic farmers and collaborate with them in designing training programs for other farmers interested in transitioning to organic or sustainable farming.

(e) The study highlights the vitality of authentic information recording and seamless sharing in the adoption of conservation practices. Lack of information results in reluctance to adopt and knowledge gaps in farming methods and professional support or training. The study advocates greater access to education, innovation, and training and engages younger generations to pursue careers in farming.

(f) The study indicates that positive family support and trust are needed in organic farming adoption. Individual farmers are empowered through the support they receive from group collective action. Women in families interviewed have shown active involvement in preparing organic manure and generating more market opportunities by selling organic harvests and allied processed commodities.

(g) The study indicates a cause-and-effect relationship between water scarcity and organic farming adoption. Water scarcity pushed farmers to adopt organic farming, additionally using various water storage methods (water storage tanks and gravity-fed irrigation systems) and conservation methods (mulching).

15.5 Conclusion

Organic farming offers promising opportunities for enhancing sustainability in dryland agriculture, but challenges such as water scarcity, economic viability, and labor shortages must be addressed. Practices like mulching and using earthworms to improve water retention can help mitigate water scarcity, while investments in efficient irrigation and water storage are essential for long-term success. The economic transition to organic farming often requires financial sacrifices, with farmers resorting to loans or land sales. Economic constraints, fragmented landholdings, and limited market access through Agricultural Produce Market Committees (APMCs) further impact farmers' incomes and sustainability. Addressing rural infrastructure gaps, improving irrigation efficiency, and promoting climate-resilient farming are crucial for enhancing agricultural sustainability in these semi-arid regions.

These barriers require targeted financial support, market incentives, and affordable organic inputs. Addressing labor shortages, especially with less labor-intensive crops like mangoes, necessitates policies promoting fair wages, better working conditions, and labor-saving technologies. Government policies, local initiatives, and institutional support, including agricultural education, KVKs, and FPOs, are crucial for fostering informed decision-making, empowering farmers, and attracting the younger generation to organic farming. In conclusion, organic farming offers a viable pathway toward sustainable dryland agriculture in Maharashtra and can create a resilient, sustainable agricultural system that benefits both farmers and the environment.FundingThis work was funded by the Indian Council of Social Sciences Research (ICSSR) under the grant No. ICSSR/RPD/MN/2023–24/G/131.

References

Ahamad F, Tyagi SK, Singh M, Sharma AK (2023) Groundwater in arid and semi-arid regions of India: a review on the quality, management, and challenges. In: Ali S, Armanuos AM (eds) Groundwater in arid and semi-arid areas: monitoring, assessment, modelling, and management. Springer Nature, Berlin, pp 11–52. https://doi.org/10.1007/978-3-031-43348-1_2

Anand A, Mudit M, Vivek G, Pradyumna J, Manoj C (2009) Agro-climatic zonation of Maharashtra State using GIS. Trans Inst Indian Geogr 31:25–36

Anand T, Raj D, Sudarshan B, Digvijay K, Usharani K, Swati J, Ranjeet S (2022) Performance evaluation of soil and water conservation constructed structures in drought prompt areas of Satara, India. J Perform Constr Facil 36:4022050. https://doi.org/10.1061/(ASCE)CF.1943-5509.0001762

Bagwan WA (2023) Nature-based solutions (NbS) for dryland agriculture in semi-arid regions of Maharashtra, India: a short review with possible approaches for building climate resilience. In: Chatterjee U, Shaw R, Kumar S, Raj AD, Das S (eds) Climate crisis: adaptive approaches and sustainability. Springer Nature, Berlin, pp 277–291. https://doi.org/10.1007/978-3-031-44397-8_15

Beltagi MS, Ismail AM, El-Mogy MM (2022) Natural farming techniques for water conservation and soil health. J Sustain Agric 45(3):123–135. https://doi.org/10.3390/agronomy12081881

Central Research Institute for Dryland Agriculture (CRIDA) (2011a) Agriculture contingency plan for district: Pune, Maharashtra. ICAR-CRIDA. https://www.icar-crida.res.in/CP/Maharastra/MPKVV%2C%20Rahuri/MH8-%20PUNE%2031.03.2011.pdf

Central Research Institute for Dryland Agriculture (CRIDA). (2011b Agriculture contingency plan for district: Satara, Maharashtra. ICAR-CRIDA. https://www.icar-crida.res.in/CP-2012/statewiseplans/Maharastra%28Pdf%29/MPKVV%2C%20Rahuri/MH10-SATARA%2031.03.2011.pdf

Central Research Institute for Dryland Agriculture (CRIDA). (2011c) Agriculture contingency plan for district: Solapur, Maharashtra. ICAR-CRIDA. https://www.icar-crida.res.in/CP/Maharastra/MPKVV%2C%20Rahuri/MH1-Solapur%203.2.2011.pdf

Chatzimichael K, Genius M, Tzouvelekas V (2014) Informational cascades and technology adoption: evidence from Greek and German organic growers. Food Policy 49:186–195

Creswell JW (2015) Educational research: Planning, conducting, and evaluating quantitative and qualitative research. Pearson Inc.

Dash RK, Chandiramani J, Miglani V, Mukherjee S, Ghosh BK (2023) The status of rural infrastructure: a district level study of Maharashtra. J Rural Dev 41(3):318–339

Demo AH, Asefa Bogale G (2024) Enhancing crop yield and conserving soil moisture through mulching practices in dryland agriculture. Front Agron 6:1361697

Dhale OS, Pawar G, Bendre N, Bhoite S (2024) Entomofauna diversity and prey-predator relationship associated with organic paddy field at Mahadare conservation reserve, Satara District, Maharashtra. J Entomol Zool Stud 12:20–28. https://doi.org/10.22271/j.ento.2024.v12.i3a.9314

Dound RV, Suramwad SR, Kolgane BT (2018) Study of socio-economic characteristics of pomegranate growing farmers in Solapur district of Maharashtra state. J Pharm Phytochem 7(1S):2959–2962

El-Beltagi HS, Basit A, Mohamed HI, Ali I, Ullah S, Kamel EA et al (2022) Mulching as a sustainable water and soil saving practice in agriculture: a review. Agronomy 12(8):1881

Gachango FG, Andersen LM, Pedersen SM (2015) Adoption of voluntary water pollution reduction technologies and water quality perception among Danish farmers. Agric Water Manag 158:235–244

Genius M, Pantzios CJ, Tzouvelekas V (2006) Information acquisition and adoption of organic farming practices. J Agric Resour Econ 31:93–113

Giovanopoulou E, Nastis SA, Papanagiotou E (2011) Modeling farmer participation in agri-environmental nitrate pollution reducing schemes. Ecol Econ 70:2175–2180

Gnanaprakasam PD, Vanisree AJ (2022) Recurring detrimental impact of agrochemicals on the ecosystem, and a glimpse of organic farming as a possible rescue. Environ Sci Pollut Res 29:75103–75112. https://doi.org/10.1007/s11356-022-22750-1

Government of Maharashtra (2017) Maharashtra Vision 2030. Retrieved from https://plan.maharashtra.gov.in/Sitemap/plan/pdf/final_Vision_Eng_Oct2017.pdf

Jothiprakash V, Marimuthu G, Muralidharan R, & Senthilkumar N (2003) Delineation of potential zones for artificial recharge using GIS. Journal of the Indian Society of Remote Sensing, 31(1):37–47. https://doi.org/10.1007/BF03030750

Kale RB, Gavhane AD, Thorat VS, Gadge SS, Wayal SM, Gaikwad SY, Singh S, Khandagale KS, Bhat R, Mahajan V (2024) Efficiency dynamics among onion growers in Maharashtra: a comparative analysis of drip irrigation adopters and non-adopters. BMC Plant Biol 24:237. https://doi.org/10.1186/s12870-024-04875-2

Knowler D, Bradshaw B (2007) Farmers' adoption of conservation agriculture: a review and synthesis of recent research. Food Policy 32(1):25–48. https://doi.org/10.1016/j.foodpol.2006.01.003

Koesling M, Flaten O, Gudbrand L (2008) Factors influencing the conversion to organic farming in Norway. Int J Agric Resour Gov Ecol 7(1):78–95. https://doi.org/10.1504/IJARGE.2008.016981

Kumar R, Kumar S, Yashavanth BS, Venu N, Meena PC, Dhandapani A, Kumar A (2023) Natural farming practices for chemical-free agriculture: implications for crop yield and profitability. Agriculture 13. https://doi.org/10.3390/agriculture13030647

Läpple D, Kelley H (2015) Spatial dependence in the adoption of organic drystock farming in Ireland. Eur Rev Agric Econ 42:315–337

Läpple D, Van Rensburg T (2011) Adoption of organic farming: are there differences between early and late adoption? Ecol Econ 70(7):1406–1414. https://doi.org/10.1016/j.ecolecon.2011.03.002

Nautiyal CS, Chauhan PS, Bhatia CR (2010) Changes in soil physico-chemical properties and microbial functional diversity due to 14 years of conversion of grassland to organic agriculture in semi-arid agroecosystem. Soil Tillage Res 109:55–60. https://doi.org/10.1016/j.still.2010.04.008

Pande CB, Moharir KN, Varade A (2023) Water conservation structure as an unconventional method for improving sustainable use of irrigation water for soybean crop under rainfed climate condition. In: Pande CB, Moharir KN, Singh SK, Pham QB, Elbeltagi A (eds) Climate change impacts on natural resources, ecosystems, and agricultural systems. Springer International Publishing, pp 629–641. https://doi.org/10.1007/978-3-031-19059-9_28

Paradkar V, Firake NN, Gorantiwar SD, Singh AK (2023) Assessing the productivity and economic sustainability of banana cultivation under diverse colored plastic mulches in Western Maharashtra region of India. Int J Plant Soil Sci 35:980–991

Paul NC, Nangare DD (2024) Trend analysis of area, production, and productivity of nutri-cereals (pearl millet and sorghum) in Maharashtra, India: navigating challenges for food security and the way out. Natl Acad Sci Lett 47:613. https://doi.org/10.1007/s40009-024-01417-0

Prăvălie R (2016) Drylands extent and environmental issues. A global approach. Earth Sci Rev 161:259–278. https://doi.org/10.1016/j.earscirev.2016.08.003

Raina VS, Singh AP, Kour A, Tyagi B (2023) Understanding linkages between livestock sensitivity and climate variability in drylands for developing appropriate management strategies. In: Naorem A, Machiwal D (eds) Enhancing Resilience of Dryland Agriculture Under Changing Climate: Interdisciplinary and Convergence Approaches. Springer Nature, Singapore, pp 563–574. https://doi.org/10.1007/978-981-19-9159-2_27

Saldaña J (2021) The coding manual for qualitative researchers. Inc

Sarkar S, Baswas P (1980) Climatological data of the arid and semi-arid regions of India. Indian Agricultural Statistics Research Institute. Retrieved from http://ecoursesonline.iasri.res.in/mod/page/view.php?id=97675

Satyavathi CT, Ambawat S, Khandelwal V, Srivastava RK (2021) Pearl millet: a climate-resilient nutricereal for mitigating hidden hunger and providing nutritional security. Front Plant Sci 12. https://doi.org/10.3389/fpls.2021.659938

Shelar R, Nandgude S, Tiwari M, Gorantiwar S, Atre A (2023) Impact assessment of soil and water conservation measures on carbon sequestration: a case study for the tropical watershed using advanced geospatial techniques. Sustainability. https://doi.org/10.3390/su15010531

Singh R, Bindal S, Gupta AK, Kumari M (2022) Drought frequency assessment and implications of climate change for Maharashtra, India. In: Phartiyal B, Mohan R, Chakraborty S, Dutta V, Gupta AK (eds) Climate Change and Environmental Impacts: Past, Present and Future Perspectives. Springer International Publishing, pp 369–381. https://doi.org/10.1007/978-3-031-13119-6_22

Srivastava A, Chinnasamy P (2021) Developing village-level water management plans against extreme climatic events in Maharashtra (India) – a case study approach. In: Vaseashta A, Maftei C (eds) Water safety, security and sustainability: threat detection and mitigation. Springer International Publishing, pp 615–635. https://doi.org/10.1007/978-3-030-76008-3_27

Thawale U, Kuchanwar O, Chopde N, Bhoyar K, Nakade T, Mairan N, Gopal K (2021) Impact of organic inputs on fertility status of soil under certified organic farms in Nagpur district. Int J Chem Stud 9:14–18

Todmal RS (2023) Assessment of hydro-climatic trends in a drought-prone region of Maharashtra (India) with reference to rainfed agriculture. Reg Environ Chang 23:62. https://doi.org/10.1007/s10113-023-02044-z

Upwanshi M, Damry K, Pathak D, Tikle S, Das S (2023) Delineation of potential groundwater recharge zones using remote sensing, GIS, and AHP approaches. Urban Clim 48:101415. https://doi.org/10.1016/j.uclim.2023.101415

Venugopalan M, Rajendran TP, Chandran P, Goswami SN, Challa O, Damre PR (2010) Comparative evaluation of organic and non-organic cotton Gossypium hirsutum production systems. Indian J Agric Sci 80:287–292

Wollni M, Lee DR, Thies JE (2010) Conservation agriculture, organic marketing, and collective action in the Honduran hillsides. Agric Econ 41(3–4):373–384

Zhang T, Lang L, Zhao N, Lu Q, Sun B (2024) Perceived organizational support, inter-temporal choice, and farmer conservation tillage adoption. Agriculture 14(5):667

Chapter 16
Scaling Up of Vermicomposting in the Selected Landscapes of Madhya Pradesh and Chhattisgarh for Sustainable Agriculture and Land Productivity

R. S. Rawat, A. K. Thakur, and Shilpa Gautam

Abstract The overuse of chemical fertilizers in agricultural lands is causing environmental degradation and loss of soil fertility. It has negative impacts on hydrology and ecosystem health. Transitioning to organic farming practices such as vermicomposting can improve soil fertility and ecosystem health and reduce the negative impacts of chemical fertilizers. By providing examples of sustainable land and ecosystem management in Madhya Pradesh and Chhattisgarh, the World Bank-funded Ecosystem Services Improvement Project sought to assist the National Mission for a Green India. Vermicomposting was chosen as one of the activities under this project to enhance land productivity and improve ecosystem health. It involves earthworms converting waste into compost, improving soil quality, preventing erosion, and increasing crop productivity. Vermicomposting units were set up in 5022 households in 34 villages in Madhya Pradesh and 35 villages in Chhattisgarh, benefiting 15,066 individuals. Farmers, particularly women, actively participated in vermicomposting, leading to increased crop production and savings of on average at INR 3035/– per household per year due to a cutdown on chemical inputs. The project emphasized the importance of healthy soil for sustainable land productivity. Overall, vermicomposting proved to be a nature-based approach that promotes healthy ecosystems and land productivity. The article highlights the success of the practice of vermicomposting for improving land productivity, including the challenges faced and the strategy used for its successful implementation and sustainability.

Keywords Vermicomposting · Sustainable agriculture · Nature-based solution · Sustainable development goals

---

R. S. Rawat · A. K. Thakur (✉) · S. Gautam
Biodiversity and Climate Change Division, Indian Council of Forestry Research and Education, Dehradun, Uttarakhand, India

Y. D. I. Khan et al. (eds.), *Land Restoration Through Ecosystem-Based Approach*, Disaster Resilience and Green Growth, https://doi.org/10.1007/978-981-95-0389-6_16

16.1 Introduction

India has a long history of implementing programs to combat desertification, land degradation, and drought. The most significant process of desertification/land degradation in Madhya Pradesh state is vegetation degradation (8.23% in 2018–19, 8.19% in 2011–13, and 8.16% in 2003–05), which is followed by water erosion (3.66% in 2018–19, 3.65% in 2011–13, and 3.63% in 2003–05), as per the government report, which accounts for 12.52% (3.85 million ha) of the entire geographical area (SAC, 2021). The same report states that 17.06% (2.30 million hectares) of the state's total land area is experiencing desertification or land degradation (DLD) between 2018 and 2019 in Chhattisgarh. Vegetation degradation (10.03% in 2018–19, 9.97% in 2011–13, and 2003–05) is the state's most important desertification/land degradation process, followed by water erosion (6.32% in 2018–19, 5.80% in 2011–13, and 5.70% in 2003–05).

Approaches like Watershed Development Programmes since the 1970s (https://www.wri.org/research/watershed-development-india) were instrumental in improving the productivity of rainfed areas. Ecosystem approaches are highly context-specific and flexible to address management issues in different social contexts, and the landscape-based approach, such as the National Mission for a Green India, Forest Landscape Restoration Programmes, etc aimed at restoring degraded ecosystems, enhancing biodiversity, improving livelihoods, and building climate resilience through participatory and integrated land-use planning. Global hunger has been increasing since 2015, exacerbated by the pandemic, conflict, climate change, etc. Sustainable Development Goal 2 aims to end hunger by 2030. India faced food shortages post-independence due to population growth, politics, and farmer conditions until the 1960s. The green revolution in the 1960s transformed India from a food-deficient to a food-surplus nation (Somvanshi et al. 2020).

The green revolution led to high crop productivity through adaptations like (1) increased farming area, (2) planting two crops annually, (3) adoption of high-yielding variety seeds (4) increased use of chemical fertilizers and pesticides, (5) improved irrigation facilities, and (6) improved farm implements and crop protection measures (Singh 2000; Brainerd and Menon 2014) and modifications in farm equipment. Between 1961 and 2022–2023, the amount of food produced rose from 72.3 million MT to 329.69 million MT (Parliament of India. (n.d.). AU1446. Retrieved Oct 5, 2024, from https://sansad.in/getFile/annex/262/AU1446.pdf?source=pqars). Food security will continue to be one of the major challenges confronting the world, including India, as the country faces the challenge and pressure to feed more than 1 billion people today. Thus, India continues the green revolution program through schemes promoting plant breeding, irrigation, and agrochemical financing.

Numerous agencies have worked on analysis, measurement, and interventions for land degradation and ecosystem conservation. With an emphasis on flood-prone areas in nine states, the Centrally Sponsored Scheme (CSS) for River Valley Projects was implemented in 31 catchments spanning 18 states. Another CSS targeted soil reclamation and development in Haryana, Punjab, and Uttar Pradesh during the 7th Five-Year Plan. The Watershed Development Programme was introduced in shifting cultivation areas in the northeastern region and states like Andhra Pradesh and Odisha, along with the National Watershed Development Project for Rainfed Areas in multiple states. Initiatives like the Drought Prone Area Programme, the Desert Development Programme, and the Integrated Wasteland Development Projects Scheme aimed at addressing issues like desert development and wasteland rehabilitation, often with support from external funding agencies, such as the World Bank, Swiss Agency for Development and Cooperation, Danish International Development Agency, Department for International Development, and the KfW (Bhan 2013).

Benefiting from the green revolution is a continual process, according to the Government of India's Ministry of Agriculture and Farmers Welfare. The ministry is assisting states through two umbrella schemes, Krishonnati Yojana and Rashtriya Krishi Vikas Yojana, which incorporate green revolution components. The schemes include national initiatives like the National Food Security Mission, the National Project on Organic Farming, Organic Value Chain Development for the North East Region, the National Project on Soil Health and Fertility, Rainfed Area Development, Paramparagat Krishi Vikas Yojana, and more, covering agriculture, horticulture, and forestry sectors. Farmers receive assistance through these programs in organizing cluster demonstrations on paddy and wheat, seed production and distribution, nutrient management and soil ameliorants, integrated pest management, cropping system-based training, asset building such as farm machinery and implements, irrigation devices, site-specific activities, and post-harvest and marketing support, among other things (Parliament of India. (n.d.). AU1446. Retrieved Oct 05, 2024, from https://sansad.in/getFile/annex/262/AU1446.pdf?source=pqars). The Government of India recommends soil test-based balanced and integrated nutrient management through the concurrent use of both inorganic and organic sources (manure, bio-fertilizers, in situ crop residue recycling, etc.) of plant nutrients with the 4Rs approach, i.e., right quantity, right time, right mode, and right type of fertilizer, for the judicious use of chemical fertilizers and to reduce their use. Furthermore, split application, the use of slow-releasing fertilizers such as neem-coated urea, and the cultivation of leguminous crops are recommended (Parliament of India. (n.d.). AU1446. Retrieved Oct 05, 2024, from https://sansad.in/getFile/annex/262/AU1446.pdf?source=pqars).

When used sparingly and in moderation, chemical fertilizers do not negatively impact soil health. Investigations conducted over five decades under the All India Coordinated Research Project on "Long-Term Fertilizer Experiments" have revealed that continuous/overuse of chemical fertilizers can result in multi-nutrient deficiencies that affect soil health and crop productivity (Press Information Bureau, accessed Oct 5, 2024). The *government announces new initiatives for sustainable development*. Press Information Bureau. https://pib.gov.in/PressReleasePage.aspx?PRID=1806181). Recommended nitrogen, phosphorus, and potassium dosages combined with organic manure promote soil health and crop productivity. Excessive nitrogen fertilizer use may lead to nitrate pollution in groundwater exceeding the 10 mg NO_3-N/L limit, affecting human/animal health when consumed (Press Information Bureau, accessed Oct 5, 2024). The *government announces new initiatives for sustainable development*. Press Information Bureau. https://pib.gov.in/Pressreleaseshare.aspx?PRID=1696465).

The Indian Council of Agricultural Research (ICAR) has developed technology to produce various organic manures like phosphor-compost, vermicompost, and bio-enriched compost from different organic wastes. ICAR also provides training and demonstrations to educate farmers about these practices. Through the Network Project on Organic Farming scheme, ICAR is researching location-specific organic farming practices in 20 centers across 16 states (https://basai.org/wp-content/uploads/2023/08/Use-of-Fertilizers.pdf).

The Government of India has been promoting the use of bio-fertilizers, organic fertilizers, and bio-pesticides through the Paramparagat Krishi Vikas Yojana and Mission Organic Value Chain Development in the North East Region since 2015–16. Farmers are encouraged to practice organic cultivation and are provided end-to-end support, including training on producing organic fertilizers. Subsidies of INR 15000 per hectare for 3 years are being given for organic inputs (Press Information Bureau. Accessed Oct 05, 2024. The government announces new economic reforms to boost industry. Press Information Bureau. https://pib.gov.in/PressReleasePage.aspx?PRID=2037424). Under the Soil Health Card scheme, states receive assistance to promote integrated nutrient management by using chemical fertilizers along with organic manure and bio-fertilizers. Training, demonstrations, and farmer events (Kisaan Mela) are also organized (Press Information Bureau. Accessed Oct 05, 2024. The government of India launches new digital initiatives to enhance connectivity. Press Information Bureau. https://pib.gov.in/PressReleasePage.aspx?PRID=1942481) for creating mass awareness.

16.2 Fertilizer Uses in India

As per disclosure by the Department of Agriculture and Farmers Welfare, the total fertilizer consumption in India was 63644.37 (in '000 MTs) in 2021–22, while it was 63637.11 (in '000 MTs) in 2022–23, though the Government of India is promoting the biofertilizer, organic fertilizer as alternate sources of nutrients (Press Information Bureau, accessed Oct 5, 2024). *The government announces new initiatives for sustainable development*. Press Information Bureau. https://sansad.in/getFile/loksabhaquestions/annex/1714/AU1388.pdf?source=pqals). The total consumption of fertilizers (N + P2O5 + K2O) in India was 69.8 thousand tons, which has risen to 29,843.6 thousand tons as of 2022–23 (provisional figure), as per a report (The Fertiliser Association of India, accessed Oct 02, 2024. https://www.faidelhi.org/general/con-npk.pdf). Also, the fertilizer response ratio in India has decreased from 15 kg of grain per kg of NPK in 1974–79 to merely 5 kg in 2015 (Chaudhari et al. 2015).

16.3 Impact of Excessive Use of Fertilizers

Numerous studies demonstrate the beneficial effects of chemical fertilizers on plant growth and development as well as crop yield across a range of crops. However, long-term use of chemical fertilizers is detrimental to the health of the soil and other beneficial microorganisms in the environment. They reduce soil fertility due to the presence of toxic components. The environment and human health are adversely affected by the ongoing overuse of chemical fertilizers (Wang and Zhao 2019). According to Erisman et al. (2013), reactive nitrogen is a very volatile element that spreads through both air and water. Environmental deterioration can result from related issues, including nitrate leaching and runoff to water bodies (Vitousek et al. 2009). Excessive use of chemical fertilizers has been demonstrated to have detrimental effects on the environment and public health, particularly with regard to arsenic, lead, and cadmium (Khatun et al. 2022). A certain amount of these chemical fertilizers is held in the body, fruits, grains, leaves, and other parts of the plant. These enter the human body through the food chain and have an impact on health, leading to various diseases. Thus, the available research suggests that the soil's fertility is diminished by the improper application of these chemical fertilizers and toxic chemicals (Prakash 2023). Soil salinization can also result from the overuse of chemical fertilizers and soil additions like gypsum and lime (Kumar and Sharma 2020). Excessive use of fertilizers results in severe environmental issues (such as biodiversity loss, heavy metal build-up, eutrophication of water, and toxicity to various beneficial microorganisms), as well as nitrate- and sulfur- and nitrogen-containing gas accumulation that can cause issues like global warming or the greenhouse effect (Jote 2023).

The Intergovernmental Technical Panel on Soils (ITPS) defines soil health as "the ability of the soil to sustain the productivity, diversity, and environmental services of terrestrial ecosystems." In managed systems, soil health can be maintained, promoted, or recovered through the implementation of sustainable soil management practices (FAO 2020). The main causes of soil degradation and, consequently, the main threats to its ecological functions are erosion, organic matter decline, loss of biodiversity, compaction, sealing, point-source and diffuse contamination, pollution, and salinization (Montanarella 2007). There are numerous ways to improve the quality and productivity of soil, and one of them is vermicomposting. Seven of the ten FAO-recommended agroecological elements—which focus on increasing soil organic matter through soil biotic activities, nutrient recycling, biological interactions among different organisms, preserving biodiversity above and below ground, and avoiding synthetic fertilizers—are met by vermicomposting (Gupta et al. 2021). As per a report in 2010, 3.5 million hectares of estimated area were covered with vermicomposting in 19 states (NCOF 2010). Vermicomposting, which is more common among small and marginal landholding farmers and used in numerous crops, is thought to have been embraced by about 1.5 million farmers. It is more economical in high-value horticultural crops (Gupta et al. 2021).

16.4 Ecosystem Services Improvement Project

The Global Environment Facility/World Bank-funded Ecosystem Services Improvement Project (ESIP) supported the goals of the National Mission for a Green India by demonstrating models for adaptation-based mitigation through sustainable land and ecosystem management and livelihood benefits. The objective of the project was to improve the forest quality, land management, and non-timber forest product benefits for forest-dependent communities. One of the components of the project was scaling up sustainable land and ecosystem management in selected landscapes in the states of Madhya Pradesh and Chhattisgarh (Fig. 16.1). ESIP introduced new tools, techniques, and practices for sustainable management of natural resources, including biodiversity and carbon stocks. ESIP was implemented in the selected landscapes in Madhya Pradesh and Chhattisgarh by the Indian Council of Forestry Research and Education and the Madhya Pradesh and Chhattisgarh State Forest Departments under the overall direction of the Ministry of Environment, Forest, and Climate Change, Government of India. A total area of about 25,000 ha was covered under the scaling up of Sustainable Land Ecosystem Management (SLEM) best practices in common property resources, private lands, and government lands. More than 25,000 people, comprising forest dwellers, small landholders, marginal farmers, etc., were the direct beneficiaries of the project at the community level in the states of Madhya Pradesh and Chhattisgarh. The indirect beneficiaries of the project comprised a larger population of these states, which benefited from the improved forest quality and ecosystem services. The project was implemented in 92 villages spreading across five forest ranges (Banapura, Bhaura, Itarasi, Sukhtawa, and Budhni) in Madhya Pradesh and five forest ranges (Pali, Pandaria West, Raghunathnagar, Marwahi, and Narharpur) in Chhattisgarh, covering a total area of 63214.67 ha. The scaling up of vermicomposting was undertaken in 69 villages across all forest ranges except Naraharpur Forest Range in Chhattisgarh (ICFRE 2023).

16.5 Scaling Up of Vermicomposting in Project Areas of Madhya Pradesh and Chhattisgarh

Vermicomposting, as mentioned in the previous sections, is being encouraged by several government schemes. Literature also showed that it is an eco-friendly approach to sustainable agriculture (Tripathi et al. 2005) and recommended as a nature-based solution for environmental degradation, climate change, and food security (Gill et al. 2023). As an agrarian economy, the project areas in Madhya Pradesh and Chhattisgarh had a high potential for adopting vermicomposting. Madhya Pradesh is one of the top three states in the nation for producing oilseeds, pulses, and soybeans. Chhattisgarh is known as the "Rice Bowl" of central India. In addition to rice, the main crops grown in the state of Chhattisgarh are maize, jowar,

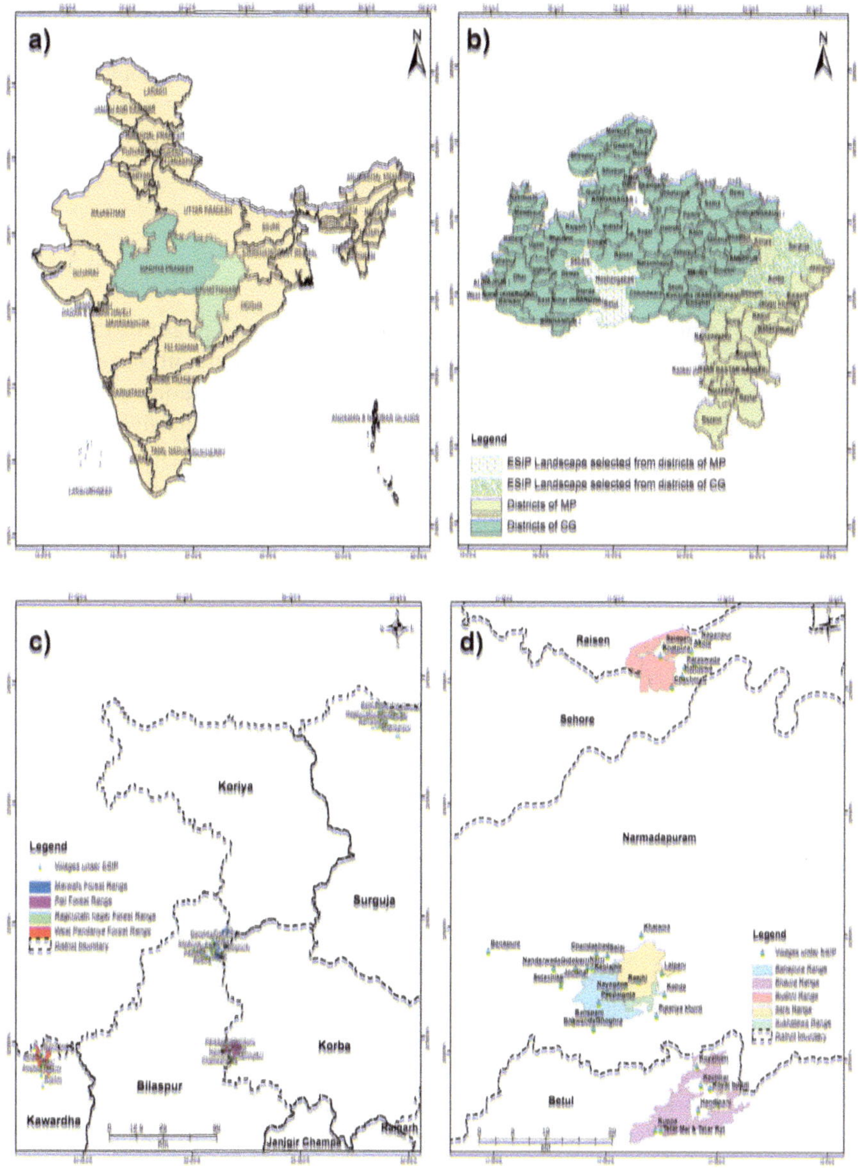

Fig. 16.1 Project area under ESIP

groundnut, gram, and wheat. Madhya Pradesh stood at the 3rd rank, while Chhattisgarh is at the 14th rank, according to the share of consumption to All India consumption in 2021–22 and 2022–23 (The Fertiliser Association of India, accessed Oct 02, 2024). https://www.faidelhi.org/member/AR-22-23.pdf). Therefore, vermicomposting could be an option for lowering these two states' reliance on chemical fertilizers, as both are known to consume large amounts of the fertilizer and are agrarian economies (ICFRE 2023). The health of the soil and the farmers' ability to subsist have been impacted by fragmented land holdings, rainfed agriculture, monoculture-based farming techniques, and a lack of knowledge about integrated farming, in addition to decreasing agricultural output. To address these problems with Integrated Farm Development practice, vermicomposting was promoted.

16.6 Why Vermicomposting?

Surveys of the project villages were conducted in the initial phase of the project from 2018 to 2019. This involved a focused group discussion with the elders and women of the villages regarding the feasibility of the implementation of scaling up of vermicomposting and their sustainability by the villagers themselves, and most importantly, their applicability and use by the farmers. The feasibility was looked after by asking basic questions like, "Do they have domestic animals (cows/buffalos/goat/pigs etc.), space for installation of vermibeds, a source of water, and the ability to feed earthworms through leaf litter/kitchen vegetable waste, and at least a kitchen garden if they not do not own any agricultural land?" Most of the farmers had domestic animals in their rear kitchen yard along with a few tall trees but were unable to integrate the waste of one into the other. To address these problems, the Integrated Farm Development practice of vermicomposting was promoted. Another reason was that people were somewhere more interested in consuming organic food.

16.7 Accessibility of Vermicompost Units to the Mass Community

The vermibeds selected were of HDPE 450 GSM, UV stabilized woven, and portable with net ventilation and ready-to-install pockets. The size of the bed was 12 ft. x 4 ft. x 2 ft. (Fig. 16.2) with the approximate capacity of producing 1000–1500 kg per vermibed one time (in 100 days). The U-shaped open vermiwash filter pipe—UPVC ¾ inch—was also included for the collection of vermiwash. The product goes by the standards set by the Bureau of Indian Standards, Government of India (IS 15907:2010). The bamboo poles of 2.5 ft. height and 25 mm thickness (14 poles per vermibed), along with a shade net sheet to cover vermibeds—90% shade (12 ft. × 10 ft)—were also included.

A total of 5022 vermicomposting units were established for scaling up of vermicomposting in the project areas during 2021–22 simultaneously (Table 16.1).

Fig. 16.2 Vermicompost unit installed by a beneficiary in the ESIP project area

Table 16.1 Scaling of SLEM best practice: vermicomposting for sustainable land productivity, ICFRE (2023)

State	No. of villages targeted	No. of vermicomposting units established	No. of HHs targeted	No. of direct beneficiaries	Area coverage (ha)
Madhya Pradesh	34	2562	2562	7686	1296
Chhattisgarh	35	2460	2460	7380	1213
Total	69	**5022**	**5022**	**15,066**	**2509**

16.8 Capacity-Building Programs for Communities

Local communities were educated during a training session about organic farming, vermicompost preparation, and its benefits. Farmers who owned land and livestock were encouraged to adopt vermicompost preparation in their fields, using animal dung and organic waste like vegetable and fruit scraps. They were taught about the advantages of vermicompost and how to sustain this practice to enhance land productivity in the long term. The beneficiaries were also informed of vermiwash use, a by-product of vermicompost preparation, and its use as a biopesticide. Rich in vitamins, plant growth hormones, enzymes, and other micro- and macronutrients, vermiwash strengthens crops' ability to withstand a variety of diseases and promotes crop growth and productivity (Pilli et al. 2019). According to a study, vermicompost had an impact on plant development that was comparable to that of chemical fertilizer applied either by itself or in part with vermicompost (Karmegam and Daniel 2008). Integrating vermicompost with chemical fertilizers increases the use efficiency of the latter by reducing both use and input cost of chemical fertilizers by 25% (Gupta et al. 2021). Therefore, the use of vermicompost and vermiwash is expected to reduce the dependency of beneficiaries on chemical fertilizers and pesticides (Karmegam and Daniel 2008). Farmers in the project villages were informed about the nutritional benefits of intercropping. Beneficiaries were given improved vegetable seeds to encourage intercropping. The goal was to cultivate fresh seasonal vegetables year-round by adopting Integrated Farm Development. Farmers were advised to utilize local resources effectively, such as land, water, and animal dung, to maximize their land's production potential while maintaining soil fertility. Promotion of high-yielding vegetable varieties suited to the project areas of Madhya Pradesh and Chhattisgarh was emphasized to achieve maximum yield. Increase in agricultural production as well as its nutritional quality at a bearable cost by producers is a big challenge in rural areas. Vermicompost is a low-cost organic amendment that any villager having livestock can easily manage. Thus, the beneficiaries of the project area have started using vermibeds for the preparation of vermicompost. Many of them have already extracted compost. Though it takes 3 months for the compost to be ready for use, since it does not take too much hard labor, the villagers have realized that the practice is sustainable so long as they are able to keep the earthworms alive and in a healthy state.

Knowledge products, in the form of flyers and pamphlets, were prepared to promote vermicompost preparation and its use. These products were distributed to the local communities of the project villages to create awareness and enhance the knowledge base. Wall paintings featuring slogans were created at significant locations like Panchayat Bhawan, community halls, houses near schools, and local market areas to reach a maximum audience.

16.9 Timing for Implementation

The implementation phase was initiated during the monsoon season. This was a deliberate decision to make the transportation of earthworms viable, as it was important that they survive during their transportation from the source to the beneficiaries of the project area.

16.10 Implementation Phase

To upscale vermicomposting activity, beneficiaries were provided with a set of one vermibed, a net for shade, an outlet pipe for collecting vermiwash, and 14 bamboo sticks for the installation of the vermibed. A total number of vermibed units distributed in project areas is 5022, along with 3 kg of earthworms per unit to the beneficiaries. Soon after the distribution of vermicompost units, a field survey was conducted to see the successful installation of vermibeds. After this, it took some time for farmers to fill the vermi bed with cow dung, leaf litter, and also kitchen waste (an average of 30–40 days, as observed in the field). The cow dung was dried for a few days and then used for filling the vermin bed. Soon after the vermibeds were partially filled, the earthworms were introduced into them, and farmers were instructed to water the beds twice (on a daily basis) to keep the mixture moist, which is required for the survival of earthworms. Monitoring of the implemented activity was conducted by the responsible consultant of the concerned project areas. Wherever disparities were discovered, the village chief (sarpanch) of the affected areas was consulted and dealt with appropriately. Since the activity was implemented with the consent and self-interest of the community members, the successful implementation and sustainability of the practices were to be assured by the community itself. The initial establishment of all the vermibeds was successful and was maintained during the project duration.

The beneficiaries were also apprised of the constraints associated with vermicomposting:

- Vermicomposting is a time-consuming process. It requires at least 100–120 days for the organic wastes to decompose to prepare vermicompost.
- Vermicomposting requires maintenance, such as maintaining humidity and shade and keeping earthworms alive.

Steps for scaling up vermicomposting in ESIP areas are outlined in Fig. 16.3a–o.

16.11 Outcome Expectations

Land productivity that has been declining due to the overuse of fertilizers will increase because vermicompost can regenerate the land. Every vital plant nutrient is abundant in vermicompost. Well-rotted vermicompost has a nutrient concentration of 1.2–2.5% nitrogen, 0.9–1.7% phosphorus, and 1.5–2.5% potassium (Yadav et al. 2023). Additionally, the villagers are more likely to gather the cow dung that was previously dispersed throughout neighborhoods and streets and use it for vermicomposting, which reduces land pollution. If the beneficiary takes advantage of the opportunity to produce and sell in the market (Rs 20–30 per kg in the nursery), it can also completely transform his financial situation. Vermicomposting on a large scale has the potential to spread awareness of the necessity of sustainable agriculture. Though continuous and regular maintenance is necessary for functional vermicomposting units, in the long run, it aims to enhance land productivity along with maintaining the ecological balance. Vermicomposting promotes the utilization of locally available resources and is one of the preparation processes of organic manure in which earthworms convert waste materials into compost (ICFRE 2023).

Fig. 16.3 (**a**) Awareness about project activities, (**b**) capacity building, (**c**) exposure visit, (**d**) brochure about vermicompost—knowledge product, (**e**) wall painting for awareness and campaign, (**f** and **g**) distribution of brochure among villagers, (**h**) distribution of vermibeds, (**i**) demonstration of vermibed installation, (**j**) beneficiary taking care of fully installed vermibed and filled in cow dung and making it ready for worm introduction, (**k**) distribution of earthworms, (**l**) beneficiary watering the vermibed, (**m**) extraction of vermicompost through sieving, (**n**) collection of vermicompost, and (**o**) applying vermicompost into the saplings

16.12 Monitoring and Evaluation of the Practice

After 6 months of the implementation of the vermicomposting, the monitoring task for assessing the success of the practices was done through the engagement of a third party as an independent consultant for the evaluation of the overall impacts of the project activities in accordance with the project objectives and indicators set out in its results framework (ICFRE 2023). They undertook a physical evaluation of the interventions carried out during the project on the local ecology of a specific landscape, such as upscaling of vermicomposting among the community members of the project area, etc. They also evaluated the socio-economic benefits generated for beneficiary communities in terms of skill enhancement and livelihood enhancement. They have reported that 79.3% of the respondents in MP and 71.2% in Chhattisgarh have received training on vermicomposting.

16.13 Challenges That Come Across and Overcoming These Challenges

Whenever community-based work is implemented on the ground, challenges inevitably arise. Upon initiating the project in the field, numerous obstacles were encountered. These challenges include engagement of stakeholders, involvement of local governance (gram panchayat, Self Help Group, Joint Forest Management Committees, etc.), capacity constraints like people who have previously worked with either NGOs or state government projects with different types of trainers, assessing their needs and understanding the fruitful impact, and, most importantly, collaboration with different categories of people belonging to different castes, religions, and beliefs.

Effective engagement begins with understanding stakeholders' perspectives, needs, and potential contributions. We talked with the existing local governance about various possibilities of implementing best practices in accordance with community needs. Capacity-building programs were launched in individual villages separately so that maximum participation appeared. The local governance, like Gram Pradhan, Self Help Group members, etc., was involved to enhance the participation of the rest of the community members. Members of different ethnic groups were brought under the same platform through regular conversation.

16.14 Conclusions

Vermicomposting is an organic process that is environmentally friendly and easy to operate. It can benefit small and marginal farmers, particularly in the face of environmental degradation and increasing food demand. While vermicomposting alone may not meet all food demand, integrating it with chemical fertilizers can produce sustainable food. In the ESIP project area, the adoption rate of vermicomposting was satisfactory, with rural women actively participating and benefiting from it. Farmers, especially women beneficiaries, reported increased crop production and savings of INR 3035 per household per year using vermicompost instead of chemical inputs (ICFRE 2023). Therefore, more farmers need to be educated about vermicomposting and its benefits by appointing more extension workers to promote sustainability in agriculture.

According to published research, it can assist in accomplishing the eight sustainable development goals listed below:

1. No poverty (Sinha et al. 2010)
2. Zero hunger (Creegan and Flynn 2020; Devi and Khwairakpam 2022)
3. Good health and well-being (Devi and Khwairakpam 2022)
4. Decent work and economic growth (Local Government Initiative on Climate Change (LoGIC) Project, Vermicomposting—a source of alternative income for women in agriculture. https://www.logicbd.org/ Accessed on 30 Oct 2024)
5. Sustainable cities and economies (Saha et al. 2022)
6. Responsible consumption and production (Saha et al. 2022)
7. Climate action (potential role of vermicompost and its extracts in alleviating climatic impacts on crop production, Vambe et al. 2023; reduces greenhouse gas emissions, Panda et al. 2022)
8. Life on land (in situ soil application of organic materials engenders soil carbon sequestration—a climate change mitigation mechanism—and overall soil health and soil water conservation, Creegan and Flynn 2020)

Vermicomposting in agriculture indirectly contributes to other SDGs, like quality education as it can equip people with skills needed for farming and related industries; gender equality since women play a significant role in agriculture, particularly in developing countries; clean water and sanitation since sustainable agricultural practices, efficient irrigation and responsible water management can reduce water wastage and ensure water availability for farming and local communities; affordable and clean energy: improved agriculture can contribute to renewable energy sources, such as bioenergy from crops or organic waste; decent work and economic growth: agriculture supports employment and economic growth; industry, innovation, and infrastructure as improved farming technologies ensures increased productivity and creating sustainable food systems and reduced inequalities since inclusive agricultural practices benefit marginalized groups, including smallholder farmers, indigenous peoples, and women, can help reduce inequalities within and among countries.

16.15 Recommendation

Any project that promotes vermicomposting must be carried out on a large scale. Forest degradation has persisted as a result of reliance on subsistence livelihoods in and near forests. Therefore, in order to scale up such practices, concentrated development of forest fringe villages, including areas where pattas (land on lease) have been allotted to people under the Forest Rights Act, 2006, should be taken into consideration. The recovery of degraded forests will greatly benefit from their quicker upliftment out of subsistence. It takes time for farmers to adopt this as a best practice. Only when they begin to reap financial rewards for their fertilizer expenditures will farmers sustainably embrace this practice. In order to assess the vermicomposting's results and properly document and disseminate the lessons learned, the project must also have a lengthy monitoring period. In order to reduce waste management systems and landfills, improve soil health, support sustainable agriculture, lessen air and land pollution from chemical fertilizers, improve soil carbon sequestration, lower methane emissions, and, last but not least, create economic opportunities (composting operations, vermicompost sales, etc.), vermicomposting must be scaled up.

References

Bhan S (2013) Land degradation and integrated watershed management in India. Int Soil Water Conserv Res 1(1):49–57

Brainerd E, Menon N (2014) Seasonal effects of water quality: the hidden costs of the Green Revolution to infant and child health in India. J Dev Econo 107:49–64. https://doi.org/10.1016/j.jdeveco.2013.11.004

Chaudhari SK, Islam A, Biswas PP, Sikka AK (2015) Integrated soil, water and nutrient management for sustainable agriculture in India. Indian J Fertil 11(10):51–62

Creegan EF, Flynn R (2020) SDG 2 Zero Hunger. In: Franco I, Chatterji T, Derbyshire E, Tracey J (eds) Actioning the global goals for local impact. Science for sustainable societies. Springer, Singapore. https://doi.org/10.1007/978-981-32-9927-6_3

Devi C, & Khwairakpam M (2022) Vermicompost for Sustainable Agriculture and Bioconversion of Terrestrial Weed Biomass into Vermicompost. In M. Turan & E. Yıldırım (Eds.), New Generation of Organic Fertilizers. IntechOpen. https://doi.org/10.5772/intechopen.100615

Erisman JW, Galloway JN, Seitzinger S, Bleeker A, Dise NB, Petrescu AR, de Vries W (2013) Consequences of human modification of the global nitrogen cycle. Philos Trans R Soc B: Biol Sci 368(1621):20130116

FAO (2020) CB1110EN/1/09.20. https://openknowledge.fao.org/server/api/core/bitstreams/ffb5feaf-8388-4e2f-b319-2260a9a6f5a2/content

Gill S, Iqbal M, Jabbar A, Amin B, Gill A, Ilyas M, Tauqeer H, Virk Z (2023) Vermicompost for sustainable future: nature-based solution for environmental degradation, climate change, and food security. Clim Resil Agric 1. https://doi.org/10.1007/978-3-031-37424-1_25

Gupta N, Pradhan S, Jain A, Nahya P (2021) Sustainable Agriculture in India 2021: what we know and how to scale up. New Delhi: Council on Energy, Environment and Water

ICFRE (2023) Ecosystem services improvement project: project completion report. Indian Council of Forestry Research and Education, Dehradun. https://icfre.gov.in/publications

Jote CA (2023) The impacts of using inorganic chemical fertilizers on the environment and human health. Organ Med Chem Intl J 13(3):555864

Karmegam N, Daniel T (2008) Effect of vermicompost and chemical fertilizer on growth and yield of hyacinth bean, *Lablab purpureus* (L.) sweet. Dyn Soil Dyn Plant 2(2):77–81

Khatun J, Intekhab A, Dhak D (2022) Effect of uncontrolled fertilization and heavy metal toxicity associated with arsenic (As), lead (Pb) and cadmium (Cd), and possible remediation. Toxicology 477:153274

Kumar P, Sharma PK (2020) Soil salinity and food security in India. Front Sustain Food Syst 4:533781

Montanarella L (2007) Trends in land degradation in Europe. In: Sivakumar MVK, Ndiang'ui N (eds) Climate and land degradation. Environmental science and engineering. Springer, Berlin, Heidelberg. https://doi.org/10.1007/978-3-540-72438-4_5

National Centre of Organic Farming (NCOF) (2010) Biofertilizers and organic fertilizers statistics year 2005–06 to 2009–10. NCOF. Department of Agriculture and Cooperation, Ministry of Agriculture, Govt of India Ghaziabad

Panda AK, Mishra R, Dutta J, Wani ZA, Pant S et al (2022) Impact of vermicomposting on greenhouse gas emission: a short review. Sustainability 14(18):11306

Pilli K, Jaison M, Sridhar D (2019) Vermicomposting and its uses in sustainable agriculture. Research trends in agriculture sciences. AkiNik Publications, pp 75–88

Prakash O (2023) Excessive use of chemical fertilizers, reduce the fertility power of the soil. Int J Eng Invent 12(8):116–118

Saha P, Barman A, Bera A (2022) Vermicomposting: a step towards sustainability. In: Sustainable crop production: recent advances, p 53

Singh R (2000) Environmental consequences of agricultural development: a case study from the green revolution state of Haryana, India. Agric Ecosyst Environ 82:97–103

Sinha RK, Valani D, Chauhan K, Agarwal S (2010) Embarking on a second green revolution for sustainable agriculture by vermiculture biotechnology using earthworms: reviving the dreams of Sir Charles Darwin. J Agric Biotechnol Sustain Dev 2(7):113

Somvanshi PS, Pandiaraj T, Singh RP (2020) An unexplored story of successful green revolution of India and steps towards ever green revolution. J Pharmacogn Phytochem 9:1270–1273

Tripathi YC, Hazarika P, Pandey BK (2005) Vermicomposting: an ecofriendly approach to sustainable agriculture. Verms and vermitechnology. APH Publishing Corporation, New Delhi, pp 23–39

Vambe M, Coopoosamy RM, Arthur G, Naidoo K (2023) Potential role of vermicompost and its extracts in alleviating climatic impacts on crop production. J Agric Food Res 12:100585

Vitousek PM, Naylor R, Crews T, David MB, Drinkwater LE, Holland E, Nziguheba G (2009) Nutrient imbalances in agricultural development. Science 324(5934):1519–1520

Wang Y, Zhao G (2019) Life cycle assessment of potential pollutant-induced human capital loss caused by different agricultural production Systems in Beijing, China. J Clean Prod 240:118141

Yadav SK, Babu S, Singh R, Yadav D, Rajanna GA (2023) The role of organic and natural ecosystems in the food industry. In: Sustainable development and pathways for food ecosystems. Academic Press, pp 115–128

Chapter 17
Plant Breeding Strategies for Sustainable Agriculture in Climate Change Context: A Literature Review

Ali Yetgin, Rajiv Kumar Srivastava, Siddharth Panda, and Sweta Rath

Abstract Climate change poses significant challenges to global agriculture, threatening food security through increased droughts, heatwaves, and erratic rainfall. Plant breeding offers a critical solution, as it develops crop varieties capable of withstanding these climate-induced stresses. This chapter explores recent advancements in climate-smart plant breeding, focusing on strategies to enhance drought and heat tolerance, improve water-use efficiency, and increase nutrient-use efficiency. Modern genomic tools, including marker-assisted selection, CRISPR gene editing, and high-throughput phenotyping, are revolutionizing the speed and precision of breeding efforts, allowing for the targeted development of climate-resilient crops. Additionally, innovative breeding approaches are being employed to enhance crop carbon sequestration, aiding in climate adaptation and mitigation. This chapter synthesizes key findings from recent research, highlighting the most promising genes and breeding techniques for addressing the environmental challenges posed by climate change. Special emphasis is placed on critical traits, including improved root architecture, enhanced photosynthetic efficiency, and greater resilience to abiotic stressors. Finally, the review outlines future directions for integrating breeding

A. Yetgin
Toros Agri Industry and Trade Co. Inc., Research and Development Center, Mersin, Turkey

Department of Biotechnology, Institute of Nature and Applied Sciences, Cukurova University, Adana, Turkey

R. K. Srivastava (✉)
Biological and Agricultural Engineering Department, Texas A&M University, College Station, TX, USA
e-mail: rajiv.srivastava@ag.tamu.edu

S. Panda
Department of Genetics and Plant Breeding, Siksha 'O' Anusandhan, Bhubaneswar, Odisha, India

S. Rath
Department of Agronomy, Siksha 'O' Anusandhan, Bhubaneswar, Odisha, India

Y. D. I. Khan et al. (eds.), *Land Restoration Through Ecosystem-Based Approach*, Disaster Resilience and Green Growth, https://doi.org/10.1007/978-981-95-0389-6_17

innovations into global agricultural systems, emphasizing the need for collaboration between scientists, policymakers, and farmers to ensure the widespread adoption of climate-smart crops.

Keywords Drought · Greenhouse gases · Food security · Plant breeding

17.1 Introduction

Plant breeding is instrumental in tackling the difficulties presented by climate change by enhancing the capacity of crops to absorb and sequester carbon in both plants and soil, thereby contributing to improved carbon sequestration. This method fosters sustainable farming practices and diminishes reliance on inputs like fertilizers and pesticides, thus improving resource use efficiency (Fess et al. 2011). By mitigating negative environmental impacts associated with conventional agricultural practices, plant breeding contributes to developing more sustainable food systems (Ahmad 2023). However, adopting genetic modification in plant breeding raises concerns regarding safety and ethical implications (Munawar et al. 2024), potentially affecting biodiversity and ecosystem health (Lóopez-Calcagno et al. 2020).

Modern plant breeders increasingly leverage genomics and data-driven techniques to accelerate breeding processes and develop crops that exhibit enhanced resilience to climate stressors (Varshney et al. 2021). Breeders can utilize genomic information to pinpoint genetic indicators linked to favorable characteristics like drought resilience and immunity to diseases, which aids in crafting innovative breeding approaches. This strategy is essential in developing crop varieties that can endure the effects of climate change, thereby mitigating environmental degradation associated with agriculture (Varshney et al. 2021). Through genomics, the accurate detection of genetic markers associated with particular attributes, such as drought resistance, disease immunity, and heat endurance, is possible, and advancements in gene editing and participatory breeding enable breeders to innovate and develop crops tailored to evolving climate conditions (Ceccarelli et al. 2013).

Phenomics, involving high-throughput imaging and sensing technologies, allows for quantifying plant traits like growth rate, biomass accumulation, and stress response (Rahman et al. 2015). These data-driven insights can inform predictive models that aid in identifying plant varieties best suited for specific environmental conditions (Bacha and Iqbal 2023). Bioinformatics, employing computational techniques for analyzing vast datasets including genetic and phenomic data (Marsh et al. 2021), empowers breeders to expedite the selection of promising plant varieties and develop innovative breeding methodologies (Yoosefzadeh Najafabadi et al. 2023). The integration of these technologies underpins smart breeding approaches, which combine data-driven methodologies with traditional breeding practices (Sharma et al. 2023). This synergistic approach holds promise in enhancing food security, reducing agriculture's environmental footprint, and improving farm profitability (Gerullis et al. 2023).

Climate-smart plant breeding is a strategic approach focused on creating new crop varieties equipped to prosper under variable weather patterns, simultaneously reducing greenhouse gas emissions and adjusting to the repercussions of climate change (Rai 2022). Moreover, it contributes to agriculture's carbon neutrality by enhancing plants' capacity to sequester carbon in soil and biomass (Harvey et al. 2014). This approach also supports ecosystem-based land restoration efforts by promoting biodiversity, improving soil health, and increasing the resilience of agricultural landscapes (IUCN 2022). Therefore, this chapter encompasses applying advanced breeding technologies and methodologies to foster the development of climate-resilient crops, offering increased yields with reduced inputs, improved water and nutrient efficiency, and enhanced adaptability to climate variability.

17.2 Climate-Smart Plant Breeding Strategies

Climate-smart plant breeding is an all-encompassing methodology that acknowledges the synergy between agriculture and the environment, aiming to enhance efficiency while minimizing the adverse impacts of agricultural methods (Rai 2022). The aim is to cultivate crop varieties that are more resilient to prevailing climatic conditions and capable of enduring the effects of climate change, including drought, extreme heat, and shifting rainfall trends (Sharma et al. 2023). The plant breeding process and strategies employed to develop crops tailored to various challenges (Fig. 17.1). Additionally, Table 17.1 summarizes the reviewed papers discussing different plant breeding strategies for coping with climate change.

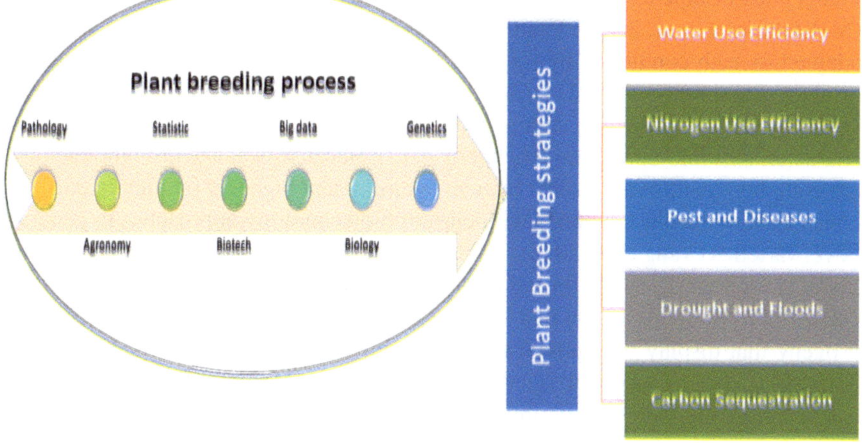

Fig. 17.1 A layout of the plant breeding process and strategies

Table 17.1 Literature review on plant breeding for different strategies

Plant breeding approaches	Modeling/ experiment	Plant types	Model types	Major findings
Developing heat- and drought-tolerant varieties	Experiment	Numerous crops, such as rice and sorghum, among others	N/A	Enhanced resilience to elevated temperatures and limited water availability (López-Calcagno et al. 2020; Mellacheruvu et al. 2019; Yang et al. 2022)
Breeding for improved water-use efficiency	Experiment	Numerous crops, such as rice and sorghum, among others	N/A	Enhanced efficiency in water use for plant growth and development (Ahmad 2023; Hatfiel and Dold 2019)
Creating crop varieties that utilize nutrients more efficiently	Modeling	Numerous crops, such as maize, wheat, rice, and others	Mathematical and computational models	Increased efficiency in nutrient uptake and utilization (Temme et al. 2024; Qaim 2020)
Breeding for resilience against diseases and pests	Modeling	Various crops, including tomato, potato, and others	Mathematical and computational models	Enhanced resilience against diseases and pests (Karlsson Green et al. 2020; Kavhiza et al. 2022; Skendžić et al. 2021)
Developing varieties with improved carbon sequestration	Modeling	Various crops, including grasses and trees	Mathematical and computational models	Increased capacity for carbon sequestration in both soil and plant biomass (Hadebe et al. 2017; Nerva et al. 2022; Selin 2023)

17.2.1 Developing Heat and Drought-Tolerant Varieties

Plants must be able to endure rising temperatures and droughts in the coming years (Mellacheruvu et al. 2019). Plant breeders can develop crop strains that are better adapted to such conditions by focusing on traits that confer resistance to heat and drought (Varshney et al. 2021). This technique entails choosing plants with characteristics that enable them to endure heat and drought stress and then breeding these characteristics into new types (Shepherd 2017). Because heat and drought tolerance are complex qualities controlled by many genes, plant breeders frequently combine conventional breeding methods with current biotechnology approaches to generate novel varieties (Yang et al. 2022). Thus, the breeding method followed can be traditional or modern (Breseghello and Coelho 2013). Conventional breeding techniques consist of hybridizing plants that exhibit favorable characteristics and choosing the progeny that display the optimal mix of these traits across successive generations (Bradshaw 2017). Conversely, contemporary biotechnological methods like marker-assisted selection and genetic modification can expedite this process by enabling breeders to pinpoint and incorporate specific genes that confer tolerance to heat and drought (Parmar et al. 2017).

Additionally, the development of plants with more extensive root systems, which enable them to access water from deeper soil layers, is one of the traits that can assist in withstanding the stresses of heat and drought. Other characteristics that can help plants to endure heat and drought stress are improvement in water-use efficiency, allowing plants to utilize water more efficiently; increased leaf wax content, which lowers leaf water loss (Shepherd 2017); increased synthesis of osmoprotectants, which aids in the stabilization of cell membranes and the maintenance of water balance (Mellacheruvu et al. 2019); and improved photosynthetic efficiency, which enables plants to produce more energy while using less water (Mellacheruvu et al. 2019).

Developing these crop varieties is particularly vital in regions already challenged by extreme heat and limited water supply, especially in dry regions (Shepherd 2017). Some crop types naturally can withstand drought and heat stress, and breeders can aim to discover and integrate these resilient traits into new strains (Shepherd 2017). Moreover, genetic engineering can be utilized to instill new characteristics associated with tolerance to heat and drought in crop varieties. (Parmar et al. 2017). For example, scientists have developed genetically modified crops that produce more specific proteins, such as heat shock protein (HSP), which helps plants withstand high temperatures (Singh and Grover 2008). Creating crop varieties that can thrive under high temperatures and water scarcity conditions is a significant strategy for dealing with climate change and reducing its impact (Srivastava et al. 2023). Food security may be ensured, and the negative environmental consequences of agriculture reduced by developing heat and drought-resistant cultivars (Kumar 2016).

17.2.1.1 Key Genes and Traits

HSP (Heat Shock Protein) Genes

These genes help plants cope with heat stress by ensuring that proteins retain their functional three-dimensional shape and that cellular membranes remain intact and functional (Elmyhun et al. 2024).

DREB1 and DREB2

These regulatory proteins play a key role in a pathway that does not involve abscisic acid (ABA) and are essential for activating genes that respond to environmental stress (ISAAA 2023).

EF-Tu Factor

The EF-Tu element is linked to a plant's ability to withstand high heat and plays a crucial role in the plant's adaptive mechanisms when faced with increased thermal conditions (Simova-Stoilova et al. 2016).

MAPK Gene

The MAPK gene has been linked to heat tolerance, playing a vital role in the plant's ability to withstand high-temperature stress (Simova-Stoilova et al. 2016).

P5CS Genes (Δ1-Pyrroline-5-Carboxylate Synthetase)

These genes enhance synthesis of proline, an osmoprotectant that stabilizes cell membranes and maintains water balance during drought stress (Mellacheruvu et al. 2019).

RD22 Genes (Responsive to Dehydration 22)

These genes help to manage and direct various functions in response to dehydration and enhance drought tolerance in plants (Shepherd 2017).

RAB18 Genes (Responsive to ABA 18)

These genes help in response to abscisic acid (ABA), regulating stomatal closure and improving water-use efficiency (drought conditions) (Shepherd 2017).

17.2.1.2 Case Studies

Heat Tolerance in Rice Through HSP Genes

Jagadish et al. (2020) explored the development of heat-tolerant rice varieties at the International Rice Research Institute (IRRI). The research focused on identifying and incorporating heat shock protein (HSP) genes. These proteins are recognized for their important function in safeguarding plants from the effects of high temperatures. They serve as molecular guardians, ensuring the stability of other proteins and preventing their denaturation during high-temperature episodes. The study successfully introduced HSP genes into rice varieties, resulting in improved thermal tolerance and the maintenance of grain quality under heat stress conditions. This advancement particularly benefits rice farmers in tropical regions where high temperatures can severely impact crop yields.

Drought and Heat Tolerance in Maize Using DREB Genes

Another significant study highlighted by Lamaoui et al. (2018) investigated how DREB genes could improve maize's resistance to dry and hot conditions. The DREB1 and DREB2 genes, which are transcription factors, control the activation of a wide array of genes that respond to stress. By manipulating these genes, researchers were able to develop maize varieties that exhibit increased resilience to both drought and heat. The study demonstrated that these genetically modified maize plants could maintain higher yields under stress conditions than non-modified plants. This research provides a promising avenue for crop varieties that can withstand the dual challenges of drought and heat, which are increasingly prevalent due to climate change.

17.2.2 Breeding for Improved Water-Use Efficiency

Breeding crops that require less water to achieve optimal yields is essential for mitigating drought impacts (Breseghello and Coelho 2013). This approach selects traits that enhance water-use efficiency, including deeper root systems, improved stomatal regulation, and enhanced photosynthetic efficiency (Wasson et al. 2012). Particularly critical in water-scarce regions and those vulnerable to increasing water stress from climate change (Raza et al. 2019), this strategy aims to maintain agricultural productivity while conserving water resources.

Breeding for improved water-use efficiency combines traditional methods of crossbreeding for desirable traits over generations with modern biotechnological tools such as marker-assisted breeding (Breseghello and Coelho 2013). Marker-assisted breeding accelerates the identification and transfer of genes associated with water conservation traits, aiding in developing drought-resistant crops (Singh et al. 2020). This can also assist in boosting agricultural resistance to drought and other kinds of water stress by generating the exact yield with less water, making crops more resilient to periods of low water supply (Hatfiel and Dold 2019).

Advanced genome editing technologies such as CRISPR offer new avenues to enhance water-use efficiency by targeting genes involved in photosynthesis and stomatal conductance (Elrys et al. 2022). These innovations promise to create crops that thrive under limited water conditions, thereby reducing the need for fertilizers and other inputs while ensuring sustainable agriculture (Hatfiel and Dold 2019; Ahmad 2023).

17.2.2.1 Key Genes and Traits

Key genes associated with improved water-use efficiency include the following:

RWC1 (Root Water Content 1)

It enhances root architecture and water uptake efficiency under drought conditions (Elrys et al. 2022).

AQP1 (Aquaporin 1)

It facilitates water transport across membranes, improving plant water-use efficiency (Singh et al. 2020).

PPDK (Phosphoenolpyruvate Carboxylase Kinase)

It enhances photosynthetic efficiency by optimizing carbon dioxide fixation, thereby reducing water loss (Ahmad 2023).

17.2.2.2 Case Studies

Drought-Resistant Maize in Africa (Sub-Saharan)

Scientists at the International Maize and Wheat Improvement Center (CIMMYT) have successfully developed drought-tolerant maize varieties through marker-assisted breeding. These varieties exhibit improved water-use efficiency by incorporating genes that regulate stomatal conductance and enhance root depth, allowing them to maintain yields even under limited water availability (Cairns et al. 2013).

Water-Efficient Rice Varieties in Asia

Traditional rice varieties require substantial water inputs in water-scarce regions of Asia, such as parts of India and Southeast Asia. The International Rice Research Institute (IRRI) has introduced rice cultivars with enhanced water-use efficiency traits, such as low transpiration rates and improved root systems. These varieties conserve water and maintain high productivity, supporting sustainable rice production in regions prone to drought (Pandey et al. 2017).

17.2.3 Enhancing Nutrient-Use Efficiency with New Varieties

With the ongoing progression of climate change, it is anticipated that the presence and allocation of nutrients in the soil will experience shifts. These shifts can greatly influence crops' yield and how well they utilize nutrients (Mirzabaev et al. 2023). Therefore, breeding programs to enhance nutrient uptake and utilization are crucial for sustaining long-term food production under evolving environmental conditions (Yadav et al. 2023).

Breeding for improved nutrient-use efficiency combines genetic insights with agricultural practices to reduce reliance on chemical fertilizers. This approach enhances crop resilience to nutrient stresses and mitigates environmental impacts associated with fertilizer use, such as eutrophication and greenhouse gas emissions (Cooper and Messina 2023).

Collaboration among researchers, farmers, and stakeholders is essential to tailoring nutrient-efficient varieties to local agricultural conditions and meeting the needs of diverse farming systems (Baligar et al. 2001). By enhancing the way plants absorb and use nutrients, these breeding approaches aid in achieving sustainable farming practices by increasing the efficiency of resource use and reducing harm to the environment (Qaim 2020).

17.2.3.1 Key Genes and Traits

Key genes associated with improved nutrient-use efficiency include the following:

PHR1 (Phosphate Starvation Response 1)

It regulates phosphate uptake and allocation, improving crop phosphorus-use efficiency (Heuer et al. 2017).

NRT1.1 (Nitrate Transporter 1.1)

It enhances nitrate uptake efficiency in plants, reducing dependence on external nitrogen fertilizers (Wang et al., 2020).

NAS (Nicotianamine Synthase)

This enzyme facilitates iron and other micronutrient uptake, which is crucial for enhancing nutrient-use efficiency in plants (Zhao et al., 2019).

17.2.3.2 Case Studies

Nitrogen-Use Efficient Wheat in Australia

This study from Australia has developed nitrogen-efficient wheat varieties through traditional breeding and marker-assisted selection. These varieties incorporate genes like NRT1.1 to exhibit improved nitrogen uptake and utilization efficiency, enabling high yields with reduced nitrogen fertilizer inputs (Garnett et al. 2015).

Phosphorus-Efficient Maize in Africa (Sub-Saharan)

In pursuing agricultural advancement, efforts have been made to breed maize varieties that exhibit greater efficiency in phosphorus utilization. These specialized varieties leverage genes like PHR1 to enhance the uptake of phosphorus from soils that are deficient in this essential nutrient, thereby ensuring consistent yields of maize even in conditions where phosphorus is limited (Mutale 2020).

17.2.4 Breeding for Disease and Pest Resistance

This approach is essential for addressing and mitigating the impacts of climate change, as shifts in climate can result in changes to pest and disease dynamics, complicating the task of crop protection (Newton et al. 2011). Plants endowed with these traits are more resistant to diseases and pests, which helps reduce agricultural losses and increase productivity (Donatelli et al. 2017). Resistance to diseases and pests in plants is typically developed through a mix of conventional breeding techniques and modern biotechnological methods (Srivastava and Yetgin 2024). Traditional breeding involves hybridizing plants with desired traits and selecting progeny that exhibit the optimal traits over successive generations (Kavhiza et al. 2022). Contemporary biotechnological strategies, such as marker-assisted selection and genetic engineering, can expedite this process by enabling the identification and incorporation of specific genes linked to resistance (National Research Council 2004). Additionally, cutting-edge gene editing tools like CRISPR-Cas9 have the potential to create crops that are resistant to diseases and pests without introducing foreign DNA (Skendžić et al. 2021). The use of wild plant species in breeding programs is also advantageous, as they often possess beneficial traits such as resistance to disease and tolerance to stress, which can improve the performance of cultivated crops (Scheben et al. 2017).

While developing disease and pest-resistant plants is an important strategy, it is not a panacea for the challenges posed by climate change due to its controversial nature (Rezv et al. 2023). It's essential to consider a range of adaptation and mitigation strategies that work together to ensure the sustainability and robustness of our food systems amid climate change (Karlsson Green et al. 2020). However, concerns about the potential risks and ethical issues related to these strategies must be considered (Kavhiza et al. 2022). A careful assessment of each method's benefits and potential downsides is crucial. A notable obstacle of this approach is the ability of pests and diseases to adapt and circumvent plant defenses quickly (Nelson et al. 2018), necessitating ongoing vigilance and flexibility to maintain the effectiveness of these breeding methods. Additionally, the application of genetic engineering in creating disease and pest-resistant crops has sparked debate over the safety and ecological consequences of genetically modified organisms (Phillips 2008).

17.2.4.1 Key Genes and Traits

Key genes associated with disease and pest resistance include the following:

RGA1 (Resistance Gene Analog 1)

It provides resistance against a broad spectrum of pathogens by recognizing pathogen avirulence proteins (Kavhiza et al. 2022).

Bt Toxin (Bacillus Thuringiensis Toxin)

It confers resistance to insect pests by producing proteins toxic to specific insect species (National Research Council 2004).

CRISPR-Cas9

It enables precise genome editing to introduce or enhance disease resistance genes like R genes or induce targeted mutations for pest resistance (Skendžić et al. 2021).

17.2.4.2 Case Studies

Indian Bt Cotton

The advancement of Bt cotton varieties by the Maharashtra Hybrid Seeds Company (Mahyco), in partnership with Monsanto, has been a game-changer for India's cotton industry. These varieties are engineered to produce the Bt toxin, which shields them from the harmful cotton bollworm, leading to a substantial reduction in the reliance on insecticides and an increase in cotton yields (Kranthi et al. 2002).

Late Blight-Resistant Potatoes in Europe

The International Potato Center (CIP) has developed late blight-resistant potato varieties using conventional breeding combined with marker-assisted selection. These varieties incorporate R genes from wild potato relatives, such as *Solanum bulbocastanum*, providing durable resistance to Phytophthora infestans, the causative agent of late blight (Grunwald et al. 2017).

17.2.5 Developing Varieties with Improved Carbon Sequestration

Plants act as organic carbon absorbers, drawing in carbon dioxide from the air via photosynthesis (Poffenbarger et al. 2023). Breeders are able to cultivate plant varieties that excel at capturing carbon, whether through enhanced biomass growth or by creating varieties that accumulate carbon within their root systems (Eckardt et al. 2023). This process contributes to the reduction of CO_2 in the atmosphere, a major factor in climate change, thereby helping to lessen its effects (Srivastava and Yetgin 2024). Through the process of photosynthesis, plants convert atmospheric CO_2 into biomass (Zhu et al. 2008). By breeding crops with an increased ability for carbon capture, the volume of carbon stored in both plants and the soil can be raised (Ontl and Schulte 2012).

The process of breeding to boost carbon capture often focuses on selecting characteristics that lead to increased biomass and the development of deeper, more expansive root systems, enhancing the soil's carbon storage capacity (Srivastava and Yetgin 2024). Plants with such root systems are able to tap into more soil carbon, thereby augmenting the soil's carbon content (Guo and Giffor 2002). Additionally, plants designed with a less dense canopy allow for better sunlight capture, which can improve photosynthesis and, consequently, carbon sequestration (Riedesel et al. 2022a). On the other hand, drought conditions can impede plant growth and biomass production, negatively affecting carbon capture (Eziz et al. 2017). Therefore, creating varieties of crops that can tolerate drought conditions is crucial to sustaining biomass production and carbon sequestration, even in environments with limited water availability (Hadebe et al. 2017).

Plant breeders must utilize a range of breeding approaches to boost photosynthetic capacity, such as deeper root systems, bigger leaf surfaces, and increased nitrogen-use efficiency (Riedesel et al. 2022b). These breeding activities have the potential to boost carbon sequestration as well as yields, drought tolerance, and nutritional value, among other benefits (Chavan et al. 2009). Enhancing carbon capture in crops is vital for sustainable farming and climate change mitigation (Fawz et al. 2020). Yet, it is important to balance this with other crop traits like yield, disease resistance, and nutrient efficiency (Dwivedi et al. 2021). Carbon capture extends beyond plant biomass to the soil via roots and remains (Ontl and Schulte 2012). Traits that boost carbon capture can also improve nutrient mobilization, affecting plant growth and biomass (Selin 2023). This enhances soil health, fertility, and agricultural productivity and resilience (Nerva et al. 2022) and boosts soil organic content, structure, moisture retention, nutrient access, and microbial life (Abrol 2009).

17.2.5.1 Key Genes and Traits

Key genes associated with improved carbon sequestration include the following:

Rubisco (Ribulose-1,5-Bisphosphate Carboxylase/Oxygenase)

It enhances photosynthetic efficiency and carbon fixation (Zhu et al. 2008).

SUSIBA2 (Sugarcane-Expressed Isoform 2 of Barley Sucrose Synthase)

It facilitates enhanced biomass accumulation and carbon storage in plant tissues (Bernhard 2010).

NAC (NAM, ATAF1/2, CUC2) Transcription Factors

They regulate root development and architecture, influencing carbon allocation to roots and soil (Guo and Giffor 2002).

17.2.5.2 Case Studies

Bioenergy Sorghum in the United States

Researchers at the University of Illinois, USA, have developed bioenergy sorghum varieties with enhanced carbon sequestration potential. These plant varieties make more effective use of the C4 photosynthetic process, resulting in greater production of biomass and enhanced carbon storage in both the plant structure above the soil and the root system (Riedesel et al. 2022a, b).

Carbon Farming in Australia

Australian farmers are adopting carbon farming practices with perennial grasses like kangaroo grass (*Themeda triandra*), which has deep root systems that enhance soil carbon sequestration. These grasses are well-adapted to drought conditions, ensuring continuous carbon storage even in arid regions (Eziz et al. 2017).

17.3 Implementation, Adoption, and Barriers

17.3.1 Implementation and Adoption

Figure 17.2 highlights the importance of cooperation between breeders, farmers, and policymakers for the effective implementation of climate-smart breeding programs (Aggarwal et al. 2018). Such partnerships ensure that breeding aligns with local needs and conditions. Collaboration allows breeders and farmers to test new crops in actual farming situations, confirming their adaptability and performance (Behera and France 2023). Policymakers play a key role by providing support through funding and regulations that encourage the use of climate-adaptable crops (Morris and Bellon 2004). Moreover, education and outreach are crucial in encouraging the use of these breeding strategies. They help raise awareness about the challenges of climate change and the advantages of eco-friendly farming. Educational initiatives like training sessions, workshops, and field schools are key in spreading knowledge and fostering support for new technologies and policies (Aggarwal et al. 2018).

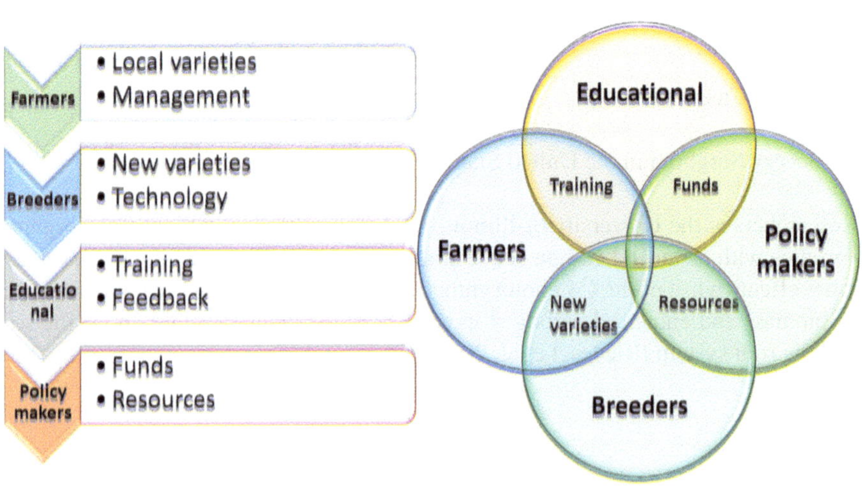

Fig. 17.2 Implementation and adoption of climate-smart plant breeding

17.3.2 Barriers

Numerous obstacles impede the adoption of climate-smart breeding methods. A key issue is the lack of farmer awareness regarding the advantages of such crops and techniques (Fasoula et al. 2020). Educational efforts, including outreach programs and model farms, are vital for demonstrating the benefits of resilient varieties. Resource accessibility, like seeds, fertilizers, and financial support, is another hurdle, especially for smallholder farmers (Muzekenyi et al. 2023). Programs like seed banks and subsidies are essential to provide fair access to these necessary resources.

Additionally, policy and bureaucratic challenges, such as complex seed certification procedures, can restrict access to innovative climate-smart crops (Steenwerth et al. 2014). Policy reform is crucial to facilitate the introduction and use of these crops. Moreover, concerns about costs and profitability also prevent farmers from implementing climate-smart methods. Financial incentives and rewards for environmental services can alleviate these worries and promote broader adoption (Jamil et al. 2021).

17.4 Sustainable Rice Ecosystem in the Changing Climate Scenario: A Case Study

Rice is a staple food in many Asian countries, providing essential carbohydrates and significantly influencing their economies and livelihoods. However, the prevalent transplanted rice cultivation system contributes to the climate crisis, consuming an estimated 60–70% of the world's agricultural water. Rice accounts for about 50% of irrigation water in Asia (Tuong and Bouman 2003), and it is projected that up to 2 million hectares of irrigated dry-season rice and 13 million hectares of irrigated wet-season rice could face water scarcity (Bouman et al. 2002). The transplanted rice system requires a constant water depth of 5–10 cm, using approximately 3000–5000 liters to produce one kilogram of rice (Joshi et al. 2009). Excessive fertilizer use further degrades soil health and fertility, with the non-renewable resource of rock phosphate for inorganic phosphorus fertilizers making this inefficiency concerning.

These challenges drive the search for more sustainable rice cultivation methods. Aerobic rice cultivation through direct seeding is a promising alternative. It eliminates the need for standing water, keeping the soil aerobic, particularly beneficial in upland and rainfed areas with limited water. This method requires 36–41% less water than the transplanted system (Priyanka et al. 2012) and has shown grain yield potentials of 6–7 tons per hectare (Huaqi et al. 2002).

Direct seeding techniques (DSR) in aerobic rice fall into three categories: wet DSR, where sprouted seeds are sown on wet soil; and dry DSR, where dry seeds are drilled or broadcasted on dry soil. Breeding for aerobic conditions involves enhancing traits like uniform emergence, seedling vigor, nutrient and moisture efficiency,

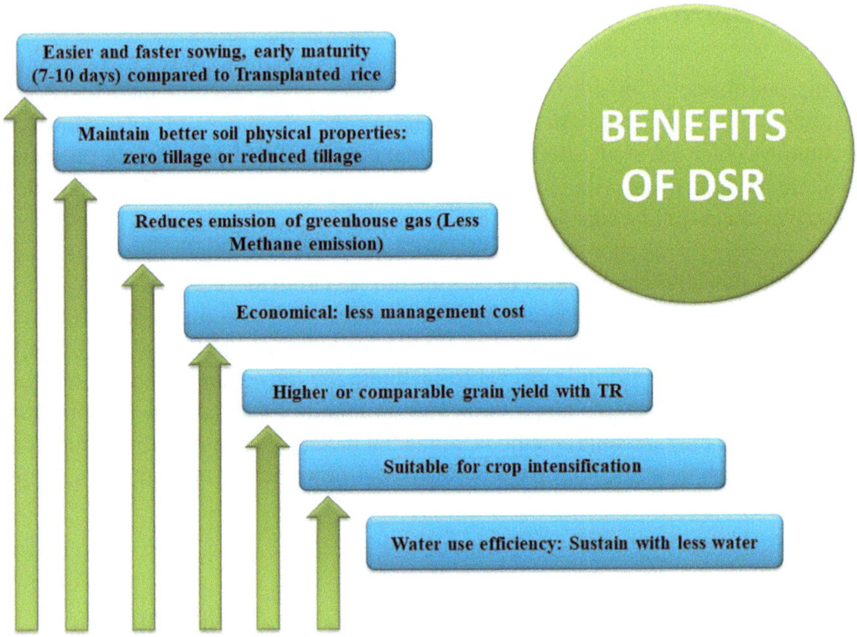

Fig. 17.3 Advantages of DSR over transplanted rice cultivation

and resistance to lodging and environmental stresses. Figure 17.3 shows the comparative analysis of DSR over the conventional methods. Recent genetic research has identified multiple QTLs and genes related to these traits (Medina-Lozano and Díaz 2022), which should be incorporated into breeding programs to develop varieties suited to changing climates. Table 17.2 mentions the different QTLs and genes related to traits governing an aerobic rice variety.

17.5 Future Scope

As climate challenges intensify, there is an urgent need for faster and more efficient breeding processes. CRISPR-Cas9 gene editing offers unparalleled precision in introducing beneficial traits into crop genomes (Elrys et al. 2022). In the future, CRISPR could be used to rapidly develop crops that combine multiple stress-resilient traits, such as heat tolerance, water-use efficiency, and improved nutrient uptake (Skendžić et al. 2021). This would significantly reduce the time required to develop new crop varieties compared to traditional breeding methods.

Furthermore, stakeholder collaboration remains essential in shaping robust, localized food systems (Mustafa et al. 2021). Cutting-edge technologies like genomic selection, high-throughput phenotyping, CRISPR/Cas9 gene editing, and remote sensing are revolutionizing the field, expediting the development of

Table 17.2 List of different QTLs and genes regulating various abiotic stresses

Sl. no.	Trait	Gene/QTL	Chr.	References
1.	Anaerobic germination	qAG9.1 and qAG9.2	9	Angaji et al. (2010)
2.	Submergence	SUB1A	9	Septiningsih et al. (2009)
3.	Uniform emergence and establishment	qEUE11.1	11	Dixit et al. (2015)
		qEMM11.1	11	
4.	Root growth	PSTOL1	12	Gao et al. (2014)
		SOR1	4	Hanzawa et al. (2013)
		qRS8b, qRDW8, and qRN8b	8	Anandan et al. (2022)
		qRL6.1, qTRL6.1	6	
5.	Nutrient uptake	OsPT1	3	Jia et al. (2011)
		OsPT8	10	
		AMT1;3	2	Sonoda et al. (2003)
		NRT2.3	1	Feng et al. (2011)
		YSL6	4	Nozoye et al. (2011)
		OsZIP1	1	Krithika and Balachandar (2016)
		OsZIP4	4	
		DRO1	9	Uga et al. (2015)
6.	Lodging resistance	qLDG3.1	3	Dixit et al. (2015)
		$qLDG_{4.1}$	4	
		APO1	6	Ookawa et al. (2010)
7.	Drought tolerance	qDTY1.1	1	Vikram et al. (2011)
		qDTY2.1	2	Ramaiah Venuprasad et al. (2009)
		qDTY3.2	3	Dixit et al. (2012)
8.	Salinity tolerance	Saltol	1	Kumar (2016)

climate-resilient crops (Awais et al. 2022). These tools empower breeders to select and cultivate crop varieties optimized for fluctuating climates, securing sustainable food production and global food security.

17.6 Conclusion

Climate-smart plant breeding is a critical tool in the fight against climate change. By developing crops that can withstand the increasingly severe environmental stresses caused by rising temperatures, droughts, and unpredictable rainfall, plant breeders can help secure global food systems. This chapter has highlighted key advances in breeding for drought and heat tolerance, improved water and nutrient efficiency, and enhanced carbon sequestration. It has also emphasized the importance of cutting-edge technologies, such as CRISPR and AI-driven breeding, in accelerating the development of climate-resilient crops. Moreover, these plant breeding technologies contribute to sustainable land restoration by promoting biodiversity, improving soil health, and increasing the resilience of agricultural landscapes. However, the slow adoption of these innovations, particularly in developing regions, remains a major challenge. Moving forward, a stronger commitment to collaborative research, farmer education, and supportive policies is needed to ensure that climate-smart crops are widely adopted. As the climate continues to change, plant breeding must remain at the forefront of global efforts to build a more resilient and sustainable agricultural system.

Statements and Declarations The authors affirm that there are no financial interests or personal relationships, which could be construed as potential conflicts of interest affecting this chapter.

Author Contributions Ali Yetgin: Writing—original draft; Rajiv Kumar Srivastava: Writing—review, editing; Siddharth Panda: Review and editing; Sweta Rath: Review.

References

Abrol IP (2009) Conservation agriculture as an adaptive and mitigation strategy to combat climate change. In: Brainstorming workshop on climate change, soil quality and food security, pp 56

Aggarwal PK, Jarvis A, Campbell BM et al (2018) The climate-smart village approach: framework of an integrative strategy for scaling up adaptation options in agriculture. Ecol Soc 23(1). https://doi.org/10.5751/ES-09844-230114

Ahmad M (2023) Plant breeding advancements with "CRISPR-Cas" genome editing technologies will assist future food security. Front Plant Sci 14

Anandan A, Ranjitkumar N, Selvaraj S et al (2022) Multi-trait association study identifies loci associated with tolerance of low phosphorus in Oryza Sativa and its wild relatives. Sci Rep 12(1):1–24

Angaji S, Abdolhamid EM, Septiningsih DJ et al (2010) QTLs associated with tolerance of flooding during germination in rice (Oryza Sativa L.). Euphytica 172(2):159–168. https://doi.org/10.1007/s10681-009-0014-5

Awais M, Li W, Cheema MJM, Zaman QU et al (2022) UAV-based remote sensing in plant stress imagine using high-resolution thermal sensor for digital agriculture practices: a meta-review. Int J Environ Sci Technol:1–18

Bacha SAS, Iqbal B (2023) Advancing agro-ecological sustainability through emerging genetic approaches in crop improvement for plants. Funct Integr Genomics 23(2):145

Baligar VC, Fageria NK, He ZL (2001) Nutrient use efficiency in plants. Commun Soil Sci Plant Anal 32(7–8):921–950

Behera UK, France J (2023) Farming systems research: concepts, design and methodology. Adv Agron 177:1–49

Bernhard M (2010) The role of SUSIBA2 in biosynthesis and carbon sequestration in transgenic sugarcane. Proc Natl Acad Sci 107(44):17576–17581

Bouman B, Xiaoguang Y, Huaqi W et al (2002) Aerobic Rice (Han Dao): a new way of growing rice in water-short areas. In:12th international soil conservation organization conference, pp 175–81

Bradshaw JE (2017) Plant breeding: past, present and future. Euphytica 213:1–12

Breseghello F, Coelho ASG (2013) Traditional and modern plant breeding methods with examples in rice (Oryza sativa L.). J Agric Food Chem 61(35):8277–8286

Cairns J, Crossa J, Zaidi PH (2013) Identification of drought, heat, and combined drought and heat tolerant donors in maise. Crop Sci 53(4):1335–1346

Ceccarelli S, Galiè A, Grando S (2013) Participatory breeding for climate change-related traits. In: Genomics and breeding for climate-resilient crops: Vol. 1 concepts and strategies. Springer, Berlin, Heidelberg, pp 331–376

Chavan, U.D., Patil JV, Shinde MS (2009) Nutritional and roti quality of sorghum genotypes. https://media.neliti.com/media/publications/62849-nutritional-and-roti-quality-of-sorghum-4b68b7e2.pdf

Cooper M, Messina CD (2023) Breeding crops for drought-affected environments and improved climate resilience. Plant Cell 35(1):162–186

Dixit S, Mallikarjuna Swamy BP, Prashant V et al (2012) Increased drought tolerance and wider adaptability of QDTY 12.1 conferred by its interaction with QDTY 2.3 and QDTY 3.2. Mol Breed 30(4):1767–1779

Dixit S, Alexandre G, Cheng RL et al (2015) Understanding rice adaptation to varying agro-ecosystems: trait interactions and quantitative trait loci. BMC Genet 16:86. https://doi.org/10.1186/s12863-015-0249-1

Donatelli M, Magarey RD, Bregaglio S et al (2017) Modelling the impacts of pests and diseases on agricultural systems. Agric Syst 155:213–224

Dwivedi S, Reynolds MP, Ortiz R (2021) Mitigating tradeoffs in plant breeding. Iscience 24(9):102965

Eckardt NA, Ainsworth EA, Bahuguna RN et al (2023) Climate change challenges, plant science solutions. Plant Cell 35(1):24–66

Elmyhun M, Abate E, Abate A (2024) Genetic analysis of tolerance to combined drought and heat stress in tropical maise. PLoS One 19:e0302272. https://doi.org/10.1371/journal.pone.0302272

Elrys AS, Elnahal AS, Abdo AI (2022) Traditional, modern, and molecular strategies for improving the efficiency of nitrogen use in crops for sustainable agriculture: a fresh look at an old issue. J Soil Sci Plant Nutr 22(3):3130–3156

Eziz A, Yan Z, Tian D et al (2017) Drought effect on plant biomass allocation: a meta-analysis. Ecol Evol 7(24):11002–11010

Fasoula D, Ioannide IM, Omirou M (2020) Phenotyping and plant breeding: overcoming the barriers. Front Plant Sci 10:1713

Fawz S, Osman AI, Doran J et al (2020) Strategies for mitigation of climate change: a review. Environ Chem Lett 18:2069–2094

Feng H, Ming Y, Xiaorong F et al (2011) Spatial expression and regulation of rice high-affinity nitrate transporters by nitrogen and carbon status. J Exp Bot 62:2319. https://doi.org/10.1093/jxb/erq403

Fess TL, Kotcon JB, Benedito VA (2011) Crop breeding for low input agriculture: a sustainable response to feed a growing world population. Sustainability 3(10):1742–1772

Gao S, Jun F, Fan X et al (2014) CYTOKININ OXIDASE/DEHYDROGENASE4 integrates cytokinin and auxin signaling to control rice crown root formation. Plant Physiol 165(3):1035. https://doi.org/10.1104/pp.114.238584

Garnett T, Plett D, Heuer S, Okamoto M (2015) Genetic approaches to enhancing nitrogen-use efficiency (NUE) in cereals: challenges and future directions. Functional plant biology 42(10):921–941

Gerullis M, Pieruschka R, Fahrner S et al (2023) From genes to policy: mission-oriented governance of plant-breeding research and technologies. Front Plant Sci 14:1235175

Grunwald NJ, Goss EM, Press CM (2017) Late blight of potato and tomato in the genomics era. Plant Dis 101(8):1084–1097

Guo LB, Giffor RM (2002) Soil carbon stocks and land use change: a meta analysis. Glob Chang Biol 8(4):345–360

Hadebe ST, Modi AT, Mabhaudhi T (2017) Drought tolerance and water use of cereal crops: a focus on sorghum as a food security crop in sub-Saharan Africa. J Agron Crop Sci 203(3):177–191

Hanzawa E et al (2013) Isolation of a novel mutant gene for soil-surface rooting in rice (Oryza Sativa L.). Rice 6(1):1–11

Harvey CA, Chacon M, Donatti CI et al (2014) Climate-smart landscapes: opportunities and challenges for integrating adaptation and mitigation in tropical agriculture. Conserv Lett 7(2):77–90
Hatfiel JL, Dold C (2019) Water-use efficiency: advances and challenges in a changing climate. Front Plant Sci 10:103
Heuer S, Gaxiola R, Schilling R et al (2017) Improving phosphorus use efficiency: a complex trait with emerging opportunities. Plant J 90(5):868–885
Huaqi W, Bouman BAM, Dule Z et al (2002) Aerobic rice in northern China: opportunities and challenges, water-wise rice production. International Rice Research Institute, Los Baños, pp 143–154
ISAAA (2023) Biotechnology for the development of drought-tolerant crops. https://www.isaaa.org/resources/publications/pocketk/32/default.asp
IUCN (2022) Ecosystem restoration. IUCN
Jagadish SVK, Pal M, Sukumaran S et al (2020) Heat stress resilient crops for future hotter environments. Plant Physiol Reports Rep 25:529–532. https://doi.org/10.1007/s40502-020-00559-9
Jamil I, Jun W, Mughal B et al (2021) Does the adaptation of climate-smart agricultural practices increase farmers' resilience to climate change? Environ Sci Pollut Res 28:27238–27249
Jia H, Hongyan R, Mian G et al (2011) The phosphate transporter gene Ospht1;8 is involved in phosphate homeostasis in Rice. Plant Physiol 156:1164. https://doi.org/10.1104/pp.111.17524
Joshi R, Mani SC, Shukla A et al (2009) Aerobic rice: water use sustainability. Oryza 46(1):1–5
Karlsson Green K, Stenberg JA, Lankinen Å (2020) Making sense of Integrated Pest Management (IPM) in the light of evolution. Evol Appl 13(8):1791–1805
Kavhiza NJ, Zargar M, Prikhodko SI (2022) Improving crop productivity and ensuring food security through the adoption of genetically modified crops in sub-Saharan Africa. Agronomy 12(2):439
Kranthi KR, Naidu S, Dhawad CS et al (2002) Bt Cotton in India: A Country Profile. International Cotton Advisory Committee (ICAC)
Krithika S, Balachandar D (2016) Expression of zinc transporter genes in rice as influenced by zinc-solubilizing Enterobacter Cloacae strain ZSB14. Front Plant Sci 7(APR2016):1–9. https://doi.org/10.3389/fpls.2016.00446
Kumar M (2016) Impact of climate change on crop yield and role of model for achieving food security. Environ Monit Assess 188:1–14
Lamaoui M, Jemo M, Datla R et al (2018) Heat and drought stresses in crops and approaches for their mitigation. Front Chem 6:26
López-Calcagno PE, Brown KL, Simkin AJ et al (2020) Stimulating photosynthetic processes increases productivity and water-use efficiency in the field. Nat Plants 6(8):1054–1063
Marsh JI, Hu H, Gill M et al (2021) Crop breeding for a changing climate: integrating phenomics and genomics with bioinformatics. Theor Appl Genet 134:1677–1690
Medina-Lozano I, Díaz A (2022) Applications of genomic tools in plant breeding: crop biofortification. Int J Mol Sci 23(6):3086
Mellacheruvu S, Tamang P, Lata C et al (2019) Genetic engineering for drought resistance in crops. In: Plant abiotic stress tolerance. Springer, Cham, pp 323–349
Mirzabaev A, Kerr RB, Hasegawa T et al (2023) Severe climate change risks to food security and nutrition. Clim Risk Manag 39:100473
Morris ML, Bellon MR (2004) Participatory plant breeding research: opportunities and challenges for the international crop improvement system. Euphytica 136(1):21–35
Munawar N, Ahsan K, Ahmad A (2024) CRISPR-edited plants' social, ethical, policy, and governance issues. In: Global regulatory outlook for CRISPRized plants. Academic Press, pp 367–396
Mustafa MA, Mabhaudhi T, Massawe F (2021) Building a resilient and sustainable food system in a changing world–a case for climate-smart and nutrient dense crops. Glob Food Sec 28:100477
Mutale CE (2020) Genetic analysis in tropical maize for phosphorus utilization in phosphorus-limited soil (Doctoral dissertation, The University of Zambia). Url: https://dspace.unza.zm/server/api/core/bitstreams/92650345-c170-40e5-a493-df54a4f34ab6/content

Muzekenyi M, Zuwarimwe J, Kilonzo BM (2023) Economic challenges limiting small-scale commercial farming development in rural areas of South Africa. Afr J Food Agric Nutr Dev 23(4):23026–23038
National Research Council (2004) Biological confinement of genetically engineered organisms. National Academies Press
Nelson R, Wiesner-Hanks T, Wisser R, Balint-Kurti P (2018) Navigating complexity to breed disease-resistant crops. Nat Rev Genet 19(1):21–33
Nerva L, Sandrini M, Moffa L et al (2022) Breeding toward improved ecological plant–microbiome interactions. Trends Plant Sci 27:1134
Newton AC, Johnson SN, Gregory PJ (2011) Implications of climate change for diseases, crop yields and food security. Euphytica 179:3–18
Nozoye T, Seiji N, Takanori K et al (2011) Phytosiderophore efflux transporters are crucial for iron acquisition in Graminaceous plants. J Biol Chem 286:5446. https://doi.org/10.1074/jbc.M110.180026
Ontl TA, Schulte LA (2012) Soil carbon storage. Nat Edu Know 3(10)
Ookawa T et al (2010) New approach for rice improvement using a pleiotropic QTL gene for lodging resistance and yield. Nat Commun 1(1):1–11
Pandey V, Shukla A, Pandey D (2017) Sustainable agriculture through drought mitigation and enhanced water use efficiency in rice. In: Advances in rice research for abiotic stress tolerance. Springer, Singapore, pp 143–160
Parmar N, Singh KH, Sharma D et al (2017) Genetic engineering strategies for biotic and abiotic stress tolerance and quality enhancement in horticultural crops: a comprehensive review. 3 Biotech 7:1–35
Phillips T (2008) Genetically modified organisms (GMOs): transgenic crops and recombinant DNA technology. Nat Edu 1(1):213
Poffenbarger H, Castellano M, Egli D et al (2023) Contributions of plant breeding to soil carbon storage: retrospect and prospects. Crop Sci 63(3):990–1018
Priyanka S, Baghel J, Subhash B (2012) Aerobic rice, a new approach of rice cultivation. Int J Res BioSci 1(1):1–6
Qaim M (2020) Role of new plant breeding technologies for food security and sustainable agricultural development. Appl Econ Perspect Policy 42(2):129–150
Rahman H, Ramanathan V, Jagadeeshselvam N et al (2015) Phenomics: technologies and applications in plant and agriculture. In: PlantOmics: the omics of plant science, pp 385–411
Rai KK (2022) Integrating speed breeding with artificial intelligence for developing climate-smart crops. Mol Biol Rep 49(12):11385–11402
Raza A, Razzaq A, Mehmood SS et al (2019) Impact of climate change on crops adaptation and strategies to tackle its outcome: a review. Plants 8(2):34
Rezv HUA, Tahjib-Ul-Arif M, Azim MA et al (2023) Rice and food security: climate change implications and the future prospects for nutritional security. Food Energy Secur 12(1):e430
Riedesel L, Laidig F, Hadasch S et al (2022a) Breeding progress reduces carbon footprints of wheat and rye. J Clean Prod 377:134326
Riedesel S, Liebhard P, Riedesel K (2022b) Bioenergy sorghum breeding: success story or brave new world? Crop J 10(5):715–722
Scheben A, Wolter F, Batley J et al (2017) Towards CRISPR/Cas crops–bringing together genomics and genome editing. New Phytol 216(3):682–698
Selin NE (2023) Carbon sequestration. Encyclopedia Britannica, https://www.britannica.com/technology/carbon-sequestration
Septiningsih EM, Alvaro MP, Darlene LS et al (2009) Development of submergence-tolerant rice cultivars: the Sub1 locus and beyond. Ann Bot 103(2):151–160. https://doi.org/10.1093/aob/mcn206
Sharma N, Raman H, Wheeler D, Kalenahalli Y, Sharma R (2023) Data-driven approaches to improve water-use efficiency and drought resistance in crop plants. Plant Sci 336:111852
Shepherd T (2017) Wax pathways. In: Encyclopedia of applied plant sciences. Elsevier, pp 1–6. https://doi.org/10.1016/B978-0-12-394807-6.00100-9

Simova-Stoilova L, Vassileva V, Feller U (2016) Selection and breeding of suitable crop genotypes for drought and heat periods in a changing climate: which morphological and physiological properties should be considered? Agriculture 6(2):26

Singh A, Grover A (2008) Plant Hsp100/ClpB-like Proteins. In: Vivo functions and molecular chaperone activity, in plant molecular biology. Springer, Berlin, Heidelberg, pp 215–225

Singh S, Roychowdhury R, Ghosh M (2020) Aquaporin: a potential candidate for drought and salt stress management in plants. In: Mechanistic approaches in plant science. Springer, Singapore, pp 231–246

Skendžić S, Schmitz M, Widera D (2021) CRISPR-Cas9 genome editing: from basic research to application. Trends Biotechnol 39(5):491–509

Sonoda Y, Akir I, Satomi S et al (2003) Distinct expression and function of three ammonium transporter genes (OsAMT1;1-1;3) in Rice. Plant Cell Physiol 44:726. https://doi.org/10.1093/pcp/pcg083

Srivastava RK, Yetgin A (2024) An overall review on influence of root architecture on soil carbon sequestration potential. In: Theoretical and experimental plant physiology, pp 1–14

Srivastava RK, Panda RK, Chakraborty A et al (2023) Drought tolerance agronomic management strategies for Rainfed and irrigated maize crop in eastern India. In: Integrated drought management, vol 2. CRC Press, pp 593–606

Steenwerth KL, Hodson AK, Bloom AJ et al (2014) Climate-smart agriculture global research agenda: scientific basis for action. Agric Food Secur 3(1):1–39

Temme AA, Kerr KL, Nolting KM et al (2024) The genomic basis of nitrogen utilisation efficiency and trait plasticity to improve nutrient stress tolerance in cultivated sunflower. J Exp Bot:erae025

Tuong TP, Bouman BAM (2003) Rice production in water scarce environments, in 'water productivity in agriculture: limits and opportunities for improvement. CABI Publishing, Wallingford

Uga Y, Yuka K, Eiji Y et al (2015) A QTL for root growth angle on rice chromosome 7 is involved in the genetic pathway of DEEPER ROOTING 1. Rice 8:8. https://doi.org/10.1186/s12284-015-0044-7

Varshney RK, Barmukh R, Roorkiwal M et al (2021) Breeding custom-designed crops for improved drought adaptation. Adv Genet 2(3):e202100017

Venuprasad R, Dalid CO, Valle M et al (2009) Identification and characterisation of large-effect quantitative trait loci for grain yield under lowland drought stress in rice using bulk-Segregant analysis. Theor Appl Genet 120(1):177–190

Vikram P, Swamy BP, Dixit S et al (2011) QDTY 1.1, a major QTL for Rice grain yield under reproductive-stage drought stress with a consistent effect in multiple elite genetic backgrounds. BMC Genet 12(1):1–15

Wang W, Hu B, Li A, Chu C (2020) NRT1. 1s in plants: functions beyond nitrate transport. Journal of Experimental Botany 71(15):4373–4379

Wasson AP, Richards RA, Chatrath R et al (2012) Traits and selection strategies to improve root systems and water uptake in water-limited wheat crops. J Exp Bot 63(9):3485–3498

Yadav MR, Kumar S, Lal MK et al (2023) Mechanistic understanding of leakage and consequences and recent technological advances in improving nitrogen use efficiency in cereals. Agronomy 13(2):527

Yang Y, Yu J, Qian Q et al (2022) Enhancement of heat and drought stress tolerance in rice by genetic manipulation: a systematic review. Rice 15(1):1–18

Yoosefzadeh Najafabadi M, Hesami M, Eskandari M (2023) Machine learning-assisted approaches in modernised plant breeding programs. Genes 14(4):777

Zhao RR, Qu BY, Tong YP, Zou CQ (2019) Iron and zinc accumulation in winter wheat regulated by NICOTIANAMINE SYNTHASE responded to increasing nitrogen levels. J Plant Nutri 42(14):1624–1636

Zhu XG, Long SP, Ort DR (2008) Improving photosynthetic efficiency for greater yield. Annu Rev Plant Biol 61:235–261

Part IV
Cutting-Edge Innovations, Digital Tools, and Co-creation Strategies

Chapter 18
Utilizing Decision Support Tools to Optimize Resource Allocation for Different Categories of Farmers in Northern Ghana

Dilys S. MacCarthy, Samuel G. K. Adiku, Bright S. Freduah, Edward E. Onumah, Collins N. O. Larkai, Pierre S. Traore, and Helen Greatrex

Abstract Based on resource endowments, West African farmers can be classified into small scale, medium scale, or large scale. Each classification faces a decision-making challenge when choosing crops and allocating resources for maximum returns. This is more difficult due to frequent variabilities in weather, agro-input, and commodity prices. This paper employs the Decision Support System for Agro-technological Transfer (DSSAT) crop model to simulate the yields of six crops (maize, cowpea, soybean, rice, groundnut, and millet) in the Northern and the Upper East regions of Ghana for a period of 20 years (1990–2010) for each farmer classification. The simulated yields were input to a linear programming (LP) framework that optimized the resource allocation based on farmgate prices and production costs. We also employed the ENSO to explore the value of seasonal weather

D. S. MacCarthy (✉) · B. S. Freduah
Soil and Irrigation Research Centre, School of Agriculture, College of Basic and Applied Sciences, University of Ghana, Kpong, Ghana
e-mail: dmaccarthy@ug.edu.gh

S. G. K. Adiku
Department of Soil Science, School of Agriculture, College of Basic and Applied Sciences, University of Ghana, Kpong, Ghana

E. E. Onumah · C. N. O. Larkai
Department of Agriculture Economics and Agribusiness, School of Agriculture, College of Basic and Applied Sciences, University of Ghana, Accra, Ghana

P. S. Traore
International Crops Research Institute for the Semi-arid Tropics (ICRISAT), Bamako, Mali

Manobi Africa PLC, Port-Louis, Mauritius

H. Greatrex
Department of Geography, Department of Statistics and the Institute for Computational and Data Sciences, The Pennsylvania State University, University Park, PA, USA

Y. D. I. Khan et al. (eds.), *Land Restoration Through Ecosystem-Based Approach*, Disaster Resilience and Green Growth, https://doi.org/10.1007/978-981-95-0389-6_18

forecasting on farm plan optimization. Our results indicate that the weather variability has impact on crop yields as captured by the DSSAT model. The optimized resource allocation differed year to year, suggesting an annual update of the optimized farm plan for each farmer type. Combining a dynamic crop and economic optimization models provides a good basis for enhancing agricultural planning and hence, should be promoted for use to support farmers decision-making.

Keywords Cereals · Climate variability · DSSAT · Farm typology · Legumes · Linear programming

18.1 Introduction

West African farmers, on year-to-year basis, make decisions to maximize returns to their cropping investments (Assouto and Houensou 2020; Wahab et al. 2022). Decision determinants include crop mix choices, planting times, inputs (e.g., fertilizer, other soil management options, and labor), among others. Decisions to expand or increase investment levels depend on farmer type, which can be classified into small, medium, or large scale (commercial farming). This classification is based on their resource endowment and capital. As no 2 years are the same in terms of weather and given that different crops respond differently to weather conditions and other production practices, an optimized crop choice, crop mix, and resource allocation for 1 year may not be valid for another year.

Indeed, there is evidence that weather patterns have changed in many locations in northern Ghana (Adu-Prah et al. 2019). The data suggests that though the seasonal or annual rainfall over the past 50 years has not changed significantly, it is the within-season distribution of the rainfall that has shown significant changes (Adiku 2019). In the Upper West Region, where the season often begins in April/May, there is a significant reduction in the early season rainfall, favoring late season (August/September) rainfall. These shifts in the within-season weather pattern introduce an early season drought and late season flooding. Field observations indicate that the season onset has shifted from March/April to May/June, implying that traditional cropping calendars that relied on an even rainfall distribution within the season can no longer be relied upon (Gbangou et al. 2019; Atiah et al. 2021; Kumi et al. 2023).

Responding to climate variability and change requires medium- to long-term sequence of information on crop yields, investment, and other variables. Except for aggregated district yield data published as Agricultural Statistics by the Ministry of Food and Agriculture, (MoFA), farm-level documentation of crop yields and performance over time is poor. Therefore, the memory of weather variability impact on farm decisions is not immediately available at the farm level.

Dynamic crop models such as the Decision Support System for Agro-Technological Transfer (DSSAT) (Jones et al. 2003) have been shown to reliably simulate crop response to soil, weather, and management factors. Over the past 10 years, crop models have been validated for a number of crops including maize, millet, rice, sorghum, cowpea, and soybean (MacCarthy et al. 2009; Narh et al. 2014; Adiku et al. 2005). Conceivably, then, crop models can be used to fill the response information gap for several locations and time scales.

However, for farm planning purposes, it is the economic returns to the farmer that is of interest. The derivation of an optimized resource allocation thus requires the conversion of the crop yield to economic indicators, such as revenue. Furthermore, the derivation of an optimized farm plan can be facilitated using models such as linear programming (LP), which offers the possibility of selecting the best alternative from a number of available courses of action (Sowell and Ward 1998). In agriculture, linear programming techniques have been applied to farm machinery usage, planning of crop irrigation systems and determination of optimum crop mix (Sowell and Ward 1998), among others. In most studies, linear programming has been applied for cross-sectional studies involving many farmers but at a given point of time (e.g., in a given year) instead of over time scales. Drafor et al. (2013) applied linear programming to investigate how rural communities respond to food shortage in the lean season, using a base data of 2005. Another study used linear programming to select the optimum crop for farmers drawn from 32 operation areas in the Eastern Region of Ghana (Otoo et al. 2015).

Given that no 2 years are the same in terms of weather, agro-input and production costs, as well as market prices, an optimized farm plan for a given year may not be valid for another year. The question that lingers is whether an optimized farm plan for a given year would change over the course of time. As indicated, long-term temporal field-scale yield data required as input into LP is scant and not readily available. We conjecture, therefore, that a combination of dynamic crop modelling (DSSAT) with economic optimization models (LP) provides an innovative framework for generating relevant outputs for assessing the impact of weather variability and climate change, and commodity price changes on farmers' returns and profitability. One question of interest is whether some fore-knowledge of the coming seasons' potential can influence farmer decision-making, e.g., in terms of cop choice and optimized crop mix, among others. For this, we relied on the previous studies by Adiku and Stone (1995) and MacCarthy et al. (2017), who showed that the Pacific (NiNO3 region) October to December (OND) ENSO (El Niño-Southern Oscillation) has an impact on the April to July rainfall in many parts of Ghana. Positive or warm phase OND ENSO often resulted in below-average rainfall in Ghana. ENSO-based rainfall categories are published in the literature. Therefore, for each of the simulation years, the historical weather data was replaced by the appropriate ENSO-based analogue for the yield simulations.

The aim of this study is two-fold: First, we evaluate the economic returns to different farmer types (small, medium, and large scale) over a period of 20 years (1990–2010) using the DSSAT model and subsequently input the simulated yields into the LP model for farm plan optimization; second, we explored the value of seasonal forecasting in the optimization of farm plans.

18.2 Materials and Methods

18.2.1 Sites

Two locations in Northern Ghana, namely Tamale in the Northern region and Garu-Tempane in the Upper East region, were selected in this study (Fig. 18.1). The two locations are characterized by one main rainy season with onset in April/May with the rains ceasing in September/October (Fig. 18.2).

The annual rainfall ranges between 695 and 1580 with a mean annual average of 1066 mm and a coefficient of variation (CV) of 18% at Tamale whereas that for Garu-Tempane ranged between 671 and 1563 with an average of 957 mm and a CV of 20%. The average minimum and maximum temperatures are 22.8 and 34.4 at Tamale and 22.9 °C and 35.2 °C, at Garu-Tempane, respectively. Tamale falls within the Guinea savannah agro-ecological zone, whereas Garu-Tempane falls within the Sudan savannah agro-ecology, and both locations cultivate crops including maize, millet, rice, sorghum, cowpea, etc. The soils at the two locations are generally coarse textured, low in soil organic carbon, generally low in nutrients and shallow in depth.

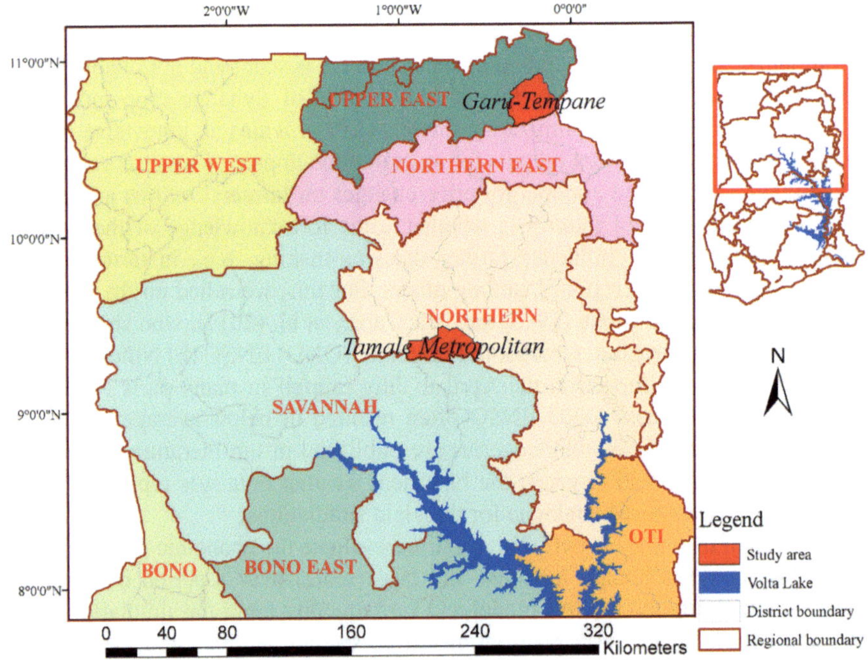

Fig. 18.1 Map of Northern Ghana showing the two study sites

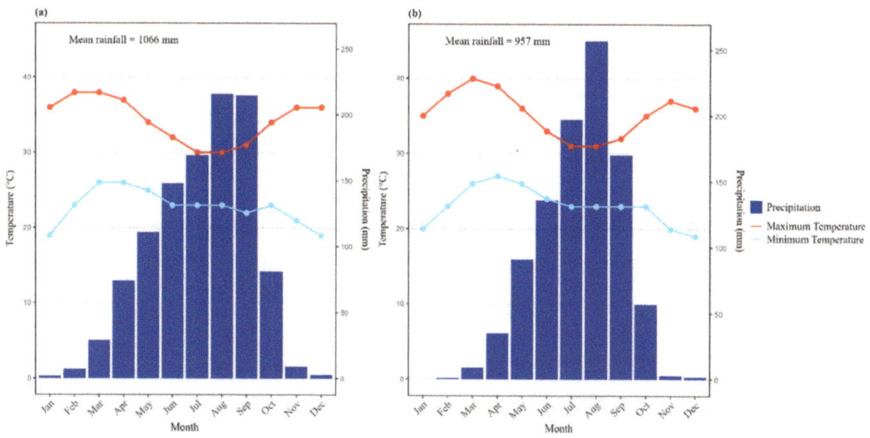

Fig. 18.2 Average total monthly total rainfall, minimum and maximum temperature for Tamale and Garu-Tempane for 1990 to 2020

18.2.2 Farm Survey and Resource Outlay

A field survey was carried out on 300 farmers in Tamale and Garu-Tempane in the year 2018. Farmers were interviewed using structured questionnaires to document their resource endowments with respect to farm size, crops grown, fertilizer and other input usage levels, capital outlay, yields, farm gate prices, and labor availability. Out of the 300 respondents, 100 respondents were from Garu-Tempane in the Upper East Region while 200 respondents were from the Tamale in the Northern Region. The following districts were purposively selected for the random sampling of the respondents: Savelugu District, Tolon District, Nanton District and Kumbungu District in Northern Regions and the Garu-Tempane District of the Upper East region. A categorization of the farmers was made based on the farm sizes cultivated. Three categories were obtained, namely small-scaled farmers (less than 2 hectares), medium-scaled farmers (between 2 hectares and 4 hectares), and large-scaled farmers (more than 4 hectares).

Farmer types varied in the survey locations. As shown in Table 18.1, 47% of farmers held less than 2 ha, 38% held between 2 and 5 ha, and the remaining 15% held between 5 and 10 ha. This distribution mirrors closely that previously published by the Ghana Livelihood Survey. Based on acreages, farmers were grouped into three categories, namely those holding up to 2 ha as small scale, between 2 and 5 ha as medium scale, and between 5 and 10 ha as large scale (Table 18.1).

Labor and capital availability also varied among farmer types (Table 18.1). Typically, small-scale farmers have a maximum of 24 man-days, medium-scale farmers have 36 man-days, and large-scale farmers have 63 man-days. The capital available to each farmer type for investment also varies, with the small-scale farmers capable of purchasing 30 kg N/ha, medium-scale farmers 60 kg N/ha and large-scale farmers 90 kg N/ha. Observations from the survey indicated that small-scale

Table 18.1 Land acreage and resource levels of the three farm categories

Farmer classification	Acreage	Inputs	Labor	Capital
Small scale	2.0 ha	30 kg N/ha to cereals	24 man-days	GH¢ 695
Medium scale	3.6 ha	60 kg N/ha to cereals	36 man-days	GH¢ 1190
Large scale	6.8 ha	90 kg N/ha to cereals	63 man-days	GH¢ 1479

farmers apply not more than 30 kg N/ha to maize and millet. The medium- and large-scale farmers applied 60 and 90 kg N/ha, respectively, to maize, rice, and millet (Table 18.1).

18.2.3 Crop Modelling

We employed the DSSAT crop model (Jones et al. 2003) to simulate the yield of each of the six crops (maize, cowpea, soybean, rice, groundnut, and millet) for each location and farmer type for a period of 20 years (1990–2010). A detailed description of the DSSAT model is available from Jones et al. (2003). The DSSAT suite has a number of crop models. Plant growth processes are based on carbon capture, coupled to water and nutrient uptake from the soil submodule. Hydrologic and nutrient dynamics are simulated in the soil sub-model. Inputs to the model include weather, planting date, fertilizer and manure application rates and time, soil depth, and water holding parameters, among others. As indicated, the various crop-specific models in the suite have been previously validated for the northern Ghana soil management and weather conditions (Adiku et al. 2005; MacCarthy et al. 2018).

This study obtained weather data from the Ghana Meteorological Agency, Accra. Observations from the survey indicated that small-scale farmers applied about 30 kg N/ha to maize and millet, whilst the medium- and large-scale farmers applied 60 and 90 kg N/ha, respectively, to maize, rice, and millet (Table 18.1).

18.2.4 Economic Modelling

This study used the linear programming (LP) framework available in Excel to determine optimum resource allocation to farm enterprises. The LP model was solved using Simplex method available in the *Solver* Add-in (in Microsoft Excel).

Input requirements such as land size, agro-inputs, and other labor costs and market prices were derived from the survey conducted. For a given year, the simulated yield for each crop and farmer type by DSSAT is converted to revenue using the farmgate price.

The implementation of LP model requires the formulation of the farm enterprise as a model (Eq. 18.1), whose objective is to optimize resource allocation for maximization of returns or minimization of costs. In this study, the constraints considered include farm size, capital, and labor availability.

$$f = c_1 x_1 + \ldots + c_n x_n \tag{18.1}$$

Subject to a set of constraints expressed as inequalities:

$$a_{11} x_1 + \ldots + a_{1n} x_n \le b_1$$

$$a_{m1} x_1 + \ldots + a_{mn} x_n \le b_m \text{ with } x_i \ge 0 \tag{18.2}$$

The *a*'s, *b*'s, and *c*'s are constants determined by the capacities, needs, costs, profits, productivity, and other requirements and restrictions of the problem.

The study first defines decision variables, namely the number of hectares of land planted to each crop type, x_j, with x_1= maize, x_3 = millet, x_3 = rice, x_4= groundnut, x_5 = cowpea and x_6 = soybean. Secondly, we specify the constraints, bj, based on the survey data with b_1 for land, b_2 for capital b_3 for labour. Thirdly, we determined for each given year the gross margin returns for each crop, c_j, as the product of the simulated yield and CPI adjusted farm gate price, less the CPI adjusted cost of production.

Mathematically, we write:

$$Maximize\ z = \sum_{j=1}^{6} c_j x_j \tag{18.3}$$

with

$$c_j = y_j \times p_j - p_{cj} \tag{18.4}$$

where for crop j, y_j is the simulated yield (kg/ha), p_j is the CPI-adjusted farm gate price (GHC/kg), and p_{cj} is CPI-adjusted the production cost (GHC/ha).

The Simplex Tableau was set as follows:

$$a_{1,1}x_1 + a_{1,2}x_2 + a_{1,3}x_3 + a_{1,4}x_4 + a_{1,5}x_5 + a_{1,6}x_6 \le b_1 \left(\text{Land}\right) \tag{18.5}$$

$$a_{2,1}x_1 + a_{2,2}x_2 + a_{2,3}x_3 + a_{2,4}x_4 + a_{2,5}x_5 + a_{2,6}x_6 \le b_2 \left(\text{Labour}\right) \tag{18.6}$$

$$a_{3,1}x_1 + a_{3,2}x_2 + a_{3,3}x_3 + a_{3,4}x_4 + a_{3,5}x_5 + a_{3,6}x_6 \le b_3 \left(\text{Capital}\right) \tag{18.7}$$

and the non-negativity ...

$$x_j >= 0. \tag{18.8}$$

where a_j is the amount of resource required to produce one unit of crop j. Resource levels and constraints information for each farmer-type were derived from the survey, as summarized in Table 18.1. The values of a_j depend on crop type and farmer type, as investments per ha vary (Table 18.1). Also, the capital outlay varies with farmer type as summarized in Table 18.1.

Commodity prices and production cost obtained for the survey year (2018) were used as the base to estimate those of each of the 20 simulation years using the Consumer Price Index (CPI) data published by the Ghana Statistical Services Department. The values of the activity coefficients a_1, a_2,etc., and the returns to each crop, c_j, were derived from the survey information as summarized in Table 18.1. Crop yield for any farmer type and year were derived from crop model simulations.

For a given farmer type, three LP scenarios were executed. First, the "best" crop(s) and acreages that will maximize the returns for the given constraint levels for a given year were determined. This assumes that the farmer is purely commercially oriented. Second, a condition was imposed to include at least one cereal crop as a subsistence measure with a minimum of 0.25 ha. Third, the farmer can access credit to the tune that can fully utilize his total land while still maintaining the subsistence cereal production.

In all, 57 outputs were generated for each farmer type (19 years × 3 scenarios), leading to 342 outputs for all farmer types, scenarios, and two locations.

Resource levels and constraints information for each farmer type were derived from the survey, as summarized in Table 18.1. Commodity prices and production costs obtained for the survey year (2018) were used as the base to estimate those of each of the 19 simulation years using the Consumer Price Index (CPI) data published by the Ghana Statistical Services Department. The values of the activity confidence a1, a2, etc., and the returns to each crop, cj, were derived from the survey information. Crop yield for any farmer type and year were derived from crop model simulations.

18.2.5 Description of Scenarios

Scenario analysis involved the input of simulated yield using both historical and forecast seasonal weather, based on ENSO. Instead of simulation yields for each year based on the forecast, we adopt a simple approach whereby investments are doubled if the forecast predicts good wet rainfall years but are unchanged for normal years. For a given farmer type, the three investment scenarios executed are summarized in Table 18.2.

In the first scenario, the highest yielding crop(s) with good price and acreages that will maximize returns under the current farmer constraint levels (land, capital, labor) was determined for each simulation year.

For the second investment scenario, a condition was imposed to enable small- and medium-scale farmers to cultivate at least 0.25 ha of one cereal crop (maize, millet, and rice) as a subsistence crop under the same labor and land and capital constraints but assuming that the farmer can double his capital. Scenario 2 was not applied to large-scale farmers as they are not constrained by capital for investment.

Table 18.2 Description of investment scenarios

Scenario	Condition	Capital investment	Labor, land constraints	Farm scale	Note
1	Best crop and land allocation for max. returns	Normal	Fixed (based on constraints)	SS, MS, LS	No additional capital
2	Subsistence[a]	Double for all years	Fixed (land, labor)	SS, MS	
3	Subsistence[a]	Double for normal, triple for good years		SS[b], MS, LS	MS & LS respond to weather forecast

SS small scale, *MS* medium scale, *LS* large scale
[a]Cultivate at least 0.25 ha of one of the cereals crops (maize, millet and rice)
[b]SS double investment in both normal and good years

In the third investment scenario, it is assumed that the farmer (medium and large scale) responds to seasonal weather forecasts by increasing investments in good years. For this third scenario, the medium- and large-scale farms relied on weather forecasts and increased their investment capital three times in good weather years, while investment is doubled for the remaining normal weather years. Small-scale farmers under scenario 3 did not access weather forecasts but doubled investments all years.

18.3 Results and Discussions

18.3.1 Crop Model Results

The yields of the various crops vary from year to year. Over the 19 years, the average crop yields are summarized in Fig. 18.3. Yields of the cereals were generally higher than those of the legumes, a situation that may be attributed to the application of fertilizers to the cereals. In Tamale, the simulated yields obtained for the different categories of farms (small, medium, and large) for the cereals were consistently lowest with the small-scale farmers and highest with large-scale farmers. This is mainly in response to the amount of fertilizer used. Inter-annual variability in yields was highest with rice among the cereals. Variability in yields was generally below 25%, a situation that could be explained by the averaging of yields over different planting dates, thus probably buffering year-to-year variation in yields. With the legumes, differences in yields between the farm categories were much smaller compared to the cereals, mainly because fertilizers were not used and the main difference between categories was plant spacing. Inter-annual variabilities in yields were generally similar to those of the cereals.

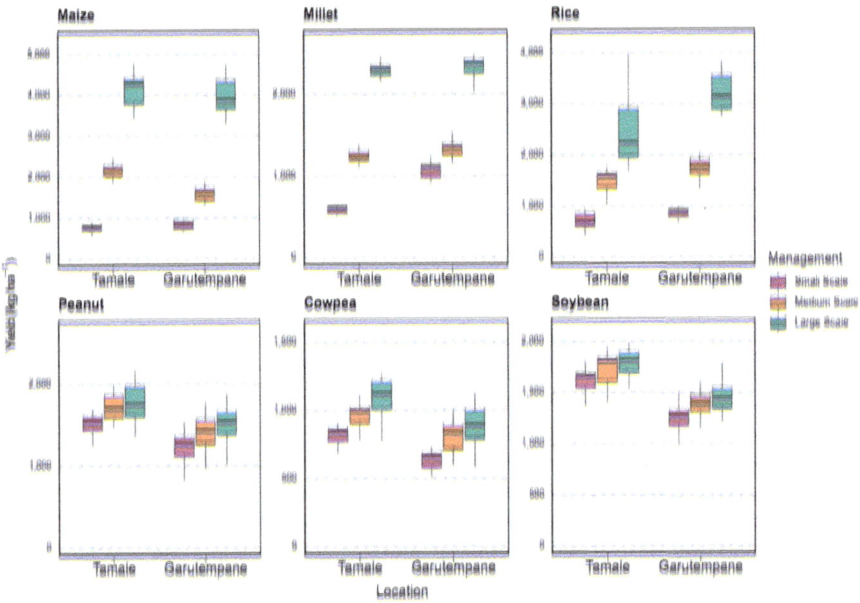

Fig. 18.3 Simulated yield for different farm categories at Tamale and Garu-Tempane. Each box is comprised of yield from 1990 to 2009

18.3.2 Economic Modelling Output

18.3.2.1 Optimum Land Allocation to Crops at Tamale

The results from the LP indicated that not all the crops contribute to the overall returns to the farmers. For small-scale farmers with the current resource outlay (Table 18.2), the yearly cultivation of soybean at 0.51 ha would maximize the returns over 18 out of 19 times (Fig. 18.4). Additionally, 1.34 ha of soybean would maximize returns once out of the 19 years. Detailed results (not shown) indicated that all the capital would be used, but land and labor would remain in excess. Under the second scenario, the inclusion of a staple crop would reduce the land allocated to soybean to 0.28 ha with the additional capability of 0.25 ha of rice in 18 out of 19 years and to 1.11 ha once out of 19 years. The inclusion of a staple crop in scenario 2 is more realistic since the survey showed that no farmer cultivated only a single crop, as scenario 1 would imply. Under scenario 3, where capital is doubled, land allocation to soybeans increased to 0.79 ha in 18 out of 19 years and to 1.85 ha in 1 out of 19 years.

The output for medium-scale farmers shows that soybean was again selected for maximum returns in scenario 1 at 0.87 ha land area for 15 out of 19 years as the investment capital outlay was also higher. Additionally, 0.66 ha and 0.33 ha of land were allocated to rice at 3 out of 19 years and 1 out of 19 years, respectively. For scenario 2, however, the output suggests that millet should be cultivated at an

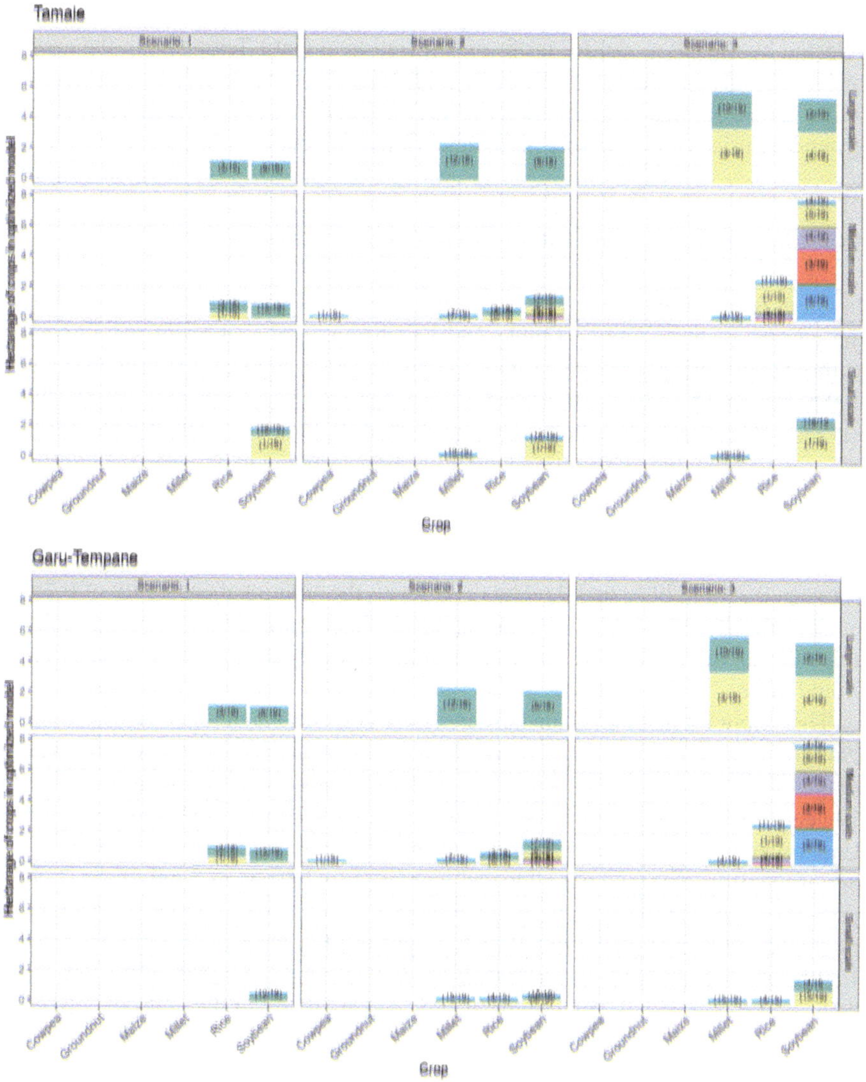

Fig. 18.4 Optimized land allocation to crops for different farmers and investment scenarios

acreage of 0.25 ha 7 out of 19 years (or 38% of the time), rice at 0.5 ha (61%) of the time and soybean at 1.26 ha all the time to maximize returns.

For the large-scale farmer, it was assumed that the pure commercial orientation removes the need for a subsistence crop. Therefore, the model is allowed to select any crop(s) combinations to maximize the returns. The model selected for scenario 1 an increased rice cultivation of 1.2 ha (72% of the time) to be complemented with soybean at 1.09 ha of soybean (38% of the time). For scenario 2, the model selected millet at 2.35 ha (67% of the time) and soybean at 2.2 ha for the remaining times.

It is worth noting that in all the analysis, maize, groundnut, and cowpea were not selected by the LP model for profit maximization at Garu-Tempane. In the case of maize, though the yields are relatively high (Fig. 18.3), the price is relatively low, while fertilizer requirements increase the production cost. Though cowpea and groundnut prices are relatively high, their yields are low.

18.3.2.2 Optimum Land Allocation to Crops at Garu-Tempane

In Garu-Tempane farming systems, an initial analysis of small-scale farmers with the current resource outlay (Table 18.2) focusing solely on profit maximization without considering subsistence showed that the yearly cultivation of soybean alone at 0.51 ha would maximize the returns as was the case for Tamale (Fig. 18.4). Under the second scenario, a minimum amount of land (0.25 ha) was allowed for one staple crop to be included, resulting land allocation reducing to between 0.18 (21% of years) and 0.28 ha (79% of the years). Detailed results (not shown) indicated that all the capital would be used, but land and labor would remain in excess. In scenario three, doubling investment capital would increase the land allocated to soybean to between 0.79 ha (79%) and 0.69 ha (21%) with the additional capability of 0.25 ha of rice or millet. The inclusion of a staple crop in scenario 2 and 3 is more realistic since the survey showed that no farmer cultivated only a single crop as scenario 1 would imply.

The output for medium-scale farmer selected rice for maximum returns at 0.66 ha (95% of the time) and 0.14 ha (5% of the years). For scenario 2, however, the output suggests that 0.47 ha of rice and 0.25 ha of soybean should be cultivated 18 out of 19 years (or 95% of the time) and cowpea one out of 19 years to maximize returns. Under scenario 3, when farmers took advantage of forecast to triple investment in 7 out of the 19 years, the land allocated increased to 2.39 ha in 3 out of the 7 good years (0.25 ha of millet), 2.29 ha in another 3 out of the 7 good years (0.25 ha of rice), and 1.8 ha rice once (with 0.25 ha soybean). For the remaining normal years, 1.14 ha was allocated to rice (with 0.25 ha soybean) for about 92% of the times and 0.16 ha of millet with 0.25 ha soybean allocated about 8% of the times to maximize returns.

For the large-scale farmer, it was assumed that the pure commercial orientation removes the need for a subsistence crop. Therefore, the model is allowed to select any crop(s) combinations to maximize the returns. The model selected for scenario 1 soybean cultivation at 1.09 ha (42% of the time) to be complemented with soybean at 1.17 ha of millet (37% of the time) and 0.9 ha of maize for 21% of the time. For scenario 2, the model selected millet at 2.35 ha (67% of the time) and soybean at 2.2 ha for the remaining times.

It is worth noting that in all the analyses, maize, groundnut, and cowpea were not selected by the LP model for profit maximization at Garu-Tempane. In the case of maize, though the yields are relatively high (Table 18.2), the price is relatively low, while fertilizer requirements increase the production cost. Though cowpea and groundnut prices are relatively high, their yields are low.

18.3.3 Responding to Weather Forecast

One major objective of this study is to explore the extent to which seasonal weather forecast can impact farm decision-making. Several works in Ghana have attempted to forecast the coming season and its impact on farm management decision-making, such as the choice of planting date, manure and fertilizer application rates, among others (Adiku et al. 2007; MacCarthy et al. 2017). In particular, it has been shown the ENSO can be useful in this regard, with the La Nina and some Neutral phases indicating good years (seasonal rainfall falls within the upper quartile of the probability distribution. Thus, based on the pre-season ENSO phase, the following years were selected as good years at Tamale and Garu-Tempane within the simulation period of 1991 to 2009: 1991, 1994, 1996, 1999, 2000, 2008, 2009.

The LP model output for the medium-scale farmer indicated that if the farmer responded to the expected good season by tripling the capital investment, millet will be selected at 0.25 ha in 3 out of the 7 years, rice at 0.64 ha in 4 out of 7 years combined with soybean cultivation at 2.04 ha in all the 7 good years (Fig. 18.5). In the case of the large-scale farmer, the selected crops were millet (3.51) in 3 out of the 7 good years and soybean at 3.26 ha in the remaining 4 good years. Total acreage utilized would still fall short of the total land of 10 ha available.

In the case of Garu-Tempane, if the medium-scale farmer responded to the forecasted good seasons by tripling initial investment, 1.80 ha of rice in combination with 0.25 ha of soybean was selected by the LP to maximize returns in all 7 years. For the large-scale farmer, 3.51 ha of millet was selected in 4 out of the 7 good years, 2.70 ha of maize was selected for 2 years out of 7 good years, and 3.25 ha of soybean was selected in a year to maximize returns. As with the Tamale site, land

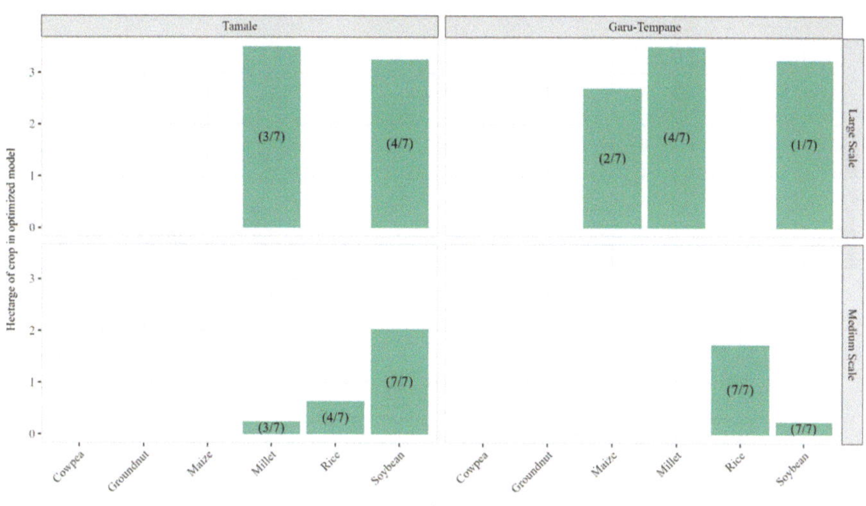

Fig. 18.5 Optimized allocation to crops for different farmers and investment scenarios in good years

utilized was less than what is available for each category. In effect, capital continued to be the major limitation to maximizing returns to the farmers.

18.3.4 Economic Returns at Tamale and Garu-Tempane

The economic returns (profits) vary depending on the farmer's type and investment scenario. For the current capital outlay, small-, medium-, and large-scale farmers would obtain 529, 895, and 1335 profits from cultivating soybeans only in Tamale. In Garu-Tempane, the profits were 362, 193, and 555 for the small-, medium-, and large-scale farmers, respectively, when only soybean was selected to be cultivated. The inclusion of a subsistence cereal with the capital investment remaining same as in scenario 1 in Tamale resulted in the reduction in overall profit for small- and medium-scale farmers to 320 and 858 Ghana Cedis, respectively (Fig. 18.6). In Garu-Tempane, the profit margins for the small and medium scales were 193 and 882, respectively. Scenario 3, which involves doubling of investment for small-scale farmers with the capital of the medium- and large-scale farmers tripled, resulted in profit margins of 817, 2051, and 3177 for the small-, medium-, and large-scale farmers, respectively, in Tamale. Generally, the variability of the profit margin for the small-scale farmers was lower than those of the medium- and large-scale farmers (Fig. 18.6) at both sites.

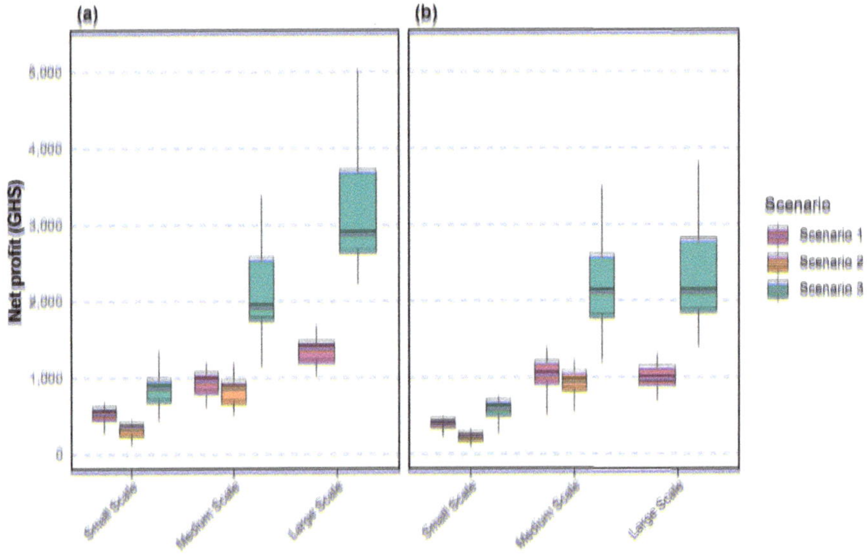

Fig. 18.6 Interannual variability in net profit of three farm categories (small, medium, and large scale) under different management scenarios for the (**a**) Tamale and (**b**) Garu-Tempane sites (1 GHC = 0.18 US$)

Furthermore, if medium- and large-scale farmers respond to weather forecasts by tripling their capital investment, profits would reach 2909 and 4115 Ghana Cedis, respectively. Clearly, capital seems to be the major constraint to maximizing profits. However, whether farmers can access credit would depend on their financial strength and collateral, which are often weak.

The result also suggested that responding to weather forecasts leads to increased production and net returns. In the main, large-scale farmers seem to benefit from seasonal forecast technology. Yet, little progress has been made in improving the predictive capacity of seasonal forecasting. This aspect requires further research attention.

18.4 Conclusion

The LP framework was successfully utilized to optimize land allocation with the aim of maximizing returns. Optimal crop combination obtained from the LP framework involves soybeans, millet, rice, and maize. Increases in the capital holdings of a farmer increases the optimal values of the decision variables as well as the maximum value of the objective function. A major constraint to maximizing returns in the study area is access to capital to invest in the farming enterprise. The returns on investments varied significantly among the three category of farmers, with the large scale farmers benefitting most. Knowledge of pre-season weather forecasts, which allows farmers to increase their capital investment, has the potential to further maximize returns on investments.

References

Adiku SGK (ed) (2019) Climate change: variability and biophysical landuse in Ghana. Afram Publications (Ghana) Limited, Accra

Adiku SGK, Stone RC (1995) Using the southern oscillation index for improving rainfall prediction and agricultural water management in Ghana. Agric Water Manag 29(1):85–100

Adiku SG, MacCarthy DS, Hathie I, Diancoumba M, Freduah BS, Amikuzuno J et al (2005) Climate change impacts on west African agriculture: an integrated regional assessment (CIWARA). In: Rosenzweig C, Hillel D (eds) Handbook of climate change and agrosystems: the agricultural model Intercomparison and improvement project integrated crop and economic assessments, part 2. World Scientific, pp 25–73

Adiku S, Mawunya F, Jones J, Yangyouru M (2007) Can ENSO help in agricultural decision-making in Ghana? In: Sivakumar MVK, Hansen J (eds) Climate prediction and agriculture: advances and challenges. Springer, Berlin, Heidelberg, pp 205–212. https://doi.org/10.1007/978-3-540-44650-7_20

Adu-Prah S, Appiah-Opoku S, Aboagye D (2019) Spatiotemporal evidence of recent climate variability in Ghana. Afr Geogr Rev 38(2):172–190. https://doi.org/10.1080/19376812.2017.1404923

Assouto AB, Houensou DA, Semedo G (2020) Price risk and farmers' decisions: a case study from Benin. Scient Afr 8:e00311. https://doi.org/10.1016/j.sciaf.2020.e00311

Atiah WA, Muthoni FK, Kotu B, Kizito F, Amekudzi LK (2021) Trends of rainfall onset, cessation, and length of growing season in northern Ghana: comparing the rain gauge, satellite, and farmer's perceptions. Atmos 12(12):1674. https://doi.org/10.3390/atmos12121674

Drafor I, Kunze D, Sarpong DB (2013) Food security: how rural Ghanaian households respond to food shortages in lean season. Inter J Agric Manag 2(4):199–206. https://doi.org/10.22004/ag.econ.175073

Gbangou T, Ludwig F, van Slobbe E, Hoang L, Kranjac-Berisavljevic G (2019) Seasonal variability and predictability of agro-meteorological indices: tailoring onset of rainy season estimation to meet farmers' needs in Ghana. Clim Serv 14:19–30. https://doi.org/10.1016/j.cliser.2019.04.002

Jones JW, Hoogenboom G, Porter CH, Boote KJ, Batchelor WD, Hunt L et al (2003) The DSSAT cropping system model. Eur J Agron 18(3–4):235–265

Kumi N, Adeliyi TE, Asante VA, Abiodun BJ, Lamptey BL (2023) Impact of rainfall onset date on crops yield in Ghana. Front Sustain Food Syst 7:1176385. https://doi.org/10.3389/fsufs.2023.1176385

MacCarthy DS, Sommer R, Vlek PL (2009) Modeling the impacts of contrasting nutrient and residue management practices on grain yield of sorghum (Sorghum bicolor (L.) Moench) in a semi-arid region of Ghana using APSIM. Field Crop Res 113(2):105–115

MacCarthy DS, Adiku SG, Freduah BS, Gbefo F, Kamara AY (2017) Using CERES-maize and ENSO as decision support tools to evaluate climate-sensitive farm management practices for maize production in the northern regions of Ghana. Front Plant Sci 8:31

MacCarthy DS, Kihara J, Masikati P, Adiku SGK (2018) Decision support tools for site-specific fertilizer recommendations and agricultural planning in selected countries in Sub-Sahara Africa. In: Bationo A, Ngaradoum D, Youl S, Lompo F, Fening JO (eds) Improving the profitability, sustainability and efficiency of nutrients through site specific fertilizer recommendations in West Africa agro-ecosystems, vol 2. Springer, pp 265–289

Narh S, Boote KJ, Naab JB, Abudulai M, M'Bi Bertin Z, Sankara P et al (2014) Yield improvement and genotype× environment analyses of peanut cultivars in multilocation trials in West Africa. Crop Sci 54(6):2413–2422

Otoo J, Ofori JK, Amoah F (2015) Optimal selection of crops: a casestudy of small scale farms in Fanteakwa district, Ghana. Int J Sci Technol Res 4(05):142–146

Sowell RS, Ward R (1998) Modeling processes and operations with linear programming. In: Peart RM, Curry RB (eds) Agricultural systems modeling and simulation. 270 Madison avenue. Marcel Dekker Inc, New York, pp 112–156. https://doi.org/10.1201/9781482269765

Wahab I, Hall O, Jirström M (2022) The maize is the cost of the farming, and the cassava is our profit: smallholders' perceptions and attitudes to poor crop patches in the eastern region of Ghana. Agric Food Secur 11(1):14. https://doi.org/10.1186/s40066-022-00361-w

Chapter 19
Promoting Ecosystem-Based Adaptation to Climate Change in a Fragile Ecological Context of the North Region of Cameroon: A contribution of Innovation Niche Approach

Thierry Gaitan Seutchueng Tchuenga, Émile Blaise Siewe Pougoue, and Fréderic Saha

Abstract Agrosylvopastoral areas in arid and semi-arid regions play an important role in the survival of rural populations. However, the intensification of drought linked to climate change (CC), land degradation, and deforestation is having a considerable impact on the resilience of these areas. This study aims to demonstrate the role of the innovation niches approach in the adaptation of fragile ecosystems to climate change in North region of Cameroon. Data was collected from stakeholders in agro-sylvopastoral activities and analyzed using a harmonized framework for classifying and selecting innovation niches. The results identified two key innovations: the Agricultural Innovation Platform (AIP), which promotes climate-smart agricultural practices, and the Watershed Approach (WA), which integrates landscape management and forest protection. These approaches strengthen ecological resilience, improve soil restoration, and enhance community well-being. From a scientific point of view, it highlights the conceptual relationship between innovation niche and ecosystem-based adaptation (EbA) for the promotion of climate change adaptation and mitigation. In practical terms, it provides levers that can be mobi-

T. G. Seutchueng Tchuenga (✉)
Faculty of Arts, Letters and Humanities, University of Yaoundé 1, Yaoundé, Cameroon

CIFOR-ICRAF, Project ReSI-NoC, Garoua, Cameroon

Agricultural Research Institute for Development, Yaoundé, Cameroon

É. B. Siewe Pougoue
National Polytechnic School, Department of Agriculture, Livestock and Derived Products (AGEPD). Bio-resources and Food Technology Laboratory, University of Maroua, Maroua, Cameroon

F. Saha
Faculty of Arts, Letters and Humanities, University of Yaoundé 1, Yaoundé, Cameroon

Y. D. I. Khan et al. (eds.), *Land Restoration Through Ecosystem-Based Approach*, Disaster Resilience and Green Growth, https://doi.org/10.1007/978-981-95-0389-6_19

lized to multiply and accelerate innovation processes and thus contribute to the sustainability of agricultural systems in drylands of Cameroon.

Keywords Innovation niche · Ecosystem · Watershed · Natural resource

19.1 Introduction

Since the early 1990s, the Intergovernmental Panel on Climate Change has provided evidence of accelerating global warming and climate change (CC) (IPCC 2013). Nevertheless, the Sahelian zone of Africa experienced a general downward trend in total annual rainfall and the number of consecutive rainy days between 1968 and 1998 (Ly et al. 2013). North region of Cameroon is a part of the Sahel region that is particularly vulnerable to the impacts of climate change due to phenomena such as desertification, prolonged droughts, and erratic rainfall variations (Awazi et al. 2023; Lombe et al. 2024). These phenomena have led to severe soil degradation, lower agricultural productivity, and reduced food security for rural communities, who depend mainly on rainfed agriculture and pastoralism (Saha et al. 2020; Ly et al. 2013). The consequences of these impacts are particularly serious in a region where agriculture is highly dependent on natural and climatic resources (Muluneh 2021). For example, the region's high dependence on rainfall makes it particularly vulnerable to climate variations, with serious consequences for food security and economic stability (Geneletti et al. 2016). Ecosystem-based Adaptation (EbA) represents a strategic response to these challenges, relying on nature-based solutions to restore degraded ecosystems, protect biodiversity, and improve community resilience (Calliari et al. 2022). However, although practices such as conservation agriculture, agroforestry, and sustainable water management have demonstrated their effectiveness, their implementation remains limited by factors such as the lack of coordination between stakeholders and the need for an approach adapted to local realities (Klein et al. 2007; Matos et al. 2022). Yet, there is limited evidence of cases involving local governments and constellations with diverse actor types, such as governance actors, the private sector, or civil society actors, and individuals or households (Petzold et al. 2023). In this context, the innovation niche approach (INA) offers an opportunity to promote innovative solutions for adapting to climate change (Vasseur 2021; Tchuenga et al., 2024) based on a multi-stakeholder approach. The innovation niche approach (INA) offers a structured way to promote collaboration, focusing on experimentation in protected areas to test and refine sustainable practices (Schot and Geels 2008; Hekkert et al., 2007). This approach makes it possible to strengthen the resilience of ecosystems while encouraging stakeholder involvement. The aim of this study is to examine the INA's contribution to promoting EbA in North region of Cameroon. Based on two experimental case studies, such as innovation platforms for resilient agriculture and sustainable water management. This research assesses the effectiveness of innovative approaches in ecological restoration and improving the livelihoods of local communities. This study is significant in exploring the practical application of INA in a fragile context, providing avenues for sustainable and scalable climate adaptation, adapted to local realities and capable of influencing regional and national climate change adaptation policies (Rodima-Taylor 2012; Cortinovis et al., 2022).

19.2 Research Framework

19.2.1 Innovation Niche Approach

Agrinatura and FAO (2020) consider an innovation niche as a place where innovation can be learned from and/or experimented with and which, if strategically managed, can lead to a sustainable transformation. It is an enabling space where innovation is developed or co-constructed with a view to driving sustainable change in relation to the promotion of agriculture (TAP 2017). The notion of innovation niche has been placed at the heart of transition theory (Mathé et al., 2016) through the multi-actor model developed by Geels (2006). This model combines several concepts relating to the evolution of the economy (trajectories, regimes, niches, speciation, dependency path, routines), including that of the niche. Transition results from the interaction between three levels of reality: the socio-technical landscape at the macro level, the socio-technical regime at the meso level, and niche innovations at the micro level. Geels (2006) considers niches as protected spaces in which radical innovations can be developed. They are described as stable spaces protected by precise rules, in which producers (R&D scientists) sponsored by national or private actors try to stabilize an innovation by various means (scientific networks, publications, seminars, etc.). Mathé et al., 2016). According to Toillier et al. (2020), an IN should be based on a specific object, which may be a commodity chain (mango, rice, coffee, fish, etc.) or a thematic area [local land tenure, rural entrepreneurship, small and medium-sized enterprises (SMEs), ecosystem-based adaptation (EbA), etc.]. In this way, the stakeholders involved are configured in a network constituting an innovation system (IS) (Geels 2004; Temple et al., 2017; Touzard et al. 2015). Based on this definition of Toillier (Toillier and de Bellaire 2016), two innovation niches were chosen with a thematic area close to EbA practices, such as innovation platforms for the promotion of agroecological practices and watershed approaches for the best practices of water management. It is the case of the platforms for the promotion of agroecological practices and sustainable water management. The potential of those case studies has been demonstrated in combating climate change impact. Ngondjeb (2013) underscores the potential of EbA in mitigating drought impacts and restoring degraded ecosystems, particularly in the Sahelian zones. Research by Chomba et al. (2020) highlights that EbA approaches such as agroforestry and watershed management enhance agricultural productivity and reduce vulnerability in semi-arid regions. Recent studies by Doswald et al. (2020) and Locatelli et al. (2021) emphasize the role of EbA in addressing challenges such as water scarcity, soil degradation, and loss of biodiversity in semi-arid and arid regions.

19.2.2 Ecosystem-Based Adaptation (EbA) and Innovation Niche: Interconnections and Implications for Climate Resilience

In the context of climate change adaptation, particularly in locations such as the north region of Cameroon, the interaction of EbA and innovation niches creates a holistic framework for sustained development (Pougoue et al. 2024). This framework enables local communities to leverage natural resources, experiment with creative solutions, and incorporate these solutions into bigger policy and economic institutions (Danielsen et al. 2022). The interconnection of EbA and innovation niches creates a dynamic framework capable of mitigating susceptible regions such as the north region of Cameroon. Innovation niche, as an integrated approach, promotes resistance to climate change while also supporting natural restoration (Bas et al. 2022). EbA practices frequently develop from innovation niches in which communities can try nature-based solutions, serving as the central adaptation approach, providing a nature-based framework for tackling climate change concerns in vulnerable ecosystems (Nyathi et al. 2024). Innovation niches create protected environments for experimenting with small-scale, community-driven initiatives. Innovation niches emphasize participatory approaches, involving local stakeholders in designing and implementing projects. This fosters community ownership and enhances the sustainability of EbA interventions (CIFOR 2023). These niches enable the development, testing, and refinement of EbA practices, such as agroforestry and watershed management, before scaling them to broader systems (Geels et al. 2020).

In North region of Cameroon, innovation niches have been instrumental in tailoring EbA strategies to the region's unique environmental and socioeconomic conditions, enabling more targeted and effective interventions (CIFOR 2023). These niches empower local communities by involving them in the co-design and implementation of projects, ensuring that solutions align with their cultural and economic realities. Additionally, innovation niches provide a framework for scaling successful practices, supported by evidence of their efficacy and adaptability in real-world settings (Zeratsion et al. 2024). This approach equips communities to thrive in dynamic and challenging environments, driving policy alignment and enabling the upscaling of proven solutions. EbA and innovation niches together represent a pathway to climate resilience for regions like North region of Cameroon (Rawlins et al. 2023).

19.3 Methodology

19.3.1 The Study Area

The study area is in the north region of Cameroon (Fig. 19.1). The North Region of Cameroon lies between 8° and 10° North latitude and 13° and 14°05′ East longitude. It is bordered to the north by the extreme north region, to the south by the Adamaoua Region, to the west by the Republic of Nigeria, and to the east by the Republic of Chad and the Central African Republic. The region is in the Sudano-Sahelian region, where rainfall is low and variable (Saha 2019). The region is experiencing different levels of changes in land use and ecosystems. The human modification index indicates moderate to high levels of modification. Land and ecosystems, such as national parks, are highly degraded (Wolf et al., 2023). It is one of the most underdeveloped and rural regions in the country. This situation is exacerbated because almost half of the territory is occupied by protected areas and zones of hunting interest, which considerably reduces the space dedicated to agriculture and livestock farming and puts increased pressure on natural resources. In the past (1990–2000), excluding local populations from resource management has led to the failure of the co-management approach of protected areas. When defining the spatial extension of protected areas, assessing their basic needs has hardly been

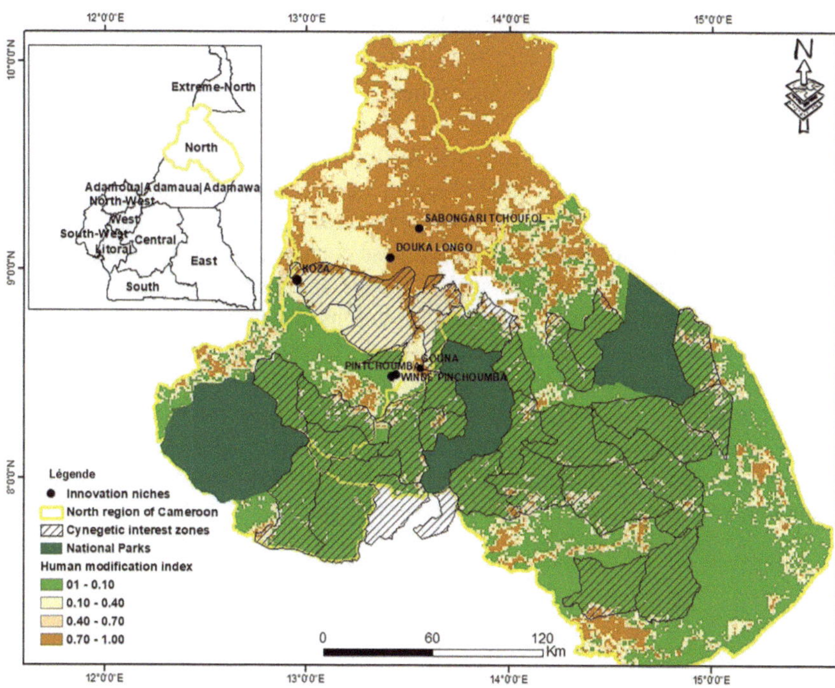

Fig. 19.1 The study area. (Source: Adapted from NASA, Socio-economic data)

considered. So, their only alternative is to operate illegally by deforesting and degrading protected areas. From 1970 to 1980 and from 1980 to 1985, the population of the underpopulated northern region grew considerably as people from the far north of the country began to develop their farmland (Fedreheim and Blanco 2017). The drought of 1983–1984 accelerated the rate of migration. The consequences include the disappearance of scrubland (vegetation) and the increasing scarcity of arable land (Palou 2011). The average cultivated area per farm is between 1.5 and 3 ha, and in some areas, it is no longer possible to follow (Djiongo et al., 2015). Deforestation quickly leads to an immediate deterioration in the physical and chemical quality of the soil, which impairs soils that are less permeable and highly susceptible to (Fearnside 2005; Laurance et al., 2013; Tolimir et al., 2020). Deforestation also leads to rapid negative changes in soil microbial communities (Kroeger et al. 2018) and soil physical properties (soil density and permeability), which modifies the exchange of greenhouse gases between the soil, plants, and the atmosphere (Medina et al. 1980; Van der Werf et al. 2009). The deforestation triggers atmospheric warming (IPCC 2013; Longobardi et al. 2016), which in turn leads to desertification, land degradation, and the drying out of surface waters (MINEPDED 2015).

19.3.2 The Qualitative Research Paradigm, a Key Research Design for Innovation Niche

Innovation niches, as a protected space where new technologies, practices, or ideas are cultivated, often involve complex social, cultural, and contextual dynamics. Qualitative methods offer the depth and flexibility needed to explore these complexities. This paradigm has facilitated exploring the intervention of various stakeholders and the local community context (Creswell and Plano Clark 2018). It allowed the comprehensive exploration of these contexts, capturing factors such as community or local knowledge and stakeholder interactions (collaboration). Stakeholders often collaborate to cocreate solutions. Some key methods of qualitative research, such as case studies, were used. It has provided rich insights into the specific environments where innovations were developed and tested. The paradigm has facilitated participatory approaches, enabling this research to study the engagement of communities and stakeholders in fostering mutual learning and empowerment. The mobilization of inductive reasoning helps in understanding how and why certain innovations succeed or fail within specific niches.

19.3.3 Data Types and Sources

Qualitative data were used in this research. The qualitative data was captured through primary data sources through interviews, focus group discussions, and multi-actor workshop consultations. Also, some data from this study were derived from secondary data sources. We have consulted research and development projects (AFD/DPGT/PRASAC) on land management, water supply and management, land tenure, natural resource management, and the implementation of agroecological practices. The documentary research enabled us to achieve the objective of identifying the organizations in the region and some innovation niches.

19.3.4 Population and Sampling Procedure

The samples for this study were defined according to the spatial scales of the organization's intervention. In terms of geography, the organization's activities were framed at the regional and local level, leaving aside the national level (Nelson and Rosenberg 1993; Lundvall 1992). The local scale is the level at which innovations are implemented and tested (niche). This scale is associated with Producer Organizations (POs), which are generally the beneficiaries of other organizations at a macro or regional level (Fig. 19.2).

The micro level is the village level, where the innovation is demonstrated (field school) and transferred by one or more support service providers (Fofiri Nzossie et al. 2016; Mathé et al. 2019; Faure et al. 2019). The regional scale represents all the organizations involved in projects and programs in the region, including the

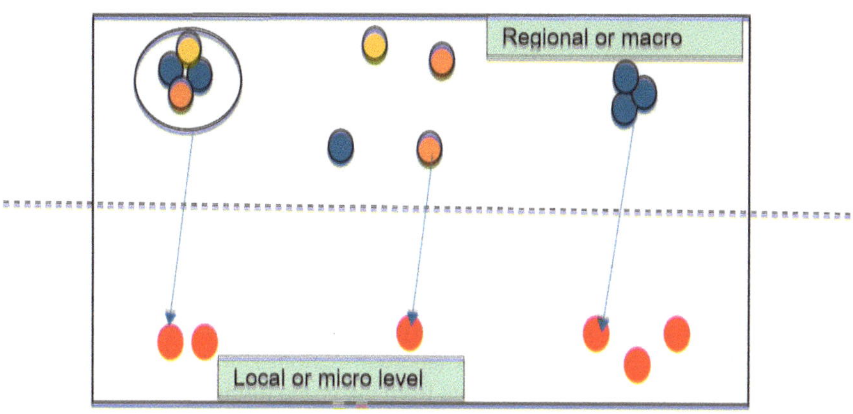

Fig. 19.2 Scale of organizations sample
The variation of scales between organizations highlights the relationship in two dimensions. The circles, shown in different colors, illustrate the diversity of organizations that support local organizations. The local level shows that innovation niche (IN) is an entity in the innovation system

state, international cooperation, and nongovernmental organizations (NGOs) (Ndah et al. 2020). Two innovation niches based on the EbA approach have been used to illustrate the effectiveness of their experimentation because the innovation niche approach (INA) involves structured processes for testing new practices or technologies in real-world contexts (Hekkert et al. 2007).

19.3.5 Data Collection and Procedure

A mixed method was used to combine participatory, multi-actor (consultative and synergistic workshops), and systemic (Ndah et al. 2020) approaches. The participatory approach made it possible to consider the opinions, viewpoints, and field experiences of the organization involved in EbA practices. This participatory approach integrated local knowledge and scientific inquiry into developing and testing EbA strategies. Data was collected in three main phases and in an iterative form. It involved field surveys, stakeholder interviews, and ecological assessments across pilot villages in the north region of Cameroon, such as Koza2, Douka Longo, Pintchoumba, and Gouna (Fig. 19.1). The pilot villages were selected based on the effectiveness and clarity of the innovation tested in the villages, as well as the number of beneficiaries. Also, the data from the baseline survey on innovation niche and innovation support services carried out within the framework of the ReSI-NoC project in 34 villages were used in the present study. The villages were sampled according to three thematic interventions, namely: "degraded areas around Garoua," "areas near protected areas," and "cotton frontier areas." We employed a niche innovation framework, selecting small-scale interventions within localized environments that have the potential for broader application. The study emphasizes flexibility and community participation in testing adaptation strategies, ensuring local stakeholders have ownership over the processes.

19.3.5.1 Consultative Workshops

The consultative workshop aimed to identify the initiatives promoted in the development process in connection with agriculture and forestry in the North region of Cameroon region. The workshop provides a list of stakeholders ($N = 40$) and innovative initiatives ($N = 60$) they promote in the region. Supplementary information was collected during the inter-project synergies workshop with a view to capitalizing on innovative initiatives and fostering synergy of action between the stakeholders operating in the North region of Cameroon region. During the workshops, a matrix was drawn to summarize the data from the group discussions (description of each action, nature and primary objective of the innovation, obstacles to implementation, area of implementation, etc.).

19.3.5.2 Selection of Innovation Niche for This Study

After identifying innovative initiatives, 12 were selected under the framework of the ReSI-NoC project. The selection was based on the "scoring/notation grid" for the cases identified (Toillier et al. 2020) to judge their relevance and consistency (Siéwé et al. 2020). The selection criteria were based on the characteristics of the innovation cases with the research questions, the accessibility of the data on innovation, and the potential for supporting this innovation case. The grid was adapted to take account of the support and impact objectives of the ReSI-NoC project, considering economic profitability, social equity, and, above all, ecological sustainability. Understanding the links between well-being (and socioeconomic sectors) and ecosystem services can help to identify contextually appropriate physical and ecosystem interventions (World Bank and WWF 2014). A coefficient of 3 was assigned to the characteristics linked to the significant challenges, such as well-being and the environment. For well-being, the assessment focused on the impact of the innovation on the fundamental pillars such as food security, income, social cohesion, and employment. For the environment, the assessment was applied to the impact of innovation on climate change, biodiversity, and natural resources. The present study chose two cases of innovation niches among the 12 related to EbA practices (agroforestry and sustainable water management).

19.3.5.3 Key Informant Interview

Stakeholders identified during workshops whose activities are circumscribed around agroforestry components and water management were selected for an interview. Twenty of the 40 actors identified at workshops were interviewed because of their openness and collaboration during the survey. The interview guide was divided into three parts. The first part was identifying the structure/organization (name, location, type of structure, area(s) of intervention, role, approach, mission assigned to the structure, support offered, etc.). The second was looking at innovative initiatives (areas of intervention, innovative practices, etc.), and the third at the elements characterizing the innovative initiatives (type of innovation, objective, beneficiaries, impact on food security, income, employment, social cohesion, biodiversity/natural resources, and climate change).

19.3.5.4 Field Research

The statements of actors' achievements reported during interviews are not always credible (Dunwoodie et al. 2023). A field exploration was made to observe the gap between what actors state and what is achieved on the ground. Exploratory field visits (Sawadogo 2021) are essential in action research because they provide more tangible arguments and irrefutable evidence about a specific situation, fact, or achievement. It was possible to verify the effectiveness of the achievements made by project leaders in two case studies based on EbA practices. This was based on verifying the accuracy of the information obtained from interviews (upstream) with project owners.

Focus groups were also held in the field where EbA initiatives had been implemented. The optimal size of participants was between 6 and 15, creating a large enough group to facilitate discussion without preventing balanced participation (Jayasekara 2012; Li et al. 2015). Participants were invited to participate based on their experiences related to EbA practices and learning (Domecq et al. 2014; Jayasekara 2012). The focus groups were organized in comfortable, relaxed, and non-threatening public places (Kitzinger 1995). The focus groups were aimed more at two categories of groups. These were the direct beneficiaries of the interventions of the project actors and the community in which EbA has been promoted.

Also, semi-structured interviews were conducted with community leaders and those from village organizations. It was an open-ended question focusing on the characteristics of the innovation being promoted, which had already been identified with the stakeholders (origin of the idea that led to the innovation being introduced, date on which the innovation was introduced, cost of the innovation, successes recorded, etc.), the constraints and avenues for improvement (problems encountered in implementing the innovation, solutions envisaged, and recommendations for overcoming the constraints identified, etc.), and the adoption of those innovations.

On the field, photography was used as proof of information and content of EbA practices. This study essentially illustrates our arguments and justifies the effectiveness of data collection on EbA practices. Two types of photography were used: aerial images taken by drone and those taken with a Canon EOS 4000D digital camera. These two instruments were used to give or clarify the content of the EbA innovation niches.

19.3.6 Procedure for Data Analysis

Using an approach based on content analysis, each data collected on innovation niche, EbA, and organizations was filled in an Excel sheet (2016 version). These qualitative data were analyzed for each actor's comments (Fraenkel and Wallen 2009). Data were constructed based on the interview and focus group guide by paying more attention to the occurrences of meaning in the participants' addresses. The analysis used quantitative statistical analysis from the content analysis output data. Each category of analysis in the quantitative aspect was supported by the discourses of the data collected during the survey. Several researchers consider a combination of quantitative and qualitative approaches effective and relevant. Moreover, according to Savoie-Zajc and Karsenti (2004), "beyond the recognized advantages of the quantitative approach, there is always the possibility of diversifying the means of collection to bring out fuller and richer explanations by using the complementarity with the qualitative approach."

19.4 Results

19.4.1 Approach to Promoting EbA Practices in Innovation Niches

The innovation niche shows diverse support provided by actors and received by households. The approach to promoting ecosystem-based adaptation (EbA) practices is based on actors' intervention. Different actors intervene in the innovation niche in promoting EbA practices, as observed in Fig. 19.2. The diversity of actors (group, community, association, NGO, etc.) involved in EbA promotion includes project leaders. They belong to different categories, as shown in Fig. 19.3. Around 38% of the actors belong to the public sector, while 34% fall in the parapublic organizations. They are followed by international and local NGOs (10% and 8%, respectively). We also notice the presence of cooperation organizations (6%) and producer organizations (4%).

This innovative niche approach seeks to promote collective EbA practices. This match between the proposed innovations and the population's needs is confirmed by the level of satisfaction expressed by the population. Significant contributions to EbA can be observed during the actor's intervention, experimenting with nature-based solutions to face climate change. Actors have used different approaches in promoting EbA practices (Fig. 19.4). Learning and knowledge sharing through training and awareness are the best ways to support nature-based solutions in the innovation niche. Niche innovation case studies have been used to analyze how EbA practices are conducted and implemented. This can be shown in two case studies.

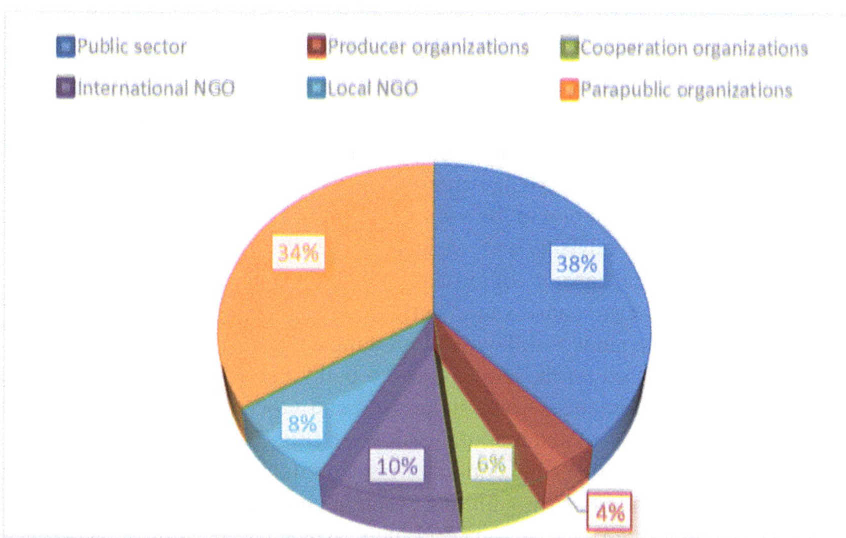

Fig. 19.3 Different types of actors in the innovation niche

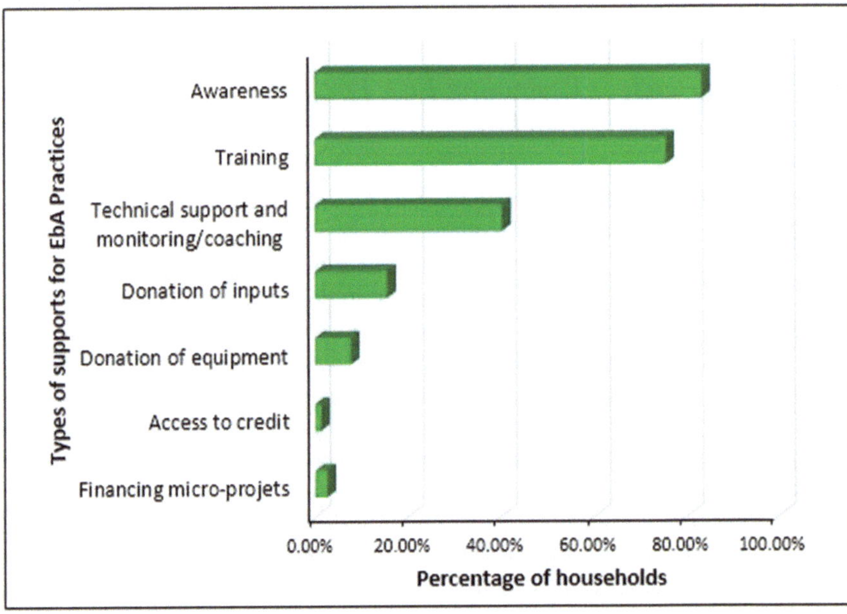

Fig. 19.4 Approaches used by actors in promoting EbA practices

19.4.1.1 Innovation Platforms for the Promotion of Agroecological and Agroforestry Practices

The innovation platforms were promoted in the north region of Cameroon between 2016 and 2020 under the Program for the Security and Integrated Management of Agro-Pastoral Resources (namely PASGIRAP). The platform was just for cotton growers to combat the decline in soil fertility. Also, the innovation platforms were set up in response to the problems of ongoing degradation of land and ecosystems and climate change, which over time have led to a decline in agricultural productivity in North region of Cameroon and impoverished populations, especially in rural areas. The learning and knowledge-sharing strategies used within the platforms were on-site in a common plot in the village (Fig. 19.5).

An agroecological advisor was assigned to each platform to support and strengthen the technical capacity of farmers. The agroecological practices promoted in the common plots were conservation agriculture based on crop rotation and crop associations, use of organic fertilizer, and growing crops on biomass. The agroforestry practice was based on the use of trees to build a strip around farmers' plots (Fig. 19.5). On-field practices were supervised and co-constructed with farmers, and their participation has increased their adoption (Fig. 19.6). Adoption here concerns both platform members and communities. Communities learn how to manage and sustain their plantations through tree planting and soil restoration techniques. Vulnerable groups, such as small-scale farmers, women, and indigenous peoples, were empowered, ensuring that EbA practices meet their specific needs and

Fig. 19.5 Experimental plot

challenges. The participation of communities in the design and implementation of ecosystem-based solutions, as stipulated by the innovation niche approach, is more likely to increase adoption of practices and technologies and maintain them over time, contributing to long-term resilience.

Among several practices and techniques disseminated through the platforms, the most widely adopted of these are improved seeds (new rice varieties), grass strips to demarcate plots, fallows, hedgerows, and crop rotation and association. Appiah-Adu et al. (2016) showed that platforms have considerable potential to increase the adoption of technologies in Ghana, particularly crop varieties and the practices and technologies developed. They have enormous potential for contributing to the adaptation and mitigation of climate change and variability. More than 75% of farmers (Fig. 19.6) recognized that these potentials are high or even very high. So, researchers have indicated some practices and techniques as contributing most to adaptation to climate change. These include mulching the soil with biomass residues, which helps limit water evaporation and stabilize soil moisture for good plant growth. There are also grass and tree strips that, as mentioned above, make an effective and

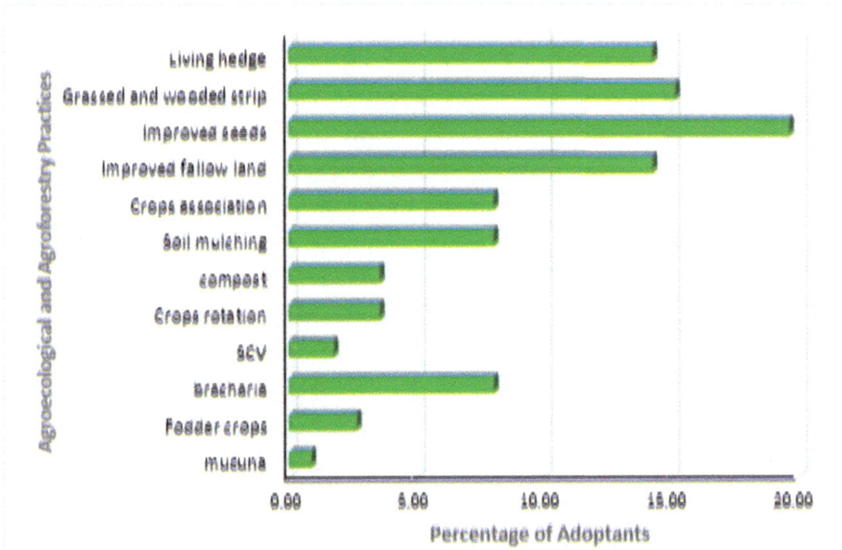

Fig. 19.6 Farmer's adoption of agroecological and agroforestry practices

significant contribution to reducing water erosion in crop plots. This practice also helps to increase soil fertility by retaining organic matter, which is one of the fertilizing elements in agricultural plots. Living fences and improved seeds are climate-smart innovations with great potential for adaptation and mitigation. Shelterbelts protect crops from strong winds during the growing season. Depending on the type of ligneous plant used, the communities that practice them reap the benefits of the ecosystemic role they play. Regarding seeds, long-time producers were under the yoke of traditional seeds, generally from previous harvests. The introduction of improved seeds by the platforms has been appreciated by farmers, who recognize their effectiveness in the current climate change context. For example, in the past, rice seed was only available for lowland production. Today, farmers have rice seeds that can be produced in every terrain. These include rainfed rice, Nérica 1 and 2. By experimenting with agroforestry and agroecology practices in platforms considered innovation niches, communities and stakeholders have tested the effectiveness of EbA techniques in addressing specific local challenges like drought and land degradation. This was done in small pilot sites before being expanded in the north region, allowing communities to learn and adapt practices that fit their environmental needs. Initial results showed significant improvements in ecosystem health, with increased vegetation cover, improved soil quality, and enhanced water retention in reforested zones. Agroforestry interventions in plotted villages resulted in a 20% reduction in soil erosion and a notable increase in crop yields due to improved soil fertility and moisture retention.

19.4.1.2 Sustainable Water Management Approach

The watershed is an integrated approach to the sustainable management of landscapes in the Sudano-Sahelian zone of Cameroon. It is an innovation niche developed in the south of Garoua (Fig. 19.1). In the watershed innovation niche, our attention is based on sustainable irrigation and wetland restoration to improve water security. This is an ecosystem-based solution (EBS). The watershed approach effectively took shape with PASGIRAP since 2016. Still, it started in the study area in 2018 with the ABC-PADER project (namely, Supporting Change in Cameroon's Cotton-Producing Basin, Cameroon—Rural Development Support Program Development) and other nongovernmental organizations such as the NGO "Actions pour la Biodiversité et Gestion des Terroirs" (ABIOGeT). The watershed approach aims to improve food security and support fragile and degraded ecosystems to adapt to climate change. This approach has been put into practice in the Mayo-Douka watershed by the agreement between the municipality and the ABC-PADER project for the benefit of five villages (Fig. 19.1) surrounding the watershed. Several actions have been completed through the ABC-PADER projects. However, our focus is on Smart Valleys and village water supply systems. This innovative nature-based solution is used as an adaptation strategy to climate change. The rivers in the region are vital for the local population. However, prolonging the dry season means that surface water is drying up completely, considerably impacting agro-pastoral activities. It will therefore be necessary to invest in measures and technologies to save water so that it can be used appropriately in the future. Water-saving measures will likely become increasingly important due to rapid population growth and the prevailing drought. With this in mind, the Smart Valley and the village water supply system have been promoted as innovative nature-based technologies to combat climate change. The village water supply system is a temporary micro-dam designed for water retention (Fig. 19.7).

They are built to stop water from flowing at one point to create a permanent or temporary surface water reserve and to slow river water down at one point to force it to infiltrate and replenish the water table (Djombaye 2005). It was for this purpose that they were built in the region. These are devices built downstream of watercourses to limit the watercourse flow. The population was trained on the process. During the survey, the technical skills of the farmers were improved. Also, for the long-term use and management of this technology, a farmer's organization was created to share the knowledge with the community.

The Smart Valley approach was selected as a type of climate change resilience for African smallholder rice farmers at the Conference of the Parties (COP 22) on climate change, held in 2016 in Marrakech, Morocco (Delphine et al. 2019). Smart Valleys is a participatory, sustainable, low-cost lowland development approach for rice-based production systems in Sub-Saharan Africa. Smart Valley is a low-cost, sustainable, and participatory approach to developing inland valleys. It has been implemented in the northern region, in the Ngong subdivision. To ensure the sustainability of agro-pastoral and agro-forestry systems, the Smart Valley approach facilitates access to water and its sustainable management for rice production and

Fig. 19.7 The village water supply system (micro-dam). (Author, 2022)

Fig. 19.8 Smart Valley at Mayo Bangai in Ngong Subdivision. (Author, 2023)

market gardening, which are at the heart of income generation, particularly among women (Fig. 19.8). It is a new and innovative technology introduced in the watershed of Mayo-Douka. The community was trained on detecting the water flow, soil sample studies, and the line drawing in the same direction as the water flow. Farmers were taught by an ABC-PADER researcher how to build a small square to keep water for the rice growth. On a surface of 2500 m^2, 32 bags of 106 kg were produced. Before the implementation of Smart Valley, production in the village was very low (less than 10 bags on the same surface).

Farmers in Benin and Togo, who have adopted the Smart Valleys approach, have seen their rice yields under rainfed conditions increase from 1.5–2 tons per hectare to 5–4.5 tons per hectare, especially if accompanied by good crop management practices (Aminou and Ka 2017). Communities reported improved livelihoods through diversified income streams, such as the sale of tree products (fruit, fodder, and wood) and better agricultural yields. The water harvesting interventions reduced seasonal migration in search of water, stabilizing local economies.

19.4.2 Challenges of Ecosystem-Based Adaptations in Innovation Niche

19.4.2.1 Favorable Environment and Scalability of Ecosystem Base Adaptation Practices

Despite the growing recognition of ecosystem-based adaptation (EbA) as a sustainable approach to climate resilience, several challenges hinder its scaling. The innovation niche approach has successfully promoted EbA practices, but the lack of favorable environments remains a significant obstacle (Schmidt and Walz 2021; Selje et al. 2024). For instance, administrative decentralization, a key factor in empowering local governments, has been cited as ineffective by 70% of stakeholders involved in EbA projects (Ibrahim 2024). As one stakeholder emphasized, "*The slowness of the administrative process when having cooperation with the government to implement a project. Local authorities don't have power in decision-making; only the central government. Consequently, the local community's needs are often delayed or misaligned because they require approval from a central authority that is disconnected from on-the-ground realities.*" Studies by Nguyen et al. (2020) and Patel and Sharma (2017) highlight that this centralized control often marginalizes vulnerable regions, limiting their ability to implement EbA initiatives effectively. Additionally, agroecological and agroforestry practices require funding. Due to the local community's precariousness and lack of support, it has been challenging to sustain agroforestry practices because of the high cost of implementing agroecological activities. The lack of awareness among organizations and local governments regarding funding opportunities at national and regional levels (Schmidt and Walz 2021) created a gap and restricts access to financial resources needed for adaptation at the local level (Selje et al. 2024).

Centralized systems struggle to incorporate traditional knowledge and adapt to local ecological conditions. As Anderson et al. (2019) noted, participatory approaches that engage local communities are vital for aligning adaptation strategies with each area's unique vulnerabilities and strengths. Without such inclusivity, the effectiveness of EbA practices is significantly compromised. Sociocultural beliefs act as a brake on the innovation support process. It limits support for innovation and participation in EbA initiatives. There are also sociocultural beliefs that hinder the process of providing support for the promotion of EbA. It is the case of beliefs against agroforestry practices. Also, traditional gender roles limit women's participation in formal decision-making, which can restrict the adoption of EbA practices that might otherwise benefit from their involvement. The inclusion is linked to considering the different social components.

In the various EbA processes. The fact that the head of the household is the primary contact for the initiatives hinders the dissemination of knowledge. It is important to have specific activities for the other members. In the innovation niches, the stakeholders' initiatives (learning, experimentation) involve households. However,

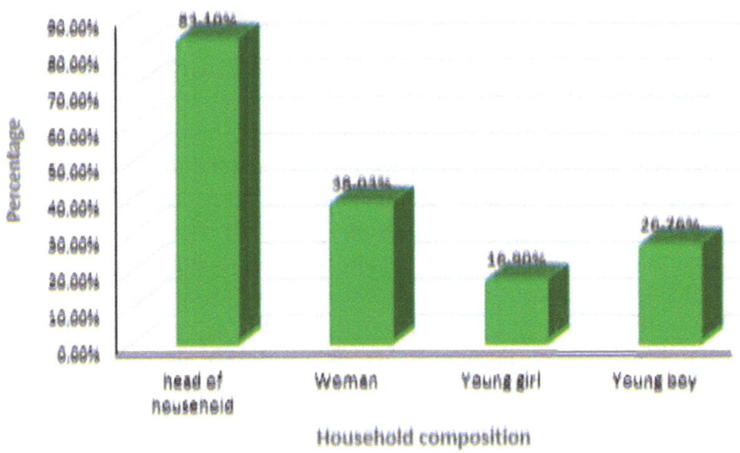

Fig. 19.9 Inclusion of different social components in the various EbA processes

heads of household are most involved in these initiatives (58%). There is also a low level of participation among young girls and boys (Fig. 19.9).

19.4.2.2 Coordination in Supporting Ecosystem-Based Adaptation Practices

Since different types of actors are involved in supporting ecosystem-based adaptation practices, the present study has addressed the territorial anchoring temporality of those actors. Ecosystem-based adaptation support is mainly based on four services provided by actors. The first is capacity building. This support includes the provision of formal training and experiential learning processes. The second is awareness-raising and knowledge exchange, which contribute to knowledge awareness and the dissemination of scientific or technical information. The third is advice, consultancy, and support activities aimed at solving problems and co-constructing solutions at local stakeholders' request. For example, support for conservation, improvement of the production process, and support for agroecological development. The last is the services to facilitate the acquisition of resources for the EbA innovation process (access to inputs, facilities, equipment, and financing). For example, financial support, distribution of seedlings, reforestation, and supply of conservation equipment.

As the ecosystem-based adaptation practices comprise stakeholders, it was necessary to examine this support over time in the innovation niche within the region. The territory aspect of the innovation niche becomes a crucial factor in determining the capacities of actors through the permanence of their interventions and territorial anchoring. The reason for emphasis on territorial anchoring is that it helps to explain

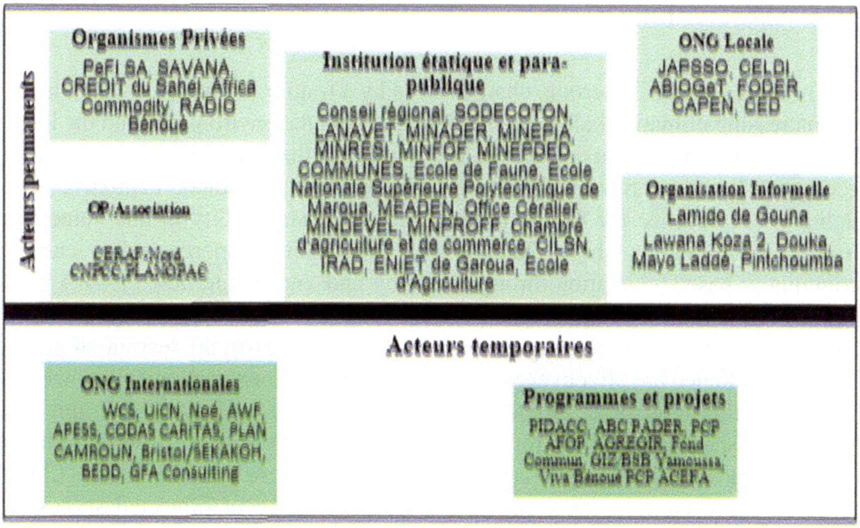

Fig. 19.10 Temporality of the actor in supporting EbA practices

the importance of the ifs and buts of spatial contexts. The anchoring and mobility coincide with the reality of actors in the North region of Cameroon region in supporting EbA practices. Some are anchored in the territory (permanent), while others are mobile (temporary).

After their interventions, some actors go elsewhere (outside the North region of Cameroon region) to continue their interventions, while others remain established in the region. Consequently, some variations in the territorial anchoring of actors' interventions in supporting EbA practices (Fig. 19.10). Territorial anchoring and mobility raise questions about the permanence of stakeholders and their EbA innovation support initiatives. In the case of the north region of Cameroon, innovation is nurtured through projects and programs. They have a limited time (1 to 5 years) to support innovation. At the end of projects and programs, the departure of actors to other areas of interest causes the problem of lacking the follow-up of EbA practices, the non-continuity of their support by new actors, their weak scaling up, and their very weak institutionalization. Therefore, the absence of effective intermediary services that can coordinate and link past and present stakeholders' actions on the regional territory contributes to the fragmentation of the support provided for EbA practices over time, making them unsustainable. For example, a fragmented agricultural service system with many competing service providers may leave ample room for innovation if strong coordination is not required. As was the case in England, the integration of synergies between actors in the North region of Cameroon region would make it possible to eliminate differences in the geographical coverage and content of EbA support services.

19.5 Discussion

The north region of Cameroon, characterized by a fragile ecological context, is facing increasing climatic challenges that threaten both the environment and the livelihoods of its inhabitants. This study explored the potential for promoting ecosystem-based adaptation (EbA) through the innovation niche approach, highlighting its relevance for fostering resilience and sustainability in this vulnerable region. The findings show that the Innovation niche approach has enabled community-based adaptation through learning and knowledge flow on EbA practices to limit climate change impact. The learning is focused on technical problem-solving, linked to climate change impact (Raven et al. 2010). So, learning has been shown to enhance the effectiveness of integrating traditional ecological knowledge with scientific approaches (Reed et al. 2021). It supported robust learning and knowledge-sharing processes for a transformative framework in fostering climate resilience and sustainability. Existing studies, such as those by Hekkert et al. (2007) and Schot and Geels (2008), emphasize the importance of protected spaces for experimenting with innovative solutions, engaging stakeholders, and refining interventions before broader application. Research in Sub-Saharan Africa, including Cameroon, demonstrates the utility of pilot community-driven agroecological practices that balance livelihood needs with environmental conservation (Tiani et al. 2015; Shikuku et al. 2021).

Considering engaging stakeholders as mentioned by Hekkert et al. (2007) and Schot and Geels (2008), diverse actors with multiple approaches operate in the innovation niche to support and foster EbA practices. Audouin et al. (2021) mention that innovation is supported by different stakeholders. However, the diversity of actors in the innovation niche is a function of their intervention's duration. In literature, the temporal dimension of actors and their intervention has been largely absent from studies on innovation ecosystems, or even innovation systems (Palmer et al. 2017). When temporality is considered, the focus is often on how ecosystems and networks evolve through a life cycle of phases and how EbA innovations can be nurtured and sustained over time in an innovation niche (Baker and Nenonen 2020).

Moreover, there is a need for coordination between service providers and other actors to fully support EbA innovation practices (Toillier et al. 2018). Its absence can lead to confusion among farmers who do not know where to find information on EbA practices. Also, without coordination among actors in the innovation niche, duplication of action and geographical imbalance in support provision to local organizations are observed (Garforth et al. 2003). This can be linked to the inadequate distribution of support services among stakeholders (Isaksen et al. 2018).

More studies are needed to better understand the long-term effects of INA-facilitated EbA on ecosystems and community resilience. Furthermore, looking at the socioeconomic aspects of EbA, such as gender dynamics and equality concerns, might provide a more complete picture of its advantages and drawbacks. The combination of EbA and INA is a revolutionary approach to climate adaptation, potentially strengthening some of the world's most vulnerable ecosystems and populations.

19.6 Conclusion

This study aimed to investigate the potential of ecosystem-based adaptation (EbA) for promoting climate resilience in vulnerable ecological situations using the innovation niche approach. The findings showed that the north region benefited from promotions of EbA techniques, notably through agroforestry and watershed management. These practices and techniques enable communities to adapt to climate unpredictability. However, multiple stakeholders, such as local communities, researchers, politicians, and nongovernmental organizations (NGOs), collaborate in developing adaptable solutions. Though the study identifies limitations such as favorable intervention environment, disparities in community engagement, and centrality in policy and decision-making. Nonetheless, the study highlights the adoption of EbA practices, which is mostly attributable to actor interaction. This research is important because it provides actionable insights into how natural systems might be used to build resilience in tackling the impact of climate change. It demonstrates the possibility of cost-effective, sustainable solutions to ameliorate climate impacts in resource-constrained locations by investigating its EbA application in fragile ecosystems. The innovation niche approach empowers local communities, particularly women and adolescents, by increasing their adaptive ability and creating sustainable livelihoods. Furthermore, by integrating multiple stakeholders in innovation niches, the research guarantees that adaptation methods are locally relevant and inclusive, thereby lowering inequities and strengthening community ownership. However, restrictions such as context-specific issues, reliance on transitory actors, resource limits, and knowledge gaps must be recognized. Addressing these constraints through long-term engagement, capacity building, and institutional strengthening is critical for increasing the efficacy and sustainability of EbA programs within INA frameworks.

Acknowledgements This work was carried out as part of the "Strengthening Innovation Systems in Northern Cameroon" (ReSI-NoC) project, a four-year project (2020–2024). This project is being implemented by a consortium of four partner organizations: World Agroforestry (ICRAF), the Agricultural Research Institute for Development (IRAD), the Center for International Cooperation in Agricultural Research for Development (CIRAD), and the Center for International Forestry Research (CIFOR). It was funded by the European Union (EU) under contract number FOOD/2020/416-105.

References

Agrinatura et FAO (2020) Partenariats de niche d'innovation: un guide du processus de coaching. CDAIS, Paris, and FAO, Rome, pp 32

Aminou A, Ka AA (2017) Impact de la technologie smart-valley pour l'aménagement des bas-fonds sur le revenu et le rendement des petits producteurs de riz au bénin. ISSN: 1840-703X, Cotonou (Bénin)

Anderson C, Johnson M et al (2019) Participatory approaches to local adaptation: engaging communities for sustainable solutions. Environ Policy Gov 29(4):214–226. https://doi.org/10.1002/eet.1892

Appiah-Adu K, Okpattah BK, Djokoto JG (2016) Technology transfer, outsourcing, capability and performance: a comparison of foreign and local firms in Ghana. Technol Soc 47:31–39. https://doi.org/10.1016/j.techsoc.2016.07.002

Audouin S, Dugué P, Randrianarisona N, Tim Ndah HT, Ratsimbazafy T, Andriamaniraka H, Noharin-janaharya ES, Ralisoa N et al (2021) Quelle place du conseil agricole dans les services support à l'innovation à Madagascar ? Cah Agric 30:29. https://doi.org/10.1051/cagri/2021017

Awazi NP, Ambebe TF, Ngwabie NM, Fonteh MF (2023) Droughts and desertification in the era of anthropogenic climate change: manifestations, impacts, and nature-based solutions in the Guinea high Savannah and Sudano-Sahelian regions of Cameroon. In: Handbook of nature-based solutions to mitigation and adaptation to climate change. Springer International Publishing, pp 1–18. https://doi.org/10.1007/978-3-030-73200-3_50

Baker JJ, Nenonen S (2020) Collaborating to shape markets: emergent collective market work. Ind Mark Manag 85:240–253. https://doi.org/10.1016/j.indmarman.2019.12.006

Bas TG, Gagnon J, Gagnon P, Contreras A (2022) Analysis of agro alternatives to boost Cameroon's socio-environmental resilience, sustainable development, and conservation of native forests. Sustainability 14(14):8507. https://doi.org/10.3390/su14148507

Calliari E, Castellari S, Davis M, Linnerooth-Bayer J, Martin J, Mysiak J et al (2022) Building climate resilience through nature-based solutions in Europe: a review of enabling knowledge, finance and governance frameworks. Clim Risk Manag 37:100450. https://doi.org/10.1016/j.crm.2022.100450

Chomba S, Sinclair F, Savadogo P, Bourne M, Lohbeck M (2020) Opportunities and constraints for using farmer-managed natural regeneration for land restoration in sub-Saharan Africa. Front Forests Global Change 3:571679. https://doi.org/10.3389/ffgc.2020.571679

CIFOR (2023) Innovation niches approach fosters sustainable agrosilvopastoral efforts in Northern Cameroon. Retrieved from https://forestsnews.cifor.org

Cortinovis C, Olsson P, Boke-Olén N, Hedlund K (2022) Scaling up nature-based solutions for climate-change adaptation: potential and benefits in three European cities. Urban For Urban Green 67:127450. https://doi.org/10.1016/j.ufug.2022.127450

Creswell JW, Plano Clark VL (2018) Designing and conducting mixed methods research, 3rd edn, Sage

Danielsen F et al (2022) Community monitoring of natural resource systems and the environment. Annu Rev Environ Resour 47(1):637–670. https://doi.org/10.1146/annurev-environ-112421-023116

Delphine BNA, Dial N, Mahama G, Ibrahima O, Bassirou B, Vanessa ENT (2019) Adaptation and resistance of smart valleys bunds in Sudanian's lowland: experience in four lowlands in Burkina Faso. Am J Water Resour 7(2):58–61. https://doi.org/10.12691/ajwr-7-2-2

Djiongo C, Mpong SM, Monga O (2015) Estimation of aboveground biomass from satellite data using quaternion-based texture analysis of multi-chromatic images. In: In 2015 11th international conference on signal-image Technology & Internet-Based Systems (SITIS). IEEE, pp 68–75. https://doi.org/10.1109/SITIS.2015.15

Djombaye B (2005) Normes de construction des biefs, évaluation des effets et estimation d'impacts dans les bas-fonds de Kaélé, Province de l'Extrême-Nord Cameroun [Master's thesis, Université de Dschang]. Université de Dschang

Domecq JP, Prutsky G, Elraiyah T, Wang Z, Nabhan M, Shippee N et al (2014) Patient engagement in research: a systematic review. BMC Health Serv Res 14:1–9. https://doi.org/10.1186/s12913-014-0509-5

Doswald N, Sánchez Torrente L, Reumann A, Leppert G, Moull K, Rocío Pérez JJ et al (2020) Evidence gap and intervention heat maps of climate change adaptation in low- and middle-income countries. DEval Discussion Papers, 02/2020. Deutsches Evaluierungsinstitut der Entwicklungszusammenarbeit (DEval). https://nbn-resolving.org/urn:nbn:de:0168-ssoar-71447-8

Dunwoodie K, Macaulay L, Newman A (2023) Qualitative interviewing in the field of work and organisational psychology: benefits, challenges and guidelines for researchers and reviewers. Appl Psychol 72(2):863–889. https://doi.org/10.1111/apps.12377

Faure G, Knierim A, Koutsouris A, Ndah HT, Audouin S, Zarokosta E, Wielinga E, Triomphe B, Mathé S, Temple L, Heanue K (2019) How to strengthen innovation support services in agriculture with regard to multi-stakeholder approaches. J Innov Econ Manag 28(1):145–169. https://doi.org/10.3917/jie.028.0145

Fearnside PM (2005) Deforestation in Brazilian Amazonia: history, rates, and consequences. Conserv Biol 19(3):680–688. https://doi.org/10.1111/j.1523-1739.2005.00646.x

Fedreheim GE, Blanco E (2017) Co-management of protected areas to alleviate conservation conflicts: Experiences in Norway. Int J Coms 11(2). https://doi.org/10.18352/ijc.749

Fraenkel JR, Wallen NE (2009) The nature of qualitative research. In: How to design and evaluate research in education. McGraw-Hill, pp 420–438

Fofiri Nzossie EJ, Ndame JP, Temple L, Simeu Kamdem M (2016) L'innovation agricole dansla zone soudano-sahélienne du Cameroun : acteurs et politique d'intervention. In Pour unegéographie rurale de l'action. (Eds) Ngo Balepa ASF, Moupou M, Mougoue B., Nji Fogwe Z,Tchawa P. Edition Clé Yaoundé, pp 411–425

Garforth C, Angell B, Archer J, Green K (2003) Fragmentation or creative diversity? Options in the provision of land management advisory services. Land Use Policy 20(4):323–333. https://doi.org/10.1016/S0264-8377(03)00035-8

Geels F (2004) From sectoral systems of innovation to socio-technical systems. Res Policy 33(6–7):897–920. https://doi.org/10.1016/j.respol.2004.01.015

Geels FW (2006) Multi-level perspective on system innovation: relevance for industrial transformation. In: Understanding industrial transformation: views from different disciplines. Springer, Dordrecht, pp 163–186

Geels FW, Sovacool BK, Schwanen T, Sorrell S (2020) Sociotechnical transitions for deep decarbonization. Science 357(6485):1242–1244

Geneletti D, Zardo L, Cortinovis C (2016) Promoting nature-based solutions for climate adaptation in cities through impact assessment. In: Handbook on biodiversity and ecosystem services in impact assessment. Edward Elgar Publishing, pp 428–452

Hekkert MP, Suurs RAA, Negro SO, Kuhlmann S, Smits REHM (2007) Functions of innovation systems: a new approach for analysing technological change. Technol Forecast Soc Chang 74(4):413–432. https://doi.org/10.1016/j.techfore.2006.03.002

Ibrahim AH (2024) Decentralization and its impact on improving public services. Int J Social Sci 7(2):45–53

IPCC (2013) Abstract for Policymakers. Climate Change (2013) The physical science basis. In: Stocker TF, Qin D, Plattner G-K, Tignor M, Allen SK, Boschung J (eds) Contribution of working Group I to the fifth assessment report of the intergovernmental panel on climate change. University Press, Cambridge

Isaksen A, Tödtling F, Trippl M (2018) Innovation policies for regional structural change: combining actor-based and system-based strategies. In: Innovation and regional development. Springer International Publishing, pp 221–238. https://doi.org/10.1007/978-3-319-74429-0_12

Jayasekara RS (2012) Focus groups on nursing research: methodological perspectives. Nurs Outlook 60(6):411–416. https://doi.org/10.1016/j.outlook.2012.05.004

Klein RJ, Huq S, Denton F, Downing TE, Richels RG, Robinson JB, Toth FL (2007) Inter-relationships between adaptation and mitigation. In: Climate change 2007: impacts, adaptation and vulnerability. Contribution of working group II to the fourth assessment report of the inter-governmental panel on climate change. Cambridge University Press, pp 745–777

Kitzinger J (1995) Introducing focus groups. Br Med J 311:299–302. https://doi.org/10.1136/bmj.311.7000.299

Kroeger ME, Delmont TO, Eren AM, Meyer KM, Guo J, Khan K, Nüsslein K (2018) New biological insights into how deforestation in Amazonia affects soil microbial communities using metagenomics and metagenome-assembled genomes. Frontiers in microbiology 9:1635. https://doi.org/10.3389/fmicb.2018.01635

Laurance WF, Sayer J, Cassman KG (2013) Agricultural expansion and its impacts on tropical nature. Trends Ecol Evol 29(2):107–116. https://doi.org/10.1016/j.tree.2013.12.001

Li TC, Villarrubia G, Sun SD, Corchado JM, Bajo J (2015) Resampling methods for particle filtering: identical distribution, a new method, and comparable study. Front Inf Technol Electron Eng 16(11):969–984. https://doi.org/10.1631/FITEE.1500064

Locatelli B, Pavageau C, Pramova E, Di Gregorio M (2021) Climate change adaptation through watershed management and agroforestry: evidence from developing countries. Curr Opin Environ Sustain 15:1–8. https://doi.org/10.1016/j.cosust.2021.05.003

Lombe P, Carvalho E, Rosa-Santos P (2024) Drought dynamics in Sub-Saharan Africa: impacts and adaptation strategies. Sustainability 16(22):9902. https://doi.org/10.3390/su16229902

Longobardi A, Buttafuoco G, Caloiero T, Coscarelli R (2016) Spatial and temporal distribution of precipitation in a Mediterranean area (southern Italy). Environ Earth Sci 75:1–20. https://doi.org/10.1007/s12665-016-5562-6

Lundvall BA (1992) National systems of innovation: Towards a theory of innovation and interactive learning. Francis Printer. (DOI and ISSN not applicable; this is a book)

Ly M, Traore SB, Alhassane A, Sarr B (2013) Evolution of some observed climate extremes in the West African Sahel. Weather Clim Extr 1:19–25. https://doi.org/10.1016/j.wace.2013.07.005

Mathé S et al (2016) Diversité des trajectoires vers l'agriculture biologique dans les pays en développement: Le cas du Cameroun. 10ème journée de recherche en sciences sociales à Paris la Défense (IESIEG), 8 et 9 Déc. 2016

Mathé S, Audouin S, Fongang G, Traoré O, Gerster-Bentaya M, Knierim A, Randrianarison N, Ndah HT, Toillier A (2019) Designing frameworks for characterizing and assessing innovation support services and innovation support providers: SERVInnov project. ESEE 2019, Acireale, p 26. Retrieved from http://agritrop.cirad.fr/593370/7/Math%C3%A9%20et%20al_Designing%20framework_ESEE2019_20190601_VF.pdf

Matos S, Viardot E, Sovacool BK, Geels FW, Xiong Y (2022) Innovation and climate change: a review and introduction to the special issue. Technovation 117:102612. https://doi.org/10.1016/j.technovation.2021.102612

Medina E, Klinge H, Jordan C, Herrera R (1980) Soil respiration in Amazonian rain forests in the Rio Negro basin. Flora 170(3):240–259. https://doi.org/10.1016/S0367-2530(17)31209-4

MINEPDED (2015) Plan national d'adaptation aux changements climatiques pour le Cameroun (PNACC)-Cameroun. Draft, Yaoundé

Muluneh MG (2021) Impact of climate change on biodiversity and food security: a global perspective EN a review article. Agric Food Secur 10(1):1–25. https://doi.org/10.1186/s40066-021-00280-7

Ndah HT, Knierim A., Randrianarison N., Mathé S., Audouin S., Toillier A., Crestin-Billet S., Soule Adam N., Noharinjanahary E. S., Traoré O., Michel T., Temple L., Tiatite N., Goumbri JDD., Gerster-Bentaya M., Fongang G., Andriamaniraka J. H., Andriamparany R. R (2020) Co designed Methodological Fram-work and Guidelines for in-depth Case Study Analysis, SERVInnov project, Deliverable 1.3, Universität Ho-henheim, Stuttgart, Germany

Nelson RR, Rosenberg N (1993) Technical innovation and national systems. In: Nelson RR (ed) National innovation systems: a comparative analysis. Oxford University Press, pp 1–18

Ngondjeb YD (2013) Agriculture and climate change in Cameroon: an assessment of impacts and adaptation options. Afr J Sci Technol Innov Dev 5(1):85–94. https://doi.org/10.1080/20421338.2013.782147

Nguyen X, Tran Y, Le Z (2020) The impact of centralized control on rural development: evidence from Southeast Asia. J Rural Stud 38(2):45–56. https://doi.org/10.1016/j.jrurstud.2020.02.003

Nyathi D, Ndlovu J, Hadebe S (2024) Challenges of implementing urban ecosystem-based adaptation to climate change in developing countries: a systematic review. In: Ghosh S, Majumdar S, Cheshmehzangi A (eds) Cities of tomorrow: urban resilience and climate change preparedness. Urban sustainability. Springer, Singapore. https://doi.org/10.1007/978-981-97-9658-8_13

Palmer M, Medway D, Warnaby G (2017) Theorizing temporary spatial clusters and institutional boundary-work in industrial marketing. Ind Mark Manag 61:104–113. https://doi.org/10.1016/j.indmarman.2016.08.010

Palou OM (2011) Déterminants socioéconomiques de la gestion paysanne des plantations d'Acacia senegal et de la production de la gomme arabique au Nord Cameroun. Thèse en vue de l'obtention du doctorat à l'Université de Toulouse. Spécialité : Ecologie végétale, p 275

Patel A, Sharma K (2017) Marginalization of vulnerable regions under centralized governance: a case study of agricultural economies. Int J Dev Stud 24(3):122–135. https://doi.org/10.1080/02589001.2017.1348520

Petzold J, Hawxwell T, Jantke K, Gonçalves Gresse E, Mirbach C, Ajibade I, Garschagen M (2023) A global assessment of actors and their roles in climate change adaptation. Nature climate change 13(11):1250–1257. https://doi.org/10.1038/s41558-023-01824-z

Pougoue EBS, Tchuenga TG, Mathé S, Degrande A, Ijang PTN, Kemajou SHB (2024) Climate change and integrated farmland Management in Northern Cameroon: highlighting pro-accessibility and pro-restoration strategies: climate change and land management strategies. Afr J Land Policy Geospat Sci 7(6):1700–1723

Raven RPJM, Van den Bosch S, Weterings R (2010) Transitions and strategic niche management: towards a competence kit for practitioners. Int J Technol Manag 51(1):57–74

Rawlins J, Monteith S, Kalaba FK, Clements HS (2023) Ecosystem-based adaptation in Africa: integrating mitigation and adaptation. Reg Environ Chang 23(1):46. https://doi.org/10.1007/s10113-023-01984-8

Reed MS, Duncan S, Manners P et al (2021) The role of stakeholder collaboration in scaling up ecosystem-based adaptation. Sustain Sci 16(2):587–598. https://doi.org/10.1007/s11625-020-00859-7

Rodima-Taylor D (2012) Social innovation and climate adaptation: local collective action in diversifying Tanzania. Appl Geogr 33:128–134. https://doi.org/10.1016/j.apgeog.2011.10.005

Saha F (2019) Adaptation aux risques naturels et incertitudes climatiques en milieu soudano-sahélien au Cameroun. NAAJ Revue africaine sur les changements climatiques et les énergies renouvelables 1(1):157–173. https://doi.org/10.46711/naaj.2019.1.1.8

Saha F, Tchindjang M, Dzana JG, Nguemadjita D (2020) Risques naturels dans la région de l'Extrême-Nord du Cameroun et dynamique des extrêmes hydrologiques du système Chari-Logone. Physio-Géo. Géographie physique et environnement, (Volume 15):69–88

Savoie-Zajc L, Karsenti T (2004) La méthodologie. In T. Karsenti et L. Savoie-Zajc (dir.), La recherche en éducation : étapes et approches. Éditions du CRP, Sherbrooke, pp 109–121

Sawadogo HP (2021) Saturation, triangulation et catégorisation des données collectées. Guide décolonisé et pluriversel de formation à la recherche en sciences sociales et humaines

Schmidt K, Walz A (2021) Ecosystem-based adaptation to climate change through residential urban green structures: co-benefits to thermal comfort, biodiversity, carbon storage and social interaction. One Ecosyst 6:e65706. https://doi.org/10.3897/oneeco.6.e65706

Schot J, Geels FW (2008) Strategic niche management and sustainable innovation journeys: theory, findings, research agenda, and policy. Tech Anal Strat Manag 20(5):537–554. https://doi.org/10.1080/09537320802292651

Selje T, Schmid LA, Heinz B (2024) Community-based adaptation to climate change: Core issues and implications for practical implementations. Climate 12(10):2225–1154

Shikuku KM, Thornton PK, Förch W, Winowiecki L, Herrero M (2021) Exploring the socio-economic and environmental impacts of agroecological innovations in Sub-Saharan Africa. Agric Syst 191:103141. https://doi.org/10.1016/j.agsy.2021.103141

Siéwé Pougoué EB et al (2020) Production d'Œufs et Bien-être des Aviculteurs au Sud du Bénin. Revue Africaine d'Environnement et d'Agriculture 3(4):37–48

TAP-Plate-forme pour l'agriculture tropicale (2017) Cadre commun de travail pour le renforcement des capacités des systèmes d'innovation agricole : Fondements conceptuels. CAB International, Walling-ford, Royaume-Uni

Temple L, Machicou Ndzesop N, Fongang Fouepe GH, Ndoumbe Nkeng M, Mathé S (2017) Système National de Recherche et d'Innovation en Afrique: le cas du Cameroun 1. Innovations 2:41–67. https://doi.org/10.3917/inno.053.0041

Tchuenga GTS, Tchindjang M, Ngome PIT, Degrande A, Basga SD, Saha F (2024) Agricultural Innovations and Adaptations to Climate Change in the Northern Cameroon Region. Sustainability 16(22):10096

Tiani AM, Besa MC, Devisscher T, Pavageau C, Butterfield R, Bharwani S, Bele MY (2015) Assessing current social vulnerability to climate change: a participatory methodology, p 32. CIFOR Working Paper No. 169. Bogor, Indonesia: Center for International Forestry Research (CIFOR). https://doi.org/10.17528/cifor/

Toillier A, de Bellaire LDL (2016) Contributions de la recherche à l'innovation dans les filières. In Développement durable et filières tropicales (Chapitre 7). Éditions Quæ, pp 95–108

Toillier A, Faure G, Chia E (2018) Designing and organizing support for collective innovation in agriculture. In : Innovation and development in agricultural and food systems. Faure Guy (ed.), Chiffoleau Yuna (ed.), Goulet Frédéric (ed.), Temple Ludovic (ed.), Touzard Jean-Marc (ed.). Versailles : Ed. Quae, pp 108–121 (Synthèses : Quae)

Toillier A, Kola PN, Traore O, Noufé T (2020) Analysis and evaluation of selected case studies: methodological guide for Burkina WP3 (draft). CIRAD, CEDRES, Burkina Faso

Tolimir M, Kresović B, Životić L, Dragović S, Dragović R, Sredojević Z, Gajić B (2020) The conversion of forestland into agricultural land without appropriate measures to conserve SOM leads to the degradation of physical and rheological soil properties. Sci Rep 10:13668. https://doi.org/10.1038/s41598-020-70464-6

Touzard JM, Temple L, Faure G, Triomphe B (2015) Innovation systems and knowledge communities in the agriculture and agrifood sector: a literature review. J Innov Econ Manag 17(2):117–142

Van der Werf GR, Morton DC, DeFries RS, Olivier JGJ, Kasibhatla PS, Jackson RB, Collatz GJ, Randerson JT (2009) CO2 emissions from forest loss. Nat Geosci 2:737–738. https://doi.org/10.1038/ngeo671

Vasseur L (2021) How ecosystem-based adaptation to climate change can help coastal communities through a participatory approach. Sustainability 13(4):2344

Wolf ID, Sobhani P, Esmaeilzadeh H (2023) Assessing changes in land use/land cover and ecological risk to conserve protected areas in urban–rural contexts. Land 12(1):231. https://doi.org/10.3390/land12010231

World Bank and WWF (2014) Operational framework for ecosystem-based adaptation implementing and mainstreaming ecosystem-based adaptation responses in the greater mekong sub region, p 88. https://awsassets.panda.org/downloads/wwf_wb_eba_project_2014_gms_ecosystem_based_adaptation_general_framework.pdf

Zeratsion BT, Manaye A, Gufi Y, Tesfaye M, Werku A, Anjulo A (2024) Agroforestry practices for climate change adaptation and livelihood resilience in drylands of Ethiopia. For Sci Technol 20(1):47–57

Chapter 20
Soil and Water Conservation Technologies for Ecosystem Restoration in Drylands

Stephen Mureithi, Samuel Mwendwa, and Yonathan Ayalew

Abstract Drylands cover over 41% of the earth's surface and are home to over 2 billion people, primarily in developing nations. Land degradation in these regions stands as one of the most significant environmental challenge of the twenty-first century, affecting grazing resources, food and nutrition security, water availability, biodiversity, and overall environmental health. This degradation is driven by a combination of natural factors, such as climate variability and human activities, including overgrazing and deforestation. The socioeconomic implications are profound, exacerbating poverty and food insecurity while undermining sustainable development. The urgency of addressing land degradation is highlighted by the United Nations' Decade on Ecosystem Restoration (2021–2030), emphasizing the need for effective restoration strategies in dryland ecosystems. We reviewed various soil and water conservation technologies aimed at ecosystem restoration in drylands, highlighting several key areas of focus. Active involvement of local communities in restoration efforts has proven essential for enhancing the sustainability and effectiveness of conservation practices, as communities that take ownership of these initiatives are more likely to engage in and support long-term activities. Additionally, the implementation of integrated watershed management strategies has shown significant promise in enhancing water retention and availability in targeted areas, which is critical for boosting agricultural productivity and ensuring that local livestock keepers and farmers can sustain their livelihoods. Furthermore, restoration practices that incorporate the replanting of native vegetation have led to notable increases in local biodiversity metrics, crucial for building ecosystem resilience, as diverse ecosystems are better equipped to withstand environmental changes and stressors. Lastly, the adoption of these technologies has contributed to improvements

S. Mureithi (✉) · S. Mwendwa
Department of Land Resource Management and Agricultural Technology (LARMAT), University of Nairobi, Nairobi, Kenya
e-mail: stemureithi@uonbi.ac.ke

Y. Ayalew
World Food Programme (WFP) Climate and Resilience Service, Rome, Italy

Y. D. I. Khan et al. (eds.), *Land Restoration Through Ecosystem-Based Approach*, Disaster Resilience and Green Growth, https://doi.org/10.1007/978-981-95-0389-6_20

in agricultural productivity, thereby supporting food security and enhancing the livelihoods of communities reliant on these fragile ecosystems. Overall, our analysis underscores the importance of combining traditional knowledge with modern conservation techniques, as this integrated approach can significantly mitigate land degradation and bolster ecosystem resilience, paving the way for sustainable development in dryland regions.

Keywords Drylands · Land degradation · Ecosystem restoration · Watershed management · Traditional knowledge

Nomenclature

Drylands	Regions characterized by arid or semi-arid conditions with limited water availability.
Ecosystem restoration	Assisting in the recovery of ecosystems that have been degraded or destroyed, as well as conserving intact ecosystems.
Land degradation	Deterioration of land quality, often resulting from human activities such as deforestation or overgrazing.
Resilience	Ability of ecosystems to withstand and recover from disturbances.
Soil and water conservation	Practices designed to prevent soil erosion, enhance soil fertility, and maintain water quality, aiming to sustainably manage land and water resources, ensuring their availability for future generations while promoting ecosystem health.
Water harvesting	Collection and storage of rainwater for beneficial use.

20.1 Introduction

Drylands cover over 41% of the Earth's land surface and sustain more than 2 billion people, mostly in developing regions, and are faced with major environmental challenges. For instance, Bristol-Alagbariya (2023) highlights the UN Convention to Combat Desertification's role in promoting sustainable land management in drylands; Clegg et al. (2024) discuss how innovative space technologies can aid in achieving sustainable development goals in drylands; and Naorem (2023) introduces the unique challenges of drylands and the need for resilience-building strategies in agriculture. Chief among these challenges is the pervasive issue of land degradation, which imperils not only local ecosystems and communities but also global efforts toward sustainable development. The United Nations' declaration of the Decade on Ecosystem Restoration 2021–2030 underscores the urgency of addressing this pressing concern, emphasizing the need to rehabilitate degraded landscapes while safeguarding those still relatively intact (von Holle et al. 2020; Ament et al. 2023).

The impacts of land degradation in drylands are particularly acute as also reported by de Souza Oliveira Filho et al. (2019) who assesses the benefits of grazing exclusion on soil fertility in semi-arid regions; Jiang et al. (2024) who identifies global hotspots of land degradation needing targeted interventions; and Wang et al. (2023) who examines trends in land cover change in drylands and their ecological impacts. Therefore, the interplay between soil and water conservation technologies and vegetation planting works assumes paramount importance in the process of ecosystem restoration. In the context of research for development, such technologies can also be referred to as livelihood assets and the act of installing them as livelihood asset creation. Despite advancements in research and implementation of livelihood assets in drylands, significant gaps persist in our understanding and application of effective strategies for mitigating environmental decline and fostering resilience in these vulnerable regions.

Land degradation is driven by a combination of natural and human-induced factors, as also identified by Wang et al. (2024a, b), who discuss strategies to improve soil health by optimizing the plant-soil-miocrobe-anthropogenic activity nexus, and Yan et al. (2024) on combating land degradation through human efforts: ongoing challenges for sustainable development of global drylands. Natural hazards, such as steep slopes, landslides, floods, high winds, and intense rainfall, contribute to degradation, as also observed by Badapalli et al. 2023a, b, studies that explore the visual and ecological consequences of desertification in semi-arid regions, and causes and mitigation strategies for land degradation and desertification, respectively. Sarma et al. (2024) also discuss the impact of natural hazards on land degradation exacerbated by climate change, while Weeraratna 2022b identifies the key factors contributing to land degradation for effective management, exacerbated by climate change effects. Human activities, including agricultural practices, deforestation, overgrazing, and improper soil and water management, intensify degradation (Khan et al. 2024). Additionally, poverty forces rural households to rely on small-scale farming, increasing pressure on arable land. Population growth further strains resources, highlighting the complex interplay of socio-economic, environmental and climatic factors.

The consequences of land degradation in drylands pose a severe threat to biodiversity, water availability, food security, and environmental quality (Habtu 2024), disproportionately impacting the livelihoods of millions of people who depend on these fragile ecosystems. Addressing this issue is justified on two fronts: first, the socioeconomic implications of land degradation in drylands are profound, exacerbating poverty, feed, food, and nutrition insecurity while undermining efforts to achieve sustainable development goals. Second, the ecological and hydrological significance of dryland ecosystems extends far beyond their immediate boundaries, with implications for global climate stability, biodiversity conservation, and provision of ecosystem services.

The effectiveness of soil and water conservation asset creation activities hinges on robust design and participatory planning processes that are deeply rooted in local contexts and dynamics. Well-prepared community-based participatory plans and quality interventions are more likely to yield tangible benefits, gain community support, ensure asset maintenance, encourage replication, and attract government attention for widespread adoption (Sahal and Tilahun, 2023). Conversely, participatory planning that lacks strong technical knowledge and understanding can lose its effectiveness. Therefore, the planning and execution of soil and water conservation activities must adhere to strict technical standards, considering factors, such as rainfall patterns, soil types (Mureithi et al. 2024a), land use, and slope. Activities should be integrated with other components, such as access to markets, financial services, and value chains (particularly for fodder and livestock), while thoroughly mainstreaming gender and protection considerations. Finally, these initiatives should be implemented at the catchment scale to achieve sufficient impact and yield multiplier effects, thereby strengthening the sustainability of interventions.

Against this backdrop, this chapter critically examines the role of soil and water conservation technologies in promoting ecosystem restoration and resilience in drylands. By synthesizing existing research findings, identifying key research gaps, and proposing ideas for future investigation, our goal is to enhance the understanding of the complex dynamics shaping dryland ecosystems and inform evidence-based interventions for sustainable land management. Furthermore, we explore innovative approaches and case studies from diverse dryland restoration settings to elucidate the potential of soil and water conservation measures in mitigating land degradation, enhancing ecosystem services, and improving the livelihoods of communities reliant on these landscapes. Ultimately, our goal is to catalyze action and inspire collaboration among researchers, policymakers, and practitioners to advance watershed resilience and sustainable development in arid regions.

20.2 Land Degradation

Land degradation is characterized by the progressive deterioration of the land's productive potential across diverse landscapes, encompassing rain-fed areas, arable land, irrigated zones, rangelands, and forests, each crucial in sustaining various ecosystems and human activities (Drobitko et al. 2023; Gunawardena 2022; Jiang et al. 2024; Mureithi et al. 2014). This degradation extends beyond mere loss of fertility or erosion, encompassing the intricate interplay between environmental factors and human interventions that compromise the land's ability to support life and provide essential resources. From smallholder subsistence farming to large-scale agricultural operations, the complex web of land use practices intertwines with degradation, accentuating its significance in shaping farming systems and livelihoods. Moreover, the economic ramifications of land degradation reverberate widely (Pei et al. 2024), as diminished land productivity translates into decreased agricultural yields, reduced biodiversity, increased vulnerability to natural disasters, and heightened socio-economic disparities. Consequently, recognizing the intrinsic linkages between land degradation and its impacts on land utilization serves as a cornerstone in the formulation of effective strategies aimed at mitigating environmental degradation, enhancing land resilience, and fostering sustainable development pathways for present and future generations. This section examines the effects of land degradation on ecosystem structure, function, and services provision, highlighting the cascading consequences for biodiversity, soil fertility, water resources, and climate regulation.

20.2.1 Biodiversity Decline

Land degradation poses a significant threat to biodiversity (Jiang et al. 2024) by altering habitat quality, reducing ecosystem resilience, and diminishing the abundance and diversity of plant and animal species. As vegetation cover declines and ecosystems become fragmented and degraded, species populations decline, habitats become increasingly fragmented, and ecological interactions are disrupted. The decline in biodiversity not only disrupts ecosystem functioning but also jeopardizes the delivery of crucial ecosystem services, including seed dispersal, pollination, and pest control, resulting in significant consequences for both ecosystem resilience and human well-being. Furthermore, the loss of biodiversity can lead to a reduction in genetic diversity, which is essential for species adaptability to changing environmental conditions. This genetic erosion limits the ability of species to respond to stressors, such as climate change, disease, and habitat loss, ultimately increasing the risk of extinction for vulnerable species. Moreover, the decline in biodiversity can have cascading effects on food webs and ecosystem stability, as the loss of keystone species can disrupt trophic interactions and lead to further declines in other species (Ripple et al. 2024). Therefore, preserving biodiversity is not only vital for maintaining ecosystem integrity but also for ensuring the resilience of ecosystems in the face of environmental changes.

20.2.2 Soil Degradation and Fertility Loss

Land degradation contributes to soil degradation and fertility loss (Weeraratna 2022a), through processes such as erosion, compaction, salinization, and nutrient depletion. As fertile topsoil is eroded, soil structure deteriorates, nutrient cycling is disrupted, and soil fertility declines, leading to reduced agricultural productivity and increased vulnerability to drought and other environmental stressors. Moreover, compaction and salinization further degrade soil quality, inhibiting plant growth and compromising the ability of ecosystems to support diverse and productive plant communities. The implications of soil degradation extend beyond immediate agricultural impacts; they also affect food security and rural livelihoods, as farmers face declining yields and increased costs for fertilizers and soil amendments (Gomiero 2016). Additionally, degraded soils are less capable of retaining water, exacerbating drought conditions and leading to increased reliance on irrigation, which can further strain local water resources. This cycle of degradation not only threatens agricultural productivity but also undermines the resilience of farming systems, making it imperative to implement sustainable soil management practices (Gunawardena 2022). Strategies, such as crop rotation, cover cropping, and organic amendments, can help restore soil health, improve fertility, and enhance the capacity of soils to support diverse plant communities.

20.2.3 Water Scarcity and Quality Decline

Land degradation exacerbates water scarcity and quality decline (Ghosh and Mandal 2023) by reducing water infiltration, increasing surface runoff, and impairing water storage and purification capacity. As vegetation cover declines and soil structure deteriorates, water infiltration rates diminish, leading to increased runoff and erosion. This not only exacerbates soil erosion but also reduces groundwater recharge, exacerbating water scarcity in affected areas. Additionally, sediment runoff and nutrient leaching further degrade water quality, impairing aquatic habitats, reducing biodiversity, and compromising human health and livelihoods. The deterioration of water quality poses significant risks to both ecosystems and communities, as polluted water can lead to health issues and reduce the availability of clean drinking water. Furthermore, the combined effects of water scarcity and quality decline can lead to increased competition for water resources, heightening tensions among communities and exacerbating social conflicts. Effective management strategies, such as implementing watershed management practices and restoring riparian buffers, are essential for improving water quality and enhancing water availability, as also recommended by Stutter et al. (2019). By addressing the root causes of land degradation, these strategies can help mitigate water scarcity and promote sustainable water resource management in affected regions.

20.2.4 Climate Change Amplification

Land degradation contributes to climate change (Nguyen et al. 2023) by releasing stored carbon from degraded soils, reducing carbon sequestration capacity, and altering regional climate patterns. As soils degrade and vegetation cover declines, the capacity of ecosystems to sequester carbon diminishes, leading to the release of carbon dioxide and other greenhouse gases into the atmosphere. Additionally, changes in surface albedo, evapotranspiration rates, and atmospheric moisture fluxes associated with land degradation can further exacerbate climate change by altering regional climate patterns, amplifying temperature extremes, and disrupting precipitation regimes, with cascading effects on ecosystem dynamics and human societies. The feedback loop created by land degradation and climate change poses significant challenges for global efforts to mitigate climate impacts, as degraded landscapes are often less resilient to extreme weather events, leading to increased vulnerability for local communities. Moreover, the loss of carbon-rich soils not only contributes to atmospheric greenhouse gas concentrations but also diminishes the potential for natural climate solutions that rely on healthy ecosystems for carbon sequestration. Therefore, integrating land restoration and sustainable land management practices into climate change mitigation strategies is critical for addressing both land degradation and climate change simultaneously, similar to the recommendations of Rapiya et al. (2024). Initiatives, such as afforestation, reforestation, and sustainable agricultural practices, can enhance carbon storage, improve ecosystem resilience, and contribute to global climate goals.

20.2.5 Desertification: Land Degradation in the Drylands

Desertification, defined as the process of land degradation in arid, semi-arid, and dry sub-humid areas (AbdelRahman 2023; Roy et al. 2024) resulting from various factors, including climatic variations and human activities, represents a significant environmental challenge with far-reaching socio-economic implications. This section delves into the multifaceted dimensions of desertification, examining its causes, manifestations, impacts, and mitigation strategies within the context of dryland ecosystems.

20.2.5.1 Causes of Desertification

Desertification is driven by a complex interplay of natural and anthropogenic factors (Rodrigues do Nascimento 2023; Oswald and Harris 2023), which can act synergistically to exacerbate land degradation processes. Primary causes include climate variability and change (Ahmed et al. 2024), where dryland regions are inherently vulnerable to climatic fluctuations, such as prolonged droughts, erratic rainfall patterns, and increased temperatures, accelerating soil erosion, reducing vegetation cover, and compromising ecosystem resilience. Additionally, land use and land cover changes (Ontel et al. 2023) stemming from unsustainable practices, such as continuous overstocking leading to overgrazing in the rangelands, which is especially severe around watering points, deforestation, inappropriate agricultural methods, and uncontrolled urbanization, further degrade soil fertility, deplete natural vegetation, and disrupt hydrological cycles, thereby leading to land degradation. Soil degradation (Mureithi et al. 2024b), marked by erosion, salinization, compaction, and nutrient depletion, contributes significantly to the degradation of drylands, rendering the land less productive and more susceptible to desertification processes. Furthermore, water scarcity and mismanagement of available water (Abdulhafedh

Fig. 20.1 Severe rangeland degradation at the Jerusalem Catchment in Wamba, Samburu County in northern Kenya. (Photo: S. Mureithi, Oct. 2023)

2024) exacerbate desertification by limiting water resources, thereby reducing soil moisture, impeding vegetation growth, and heightening competition for water resources among different land uses and stakeholders. These interrelated factors highlight the urgency of addressing desertification through comprehensive strategies that address both natural and human-induced drivers of land degradation.

20.2.5.2 Manifestations of Desertification

Desertification manifests through a range of observable phenomena, each of which contributes to the degradation of dryland ecosystems:

Loss of vegetative cover Reduced vegetation cover, characterized by the depletion of native plant species and the proliferation of invasive, invader, or increaser species, is a hallmark of desertification in drylands. This loss of vegetation not only diminishes the aesthetic appeal of the landscape (Fig. 20.1) but also disrupts ecological and hydrological balance, reduces biodiversity, and compromises the ability of ecosystems to provide vital services, such as soil stabilization and carbon sequestration (Akanwa et al. 2024). For instance, while native species are adapted to local conditions and contribute to soil health, invasive species often outcompete them, leading to monocultures that lack ecological resilience. Furthermore, studies show that areas with reduced vegetation cover experience increased soil erosion and desertification rates, exacerbating the loss of fertile land. In contrast, regions that implement reforestation and sustainable land management practices demonstrate improved vegetation recovery, showcasing the potential for restoration efforts to mitigate these losses. Thus, addressing the loss of vegetative cover requires a multifaceted approach that includes promoting biodiversity, controlling invasive species, and restoring native plant communities, as also suggested by Khattak et al. (2024).

Soil Erosion Accelerated soil erosion is another pervasive consequence of desertification, manifesting through sheet and rill erosion, wind-driven sediment transport, and gully formation. This erosion leads to the loss of fertile topsoil, which is crucial for sustaining agricultural productivity. Studies reveal that soil erosion can reduce crop yields by up to 50% in severely affected areas, highlighting the economic implications of this degradation. Moreover, as soil fertility declines, the capacity for crops to thrive diminishes, making agricultural systems increasingly vulnerable to environmental stressors such as drought. Research by Valentina-Daniela and Diana-Mura (2023) emphasizes that sediment runoff from eroded areas degrades water quality in downstream ecosystems, impairing aquatic habitats and exacerbating flooding. This degradation creates a feedback loop; as agricultural productivity declines, communities may resort to unsustainable practices, further exacerbating soil erosion. Conversely, sustainable practices, like contour farming and agroecological techniques, have shown promise in reducing erosion rates and enhancing soil health. Therefore, a comprehensive strategy that combines soil conservation techniques with community engagement is essential for reversing soil erosion trends (Albaladejo et al. 2021).

Decline in Soil Quality Soil degradation processes, such as compaction, salinization, and nutrient depletion, compromise soil quality and resilience, posing significant challenges for sustainable land management and ecosystem restoration efforts. Compacted soils exhibit reduced porosity and infiltration rates, exacerbating runoff and erosion, while salinization diminishes soil fertility and inhibits plant growth (Mureithi et al. 2024b). Research indicates that nutrient depletion can lead to a decline in soil organic matter, which is crucial for maintaining soil health and fertility. This decline necessitates the use of soil amendments and fertilizers to maintain agricultural yields, creating a cycle of dependency that can further degrade soil health over time. In contrast, practices such as organic farming and the use of cover crops have been shown to enhance soil quality and resilience, promoting the restoration of degraded lands. Furthermore, integrated approaches that combine agroecological practices with traditional knowledge can significantly improve soil health and agricultural productivity, demonstrating the importance of a holistic view in soil management strategies, as also posited by Raupp et al. (2024).

Water Scarcity Vegetation cover removal leaves the soils bare, and compacted soils hamper rainwater infiltration, leading to more water running overland, carrying soil sediments with it, resulting in soil loss. At catchment scale, this leads to declining water availability, exacerbated by climate change and unsustainable water use practices, heightening water scarcity in drylands, exacerbating competition for limited water resources, and hindering agricultural productivity and livelihoods (Salvia et al. 2019, 2022). As water becomes increasingly scarce, communities reliant on agriculture and livestock grazing face mounting challenges in meeting their water needs, leading to conflicts over water allocation, reduced crop yields, and livestock losses, especially during protracted dry seasons and droughts. Moreover, water scarcity undermines ecosystem functioning, diminishing habitat quality for aquatic and terrestrial species and exacerbating the risk of ecological collapse.

Together, these signs of desertification highlight the critical necessity for coordinated efforts to tackle the fundamental causes of land degradation and encourage sustainable land management practices within dryland ecosystems. By addressing the root causes of desertification and implementing targeted interventions to restore ecosystem function and resilience, we can mitigate the impacts of land degradation, safeguard livelihoods, and ensure the long-term sustainability of dryland landscapes. A similar recommendation has been given by Jain et al. (2024).

20.2.5.3 Impacts of Desertification

Desertification exerts profound socio-economic and environmental impacts, each of which has far-reaching consequences for both human well-being and ecosystem health:

Loss of Livelihoods Desertification has profound socio-economic impacts, notably the erosion of livelihoods for millions of people who depend on dryland ecosystems for their survival. As fertile land becomes degraded and water resources dwindle, agricultural productivity declines, grazing lands shrink, and access to essential resources becomes increasingly scarce. This degradation exacerbates poverty and food insecurity, pushing vulnerable communities deeper into cycles of deprivation and dependency (Singh and Chaudhary 2023). As fertile land becomes degraded and water resources diminish, agricultural productivity declines, grazing lands shrink, and access to essential resources becomes increasingly scarce. This exacerbates poverty and food insecurity, pushing vulnerable communities further into deprivation and perpetuating cycles of poverty and dependency. The loss of livelihoods not only threatens individual and community well-being but also has broader implications for social stability and economic development. Furthermore, communities facing desertification are often forced to migrate in search of better opportunities, leading to increased urbanization and strain on city resources. Regions with proactive land management policies are better equipped to mitigate these impacts, emphasizing the importance of sustainable practices in preserving livelihoods. Therefore, addressing the socio-economic impacts of desertification requires a comprehensive approach (Jain et al. 2024) that includes enhancing adaptive capacities, promoting alternative livelihoods, and ensuring equitable access to resources.

Biodiversity Loss Desertification poses a significant threat to biodiversity by degrading habitat quality, reducing ecosystem resilience, and diminishing the abundance and diversity of plant and animal species (Badapalli et al. 2023a). As vegetation cover declines and ecosystems become fragmented and degraded, species populations decline, habitats become increasingly fragmented, and ecological interactions are disrupted. The degradation of biodiversity disrupts not only ecosystem functioning but also the provision of crucial ecosystem services, such as pollination, soil fertility, and pest control, leading to significant repercussions for human well-being and the health of ecosystems. Additionally, the loss of biodiversity has dire repercussions for human well-being, as it undermines food security and the health of ecosystems that support livelihoods and cultural practices (Sahoo et al. 2023). Thus, conservation efforts aimed at preserving biodiversity in dryland ecosystems must be prioritized, including the restoration of habitats and the promotion of sustainable land-use practices that enhance ecological integrity.

Climate Change Desertification contributes to climate change by releasing stored carbon from degraded soils, reducing carbon sequestration capacity, and altering regional climate patterns (Rodrigues do Nascimento, 2023). As soils degrade and vegetation cover declines, the capacity of ecosystems to sequester carbon diminishes, leading to the release of carbon dioxide and other greenhouse gases into the atmosphere. Additionally, changes in surface albedo, evapotranspiration rates, and atmospheric moisture fluxes associated with desertification can further exacerbate climate change by altering regional climate patterns, amplifying temperature extremes, and disrupting precipitation regimes, with cascading effects on ecosystem dynamics and human societies. Regions experiencing desertification may see an increase in extreme weather events, further complicating efforts to adapt to climate change (Owusu et al. 2024). Therefore, addressing the intersection of desertification and climate change requires integrated strategies that promote sustainable land management, enhance carbon sequestration, and foster community resilience in the face of environmental challenges.

Social Conflict and Migration Desertification can exacerbate social tensions and trigger migration as communities compete for dwindling natural resources (Al-Obaidi et al. 2022). As access to land, water, and other essential resources becomes increasingly scarce, competition among different land uses and stakeholders intensifies, leading to conflicts over resource allocation, land tenure disputes, and social unrest. Moreover, as livelihoods become increasingly precarious and ecosystems degrade, vulnerable communities may be forced to migrate in search of alternative livelihood opportunities, exacerbating social dislocation, straining social support systems, and undermining social cohesion in affected areas.

The impacts of desertification on socio-economic conditions and the environment are deep and complex, affecting human well-being, ecosystem health, and global sustainability. By tackling the root causes of desertification and executing specific interventions aimed at restoring ecosystem function and resilience, we can lessen the effects of land degradation, protect livelihoods, and promote the long-term sustainability of dryland ecosystems for future generations. Regions with effective governance and conflict resolution mechanisms are better equipped to manage the social impacts of desertification, emphasizing the importance of inclusive policies that address resource management and community needs (Adebayo 2024). This underscores the necessity of developing adaptive strategies that not only address environmental challenges but also promote social cohesion and stability in affected communities.

20.2.5.4 Mitigation and Adaptation Strategies Against Desertification

Effective desertification mitigation and adaptation strategies encompass a diverse array of interventions tailored to address the complex and interrelated drivers of land degradation in dryland ecosystems and simultaneously build the resilience of land and livelihoods as well:

Sustainable Land Management Implementing sustainable land management practices is paramount for mitigating desertification and restoring ecosystem health (Salvia et al. 2019). Approaches such as conservation agriculture, agroforestry, soil and water conservation measures, and rangeland rehabilitation can enhance soil fertility, restore vegetation cover, and improve water retention capacity. By promoting regenerative agricultural practices and ecosystem-based land management techniques, sustainable land management fosters ecosystem resilience and supports sustainable livelihoods in dryland areas. These practices can lead to increased biodiversity, improved soil structure, and enhanced carbon sequestration, which are critical for combating climate change. For example, agroforestry systems not only provide food and timber but also improve soil health and water retention, creating a more resilient agricultural landscape (Vidadala 2024). Additionally, integrating crop-livestock systems can optimize resource use and reduce land degradation, highlighting the need for a systems-based approach to land management. Thus, sustainable land management is a vital component of a broader strategy to combat desertification, requiring collaboration among farmers, researchers, and policymakers to ensure effective implementation and adaptation to local conditions.

Integrated Water Resource Management Adopting integrated water resource management approaches is essential for addressing water scarcity and enhancing ecosystem resilience in dryland regions (Kalogiannidis et al. 2023). Strategies such as rainwater harvesting, watershed management, and water-use efficiency measures can alleviate water stress and support ecosystem functioning. By optimizing water allocation, enhancing water storage capacity, and promoting sustainable water use practices, integrated water resource management enhances the adaptive capacity of communities and ecosystems to cope with changing environmental conditions. Moreover, the integration of social equity considerations into water management practices is crucial for ensuring that all community members have access to water resources. Comparative studies have shown that communities practicing integrated water management experience improved agricultural productivity and reduced conflict over water resources (Dungumaro and Madulu 2003). For instance, participatory approaches that involve local stakeholders in water management decisions can lead to more sustainable outcomes and increased resilience to climate variability. Therefore, integrated water resource management not only addresses immediate water scarcity issues but also contributes to long-term sustainability and resilience in dryland ecosystems.

Policy and Governance Interventions Strengthening policy frameworks, promoting land tenure security, and enhancing governance mechanisms are critical for facilitating coordinated action and supporting community-based initiatives for desertification control and sustainable land management (Grigorieva et al. 2023). By fostering collaboration among stakeholders, enhancing institutional capacity, and promoting participatory decision-making processes, policy and governance interventions can create an enabling environment for effective desertification mitigation and adaptation efforts. Research indicates that secure land tenure is associated with improved land management practices and increased investments in sustainable agriculture (Bayisa et al. 2024). Furthermore, the role of governance in managing natural resources cannot be overstated; effective governance structures that promote transparency, accountability, and inclusivity are essential for addressing the complex challenges posed by desertification. Comparative analyses highlight that countries with robust governance frameworks are better positioned to implement successful land management policies, demonstrating the importance of effective governance in addressing desertification (Jain et al. 2024). Thus, strengthening policy and governance interventions is vital for creating a supportive environment for sustainable land management and desertification mitigation.

Climate Change Adaptation Integrating climate change adaptation measures into desertification mitigation efforts is essential for enhancing the adaptive capacity of dryland communities and ecosystems (Roy et al. 2024; Salinas and Mendieta 2013). Strategies such as promoting climate-smart agriculture, enhancing ecosystem-based adaptation strategies, and fostering community resilience can help communities cope with the impacts of climate change while simultaneously addressing land degradation and desertification. By mainstreaming climate change considerations into land use planning, resource management, and development policies, climate change adaptation measures contribute to building resilience and promoting sustainable development in dryland ecosystems. Moreover, the adoption of climate-smart practices has been shown to improve food security and livelihoods while reducing greenhouse gas emissions, highlighting the dual benefits of these strategies. For instance, practices such as intercropping and agroecological farming not only enhance productivity but also improve soil health and biodiversity, making agricultural systems more resilient to climate variability (Otieno 2024). Additionally, investing in education and capacity-building initiatives can empower communities to implement adaptation strategies effectively, ensuring that they are equipped to face the challenges posed by climate change and desertification. Therefore, integrating climate change adaptation into desertification mitigation strategies is crucial for fostering resilience and promoting sustainable development in vulnerable dryland regions.

20.3 Reversing Land Degradation Through Ecosystem Restoration and Effective Water Management Strategies

Ecosystem restoration offers a promising approach to reversing land degradation and restoring ecological function in degraded landscapes (Young et al. 2022). This section explores the principles, methodologies, and benefits of ecosystem restoration, highlighting its potential to enhance biodiversity, improve soil fertility, restore water resources, and mitigate climate change. Strategies tailored for arid and semi-arid regions play a pivotal role in mitigating the challenges posed by limited water resources and harsh environmental conditions. One such strategy involves the encouragement of rainwater storage, emphasizing the retention of moisture or conservation (Rojano-Cruz et al. 2023). This approach aims to prevent soil erosion and keep water from the rain within the area where it rains, thereby enhancing productivity while averting erosion and conserving moisture. To achieve this, various physical structures along contours, such as deep trenches (DTs), soil or stone-faced lines, and banquettes, eyebrows, half-moons, are implemented, alongside practices like compost making, mulching, contour cultivation, tie ridging, manuring, and planting of drought-tolerant species (Patra 2022). The selection and design of these structures depend on several factors, including climate, rainfall characteristics, soil properties (Mureithi et al. 2024a), past erosion patterns, slope steepness, cropping patterns, farm size and system, labor availability, and local knowledge. Each measure is intricately linked to food, fodder, or tree crop management and soil fertility stabilization, underscoring the interconnectedness of physical interventions with agricultural practices. Another vital strategy for water management in arid and semi-arid regions is water harvesting, which involves capturing and storing rainwater (Fig. 20.2) for various purposes.

The water harvesting structures may include contour bunds, check dams, farm ponds, rooftop rainwater harvesting systems, and subsurface dams. By harnessing rainwater, water harvesting contributes to soil moisture replenishment, groundwater recharge, and agricultural production, thereby enhancing resilience to drought and

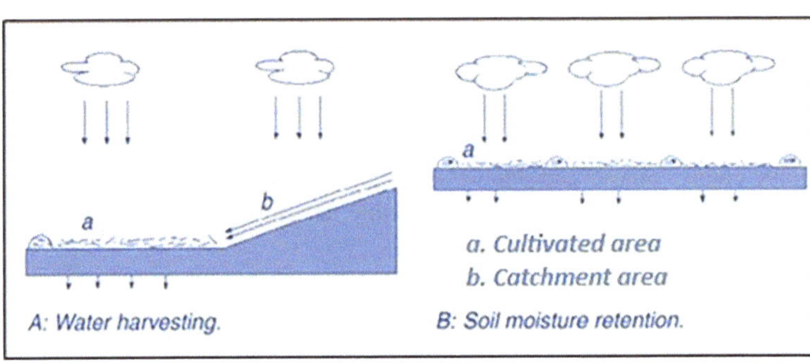

Fig. 20.2 2D illustration of in situ water harvesting with a catchment and cultivated area

promoting sustainable water use (Gebru et al. 2021). Additionally, water harvesting can support multiple livelihood activities, including agriculture, livestock rearing, and household water supply, thereby improving socio-economic conditions in arid and semi-arid regions (Lutta et al. 2020). Through the adoption of these strategies, arid and semi-arid regions can enhance water security, improve agricultural productivity, and build resilience to climate variability, ultimately fostering sustainable development and livelihoods in these water-stressed environments.

20.3.1 Principles of Ecosystem Restoration

Ecosystem restoration is guided by several key principles aimed at promoting the recovery of degraded ecosystems and enhancing their resilience to future disturbances (Young et al. 2022). These principles include prioritizing ecological integrity by fostering the recovery of native vegetation communities, reinstating natural ecological processes, and promoting diverse and resilient ecosystems. Additionally, ecosystem restoration employs adaptive management approaches that facilitate learning, experimentation, and adjustment based on monitoring and feedback, allowing for iterative improvements in restoration outcomes over time. Restoration initiatives have a higher chance of success when they are well-planned and actively engage multi-stakeholders, including local communities, landowners, governments, and relevant practitioners, in planning, implementation, and monitoring processes to ensure the integration of diverse perspectives, knowledge, and values (Higgs et al. 2018). Moreover, it fosters collaboration and partnerships among diverse stakeholders, including government agencies, non-governmental organizations, research institutions, and local communities, to leverage resources, expertise, and collective action towards shared restoration goals.

20.3.2 Methodologies for Ecosystem Restoration

Ecosystem restoration employs a variety of methodologies and techniques tailored to the specific context and objectives of restoration projects. These methodologies may include restoring native vegetation communities through reforestation, afforestation, and revegetation efforts (Grigoriadis et al. 2023), aimed at enhancing rangeland health, habitat quality, biodiversity, and ecosystem resilience while mitigating soil erosion, improving soil fertility, and promoting carbon sequestration (Liu et al. 2023). Additionally, ecosystem restoration involves rehabilitating hydrological processes (Cui et al. 2021) such as rangeland or wetland restoration, stream channel reconstruction, and groundwater recharge enhancement to restore water resources, improve water quality, and enhance aquatic habitat suitability while reducing the risk of floods and droughts (Verdonschot and Verdonschot 2023). Implementing soil conservation and restoration techniques, such as soil amendment, erosion control through rainwater harvesting, and soil bioremediation, aims to improve soil fertility, structure, and resilience, thereby enhancing agricultural productivity, supporting vegetation establishment, and mitigating land degradation. Moreover, integrating climate-smart principles into restoration efforts, such as promoting climate-resilient species, enhancing carbon sequestration, and adapting to changing climatic conditions, aims to enhance the adaptive capacity of restored ecosystems and contribute to climate change mitigation and adaptation efforts (Fatorić and Daly 2023).

In addition to these methodologies, participatory approaches that engage local communities in the restoration process are increasingly recognized as vital for successful ecosystem restoration. Involving stakeholders not only fosters a sense of ownership and responsibility but also ensures that restoration efforts align with local needs and knowledge (Ngulube et al. 2024). Community-led initiatives can enhance the effectiveness of restoration projects by incorporating traditional ecological knowledge and practices, which often provide valuable insights into local ecosystems and their dynamics. Furthermore, monitoring and adaptive management are essential components of restoration methodologies, allowing practitioners to assess the effectiveness of interventions and adjust strategies based on observed outcomes (Nelson et al. 2024). By employing a flexible approach that incorporates feedback and continuous learning, restoration projects can adapt to changing environmental conditions and improve long-term sustainability, ultimately leading to more resilient ecosystems.

Another critical aspect of ecosystem restoration methodologies is the use of technology and innovative practices to enhance restoration outcomes. Remote sensing and geographic information systems (GIS) enable detailed mapping and monitoring of ecosystems, facilitating data-driven decision-making and efficient resource allocation (Raihan 2024). Technologies such as drones and satellite imagery can provide real-time insights into vegetation health, soil conditions, and hydrological changes, thereby improving the effectiveness of restoration interventions (Wang et al. 2024a, b). Additionally, the application of ecological engineering principles, such as the design of artificial wetlands or green infrastructure, can create multifunctional landscapes that support biodiversity, enhance ecosystem services, and improve resilience to climate change (Makrickas 2024). By integrating technological advancements with traditional restoration practices (Haq et al. 2023), practitioners can develop more effective and scalable methodologies that address the multifaceted challenges of ecosystem degradation and contribute to the overall health of the environment.

20.3.3 Benefits of Ecosystem Restoration

Ecosystem restoration offers a range of multiple benefits for both ecosystems and human well-being (Hernández-Blanco et al. 2022). Restored ecosystems provide habitat for native plant and animal species, enhance habitat connectivity, and promote species recovery, thereby supporting biodiversity conservation and ecosystem resilience (Piczak et al. 2023). Furthermore, ecosystem restoration improves soil fertility, reduces soil erosion, enhances water infiltration and retention, and promotes groundwater recharge, mitigating land degradation, improving water quality, and enhancing water availability for ecosystems and human communities (Khan et al. 2024). Additionally, restored ecosystems sequester carbon, mitigate greenhouse gas emissions, and enhance carbon storage capacity, contributing to climate change mitigation efforts and enhancing the resilience of ecosystems and human societies to climate-related impacts (Rajasugunasekar et al. 2023). Moreover, ecosystem restoration generates employment opportunities, supports sustainable livelihoods, enhances ecosystem services provision, and promotes cultural and recreational values, thereby contributing to poverty alleviation, social cohesion, and human well-being (Wairore et al. 2016).

20.4 Soil and Water Conservation Technologies Useful in Ecosystem Restoration

Arid and semi-arid regions are characterized by distinct climatic and environmental features (Huang et al. 2016), such as high aridity, high evapotranspiration generally low annual rainfall. With mean maximum temperatures exceeding 30 °C during the dry season months and mean annual rainfall typically falling below 600 mm, these regions experience harsh and arid conditions. Rainstorms are often brief yet intense and heavy, leading to limited water availability and high rates of evaporation, particularly during the growing season. The prevalence of long dry seasons punctuated by short humid periods further exacerbates water and pasture scarcity challenges. Consequently, arid and semi-arid regions face escalating land degradation and desertification, compounded by intensive rainfall events that result in significant surface runoff. Given these conditions, it becomes imperative to harvest, store, and utilize the minimal rainfall available as efficiently as possible to sustain ecosystems and support livelihoods in these water-stressed environments.

20.4.1 Strategy 1: Soil Moisture Retention or Conservation

Various soil moisture retention and erosion control measures are utilized in agricultural and environmental management (Luo et al. 2024), and they can also be customized for the drylands under restoration. These include hillside terraces (HTs), stone bunds, stone-faced soil bunds, and simple soil bunds, all designed to reduce runoff, trap sediment, and enhance water infiltration (Czigány et al. 2023). Fanya juu, for instance, a soil and water conservation technique that has been adapted and widely used in Africa, especially in Kenya, Tanzania, Uganda, and Ethiopia, combines soil and stone to create barriers that slow water flow (Gemeda et al. 2024).

Deep trenches capture surface runoff, while gully control measures stabilize and rehabilitate eroded areas. These measures collectively contribute to sustainable land management, improving soil moisture retention, reducing erosion, and enhancing agricultural productivity.

20.4.1.1 Hillside Terraces

Hillside terraces (HTs) (Ajith et al. 2024) are physical structures built along contours, particularly effective in steep, degraded slopes and shallow soils (Fig. 20.3). They support rangeland grass reseeding and tree planting efforts, control runoff, and shield downstream fields from flooding. HT complements reforestation and area closure initiatives, making them suitable for semi-arid and arid regions as well as areas with medium rainfall. These terraces typically have a height of 50–75 cm, and controlled grazing is a precondition. By capturing runoff and promoting water infiltration, these terraces not only protect downstream fields from flooding but also create microclimates that support tree planting and vegetation growth. Their implementation is particularly beneficial in semi-arid and arid regions, where they can significantly improve soil moisture retention and reduce erosion. Previous efforts have demonstrated that areas utilizing hillside terraces experience increased agricultural productivity and enhanced biodiversity, making them an essential tool for restoring degraded landscapes and promoting ecological resilience (Socci et al. 2019).

Fig. 20.3 Hillside terraces formed using stone bunds placed along the contour in an arid environment. (Source: WFP)

Fig. 20.4 Deep trench constructed on a slope (left), and an illustration where deep trench should be installed (right). (Source: Y. Ayalew I WFP)

20.4.1.2 Deep Trenches

Deep trenches (DTs) are constructed as level channels designed to capture and retain surface runoff from elevated areas or slopes, thereby protecting downstream areas (Maingi 2024). The objectives of deep trenches include catching runoff on hillsides, grazing lands, and gully offsets to enhance soil moisture for biomass production while reducing soil erosion and gully expansion (Palmer 2024). Additionally, they contribute to replenishing groundwater and increasing the discharge of wells and springs, as well as improving the survival rate of seedlings and supporting natural regeneration. Deep trenches are ideally located in areas with soil depth greater than 1 m and well-drained soils, though they are not recommended for heavy clay (Vertisol) or sandy soils. Technical standards for deep trenches include a depth of 1 m, a width of 1–3 m, an embankment wall height of 0.75–1.2 m, a top width of 0.6–1.2 m, a bottom width of 2–3.5 m, and a wall length ranging from 50 to 70 m (Fig. 20.4).

Beyond their technical specifications, deep trenches serve as an effective strategy for managing water resources and enhancing agricultural productivity (Deng et al. 2006). By capturing and retaining surface runoff, they not only mitigate soil erosion and prevent gully expansion but also promote soil moisture retention, which is crucial for biomass production. Implementing deep trenches in suitable locations can lead to significant benefits, such as improved groundwater recharge, increased well and spring discharge, and higher seedling survival rates. Successful case studies have shown that areas utilizing deep trenches experience enhanced natural regeneration and overall ecosystem resilience, making them a vital component of sustainable land management practices.

Design and Specification

Determination of runoff volume is done using (Eq. 20.1)

$$\text{Runoff volume}\left(m^3\right) = K \times R\left(\text{rainfall depth}\right) - m \times A\left(\text{area of catchments}\right) - m^2 \tag{20.1}$$

$$\text{Volume of deep trench} = d\left(\text{effective water depth}\right) \times \text{width of the trench} \times l\,\text{is length of the trench}$$

d = 0.85 m (0.15 m is Antecedent soil moisture and siltation), l = 1 m, w =?

$=0.84m^3 = 0.85\ m \times 1\ m \times w$

$w = 0.84\ m^3/0.85m^2 = 0.988\ m = 1\ m$

20.4.1.3 Soil Bunds (Swales)

Soil bunds, or swales, are essential for soil conservation and water harvesting in farmland (Teale 2023; Twomey et al. 2024). Adhering to specific technical standards ensures their effectiveness: ditches are constructed with widths of 40 cm and depths of 60 cm for stable soil, while for unstable soil, the bottom width is 30 cm and the top width is 50 cm, with a consistent depth (Fig. 20.5). Berms, crucial for stability, measure 20 cm in width for both soil types, and tie points are integrated every 6–10 m for plantation purposes. Embankments, vital for containment, have a top width of at least 30 cm and a bottom width ranging from 1 to 1.5 m, with proper layering, leveling, and compaction allowing the height to reach up to 60 cm. Spacing between bunds depends on slope gradient, with a recommendation of 1 m VI for slopes up to 15% and 1.5–2 m VI for slopes between 15 and 20%. Modifications may be needed for swales crossing depression points. These structures not only enhance soil conservation and water harvesting but also improve crop resilience and productivity by promoting water infiltration and reducing erosion. Successful implementation strategies involve tailoring the design to local conditions, ensuring that bund dimensions and spacing are adapted to the specific slope gradients and soil types. Previous efforts have demonstrated that well-designed soil bunds can lead to significant improvements in agricultural yield and ecosystem health, making them a valuable tool for sustainable land management (Tefera et al. 2024).

20.4.1.4 Fanya Juu

Fanya juu, a soil conservation technique, plays a vital role in both soil preservation and water harvesting within farmland settings (Bayle and Muluye 2023). Adhering to minimum technical standards ensures its effectiveness (Fig. 20.6). Ditches, rectangular in shape, are constructed with specific widths (width = 40 cm for stable soil, bottom width = 30 cm for unstable soil) and depths (depth = 60 cm for both stable and unstable soil), with berms added for stability (20 cm for both) and tie points

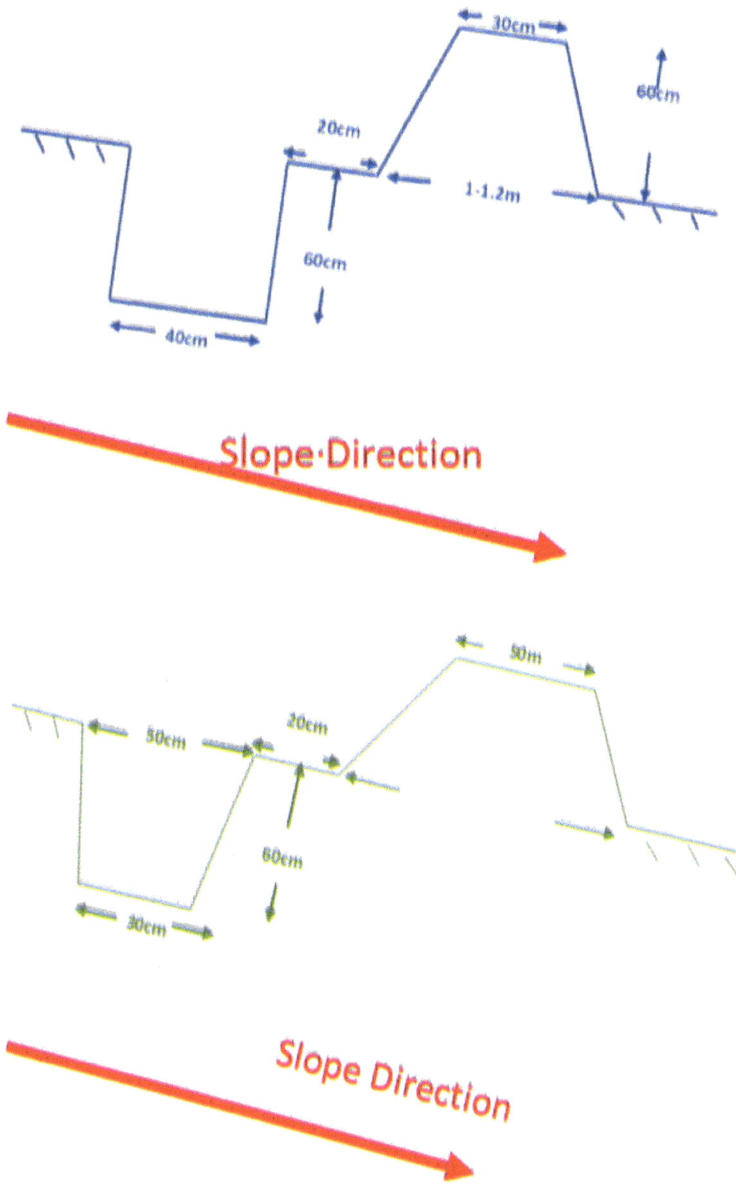

Fig. 20.5 Swales profiles for stable (top) and unstable (bottom) soils. (Source: Y. Ayalew | WFP)

integrated for plantation purposes. Embankments, positioned on the top side of slopes, are layered and compacted to achieve optimal height (top width min. 30 cm, bottom width 1–1.5 m, height up to 60 cm). Spacing between fanya juu structures is determined by slope gradient (1 m VI for slopes up to 8%) and often combined with

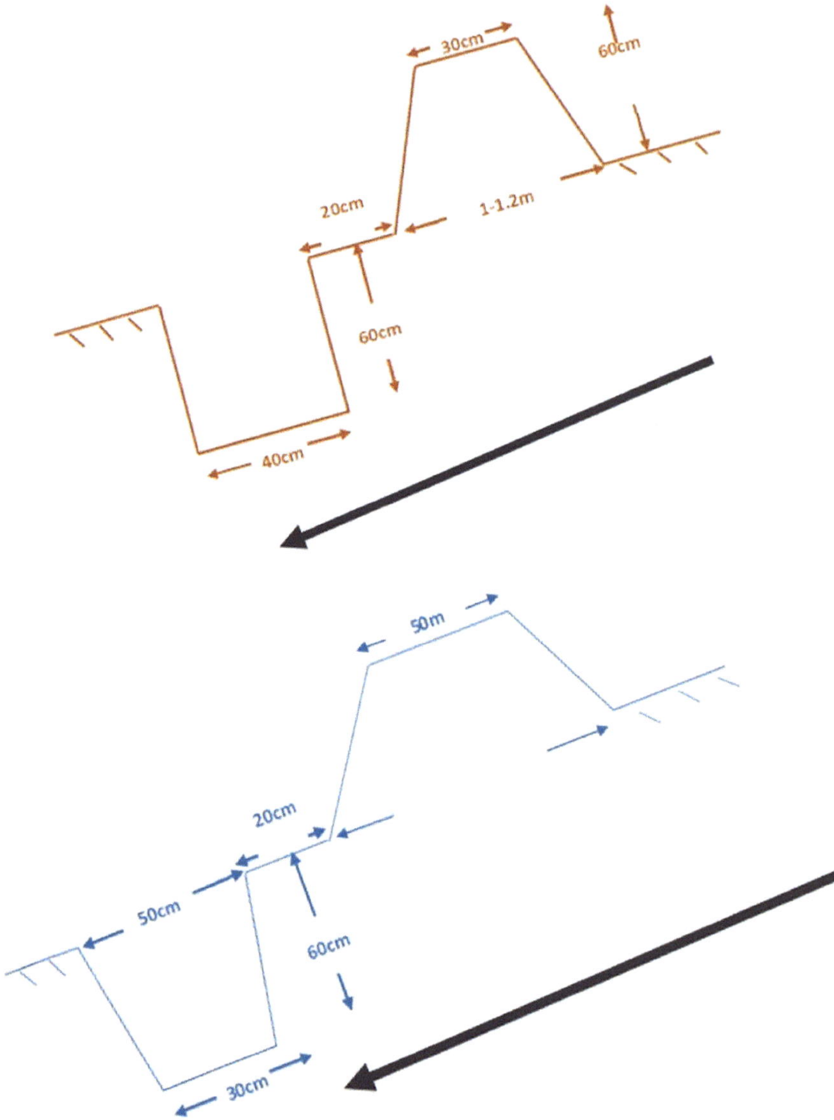

Fig. 20.6 Fanya juu profiles for stable (top) and unstable (bottom) soil. (Source: Y. Ayalew | WFP)

stabilization measures, such as grasses or pigeon pea. Modifications may be necessary for fanya juu structures that should not cross depression points, requiring buffer zones and treatment of depressions with check dams.

Fanya juu serves as an effective soil conservation and water harvesting technique that significantly enhances agricultural sustainability (Dangiso and Wolka 2024). By capturing and retaining runoff, these structures improve soil moisture and fertility, thereby promoting healthier crop growth and reducing erosion. The implementation of fanya juu is particularly beneficial in areas prone to soil degradation, as it fosters better water management and supports the establishment of vegetation through integrated tie points (Masha et al. 2023). Successful case studies have shown that farms employing fanya juu experience increased yields and improved resilience to climate variability, making it a valuable practice for restoring degraded lands and enhancing food security (Bayle and Muluye 2023).

20.4.1.5 Stone Bund/Lines

Stone lines or bunds serve as effective soil conservation methods on slopes exceeding 5% (bund) and as water harvesting solutions on gently sloping plains (less than 5%) in semi-arid regions (line) (Fig. 20.7). Consequently, these stone lines incorporate aspects of both macro-catchment and micro-catchment technologies, depending on whether they collect runoff from an external catchment or from within the field.

Typically, a stone line measures about 25 cm in height and has a base width ranging from 35 to 40 cm. Constructed using a combination of small and large stones, these lines are positioned along the contour of the land. Smaller stones are placed on the upslope side, while larger stones are positioned beneath them to help slow runoff, capture fertile soil sediment, and improve water infiltration. The spacing between the lines is determined by the slope and the availability of stones. For slopes under 1%, a distance of 20 m between lines is recommended, while for slopes between 1% and 2%, the spacing should be 15 m.

Stone lines are both easy and inexpensive to build, especially when stones are readily accessible. They are prevalent across Africa, found in both dry and humid regions, and are utilized wherever loose stones are present in the fields. In the Sahel, particularly in Burkina Faso and Niger, these lines tend to be smaller, usually consisting of no more than three stones in width and one or two stones in height (Mekdaschi and Liniger 2013).

Fig. 20.7 Stone lines installed along the contour. (Source: WFP)

The stone bund serves as a crucial structure in water management, primarily by slowing down the speed of runoff water, allowing for better infiltration and limiting the transport of solid particles and sediments upstream (Bedadi et al. 2023). Additionally, it plays a role in reducing slope gradients and facilitating the gradual formation of bench terraces, which are essential for soil conservation and water retention (Fig. 20.7). Moreover, the implementation of stone bunds contributes to the restoration of degraded areas, aiding in the regeneration of vegetation and overall ecosystem rehabilitation. A stone bund typically has a foundation of 30 cm and a height of 65 cm (upper side) and 100 cm (lower side).

20.4.1.6 Stone-Faced Soil Bund with Spillway

Stone-faced soil bunds with spillways are implemented on farmlands characterized by shallow soil and ample availability of stones (Leykun et al. 2023). They are particularly suitable for reclaiming land with slopes ranging from 3% to 15%, featuring sandy loam or clay loam sandy soil with a crusted and compacted surface. The vertical interval between consecutive terraces is typically set at 1 m (Fig. 20.8). With proper maintenance, these terraces gradually form due to the accumulation of soil above the bund, enhancing soil conservation and water retention efforts over time. Stone-faced soil bunds with spillways play a crucial role in sustainable land management by effectively capturing runoff and promoting soil moisture retention (Tamene et al. 2021). These bunds not only prevent soil erosion on sloped farmlands but also facilitate the gradual accumulation of soil, leading to improved fertility and enhanced agricultural productivity over time. Their implementation is especially beneficial in areas with shallow soils and abundant stones, as they provide a cost-effective solution for land reclamation. Previous experiences have shown that farms utilizing stone-faced bunds often report increased crop yields and better resilience

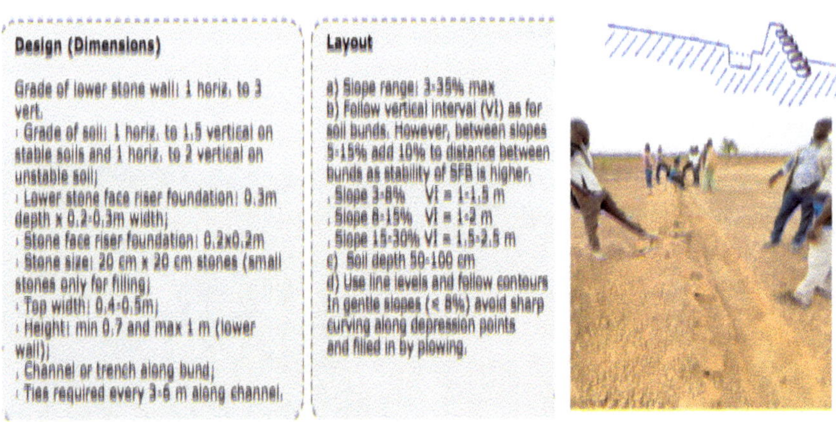

Fig. 20.8 Design and layout specifications of a stone-faced soil bund with spillway. (Source: Y. Ayalew | WFP)

against adverse weather conditions, making them a vital practice for enhancing soil conservation and supporting sustainable agriculture.

Layout and Construction

20.4.1.7 Strategy 2: Water Harvesting

Water harvesting, defined as the collection of runoff for productive use, entails various techniques aimed at harnessing runoff from different sources (Shaikh and Birajdar 2024). This runoff, obtained from roofs, ground surfaces, and seasonal streams, serves as a valuable resource for agricultural, domestic, and other purposes. Two primary types of water harvesting systems are commonly employed:

(i) Rainwater harvesting (Tholibon et al. 2024): This method focuses on capturing runoff from roofs or ground surfaces. Rainwater harvesting systems are designed to channel rainfall into storage facilities, such as tanks, cisterns, or reservoirs, where it can be stored and utilized as needed. Rainwater harvesting is a sustainable practice that effectively utilizes precipitation as a critical water resource, particularly in regions experiencing water scarcity. By capturing runoff from roofs and impermeable surfaces, this method not only reduces surface runoff and erosion but also enhances groundwater recharge. Systems can vary in complexity, ranging from simple rain barrels to elaborate underground cisterns equipped with filtration systems. The stored rainwater can be used for various purposes, including irrigation, domestic use, and even livestock watering, significantly reducing dependency on conventional water sources. Implementing rainwater harvesting systems contributes to improved water security and promotes environmental sustainability by maximizing the use of available rainfall (Rao et al. 2023; Ssekyanzi et al. 2024).

(ii) Floodwater harvesting: Floodwater harvesting systems are designed to capture runoff from seasonal streams or temporary watercourses during periods of heavy rainfall or flooding. These systems typically involve the construction of structures, such as check dams, retention ponds, or diversion channels, to capture and store excess water for later use (Hamlat et al. 2024). Floodwater harvesting is a strategic approach designed to manage excess water during periods of heavy rainfall, transforming potential hazards into valuable resources. This technique focuses on capturing and storing runoff from seasonal streams and flood-prone areas, utilizing structures like check dams, retention ponds, and diversion channels to direct water flow. By effectively managing floodwaters, these systems not only mitigate the risks of flooding but also enhance soil moisture levels, supporting agricultural activities during dry spells. Floodwater harvesting can significantly improve crop yields and promote biodiversity by replenishing local aquifers and creating wetland habitats (Haubrock et al.

2025). This method is particularly advantageous in arid and semi-arid regions, where water scarcity is a pressing challenge, enabling communities to harness natural water resources for sustainable development.

20.4.1.8 Principle of Water Harvesting

The principle of water harvesting involves deliberately leaving a portion of land uncultivated, known as the catchment area, to allow rainwater runoff to flow to the cultivated zone where crops are grown (Vetter et al. 2014). Here, soil moisture conservation methods, such as earth or stone structures, are utilized to pond the runoff, facilitating its infiltration into the soil for crop root absorption. This approach optimizes water availability for agricultural purposes, enhancing crop productivity and promoting sustainable land management practices.

The principle of water harvesting is centered around the strategic management of rainwater to maximize its availability for agricultural use (Ertop et al. 2023). By intentionally leaving a portion of land uncultivated, referred to as the catchment area, rainwater runoff is directed toward cultivated zones where crops are grown. This method not only allows for the natural flow of water but also enhances soil moisture conservation through the use of earth or stone structures that pond the runoff. These structures facilitate the infiltration of water into the soil, making it accessible for crop roots, thereby optimizing water availability and improving crop productivity.

In the context of rainwater harvesting (Fig. 20.9), the principle is similar but focuses specifically on capturing precipitation directly from surfaces such as roofs and paved areas. Systems are designed to channel this runoff into storage facilities,

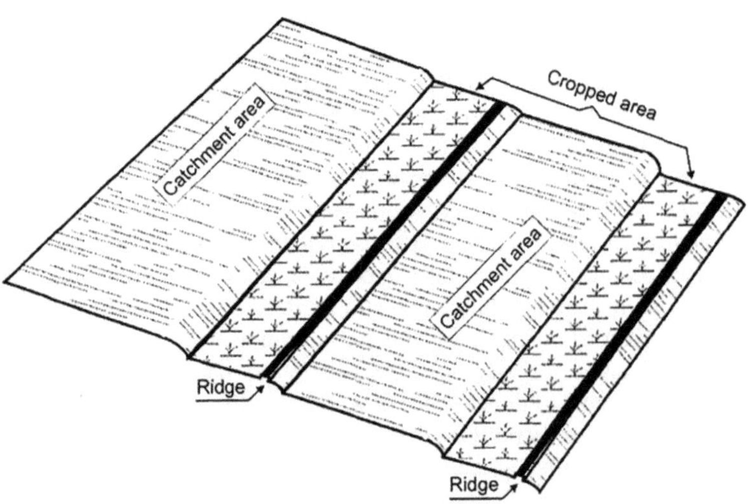

Fig. 20.9 Schematic of the principle of runoff water harvesting. (Source: Theib Oweis, 2004)

ensuring that the collected rainwater can be utilized during drier periods. By integrating both principles, water harvesting systems create a comprehensive strategy for sustainable land management, enabling farmers to enhance water security, reduce reliance on external water sources, and promote resilient agricultural practices. This holistic approach not only supports crop growth but also contributes to the overall health of the ecosystem by maintaining soil moisture and reducing erosion.

20.4.1.9 Conditions for Water Harvesting

Climate

Water harvesting is especially effective in semi-arid regions (Ammar et al. 2016) with an annual average rainfall (AAR) ranging from 300 to 700 mm. It is also practiced in some arid areas with lower rainfall levels between 100 and 300 mm AAR. In these areas, summer is typically the main rainfall period, but it coincides with high evaporation rates, increasing the likelihood of crop failure due to water scarcity.

Slope

The suitability of water harvesting depends significantly on the slope of the terrain (Al-Hasani et al. 2023). In general, steep slopes generate more runoff compared to gentle slopes; however, water harvesting is discouraged on slopes steeper than 5% because of challenges like uneven runoff distribution, increased soil erosion, and the high costs associated with necessary structures. In areas with steep slopes exceeding 5%, specific rainwater harvesting measures can still be implemented to manage runoff effectively. Techniques such as constructing terraces or contour bunds can help slow down water flow, allowing for better infiltration and reducing soil erosion. Additionally, utilizing smaller, strategically placed catchment areas can capture and direct runoff more efficiently, mitigating the risks associated with steep terrain while maximizing water retention.

Soil Type, Texture, Depth, and Structures

The effectiveness of water harvesting largely depends on soil characteristics. In cultivated areas, soils need to be deep enough to retain adequate moisture and fertile enough to support optimal crop growth. On the other hand, soils in the catchment area should have a low infiltration rate to enhance runoff. Water infiltration capacity is influenced by soil texture and structure, with sandy soils exhibiting higher infiltration rates compared to clay soils.

Table 20.1 Reference evapotranspiration (ETo) for dryland climatic zones

Climate zone	Mean daily temperature		
	Low(<15 °C)	Medium (15–20 °C)	High (>25 °C)
	ETo (mm/day)	ET (mm/day)	ETo (mm/day)
Desert/arid	4–6	7–8	9–10
Semi-arid	4–5	6–7	8–9
Moist (sub-humid)	3–4	5–6	7–8
Humid	1–2	3–4	5–6
Crop		Average Kc	
Cotton		0.82	
Groundnuts		0.79	
Legumes		0.79	
Maize		0.82	
Millet		0.79	
Sorghum		0.78	

Crops

Water harvesting practices should consider the suitability of crops for the local environment (Kone et al. 2024). Different crops have varying water requirements and adaptability to water distribution patterns. Perennial crops, such as trees, require concentrated water sources, while annual crops benefit from more evenly distributed water. Adjustments in crop selection and cultivation techniques, such as leveling the cultivated area, can optimize water use efficiency and crop productivity.

Calculation of C:CA Ratio

Calculation of the crop to catchment area ratio (C:CA) (Frezghi et al. 2019) entails assessing the water requirements of crops and comparing them with the design rainfall for water harvesting systems tailored to various climatic zones (see Table 20.1). The crop evapotranspiration (ETcrop) is determined by multiplying the crop coefficient (Kc) by the reference evapotranspiration (ETo), measured in millimeters per day (mm/day).

The crop water requirement (CWR) can be calculated using the formula:

$$\text{CWR} = \text{ETcrop} \times \text{total days in growing season}$$

$$\text{CWR} = \text{ETcrop} \times \text{total days in growing season} \quad (20.2)$$

where

ETcrop = Kc × EToETcrop = $Kc \times ETo$
KcKc = Average crop coefficient
EToETo = Average reference evapotranspiration (mm/day)

The Formula to Calculate the C:CA Ratio

1. WN in the cultivated area (CA) = WH in the catchment area (C).

2. WN in the cultivated area (CA) = [CWR–design rainfall] × CA (m^2) and WH in catchment area (C) = R × DR× efficiency factor × C (m^2).
3. Therefore: [CWR–DR] × CA = R × DR× efficiency factor × C.

$$C : CA = \frac{\text{CWR} - \text{Design rainfall}}{\text{R} \times \text{Design rainfall} \times \text{Efficiency factor}} \tag{20.3}$$

In the context of designing a water harvesting system, the "design rainfall" serves as the benchmark for determining the necessary infrastructure. The runoff factor (R) is pivotal in assessing runoff volume, typically ranging between 0.20 and 0.30 on gentle slopes and increasing to approximately 0.50 in rocky catchment areas. Efficiency, denoted by an efficiency factor, falls within the range of 0.5–0.75, reflecting the system's ability to capture and utilize rainfall effectively. Additionally, the recommended size for micro catchment areas per tree varies from 10 to 100 m^2, influenced by factors such as climate and tree species. For seasonal crops, a conventional rule of thumb suggests a crop to catchment area (C:CA) ratio of 3:1, implying that the catchment should be three times larger than the cultivated area to optimize water management.

20.4.2 Gully Rehabilitation Techniques

Gully rehabilitation contributes significantly to water management by reducing sediment load in waterways, improving runoff regulation, and enhancing groundwater recharge in certain contexts (Zhuang et al. 2024). These benefits demonstrate its complementary role in integrated water resource management. While not a water management strategy per se, gully rehabilitation has important water-related benefits, making it a valuable component of holistic watershed management approaches. Gully rehabilitation techniques are essential for mitigating the detrimental effects of gully erosion (Wilkinson et al. 2024), a process characterized by the concentration of water in narrow channels that swiftly removes soil, resulting in channels larger than rills. Often, gully formation signifies land misuse, emphasizing the importance of proactive land management practices. Prevention strategies are preferable to remediation efforts, highlighting the significance of good land husbandry. Effective gully control involves several measures, including treating gully catchments to reduce and regulate runoff volume and peak rates, diverting runoff water upstream of the gully area, and stabilizing gullies through structural interventions and accompanying re-vegetation initiatives. These approaches collectively contribute to the restoration and preservation of soil and landscape integrity, promoting sustainable land use practices. The following are key gully rehabilitation technologies.

20.4.2.1 Gully Wall Reshaping

Gully wall reshaping entails the modification of steep slopes along active gully flanks into gentle slopes (Ebong et al. 2023), typically with a minimum gradient of 45 degrees (Figs. 20.10 and 20.11). This technique involves cutting into the gully's depth by up to two-thirds and creating small trenches along contours to facilitate the re-vegetation of the slanted parts of the gully walls and beds. Technical procedures for gully wall reshaping include controlling or diverting the flow above the gully head, managing the gully head and side banks through cutting and filling processes, and constructing supporting check dams and retaining walls where necessary. Additionally, measures are implemented to protect against livestock interference, often incorporating the cut and carry system to maintain stability and promote vegetation growth.

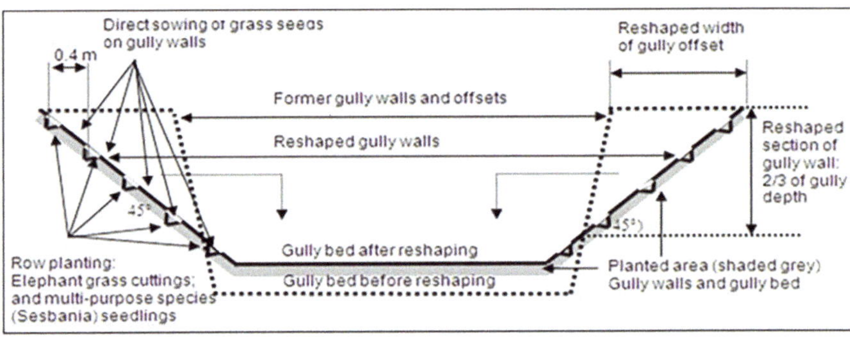

Fig. 20.10 Gully reshaping cross-section. (Source: Desta and Adugna 2012)

Fig. 20.11 Reshaped gulley with fresh plantings. (Source: Y. Ayalew | WFP)

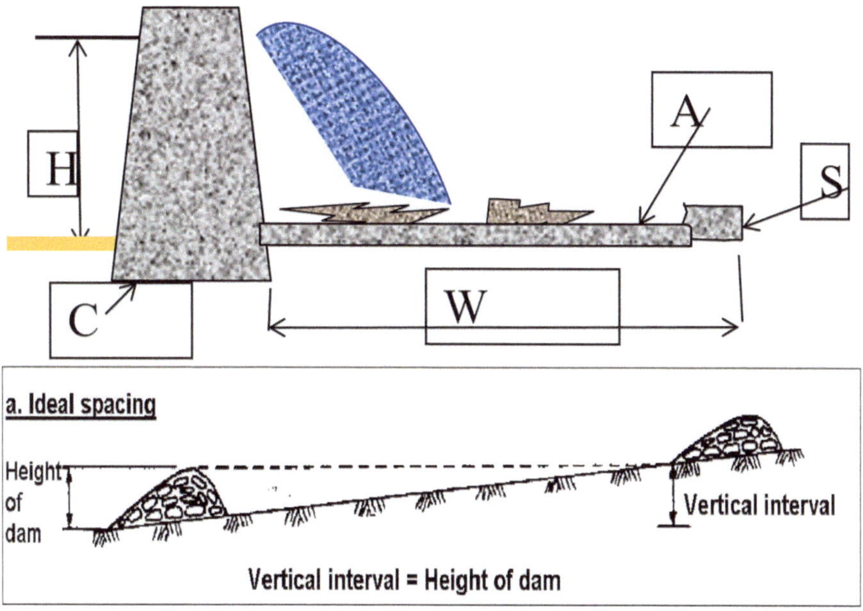

Fig. 20.12 Cross-section dimension of a loose stone check dam. (Source: WFP)

Fig. 20.13 Loose stone check dams. (Source: WFP)

20.4.2.2 Loose Stone Check Dam

The loose stone check dam is a structure crafted from relatively small rocks and positioned across the gully or small stream (de Haas et al. n.d.) (Figs. 20.12 and 20.13). These dams help slow down the velocity of runoff, effectively preventing the deepening and widening of the gully. The sediments that accumulate behind the check dam can be used for planting crops, trees, shrubs, or grasses, providing farmers with additional income opportunities. The spacing between loose stone check dams is determined by multiplying the dam's height by 1.2 and dividing by the gully bed slope, ensuring optimal effectiveness in water flow regulation and erosion control (Eq. 20.4).

$$\mathrm{A} = \mathrm{apron}, \mathrm{S} = \mathrm{sill}, \mathrm{W} = \mathrm{Apron\ width}, \mathrm{C} = \mathrm{Check\,dam}, \mathrm{H} = \mathrm{dam\,height} \quad (20.4)$$

$$\mathrm{S}(\mathrm{spacing}) = \frac{\mathrm{Height}(\mathrm{m}) \times 1.2}{\mathrm{Gully\,bed\,slope}(\mathrm{decimals})} \quad (20.5)$$

The number of check dams can also be estimated can also be estimated from gully length and spacing (Eq. 20.6).

$$N = L / S \quad (20.6)$$

Where

N = number of check dams
L = gully length
S = spacing (s)

20.4.2.3 Gabion Check Dam

Gabion check dams consist of rectangular boxes, typically constructed from mesh galvanized steel wire, of varying sizes (Adimassu et al. 2022). These boxes are connected by wire and filled with stones or soil materials, functioning as the foundational components of the structure. Gabion check dams are frequently deployed to manage large and medium gullies on highly eroded grazing and cultivated lands (Blanco and Lal 2023), often in conjunction with catchment treatment and protection measures (Figs. 20.14, 20.15, 20.16, and 20.17). By leveraging gabion technology, erosion is effectively mitigated, and landscapes are stabilized, contributing to the sustainable management of land resources and environments.

Fig. 20.14 Gabion check dam. (Source: WFP)

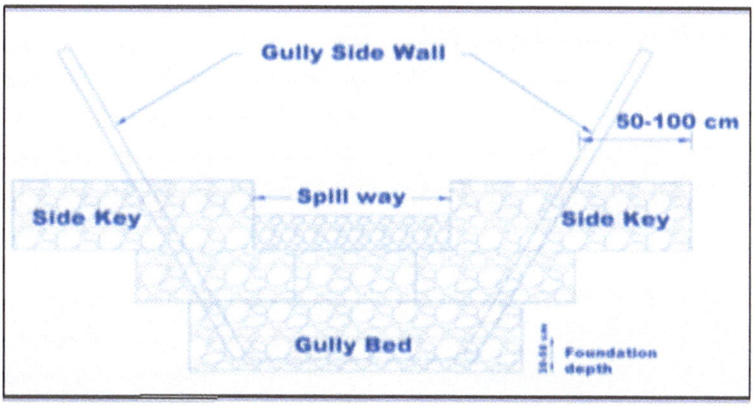

Fig. 20.15 The cross-sectional layout of a gabion check dam. (Source: WFP)

20.4.2.4 Gabion Pond

20.4.2.5 Stabilizing Gullies and Gully Control Structures Using Vetiver Grass

Soil erosion represents a major environmental challenge for many developing countries, resulting in the significant loss of valuable soil crucial for agricultural productivity. Although management strategies can often be expensive and less effective, using vegetative methods like grasses presents a cost-effective solution for rehabilitating degraded landscapes. One notable option is vetiver grass, now reclassified as

Fig. 20.16 Gabions installed to control gullies. (Source: WFP)

Fig. 20.17 A gabion pond created on a reshaped gully in Ethiopia. (Source: WFP)

Chrysopogon zizanioides (Figs. 20.18, 20.19, and 20.20) (Nyaga et al. 2022), which is recognized for its affordability and eco-friendliness in combating soil erosion and promoting water conservation. Its effectiveness is enhanced when integrated with other soil conservation practices, such as gabions or reshaped gully embankments, which create biophysical barriers to sediment loss while facilitating water infiltration.

Characterized by its tall, tufted growth and a robust, complex root system, vetiver plays a vital role in watershed protection. It helps slow and spread runoff

Fig. 20.18 *Chrysopogon* zizanioides (L. Roberty) commonly known as vetiver grass

Fig. 20.19 Vetiver splits planted directly at the base of the gullies in Chepareria, West Pokot County, Kenya. M Nyaga | Drylands Transform

Fig. 20.20 Vetiver splits planted in the horizontally laid bags along the contour lines and at the gully mouth at Chepareria, West Pokot County, Kenya. M Nyaga | Drylands Transform

harmlessly across farmland, recharge groundwater, reduce siltation in drainage systems and water bodies, and limit the input of agrochemicals into aquatic ecosystems. This resilient grass can thrive in conditions that many other plants cannot due to its unique physiological and morphological characteristics. These include a fine-structured root system that resists pests, diseases, and fire, high efficiency in absorbing nutrients and heavy metals from contaminated water, and a quick recovery rate following adverse conditions.

In carrying out degraded rangeland restoration in West Pokot County in Kenya under the Drylands Transform research project (Figs. 20.19 and 20.20), the rest of the bare areas were ripped followed by broadcasting of a mixture of seeds from different grasses at a rate of 14.4 kg/ha. The mixed grass seeds used were *Cenchrus ciliaris* (African foxtail), *Eragrostis superba* (Maasai lovegrass), and *Chloris roxburghiana* (African horsetail). These are all indigenous perennial grasses that are robust, establish well from seed, and are drought, grazing, and fire resistant. Seeds of forage legumes *Clitoria ternatea, Crotalaria juncea, Neonotonia wightii* (Glycine), and *Macroptilium atropurpureum* (Siratro) were also incorporated in the mixture to improve the quality of the pasture (Nyaga et al. 2022).

20.4.3 Selected Runoff Farming Technologies

These technologies collectively enhance water management strategies in agricultural settings, particularly in areas prone to low rainfall and soil erosion. Each method has unique applications and benefits, contributing to sustainable land and water management practices.

Fig. 20.21 Eyebrow basins. (Source: WFP)

20.4.3.1 Eyebrow Basin

Eyebrow basins are semi-circular stone-faced structures constructed along the contours (Ertop et al. 2023). These basins are designed for moisture harvesting and tree cultivation in regions with low rainfall and can be built on slopes of up to 30% for spot planting. They effectively manage runoff and aid in replenishing groundwater tables, making them ideal for shallow, stony soils found in dry areas (Fig. 20.21).

20.4.3.2 Herring Bones

Herring bones (Bashir n.d.) are small trapezoid structures, also known as A structures, designed for planting trees and other species (Fig. 20.22). Most effective in areas with medium to low rainfall (500–900 mm), they are suitable for degraded areas with slopes <5%, soils at least 50 cm deep, small plateaus with gentle slopes, abandoned lands for restoration, and portions of forest land requiring enrichment. Additionally, they can be utilized around homesteads for valuable trees.

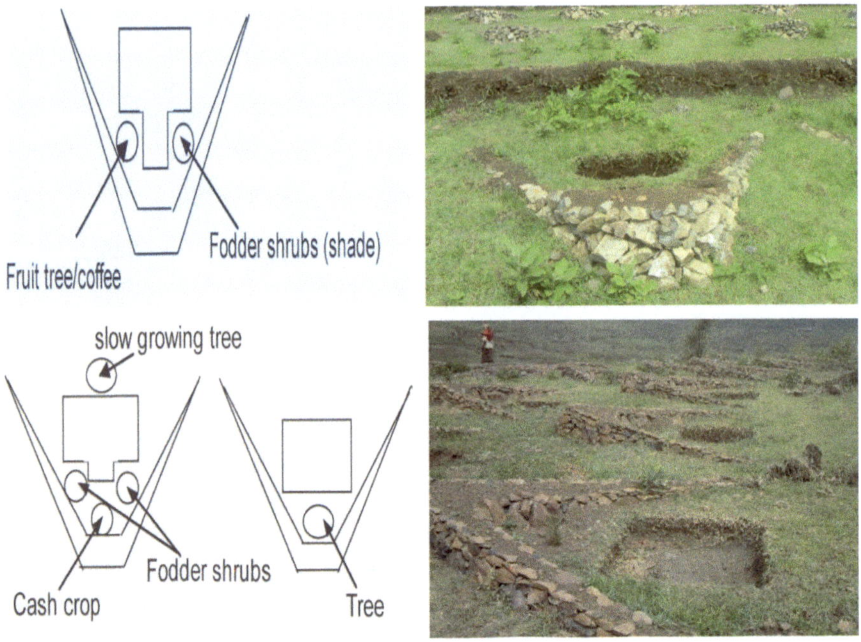

Fig. 20.22 Herring bones. (Source: WFP)

20.4.3.3 Multifunctional Halfmoons (Demilunes)

Demilune is a traditional Sahel planting method creating contours and water retention trenches to mitigate rainwater runoff (Eshete et al. 2023). These structures improve water infiltration, creating favorable micro-climate conditions for seedlings to flourish. Various types exist with different land uses, densities, shapes, and sizes, including demilune-Agricole, agro-silvo-pastoral, and silvo-pastoral halfmoons (Figs. 20.23, 20.24, 20.25, 20.26, 20.27, 20.28, and 20.29). Technical specifications include:

- Half-moons with 4 m diameter.
- Composed of 2 m long water collection trench, 0.5 m width, and 0.5 m depth.
- Slice swallows: 4 m half-moon length, 0.30 m width, and 0.3 m depth.
- Internal and external planting Zai pits (0.3 m diameter, 0.2–0.3 m depth).
- Waiting for the pit for end-of-season rain crops (e.g., watermelons, etc.)
- Millet plantation at the foot of the bulge—platform for forest tree/forage.
- Ratio runoff zone (runoff area) and receiving area (run-on area); C:CA = 3–4:1.
- Interval between half-moons on level curve lines: 2–4 m.
- Distance between level curve lines: 6 m.
- Vertical spacing: 2–3 m.

20.4.3.4 Demilune Classic

Fig. 20.23 Demilune classic. (Source: WFP/RBD)

20.4.3.5 Demilune with External Zai

Fig. 20.24 Demilune with external Zai pits. (Source: WFP/RBD)

20.4.3.6 Demilune with Internal and External Zai

Fig. 20.25 Demilune with internal and external Zai pits. (Source: WFP/RBD)

20.4.3.7 Improved Demilune

Fig. 20.26 Improved demilune. (Source: WFP/RBD)

Fig. 20.27 A composite of improved demilunes in Dosso region, Niger. WFP/RBD/Mar 2023

Fig. 20.28 Catchment scale restoration using half-moons in Maladi, Niger. (Source: WFP/RBD)

20.4.3.8 Demilune Forest

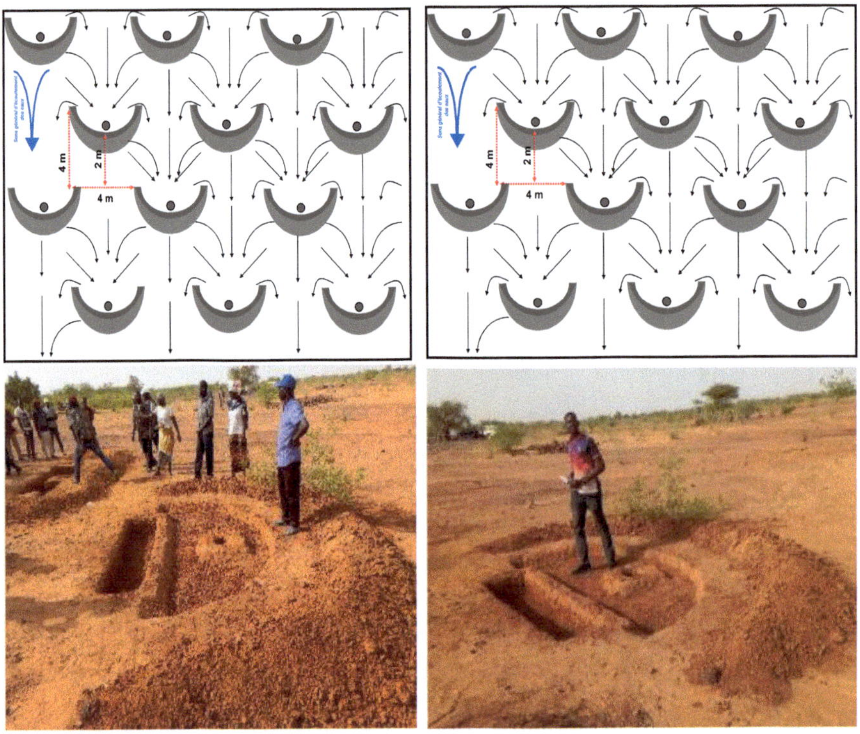

Fig. 20.29 Demilune Forest. (Source: WFP)

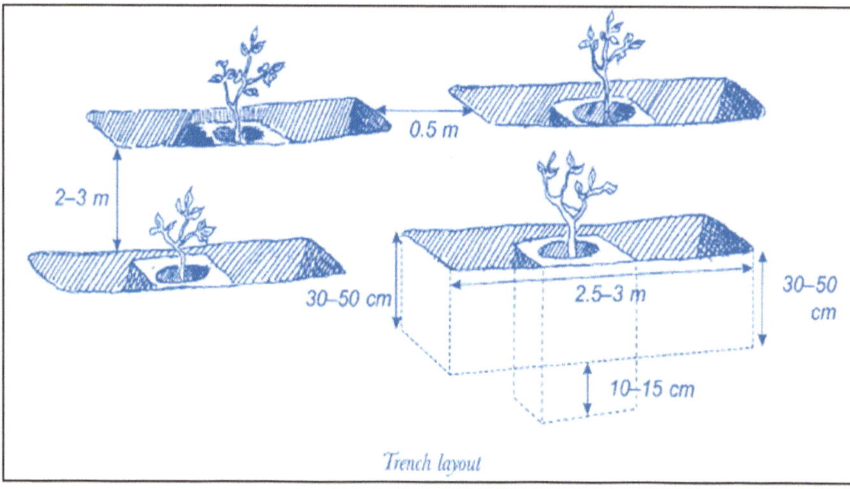

Fig. 20.30 Micro-trenches suitable for tree planting in water scarce dryland zones. (Source: WFP)

Fig. 20.31 Micro-trenches. (Source: WFP)

20.4.3.9 Micro-trench

Micro-trenches are extensive, deep pits that are built along the contours of the land, serving to collect and store rainfall water to support tree and fruit growth in dry zones (Figs. 20.30 and 20.31) while controlling erosion (Abrha et al. 2024). They act as part of a rainfall multiplier system, enhancing water harvesting effects.

Fig. 20.32 Cylindrical shape technique used for tree planting. (Source: WFP/RBD)

Fig. 20.33 Zai pits collect more water for crops. (Source: WFP)

20.4.3.10 Cylindrical Shape

The cylindrical shape technique (Walia et al. 2024) involves creating cylindrical structures along the contours to capture and retain rainwater. These structures act as reservoirs for storing runoff, facilitating groundwater recharge, and providing a water source for vegetation during dry periods. Typically constructed using natural materials such as stones or compacted soil, cylindrical shapes are effective in both hilly and flat terrains, contributing to soil moisture conservation and erosion control (Fig. 20.32).

Fig. 20.34 A combination of several technologies (deep and micro-trenches, eye-brows, Zai pits, and vegetation planting) in a catchment in Ethiopia (Photo: Y. Ayalew | WFP)

20.4.3.11 Zai Pits

The Zai or water pocket technique involves digging holes in blocked surface areas to collect runoff and infiltrate it (Figs. 20.33 and 20.34), aiding in moisture retention and enhancing soil fertility for agricultural purposes (Fita et al. 2023).

20.4.3.12 Water Pond

Water ponds are man-made depressions (Fig. 20.35) designed to capture and store rainwater for various purposes, including irrigation, livestock watering, and domestic use (Rao et al. 2024). These ponds can range in size from small-scale household ponds to larger community or agricultural reservoirs. By collecting runoff from surrounding areas, water ponds help mitigate flooding, replenish groundwater, and provide a reliable water source during dry seasons. Proper pond design considers factors such as slope, soil type, and water table depth to optimize water retention and minimize evaporation. Additionally, water ponds can support aquatic ecosystems and biodiversity, enhancing overall landscape resilience and productivity.

Excavated water ponds are a type that encompasses various types of water bodies formed through different methods, including embankment, excavation, or a combination of both (Hazarika and Hazarika n.d.). Embankment ponds are typically created by constructing a dam or embankment across a stream or depression in the landscape, allowing for the storage of significant water depths. These ponds can vary in size and are suitable for a range of land slopes, from gentle to steep. In

Fig. 20.35 A water pond. (Source: WFP)

contrast, excavated ponds are formed by digging a pit or dugout in level terrain, ideal for situations where only a small water supply is required. These ponds are often sustained by rainfall, surface runoff, or groundwater, offering a reliable water source for various purposes. Additionally, cut-and-fill ponds are crafted through a combination of excavation and embankment techniques on sloping terrain, providing a versatile solution for water storage and management.

Pond Site Selection

Drainage Area

When selecting a site for a pond, careful consideration must be given to the drainage area to ensure optimal functionality and longevity of the water body (Alene et al. 2022). It is imperative that the catchment area chosen does not include runoff from sources such as farmsteads, feedlots, sewage lines, urban sites, or other areas prone to contamination. This precaution is necessary to safeguard the pond from pollutants and maintain water quality. Additionally, to preserve the desired depth and capacity of the pond, it is essential that inflow from the catchment is relatively free of silt. If suitable land free from potential contaminants is not readily available, it is recommended to implement proper conservation practices in the watershed before pond construction begins. Initiating measures to protect the drainage area should be prioritized as soon as the decision to build a pond is made, ensuring the long-term viability and effectiveness of the water resource.

Pond Area

When determining the area for a pond, several key considerations must be considered to ensure its effectiveness and safety. Ideally, ponds should be positioned at low points within the catchment area, allowing rainwater runoff to flow naturally towards the excavated pond by gravity (Kelly and Kohler 2017). Additionally, the land surface where the pond is situated should be relatively flat, with an ideal slope of no more than 4% (equivalent to 4 cm per meter). It's advisable to locate the pond as close as possible to the primary water usage area to minimize transportation needs. One should avoid locating the pond in areas where the spillway could pose risks such as loss of life, injury to humans or livestock, or damage to nearby residences. Furthermore, sites near unstable ground, such as gullies or landslides, or in close proximity to deep-rooted trees, should be avoided to prevent potential hazards and ensure the structural integrity of the pond over time.

Soil Type

When selecting a site for a pond, the soil type plays a critical role in determining its suitability and water retention capacity. Ideally, soils with minimal seepage and percolation are preferred to ensure effective water retention (Mbilinyi et al. 2005). Deep clay or silty soils are particularly well-suited for this purpose. Sites with underlying strata of sand, gravel, limestone, or fractured rock at shallow depths should be avoided as they may lead to high seepage losses. Additionally, areas with soils containing outcrops and stones should be excluded from consideration. It is essential to investigate the soil profile depth before initiating pond excavation. Soils with a depth greater than 1 m, devoid of stones, and with a low pH level may be deemed suitable for dugout pond construction. Deeper ponds are necessary in cases where a permanent or year-round water supply is essential or when seepage and evaporation losses are anticipated to be high.

Runoff Estimation

Excavated ponds come in three primary types, each with distinct water sources: those fed by surface runoff, those supplied by groundwater aquifers, typically found in layers of sand and gravel, and those receiving water from both sources. The volume of runoff, crucial for filling surface-fed ponds, depends on various interconnected factors. These include the physical characteristics of the area, such as relief, soil infiltration capacity, vegetation cover, and the size and shape of the catchment, as well as factors like land use and surface storage (Fig. 20.36). Storm attributes, including the amount, intensity, and duration of rainfall, also significantly influence runoff. Estimating the runoff volume from a specific catchment area can be accomplished using a formula:

$$\text{Runoff volume}\left(\text{m}^3\right) = \text{K} \times \text{Design rainfall}\left(\text{m}\right) \times \text{Area of catchments}\left(\text{m}^2\right)$$

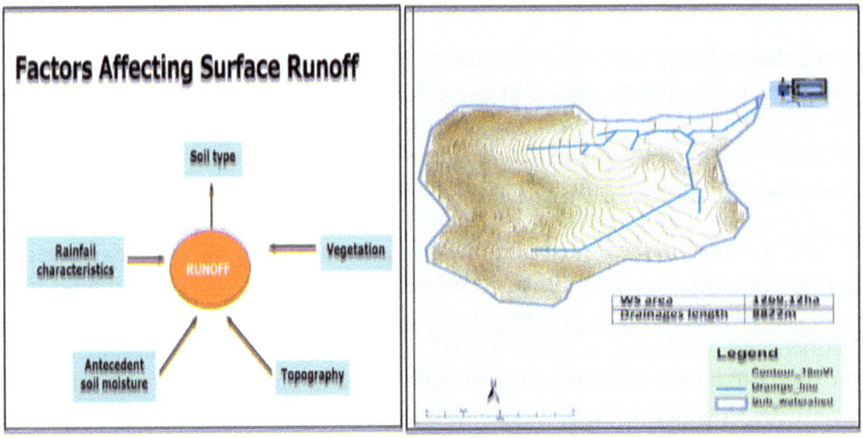

Fig. 20.36 Factors affecting surface runoff. (Source: WFP)

Here, "K" represents the runoff coefficient, calculated as the ratio of runoff (r) to precipitation (p).

Pond Capacity and Yield

Determining the capacity of a dugout pond involves considering both the water demand and the expected inflow over a specific period (Akandi 2013). This capacity is influenced by factors such as catchment size and the characteristics affecting water yield. For instance, a conservative estimate suggests that about 20% of seasonal rainfall may contribute to runoff in black soil areas, while in red soil areas with mild to medium slopes, this runoff may be around 10%. Additionally, storage losses like seepage and percolation losses, as well as soil type in the catchment area, play a role in determining the actual storage capacity. Siltation, influenced by soil type, must also be considered as it impacts the overall storage capacity of the pond. Water loss in consideration when estimating storage capacity is presented in Fig. 20.37.

$$\text{Storage capacity required} = \text{Water demand} + \text{Losses in the reservoir} + \text{Siltation}$$

Water Demand for Livestock

Water demand for selected livestock is calculated using Eq. 20.7 and is presented in Table 20.2. Water demand for domestic use is calculated using Eq. 20.8.

$$WL = \frac{NL \times Ac \times T}{1000} \tag{20.7}$$

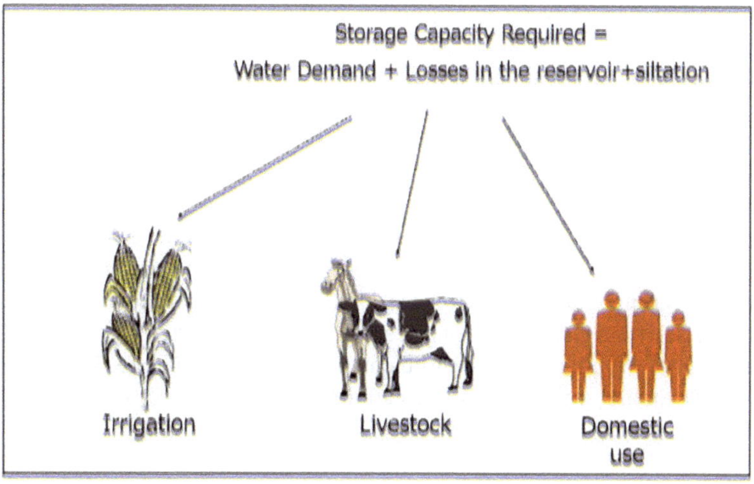

Fig. 20.37 Illustration of the water in the reservoir loses consideration when estimating storage capacity. (Source: WFP)

Table 20.2 Water consumption per domestic animal per day

Animal	Consumption liters/day
Camel	50
Cattle	25–35
Sheep	5–15
Goat	5–15
Donkey	16–20
Chicken	15–20/100 heads

where

WL = Water needed for livestock during the critical period in cubic meters
NL = Number of animals to be watered from the reservoir
Ac = Average rate of animal water consumption in liters per day per animal 25–60 l/animal/day
T = Duration of the critical period in days

Water Demand for Domestic Use

$$\mathrm{Wd} = \frac{\mathrm{Po} \times \mathrm{Dc} \times \mathrm{T}}{1000} \tag{20.8}$$

where

Wd = Domestic water supply during a critical period in cubic meters
Po = Users of the reservoir

Dc = Average rate of water consumption in liters per day per person 20–40 l/person/day
T = Duration of the critical period in days

Irrigation Water Demand

The calculation of irrigation water demand is influenced by several interconnected factors, including the water needs of the crops being irrigated, the anticipated effective rainfall throughout the growing season, the efficiency of the chosen irrigation method, losses from evaporation and seepage, and the expected inflow to the pond. The capacity of the pond must be sufficient to satisfy crop water needs and compensate for inevitable water losses. Irrigation water demand is determined using Eq. 20.9.

$$Ir\left(m^3\right) = \frac{10 \times ETcrop\left(mm\right) \times Ca\left(ha\right)}{Eff} \tag{20.9}$$

where

Ir = Irrigation water requirements (m^3) for the entire dry period.
Etcrop = Crop water requirement (mm) during the dry period
Ca = Area irrigated with water from the reservoir (ha)
Eff = Overall water application efficiency

Losses: Evaporation and Seepage
Evaporation losses: Can be calculated or measured using Pan A.
Seepage losses:

- Difficult to assess as it depends on permeability of the prevailing soil.
- As a rule of thumb can be assumed equal to ETo losses.
- Consider siltation and other losses.

Total losses = evaporation + seepage + siltation.

20.4.3.13 Diversion Channel

For a diversion channel, if a natural watercourse does not exist, one or two trenches should be excavated to direct water into the silt trap and the pond (Ward and Pulido-Velazquez 2008). These trenches can be constructed from compacted earth, stone pavement, or lined with cement to ensure durability. Additionally, it is crucial to maintain a very gentle gradient in the channel to prevent damage.

Fig. 20.38 A silt trap. (Source: Y. Ayalew)

20.4.3.14 Silt Trap

Silt traps are essential for collecting sediment and reducing silt accumulation in the pond (Robotham et al. 2021). The design capacity of these traps should consider the surface condition and sediment yield of the catchment area. Ideally, they should be positioned some distance away from the reservoir, typically between 5 and 20 m, based on the topography. Silt traps must be sufficiently sized and deep, typically ranging from 1 to 2 m, to accommodate the anticipated rate of siltation. Figure 20.38 illustrates a silt trap.

Spillway Design

The technical considerations for a spillway design should be considered when installing a silt trap. The determination of peak discharge—Rational Formula—is presented in Eq. 20.10.

$$Qp = \frac{Kr \times I \times C}{3.6} \tag{20.10}$$

where

Qp = peak flood discharge (m^3/s)
Kr = rational runoff coefficient (different from K, annual or seasonal runoff coefficient)
I = rainfall intensity (mm/hr)
C = area of the catchment (km^2)

20.5 Key Considerations in Water Interventions

Key considerations in water interventions emphasize the need for holistic approaches beyond mere technological solutions, emphasizing integration with environmental and agroecological factors. Understanding the context is crucial, requiring assessments of various elements such as topography, geology, weather patterns, livelihoods, social norms, land tenure, policies, and potential conflicts. Recognizing that water management begins with people, community involvement is prioritized to ensure sustainability and social cohesion. Acknowledging that water flows across landscapes and sectors, a multi-sectoral approach is advocated. Given the complexity of water challenges, specialized and collaborative solutions are essential, highlighting the importance of partnerships.

20.5.1 Future Research Directions in Ecosystem Restoration in Drylands

Future research in ecosystem restoration in drylands should focus on developing innovative, context-specific strategies that address the unique challenges of these environments. This includes exploring the integration of traditional ecological knowledge with modern scientific approaches to enhance restoration effectiveness. Research should also prioritize the identification of native plant species that are resilient to climate change and can support biodiversity and encourage water infiltration while improving soil health. Additionally, investigating the socio-economic impacts of restoration initiatives on local communities is vital to ensure that restoration efforts provide tangible benefits, such as improved livelihoods and better food and nutrition security. Furthermore, the development of advanced monitoring technologies, such as remote sensing and data analytics, can facilitate real-time assessment of restoration progress and ecosystem health. Collaborative research initiatives that engage local stakeholders, policymakers, and both biophysical and social scientists will be essential for fostering transdisciplinary knowledge exchange and building adaptive management frameworks. By focusing on these areas, future research can significantly contribute to scaling up ecosystem restoration efforts, ultimately leading to more resilient dryland ecosystems.

20.6 Conclusions and Recommendations

In conclusion, combating desertification and land degradation in dryland ecosystems necessitates a comprehensive and coordinated approach that addresses the root causes and implements effective solutions. The intricate complexities of these challenges have been examined, highlighting the various causes, manifestations, and impacts on both ecological and socio-economic systems. From unsustainable land use practices to climate change-induced stressors, the drivers of desertification and land degradation are diverse and interconnected, requiring proactive measures to mitigate their adverse effects. It is evident that ecosystem restoration, grounded in principles of biodiversity conservation, ecosystem services enhancement, and community engagement, holds promise as a cornerstone in restoring degraded landscapes and building resilience against environmental degradation. Furthermore, the integration of soil and water conservation technologies, such as soil moisture retention and water harvesting, offers practical solutions to improve soil fertility, reduce erosion, and enhance water availability in arid and semi-arid regions. However, to effectively address these challenges, concerted efforts are needed at multiple levels, from local grassroots initiatives to national policies and international cooperation. Priority should be given to raising awareness and fostering understanding among policymakers, stakeholders, and local communities, while also promoting collaboration and knowledge-sharing to mobilize resources and support innovative approaches. By prioritizing combating desertification and land degradation on the environmental agenda, integrating management strategies into broader development frameworks, and investing in sustainable practices, we can pave the way for a more resilient and sustainable future for dryland ecosystems and the communities that depend on them.

Recommendations

1. Prioritize awareness and education of soil and water conservation practices through targeted educational initiatives for stakeholders and communities.
2. Develop holistic land management approaches that prioritize soil and water conservation, such as contour farming and agroforestry.
3. Invest in the adoption and promotion of soil and water conservation technologies, like rainwater harvesting and check dams to improve soil moisture retention and reduce erosion.
4. Foster multi-stakeholder partnerships and collaboration to mobilize resources and share best practices in soil and water conservation.
5. Integrate desertification and land degradation management into national policies and development agendas, allocating sufficient resources and establishing robust regulatory frameworks to support implementation efforts.

Acknowledgments The authors acknowledge the anonymous reviewers for their technical advice to improve this work.

Conflict of Interest The authors declare no conflict of interest.

Example: CWR Calculation

Crop: Sorghum
Length of Growing Season: 4 months
Average Crop Coefficient (Kc): 0.78

Average Reference Evapotranspiration (ETo):

$$\text{ETo} = (9 + 8.5 + 8 + 8)4 = 8.4\ \text{mm/day}$$
$$\text{ETo} = 4(9 + 8.5 + 8 + 8) = 8.4\ \text{mm/day}$$

Crop Evapotranspiration (ETcrop):

$$\text{ETcrop} = \text{Kc} \times \text{ETo} = 0.78 \times 8.4 = 6.55\ \text{mm/day}$$
$$\text{ETcrop} = \text{Kc} \times \text{ETo} = 0.78 \times 8.4 = 6.55\ \text{mm/day}$$

Average CWR for Growing Season:

$$\text{Average CWR} = \text{ETcrop} \times \text{Total Days in Growing Season} = 6.55\ \text{mm/day} \times 120\ \text{days} \approx 790\ \text{mm}$$
$$\text{Average CWR} = \text{ETcrop} \times \text{Total Days in Growing Season} = 6.55\ \text{mm/day} \times 120\ \text{days} \approx 790\ \text{mm}$$

Example of Calculation of the C:CA Ratio for Crops

Climate: Semi-arid
Water Harvesting Method: Small-scale, such as contour ridges
Crop Type: Sorghum
Water Requirement for Crop: 550 mm
Design Rainfall: 320 mm
Runoff Coefficient (R): 0.50
Efficiency Factor: 0.70
C:CA = (550–320)/(320 × 0.50 × 0.70) = 2.05

Conclusion The catchment area must be approximately 2 times larger than the cultivated area.

References

AbdelRahman MAE (2023) An overview of land degradation, desertification and sustainable land management using GIS and remote sensing applications. Rendiconti Lincei Scienze Fisiche e Naturali 34:1–42

Abdulhafedh A (2024) Global water scarcity and watershed management planning: a case study of clear lake, IOWA, USA. Open Access Library Journal 11(1):1–14

Abrha H, Yaebiyo G, Abraha B (2024) Investigation of moisture collection structures and introduction of resistant tree species in the semi-arid slopes of tigray, Ethiopia. Asian J Res Agric Forestry 10(1):11–23

Adebayo WG (2024) Resilience in the face of ecological challenges: Strategies for integrating environmental considerations into social policy planning in Africa. Sustain Dev

Adimassu Z, Tamene LD, Kibret G, Assefa A, Dawit M (2022) Catalogue of sustainable land management and climate smart agriculture practices in Ethiopia
Ahmed A, Ishak MY, Umar NK, Zangina AS, Imam MN (2024) Climate change and desertification syndrome: the interaction of drivers in the drylands of Nigeria and the possibility of reducing adverse changes. Arab J Geosci 17(3):1–19
Ajith A, Francis KA, Pillai RJ (2024) Evaluation of pore-pressure variation and slope stability on terraced cultivation using physics-based landslide susceptibility model. Geomorphology 450:109081
Akandi AJ (2013) Impact of small reservoirs and dugouts in Ghana on hydrology and water allocation in the black Volta basin
Akanwa AO, Iko-ojo IV, Ezeomedo IC, Ikegbunam FI, Igwe PU, Muoghalu LN, Okeke SO, Okonkwo AU, Odimegwu CN, Nkwocha KF (2024) Effects of climatic risks on soil erosion/desertification in Southern and Northern Nigeria using GIS/remote sensing analysis. In: Climate crisis: adaptive approaches and sustainability. Springer, pp 151–170
Albaladejo J, Díaz-Pereira E, de Vente J (2021) Eco-holistic soil conservation to support land degradation neutrality and the sustainable development goals. Catena 196:104823
Alene A, Yibeltal M, Abera A, Andualem TG, Lee SS (2022) Identifying rainwater harvesting sites using integrated GIS and a multi-criteria evaluation approach in semi-arid areas of Ethiopia. Appl Water Sci 12(10):238
Al-Hasani B, Abdellatif M, Carnacina I, Harris C, Al-Quraishi A, Maaroof BF, Zubaidi SL (2023) Integrated geospatial approach for adaptive rainwater harvesting site selection under the impact of climate change. In: Stochastic environmental research and risk assessment, pp 1–25
Al-Obaidi JR, Yahya Allawi M, Salim Al-Taie B, Alobaidi KH, Al-Khayri JM, Abdullah S, Ahmad-Kamil EI (2022) The environmental, economic, and social development impact of desertification in Iraq: a review on desertification control measures and mitigation strategies. Environ Monit Assess 194(6):440
Ament R, Clevenger A, van der Ree R (2023) Addressing ecological connectivity in the development of roads, railways and canals. IUCN WCPA technical report series
Ammar A, Riksen M, Ouessar M, Ritsema C (2016) Identification of suitable sites for rainwater harvesting structures in arid and semi-arid regions: a review. Int Soil Water Conserv Res 4(2):108–120
Badapalli PK, Kottala RB, Pujari PS (2023a) Impact of desertification in semi-arid regions. In: Aeolian desertification: disaster with visual impact in semi-arid regions of Andhra Pradesh, South India. Springer, pp 95–100
Badapalli PK, Kottala RB, Pujari PS (2023b) Land degradation and desertification. In: Aeolian desertification: disaster with visual impact in semi-arid regions of Andhra Pradesh, South India. Springer, pp 13–49
Bashir N (n.d.) Techniques to control wind Erosion in Libya
Bayisa H, Kebede B, Benti F (2024) Factors influencing the implementation and adoption of sustainable land management practices on wacaca mountain in Central Highlands of Ethiopia. In: Environmental modeling & assessment, pp 1–18
Bayle D, Muluye K (2023) Review on successful soil conservation methods in Ethiopia. Cogent Food Agric 9(2):2274171
Bedadi B, Beyene S, Erkossa T, Fekadu E (2023) Soil management. In: The soils of Ethiopia. Springer, pp 193–234
Blanco H, Lal R (2023) Mechanical structures and engineering techniques. In: Soil conservation and management. Springer, pp 299–329
Bristol-Alagbariya ET (2023) UN convention to combat desertification as an international environmental regulatory framework for protecting and restoring the world's land towards a safer, more just and sustainable future. Int J Energy Environ Res 11(1):1–32
Clegg SR et al (2024) Tackling sustainable development goals through new space. Project Leadership Soc 5:100107
Cui W, Liu J, Jia J, Wang P (2021) Terrestrial ecological restoration in China: identifying advances and gaps. Environ Sci Eur 33:1–14

Czigány S, Sarkadi N, Lóczy D, Cséplő A, Balogh R, Fábián SÁ, Ciglič R, Ferk M, Pirisi G, Imre M (2023) Impact of agricultural land use types on soil moisture retention of loamy soils. Sustainability 15(6):4925

Dangiso Y, Wolka K (2024) Fanya Juu terraces improve soil properties of cultivated land in erosion-prone semi-arid area of Ethiopia. Sci Rep 14(1):25485

de Haas T, Roelofs L, Conway S, McElwaine J, Merrison J, Patel M, Sylvest M (n.d.) EGU general assembly 2023

de Souza Oliveira Filho J, Vieira JN, da Silva EMR, de Oliveira JGB, Pereira MG, Brasileiro FG (2019) Assessing the effects of 17 years of grazing exclusion in degraded semi-arid soils: evaluation of soil fertility, nutrients pools and stoichiometry. J Arid Environ 166:1–10

Deng X-P, Shan L, Zhang H, Turner NC (2006) Improving agricultural water use efficiency in arid and semiarid areas of China. Agric Water Manag 80(1–3):23–40

Desta L, Adugna B (2012) A field guide on gully prevention and control. Nile Basin Initiative Eastern Nile Subsidiary Action Program (ENSAP), Addis Ababa, Ethiopia, 251(March), 74

Drobitko A, Markova N, Tarabrina A-M, Tereshchenko A (2023) Land degradation in Ukraine: retrospective analysis 2017-2022. Int J Environ Stud 80(2):355–362

Dungumaro EW, Madulu NF (2003) Public participation in integrated water resources management: the case of Tanzania. Phys Chem Earth Parts a/b/C 28(20–27):1009–1014

Ebong ED, Urang JG, Melouah O, Ugi AU, Sule A (2023) Near-surface geophysical characterization of gully erosion hazard-prone area in Calabar, southern Nigeria. Acta Geophys 72:1–12

Ertop H, Kocięcka J, Atilgan A, Liberacki D, Niemiec M, Rolbiecki R (2023) The importance of rainwater harvesting and its usage possibilities: Antalya Example (Turkey). Water 15(12):2194

Eshete A, Teshome S, Edaso A, Yilma G, Furo G (2023) Selection of soil and water conservation technologies and native tree species for rehabilitation of degraded arid lands in Southeast Ethiopia. Int J For Res 2023:1

Fatorić S, Daly C (2023) Towards a climate-smart cultural heritage management. Wiley Interdiscip Rev Clim Chang 14(6):e855

Fita T, Getu E, Wakgari M, Woldetsadike K (2023) Host preference of Aulacaspis tubercularis Newstead (Homoptera: Diaspididae) to different mango cultivars in East Wollega and East Shewa Zones, Ethiopia. Am J Entomol 7(2):70–82

Frezghi H, Singh B, Ogbazghi W (2019) Ratio of catchment to cultivated area for different slopes under runoff farming of sorghum at Hamelmalo, Eritrea. J Soil Water Conserv 18(1):52–58

Gebru TA, Brhane GK, Gebremedhin YG (2021) Contributions of water harvesting technologies intervention in arid and semi-arid regions of Ethiopia, in ensuring households' food security, Tigray in focus. J Arid Environ 185:104373

Gemeda FT, Gemeda BS, Sori T (2024) Watershed management in Ethiopia and its effects on soil quality and productivity. Eur J Sustain Dev Res 8(1)

Ghosh D, Mandal A (2023) Climate change: its impact on land degradation and plant nutrients dynamics. In: Climate change impacts in India. Springer, pp 189–209

Gomiero T (2016) Soil degradation, land scarcity and food security: reviewing a complex challenge. Sustainability 8(3):281

Grigoriadis V, Gold E, Hutchinson G, Frewer LJ, Brereton P, Flannery D, Byrne KA, Garvey J (2023) How financial mechanisms can incentivize provision of ecosystem services from land restoration: a systematic review protocol. PLoS One 18(7):e0289120

Grigorieva E, Livenets A, Stelmakh E (2023) Adaptation of agriculture to climate change: a scoping review. Climate 11(10):202

Gunawardena UADP (2022) Soil degradation: causes, consequences, and analytical tools. In: Land degradation neutrality: achieving SDG 15 by forest management. Springer, pp 155–170

Habtu S (2024) Revitalizing key conditions and integrated watershed management to mitigate land degradation and sustain water availability for agriculture in semi-arid regions: a case study of Ethiopia. Qeios

Hamlat A, Meharzi S, Guidoum A, Sekkoum M, Mokhtari Y, Kadri CB (2024) GIS-based multicriteria analysis for flood hazard areas mapping of M'zab wadi basin (Ghardaia, North-Central Algeria). Arid Land Res Manag 38(1):1–25

Haq SM, Pieroni A, Bussmann RW, Abd-ElGawad AM, El-Ansary HO (2023) Integrating traditional ecological knowledge into habitat restoration: implications for meeting forest restoration challenges. J Ethnobiol Ethnomed 19(1):33

Haubrock PJ, Stubbington R, Fohrer N, Hollert H, Jähnig SC, Merz B, Pahl-Wostl C, Schüttrumpf H, Tetzlaff D, Wesche K (2025) A holistic catchment-scale framework to guide flood and drought mitigation towards improved biodiversity conservation and human wellbeing. Wiley Interdiscip Rev Water 12(1):e70001

Hazarika BB, Hazarika BB (n.d.) A comprehensive review of traditional and modern soil and water conservation practices

Hernández-Blanco M, Costanza R, Chen H, DeGroot D, Jarvis D, Kubiszewski I, Montoya J, Sangha K, Stoeckl N, Turner K (2022) Ecosystem health, ecosystem services, and the well-being of humans and the rest of nature. Glob Chang Biol 28(17):5027–5040

Higgs E, Harris J, Murphy S, Bowers K, Hobbs R, Jenkins W, Kidwell J, Lopoukhine N, Sollereder B, Suding K (2018) On principles and standards in ecological restoration. Restor Ecol 26(3):399–403

Huang J, Ji M, Xie Y, Wang S, He Y, Ran J (2016) Global semi-arid climate change over last 60 years. Clim Dyn 46:1131–1150

Jain S, Srivastava A, Khadke L, Chatterjee U, Elbeltagi A (2024) Global-scale water security and desertification management amidst climate change. Environ Sci Pollut Res 31:1–25

Jiang K, Teuling AJ, Chen X, Huang N, Wang J, Zhang Z, Gao R, Men J, Zhang Z, Wu Y (2024) Global land degradation hotspots based on multiple methods and indicators. Ecol Indic 158:111462

Kalogiannidis S, Kalfas D, Giannarakis G, Paschalidou M (2023) Integration of water resources management strategies in land use planning towards environmental conservation. Sustainability 15(21):15242

Kelly AM, Kohler CC (2017) Climate, site, and pond design. In: Dynamics of pond aquaculture. CRC Press, pp 109–133

Khan S, Masoodi TH, Islam MA, Arjumand T, Raja A, Parrey AA, Pallavi A, Bhat JH (2024) Ecosystem degradation to restoration: a challenge. In: Climate crisis: adaptive approaches and sustainability. Springer, pp 19–33

Khattak WA, Sun J, Hameed R, Zaman F, Abbas A, Khan KA, Elboughdiri N, Akbar R, He F, Ullah MW (2024) Unveiling the resistance of native weed communities: insights for managing invasive weed species in disturbed environments. Biol Rev 99(3):753–777

Kone S, Balde A, Zahonogo P, Sanfo S (2024) A systematic review of recent estimations of climate change impact on agriculture and adaptation strategies perspectives in Africa. Mitig Adapt Strateg Glob Chang 29(2):18

Leykun S, Teklay A, Gurebiyaw K, Dile YT, Bayabil HK, Ashenafi M (2023) Impacts of soil and water conservation measures on soil physicochemical properties in the Jibgedel Watershed, Ethiopia. Environ Mon Assess 195(4):447

Liu J, Gao W, Liu T, Dai L, Wu L, Miao H, Yang C (2023) A bibliometric analysis of the impact of ecological restoration on carbon sequestration in ecosystems. Forests 14(7):1442

Luo S, Zhou B, Likos WJ, Lu N (2024) Determining capillary pore–size distribution of soil from soil–water retention curve. J Geotech Geoenviron 150(2):4023138

Lutta AI, Wasonga OV, Nyangito MM, Sudan FK, Robinson LW (2020) Adoption of water harvesting technologies among agro-pastoralists in semi-arid rangelands of South Eastern Kenya. Environ Syst Res 9(1):1–12

Maingi SM (2024) Assessment of runoff for design of farm ponds for irrigation in Maragua Watershed, Murang'a County, Kenya. JKUAT-COETEC

Makrickas E (2024) Multifunctional green infrastructure and ecosystem services for social and ecological needs: a case study on peatlands

Masha M, Bojago E, Belayneh M (2023) Assessing the impacts of soil and water conservation practices on soil physicochemical properties in contrasting slope landscapes of Southern Ethiopia. J Agric Food Res 14:100876

Mbilinyi BP, Tumbo SD, Mahoo HF, Senkondo EM, Hatibu N (2005) Indigenous knowledge as decision support tool in rainwater harvesting. Phys Chem Earth Parts a/b/c 30(11–16):792–798
Mekdaschi R, Liniger HP (2013) Water harvesting: guidelines to good practice. Centre for Development and Environment
Mureithi SM, Verdoodt A, Gachene CKK, Njoka JT, Wasonga VO, De Neve S, Meyerhoff E, Van Ranst E (2014) Impact of enclosure management on soil properties and microbial biomass in a restored semi-arid rangeland, Kenya. J Arid Land 6(5):561–570. https://doi.org/10.1007/s40333-014-0065-x
Mureithi SM, Mwendwa S, Kamau S (2024a) Soil system status, threats, utilization pattern, and sustainable management strategies in the global south: african perspectives. In: Sustainable soil systems in global south. Springer, pp 719–747
Mureithi SM, Mwendwa S, Neina D (2024b) Soil types, formation processes, and characteristics in the global south. In: Sustainable soil systems in global south. Springer, pp 3–47
Naorem A (2023) Drylands: an introduction. In: Enhancing resilience of dryland agriculture under changing climate: interdisciplinary and convergence approaches. Springer, pp 3–14
Nelson CR, Hallett JG, Romero Montoya AE, Andrade A, Besacier C, Boerger V, Bouazza K, Chazdon R, Cohen-Shacham E, Danano D (2024) Standards of practice to guide ecosystem restoration: A contribution to the United Nations Decade on Ecosystem Restoration 2021–2030. Food & Agriculture Org
Ngulube NK, Tatano H, Samaddar S (2024) Toward participatory participation: a community perspective on effective engagement in post-disaster recovery and reconstruction. Community Dev 55(6):895–915
Nguyen TT, Grote U, Neubacher F, Rahut DB, Do MH, Paudel GP (2023) Security risks from climate change and environmental degradation: implications for sustainable land use transformation in the Global South. Curr Opin Environ Sustain 63(June):101322. https://doi.org/10.1016/j.cosust.2023.101322
Nyaga MN, Mureithi SM, Wegesa J, Makokha WL (2022). Practical training on gully rehabilitation using Vetiver grass (*Chrysopogon zizanioides*) at Chepukat Livestock Café site in Chepareria Ward, West Pokot County, Kenya. Field report, University of Nairobi, Triple L, Jjaja Regeneration
Ontel I, Cheval S, Irimescu A, Boldeanu G, Amihaesei V-A, Mihailescu D, Nertan A, Angearu C-V, Craciunescu V (2023) Assessing the recent trends of land degradation and desertification in Romania using remote sensing indicators. Remote Sens 15(19):4842
Oswald J, Harris S (2023) Desertification. In: Biological and environmental hazards, risks, and disasters. Elsevier, pp 369–393
Otieno M (2024) current climate, soil, and natural vegetation. In: Sustainable agroecological practices in Sub-Saharan Africa in the face of climate change. Springer, pp 21–51
Owusu AB, Fynn IEM, Adu-Boahen K, Kwang C, Mensah CA, Atugbiga JA (2024) Rate of desertification, climate change and coping strategies: insights from smallholder farmers in Ghana's Upper East Region. Environ Sustain Ind 23:100433
Palmer T (2024) Seek higher ground: the natural solution to our urgent flooding crisis. Univ of California Press
Patra AK (2022) Introductory agroforestry. CRC Press
Pei X, Zhao X, Liu J, Liu W, Zhang H, Jiao J (2024) Habitat degradation changes and disturbance factors in the Tibetan plateau in the 21st century. Environ Res 260:119616
Piczak ML, Perry D, Cooke SJ, Harrison I, Benitez S, Koning A, Peng L, Limbu P, Smokorowski KE, Salinas-Rodriguez S (2023) Protecting and restoring habitats to benefit freshwater biodiversity. Environ Rev 32:438
Raihan A (2024) A systematic review of Geographic Information Systems (GIS) in agriculture for evidence-based decision making and sustainability. Global Sustain Res 3(1):1–24
Rajasugunasekar D, Patel AK, Devi KB, Singh A, Selvam P, Chandra A (2023) An integrative review for the role of forests in combating climate change and promoting sustainable development. Int J Environ Clim Change 13(11):4331–4341

Rao MCS, Udayana SK, Swami DV (2023) Achieving land degradation neutrality to combat the impacts of climate change. In: Enhancing resilience of dryland agriculture under changing climate: interdisciplinary and convergence approaches. Springer, pp 77–96

Rao A, Laura JS, Dhania G (2024) Water conservation for environmental sustainability. In: Water, the environment and the sustainable development goals. Elsevier, pp 85–106

Rapiya M, Truter W, Ramoelo A (2024) The integration of land restoration and biodiversity conservation practices in sustainable food systems of Africa: a systematic review. Sustainability 16(20):8951

Raupp P, Carrillo Y, Nielsen UN (2024) Soil health to enhance ecological restoration and conservation. J Sustain Agric Environ 3(4):e70022

Ripple WJ, Whalen DN, Wolf C, Cao Y, Schulte J, Swann S, Woodrich ST, Newsome T, Cairncross R, Wirsing AJ (2024) Trophic cascades and climate change. Food Webs e00362:e00362

Robotham J, Old G, Rameshwaran P, Sear D, Gasca-Tucker D, Bishop J, Old J, McKnight D (2021) Sediment and nutrient retention in ponds on an agricultural stream: evaluating effectiveness for diffuse pollution mitigation. Water 13(12):1640

Rodrigues do Nascimento, F. (2023) Agenda in addressing global environmental changes and desertification. In: Global environmental changes, desertification and sustainability. Springer, pp 75–84

Rojano-Cruz R, Martínez-Moreno FJ, Galindo-Zaldívar J, Lamas F, González-Castillo L, Delgado G, Párraga J, Ramírez-González V, Durán-Zuazo VH, Cárceles-Rodríguez B (2023) Impacts of a hydroinfiltrator rainwater harvesting system on soil moisture regime and groundwater distribution for olive groves in semi-arid Mediterranean regions. Geoderma 438:116623

Roy P, Pal SC, Chakrabortty R, Chowdhuri I, Saha A, Ruidas D, Islam ARMT, Islam A (2024) Climate change and geo-environmental factors influencing desertification: a critical review. Environ Sci Pollut Res:1–14

Sahal KM, Tilahun ZA (2023) Factors affecting farmer's adoption of soil and water conservation practice in case of Gode District, Shebelle Zone, Somali Region, Ethiopia

Sahoo G, Mishra P, Wani AM, Sharma A, Mishra D, Patra D, Mishra I, Behera M (2023) Impact of climate change on livelihood security and biodiversity–issues and mitigation strategies. In: Climate change impacts on natural resources, ecosystems and agricultural systems. Springer, pp 1–27

Salinas CX, Mendieta J (2013) The cost of mitigation strategies for agricultural adaptation to global change. Mitig Adapt Strateg Glob Chang 18:933–941

Salvia R, Egidi G, Vinci S, Salvati L (2019) Desertification risk and rural development in Southern Europe: permanent assessment and implications for sustainable land management and mitigation policies. Land 8(12):191

Salvia R, Quaranta V, Sateriano A, Quaranta G (2022) Land resource depletion, regional disparities, and the claim for a renewed 'sustainability thinking'under early desertification conditions. Resources 11(3):28

Sarma B, Sarmah N, Malakar A (2024) Degradation of water and land resources in the himalayan mountain ecosystems. In: Addressing the climate crisis in the Indian Himalayas: can traditional ecological knowledge help? pp 31–52

Shaikh MAJ, Birajdar F (2024) Water harvesting: importance and techniques for mitigating drought in solapur district. Int J Res Eng Sci Manag 7(2):74–83

Singh V, Chaudhary N (2023) Land degradation, desertification, and food security in North-East India: present and future scenarios. In: Sustainable development goals in Northeast India: challenges and achievements. Springer, pp 153–166

Socci P, Errico A, Castelli G, Penna D, Preti F (2019) Terracing: from agriculture to multiple ecosystem services. In Oxford research encyclopedia of environmental science

Ssekyanzi G, Ahmad MJ, Choi K-S (2024) Sustainable solutions for mitigating water scarcity in developing countries: a comprehensive review of innovative rainwater storage systems. Water 16(17):2394

Stutter M, Kronvang B et al (2019) Current insights into the effectiveness of riparian management, attainment of multiple benefits, and potential technical enhancements. J Environ Qual 48(2):236–247

Tamene LD, Abera W, Woldearegay K, Mulatu K, Mekonnen K (2021) Creating multifunctional climate resilient landscapes: approaches, processes and technologies
Teale J (2023) The effect of soil bunds as Natural Flood Management features on soil water chemistry and hydraulic conductivity. Durham University
Tefera ML, Awada H, Pirastru M, Kombiok JM, Adjebeng-Danquah J, Adombilla R, Asungre PA, Mahama G, Carletti A, Seddaiu G (2024) Remote sensing and field data analysis to evaluate the impact of stone bunds on rainfed agriculture in West Africa. Land 13(10):1654
Tholibon DA, Mokhtar H, Latib FWM, Ismail MI, Nujid MM, Salam R (2024) The capability of rainwater harvesting as an alternative water supply. IOP Conf Ser Earth Environ Sci 1303(1):12024
Twomey AJ, Nunez K, Carr JA, Crooks S, Friess DA, Glamore W, Orr M, Reef R, Rogers K, Waltham NJ (2024) Planning hydrological restoration of coastal wetlands: Key model considerations and solutions. Sci Total Environ 169881:169881
Valentina-Daniela B, Diana-Mura B (2023) Monitoring of soil desertification-quality parameters. In: International conference on reliable systems engineering, pp 56–65
Verdonschot PFM, Verdonschot RCM (2023) The role of stream restoration in enhancing ecosystem services. Hydrobiologia 850(12):2537–2562
Vetter T, Rieger A-K, Nicolay A (2014) Disconnected runoff contributing areas: evidence provided by ancient watershed management systems in arid north-eastern Marmarica (NW-Egypt). Geomorphology 212:41–57
Vidadala R (2024) Reimagining agroforestry: climate-resilient landscapes for regenerative agriculture. In: Agroforestry solutions for climate change and environmental restoration. Springer, pp 171–201
von Holle B, Yelenik S, Gornish ES (2020) Restoration at the landscape scale as a means of mitigation and adaptation to climate change. Curr Landsc Ecol Rep 5:85–97
Wairore JN, Mureithi SM, Wasonga OV, Nyberg G (2016) Benefits derived from rehabilitating a degraded semi-arid rangeland in private enclosures in West Pokot County, Kenya. Land Degrad Dev 27(3):532–541
Walia SS, Kaur K, Kaur T (2024) Water harvesting: importance, its techniques. In: Rainfed agriculture and watershed management. Springer, pp 125–137
Wang H, Liu Y, Wang Y, Yao Y, Wang C (2023) Land cover change in global drylands: a review. Sci Total Environ 863:160943
Wang L, Lu P, Feng S, Hamel C, Sun D, Siddique KHM, Gan GY (2024a) Strategies to improve soil health by optimizing the plant–soil–microbe–anthropogenic activity nexus. Agric Ecosyst Environ 359:108750
Wang R, Sun Y, Zong J, Wang Y, Cao X, Wang Y, Cheng X, Zhang W (2024b) Remote sensing application in ecological restoration monitoring: a systematic review. Remote Sens 16(12):2204
Ward FA, Pulido-Velazquez M (2008) Water conservation in irrigation can increase water use. Proc Natl Acad Sci 105(47):18215–18220
Weeraratna S (2022a) Effects of land degradation. In: Understanding land degradation: an overview. Springer, pp 23–38
Weeraratna S (2022b) Factors causing land degradation. In: Understanding land degradation: an overview. Springer, pp 5–22
Wilkinson SN, Rutherfurd ID, Brooks AP, Bartley R (2024) Achieving change through gully erosion research. Earth Surf Process Landf 49(1):49–57
Yan Z, Guo Y, Sun B, Gao Z, Qin P, Li Y, Yue W, Cui H (2024) Combating land degradation through human efforts: ongoing challenges for sustainable development of global drylands. J Environ Manag 354:120254
Young RE, Gann GD, Walder B, Liu J, Cui W, Newton V, Nelson CR, Tashe N, Jasper D, Silveira FAO (2022) International principles and standards for the ecological restoration and recovery of mine sites. Restor Ecol 30:e13771
Zhuang J, Peng J, Li Z, Leng Y, Zhu X, Huang W, Jia Z, Liu Y, Zhu Y (2024) The evolution of gully erosion in the Rift Valley of Kenya over the past 50 years. J Afr Earth Sci 209:105099

Chapter 21
MaxEnt's Forecast of Climate Change's Effect on Habitat Suitability of *Anacardium occidentale* L. in Togo (West Africa)

Kossi-Messan Jacques Agboka , Dodiomon Soro , Komi Agboka, and Wouyo Atakpama

Abstract Climate change affects the world's ecosystems, posing severe challenges for scientists, environmentalists, and policymakers. *Anacardium occidentale* L., a plant of the Anacardiaceae family, is both ecologically and commercially significant due to its high drought resistance. However, the features of habitat distribution and the primary environmental factors influencing its appropriateness are little understood in Togo. Using 2538 occurrence data and 17 soil and environmental parameters, HadGEM3-GC3.1-L and MIROC6 models were used to predict current and future appropriate habitats under two Shared Socioeconomic Pathways (SSP245 and SSP585) by 2070. The results show that soil, followed by yearly precipitation (bio12) and temperature seasonality (bio4), are the most important environmental variables influencing *A. occidentale*'s regional spread. MaxEnt projected that very advantageous habitats for *A. occidentale* would decrease, while less favorable, moderately favorable, and unfavorable habitat areas would increase. Furthermore, under

K.-M. J. Agboka (✉)
Ecole Supérieure d'Agronomie (ESA), Université de Lomé, Lomé, Togo

WASCAL-Graduate Research Program on Climate Change and Biodiversity, Université Felix Houphouet-Boigny, Abidjan, Côte d'Ivoire
e-mail: agboka.j@edu.wascal.org

D. Soro
Laboratoire des Milieux Naturels et Conservation de la Biodiversité-UFR Biosciences-UFHB (Côte d'Ivoire), Abidjan, Côte d'Ivoire

K. Agboka
Ecole Supérieure d'Agronomie (ESA), Université de Lomé, Lomé, Togo

West African Science Service Centre on Climate Change and Adapted Land Use (WASCAL), Université de Lomé, Lomé, Togo

W. Atakpama
Laboratoire de Botanique et d'Ecologie Végétale (LBEV), Faculté des Sciences, Université de Lomé, Lomé, Togo

Y. D. I. Khan et al. (eds.), *Land Restoration Through Ecosystem-Based Approach*, Disaster Resilience and Green Growth, https://doi.org/10.1007/978-981-95-0389-6_21

two temperature change scenarios (SSP245 and SSP585), Togo would have essentially no possible appropriate habitat for *A. occidentale* by 2070. Our findings might be used to develop a climate-resilient cashew (*Anacardium occidentale* L.) agriculture plan in Togo, helping to preserve and conserve this species.

Keywords MaxEnt · Suitable habitat · *Anacardium occidentale* L. · Climate change · Potential geographic distribution

21.1 Introduction

Climate change is one of the most serious environmental, social, and economic concerns to our planet today (Chomitz et al. 2006). In Africa, climate change and variability are expected to have the greatest impact on forestry and agriculture (Koungbanane et al. 2020). Togo's agriculture sector employs two-thirds of the population, and climate change-induced changes in weather patterns, such as droughts and heatwaves, are likely to have an impact on crop production. (Togolese Republic 2021). Climate change poses a worldwide hazard to species; thus, understanding their current and future geographical distribution is critical for implementing an effective conservation plan. One of the challenges that ecologists face is predicting how various types of ongoing environmental change will affect species' geographical distribution, global biodiversity patterns, and the ecosystem services that biodiversity provides to humans (Zhang et al. 2020). Many severe alterations in species distributions and extinctions are taking place, particularly in fragile or fragmented ecosystems (Zhang et al. 2018). Many studies have anticipated environmental changes, such as altering patterns of precipitation and temperature variations, which would undoubtedly influence habitat suitability in diverse ecosystems in the future. Xu et al. (2023). This understanding will assist managers in addressing any possible hazards that may alter species' regional range (Wang et al. 2014). When distribution data is insufficient, species distribution modeling (SDM) can be used to analyze ecological demands and forecast a species' future biogeography as well as its existing distribution region (Anderson et al. 2003). Several SDMs, including DOMIN (Carpenter et al. 1993), ecological niche factor analyses (ENFA) (Hirzel et al. 2002), the bioclimate envelope (Pearson and Dawson 2003), the genetic algorithm for rule-set production (GARP) (Pearson et al. 2006; Townsend Peterson et al. 2007), CLIMEX (Park et al. 2014), and MaxEnt (Phillips et al. 2006; Yang et al. 2022), have been used to predict the impact of climate change on the geographic distribution of species (current and future).

MaxEnt has been shown to be one of the most dependable habitat modeling tools in terms of prediction abilities (Elith et al. 2006; Khanum et al. 2013; Atakpama et al. 2023; Biaou et al. 2023; Wa Rusaati and Kang 2024). The core idea underlying MaxEnt is to choose a probability distribution that maximizes entropy while meeting the restrictions (Anderson et al. 2003). One of MaxEnt's primary features is its capacity to handle presence-only data, which makes it useful in cases when absence data is difficult to get (Xu et al. 2023).

The cashew tree (*Anacardium occidentale* L.) is a member of the Anacardiaceae family (Brunel et al. 1984). It is primarily grown for the nut, which is a key industrial and export raw resource (Akinhanmi et al. 2008). The cashew (*Anacardium occidentale* L.) is a Brazilian tree that is rich in phenolic lipids. The cashew tree may reach a height of 15 meters and has a sturdy, twisted trunk with woody branches. It may be found in tropical countries including Ghana, Côte d'Ivoire, Nigeria, Kenya, Tanzania, Mozambique, and Togo, among others (Akinhanmi et al. 2008). According to Akinhanmi et al. (2008), cashew has antioxidant, antigenotoxic, antimutagenic, antiulcerogenic, anti-inflammatory, antibacterial, antifungal, and larvicide activities, and anthocyanins, carotenoids, ascorbic acid (vitamin C), flavonoids, and other polyphenols. Ndiaye (2019) states that the cashew tree can withstand a wide range of circumstances, including drought and poor soil, but not cold. Cashew plants help to combat climate change through carbon sequestration (Ndiaye 2019). Reducing atmospheric carbon levels creates a healthy environment for human growth (Atakpama et al. 2023).

According to the IPCC's sixth assessment report (AR6), increasing greenhouse gas emissions will cause global surface temperatures to rise by 1.5–3 °C by the end of the twenty-first century (IPCC 2022). Such a change in climatic circumstances would have a significant impact on species distribution, as species are likely to modify their distribution range in response to climate change (Velasquez-Tibat et al. 2013). Projections (modeling) allow us to know how these species will react (expansion or reduction of distribution areas) to climate change. Shared Socioeconomic Paths (SSPs) represent changes in population, economic growth, education, urbanization, and the rate of technological progress that might affect future greenhouse gas emissions, while RCPs are based completely on greenhouse gas concentrations in the atmosphere (IPCC 2022).

Given the ecological and economic importance of *Anacardium occidentale*, it is critical to assess how changing climatic conditions may influence the species' appropriate habitat.

The present research will help to identify areas that will be less favorable (drylands) for cashew tree development in Togo in the future (2070) in order to put in place sustainable management strategies for these lands to make them favorable for cashew tree cultivation. Such a study will also aid in the development of effective management of soil and conservation methods for this significant tree species throughout Togo's different regions. We have thus chosen to perform this research, which seeks to anticipate the current and future appropriate habitat of *A. occidentale* in Togo using MaxEnt modeling. Specifically, this study will enable us to:

1. Model the potential geographical location of *A. occidentale* in Togo.
2. Identify important environmental factors that shape its distribution in Togo.
3. Predict the impact of climate change on the suitable habitat of cashews under future climate conditions.

21.2 Methodology

21.2.1 Study Area

Togo, located in West Africa, has a 56,600 km² land area and borders the Atlantic Ocean, Benin, Ghana, and Burkina Faso to the south, east, west, and north. The location is on the Guinean Gulf coast, between latitudes 6°06′N and 11°08’N, and longitudes 0°09 W and 1°49 W (Fig. 21.1). With an estimated population of 8,095,498 and a density of 143/km² (DGSCN 2022), the country is subdivided into five administrative and economic regions: Savannahs, Kara, Central, Plateaux and Maritime. Togo has two major climatic zones. The tropical Sudanian climate in the north (from the eighth parallel) has two seasons: rainy (May to October) and dry (November to April). In this zone, yearly rainfall ranges between 850 and 1400 mm, and plant growth lasts less than 175 days. The Guinean tropical climate in the south (south of the seventh parallel) is distinguished by two dry seasons and two rainy seasons of unequal length. The yearly rainfall ranges from 1000 to 1600 millimeters. The average temperature is quite high. It can reach 28 °C in the northern areas and 27 °C in the coastal zone (Badameli and Dubreuil 2015). The average relative humidity is high in the south (73–90%) but low in the north (53–67%). The average wind speed is 1.93 m/s, and the average sunshine length is 6 hours and 37 minutes every day. The average evapotranspiration is 1.540 mm per year (Badameli and Dubreuil 2015).

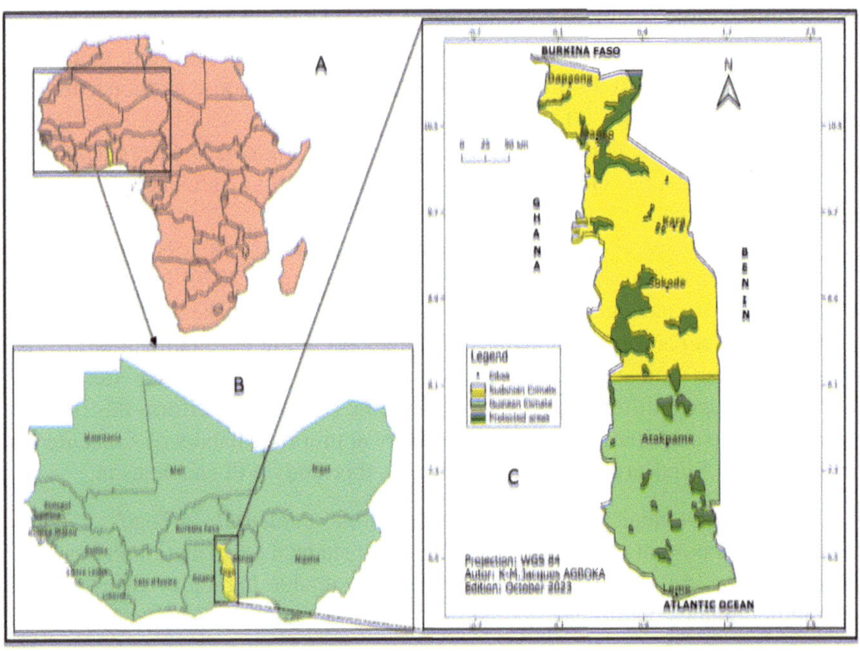

Fig. 21.1 Location of the study area (**a**) Togo in Africa, (**b**) Togo in West Africa, (**c**) Climatic zones and protected areas of Togo

21.2.2 Origin and Distribution of the Study Species

The farmed cashew is said to have started in the restinga (low vegetation found along the sandy coast of eastern Brazil) in northeastern Brazil. *Anacardium occidentale* is also likely native to the savannas of Colombia, Venezuela, and the Guianas (Mitchell and Mori 1987). It is also a dominant native in central and Amazonian Brazil's cerrados. Early travelers introduced the plant to South America west of the Andes, Central America, the West Indies, and finally to the East. Lautié et al. (2001) claim that the cashew tree expanded spontaneously and naturally in South America but that its expansion to other continents is the result of human interference. The Portuguese brought it to Mozambique and India between 1563 and 1578 (Andrighetti et al. 1994), before spreading it to Southeast Asia. It spread throughout Africa in the second half of the sixteenth century, beginning on the east coast and progressing westward. The tree then spread to Australia and portions of North America. Cashew nut cultivation in Togo began in the 1960s through collective fields managed by the Togolese Ministry of Agriculture's Société régionale d'aménagement development (SORAD) to supply the company TOGOFRUIT (Office National de promotion et de développement des cultures). Production began in colonial times. When TOGOFRUIT closed in 1982, the industry came to an end; it resumed in 2000 (MAEH 2015). Cashew nuts are currently farmed in four of the six agricultural zones (Savanes, Kara, Centrale, and Plateaux).

21.2.3 Occurrence Data Collection

From August to December 2023, incidence data for cashew trees in Togo was collected using a Global Positioning System (GPS). The research undertaken by Atakpama et al. (2023) enabled us to collect occurrence points fro.

m Benin and Ghana. GPS was used to record the locations where cashews were found. To improve the model's accuracy, it was proposed that occurrence data include as much of the area as possible where the species is affected by comparable climatic circumstances. As a consequence, fresh cashew occurrence data in West Africa was obtained from the Global Biodiversity Information Facility (GBIF, 2022) platform (https://www.gbif.org/) via the "gbif" plugin in QGIS 3.24 software1 (Fig. 21.2). These are the latest data from 1950 to 2021. After cross-checking and suppressing inaccurate and duplicated occurrences, the modeling comprised a total of 2538 occurrence points. Table 21.1 displays information on the locations of occurrence and their sources.

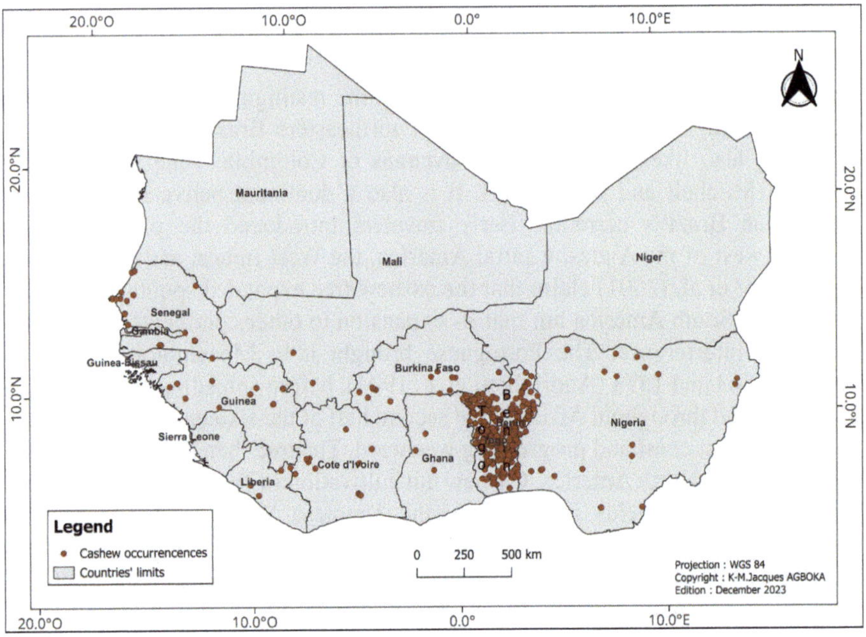

Fig. 21.2 Spatial distribution of cashew tree in West Africa

Table 21.1 Sources and number of occurrence points used for modeling

Source of occurrence points	Number of occurrence
Atakpama et al. (2023)	705
Fieldwork	833
GBIF	1000

21.2.4 Environmental Data

Environmental data such as climate, elevation, and soil was obtained. The soil data was gathered from the FAO's Harmonized Soil Database (https://www.fao.org/soils-portal/data-hub/soil-maps-anddatabases/harmonized-world-soil-database-v12/en/) (accessed on December 12, 2023). The climate and elevation data was collected from the WorldClim version 2.1 database (http://www.worldclim.org, retrieved on December 15, 2023), which includes all 19 bioclimatic variables. Data on bioclimatic variables was obtained for the present (1970–2000) and future (2070; 2061–2080) climates at a resolution of 30 arc-seconds (about 1 km^2). Previous studies reported some discontinuities in the bioclimatic variables, namely the mean temperature of the wettest quarter (Bio 8), mean temperature of the driest quarter (Bio 9), precipitation of the warmest quarter (Bio 18), and precipitation of the coldest quarter (Bio 19) in sub-Saharan Africa (Booth-Butterfield et al. 2022; Biaou et al. 2023). These discontinuities were attributed to sudden changes in the quarterly periods used to calculate these variables.

As a result, we excluded the four factors from further study. These variable layers were clipped along West African borders before being converted to ASCII files that could be used with the MaxEnt method (version 3.4). The models were run using all environmental variables (Moukrim et al. 2020; Feng et al. 2019) rather than correlative analysis and variable selection as in numerous prior research by Dimobe et al. (2020) and Atakpama et al. (2023). The potential future distributions of *Anacardium occidentale* under climate change were predicted using two Global Circulation Models (GCMs): "HadGEM3-GC3.1-LL" (Third Hadley Centre Global Environment Model in the Global Coupled Configuration 3.1) developed by the Hadley Centre, United Kingdom (Andrews et al. 2020); and MIROC6 (Model for Interdisciplinary Research on Climate) developed by the Japanese modeling community, Tatebe et al. (2019). Koo et al. (2017) forecasted future climatic conditions for the 2070 horizon (average for 2061–2080) using two IPCC-CMIP6 Shared Socioeconomic Pathways (SSP) scenarios, ssp245 and ssp585, to decrease climate uncertainty. According to Ames-Martínez et al. (2022), the ssp245 scenario has medium forcing and intermediate climate change issues, while the ssp585 scĕnario is the most chaotic if no efforts are made to mitigate the consequences of climate change.

21.2.5 Model Development and Validation

The Excel file containing cashew tree occurrence points has been converted to a CSV file for compliance with the MaxEnt model. The model was run ten times in order to improve its accuracy. Nneji et al. (2020) reported that 25% of the sites of presence were utilized to test the model, while the remaining 75% were used to train the model. To assess model accuracy, the area under the receiver operating curve (AUC) was determined using receiver operating characteristic (ROC) analysis. Swets (1988) divided AUC values into three categories: 0.5–0.7, 0.7–0.9, and ˃0.9, indicating weak, moderate, or strong model performance, respectively. AUC results near 0.5 show that the model is no better than a randomly generated model, whereas AUC values ˃0.9 reflect the model's SDM strength (Wei et al. 2018). The Jackknife technique was utilized to evaluate whether environmental factors were significant for *Anacardium occidentale* distribution (Saha et al. 2021). The real skill statistics (TSS) measure the model's ability to properly detect real presences (sensitivity) and true absences (specificity). TSS = 0 implies a random prediction, whereas values around 1 (TSS > 0.5) imply a model with strong predictive power (Allouche et al. 2006). The TSS was calculated according to Allouche et al. (2006):

True skill statistic (TSS) = Sensitivity + Specificity-1

The cashew potential, present, and future growth zones were mapped using the QGIS 3.24 software (1). According to Atakpama et al. (2023), two types of habitats were first identified using the 10-percentile criterion: unfavorable habitat (habitats with a probability below the threshold) and advantageous habitat (habitats with a probability above the threshold). The favorable habitat was then separated into three categories: less favorable, fairly favorable, and extremely favorable. The extent of each habitat, as well as its dynamics, was determined for each scenario.

21.3 Results

21.3.1 Contribution of Variables and Model Performance

The average AUC and true skill statistic (TSS) values were 0.949 (Fig. 21.3) and 0.88, indicating that habitat prediction was accurate. The soil variable (hwsd) contributed the most to the models, followed by yearly precipitation (bio12) and temperature seasonality (bio4). The altitude (elev) was the least contributing variable, as seen in Fig. 21.4.

21.3.2 Current and Future Potential Cropping Area of A. occidentale *in 2070*

Current forecasts for prospective suitability (less favorable, moderately favorable, and very suitable) regions identify 89.14% of Togo's national territory, while possibly unfavorable areas account for 10.86% (Figs. 21.5 and 21.6). The present

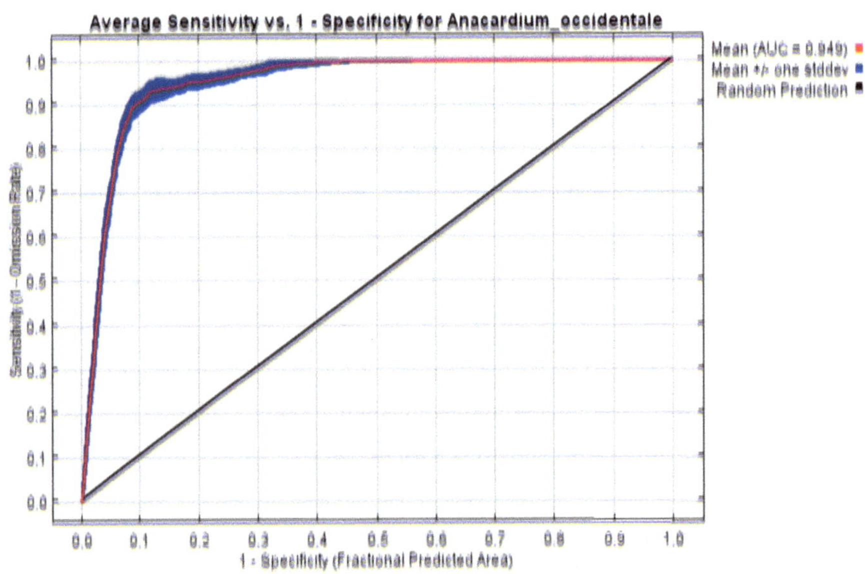

Fig. 21.3 The cross-validated areas under the receiver operating characteristic curve (AUC)

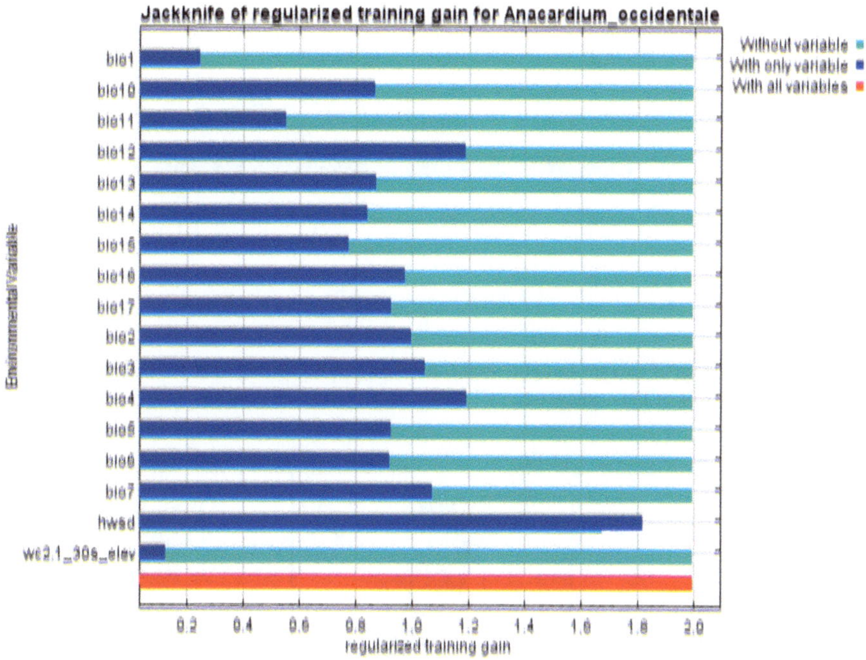

Fig. 21.4 Contribution of variables in the modeling according to the Jackknife test

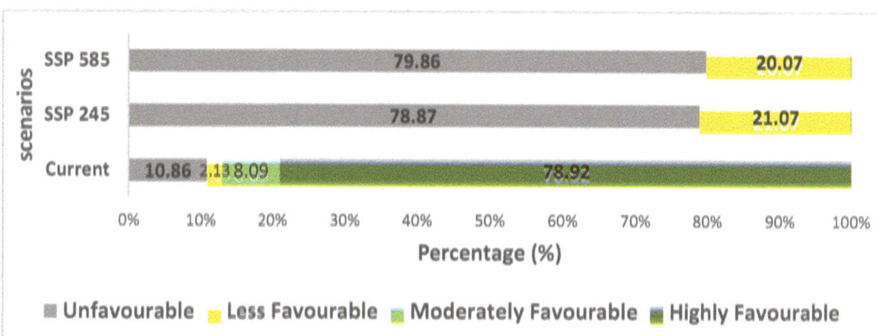

Fig. 21.5 Proportion of current and potential future habitats according to MIROC6 under ssp245 and ssp585 scenarios of the cashew tree in Togo

situation indicates that *A. occidentale* might be found in the eastern portion of the Plateaux region, the majority of the Central region, the whole Kara region, and the majority of the Savanes region (Fig. 21.7).

Figure 21.7 depicts the projected climate change consequences on the possible distribution of *A. occidentale* in Togo using the HadGEM3-GC3.1-LL and MIROC6 models for ssp245 and ssp585 scenarios. Models employing ssp245 and ssp585 data demonstrated strong prediction performance for the 2070 era, with AUC and TSS values of 0.949 and 0.88, respectively. These Global Circulation Models revealed a

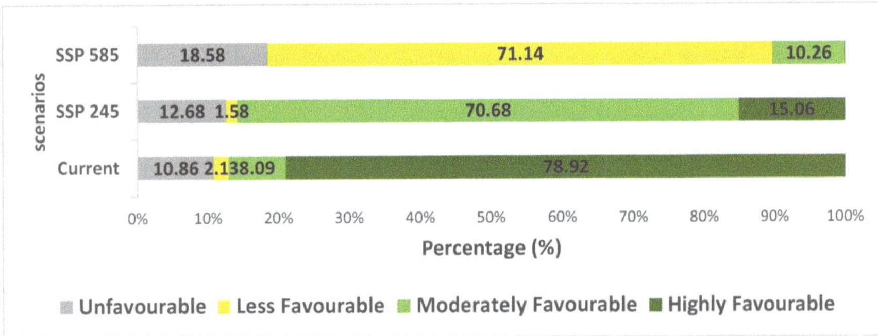

Fig. 21.6 Proportion of current and potential future habitats according to HadGEM under ssp245 and ssp585 scenarios of the cashew tree in Togo

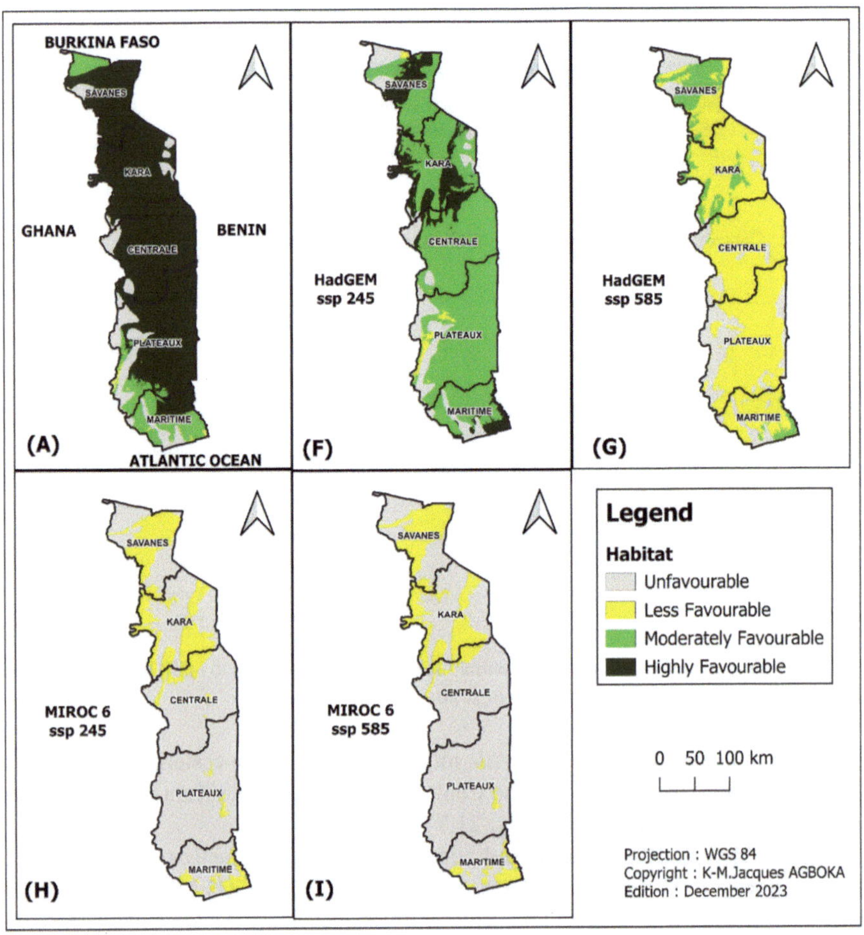

Fig. 21.7 Current habitat suitability according to economic regions; future potential distribution in 2070 for *Anacardium occidentale* in Togo, according to GCM-HadGEM A, scenarios; F, ssp. 245; G, ssp585 GCM-MIROC6, scenarios; H, ssp. 245; I, ssp. 585

large geographical extension of the suitable (less favorable, moderately favorable, and unfavorable) areas when compared to existing values. Still, the extremely advantageous appropriate regions were reduced (Fig. 21.7). The HadGEM model utilizing the ssp245 and ssp585 scenarios will dramatically reduce the number of locations designated as having high suitability (Fig. 21.6). Figure 21.5 displays the findings of the MIROC6 model for the ssp245 and ssp585 scenarios. The MIROC6 model predicts that high-potential cashew agriculture regions would vanish entirely under scenarios ssp245 and ssp585. According to the HadGEM model, possible regions of high cashew cultivation would be 15.06% and 0.02% of Togolese territory under the ssp245 and ssp585 scenarios, respectively. Under the scenario ssp245, the HadGEM model anticipated a considerable extension (from the Maritime region to the Savanes region) of possibly moderately good places for *A. occidentale*; however, in the ssp585 scenario, the same broad expansion is shown for less favorable habitats (Fig. 21.7). For the MIROC6 model, the potential unfavorable area for *A. occidentale* expands significantly under ssp245 and ssp585, with changes seen from the Maritime to the Savanes regions (Fig. 21.7).

21.4 Discussion

21.4.1 Model Reliability and Variable Contribution

The strength and validity of species distribution models are determined by the input data. This study used the maximum entropy (MaxEnt) approach, which has been used by other authors in species distribution (Abdou et al. 2016; Lyam et al. 2012), to map potential cashew-producing locations in Togo. This approach's strength is its ability to combine occurrence data and environmental parameters from across the research region (Lahoz-Monfort et al. 2014). It can also run quantitative and qualitative data (Atakpama et al. 2023). The area under the curve precisely anticipated the geographic distribution of cashew trees in Togo. Currently, the less favorable climate for cashew farming is found in areas with considerable rainfall, particularly in the southern mountainous highlands and Togo's coastline zone. The Kara and Central regions are highly beneficial. This discovery is compatible with the species' current cultivation zones and the cashew tree's rainfall requirements (Ricau 2013).

Future predictions (2070) from both models and scenarios show a decrease in the appropriate habitat (extremely favorable) compared to the existing area (current >2070). Edaphic variables are one of the environmental factors that can predict and impact species distribution, particularly for plant species (Austin and Van Niel 2011; Mod et al. 2016). This study found that soil is a major driver in the geographical distribution of cashews, followed by yearly precipitation (bio12) and temperature seasonality (bio4). Soil impacts plant physiological condition; hence, its impact on the model is considerable, as Gnahoua and Louppe (2003) described the cashew's edaphic requirements. The plant cannot live on clayey, wet, lowland soils such as those found in Togo's Maritime area. This situation might be related to soil pH, which has been shown to have a major influence on plant species distribution (Chytrý et al. 2003). The pH of the soil influences resource availability and plant nutrient uptake. Furthermore, FAO (1994) states that the ideal pH for cashew development is between 4.5 and 6.5, with a lower limit as low as 3.8. Soil moisture content is determined by seasonal temperatures and precipitation (Mod et al. 2016). The impact of seasonality, temperature, and yearly precipitation supports Lyam et al. (2012) findings. These researchers revealed that temperature and precipitation have a significant influence on plant species and vegetation distribution. Both of these substances added to soil have a direct influence on cashew regional spread and may be important environmental aspects in the species' ecology (Atakpama et al. 2023).

21.4.2 Implications of Future Climate Conditions

Predicting how species will adapt to climate change is an important step toward designing biodiversity conservation and management methods, as well as making cashew nut production climate resilient. Climate predictions indicate that the range of habitat now suitable for cashew growing in Togo might change under both HadGEM and MIROC6 scenarios (ssp245 and ssp585). This might be because the cashew tree lacks the ability to adapt to climate change. Furthermore, climate change is projected to limit the geographic distribution of cashew trees. These findings contrast with those of Atakpama et al. (2023), who found that climate change will not reduce the geographical expansion of cashew trees in Togo. This disparity is due to the fact that our study used the Shared Socioeconomic Pathways (SSPs) scenarios, whereas Atakpama et al. (2023) used the Representative Concentration Pathways (RCPs).

The forecast models used for the 2070 horizon indicated that habitats suitable for cashew production will diminish. Given the evolution of habitats using HadGEM, MIROC6, and both scenarios ssp245 and ssp585, it is clear that climate change will pose a threat to cashew growing in Togo. It is thus a crop that is vulnerable to climate change. Mod et al. (2016) also shows that biotic and anthropogenic perturbations may affect the distribution of species niches. Natural disturbances such as herbivory, human settlements, and density in SDMs should be good indicators of how climate change would affect species niches (Atakpama et al. 2023). This can help us understand how temperature and precipitation patterns influence vegetation spread. According to HadGEM's forecasts under ssp245 and ssp585 scenarios, as well as MIROC6's projections under ssp245 and ssp585 scenarios, Togo's ecosystems, which are currently very advantageous to cashew, would be more altered by climate change, becoming unsuitable by 2070.

21.5 Conclusion

This study looked at how climate change would affect the spatial distribution of prospective cashew tree-producing zones in Togo by 2070. Climate change is expected to gradually diminish *Anacardium occidentale's* suitable habitat by 2070. The findings indicate that soil is the most important predictor of the geographic spread of cashew plants in Togo. Current climatic conditions suggest that 78.92% of the country's land area is suitable for cashew nut cultivation. Under two climatic scenarios (ssp245 and ssp585), the proportion of highly suitable habitats in both models (HadGEM and MIROC6) will fall significantly by 2070. *Anacardium occidentale*'s productivity is expected to decline as the suitable habitat area declines. To adapt to climate change, preventive measures such as maintaining and cultivating A. occidentale may be required. These findings will assist cashew plantation growers and the Ministry of Agriculture, Livestock, and Rural Development in developing effective ways to strengthen the population's resistance to future climatic catastrophes. Designing areas that could serve as sustainable cashew growing zones. Designating places as potential sustainable cashew growing areas can also help *Anacardium occidentale's* conservation status.

References

Abdou L, Diouf A, Inoussa MM, Mamoudou BM, Illiassou SA, Mahamane A (2016) Modelling the geographic distribution of Prosopis africana (G. and Perr.) Taub in Niger. Environ Nat Resour Res 6(2):136–144

Akinhanmi TF, Atasie VN, Akintokun PO (2008) Chemical composition and physicochemical properties of cashew nut (*Anacardium occidentale*) oil and cashew nut shell liquid. J Agric Food Environ Sci 2(1):1–10

Allouche O, Tsoar A, Kadmon R (2006) Assessing the accuracy of species distribution models: prevalence, kappa and the true skill statistic (TSS). J Appl Ecol 43(6):1223–1232

Ames-Martínez FN, Luna-Vega I, Dieringer G, Rodríguez-Ramírez EC (2022) The effect of climate change on Arcto-Tertiary Mexican beech forests: exploring their past, present, and future distribution. Ecol Evol 12(8):e9228

Anderson RP, Lew D, Peterson AT (2003) Evaluating predictive models of species' distributions: criteria for selecting optimal models. Ecol Model 162:211–232

Andrews A, Saslow DKS, Manassaram-Baptiste D, Smith RA, Fontham ET, & American Cancer Society Guideline Development Group. (2020). Human papillomavirus vaccination 2020 guideline update: American Cancer Society guideline adaptation. CA: a cancer journal for clinicians, 70(4):274–280

Andrighetti A, Booth-Butterfield S, Cooke P, Casteel B, Lang T, Pearson D, & Rodriquez B (1994). Simultaneous versus exclusive processing of persuasive arguments and cues. Communication Quarterly, 42(1):21–35

Atakpama W, Assang MD, Ahouandjinou EBO, Dimobe K, Gouwakinnou GN, Akpagana K (2023) Priority area for sustainable Cashew (iAnacardium occidentale/i L.) growing in Togo: economic implications and conservation areas. Revue Nature et Technologie 15(1):29–40

Austin MP, Van Niel KP (2011) Improving species distribution models for climate change studies: variable selection and scale. J Biogeogr 38(1):1–8. https://doi.org/10.1111/j.1365-2699.2010.02416.x

Badameli A, Dubreuil V (2015) Diagnostic du changement climatique au Togo à travers l'évolution de la température entre 1961 et 2010. In XXVIIIe Colloque de l'association internationale de Climatologie, pp 421–426

Biaou S, Gouwakinnou GN, Noulèkoun F, Salako KV, Kpoviwanou JMRH, Houehanou TD, Biaou HSS (2023) Incorporating intraspecific variation into species distribution models improves climate change analyses of a widespread West African tree species (Pterocarpus erinaceus Poir, Fabaceae). Global Ecol Conserv 45:e02538

Booth-Butterfield S, Cooke P, Andrighetti A, Casteel B, Lang T, Pearson D, Rodriquez B (2022) Simultaneous versus exclusive processing of persuasive arguments and cues. Commun Q 42(1):21–35

Brunel JF, Hiepko P, Scholz H (1984) Flore analytique du Togo: phanérogames. Englera:3–751

Carpenter G, Gillison AN, Winter J (1993) DOMAIN: a flexible modelling procedure for mapping potential distributions of plants and animals. Biodivers Conserv 2:667–680

Chomitz KM, Buys P, DeLuca G, Thomas TS, Wertz-Kanounnikoff S (2006) At Loggerheads? Agricultural expansion, poverty reduction, and the environment in the tropics. The World Bank, Washington, p 284

Chytrý M, Tichý L, Roleček J (2003) Local and regional patterns of species richness in Central European vegetation types along the pH/calcium gradient. Folia Geobot 38:429–442. https://doi.org/10.1007/BF02803250

DGSCN (2022) General Census of Population and Housing final results. Togo: Togolese Republic Ministry to the President of the Republic, Head of Planning, Development and Regional Planning, French, p 24

Dimobe K, Ouédraogo A, Ouédraogo K, Goetze D, Stein K, Schmidt M, Thiombiano A (2020) Climate change reduces the distribution area of the shea tree (Vitellaria paradoxa CF Gaertn.) in Burkina Faso. J Arid Environ 181:104237. https://doi.org/10.1016/j.jaridenv.2020.104237

Elith J, Graham CH, Anderson RP, Dudík M, Ferrier S, Guisan A, Hijmans RJ, Huettmann F, Leathwick JR, Lehmann A (2006) Novel methods improve prediction of species' distributions from occurrence data. Ecography 29:129–151

FAO (1994) Ecocrop 1. The adaptability level of the FAO crop environmental requirements database. Version 1.0. Land Resources Management and Conservation Service, FAO, Rome

Feng X, Park DS, Liang Y, Pandey R, Papeş M (2019) Collinearity in ecological niche modelling: confusions and challenges. Ecol Evol 9(18):10365–10376. https://doi.org/10.1002/ece3.5555

Global Biodiversity Information Facility (GBIF) (2022) Angelica gigas Nakai. Available online: https://www.gbif.org/species/5537749. Accessed on 15 Dec 2023

Gnahoua GM, Louppe D (2003) Anacardier, HAL-Archives ouvertes. http://hal.cirad.fr/cirad-00429280. Accessed 19 Dec 2023

Hirzel AH, Hausser J, Chessel D, Perrin N (2002) Ecological-niche factor analysis: how to compute habitat-suitability maps without absence data? Ecology 83:2027–2036

IPCC (2022) The IPCC Sixth Assessment Report WGIII climate assessment of mitigation pathways: from emissions to global temperatures. Geoscientific Model Development, 15(24):9075–9109

Khanum R, Mumtaz AS, Kumar S (2013) Predicting impacts of climate change on medicinal asclepiads of Pakistan using Maxent modeling. Acta Oecol 49:23–31

Koo KA, Park SU, Kong WS, Hong S, Jang I, Seo C (2017) Potential climate change effects on tree distributions in the Korean peninsula: understanding model & climate uncertainties. Ecol Model 353:17–27

Koungbanane D, Zahiri PE, Totin Vodounon HS, Amoussou E, Lare LY, Koubodana DN (2020) Analyse fréquentielle et détermination des seuils pluvio-hydrologiques de risques d'inondation dans le bassin-versant de l'Oti au Togo. Afrique Sci 17(1):73–88

Lahoz-Monfort JJ, Guillera-Arroita G, Wintle BA (2014) Imperfect detection impacts the performance of species distribution models. Glob Ecol Biogeogr 23(4):504–515. https://doi.org/10.1111/geb.12138

Lautié E, Dornier M, de Souza FM, Reynes M (2001) Les produits de l'anacardier: caractéristiques, voies de valorisation et marchés. Fruits 56(4):235–248

Lyam PT, Adeyemi TO, Ogundipe OT (2012) Distribution modelling of Chrysophyllum albidum G. Don. in South-WestNigeria. J Nat Environ Sci 3(2):7–14

MAEH (2015) Programme d'investissement programme agricole, national de sécurité alimentaire et nutritionnelle (PNIASAN). MAEP

Mitchell JD, Mori SA (1987) The *A. occidentale* and its relatives (*Anacardium*: Anacardiaceae). Mem N Y Bot Gard 42:1–76

Mod HK, Scherrer D, Luoto M, Guisan A (2016) What we use is not what we know: environmental predictors in plant distribution models. J Veg Sci 27(6):1308–1322

Moukrim S, Lahssini S, Rifai N, Menzou K, Mharzi-Alaoui H, Labbaci A, Rhazi M, Wahby I, El Madihi M, Rhazi L (2020) Modélisation de la distribution potentielle de Cedrus atlantica Manetti au Maroc et impacts du changement climatique. Bois Forets des Tropiques 344:3–16. https://doi.org/10.19182/bft2020.344.a31888

Ndiaye IS (2019) Analyse de la performance d'une filière d'exportation et impact dans l'atténuation des effets des changements. In This paper was presented at the Conference on Climate Change and Food Security in West Africa co-organized by Université Cheikh Anta Diop de Dakar (UCAD) and Center for Development Research (ZEF), University of Bonn, pp 17–18. https://research4agrinnovation.org/publication/proceedings wascal-conference/. Accessed 19 Dec 2023

Nneji LM, Salako G, Oladipo S, Ayoola AO, Onadeko AB, Adedeji BE, Adeola AC (2020) Species distribution modelling predicts habitat suitability and reduction of suitable habitat under future climatic scenarios for Sclerophrys perreti: a critically endangered Nigerian endemic toad. Afr J Ecol 58(3):481–491

Park JJ, Mo HH, Lee GS, Lee SE, Lee JH, Cho K (2014) Predicting the potential geographic distribution of T hrips palmi in K orea, using the CLIMEX model. Entomol Res 44(2):47–57

Pearson RG, Dawson TP (2003) Predicting the impacts of climate change on the distribution of species: are bioclimate envelope models useful? Glob Ecol Biogeogr 12:361–371

Pearson RG, Thuiller W, Araújo MB, Martinez-Meyer E, Brotons L, McClean C, Lees DC (2006) Model-based uncertainty in species range prediction. J Biogeogr 33(10):1704–1711

Phillips SJ, Anderson RP, Schapire RE (2006) Maximum entropy modeling of species geographic distributions. Ecol Model 190(3–4):231–259

Ricau P (2013) Connaître et comprendre le marché international de l'anacarde, p 48. Available online: https://www.inter-reseaux.org/wpcontent/uploads/Guide_RONGEAD___Le_Marche_Internation al_de_l_Anacarde_v-light.pdf. Accessed on 15 Dec 2023

Saha A, Rahman S, Alam S (2021) Modelling current and future potential distributions of desert locust Schistocerca gregaria (Forskål) under climate change scenarios using MaxEnt. J Asia-Pac Biodivers 14(3):399–409

Swets JA (1988) Measuring the accuracy of diagnostic systems. Science 240(4857):1285–1293

Tatebe H, Ogura T, Nitta T, Komuro Y, Ogochi K, Takemura T, Kimoto M (2019) Description and basic evaluation of simulated mean state, internal variability, and climate sensitivity in MIROC6. Geosci Model Dev 12(7):2727–2765

Togolese Republic (2021) Togolese Republic Revised Nationally Determined Contribution Available at: Togolese Republic submission to United Nations Framework Convention on Climate Change (UNFCCC)

Velasquez-Tibat J, Salaman P, Graham CH (2013) Effects of climate change on species distribution community structure, and conservation of birds in protected areas in Colombia. Reg Environ Chang 13:235–248. https://doi.org/10.1007/s10113-012-0329-y

Wa Rusaati BI, Kang JW (2024) MaxEnt modeling for predicting the potential distribution of Lebrunia bushaie Staner (Clusiaceae) under different climate change scenarios in Democratic Republic of Congo. J Asia-Pac Biodivers 17(1):1–6

Wang JJ, Cao B, Bai CK, Zhang LL, Che L (2014) Potential distribution prediction and suitability evaluation of Fritillaria cirrhosa D. Don based on Maxent modeling and GIS. Bull Bot Res 34(5):642

Wei B, Wang R, Hou K, Wang X, Wu W (2018) Predicting the current and future cultivation regions of Carthamus tinctorius L. using MaxEnt model under climate change in China. Global Ecol Conserv 16:e00477

Xu C, Zhang L, Zhang K, Tao J (2023) MaxEnt modeling and the impact of climate change on Pistacia chinensis Bunge habitat suitability variations in China. Forests 14(8):1579

Yang J, Huang Y, Jiang X, Chen H, Liu M, Wang R (2022) Potential geographical distribution of the edangred plant Isoetes under humanactivities using MaxEnt and GARP. Glob Ecol Conserv 38:e02186

Zhang K, Yao L, Meng J, Tao J (2018) Maxent modeling for predicting the potential geographical distribution of two peony species under climate change. Sci Total Environ 634:1326–1334

Zhang K, Liu H, Pan H, Shi W, Zhao Y, Li S, Tao J (2020) Shifts in potential geographical distribution of Pterocarya stenoptera under climate change scenarios in China. Ecol Evol 10(11):4828–4837

Chapter 22
Integrating Remote Sensing and Land-Use Change to Identify Potential Wetland Restoration Areas: A Multiyear Analysis Using Landsat Data

Jintu moni Bhuyan, Manju Tiwari, and Nitish Tiwari

Abstract Wetlands play a critical role in biodiversity conservation, climate regulation, and water purification, but human activities, including urbanization, agriculture, and industrialization, increasingly threaten them. This study aims to assess wetland loss and gain in Chhattisgarh between 2002 and 2023 using remote sensing technologies, specifically Landsat 5 TM, Landsat 8, and Sentinel-1 data. The objectives include mapping wetland dynamics over time and analyzing spatial and temporal changes in wetland areas to evaluate the effectiveness of restoration and management efforts. Using the Modified Normalized Difference Water Index (MNDWI), we mapped wetland areas and classified them into permanent and transient water bodies. The results show significant spatial heterogeneity across the region, with notable wetland gains in districts like Bilaspur (4.393 ha^2), Dhamtari (5.571 ha), and Bastar (1.668 ha, 0.01668 km^2), attributed to human interventions such as water conservation projects, improved irrigation infrastructure, and natural rejuvenation. However, areas like Dakshin Bastar Dantewada experienced minimal wetland gains (0.199 ha, 0.00199 km^2) due to challenges such as mining and deforestation. These findings emphasize the varying effectiveness of local conservation strategies and the role of both natural and anthropogenic factors in wetland dynamics. The study highlights the importance of targeted wetland restoration and management practices, particularly in areas facing anthropogenic pressures. The results underscore the potential of remote sensing tools in monitoring wetland health and

J. m. Bhuyan
Forestry and Ecology, Indian Institute of Remote Sensing, ISRO, Dehradun, Dehradun, Uttarakhand, India

M. Tiwari (✉)
SVCAET&RS,2 Indira Gandhi Krishi Vishwavidyalaya, Raipur, Chhattisgarh, India

N. Tiwari
Deparetment of Agronomy, Indira Gandhi Krishi Vishwavidyalaya, Raipur, Chhattisgarh, India

Y. D. I. Khan et al. (eds.), *Land Restoration Through Ecosystem-Based Approach*, Disaster Resilience and Green Growth, https://doi.org/10.1007/978-981-95-0389-6_22

guiding future conservation efforts. The chapter explains a model that can be used for other regions in the Global South facing similar environmental challenges.

Keywords Dryland · Wetland · Water retention · Restoration · Waterbodies

22.1 Introduction

Wetlands are areas covered by water for at least part of the year. The definition of wetlands can differ greatly among various disciplines and does not have a universally accepted scientific interpretation. The term is used by scientists across different fields, resource agencies, conservation organizations, and others to denote areas wetter than typical dry land (Kaplan et al. 2019; Liu et al. 2020; Tiner 2016).

Wetlands are recognized as some of the planet's most productive and valuable resources. They play a crucial role in regulating the global climate as well as naturally purifying water, help in reducing the impacts of floods and droughts, protect shorelines from erosion, conserve soil, and serve as important habitats for wildlife (Bhowmik 2022; Klemas 2013; Liu et al. 2020; Nayak and Bhushan 2022). Natural wetland environments serve a variety of ecological purposes, including support for soil formation and stabilization, provision of food, water, and biomass from plants, landscape and ecological tourism, cultural and recreational services, climate regulation, and carbon sequestration (Ballut-Dajud et al. 2022).

Human activity poses significant threats to wetlands. Climate change and human activities threaten the resilience of wetlands globally, especially in hot, dry regions facing rapid human expansion. Pollutants from human activity have historically altered natural rivers, causing physicochemical and biological changes. Both man-made and natural disturbances continue to impact wetlands significantly. Major issues include draining wetlands for farming, urban development, pollution, climate change, excessive water extraction, and the spread of invasive species (Newton et al. 2020; Wang et al. 2020). Lu et al. (2021) have studied the impact of anthropogenic activities that caused a decline in the wetland areas and carbon pool in China from 1980 to 2010 due to human activities. Using satellite imagery and the Human Influence Index (HII), the authors found that wetland area declined from 411,000 km^2 to 214,000 km^2 and carbon pools decreased from 15.2 PgC to 7.6 PgC. The findings highlighted the need for enhanced wetland protection to reduce carbon emissions and preserve ecosystem services, providing crucial data for climate policy and action. Ballut-Dajud et al. (2022) have reviewed various factors responsible for global wetland loss, including agriculture, urbanization, aquaculture, and industrial activities. These factors were reported to reduce wetland areas and impair their critical function in carbon sequestration, leading to more significant greenhouse gas emissions and ecosystem degradation. Agriculture was found to have the most significant impact, followed by urbanization and industry. The study covered various wetland types, highlighting coastal and marine ecosystems particularly affected by anthropogenic activities. Ostad-Ali-Askari (2022) studied the effects of human activities on wetlands in southwestern Iran, focusing on how urbanization, agricultural runoff, and industrial waste contribute to the degradation, and emphasized the need for sustainable management strategies to mitigate the compounded impact of these pressures with climate change on hydrological balance, particularly in vulnerable wetland ecosystems.

Various wetland-related studies documented the loss of native species diversity and promotion of biotic homogenization due to wetland loss. India's coastal wetlands are vital for biodiversity, storm protection, and carbon sequestration. Shan et al. (2021) investigated the status and management of India's wetlands, highlighting their important ecosystem services that account for about 50% of global values. Spanning 4.1 million hectares, the wetlands are threatened by urbanization, industrialization, and agricultural practices. Joshi et al. (2021) assessed water quality in four Delhi-NCR wetlands: Sanjay Lake, Okhla Bird Sanctuary, Indira Gandhi Stadium Lake, and Bhalswa Lake. Analysis revealed alkaline water, with Bhalswa Lake showing a $Na^+ - SO_4^{2-}$ composition due to evaporation. The Water Quality Index rated the water as poor to unsuitable for drinking, with Bhalswa Lake unsuitable for irrigation. Seasonal deterioration, particularly in summer, highlights the urgent need for restoration and protection of these wetlands. Sarkar et al. (2021) studied aquatic weeds in Indian floodplain wetlands, which cover about 0.354 million hectares and are vital for biodiversity, fisheries, and local jobs. These wetlands are shrinking due to human activity, climate change, and invasive plants. The study identified 530 aquatic plant species, including 85 for medicinal use and 30 that are edible. The authors emphasized the need for sustainable management to restore these important ecosystems. Shah and Islam (2021) assessed the recreational benefits of Dal Lake in Jammu and Kashmir, estimating a consumer surplus between Rs. 6250 and Rs. 25,000 per visitor, along with an annual value of approximately Rs. 1713 million (USD 26 million). Their findings suggest that travel costs and factors such as age and education significantly impact demand, particularly among younger, more educated individuals. However, the lake's sustainability is jeopardized by human pressures and inadequate management, highlighting the urgent need for conservation efforts. This research underscores the economic significance of natural wetlands in supporting regional livelihoods and maintaining ecological balance. Basu et al. (2021) examined the decline of the Chatra Wetland in Eastern India over 28 years using geospatial techniques. They found a 60% reduction in wetland area and a 71.9% drop in ecosystem service values (ESVs) from 1991 to 2018. If no action is taken, the wetland could disappear by 2045. The research underscores the need for sustainable urban planning and recommends alternative land-use strategies to balance development and conservation. It also calls for urgent policy measures to protect ecosystem services and biodiversity. Investigations were carried out to examine the effects of human activities on three Jammu wetlands—Mansar, Gharana, and Sanasar—using geospatial techniques to analyze land-use changes from 2008 to 2018 (Asgher et al. 2021). Significant encroachment and degradation were reported, especially in Gharana, due to agriculture and urbanization. These activities have decreased water quality and increased eutrophication, particularly in lower-altitude wetlands. The findings recommend targeted conservation strategies for sustainable management. Chakraborty et al. (2023) explored the East Kolkata Wetlands (EKW), a Ramsar site in India known for its natural waste recycling. EKW has a unique ecology and essential services like biodiversity support, flood control, and pollution mitigation. The authors examined threats such as pollution from urbanization, advocating for sustainable, community-driven conservation strategies to protect EKW's ecological and socioeconomic functions. Urbanization, pollution, and climate change contribute to habitat loss and degradation. Studies address coral reef bleaching, mangrove heavy metal contamination, and estuary molluscan biodiversity (Vincent and Owens 2021).

Although conservation policies exist, inadequate enforcement emphasizes the need for targeted management to protect these ecosystems. Excessive withdrawals of water from wetlands for residential, agricultural, or industrial use are responsible for wetland degradation. The construction of dams impedes water flow and replenishment of wetlands and creates a hazard to aquatic living organisms. So, necessary steps such as increasing wetlands and preventing the illegal swamping of wetlands should be taken to conserve the wetland biodiversity from the threats of unplanned urbanization, purify the environment, and mainstream sustainable development. Though the Ramsar Convention policy (Ramsar Convention Secretariat 2016) exists to preserve wetlands and achieve sustainable development worldwide, mass consciousness, greater participation of local people, and the use of indigenous knowledge in the management strategies are needed to protect wetlands.

Remote sensing and GIS techniques are vital for monitoring and restoring wetlands. They effectively capture spatial and temporal changes (Fan et al. 2023; Hemati et al. 2023; Klemas 2013; Mahdianpari et al. 2020; Singh et al. 2020). These technologies enable consistent, large-scale observations essential for tracking ecosystem shifts, land cover, and water quality.

For example, Mahdianpari et al. (2020) demonstrated the utility of satellite imagery and cloud platforms like Google Earth Engine in Newfoundland wetland changes over decades. Singh et al. (2020) study of the Harike wetland in India further highlights how remote sensing provides insights into land use and water quality, supporting targeted conservation against agricultural encroachment. Additionally, remote sensing aids restoration planning, as noted in Klemas's (2013) review, helping managers identify optimal sites and monitor progress with high-resolution imagery. Landsat satellite data are of great importance for monitoring and restoring wetlands, since they provide long-term time series advantages of the datasets. With free, high-resolution images available since 2008, they help researchers study long-term changes in wetlands across various regions (Demarquet et al. 2023; Fan et al. 2023; Hemati et al. 2023; Mahdianpari et al. 2020; Zhang et al. 2023). Landsat's more extended temporal coverage, dating back to 1972, is crucial for wetland analysis as it enables the monitoring of long-term changes in wetland dynamics, seasonal variations, and the impact of climate change over decades, providing invaluable insights into wetland health and resilience. Landsat's reliable data support conservation efforts and land-use planning. Researchers can capture detailed seasonal changes by combining Landsat with other satellites like Sentinel-2.

Sentinel-1 SAR data improves wetland monitoring by offering high-resolution imaging that works in all weather conditions and at any time of day or night. This capability enables the detection of subtle changes in hydrology, vegetation, and surface water extent. Additionally, when combined with Landsat data, it enhances temporal analysis and resilience assessments by overcoming cloud cover limitations (Alam et al. 2023).

The integration of modern technologies of multi-temporal satellite imagery and indices like MNDWI can be used to identify wetland degradation and restoration opportunities. MNDWI is highly effective for wetland mapping due to its enhanced capability to delineate water bodies by utilizing the green and SWIR bands, which improve water feature detection compared to the NIR bands used in LSWI. This increased sensitivity to water content allows MNDWI to more accurately distinguish between water bodies and other land cover types, making it superior for wetland identification. This is described with an illustration for Chhattisgarh, India, which hosts numerous wetlands that play a critical role in biodiversity conservation, water management, and carbon sequestration. The state includes northern hills with dense forests and coal resources, fertile central plains, and the Bastar Plateau, which is rich in flora and diverse tribal communities. These features make Chhattisgarh an ecological hotspot and an ideal candidate for investigating wetland restoration techniques and innovations, to analyze the loss of wetlands due to various types of human activities, and assessing the potential for wetland restoration utilizing remote sensing and cutting-edge technologies.

22.2 Dataset and Methodology

22.2.1 Study Area

Chhattisgarh state is situated in the central part of India. It was established on November 1, 2000, by separating from Madhya Pradesh (Fig. 22.1). The state is located at 21.278° N, 81.866° E at 285–295 m elevation above MSL. This region boasts low hills and a fertile central plain enriched by the Mahanadi River, with forests covering an impressive 44% of the land. Chhattisgarh experiences a tropical hot climate, with summer temperatures soaring to 49 °C and an average annual rainfall of 1292 millimeters (Bhuarya et al. 2018). However, despite this seemingly ample rainfall, the state grapples with significant agricultural challenges stemming from uneven distribution and frequent droughts. While irrigation is paramount, many water resources remain underutilized, as historically, the state relied on tanks for water management and has a limited number of reservoirs. The documentation of Chhattisgarh's wetlands is sparse, and strategic improvements in management could effectively tackle both agricultural and ecological hurdles.

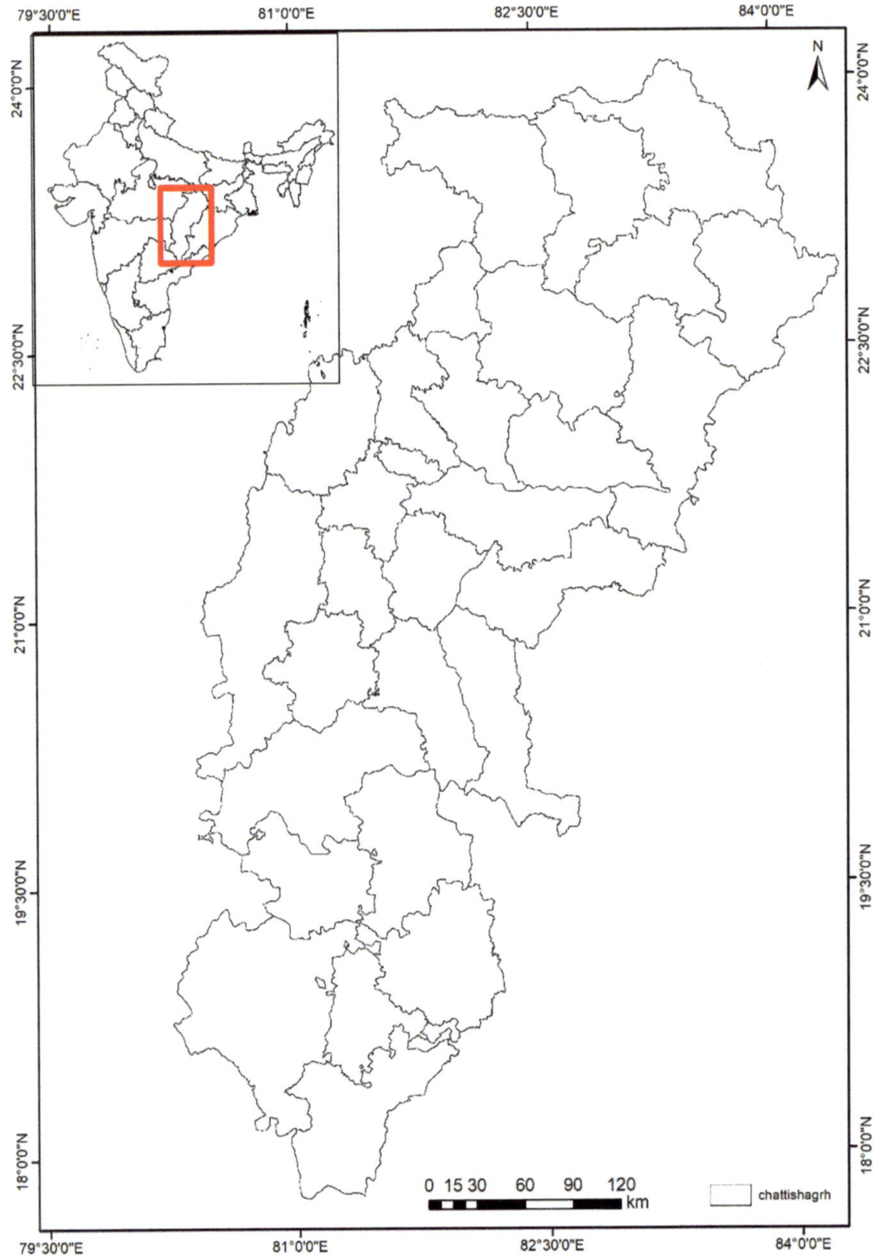

Fig. 22.1 Location map of the study area

Chhattisgarh was selected for this study due to its remarkable combination of abundant natural resources, ecological richness, and pressing environmental concerns. While often hailed as the "rice bowl of Central India," the state faces agricultural limitations due to inconsistent rainfall and recurrent droughts. However, revitalizing wetlands presents an opportunity to combat these challenges by boosting groundwater recharge, regulating floods, and enriching soil fertility.

The state houses three distinct agroecological zones: the Northern Hills, Central Plains, and the Bastar Plateau. Each offers diverse landscapes ideal for implementing innovative restoration techniques. The Bastar Plateau, celebrated for its vibrant tribal culture and biodiversity, provides an invaluable perspective on merging traditional ecological knowledge with contemporary restoration practices.

Furthermore, introducing remote sensing and GIS technologies has proven effective in pinpointing degraded wetlands and prioritizing restoration efforts. The study spans from 2002 to 2023, emphasizing crucial transformations following the state's formation, including various administrative reforms and evolving district boundaries. For clarity, the analysis incorporates a subdivision into 26 districts throughout this essential period.

22.2.2 *Dataset*

For the wetland mapping and change analysis, we utilized the Landsat series of satellite data, focusing on Landsat 8 (30-meter resolution) and Landsat 5 Thematic Mapper (30-meter resolution), both provided by the United States Geological Survey (USGS) (Table 22.1). These datasets, spanning more than three decades, enabled us to assess wetland dynamics over time and quantify the region's wetland loss. Additionally, Sentinel-1 data from the European Copernicus Program, with a spatial resolution of 10 meters, was integrated to enhance spatial accuracy and was particularly valuable for mapping inundated areas during the monsoon season. The radar capabilities of Sentinel-1 allow for reliable water detection even under cloudy conditions, making it well-suited for capturing seasonal flooding patterns in wetlands and adjacent areas.

Table 22.1 Dataset used in the study

Data	Resolution (m)	Source
Landsat 8	30	USGS
Landsat 5TM	30	USGS
District boundary		Survey of India

22.2.3 Methodology

22.2.3.1 Wetland Mapping for 2002 and 2023

Landsat 5TM and 8 OLI data were selected for this study due to their moderate spatial resolution and frequent revisit time, providing an optimal balance between detail and temporal coverage necessary for wetland monitoring. To enhance data quality, preprocessing steps involved cloud masking and surface reflectance calibration for each image within the time frame. Cloud masking relies on the Quality Assessment (QA) band (QA_PIXEL) to identify and remove pixels affected by clouds, cirrus, cloud shadows, and other distortions.

The MNDWI was calculated to highlight water bodies within the study area, as it enhances the visibility of open water features by contrasting green and short-wave infrared (SWIR) bands. The index was computed for each cloud-free image using the formula:

$$\mathrm{MNDWI} = \left(\mathrm{GREEN} - \mathrm{SWIR1}\right) / \left(\mathrm{GREEN} + \mathrm{SWIR1}\right)$$

where the green and SWIR bands represent specific spectral signatures that allow effective water discrimination from land and vegetation.

To create a robust representation of water presence over time, median and maximum MNDWI composites were generated from the filtered dataset. The median composite offers a temporally consistent view of the water bodies, reducing the effects of anomalies or outliers. In contrast, the maximum MNDWI composite captures the highest MNDWI values, thus identifying potential wetland areas. Using thresholding techniques, we classified the water bodies based on MNDWI values. Permanent water bodies, such as rivers and large reservoirs, were delineated with a threshold MNDWI >0.4. At the same time, transient wetlands, which typically experience seasonal inundation, were identified by combining the maximum MNDWI with areas marked as permanent water bodies. The classified outputs were visualized on a cloud-based platform for geospatial analysis. To provide users access to these results, we developed an interactive web application that displays the wetland and water body classifications across Chhattisgarh. This application can be accessed at https://webapp-385310.projects.earthengine.app/view/wetlandmapping

22.3 Results

22.3.1 Wetland Mapping and Loss/Gain Analysis

The wetland area changes across Chhattisgarh between 2002 and 2023 show significant spatial heterogeneity, reflecting the impact of both natural factors and human interventions (Figs. 22.2 and 22.3). In Balod, the wetland area expanded from

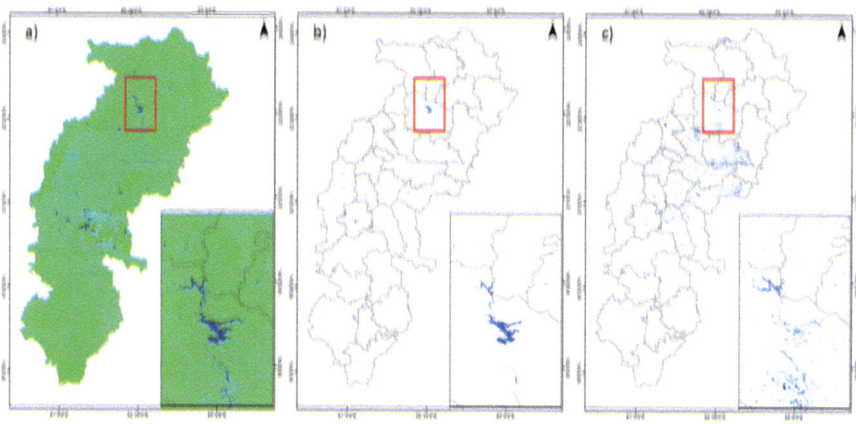

Fig. 22.2 Mapping of wetland using Landsat 8 Image for the year 2002: (**a**) MNDWI, (**b**) permanent water body, and (**c**) wetland

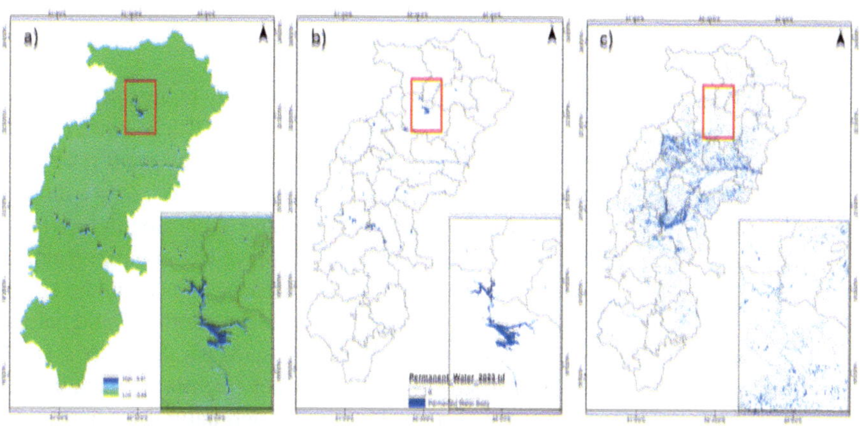

Fig. 22.3 Mapping of wetland using Landsat 8 Image for the year 2023: (**a**) MNDWI, (**b**) permanent water body, and (**c**) wetland

0.0059 km^2 to 0.03423 km^2, representing a gain of 0.02833 km^2. This increase might be attributed to developing water conservation structures such as check dams and community-driven afforestation efforts. Similarly, Baloda Bazar, with a gain of 0.02618 km^2, could have benefited from irrigation canal development and better groundwater recharge mechanisms (Fig. 22.2). These findings suggest an upward trend in water availability and management in these districts.

Balrampur showed a modest gain of 0.0028 km^2, increasing from 0.0074 km^2 to 0.0102 km^2. Being part of a largely forested and hilly area, the district may have seen limited agricultural runoff and a relatively undisturbed hydrological cycle, contributing to this minor increase. On the other hand, Bastar, which experienced a more significant gain of 0.01668 km^2, highlights the role of enhanced rainfall

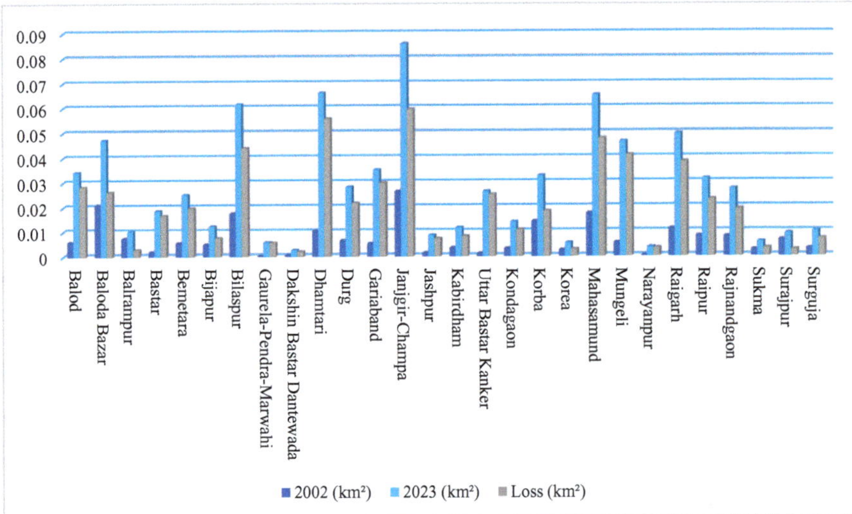

Fig. 22.4 Loss and gain of the wetland from 20,013 to 2023, district wise

patterns and potential wetland rejuvenation projects, especially in areas affected by forest degradation. Similarly, Bemetara, with a gain of 0.01965 km^2, reflects improved water storage capacity, likely due to the construction of small reservoirs and water harvesting schemes. Bijapur, a district with dense forests, saw a modest increase of 0.00744 km^2, possibly owing to natural wetland expansion in low-lying forested areas (Fig. 22.4). In contrast, Bilaspur displayed a notable gain of 0.04393 km^2, one of the highest increases among the districts. This could be attributed to anthropogenic factors, such as urban wetland restoration efforts, and natural phenomena, like enhanced precipitation. Gaurela-Pendra-Marwahi, with a small gain of 0.00557 km^2, likely benefited from isolated water retention structures and the natural hydrology of its forested terrain. Despite its rich natural resources, the district of Dakshin Bastar Dantewada showed one of the smallest gains, at 0.00199 km^2. This limited increase may result from challenges such as mining activities, deforestation, and low prioritization of wetland conservation. Conversely, Dhamtari exhibited a remarkable gain of 0.05571 km^2, reflecting the effectiveness of large-scale watershed management projects and the restoration of degraded wetlands. This district stands as an example of the positive outcomes of well-planned conservation programs.

Despite its rich natural resources, the district of Dakshin Bastar Dantewada showed one of the smallest gains, at 0.00199 km^2. This limited increase may result from challenges such as mining activities, deforestation, and low prioritization of wetland conservation. Conversely, Dhamtari exhibited a remarkable gain of 0.05571 km^2, reflecting the effectiveness of large-scale watershed management projects and the restoration of degraded wetlands. This district stands as an example of the positive outcomes of well-planned conservation programs. Durg, with a gain of 0.02169 km^2, reflects the balanced impact of urbanization and ecological management. In comparison, Gariaband showed a more substantial gain of 0.02985 km^2, likely due to its proximity to forested areas that facilitate natural water retention. The district of Janjgir-Champa stood out with the highest gain of 0.05955 km^2, emphasizing the success of extensive water management programs and likely investments in wetland restoration as part of agricultural development projects.

Jashpur, a hilly district, experienced a small gain of 0.00726 km^2, suggesting that limited anthropogenic disturbances and natural factors, such as stable rainfall, played a key role. Similarly, Kabirdham gained 0.00829 km^2, which might be due to localized water harvesting and soil moisture conservation techniques. Uttar Bastar Kanker experienced a more pronounced gain of 0.02524 km^2, highlighting the potential role of community participation in water conservation and ecological management.

Kondagaon gained 0.01087 km^2, reflecting minor improvements in water retention, while Korba, with a gain of 0.01845 km^2, could have benefited from wetland restoration projects in mining-affected areas. The Korea district, gaining 0.00284 km^2, showed minimal improvement, possibly due to the challenges of limited interventions in remote areas. On the other hand, Mahasamund, with a gain of 0.04775 km^2, demonstrated significant progress, likely due to integrated water management practices and agricultural runoff control.

Mungeli, a district with fertile agricultural lands, gained 0.04108 km^2, likely due to increased irrigation water storage and improved wetland health. Narayanpur, with a modest gain of 0.00337 km^2, might have benefited from localized community-driven conservation programs. Raigarh showed a significant gain of 0.03838 km^2, reflecting successful restoration initiatives, possibly in response to urban and industrial expansion. In Raipur, the gain of 0.02311 km^2 indicates the impact of urban wetland conservation measures, despite the pressures of rapid urbanization. Rajnandgaon displayed a moderate gain of 0.01928 km^2, reflecting a balanced approach to wetland restoration and water resource management. Sukma, with a modest gain of 0.00324 km^2, could reflect the challenges of managing wetlands in tribal and remote regions. Finally, Surajpur and Surguja, with gains of 0.00253 km^2 and 0.00729 km^2 respectively, highlight the potential of small-scale wetland restoration and conservation in hilly and forested areas. However, these minimal gains call for enhanced ecological restoration and community engagement efforts to unlock the full potential of wetland expansion in these districts.

22.3.2 *Wetland Encroached Classes*

Several Land Use/Land Cover (LULC) classes in Chhattisgarh significantly contribute to wetland loss and encroachment. Agricultural land, particularly croplands (Fig. 22.5), is one of the region's primary drivers of wetland degradation. As wetlands are drained for crop cultivation, especially for paddy fields, they lose their

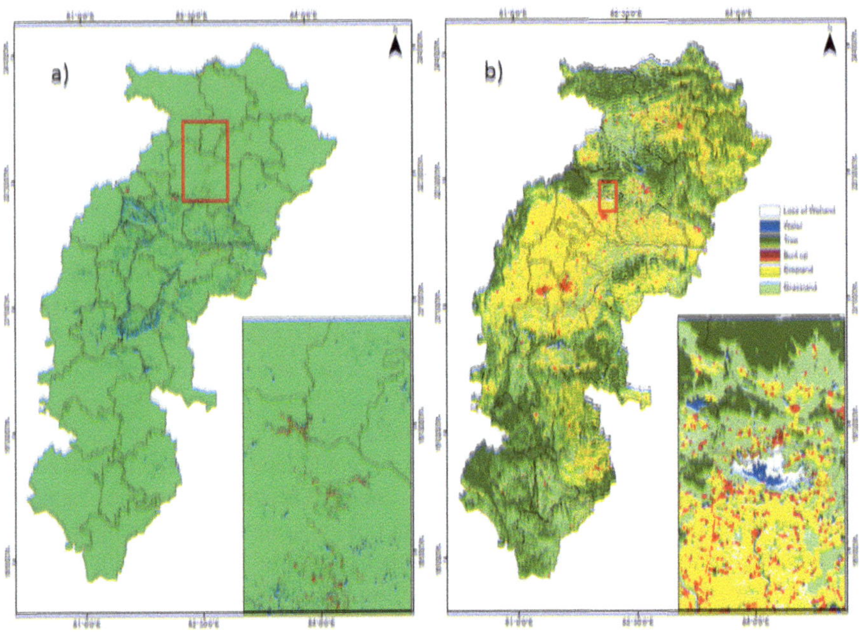

Fig. 22.5 (**a**) Loss and gain of wetland Chhattisgarh and (**b**) white area showing the loss of wetland in that area

ability to retain water, support biodiversity, and provide ecosystem services such as water purification and flood regulation. For instance, districts like Bijapur and Gaurela-Pendra-Marwahi, which have experienced major wetland loss, show a clear correlation between the expansion of agricultural land and the reduction in wetland areas. Urban and built-up areas are also a major concern, with increasing urbanization leading to the encroachment of wetlands for infrastructure development.

This is especially evident in rapidly developing areas like Raipur, where urban sprawl has contributed to significant wetland loss. Furthermore, deforestation around wetlands, such as in regions like Bastar and Surguja, exacerbates the problem by increasing sedimentation in wetland areas, altering hydrological regimes, and diminishing water retention capacity. Mining activities in areas like Korba and Jashpur, where coal extraction and other resource extractions occur, also play a significant role in wetland encroachment, either by directly filling wetland areas or altering their natural water flow. Together, these LULC classes underscore the challenges wetlands face in Chhattisgarh, emphasizing the need for focused restoration and conservation efforts to reverse the ongoing loss and degradation.

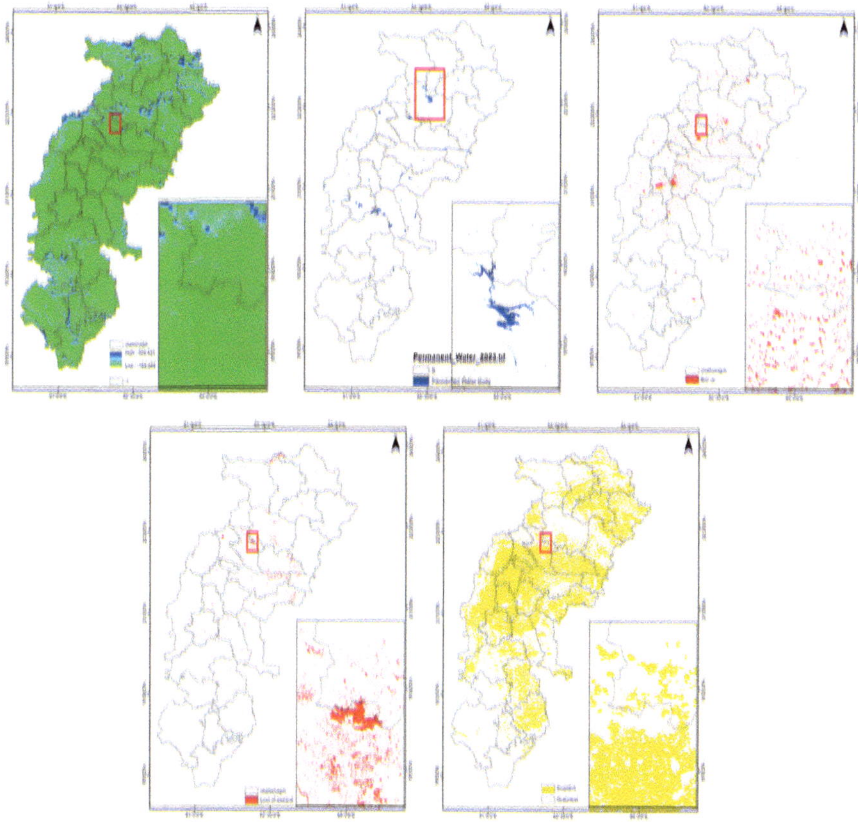

Fig. 22.6 Criteria for wetland restoration suitable area mapping

22.3.3 Potential Area of Wetland Restoration

The methodology involves a weighted multi-criteria analysis to create a wetland restoration suitability map. Five criteria are evaluated: wetland loss, water level, proximity to permanent water bodies, cropland presence, and built-up areas. Suitability conditions are assigned binary values (1 for suitable, 0 for unsuitable) based on thresholds. Weighted values are applied to each criterion (wetland loss: 40%, water level: 30%, permanent water bodies: 20%, cropland: 10%), and the final suitability map is calculated by summing these weighted scores (Fig. 22.6). Built-up areas are excluded by setting their suitability to zero. The results are normalized (0–1) for visualization and saved as a geospatial raster for further analysis.

The analysis of wetland restoration suitability across the districts was conducted using a weighted approach that integrated various environmental and land-use criteria. The primary objective was to identify the most suitable areas for wetland restoration within each district, considering factors such as water levels, proximity to permanent water bodies, cropland areas, and the condition of built-up areas

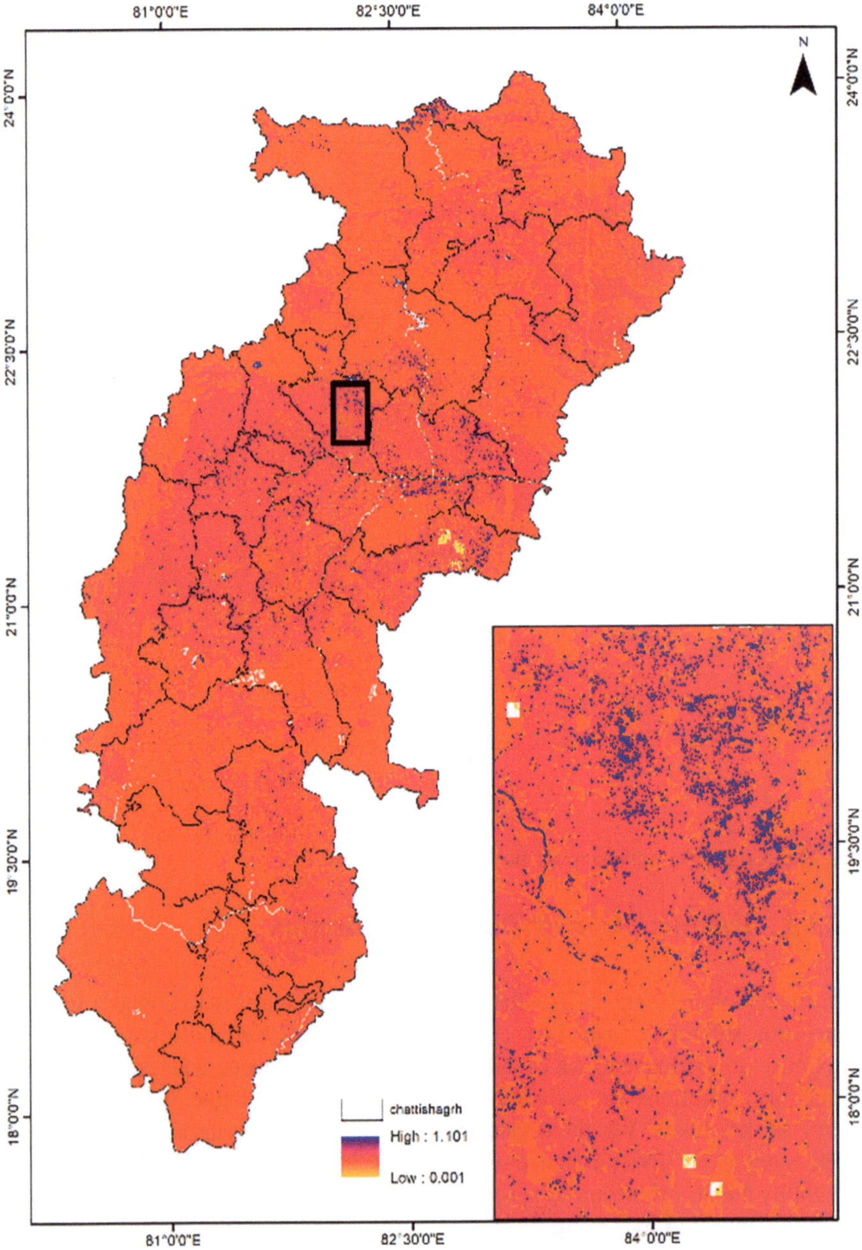

Fig. 22.7 Suitable area for wetland restoration

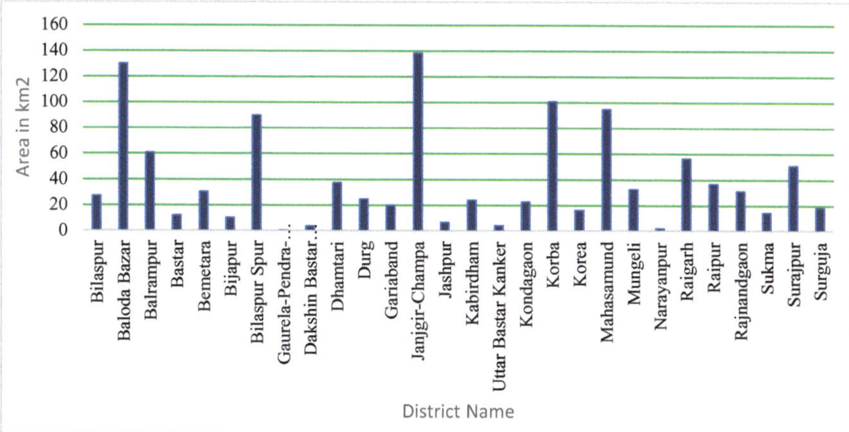

Fig. 22.8 District-wise area for suitable area for wetland restoration

(Fig. 22.7). The results provide valuable insights into the spatial distribution of suitable restoration areas, which can guide policy and decision-making for wetland conservation and restoration efforts.

The dataset reveals considerable variability in the area suitable for wetland restoration across the 28 districts, with the total suitable areas ranging from small to large districts. The district with the largest suitable area was Janjgir-Champa, covering 138.1 km^2, followed by Baloda Bazar with 130.8 km^2 and Mungeli with 94.2 km^2 (Fig. 22.8). These districts, with their expansive areas, provide substantial opportunities for wetland restoration and could serve as key targets for large-scale ecological interventions. On the other end of the spectrum, smaller districts like Gaurela-Pendra-Marwahi, with a suitable area of just 0.5472 km^2, present much more localized restoration opportunities. The limited area in such districts indicates that a more focused approach might be necessary for wetland restoration, possibly within specific wetland patches that could be restored or enhanced. The district of Nariyapur, with an area of 2.17 km^2, also falls within this category, offering a very localized yet critical opportunity for targeted restoration efforts.

The data indicates that areas such as Korba (100.467 km^2), Bilaspur (89.7237 km^2), and Mahasamund (94.5882 km^2) represent substantial land areas with a considerable portion of suitable wetland restoration zones. Due to their large areas and diverse environmental conditions, these districts would benefit from a more detailed, district-wise analysis to identify priority zones for intervention. The extent of wetland restoration suitable areas in these districts suggests a need for a comprehensive landscape-scale approach to ensure wetland ecosystems' connectivity and enhance the ecological benefits of restored areas. Interestingly, districts such as Raigarh (56.3364 km^2), Raipur (36.7083 km^2), and Sukma (14.4918 km^2), although smaller in terms of total area, still contain significant portions of land suitable for wetland restoration. These districts may require more concentrated restoration strategies to enhance local water quality, increase biodiversity, and mitigate climate impacts through ecosystem-based approaches.

Overall, the analysis underscores the spatial heterogeneity of wetland restoration potential across districts. The identified suitable areas for wetland restoration provide essential data for prioritizing restoration interventions, aligning them with environmental, economic, and social goals. This approach supports a science-driven methodology for wetland conservation, ensuring that the restoration efforts are ecologically viable and regionally relevant. Integrating this suitability map into local, regional, and national land-use planning strategies will be key to effective wetland conservation and climate change mitigation efforts. Further analysis can refine these suitability models by integrating additional environmental variables and land-use trends, ensuring that restoration initiatives are adaptive and responsive to future climate and land-use changes.

22.4 Conclusion

The wetland situation in Chhattisgarh paints a worrying picture, with wetland areas disappearing primarily due to agricultural growth, urban expansion, deforestation, and mining activities. This loss puts local wildlife at risk and harms essential ecosystem services, such as cleaning water, regulating floods, and storing carbon. However, using satellite data from Landsat and Sentinel-1 has given us valuable insights into these changes, helping to guide restoration and better management of these vital areas. The most promising sites for restoration are those once wetlands or agricultural lands that were converted from wetlands but still retain the characteristics that make wetlands so important. By focusing on these areas, we can restore crucial environmental functions, improve water quality, boost biodiversity, and even help fight climate change. Additionally, restoring wetlands can support local communities by enhancing agricultural productivity, reducing the need for irrigation, and opening doors to ecotourism. In short, wetland conservation is crucial for Chhattisgarh's future, both for its ecosystems and those who depend on them. The findings from Chhattisgarh can serve as a model for other regions in the Global South that are facing similar environmental challenges. The study offers actionable insights into policy frameworks, governance, and community participation.

References

Alam A, Bhat MS, Ahsan S, Taloor AK, Farooq H (2023) Earth observation satellite data–based assessment of wetland dynamics in the Kashmir Himalaya. Environ Monit Assess 196(1):32. https://doi.org/10.1007/s10661-023-12185-7

Asgher MS, Sharma S, Singh R, Singh D (2021) Assessing human interactions and sustainability of wetlands in Jammu, India using geospatial technique. Model Earth Syst Environ 7(4):2793–2807. https://doi.org/10.1007/s40808-020-01066-4

Ballut-Dajud GA, Sandoval Herazo LC, Fernández-Lambert G, Marín-Muñiz JL, López Méndez MC, Betanzo-Torres EA (2022) Factors affecting wetland loss: a review. Land 11(3):3. https://doi.org/10.3390/land11030434

Basu T, Das A, Pham QB, Al-Ansari N, Linh NTT, Lagerwall G (2021) Development of an integrated peri-urban wetland degradation assessment approach for the Chatra wetland in eastern India. Sci Rep 11(1):4470. https://doi.org/10.1038/s41598-021-83512-6

Bhowmik S (2022) Ecological and economic importance of wetlands and their vulnerability: a review. In: Research anthology on ecosystem conservation and preserving biodiversity. IGI Global, pp 11–27. https://doi.org/10.4018/978-1-6684-5678-1.ch002

Bhuarya HK, Sastri ASR et al (2018) Agro-climatic characterization for agro-climatic zone of Chhattisgarh. Int J Curr Microbiol App Sci 7(8):108–117. https://doi.org/10.20546/ijcmas.2018.708.013

Chakraborty SK, Sanyal P, Ray R (2023) Wetlands ecology: eco-biological uniqueness of a Ramsar site (East Kolkata wetlands, India). Springer Nature

Demarquet Q, Rapinel S, Dufour S, Hubert-Moy L (2023) Long-term wetland monitoring using the Landsat archive: a review. Remote Sens 15(3):3. https://doi.org/10.3390/rs15030820

Fan C, Yang J, Zhao G, Dai J, Zhu M, Dong J, Liu R, Zhang G (2023) Mapping phenology of complicated wetland landscapes through harmonizing Landsat and Sentinel-2 imagery. Remote Sens 15(9):9. https://doi.org/10.3390/rs15092413

Hemati M, Mahdianpari M, Shiri H, Mohammadimanesh F (2023) Comprehensive Landsat-based analysis of long-term surface water dynamics over wetlands and waterbodies in North America. Can J Remote Sens 50(1):2293058. https://doi.org/10.1080/07038992.2023.2293058

Joshi P, Siva Siddaiah N, Dixit A (2021) Urban wetlands of Delhi, India: water quality and pollution status. Chem Ecol 37(2):104–131. https://doi.org/10.1080/02757540.2020.1836164

Kaplan G, Avdan ZY, Avdan U, Kaplan G, Avdan ZY, Avdan U (2019) Mapping and monitoring wetland dynamics using thermal, optical, and SAR remote sensing data. In: Wetlands management – assessing risk and sustainable solutions. IntechOpen. https://doi.org/10.5772/intechopen.80264

Klemas V (2013) Using remote sensing to select and monitor wetland restoration sites: an overview. J Coast Res 29(4):958–970. https://doi.org/10.2112/JCOASTRES-D-12-00170.1

Liu D, Chen W, Menz G, Dubovyk O (2020) Development of integrated wetland change detection approach: in case of Erdos Larus Relictus National Nature Reserve, China. Sci Total Environ 731:139166. https://doi.org/10.1016/j.scitotenv.2020.139166

Lu M, Zou Y, Xun Q, Yu Z, Jiang M, Sheng L, Lu X, Wang D (2021) Anthropogenic disturbances caused declines in the wetland area and carbon pool in China during the last four decades. Glob Chang Biol 27(16):3837–3845. https://doi.org/10.1111/gcb.15671

Mahdianpari M, Jafarzadeh H, Granger JE, Mohammadimanesh F, Brisco B, Salehi B, Homayouni S, Weng Q (2020) A large-scale change monitoring of wetlands using time series Landsat imagery on Google earth engine: a case study in Newfoundland. GIScience Remote Sens 57(8):1102–1124. https://doi.org/10.1080/15481603.2020.1846948

Nayak A, Bhushan B (2022) Wetland ecosystems and their relevance to the environment: importance of wetlands. In: Handbook of research on monitoring and evaluating the ecological health of wetlands. IGI Global, pp 1–16. https://doi.org/10.4018/978-1-7998-9498-8.ch001

Newton A, Icely J, Cristina S, Perillo GME, Turner RE, Ashan D, Cragg S, Luo Y, Tu C, Li Y, Zhang H, Ramesh R, Forbes DL, Solidoro C, Béjaoui B, Gao S, Pastres R, Kelsey H, Taillie D et al (2020) Anthropogenic, direct pressures on coastal wetlands. Front Ecol Evol 8. https://doi.org/10.3389/fevo.2020.00144

Ostad-Ali-Askari K (2022) Review of the effects of the anthropogenic on the wetland environment. Appl Water Sci 12(12):260. https://doi.org/10.1007/s13201-022-01767-4

Ramsar Convention Secretariat (2016) An introduction to the Ramsar convention on Wetlands, 5th edn. Ramsar Convention Secretariat. https://www.ramsar.org/sites/default/files/documents/library/handbook1_5ed_introductiontoconvention_final_e.pdf

Sarkar S, Sarkar UK, Ali S, Kumari S, Puthiyotti M (2021) Status, ecological services and management of aquatic weeds of floodplain wetlands in India: an overview. Lakes Reserv Sci Policy Manag Sustain Use 26(1):76–91. https://doi.org/10.1111/lre.12353

Shah SA, Islam MS (2021) Recreational benefits of wetlands: a survey on the Dal Lake in Jammu and Kashmir of India. Int Hosp Rev 37(1):94–109. https://doi.org/10.1108/IHR-03-2021-0018

Shan V, Singh SK, Haritash AK (2021) Present status, conservation, and management of Wetlands in India. In: Al Khaddar R, Kaushika ND, Singh SK, Tomar RK (eds) Advances in energy and environment. Springer, pp 235–256. https://doi.org/10.1007/978-981-33-6695-4_22

Singh S, Bhardwaj A, Verma VK (2020) Remote sensing and GIS based analysis of temporal land use/land cover and water quality changes in Harike wetland ecosystem, Punjab, India. J Environ Manag 262:110355. https://doi.org/10.1016/j.jenvman.2020.110355

Tiner RW (2016) Wetland indicators: a guide to wetland formation, identification, delineation, classification, and mapping, 2nd edn. CRC Press. https://doi.org/10.1201/9781315374710

Vincent SGT, Owens KA (2021) Coastal wetlands of India: threats and solutions. Wetl Ecol Manag 29(5):633–639. https://doi.org/10.1007/s11273-021-09824-6

Wang X, Chen J, Zhou L (2020) Effects of human activities on the diversity of Waterbirds wintering in a shallow Lake of the middle and lower Yangtze River floodplain, China. Diversity 12(8). https://doi.org/10.3390/d12080302

Zhang M, Zhang H, Yao B, Lin H, An X, Liu Y (2023) Spatiotemporal changes of wetlands in China during 2000–2015 using Landsat imagery. J Hydrol 621:129590. https://doi.org/10.1016/j.jhydrol.2023.129590

Chapter 23
Multi-stakeholder Participatory Approaches for Sustainable Dryland Restoration in Sub-Saharan Africa

Leonard Chitongo and Kelvin Zhanda

Abstract Drylands in sub-Saharan Africa (SSA) are highly vulnerable ecosystems experiencing myriad challenges, including land degradation and climate change that threaten the region's sustainability and resilience. Over the years, several initiatives with ambitious targets, such as the African Landscape Action Plan (ALAP) and the African Forest and Landscape Restoration Initiative (AFRI100), were employed to address these challenges and restore the drylands. However, achieving the targets is a complex process that requires the participation of multiple stakeholders in designing and implementing practical and sustainable dryland restoration practices. This chapter explores participatory approaches to dryland restoration in SSA involving multiple actors. The chapter utilized a combination of scholarly and gray literature. We consider how multi-stakeholder participatory approaches might be employed to achieve dryland restoration targets in dryland ecosystems of sub-Saharan Africa, where resource constraints, among other challenges, make restoration efforts very challenging. Thematic analysis of case studies explored the efficacy of multi-stakeholder interventions in water management, forest restoration, and soil conservation practices in sub-Saharan African countries. Multi-stakeholders' direct and indirect involvement and participation have proven to be key enablers in the pursuit of ecosystem-based dryland restoration. It also emerged that multi-stakeholder approaches improve long-term sustainability and promote dryland ecological resilience and positive social and economic effects by fostering shared purpose among stakeholders such as practitioners, policymakers, and researchers. However, funding, governance, and capacity-building challenges can affect this approach. Thus, this paper recommends integrating a participatory approach to

L. Chitongo (✉)
Department of Biological and Environmental Sciences, Faculty of Natural Sciences, Walter Sisulu University, Mthatha, South Africa
e-mail: lchitongo@wsu.ac.za

K. Zhanda
Department of Architecture and Real Estate, University of Zimbabwe, Harare, Zimbabwe

Y. D. I. Khan et al. (eds.), *Land Restoration Through Ecosystem-Based Approach*, Disaster Resilience and Green Growth, https://doi.org/10.1007/978-981-95-0389-6_23

dryland restoration into national and regional policies with a focus on context-specific, adaptive, and inclusive programs.

Keywords Drylands · Ecosystem-based approach · Multi-stakeholder analysis · Restoration

23.1 Introduction

There is an unprecedented level of multiple stakeholders' interest in drylands, overlaid with a growing volume of resources, skills, and knowledge that could be required to restore dryland regions. Restoring degraded drylands has become one of the policies of governments and international organizations in recent years. Drylands are highly sensitive to human and natural disturbances (Wu et al. 2024). The extensive human dependence on natural endowments for livelihoods has intensified threats to these regions by a cycle of land degradation and poverty (The Global Centre on Adaptation 2022). With this in consideration, restoration of degraded drylands is urgently required to "prevent, stop, and reverse" ecosystem degradation (United Nations Environment Programme 2021), desertification, and safeguard the livelihoods of over two billion inhabitants in these dry areas.

The restoration of Africa's drylands is important in the context of the United Nations Decade on Ecosystem Restoration, which is time-framed from 2021 to 2030. The Decade calls on stakeholders such as businesses, the international community, politicians, civil society, and individuals to prevent, halt, and reverse ecosystem degradation worldwide. The United Nations Convention to Combat Desertification (UNCCD) pushes efforts to restore degraded land and combat desertification to attain land degradation neutrality by 2030. Africa aims to restore over 100 million hectares of land by 2030 (The Global Landscapes Forum 2021). This target aligns with the Sustainable Development Goals, African Union Agenda 2063, UNCCD's Land Degradation Neutrality targets, the Bonn Challenge, and other local and global targets. Most studies about sustainable ecosystem-based and land management for dryland restoration view restoration practices as a proxy for solving dryland resilience challenges, such as soil health, biodiversity conservation, and water recycling (Turner et al. 2022; FAO 2024).

Dryland degradation and desertification result from multifaceted interactions of numerous drivers and disturb many ecological and social functions and processes, often nonlinear and complex (Akinyemi et al. 2021; Reynolds et al. 2007). This is usually accompanied by multiple and often conflicting perspectives, values, attitudes, and primary sources of knowledge (Berkes et al. 2006). Driven by the need to bolster sustainable land management and restore drylands and degraded lands in sub-Saharan Africa over the past decades, multiple institutional, technical, and agro-environmental initiatives have been carried out across the globe (United Nations Environment Programme 2021; United Nations 2020; Akinyemi et al. 2021), yet government agencies and other top-down interventions have proven ineffective and unsustainable. In this regard, a key knowledge gap continues regarding how the multiple stakeholder participatory approaches can and cannot tackle the complex restoration in the dryland regions. Dryland restoration is a complex process (Kimmell et al. 2023). The process is characterized not only by degraded ecosystems but also by human and wildlife conflict over limited resources and trade-offs between the needs of multiple stakeholders at multiple scales. Given this multifaceted state of drylands and required measures, well-orchestrated measures and approaches are required. This calls for transdisciplinary and cohesive approaches considering the diversity and trade-offs of dryland management outcomes. Therefore, multi-stakeholders at local and national levels, including Indigenous peoples, women, and smallholder farmers, among others, have a role to play in boosting soil quality, preserving dryland forests and natural resources, and sustaining dryland communities' economies. Multi-stakeholder participatory approaches provide strategic advantages in tackling multifaceted and complex restoration processes in degraded community dryland contexts.

Participatory land restoration and rehabilitation programs are pursued to limit the impacts of climate change by storing carbon in plants and soil while often reducing people's vulnerability and enhancing their capacity to adapt (Favretto et al. 2021). This indicates that partial and disjointed strategies for restoring degraded drylands cannot be achieved in a fragmentary manner by individual and sector plans. The African Union indicated that fragmented efforts are a key challenge for ecosystem-based adaptation and resilience, mainly for low-income drylands with their inadequate resources, restricted access to entities that might facilitate coordination, and fairly limited human capital to aid many initiatives. Conceptualizing and planning climate-adapted restoration options in silos entails that collaborations are missed, and funding may become nonviable. As a result, a holistic approach in engaging all the stakeholders in adopting approaches for sustainable dryland restoration in sub-Saharan Africa becomes key.

Desk-based research explored multi-stakeholder participatory approaches in dryland restoration, focusing on sub-Saharan Africa drylands. This approach was used for an integrative and comprehensive analysis of existing scholarships, policies, reports, and case studies (Zhou and Nunes 2016). Data were collected from case study reports, peer-reviewed journal articles, publications, organizations, and governmental and nongovernmental reports. We accessed and retrieved from Google Scholar, ResearchGate, and JSTOR to gather peer-reviewed literature. Other databases, such as the Food and Agriculture Organization (FAO) and the United Nations Convention to Combat Desertification (UNCCD), were also used. Thematic analysis was used in identifying patterns and interpreting data, extracting meanings, and identifying frequent trends and frameworks in literature, including government, NGOs, women, youth, and civil society-powered initiatives, community-led programs, and private-public partnerships (Fereday and Muir-Cochrane 2006; Javadi and Zarea 2016).

The chapter reviews the role of multi-stakeholder participation in restoring African drylands. The chapter focuses on SSA drylands, where land degradation is taking a toll on household livelihoods, leading to extreme hunger and under-nutrition

(Azzarri et al. 2016). Given the importance of drylands in the SSA, millions of its inhabitants, and the pervasive food insecurity and climate change they are facing, this chapter examines how multi-stakeholder participatory approaches contribute to sustainable dryland restoration in the region through the combined efforts of local authorities, private sector partners, local communities, national governments, non-governmental institutions, and regional bodies. The chapter explores public and private stakeholders within dryland landscapes, ascertains the "winners" and "losers," and finds pathways to reconcile their differences along the restoration process. In the context of social and biophysical trends, the chapter proposes a positive vision for a future for drylands largely centered on participatory approaches.

23.2 Understanding Dryland Ecosystems

Drylands are a vulnerable part of Earth's terrestrial environments due to low water scarcity, protracted dry spells, and high exposure to degradation. Global drylands are estimated to be at 41% of the global land surface area (Prăvălie 2016; Stringer et al. 2017). More than 2.5 billion of the most marginalized, poor, and unhealthy human population reside in drylands (Stringer et al. 2017). The sub-Saharan Africa (SSA) drylands cover 13.9 million km^2 (Xie et al. 2018), which translates to more than half of the continent's surface area. The SSA region is largely covered by dryland ecosystems. Nearly two-thirds of the human population in this region depends directly on natural resources and agriculture to sustain their livelihoods.

Regardless of their harsh conditions, drylands have distinctive and diverse ecological systems that provide biodiversity and ecosystem services (BESs) important for sustainability. BESs in drylands are the benefits provided by dryland environments to human populations (Yu et al. 2024). BESs are central to sustainably provide ecological and ecosystem processes that support the well-being and survival of people in dry regions (Costanza et al. 2016). Drylands provide various BESs, including carbon sequestration, food production, sandstorm prevention, water retention, and habitat provision (Schlaepfer et al. 2017). The BESs of drylands and their interactions with the socioeconomic systems of the inhabitants are crucial determinants of the sustainability, resilience, and stability of dryland social-ecological systems (Yu et al. 2024). Ecosystem functions in dryland are highly interconnected, and changes in one function can have a cascading effect on others. Changes in vegetation cover can affect the water balance of the system and redistribution of materials, which in turn affects ecosystem productivity and nutrient cycling (Turnbull et al. 2016; Maestre et al. 2021; Mayor et al. 2019). Similarly, changes in nutrient availability can affect the diversity and productivity of vegetation and affect carbon sequestration and water balance.

The continued increase in hydro-climatic intensity, coupled with less precipitation, has exacerbated problems associated with poverty, conflicts, famine, marginalization, political instability, and migration (Prăvălie 2016). Evidence shows that the SSA drylands are the most at greater risk of food insecurity than other drylands across the globe. The drylands are remotely located, which then detaches them from governance epicenters. Due to the socio-political marginalization of dryland inhabitants, they tend to be far from political processes, which exacerbates the fragility of restoration and livelihood strategies. The severe vulnerability of drylands to climate change, coupled with extensive dependence on natural endowments for livelihoods, threatens the region with a cycle of land degradation and poverty. The critical causes of land degradation in SSA are complex and demand a collective effort that engages different and multiple actors such as nongovernmental organizations (NGOs), local communities, policymakers, and governments to promote sustainable dryland restoration practices.

23.3 The Concept of Multi-stakeholder Participation

Multi-stakeholder participatory approaches are increasingly becoming more dominant in bringing together multiple players with a prospective "stake" in a problem to sort out a solution collectively. In this chapter, the term multi-stakeholders simply refers to stakeholders working together to find a solution for particular problems, including nongovernmental players, governments, local authorities, regional groups, the private sector, and international institutions (Dodds 2015). FAO (2000) states that stakeholders are often varied and comprise many subtypes. Stakeholders are groups of people, individuals, or organizations who participate directly and indirectly. Thus, they can positively or negatively influence any course of action through decision-making, planning, and execution. These include governments, private sector entities, civil society, local communities, academia, and international organizations. Multi-stakeholder approach is cognizant that complex challenges, such as climate change, resource governance, and sustainable development, require collective action to develop long-lasting solutions. The integration of diverse perspectives, skills, expertise, inclusivity, and resources is at the center of multi-stakeholder approaches for effective actions. These approaches aim to ensure sustainable, legitimate, and effective solutions, particularly in areas like development, environmental management, and governance, by fostering consensus-building, dialogue, and shared responsibility.

In the dryland ecological management, as in other areas, the multi-stakeholder approach is commonly acknowledged as the prime way to make policy decisions for extensive projects. Concerning natural resources management, the FAO (2000) defines stakeholders as people who possess an economic, cultural, or political interest in, or influence over, a natural resource. Examples of multi-stakeholders are local government, research institutions, and community stakeholders. Details of the multi-stakeholder participation are described in Table 23.1.

Multi-stakeholder participation falls within the broad area of multi-stakeholder governance. At the same time, multi-stakeholder approaches are an established global norm in many fields (FAO 2000). They are informed by open-ended innovation (infrastructure), decentralized governance institutions (governance), and

Table 23.1 Main elements of the multi-stakeholder participation

Description	Multi-stakeholder participation aims to promote better decision-making. They ensure that the main actors' views about a certain decision are integrated at all phases through consensus-building and dialogue. The multi-stakeholder process considers that all stakeholders have a valid perspective and applicable experience and knowledge. This approach creates trust between the stakeholders and solutions that provide a win-win state (mutual benefits). The multi-stakeholder approach is community-centered, and everyone and the entity take responsibility for the outcomes. Stakeholders have a better sense of ownership of the outcomes since the inclusive and participatory approaches were engaged
Appropriate use	For resolutions that need cooperation between different stakeholders where a resolution from one stakeholder group only might not be observed by the other stakeholder groups, thus leading to defiance. The multi-stakeholders approach works smoothly when stakeholders' discussions are possible and they are willing to listen to and learn from others to resolve different interests and agree. This largely depends on the specific objectives, problems to be covered, the time, the expertise available, the participants, and the availability of other resources
Scope	It is global, national, and local, and it can be used with a wide range of structures and levels of engagement
Key output	Inclusive and transparent decision-making and strengthened stakeholder networks
Key input	The willingness of participants to learn, expertise and skills in facilitation, time to allow trust building, qualitative and quantitative information, and participation of key actors
Ease of use	The multi-stakeholder approach and the techniques used are based on common sense. Planning is a central part of ensuring a successful outcome, and time must be allowed for the design stage of the process
Training required	Several available good texts are supported by additional appropriate training
Applications	Multi-stakeholder participation has been applied and implemented in the United Nations sustainable development multi-stakeholder dialogue, the Beijing+5 global forum online discussions, the Aarhus convention process, the environment council/Shell, and the Brent spar project
Training available	Besides being still new, the multi-stakeholder participatory approach is an evolving area in dryland restoration. There is much knowledge of participatory processes at the local level compared to the national and global levels
Documentation	For information about running stakeholder engagement processes: Multi-stakeholder participation for governance and sustainability, Minu Hemmati, (2002), Earthscan, London. http://www.earthscan.co.uk/

inclusive and open processes (human) (Hemmati 2002). Participatory governance provides the basis for ensuring equitable decision-making and significant engagement. These principles include inclusivity, accountability, transparency, empowerment, and equity. Table 23.2 shows the summary of the principles of participatory governance.

Table 23.2 Principles of participatory governance

Principles	Description	Examples
Transparency	Providing open access to information and making governance processes clear and understandable	Public disclosure of budgets, policies, and decision-making criteria
Inclusivity	Ensuring the active participation of all stakeholders, particularly marginalized and underrepresented groups	Including local communities in natural resource management or urban planning
Accountability	Establishing mechanisms to hold stakeholders and decision-makers responsible for their commitments	Performance audits, feedback mechanisms, and monitoring systems
Empowerment	Building the capacity of individuals and groups to participate in governance processes effectively	Training workshops for community leaders or providing tools for citizen advocacy
Equity	Addressing power imbalances and ensuring fair treatment and representation of all stakeholders	Implementing gender-sensitive policies or ensuring equal access to decision-making platforms

The concept of multi-stakeholders brings on the idea of co-creation in dryland ecological management. While having different forms, co-creation is generally regarded as the collective integration of stakeholders' resources, networks, and knowledge (Keeys and Huemann 2017) to address challenges and/or to exploit opportunities that are impossible to independently and individually do. This implies that multiple interdependent stakeholders will be working together for a common good. The combined effect of stakeholder partnership has been suggested as a marker of the operation of the partnership. Multiple stakeholder partnerships are regarded as operational when they successfully bring resources together and assemble the required knowledge and skills of the partners. This point is reached when the efforts of all partners improve the results further than what an individual entity could achieve; hence, the whole becomes greater than the sum of the individual parts. Lasker et al. (2001) outlined a number of fundamentals for the functioning of a partnership that can affect the interaction of partners and the operation of the partnership (see Table 23.2).

It should be noted that the focus of stakeholders varies in terms of geographical scale. While some stakeholders work within national boundaries, others operate at a regional scale. Participatory instruments would therefore be highly effective by providing representation and voice to multiple stakeholders (Table 23.3).

Inclusivity in decision-making offers insights on the implications for changes in environmental policy by considering the past, present, and future interests, positions, influence, networks, and interrelations among other stakeholders' characteristics (Nonet et al. 2022; Brugha and Varvasovsky 2000). In addition, regarding the dryland restoration process, collective decisions are valuable in assessing the probability of effective implementation and modifying the restoration policy strategies for process escalation. Implementation of the ideas and strategies can be regarded as a meeting process in which many stakeholders (multi-stakeholders) regulate the important elements of the entire process (Bendtsen et al. 2021). Understanding the

Table 23.3 Key elements of multi-stakeholder participatory approaches

Critical elements	Contributions
Identification and inclusion	Indigenous groups, local communities, academic institutions, government agencies, the private sector, and NGOs are among the diverse stakeholders Inclusion of marginalized social groups (women and youth)
Integration of traditional knowledge	Incorporation of indigenous knowledge Mechanisms for balancing tradition and knowledge
Decision-making	Equal contribution to discussions and decisions Effective consultation
Policy and governance framework	Policies and governance framework to support stakeholder participatory and collaborative approaches Laws to protect the rights of stakeholders and enforce binding agreements
Conflict mmanagement	Procedures and guides to resolve stakeholders' conflicts
Long-term sustainability commitments	Commitment of pledge by all stakeholders, focusing on issues like drylands restoration policy, funding, and community participation
Assessment of outcomes	Yardstick of restoration measures in attaining soil health, water retention, biodiversity, and impact on food security and livelihoods of dryland communities

resources and interests of multiple stakeholders helps participants anticipate how they will cooperate in providing vital elements and strategies for enhancing cooperation among stakeholders. Multiple stakeholders influence policy and strategy choices (Table 23.4).

23.4 Stakeholder Identification and Engagement

This subsection looks at the main stakeholders, methods for stakeholder mapping, and strategies for promoting successful communication and engagement. Identifying, mapping, and engaging stakeholders determines the effectiveness of dryland restoration actions in SSA (Gravesen and Funder 2022). This ensures that dryland restoration initiatives harness different perspectives, expertise, and resources. Restoration of degraded dryland ecosystems is complex and diverse, which makes the participation of multiple stakeholders critical. The main stakeholders and their roles are listed in Table 23.5. Moreover, drylands comprise interrelated ecosystems and players from various sectors and backgrounds. To provide a holistic conceptual premise, this chapter proposes three categories of stakeholders in restoring degraded landscapes in SSA, involving primary stakeholders, secondary stakeholders, and external stakeholders (WHO 2000). The primary stakeholders participate in dryland restoration since they heavily depend on and influence the conditions of the land through their direct and indirect actions. The group is directly involved in the dryland restoration process and impacted by the same process. Secondary stakeholders participate in the restoration interventions by executing,

Table 23.4 Determinants of the interactions in multi-stakeholder partnership

Inputs for multi-stakeholder interaction	Factors influencing multi-stakeholder interactions
Partner relationship	Mutual respect Trust Clashes
Characteristics of partners	Level of involvement Heterogeneity
Resources	Skills Space Money Information Endorsements
Partnership characteristics	Efficiency Governance Leadership
Outside context	Community Laws and policies

Adapted from Lasker et al. (2001)

managing, governing, and supervising dryland restoration (Favretto et al. 2021). The external stakeholders involve the informal participants in dryland restoration who can influence the outcomes. For example, they provide funds for restoration or intermingle with the first and second categories. In other words, external stakeholders have a stake in the restoration process of drylands through collaborative efforts with the latter.

The multiple stakeholders outlined in Table 23.5 will certainly diverge in how they use or see dryland ecosystems and their common goods and services. Thus, their intentions and motivation for participating and engaging in a restoration process will be divergent. Reconciliation of differences and management of conflicting perspectives between stakeholder views and drives should be a pivotal component of collective drylands management to maximize social welfare and reduce costs. The issue of conflicting views and intentions underpins the importance of stakeholder mapping in understanding the interests, roles, and potential contributions of different players in dryland restoration (WHO 2000). Stakeholder mapping refers to a systematic approach for identifying, analyzing, and classifying stakeholders based on their interests, influence, and roles in a certain initiative. In dryland restoration in SSA, stakeholder mapping is vital to ensure that all relevant stakeholders are engaged. In addition, this process ensures that actors' contributions are optimized for attaining sustainable outcomes. Since dryland restoration involves complex, diverse stakeholders, highlighting various methods that cater to such complexity and the need for tailored approaches. Table 23.6 shows selected scientifically validated stakeholder mapping methods.

Table 23.5 Key stakeholders in dryland restoration activities

Stakeholders	Examples	Roles in dryland restoration	Engagement strategies
Government agencies	Ministries and departments related to environment, forestry, agriculture, land management, and water resources	Formulate and enforce policies Allocate resources Monitor restoration initiatives Collaborate through inter-agency task forces Develop policy briefs for advocacy Facilitate joint monitoring and evaluation systems	Policy advocacy Collaborative governance Data sharing capacity strengthening
Local communities	Farmers, pastoralists, indigenous peoples, and women's groups are the primary users and custodians of dryland resources	Provide traditional ecological knowledge and practices. Implementers of restoration activities Conduct participatory workshops and community consultations Use culturally appropriate communication methods Establish inclusive decision-making platforms	Participatory planning Capacity building Indigenous knowledge integration Benefit sharing
Research and academic institutions	International crops research Institute for the Semi-Arid Tropics (ICRISAT) University of Cape Town (South Africa)	Conduct studies and provide data-driven insights Develop innovative restoration technologies and models Organize knowledge-sharing platforms like webinars and conferences Partner with communities for action research projects	Collaborative research Data sharing Capacity building policy advice
Private sector	Savanna circuit tech (Kenya) and EcoPlanet bamboo (Zambia)	Invest in sustainable projects such as renewable energy and agroforestry Facilitate market access for dryland restoration-related products Develop public-private partnerships Host forums to align corporate goals with restoration objectives	Incentives and partnerships to develop public-private partnerships for private sector investment in restoration Corporate social responsibility to urge businesses to integrate restoration efforts into their CSR initiatives Market access for linkages to markets for ecosystem-based products like carbon credits, sustainable agriculture Technology transfer from corporations to bring innovative restoration technologies, such as remote sensing and drought-resistant crops

Multilateral organizations and donors	United Nations, African union, and development banks United Nations convention to combat desertification (UNCCD), green climate fund (GCF)	Provide funding, capacity-building programs, and technical expertise Align projects with global restoration frameworks such as AFR100 and Bonn challenge Foster international cooperation and accountability Regular reporting and adaptive planning	Cross-border collaboration Funding mobilization Technical assistance Global advocacy
Civil society and nongovernmental organizations	World agroforestry (ICRAF), African wildlife foundation (AWF), desert research Foundation of Namibia (DRFN)	Provide funding, technical support, and advocacy Create capacity-building programs for dryland communities Mobilize communities for participation	Partnership Co-facilitation Monitoring and evaluation Advocacy campaigns

23.5 Frameworks and Models for Participatory Approaches

23.5.1 Community-Based Restoration

Table 23.6 Overview of the relevant methods for stakeholder mapping in dryland restoration activities: methods, purpose, steps, and application

Method	Purpose	Main steps	Application in dryland restoration
Social network analysis	Maps stakeholder relationship Identifies potential collaborators and main influencers	Define stakeholder network Map relationships such as partnerships, dependencies Analyze network structure to find central or isolated actors	Identifies prominent players and opportunities for collaboration for organized action in dryland restoration
Power-interest matrix	Categorizes stakeholders based on their level of influence and interest in the project	Stakeholders' identification Assessment of power and interest levels Classify into quadrants (key players, keep informed, keep satisfied, and monitor)	Prioritizes engagement actions and focuses resources on interested stakeholders
Resource dependency mapping	Appraiser's reliance on resources by stakeholders and the possible impact of interventions	Resource-dependent groups identification Assess the needs and vulnerabilities of these groups Spatial map dependencies	The importance of including vulnerable groups (pastoralists and smallholder farmers)
Stakeholder analysis workshops	Sets a shared platform for the identification of stakeholders and definition of their roles	Expedite brainstorming sessions Use interactive tools such as mapping software and charts Form a consensus on the roles and priorities of stakeholder	Promotes collective understanding clarifies stakeholder roles for effective collaboration in dryland restoration efforts
Participatory rural appraisal (PRA)	Engages dryland communities to inclusively identify stakeholders and assess roles	Community mapping and discussions Employ participatory tools such as ranking exercises and focus groups Collaboratively compile stakeholder data	Ensures inclusivity by integrating local knowledge and addressing community needs

(continued)

Table 23.6 (continued)

Method	Purpose	Main steps	Application in dryland restoration
GIS-based mapping	Visualizes spatial distribution of stakeholders and their proximity to dryland restoration project sites	Collection of stakeholders' geographic data Map spatial overlaps like conflict zones and resource use Utilize GIS tools to visualize and analyze results	Identifies areas of collaboration and potential conflict zones. This is important in spatial planning

Community participation in dryland restoration determines the long-term sustainability of the restoration practices as the community takes ownership of the process. The increasing restoration concern is the recognition that the world cannot reverse the degradation of drylands without addressing the needs of people living in the drylands. The approach of merging community preferences and needs with plant expertise in drylands for selecting appropriate native species for large-scale natural restoration has been successful in the Sahel region as part of the Great Green Wall initiative in Africa (Gravesen and Funder 2022). An investigation carried out in Burkina Faso, Mali, and Niger indicated that the engagement of dryland villages was done through participatory diagnostic gatherings for dryland restoration planning and actions. This involved introductory group discussions, community workshops, in-depth interviews through field visits, structured questionnaires, and workshops for a consensus on plans with all the selected villages. In over 121 beneficiary communities, with over 51,000 farmers, including 52% women, participatory diagnostic gatherings were carried out. This resulted in the selection of 193 plant species. The Duwa and Sutebwo forests of the Ségou region in Mali, south of the Sahara Desert, also witnessed the effectiveness of community-based participation. The area is vulnerable to the effects of climate change coupled with forest degradation (Amadou and Diakarya 2023) and unsustainable tree felling, erosion, overgrazing, agricultural expansion, and loss of biodiversity from poaching (Nchare et al. 2024; Touré et al. 2020). Evidence indicates a development in the tree density to an average of 182 trees/ha from the previous 168 trees/ha. Over 327,000 trees were restored, thus improving soil fertility and water retention. The restoration measures have created a participatory, resilient, and inclusive community.

A good case is Tigray in Ethiopia, where the community, through decentralization policies, is actively involved in the restoration process. Although some entities contributed financial and technical support to the program, the local community always took the lead by contributing the bulk of the labor required (Gebremedihin et al. 2018). Involved organizations included the World Bank, the World Food Programme, Mekelle University, the Relief Society of Tigray, the Tigray Bureau of Agriculture and Rural Development, the Ministry of Agriculture, and German Development Cooperation (GIZ). Local bylaws were developed to guarantee equitable benefit sharing for every community group during this initiative. Successful restoration pointers included improved production of commercial crops. Tadesse et al. (2017) observed that people used the water from new hand-dug wells for irrigation, one of which can yield approximately 14,795 m^3 of water per year. Therefore, the local community farmers were able to produce at least two crops per year (Biedemariam et al. 2017), including barley, maize, and wheat; cabbage, pepper, and tomato; and many fruits. This concurs with the UN observation that community-based approaches supported by local knowledge work better and use fewer resources as long as local people obtain direct benefits.

Community-centered dryland management practices can play a key role in supporting dryland ecological functions and services. This can be done through social events, cooperatives, and certification. Social events such as festivals, seed fairs, sporting events, and climate change workshops expedite the exchange of ideas in SSA dryland communities. This goes a long way in promoting social cohesion, resulting in robust resilience to socioeconomic and dryland ecological stressors.

23.5.2 Restoration Strategic Investments

Investment is a vital strategy to improve the state of drylands and their people, coupled with using and harnessing the benefits of sustainable drylands as biodiversity stores, carbon sinks, and community food baskets (FAO 2021). The main groups in dryland restoration investment in SSA are shown in the table below (Table 23.7).

In SSA, the TerrAfrica Strategic Investment Programme (SIP) has become a beacon of investments channeled toward restoring degraded landscapes (NEPAD 2023). The SIP provided land degradation funding of $150 million from the GEF to SSA. It lured about $800 million in co-financing through its collection of 36 restoration projects addressing specific themes in 26 countries. Multi-stakeholder participation comes in to address the prospects for reversing the protracted under-investment that has constrained the restoration agenda in drylands in poor countries of SSA. A framework of stakeholders' investment needs to be framed on: (i) identifying the main groups of players (who invests?); (ii) reasons for drivers of investment in drylands (why invest?); (iii) a concise analysis of essential strategies (which investments?); and (iv) desired outcomes for incomes, well-being, natural resource management, and sustainability (what benefits?) (NEPAD 2023). The framework is very important in articulating the areas of investment in which prospects for commercial, public, household, and community sectors are identified.

Table 23.7 Dryland restoration investment groups

Type of investment	Interest group	Purpose
Government	The sectoral ministries of the state (parastatals, ministries of agriculture, transport, environment) Governance at all levels	Social, economic, and environmental benefits. Implement the restoration agendas of citizen interests
Public	Nongovernmental organizations Community-based organizations: National and international organizations Donors (multi-lateral and bilateral donor funders)	Sustain drylands public goods through programs, policies, or project interventions Create conducive dryland environments for private investments Represent a compromise between national, international, and donor priorities (Birch et al. 2010)
Private	Commercial (large-scale enterprises, corporations, companies) Households (small-scale enterprises, family farms, livelihoods, informal sector, herders)	Aim to sustain dryland livelihoods by managing human, financial, social, and natural capital

23.5.3 *Case Studies of Successful Models in Sub-Saharan Africa*

23.5.3.1 African Forest Landscape Restoration Initiative

The African Forest Landscape Restoration Initiative (AFR100) is one of the outstanding multi-stakeholder initiatives for dryland landscape restoration in SSA, which has taken the form of a multi-stakeholder platform. AFR100, as a concept in integrated landscape initiatives, has recently gained traction in the African Forest Landscape Restoration Initiative. AFR100 consists of 34 member African countries working collectively to restore over 100 million hectares of land in Africa by 2030 (Vinceti et al. 2020). The initiative contributes to a global restoration movement—the Bonn Challenge. Multi-stakeholder platforms on dryland restoration cover at least 12 of the AFR100 member countries. In SSA, this includes Cameroon, Nigeria, the Democratic Republic of Congo, Malawi, Niger, Sudan, Senegal, Ethiopia, Togo, Rwanda, and Tanzania. Multiple stakeholders and sectors are involved in these platforms. However, each of the member countries adapts its model. However, in countries like Niger and Togo, multi-stakeholder platforms operate through administrative decrees, while in other countries these platforms function as informal institutions (FAO 2024). Dryland restoration efforts in AFR100 countries differ contingent on landscape contexts. This multi-stakeholder approach has played an important role in the localization of countrywide restoration pledges and related goals into local contexts. Thus, it accelerates restoration interventions.

23.5.3.2 Trees for the Future (TREES)

As part of the restoration efforts in SSA, FAO, and UNEP, they introduced the Trees for the Future (TREES) initiative. TREES is one of seven UN World Restoration Flagships to reverse, halt, and prevent degradation. This has helped several smallholder farmers in African countries like Senegal and Tanzania to battle soil degradation, climate change, and biodiversity loss (United Nations Environment Programme 2021). Since 2014, TREES has succeeded in restoring over 41,345 hectares of degraded lands, supporting at least 50,000 households, and capturing 347 metric tons of CO2 per hectare through its model, the equivalent of emissions from over 100,000 liters of diesel consumed (UNEP 2021). The TREES initiative is projected to create 230,000 jobs by 2030 in Kenya, Senegal, Mali, Uganda, and Tanzania (UNEP 2022).

23.5.3.3 Forest Garden Program in Senegal

In 2012, TREES launched a forest garden program in Senegal, which involved working with local communities to establish sustainable agroforestry systems. The program was implemented in partnership with the Senegalese government and local NGOs. Community members were involved in planning, designing, and implementing the forest gardens, which included planting over 100,000 trees. The program resulted in increased food security, improved livelihoods, and enhanced environmental sustainability for local communities (Mentz et al. 2022).

23.5.3.4 Tree Planting and Management in the Gambia

According to Darboe (2023), TREES launched a tree planting and management program in Gambia, which involved working with local communities to plant and manage trees on their farms and in their communities. Community members were trained on tree planting, pruning, and management techniques and were provided with support and resources to establish and maintain their own tree nurseries. The program resulted in planting over 50,000 trees and improved livelihoods for local communities through the sale of tree products.

Regreening Africa

Regreening Africa (RA) is a good example of a large-scale agroforestry project in the drylands of SSA. RA is a five-year (2017–2022) restoration programme funded by the European Union and implemented by the World Agroforestry Centre (ICRAF) (United Nations Development Programme 2023). The project seeks to restore 1 million hectares of degraded land and enhance the resilience of 500,000 households across eight countries in sub-Saharan African countries, including those covered by the Great Green Wall, such as Ethiopia, Mali, Niger, and Senegal. Restoration of degraded land has been done by integrating trees into land management (United Nations Development Programme 2023). To leverage the restoration process, RA has partnered with a number of NGOs who lead the project's in-country implementation efforts. The stakeholders include the World Vision, Catholic Relief Services, CARE, OXFAM, Global Evergreening Alliance, and Sahel Eco. These implementation partners leverage their vast networks and experience to accelerate the scaling of regreening practices across the eight countries. The first phase of the project was completed in February 2023. It works with thousands of farmers to help them integrate trees to increase crop yields, improve soil fertility, and help sustain their livelihoods (Regreening Africa 2019).

Stakeholder relationships have been a critical part of the restoration process of the Regreening Africa project, and its success is attributed to the local communities and partners' participation in framing restoration strategies, data collection, and analysis. The project started with an advocacy strategy by activating interactions with various stakeholders to strengthen those that already exist. This was done through stakeholder mapping exercises like list-building, tagging supporters and partners in its content, and creating content around them. This has made them an active part of the regreening process and create a community around the regreening project.

23.5.3.5 The Global Environment Facility (GEF)

The Global Environment Facility (GEF) is an international partnership of 183 countries, the private sector, civil society organizations, and international institutions, uniquely positioned to address the intertwined issues of land degradation and poverty in sub-Saharan Africa. For 20 years, the GEF has involved communities in African countries through the UNDP and implemented the GEF Small Grants Program. The GEF Small Grants Program, implemented in over 125 countries, promotes community-based innovation, empowerment, and capacity development through sustainable development projects. Furthermore, the European Union has convened the Regreening Africa project, a five-year research-in-development project and restoration project which covers Ethiopia, Kenya, Ghana, Mali, Niger, Senegal, Rwanda, and Somalia (UNEP 2023).

23.6 Inter-Governments and Regional Frameworks

For the past two decades, Africa has generally developed a clearer voice coupled with pan-African initiatives such as the African Common Position, the African Continental Free Trade Area, the African Union, global discussions, and the Africa Climate Change Strategy (African Union 2024). All of these initiatives have had the prospect of positively influencing drylands. However, although the challenges being experienced in SSA drylands are quite similar to other dryland regions of the continent or out of the continent, the context of drylands requires dryland-specific implementation (Critchley et al. 2023). Regional and intergovernmental arrangements that specifically advocate for drylands restoration at the SSA scale and within countries are, for that reason, vital. This is a leeway of the need for regional intergovernmental authorization in decision-making to be a key effective factor in dryland restoration practices (Cordingley et al. 2015).

The TerrAfrica is a good example of a regional multi-stakeholder initiative supporting sustainable land and water management in the drylands of SSA. It is a multistakeholder partnership between the World Bank, FAO, and the New Partnership for African Development (NEPAD), among other agencies. TerrAfrica aims to mainstream and scale up sustainable land management (SLM) through assessments and monitoring of land degradation and integrated land-use planning and landscape management approaches (Cordingley et al. 2015). It usually supports this through targeted policies and funding for incentives (Critchley et al. 2023). The important aspect of TerraAfrica is the systems approach to identifying the interlinkages of climate change, biodiversity conservation, and sustainable water management with SLM instruments that are innovative and synergistic. For instance, addressing carbon in soils, pastoralism, and biodiversity in drylands of SSA (Critchley et al. 2023). This highlights the strengths of global and regional partnerships for supporting SSA drylands.

23.6.1 Women and Youth Participation

Restoration characterized by active and broad participation of groups such as women and youth has a considerable advantage, and the chance of failure is minimized. Gender-responsive dryland restoration requires effective communication, greater trust, and faster learning than when there are low levels of participation, and capacity building for stakeholders such as women, policymakers, community members, and project implementers (UNCCD 2016).

Data on women's restoration roles shows several challenges (FAO 2024). In SSA drylands, land degradation has prompted the out-migration of youth and men from dryland communities, putting women on the task of supporting the dependent population. In this regard, it is, therefore, particularly important to strengthen the participation of youth and women in dryland communities for long-term restoration and resilience. What is more challenging for women is that their gender-based tasks, such as washing, fetching water, feeding, and caring for their households, have become harder due to climate change stressors and land degradation (FAO 2024). In SSA dry communities, women bear the duty of gathering firewood and water for households, which then reduces their ability to participate in community dryland restoration initiatives. For instance, in Senegal and the Louga region, SOS Sahel educates women on how to build fuel-efficient stoves using locally available materials (Bensch and Peters 2015). Therefore, improving access to water and energy takes women out of household labor, facilitating their full participation in landscape restoration. Women in SSA need an opportunity in participatory rangeland management to contribute largely to ecosystem-based dryland restoration and management. This requires women to participate actively in discussions and meetings with other community members. Active participation of women in SSA requires gender-equitable practices and policies that empower them by promoting environmental, social, and economic rights that strengthen their voices. Mor (2019) reveals that gender-based discrimination and uneven power relations in customary systems in many societies deny women the right to plant trees, control soil degradation, and take measures to enhance soil fertility.

While young people are one of the largest demographics in SSA (Tabutin et al. 2020), they are crucial in bringing the greatest innovation, creativity, and enthusiasm. Youth in SSA dryland communities are taking critical action, bringing their landscape skills and knowledge into practice (Kemeh and Kabalan 2021), as they are gradually engaged in determining sustainable dryland regions. The Youth in Landscapes Initiative (YIL), in collaboration with Global Landscape Forum (GLF) Africa, has started gathering young people's activities to foster intergenerational dialogue on collective dry landscape governance and management and explore prospects to support the development of sustainable livelihood choices on SSA drylands (Kemeh and Kabalan 2021). Nevertheless, youth out-migration is a huge problem in SSA drylands. Actions focusing on retaining youth facilitate participation and build long-term restoration capacity. As part of measures aimed at retaining youth, for example, the Association Zoramb Naagtaaba formed the Training Centre for Rural Planners, an apprenticeship in Sahelian dryland agricultural practices for youths for a period of 3 years. This leads to rural sustainable development certification, thus providing a foundation for the establishment of proficiency in reversing land degradation across the region. According to Scaling Up Nutrition (2024), in Chad, youth have been engaged through online environmental contests, training in tree planting activities in the course of National Tree Week, and soil management. The same program created peer-to-peer learning for youths dubbed *Knowledge for Life*, in which youths, like young girls, are given knowledge and skills on how to become guides at the national tree nursery and pledge to pass on their learning to their peers (Oyekanmi 2022). Programs of this nature can broaden youth participation in dryland societies and give youth valuable knowledge for the future.

Resource mobilization and realignment are essential to the success of the multi-stakeholder platforms' restoration strategies (Table 23.8). As many countries build platforms like these, they can shape a community of restoration practice to support new platforms and share ideas.

Table 23.8 Multi-stakeholder platforms strategies and rationale

Key strategy	Rationale and objectives
Robust assessment	To understand stakeholders in existing platforms, identify networks and information flows that could be reinforced or leveraged, or any need for alignment between stakeholders and dryland restoration priorities
Dryland landscape-level interactions	To identify and address drivers of land degradation To enable direct ownership and stewardship among stakeholders to identify problems and solutions Identify power discrepancies among players and foster actions to bring inclusive exemplification and voices of marginalized players to the platforms
Long-term financial and technical resources	To maintain restoration, multi-stakeholder platforms beyond single projects
Clear direction on multi-stakeholder platforms to set up sustainable platforms	To create an exit strategy and sustainability roadmaps to ensure platform longevity

23.7 Challenges and Barriers to Multi-stakeholder Participation

Multi-stakeholder participation in dryland restoration has faced a number of challenges. Despite the relative success of participatory initiatives such as the Great Green Wall (Gravesen and Funder 2022), efforts to increase and restore degraded drylands in SSA continue to face numerous obstacles. Results reveal that several dryland communities in SSA are isolated from several critical infrastructures. These include innovations in neighbors, education, transportation, electricity, markets, sustainable water supply, and security.

In addition, governments in SSA have faced crippling opposition from the business sector, coupled with a situation where the economy is dominated by non-climate-friendly industries like forestry or crude oil. Since private sector investment in enhancing climate resilience is frequently required, a lack of private sector buy-in is especially difficult in underdeveloped nations. It makes sense that funding and expanding climate resilience initiatives would be a far more difficult and drawn-out process in such situations (Graham 2020).

Enabling actors to participate in, have a say in, and be a part of processes and decision-making is known as an inclusive process. For significant climate action, literature has emphasized the value of multi-partner collaboration, including the participation of vulnerable groups and local communities. On the other hand, research from East Africa indicates that the involvement of various players at different levels of governance is quite low. While local governments and non-state entities are frequently left out of the policy-making process, central governments take the lead in formulating policies. While some civic institutions connect with communities, Mabaya and Mafongoya (2017) found that governance structures and organizations in Zimbabwe lack openness and inclusivity in their interactions and interventions with communities (United Nations 2020).

Sibiya et al. (2023) argue that the policy integration literature has decisively cited the capacity challenge. There are continuous financial and technical capacity limitations in the SSA region. Government departments and agencies involved in climate adaptation policy lack the technical and human resources necessary to analyze and assess the information at their disposal. However, research in Kenya, Ethiopia, Nigeria, Ghana, and Uganda revealed that even when adaptation goals and initiatives are incorporated into local development plans, local governments lack the financial and human resources necessary to carry out policies or climate-resilient initiatives, despite being proclaimed self-governing by the federal government. Many studies contend that it is still challenging for African countries to meet their funding needs from bilateral and multilateral donors.

23.8 Policy Implications and Recommendations

The case studies and analysis presented in this chapter highlight the importance of participatory approaches in achieving sustainable dryland restoration in sub-Saharan Africa. The following policy implications and recommendations are drawn from the findings:

23.8.1 Policy Implications

- *Integration of participatory approaches into national policies*: Governments should integrate participatory approaches into their national policies and strategies for dryland restoration.
- *Decentralization and devolution of power*: Governments should decentralize and devolve power to local communities, enabling them to take ownership of dryland restoration initiatives.
- *Support for community-led initiatives*: Governments and development partners should provide support for community-led initiatives, including funding, technical assistance, and capacity building.
- *Recognition of land rights*: Governments should recognize and respect the land rights of local communities, ensuring that they have secure tenure and control over their lands.

23.8.2 Recommendations

- *Conduct participatory assessments and planning*: Conduct participatory assessments and planning exercises to identify the needs and priorities of local communities.
- *Establish community-led management structures*: Establish community-led management structures to oversee dryland restoration initiatives.
- *Provide training and capacity building*: Provide training and capacity-building programs for local communities, focusing on sustainable land management practices, entrepreneurship, and leadership.
- *Foster partnerships and collaborations*: Foster partnerships and collaborations among governments, NGOs, private sector entities, and local communities to leverage resources and expertise.
- *Monitor and evaluate progress*: Monitor and evaluate progress regularly, using participatory monitoring and evaluation methods to ensure that the needs and priorities of local communities are being addressed.
- By adopting these policy implications and recommendations, governments and development partners can support the scaling up of participatory dryland restoration initiatives in sub-Saharan Africa, ultimately contributing to improved livelihoods, food security, and environmental sustainability.

23.9 Conclusion and Recommendations

The chapter examined the participatory approaches of multiple actors in the restoration of sub-Saharan African drylands. Many countries in the SSA are experiencing land degradation and climate change risk. Some of the most prominent restoration initiatives are the AFR100 and the Great Green Wall. These drylands restoration initiatives have responded directly to global development initiatives such as the United Nations Sustainable Development Goals and the African Union Agenda 2063 to end the epidemics of poverty, hunger, and environmental degradation. At the forefront of the restoration efforts in the SSA are the individuals and organizations dedicated to restoring drylands. Governments, conservationists, researchers, environmental NGOs, communities of faith, and community members play a critical role in restoring degradation. From developing innovative approaches to sustainable land management to advocating for better policies and practices, the stakeholders applied their collective passion, knowledge, and resources.

The paper showed that it is imperative to engage multiple stakeholders in dryland restoration initiatives for successful and sustainable benefits. An examination of case studies has shown that sustainable and long-term restoration of drylands requires concerted approaches that involve players such as international bodies, local communities, NGOs, and governments in the region. The case examples have highlighted that multi-stakeholder involvement in dryland restoration integrates an extensive domain of science and non-science knowledge and experience, with the opportunity to introduce fresh insights and data. The impact of knowledge on subsequent actions can be further enhanced by treating conflicting perspectives fairly and openly. Therefore, including stakeholders makes it possible to connect knowledge with action. A well-coordinated, inclusive, and resourced multi-stakeholder participatory approach could generate a favorable space for fruitful discussions around conflicting restoration goals and enhance collaboration on strategic restoration determinations. Multi-stakeholder participation increases coherence, effectiveness, accountability, and reporting to both formal and informal dryland management structures.

References

African Union Development Agency – NEPAD (2023) Programme for Infrastructure Development in Africa (PIDA): First 10-year implementation report. AUDA-NEPAD. https://www.au-pida.org/wpcontent/uploads/2023/09/PIDA-Progress-Report_WEB.pdf

Amadou D, Diakarya B (2023) Agro-climatic risks analysis in climate variability context in Ségou Region. Open Journal of Statistics 13(1):170–193. https://doi.org/10.4236/ojs.2023.131009

Akinyemi FO, Ghazaryan G, Dubovyk O (2021) Assessing UN indicators of land degradation neutrality and proportion of degraded land for Botswana using remote sensing based national level metrics. Land Degrad Dev 32(1):158–172. https://doi.org/10.1002/ldr.3695

Azzarri C, Bacou M, Cox C, Guo Z, Koo J (2016) Subnational socio-economic dataset availability. Nat Clim Chang 6:115–116

Bensch G, Peters J (2015) The intensive margin of technology adoption – Experimental evidence on improved cooking stoves in rural Senegal. Journal of Health Economics 42:44–63. https://doi.org/10.1016/j.jhealeco.2015.03.006

Bendtsen EB, Clausen L.PW, Hansen SF (2021) A review of the state-of-theart for stakeholder analysis with regard to environmental management and regulation. Journal of Environmental Management 279:111773. https://doi.org/10.1016/j.jenvman.2020.111773

Berkes F (2006) From community-based resource management to complex systems: The scale issue and marine commons. Ecology and Society 11(1), Article 45. https://doi.org/10.5751/ES-01431-110145

Biedemariam M, Hadgu K et al (2017) Landscape level rehabilitation for improved agricultural productivity and ecosystem services in Abreha-we-Atsibeha, northern Ethiopia. J Arid Land 7:633–643

Birch JC, Newton AC, Aquino CA, Cantarello E, Echeverría C, Kitzberger T, Schiappacasse I, Garavito NT (2010) Cost-effectiveness of dryland forest restoration evaluated by spatial analysis of ecosystem services. Proc Natl Acad Sci 107(50):21925–21930. https://doi.org/10.1073/pnas.1003369107

Brugha RF, Varvasovszky Z (2000) Stakeholder analysis: A review. health policy and planning, 15(3):239–246. https://doi.org/10.1093/heapol/15.3.239

Cordingley J, Newton A, Rose S, Hughes K (2015) Enhancing the impact of dryland restoration initiatives. In: Reij, C., Pasiecznik, N. (Eds.), Dryland restoration successes in the Sahel and Greater Horn of Africa. ETFRN News 60, Tropenbos International, Wageningen, the Netherlands. https://www.academia.edu/117882330/

Costanza R, Daly L, Fioramonti L, Giovannini E, Kubiszewski I, Mortensen LF, Pickett KE, Ragnarsdottir KV, De Vogli R, Wilkinson R (2016) Modelling and measuring sustainable wellbeing in connection with the UN sustainable development goals. Ecol Econ 130:350–355. https://doi.org/10.1016/j.ecolecon.2016.07.009

Critchley W, Harari N, Mollee E, Mekdaschi-Studer R, Eichenberger J (2023) Sustainable land management and climate change adaptation for small-scale land users in Sub-Saharan Africa. Land 12(6):1206. https://doi.org/10.3390/land12061206

Darboe S (2023) Intertwining people and forest in the Lower River Region (LRR) of the Gambia

Dodds F (2015) Multi-stakeholder partnerships: Making them work for the Post-2015 Development Agenda. United Nations Department of Economic and Social Affairs. https://sdgs.un.org/sites/default/files/publications/16192015partnerships_background_note.pdf

FAO (2021) Climate Risk Toolbox: Guiding material for climate risk screening. https://openknowledge.fao.org/server/api/core/bitstreams/11bc3142-0997-47be-b900-759e3670093a/content

Favretto N, Shackleton S, Sallu SM, Hoffman T (2021) Editorial for special issue: "collaboration and multi-stakeholder engagement in landscape governance and Management in Africa: lessons from practice". Land 10:285. https://doi.org/10.3390/land10030285

Fereday J, Muir-Cochrane E (2006) Demonstrating rigor using thematic analysis: a hybrid approach of inductive and deductive coding and theme development. Int J Qual Methods 5(1):80–92

Food and Agriculture Organization of the United Nations (FAO) (2024) Dryland Restoration Initiative Platform (DRIP). FAO. Retrieved from https://www.fao.org/in-action/dryland-restoration-initiative-platform/en/

Gebremedihin KM, Birhane E, Tadesse T, Gbrewahid H (2018) Restoration of degraded drylands through exclosures enhancing woody species diversity and soil nutrients in the highlands of Tigray, Northern Ethiopia. Nat Conserv Res 3(1):1–20. https://doi.org/10.24189/ncr.2018.001

Graham C (2020) Managing climate change: the role of multi-stakeholder partnerships in building climate resilience in Sub-Saharan Africa. Interdiscip J Partnersh Stud 7(2)

Gravesen ML, Funder M (2022) The great green wall: an overview and lessons learnt (No. 2022: 02).DIIS Working Paper

Hemmati M (2002) Multi-stakeholder processes for governance and sustainability: Beyond deadlock and conflict. London: Routledge

Javadi M, Zarea K (2016) Understanding thematic analysis and its pitfall. J Client Care 1(1):33–39

Keeys LA, Huemann M (2017) Project benefits co-creation: shaping sustainable development benefits. Int J Proj Manag 35(6):1196–1212. https://doi.org/10.1016/j.ijproman

Kemeh SD, Kabalan AAM (2021) Youth engagement in land and ecosystem restoration. UNCCD Global Land Outlook Working Paper. Bonn

Kimmell LB, Fagan JM, Havrilla CA (2023) Soil restoration increases soil health across global drylands: A meta-analysis. Journal of Applied Ecology 60(9). https://doi.org/10.1111/1365-2664.14459

Lasker RD, Weiss ES, Miller R (2001) Partnership synergy: a practical framework for studying and strengthening the collaborative advantage. Milbank Q 79(2):179–205

Maestre FT, Benito BM, Berdugo M, Concostrina-Zubiri L, Delgado-Baquerizo M, Eldridge DJ, Guirado E, Gross N, Kéfi S, Le Bagousse-Pinguet Y, Ochoa-Hueso R, Soliveres S (2021) Tansley review: Biogeography of global drylands. New Phytologist 231(2):540–558. https://doi.org/10.1111/nph.17395

Mayor ÁG, Bautista S, Rodríguez-Mateos F, Kéfi S (2019) Connectivity-mediated ecohydrological feedbacks and regime shifts in drylands. Ecosystems 22(7):1497–1511. https://doi.org/10.1007/s10021-019-00366-w

Mentz S, Karambiri M, Smith Dumont E (2022) The great green wall initiative in Senegal-Country review

Mor T (2019) A manual for gender-responsive land degradation neutrality transformative projects and programmes, UNWomen, Global Mechanism of the UNCCD and IUCN. catalogue.unccd.int/1223_Gender_Manual.pdf

Mubaya CP, Mafongoya P (2017) The role of institutions in managing local level climate change adaptation in semi-arid Zimbabwe. Climate Risk Management 16:93–105. https://doi.org/10.1016/j.crm.2017.03.003

Nchare K, Vitouley M, & Mbih R (2024) Sustainable land management in Mali. Geography and Sustainability 5:484. https://doi.org/10.1016/j.geosus.2024.03.003

Nonet GAH et al (2022) Multi-stakeholder engagement for the sustainable development goals: introduction to the special issue. J Bus Ethics 180:945–957. https://doi.org/10.1007/s10551-022-05192-0

Oyekanmi O (2022) Climate change and environmental conflict in the Lake Chad region. Journal Administrasi Publik (Public Administration Journal) 12(2):270–279. https://doi.org/10.31289/jap.v12i2.8397

Prăvălie R (2016) Drylands extent and environmental issues. A global approach. Earth Sci Rev 161:259–278. https://doi.org/10.1016/j.earscirev.2016.08.003

Regreening Africa (2019) Regreening Africa is restoring ecosystems in 8 countries and improving the resilience of 500,000 households across sub-Saharan Africa. Retrieved from https://regreeningafrica.org/

Reynolds JF, Smith DM, Lambin EF, Turner BL II, Mortimore M, Batterbury SP et al (2007) Global desertification: building a science for dryland development. Science 316(5826):847–851. https://doi.org/10.1126/science.1131634

Schlaepfer DR, Bradford JB, Lauenroth WK, Munson SM, Tietjen B, Hall SA, Wilson SD, Duniway MC, Jia G, Pyke DA, Lkhagva A, Jamiyansharav K (2017) Climate change reduces extent of temperate drylands and intensifies drought in deep soils. Nat Commun 8(1):14196. https://doi.org/10.1038/ncomms14196

Stringer LC, Reed MS, Fleskens L, Thomas RJ, Le QB, Lala-Pritchard T (2017) A new dryland development paradigm grounded in empirical analysis of dryland systems science. Land Degradation & Development 28(7):1952–1961

Sibiya NP, Das DK, Vogel C, Mazinyo SP, Zhou L, Kalumba MA, Sithole M, Adom RK, Simatele MD (2023) Overcoming bureaucratic resistance: an analysis of barriers to climate change adaptation in South Africa. Climate 11(7):145. https://doi.org/10.3390/cli11070145

Tabutin D, Schoumaker B, Coleman H, Dutreuilh C, Reeve P, Tovey J, van Alkema BH (2020) The demography of Sub-Saharan Africa in the 21st century:transformations since 2000, outlook to 2050. Population 75(2):165–286

Tadesse L, Suryabhagavan KV, Sridhar G, Legesse G (2017) Land use and land cover changes and soil erosion in Yezat Watershed, North Western Ethiopia. International Soil and Water Conservation Research 5(2):85–94. https://doi.org/10.1016/j.iswcr.2017.05.004

Touré I, Larjavaara M, Savadogo P, Bayala J, Yirdaw E, Diakite A (2020) Land degradation along a climatic gradient in Mali: farmers' perceptions of causes and impacts. Land Degrad Dev 31(18):2804–2818. https://doi.org/10.1002/ldr.3683

Turner B, Devisscher T, Chabaneix N, Woroniecki S, Messier C. & Seddon N (2022) The role of nature-based solutions in supporting social-ecological resilience for climate change adaptation. Annual Review of Environment and Resources 47:123–148. https://doi.org/10.1146/annurev-environ-012220-010017

Turnbull LA, Isbell F, Purves DW, Loreau M, & Hector A (2016) Understanding the value of plant diversity for ecosystem functioning through niche theory. Proceedings of the Royal Society B: Biological Sciences 283(1832):20160536. https://doi.org/10.1098/rspb.2016.0536

UNCCD (2016) Turning the tide: the gender factor in achieving land degradation neutrality. www.unccd.int/sites/default/files/documents/2017_Gender_ENG.pdf

United Nations (2020) Challenges ahead: climate change in the context of weak institutions. Paper prepared for the UN Inter-Agency Expert Group Meeting on Implementation of the Third United Nations Decade for the Eradication of Poverty (2018–2027), 4–6 March 2020, Rome

United Nations Environment Programme (UNEP) (2022) Annual report Nairobi, Kenya:UNEP. Retrieved from https://www.unep.org/resources/annual-report-2022

United Nations Environment Programme (2023) The Great Green Wall is the wall we need [Speech]. UNEP. https://www.unep.org/news-and-stories/speech/great-green-wall-wall-we-needleap.unep.org+14

United Nations Development Programme (2023) Great Green Wall. Homegrown solutions to accelerate climate action and development. Retrieved from https://www.undp.org/africa/stories/great-green-wall-homegrown-solutions-accelerate-climate-action-and-development

United Nations Environment Programme (2021) Ecosystem restoration for people, nature and climate: becoming #GenerationRestoration. United Nations

Vinceti B, Valette M, Bougma AL, Turillazzi A (2020) How is forest landscape restoration being implemented in Burkina Faso? Overview of ongoing initiatives. Sustainability 12(24). https://doi.org/10.3390/su122410430

World Health Organization (2000) The world health report 2000: Health systems – Improving performance. Geneva: World Health Organization. https://www.who.int/publications/i/item/924156198X

Wu B, Smith WK, Zeng H (2024) Dryland dynamics and driving forces. In: Fu B, Stafford-Smith M (eds) Dryland social-ecological systems in changing environments. Springer, Singapore. https://doi.org/10.1007/978-981-99-9375-8_2

Xie H, Perez N, Anderson W, Ringler C, You L (2018) Can Sub-Saharan Africa feed itself? The role of irrigation development in the region's drylands for food security. Water International 43(6):796–814. https://doi.org/10.1080/02508060.2018.1516080

Yu X, Liu Y, Niu S, Zhao W, Fu C, Chen Z (2024) Structure, functions, and interactions of dryland ecosystems. In: Fu B, Stafford-Smith M (eds) Dryland social-ecological systems in changing environments. Springer Nature, pp 69–107. https://doi.org/10.1007/978-981-99-9375-8_3

Zhou L, Nunes MB (2016) Formulating a framework for desktop research in Chinese information systems. In: Martins J, Molnar A (eds) Handbook of research on innovations in information retrieval, analysis, and management. IGI Global Scientific Publishing, pp 307–325. https://doi.org/10.4018/978-1-4666-8833-9.ch011

Chapter 24
Indigenous Ecological Knowledge and Climate Change Adaptation

Samson Olusoji Afolabi, Emmanuel Babatunde Ajala, Innocent Onu Alenyi, and Olubunmi Abosede Odusanya

Abstract The peculiarity of land globally in human development circles across landscapes foregrounds that land management requires sensitive and deliberate sustainability approaches. The concern for food security with smallholder farmers as the power base and the nature of dryland regions in the Global South suggests a proactive focus due to variations in soil water content in different seasons, particularly with the unpredictable climate change scenarios. The indigenous or traditional approach to climate change adaptation, based on the existing knowledge bank of indigenous peoples globally, is a verified knowledge domain preserved by local or indigenous communities through oral traditions. In agriculture and climate change, scientific efforts and inquiry into the reliability of indigenous knowledge for climate change adaptation and agriculture are growing. In other development circles like health and medicine, verification and adoption have been attained to a very high degree. The United Nations Framework Convention for Climate Change (UNFCCC) in 2015, through Conference of the Parties (COP) 22 in Paris, promoted the inclusion of indigenous knowledge systems. The Local Communities and Indigenous Peoples' Platform (LCIPP) was the base formed for indigenous peoples' participation in the global climate body UNFCCC. Evidences abound that phenological observations in plants, insects, and the animal kingdom are natural signals of climate and weather changes, suitable for climate change adaptation globally and drylands regions of the Global South in particular. Ancestral groups also used cloud formation, wind direction, and moon phases to interpret environmental changes. However, detailed documentation beyond oral traditions is gradually growing.

S. O. Afolabi (✉) · E. B. Ajala
Glorious Vision University (Formerly, Samuel Adegboyega University), Ogwa, Nigeria
e-mail: safolabi@gvu.edu.ng; librarian@gvu.edu.ng

I. O. Alenyi
National Agency for the Great Green Wall, Abuja, Nigeria

O. A. Odusanya
Olabisi Onabanjo University, Ago Iwoye, Ogun State, Nigeria
e-mail: bunmi.odusanya@oouagoiwoye.edu.ng

Y. D. I. Khan et al. (eds.), *Land Restoration Through Ecosystem-Based Approach*, Disaster Resilience and Green Growth, https://doi.org/10.1007/978-981-95-0389-6_24

Scientific investigations with an empirical focus are increasing to ascertain and improve on the existing correlations between indigenous knowledge and scientific knowledge. In conclusion, the possibility of integration with existing adaptation strategies is very high.

Keywords Indigenous knowledge · Oral tradition · Drylands · Climate change · Phenology · Adaptation

24.1 Introduction

The peculiarities of groups in various regions globally mean that inhabitants have developed measures of managing their conditions in every sector sustainably so that they can cope with emerging issues as they evolve. The body of knowledge relevant for survival socially, economically, spiritually, physiologically, culturally, and technically exists among indigenous or traditional groups locally in various continents globally.

The growth and continuity in such regions have developed sustainably that indigenous groups, as early settlers, have gone through past centuries coexisting among themselves and relating to external groups and globally other continents using the knowledge gathered to solve their peculiar problems sustainably. Such knowledge, often described as local, indigenous, traditional, or cultural, exists in regions and continents globally, and it has become the body of adaptive knowledge to cope and survive in their natural settings.

The Global South as constituted by 134 nations globally, but in different continents like Africa, Asia, Latin America, the Caribbean, and Oceania, known for a common stand on global issues related to development generally. Brooke-Holland (2004) noted that the Global South is a voice segmentation or division that combines nations with specific, uniform agreement on contemporary global issues. Even though the term Global South is gaining prominence with the developing nations, the G7 was established in 1960 to promote the collective interest of the United Nations; it is an identity based on which the Global South nations are scattered across about six continents globally.

The term signaled a division of the territories into distinct geographical areas of North and South, but such assertions may be inappropriate on intellectual grounds. This is because several countries in the so-called "Global South or Global North" are scattered in various continents across the globe. Small countries like the Benin Republic in West Africa and the Fiji Islands in Oceania (South Pacific Ocean) are members, while Nigeria and other big nations are part of the Global South nations.

The 125 countries that participated in the first virtual meeting of the Global South nations in India in January 2023 gave an indication of countries tied to the Global South. The Global South landscapes or nationalities in sub-Saharan Africa include all Southern lands and are knitted to other Global South lands in Asia, Latin America, the Caribbean, and Oceania.

The traditional or indigenous knowledge (IK) of these nations, with its uniqueness in each locality, has peculiar features that are distinct and unique and applicable to specific environments among indigenous peoples in these nations. Although limited by the oral mode of transmission, traditional knowledge remains unique, and civilization and modernization have not been able to erode its base.

According to United Nations (2024b), the indigenous peoples are custodians of peculiar knowledge, ideas, and practices that prioritizes equality between nature and human beings. The intergenerational and community-based nature of indigenous knowledge is a fundamental characteristic. As early or pioneer settlers, the indigenous peoples were among the first to notice the early signals of climate change, as we know it today, as an issue of global concern. This is because they are very close to the environment, and all their activities as higher animals revolve around climate and the natural environment.

Climate change, with varying impacts on the globe, has profound effects reflected across the Global South's dryland landscapes. With the least contribution to climate change, the Global South bears the significant consequences. The ongoing debate on climate justice puts the Global South at the top of the need for lifeline support to diversify energy systems and adapt their infrastructure to a changing climate.

Maximum efforts are made to ensure supplementary water supply at critical periods to ensure optimum water security for the plant life and sustainable agricultural production.

Globally, indigenous knowledge and scientific strategies exist for adaptation to various climate change scenarios.

24.2 Evolution of Management and Conservation Principles for Sustainability of Drylands Using Indigenous Knowledge

Global concerns about the sustainable management of drylands have been a major agenda item at the United Nations. Sustainable resource management is needed to ensure that all lands in wet and dry regions and even large water bodies across continents are proactively focused on supporting human development and economic sustainability.

The United Nations Convention to Combat Desertification (UNCCD) was an international agreement initiated in 1994. Its membership comprises nations that agreed to support a new integrated approach to the problem of desertification. The goal was land protection and restoration to ensure a more sustainable future. The 197 member nations agreed to focus on effective ways of improving the lives of dryland inhabitants to maintain and restore land and soil resources to avert degradation and desertification IOM& UNCCD (2019). The collection, protection, improvement, and dissemination of relevant indigenous knowledge and technologies to tackle desertification are the core content of Article 18 (2) of the Convention to Combat Desertification (UNCCD 2022).

Through the Paris Agreement, the United Nations Framework Convention for Climate Change (UNFCCC) agreed that adaptation strategies should rest on the best practices of scientific findings and verified traditional knowledge. The International Panel for Climate Change Working Group II Sixth Assessment Report further acknowledges that "these knowledge systems provide useful climate change information, observations, solutions and enable improved adaptation." The Local Communities and indigenous Peoples' Platform (LCIPP) is the base for indigenous peoples' participation in the global climate body, the United Nations Framework for Climate Change (UNFCCC). The body UNFCCC recognized the role of indigenous knowledge in climate issues in terms of solutions and challenges of climate change based on traditional knowledge for inclusion in climate policies, plans, and actions (IPCC 2022; UNCCD 2022).

The promotion of indigenous knowledge systems through the Conference of the Parties 22 (COP22) in 2015 paved the way for the incorporation of indigenous knowledge systems in global climate discussions by the UNFCC. The importance of traditional knowledge in climate change management, as reaffirmed by Intergovernmental Panel on Climate Change (IPCC), was that "indigenous and traditional knowledge may prove useful for understanding the potential of certain adaptation strategies that are cost-effective, participatory and sustainable".

Promotion of indigenous knowledge on climate issues started receiving more attention from policy to research, higher levels of intervention, and global reckoning since the United Nations Framework Convention on Climate Change and Intergovernmental Panel on Climate Change prompted actions.

24.3 Drylands Landscape in the Global South

Drylands are commonly defined as regions in which the annual potential evapotranspiration (PET) greatly exceeds annual precipitation (P); that is, the ratio of the annual potential evapotranspiration is less than 0.65 (Huang et al. 2015). Specifically, more than 40% of the Earth's surface is made up of dryland ecosystems, which include rangelands, grasslands, and woodlands. These ecosystems are distinguished by their great temporal and spatial variability in rainfall (Daryanto et al. 2013). All continents contain drylands, which are regions that are hyperarid, arid, semiarid, and dry subhumid (Stringer et al. 2021). The biomass and soils of these ecosystems, grasslands, savannas, and woodlands store much of the world's terrestrial carbon and are rich in biodiversity (Gann et al. 2019). The geology, culture, and history of drylands are diverse and abundant.

The proportion of the world's population living in drylands is 38%, and the region is among the region's most vulnerable to human activity and climate change (Steenwerth et al. 2014). Out of these, two billion people live in arid and semiarid parts of Africa, Asia, Latin America, and the Middle East, around 100 nations (Stavi et al. 2021). Drylands provide a significant amount of the food and fiber consumed globally. Pastoral and agricultural groups that rely on the land for their livelihoods are frequently found in these areas (Nishi and Subramanian 2023).

The terrain, ecology, and socioeconomic circumstances of the drylands of the Global South vary diversely. Human activities like grazing, agriculture, and resource extraction have traditionally modified dryland landscapes. However, in many places, these practices and population expansion have also exacerbated desertification and land degradation. According to Brevik (2013), dryland soils are generally deficient in nutrients and organic matter, which reduces their agricultural productivity if improperly managed. Drylands of the Global South are especially susceptible to the effects of current climatic variations due to complicated dynamics that combine environmental fragility with socioeconomic difficulties.

The effect of altered rainfall patterns and soil degradation has made drylands the most impacted by climate change (Li et al. 2023), which lessens species and human capacity to adapt to dryland environments. In drylands, poverty, food insecurity, and water scarcity will probably worsen due to global climate change scenarios. Water resources are unpredictable and unreliable due to the substantial seasonal and inter-annual variability in rainfall in drylands. According to Gomiero (2016), between 20% and 35% of drylands currently experience some kinds of land degradation, which is predicted to increase dramatically under various emission scenarios.

Dryland soils store about one-third of the global organic carbon, crucial for improving soil structure and function and determining soil productivity. One of the most important causes of land degradation in drylands is soil erosion, which leads to the loss of soil organic carbon found in the soil's roots and woody components and, ultimately, a decrease in land productivity (Herbert et al. 2015). According to Diamini et al. (2016), grasslands may store up to 70 tons of soil carbon per hectare and contain almost 10% of the carbon on the Earth.

The grasslands, more than other plants, especially those with a perennial predominance, can tolerate high temperatures and high concentrations of carbon dioxide (CO_2). Due to these characteristics, grasslands offer adaptation potential and can continue to feed grazing cattle as temperatures and atmospheric CO_2 concentrations rise due to climate change. Temperatures in hot, tropical regions have already risen above the global average in drylands, and under higher emissions scenarios, temperatures are expected to rise by 2 °C to 4 °C by 2100 (Stringer et al. 2021). The worldwide greenhouse gas emissions, with over 75% and nearly 90% of all carbon dioxide emissions coming from fossil fuels like coal, oil, and gas and from these industries, are by far the most significant contributors to climate change (IPCC 2023). The greenhouse gas emission covers the earth and traps heat from the sun, thus leading to global warming, which causes climate change and its attendant effects.

The global warming rate is on the increase more than it has ever been, and the increasing temperatures are altering weather patterns and disrupting natural equilibrium. Humans and other living things on Earth are highly endangered. Climate change, drought, flooding, and land degradation frequently worsen these problems.

The significant aspects of current and future climate change throughout terrestrial ecosystems, especially in drylands, include increasing aridity, alteration in precipitation, and global warming. By increasing the frequency of extreme weather events like heat waves and droughts, ongoing climate change has increased climate variability (Li et al. 2023).

The drylands are under an unprecedented threat from climate change, which will seriously affect the environment, society, and economy. According to the IPCC report (2023), by the end of the twenty-first century, the world's drylands might increase by 10% with increased greenhouse gas emissions, and up to 80% of this growth occurs in developing nations. Land degradation can also result from converting dryland areas to other uses, particularly farming. Up to 60% of the soil's organic carbon below ground and up to 95% of its carbon above ground may be lost (Tuomas and Noora 2024). These losses lower drylands' ability to absorb carbon and directly increase atmospheric concentrations of greenhouse gases.

In addition to significantly altering rainfall patterns caused by climate change, the loss of vegetation may result in soil surface hardening, higher water runoff, and decreased groundwater recharge. Climate change is also predicted to increase the frequency and intensity of droughts, which is especially problematic in drylands where organisms and communities already have to deal with drastic climatic fluctuations (Furtak and Wolińska 2023).

Land degradation, also known as desertification in arid regions, is expected to be accelerated by global climatic variation, while the land's ecological and economic productivity decreases. Climate change accelerates arable land degradation, plant cover loss, and soil erosion. The risk of desertification rises as arid and semiarid areas get drier. Apart from decreasing agricultural output, land degradation interferes with drylands' ecosystem services, including water filtration, soil stabilization, and carbon sequestration.

The drylands of the Global South, which already struggle with water scarcity, high temperatures, and low soil fertility, are greatly impacted by climate change. Droughts are becoming more common and severe in the Global South drylands, which limit the amount of water available for agriculture and human and animal consumption (Springer et al. 2021). Food security and livelihoods are at risk, particularly for smallholder farmers and pastoralists, as crop yields and pasture quality diminish. Rising temperatures and unpredictable rainfall patterns accelerate soil erosion, vegetation loss, and land degradation.

Agricultural productivity is further restricted by the soil's reduced capacity to hold onto water and nutrients, which increases the likelihood of desertification on the land. Communities may adopt less sustainable farming techniques owing to conventional farming methods becoming less feasible due to shortened growing seasons and heat stress on plants. When precipitation decreases and evaporation rates rise, surface water and groundwater supplies deplete severely. Communities may have to compete for limited resources, which may impact access to drinking water and the irrigation required for agricultural purposes (Dao et al. 2024).

According to Weiskopf et al. (2020), changes in biodiversity result from the impact of climate change on the habitat ranges of several plant and animal species.

The ecosystem services that drylands offer, such as soil stabilization and carbon sequestration, may also be hampered by the loss of vegetation cover. Conflicts over land, water, and grazing rights might arise due to resource scarcity brought on by climate change (Godde et al. 2021). People may be compelled to relocate from impoverished areas to more fertile areas or urban centers, putting extra strain on social services and infrastructure. Crops, animals, and human health can all be directly threatened by rising temperatures and protracted heat waves. Heat stress can raise mortality rates among susceptible groups and lower labor output. Increased poverty, resource-related conflicts, and relocation are part of the socioeconomic climate change effects in arid regions. Declining production can cause social unrest and economic hardship for settlements where crop and animal husbandry are the major means of livelihood. In certain instances, this leads to migration across borders or from rural to urban regions, further taxing social services and infrastructure (Alarima 2018).

Although the Global South drylands encounter many difficulties, there are also chances to advance sustainable development and strengthen climate change resistance. The well-being of dryland communities can be enhanced and the effects of climate change can be lessened with adaptation strategies that combine traditional knowledge with contemporary methods and policy initiatives that promote sustainable land management (Herbert et al. 2015). Degraded areas can be restored, and soil fertility can be increased with the use of techniques like agroforestry, conservation agriculture, and soil and water conservation (Wolka et al. 2018). Minimizing soil erosion and introducing water retention practices like contour farming, mulching and cover crops, are put in place to boost the drought resistance capacity of dryland, as water retention efficiency is essential. Techniques like drip irrigation, rainwater harvesting, and building tiny dams can help communities better manage their limited water supplies.

Dryland agricultural output can be increased by creating and encouraging the usage of crop varieties resistant to drought. Furthermore, combining agricultural production systems with livestock management can maximize resource utilization and enhance food security. Vulnerability to climate shocks is reduced by diversifying sources of income, such as through agropastoral systems. Addressing dryland issues requires effective policies supporting climate adaptation and sustainable development. The potential for solar and wind energy in the Global South regions is substantial. Rural communities may receive sustainable electricity, lessen their dependency on fossil fuels, and open up new business prospects by investing in renewable energy infrastructure. Climate change poses serious threats to water resources, agricultural productivity, biodiversity, and socioeconomic stability in the drylands of the Global South. Nevertheless, these areas also provide chances for climate resiliency and sustainable development. The Global South can make its drylands more sustainable by using adaptive land management techniques, increasing water usage efficiency, encouraging climate-resilient agriculture, and investing in renewable energy (Erezi et al. 2023). Policies integrating environmental, social, and economic measures must be prioritized to ensure that dryland communities can prosper due to climate change peculiarities.

24.4 Impacts of Climate Change on Dryland Landscape in the Global South

Drylands, which encompass arid, semiarid, and subhumid lands, form about 41% of the global land surface with a population of over 2 billion people, many of whom are mostly inhabitants of the Global South. These regions are especially susceptible to climate change, as they already have limited water resources, high temperatures, and delicate ecosystems (UNCCD 2020). The effects of climate change on drylands in the Global South are far-reaching and diverse, significantly affecting biodiversity, agriculture, water resources, and the general livelihoods of the inhabitants.

The drylands in these areas face increasing threats from rising temperatures, unpredictable rainfall patterns, prolonged droughts, and degradation of lands, all of which profoundly impact ecosystems, agriculture, water resources, and human livelihoods. It is crucial to consider the effects of climate change on these fragile landscapes urgently. Such that we can develop sustainable strategic management principles that will foster resilience and support the well-being of affected communities.

24.4.1 Temperature and Precipitation Patterns

The direct effects of climate change on drylands in the Global South vary, but the alteration of temperature and precipitation patterns is on a large scale. Climate change has led to a significant temperature increase in dryland regions of the Global South, with some areas experiencing warming rates higher than the global average. Higher temperatures exacerbate evapotranspiration, reducing soil moisture and water availability and affecting agricultural productivity and natural vegetation (IPCC 2021).

The nations in the Sahel regions of Africa have mean annual rainfall reduced from above 800 mm to lower than 200 mm between the South and the North. Extreme rainfall also results in urban floods, which affect living conditions, particularly drinking water distribution, apart from topsoil erosion (Basutt 2019). The alteration of climatic elements contributes greatly to soil erosion, groundwater level depletion, and the depletion of surface water bodies, impacting both crop yields and biodiversity, with an increased threat to food security and livelihoods of millions who depend on crops and pastoralism.

24.4.2 Desertification and Land Degradation

Desertification, "as the process through which fertile land becomes desert," is accelerated by climate change and activities of human beings. In the Global South, drylands are particularly susceptible to desertification due to the combined effects of overgrazing, deforestation, unsustainable agricultural practices, and extreme weather events (UNEP 2019). Climate-induced factors, such as reduced rainfall and increased temperatures, worsen these conditions by intensifying soil degradation, loss of vegetation cover, and reduced land productivity.

The African nation with 45% of the world's degraded land is facing challenges of food insecurity and poverty as heightened by climate change. The increasing desertification, as escalated by climate change threats, negatively impacts agricultural activities, leading to more hunger and increased migration (AFDB 2025). The water shortage creates an opportunity for poor hygiene and unavailability of clean drinking water, thus promoting the spread of waterborne infectious diseases and malnutrition due to exacerbated food insecurity imposed by climate change hazards (WHO 2020).

The issue of desertification and land degradation is currently of great concern, given the attendant effects of climate change, particularly in the drylands of the Global South and all other parts of the world. The decision at the Conference of the Parties 16 (COP 16) on land restoration and drought resilience led to the projection of 2.6 billion dollars for the restoration of degraded lands, approximately 1 billion hectares globally by 2030. It was a noble decision to ameliorate the problem of land desertification and degradation globally (CGIAR 2025).

24.4.3 Impact on Biodiversity and Ecosystem Services

Dryland ecosystems in the Global South are home to diverse plant and animal species uniquely adapted to harsh conditions. Climate change disrupts these ecosystems, causing shifts in species distribution, changes in phenological patterns, and a decline in habitat quality. The loss of plant species due to higher temperatures and reduced soil moisture affects the entire food chain, leading to a decrease in wildlife populations and altering the ecosystem balance. In arid regions like the Kalahari Desert, temperature increases and prolonged droughts have led to shifts in vegetation types, with resilient species replacing less tolerant ones. This change in vegetation dynamics implies pastoral communities because of livestock grazing (Millennium Ecosystem Assessment 2005).

The various ecosystem services, such as fertility and soil water regulation, provision of agricultural products, carbon sequestration, and nonmaterial services such as spiritual roles common among indigenous communities, have been impaired by the negative effects of climate change.

24.4.4 Socioeconomic Impacts

Communities in the drylands of the Global South are among the most vulnerable to the socioeconomic impacts of climate change. The dependence on natural rainfall for agriculture, livestock, and general livelihoods made them highly sensitive to natural climate trends. Increased droughts and extreme weather events often result in crop failures, livestock losses, and reduced incomes, exacerbating poverty and food insecurity (FAO 2020).

Climate-induced stress has also led to increased migration, resource-based conflicts, and social instability in many regions. For example, in the Lake Chad Basin in Central Africa, diminishing water resources have contributed to tensions among farmers, herders, and fisher folk, intensifying regional conflicts (Okpara et al. 2015).

24.5 Adaptive Strategies and Mitigation Efforts

Several adaptation and mitigation strategies are required to tackle the menace of climate change in dryland landscapes of the Global South. Sustainable land management practices, like agroforestry, soil conservation techniques, and water harvesting, can help improve soil fertility, reduce erosion, and enhance water availability.

Community-based adaptation approaches that involve local populations in decision-making and resource management are also crucial for building resilience to the current challenges of frequent climatic variation. The use of indigenous knowledge and modern technologies can create effective strategies for coping with the adverse effects of climate variability in drylands (Nyong et al. 2007).

On a broader scale, international initiatives like the Great Green Wall project in Africa were projected to recover degraded lands and increase biodiversity growth across dryland regions. This large-scale effort to reforest and restore lands is essential for tackling increasing desertification and promoting sustainable development in the Global South.

In order to address these impacts, holistic approaches that include sustainable management of land resources, integrated management of water resources, and community-based adaptation strategies are essential to build resilience so as to sustain the livelihoods of millions of people and the overall stability of global ecosystems. The increasing threats of climate change are posing a significant challenge to continuous concerted efforts at the regional and global levels to mitigate its effects in order to secure a sustainable future for dryland inhabitants.

24.6 Indigenous Knowledge and Indigenous People

Indigenous knowledge (IK) "includes the expressions, practices, beliefs, understandings, insights, and experiences of indigenous groups, generated over centuries of profound interaction with a particular territory." There is uniqueness in its iterations and mechanisms with each community, or people's group, even where there is similarity across groups because of being embedded in a broader, common culture. In all local communities, indigenous knowledge is the basic structure of "indigenous governance, ecological stewardship, social, ethical, linguistic, spiritual, medical, food, and economic systems, so that the continual production and reproduction of local, land-based knowledge is the basis of indigenous identity" (Grey 2014). Indigenous knowledge has been a tool for managing the economy, social system, religious life, spiritual beliefs, food production, health, and all segments of community life.

The advent of civilization has not been able to completely erode or reduce the value and importance of indigenous knowledge. Modern religions and scientific interventions have not been able to replace indigenous beliefs and practices; rather, the integration of indigenous knowledge in modern religions has been a promotional factor to make modern religions and scientific inventions more applicable and appealing to the needs of indigenous groups.

In agriculture, which has been mankind's pioneer vocation since creation, the systematic observation of climate and weather characteristics, the soils and crops peculiar to specific communities and people groups, has become a guide to manage their activities with little or no support from external agencies.

The systematic observations over the ages have become the body of knowledge often transferred to generations. Inhabitants of various lands globally have adjusted their lifestyle based on the knowledge of their physical environment in terms of changes in seasons and the peculiar signs associated with such changes in biological features like plants and animals. Even though indigenous knowledge ideas were not written down in the early days until recent times, transmission between generations was done verbally in various forms. In many cases, direct communication and indirect communication in songs, stories, proverbs, and taboos are used to transfer indigenous knowledge information. Sometimes, physical demonstrations like soup making, farm practice, mode of dancing, mode of worship, practical ways of interpreting moon phases, sun appearance, and many more are also used to transfer indigenous knowledge information. These are the means of communication to ensure that coming generations have ideas of how their local economy is managed at the community level. It is interesting to note that indigenous knowledge has the potential to enrich climate policies and actions, particularly when integrated with meteorological scientific observations.

24.6.1 *Importance of Indigenous Knowledge to Climate Change Adaptation and Resilience*

Climate change is one of the biggest challenges facing the world today (Rodriguez and Alsop 2016), causing sudden and drastic changes, threatening the survival of humanity (Arshad et al. 2018), and impacting all societies (Heininen 2022). The impacts of climate change are more severe and intense in poor and marginalized communities (Ara and Reazul 2013), and many indigenous communities are the most vulnerable (Leal et al. 2021). Adapting to the multidimensional changes brought forth by climate change, indigenous communities rely heavily on their indigenous Knowledge (IK) systems to enhance adaptation and resilience to climate change (Apraku et al. 2018).

Indigenous knowledge refers to the repository of knowledge, practices, beliefs, and insights developed over generations within indigenous communities (Sukula 2006). Indigenous knowledge is deeply rooted in the cultural, social, and ecological contexts of specific indigenous groups and is closely tied to their relationship with the land, the environment, and the traditional ways of life (Kusumastutie et al. 2023; Radcliffe et al. 2021).

The use of indigenous knowledge, alongside scientific knowledge or meteorological services, provides an efficient solution to enhance climate change adaptation and build the resilience of indigenous communities in the face of a changing climate (Chen and Cheng 2020), which has also increased the severity and frequency of disasters (Heininen 2022).

Since time immemorial, indigenous peoples have respectfully lived with the natural world and have a deep connection to the land, water, and ecosystems that are central to their cultures, languages, and livelihoods. Through their experience and observations over generations, the indigenous people have been able to build a body of knowledge that has become codes and regulations guiding their actions and activities. Indigenous peoples, as pioneer settlers in various communities globally, have been able to adapt to their environments. Close observation of their natural environments makes them fix occupationally and in other areas. Communities in riverine areas are predominantly fisher folks. Pastoralists are peculiar to savannah and desert vegetation, and predominant crop farmers occupy forest and high forest vegetation.

24.6.2 The Role of Government and Policymakers

According to the United Nations Development Program (2024b), to promote indigenous knowledge for utilization as a means of combating climate change, policymakers through the government must act proactively on the inclusion of indigenous knowledge into strategies and action plans for climate change in the following ways.

(i) Recognition of indigenous peoples' rights to land and traditional territories as a matter of legal obligation and core to environmental sustainability. Promotion and preservation of local cultures and objects relevant to culture and historical issues as core ingredients for indigenous peoples to sustainably maintain identity, social cohesion, knowledge, rights, and resilience.
(ii) Establish frameworks with legal backing for "Free Prior and Informed Consent." Free Prior and Informed Consent (FPIC) enables the protection of people's collective rights.
(iii) The integration of indigenous knowledge into social and environmental policy frameworks, both at national and regional levels. Collaboration with indigenous peoples at the community level with respect for their knowledge, ideas, and practices.

(iv) Introduction of education, indigenous history, knowledge, and culture to ensure preservation and perpetuity as natural heritage.
(v) Provision of access to climate finance, which will be directed by indigenous people, to enhance the execution of sustainable adaptation projects and resilience (UNDP 2024b).

The inclusion of indigenous peoples as key actors in climate actions will support the development of more contextually relevant and highly adaptive solutions, which will foster innovation. It will also promote local ownership, create a sense of belonging, and ensure a greater impact of climate initiatives (UNDP 2024b). This is a strong reason why it is important to recognize, promote, and protect indigenous knowledge.

The approval of the Cacaus of indigenous Peoples and the Cacaus for Local Communities at the Conference of Parties 16 (COP 16) in 2024 was meant to ensure adequate representation of the peculiar situation of indigenous people globally for promoting indigenous knowledge inclusion in climate change adaptation and land restoration issues (CGIAR 2025).

24.7 Survey of Indigenous Knowledge on Climate Change among the Global South Drylands

24.7.1 Indigenous Knowledge and Climate Change in Africa

Africa, as one of the continents in the Global South, is a large reservoir of accumulated indigenous knowledge over the past decades on how climate and weather variables interact. The climate change effects on Africans are great to the extent that the livelihoods of 100 million people are endangered due to heat and drought. Approximately 40 million Africans migrated to the northern part of Africa and lately to Europe through a long journey that claimed lives on the sea and in the desert (Adesina 2015). The challenge of nomads migrating forcefully than ever before in this decade towards the well-watered areas signaled the pressure of advancing desert and hot temperatures that cannot be easily adapted to, except those humans and animals keep migrating.

The growth in indigenous knowledge research based on the utilization and application to manage climate change challenges in several countries of the Global South, particularly in Africa and other lands, led to policy formulation by the government for approval of the indigenous knowledge system policy as an instrument to promote the adoption of indigenous knowledge practices. That was the situation in South Africa when the government in 2004 took a step on indigenous knowledge policy formulation (Ncube and Largadien 2015). The breakthrough in the South African region came as the indigenous knowledge policy formulation was formally adopted. There was viral growth in indigenous knowledge documentation and inclusion as the South African government funded such projects through the National Research Foundation (Malferethane 2012).

Smallholder farmers in the Delta region of Nigeria found the integration of scientific and local knowledge feasible for interpreting and predicting weather and climate. They depended on organized indigenous knowledge systems to survive long-term climatic change due to poor access to meteorological services until the availability of daily forecasts broadcast in the electronic media (Fitchett and Ebhuoma 2017).

The various indigenous knowledge information in Africa related to weather changes, using plants, animals, insects, moon phases, and wind direction to interpret changes in climate and adopt adaptation strategies, are stated below.

24.7.1.1 Cloud Formation, Moon Shapes, Patterns of Star Arrangement, and Wind Direction

The patterns of cloud formation, stars arrangement, characteristic or peculiar wind and wind directions, and moon shapes have been used to predict weather patterns at the onset of season, during the season, and the transitional period between seasons. Afolabi (2022), in a study among farmers in Southwestern Nigeria, noted the use of dry winds to predict rainfall or the onset of dry season and drought periods. Tfwala et al. (2023) also reported the use of the crescent shape appearance of the moon facing up to predict dry months, while crescent shape facing sideways signifies an early warning for large amounts of rainfall among the Eswatini nation in South Eastern Africa. Depending on the region and locality, wind movement has been found useful for weather prediction.

The use of Lunar cycles, peculiar animal behavior, resurgence of certain insects, and appearance of certain animals has been found relevant as indicators of changes in weather and climate patterns. Crickets, toads, and frogs are known for emerging during rainy periods. The careful study of meteorological parameters like air temperature, cloud color formation, and cloud shapes was peculiar to communities for the interpretation and prediction of rainfall or drought (Masinde et al. 2012).

24.7.1.2 Flower Production and Fruiting in Trees

Plant phenology was affirmed by Gordo and Jose Sanz (2010) as strictly controlled by climate, and it is one of the most reliable or dependable bioindicators of climate change. Trees generally show signals of weather changes regularly, and they have become dependable over the years. Based on plant phenology, plants may indicate good rainfall or drought, onset of dry season, and peak of dry season (Mafongoya et al. 2017). In the oil palm production cycle in Nigeria, fresh fruiting commences in November in different regions, and they are signal of farmers' expectation of good yield and the change in weather to the dry season. Fruit production is more in mango and cashew; this is also used as an indicator of which area receives rainfall first. Generally, some trees like cashew produce new flowers to signal the onset of the dry season in many parts of Africa. Flamboyant tree *Delonix regia* sheds leaves to produce red flowers in the peak of the dry season (Afolabi 2022).

24.7.1.3 Animals and Insects' Response to Weather Changes

Large animals like cattle, birds, fowls, and amphibians are prediction tools for farmers; the croaking of toads and crickets and the peculiar tone of singing birds are signals of rain onset. This has been reported in Nigeria (Afolabi 2022) and in Uganda (Okonya et al. 2017). The movement in large herds of cattle from dry regions in Northern Africa towards the middle and Southern belts of sub-Saharan countries has been noticed over the years as the grasses have reduced considerably in the Northern parts of the drylands. The movement continues south to seek green grasses and avoid the increasingly harsh weather and advancing desertification in dry Northern regions.

Several animals and insects have been noted to respond to changes in weather through behavioral patterns. Ants have been noted for attempts to store food in the dry season, and termites build up the termitarium in the dry season, such that it is peculiar to see a tall new termitarium in the dry season. Croaking of toads and frogs always occurs in the wet season when rainfall is very common. The colonization of new beehives between November and February, mainly by bees to form new colonies, has been observed too.

24.7.1.4 Temperature and Winds

Periods of rainfall and drought have characteristic temperature patterns that could predict rainfall with high accuracy using local or traditional knowledge of a particular environment. The parameters used by the meteorological sciences in weather prediction are also employed by indigenous people using judgmental sense and experience of living in a particular environment for a long time. Isacc et al. (2009) noted that the commencement of the rainfall season and closure follow certain patterns in which peculiar, strange winds, cold or high, low or extremely high winds, precede rainfall with consistency over time in particular locations. Violent winds in the dry season may be an indication of the rainy season.

24.7.1.5 Blooming of New Leaves and Shedding of Old Leaves

Peculiarity of phenology in botanical sciences is synonymous with responses of plants to changes in climate and weather by leaf shedding, blooming of new leaves, and color changes in leaves as signals of adjustment to drought or temperature changes in various landscape in Global South dryland. In the tropics, sub-savannah, and temperate regions, various plants like lrvingia, fig (*Ficus sycomorus*), baobab

(*Adansonia digitata*), iroko (*Milicia excelsa*), and locust bean (*Parkia globerina*) trees show various signals with weather change. In adaptation to climate change, the relevance of these plants among others has formed a body of knowledge for indigenous, local, or traditional groups as pointers to drought, excessive rainfall, high temperature, high humidity, and other weather parameters. They are key interpreters to control, adapt, or build resilience to climate change (Afolabi 2022).

24.7.1.6 Peculiar Crops for Climatic Variation Adaptation

Several crops in various communities across the Global South landscapes are climate resilient. Such crops include Bambara, varieties of beans, and watermelon, which can adjust to reduced rainfall patterns.

The cultivation of climate-resilient crops like cowpea (*Vigna unguiculata*), Pigeon pea (*Cajanus cajan*), African yam bean (*Sphenostylis stenocarpa*), Lima bean (*Phaseolus lunatus*), and Bambara groundnut (*Vigna subterranea*) is used for adaptation to changes in climate in most parts of Africa, particularly in the drylands (Jiri et al. 2017). Some climate-resilient crops like African yam bean and Bambara groundnut are often classified as underutilized crops, but their climate-resilient property makes them peculiar.

24.7.1.7 Cropping Pattern

Multiple cropping, as peculiar to indigenous or local communities globally, has been used globally to manage climatic variation in ancient times, but has fallen into disuse in recent decades, and is coming up in the wake of unending climate change complexities. Zero-tillage farming in south-western Nigeria is an age-old cropping pattern in which trash burning is not done. Planting is done directly after manual clearing without bush burning within the dried thrashes. The advantages of direct soil fertility improvement, reduced soil erosion, and the possibility of time adjustment to adapt to unpredictable climatic variation make the zero-tillage cultural farming unique. According to Fane et al. (2024), in Burkina Faso and Mali, agroforestry systems were developed by the indigenous people to adapt to climate change. It involves planting trees with crops to reduce soil erosion and increase soil fertility. The crop diversification helps to reduce total crop failure because of the adverse effects of pests, diseases, and weather events (UNDP 2024b).

24.7.2 *Indigenous Knowledge and Climate Change in Asia, Caribbean, and Latin America*

The degree of knowledge on the environment, like other sectors, among indigenous groups globally has been an asset. The depth of knowledge based on several generations of romance with land, environment, and organisms in an ecosystem usually leads to continuous interaction, evaluation, and redefinition of such interactions. These eventually lead to the accumulation of knowledge built up consciously and unconsciously over centuries on managing challenges in a locality. The knowledge body often becomes an opportunity to combat various global challenges in localities and regions.

24.7.2.1 Pastoralists Approach

The indigenous peoples in Asian countries are tapping into indigenous local or customary knowledge peculiar to their environment for climate change adaptation. In both livestock and crop production in the Himalayas and Bangladesh, specific adaptation methods were adopted, drawing from their knowledge of the local environment and customary practice. The Himalayan regions, as the home to pastoralists, witnessed more frequent movement across pasture to give room for regeneration on a seasonal basis (UNDP 2024a).

Movement across pasture and rangelands is a common practice globally in response to natural seasonal changes and their effects on pastures. This customary practice helps the regrowth of the grasses and also breaks the cycle of pasture pests and insect reproduction. As a natural way of managing ecosystems, this practice considerably manages pest and insect populations.

24.7.2.2 Production of Tolerant Trees and Vegetables

Crop adaptation as a natural method of coping with climate change in Bangladesh involves the cultivation of saline-tolerant reeds, timber, and fruit trees to cope with flooding and salinization in areas with perennial floods without challenges of storm surges. Floating vegetable gardens are cultivated for adaptation to flooding by the saline water (UNDP 2024a).

24.7.2.3 Traditional Crops

In various parts of Latin America, including Colombia, Argentina, Brazil, Mexico, Peru, and the Republic of Venezuela, indigenous people account for 35% of the population in the forest areas. Colombia has about 102 indigenous peoples spread over 32.1 million hectares within the country; several local practices were developed over the years to cope with their environmental conditions, one of which is the cultivation of Guajira beans, *Vigna unguiculata.* The Wayum community often cultivates the Guajira beans as one of the means of adapting to changes in climate. It is naturally adaptable to the hot, dry climate, and drought conditions of the community, so the production is widespread (UNDP 2024b; UNDDR 2023).

24.8 Documentation of Indigenous Knowledge Ideas Related to Climate Change in the Global South Drylands

The current challenge globally is the transitional growth in the volumes of indigenous knowledge documentation. Ecological indigenous knowledge, sometimes called traditional or local knowledge, was developed through oral traditions over centuries in specific regions, nations, communities, and environments globally. The oral tradition, inheritance, and transfer of indigenous knowledge information transition from ancient times into the modern age. This is particularly true in most parts of the world, particularly in African and Asian countries.

In a study in Southwestern Nigeria, West Africa, one of the populous countries in the Global South, it was reported that 75% of the respondents asserted that indigenous knowledge information was relayed to them through oral traditions by adults, local leaders, chiefs, or traditionalists (Afolabi 2022). The respondents further affirmed that some of the people who are indigenous knowledge custodians shared the knowledge verbally. Since such local, traditional, or indigenous knowledge information on all segments of human economy and lifestyle, health, agriculture, education, human medicine, climate, weather, and building structure was shared through the oral transmission mode. The documentation was made based on the quantity and quality of information available offhand through human memory.

The core challenge observed in Africa and some parts of the world is the unwillingness to share traditional or indigenous knowledge information, which is often perceived as sacred or exclusive to specific people, age groups, classes, genders, or the elderly as custodians. In some African countries and local communities, gender and age discrimination exist among custodians of indigenous knowledge information. Extensive studies are ongoing in various parts of the world, including the Global South drylands countries, to document, examine, compare, and validate the veracity and accuracy of local knowledge as a potent tool.

The extensive work (Apata et al. 2009; Masinde et al. 2012; Matowayinka 2016; Mafongoya et al. 2017; Ajibade and Etche 2017; Ajayi 2018; UNFCCC 2020; UNCCD 2020; Leal et al. 2021; Afolabi 2022) that has been carried out in various parts of the world, particularly across all regions of Africa, is basic foundational research on indigenous knowledge. The studies of Iwu (2013) and NNMDA (2013) on the study and documentation of medicinal plants and traditional knowledge peculiar to Africa on human health and natural medicine were great foundational information on the value of indigenous knowledge. These studies are foundations to propel the knowledge search on the relevance of indigenous knowledge to global development, especially climate and weather adaptation. Local knowledge has been used by local indigenous groups as pioneer settlers in lands and jungles and is found relevant to present realities if carefully and scientifically examined.

The review study by Dorji et al. (2024) about the contribution of indigenous knowledge to climate change concluded that indigenous knowledge and climatic variables in the context of global climate change have multidisciplinary value due to the aggregation of authors in coaction in various domains of knowledge to share the holistic effect of climate and weather changes on the environment and all sectors.

The identification of local or traditional knowledge usage in weather prediction, preparation for disasters through early warning in agriculture and nonagricultural sectors of power generation, human health and safety, and urban and rural planning for housing, roads, and other infrastructural development has continued to gain attention. Ongoing efforts are to document indigenous knowledge's past and review, examine, improve, integrate, or adopt it for national and global intervention adaptation programs.

24.9 The Critical Role of Information Communication Technology in Documentation, Research, and Utilization of Traditional or Indigenous Ecological Knowledge

The era of information communication technology, as a contemporary mode of information dissemination through virtual space, with the world as a global village, makes the use of hard copies of documents in physical libraries and other domains a wholesome task, very costly, time-consuming, and unnecessary.

The neglect, abuse, loss, and nonrecognition of the premium value of indigenous or traditional knowledge of the early ages, as endowed by nature, since the early days, call for urgent action.

The need for critical review through research and various accredited documentation processes cannot be overemphasized, and it must be urgent. Wholesome documentation is core, but holistic evaluation and empirical comparative investigative studies as a follow-up are a major task for this age. The documentation of indigenous knowledge related to climatic variables is not just the core issue, but a holistic, multidisciplinary, collaborative, and integrative approach to indigenous knowledge systems research.

To remediate the current challenges of our age due to the lack of sustainable development of indigenous traditional knowledge in all sectors, documentation, taking advantage of digital tools, is core to redeeming the global population from the current environmental, nonenvironmental, sociocultural, human, and nonhuman challenges of this age.

24.10 Opportunities in the Global South Drylands for Digital Tools Intervention

- Auto or digitally operated gadgets for indigenous knowledge information recording.
- Nonrenewable-energy-operated recording equipment for voice recording, solar energy, bioorganic waste, and wind energy generation.

- Virtual communication within the world as a global village, affordably, in a timely manner, and with less risk, is crucial.
- Conclusively, the information and communication age are a great asset to this generation. In this context, the Global South landscape responds to the core values that brought those nations together.

The common ground was their core eagerness and concern for global issues of health, food and nutrition, security, environment, climate, peace, and conflict resolution, among others. Information and communication technology, as a digital tool, can deliver sustainable solutions if properly employed in indigenous knowledge research, documentation, and utilization.

References

Adesina A (2015) The Great Green Wall is helping African countries to mitigate climate change. Available via https://www.afdb.org. Accessed 11 Oct 2024

Afolabi SO (2022) Farmers' perception of indigenous knowledge on climate change adaptation strategies on crop production in Oyo and Ogun States of Nigeria. Dissertation, Benson Idahosa University Nigeria

African Development Bank (2025) Desertification and land degradation. Available via http://www.afdg.org/. Accessed 13 Jan 2025

Ajayi OC (2018) Aggregating indigenous and scientific knowledge for better climate adaptation. Available via http://www.cta.int/en/article. Accessed 10 Oct 2024

Ajibade LT, Etche JO (2017) Indigenous for climate change adaptation in Nigeria. In: Mafongoya PL, Ajayi OC (eds) Indigenous knowledge systems and climate change management in Africa, vol 10. CTA, pp 231–235

Alarima CI (2018) Factors influencing rural-urban migration of youths in Osun state, Nigeria. J Trop Agric Food Environ Exten 17(3):34–39

Apata TG, Samuel KD, Adeola AO (2009) Analysis of climate change perception and adaptation among arable crop farmers in Southwestern Nigeria. Paper presented at the international association of agricultural economics conference, pp 2–9

Apraku A, Akpan W, Moto P (2018) Indigenous knowledge, global ignorance? Insights from an eastern cape climate change study. S Afr Rev Sociol 49(2):1–21

Ara Parvin G, Reazul Ahsan SM (2013) Impacts of climate change on food security of rural poor women in Bangladesh. Manag Environ Qual Int J 24(6):802–814

ASEAN (2023) The voice of Global South Summit. Participation of ASEAN Member states. Association of South East Asian Nations. Accessed via http://www.indmissio.nasean.gov.in/

Basutt M (2019) Rainfall trends in the Africa Sahel: characteristics, processes and causes. In: Lewis SC, Haliman M (eds) WIRES climate change, p e591

Brevik E (2013) The potential impact of climate change on soil properties and processes and corresponding influence on food security. J Agric For 3(3):398–417. https://doi.org/10.3390/agriculture3030398

Brooke-Holland L (2004) What is the global South? Review by House of Commons Library United Kingdom. Available via http://commonslibrary.parliament.uk/. Accessed 10 Oct 2024

Chen TL, Cheng HW (2020) Applying traditional knowledge to resilience in coastal rural villages. Int J Disaster Risk Reduct 47:101564

Consultative Group International Agricultural Research (2025) UNCCD COP 16 Wrap up: Key Outcomes from negotiations. CGIAR Newsletter. Accessed via https://www.cgiar.org/news-events/all-news/. Accessed 15 Jan 2025

Dao PU, Heuzard AG et al (2024) The impacts of climate change on groundwater quality: a review. Sci Total Environ 912:169241. https://doi.org/10.1016/j.scitotenv.2023.169241
Daryanto S, Eldridge DJ, Throop HL (2013) Managing semi-and woodlands for carbon storage: grazing and shrub effects on above and below ground carbon agriculture. Ecosyst Environ 169:11
Diamini P, Chivenge P, Chaplot V (2016) Overgrazing decreases soil organic carbon stocks the most under dry climates and low soil pH: a meta-analysis shows. Agric Ecosyst Environ 221:258–269. https://doi.org/10.1016/j.agee.2016.01.026
Dorji T, Rinchen K et al (2024) Understanding how indigenous knowledge contributes to climate change adaptation and resilience: a systematic literature review. Environ Manag 74:1101–1123
Erezi E, Ogar EE, Oladosu T (2023) Promoting sustainable agriculture and climate resilience in African nations. Int J Agric Earth Sci 9(5):27–45. https://doi.org/10.56201/ijaes.v9.no5.2023.pg27.45
Fane S et al (2024) Adoption of agroforestry systems by smallholders' farmers with Sudano-Sahelian zones of Mali and Burkina Faso, in West Africa. Agrofor Syst 97:2385. https://doi.org/10.1007/s10457-02401020-8
Fitchett JM, Ebhuoma E (2017) Phenological cue intrinsic in indigenous knowledge systems for forecasting seasonal climate in the Delta state of Nigeria. Int J Biometeorol 62:1115–1119
Food and Agriculture Organization (2020) State of food security and nutrition in the world 2020. FAO, Rome. Available via https://www.fao.org/3/ca9692en/CA9692ENpdf
Furtak K, Wolińska A (2023) The impact of extreme weather events as a consequence of climate change on the soil moisture and on the quality of the soil environment and agriculture – a review. Catena 231:107378. https://doi.org/10.1016/j.catena.2023.107378
Gann GD, McDonald T, Walder B, Aronson J, Nelson CR, Jonson J, Hallett JG, Eisenberg C, Guariguata MR, Liu J, Hua F, Echeverría C, Gonzales E, Shaw N, Decleer K, Dixon KW (2019) International principles and standards for the practice of ecological restoration. Restor Ecol 27(5s1). https://doi.org/10.1111/rec.13035
Godde CM, Mason-D'Croz D, Mayberry DE, Thornton PK, Herrero M (2021) Impacts of climate change on the livestock food supply chain; a review of the evidence. Glob Food Sec 28:100488. https://doi.org/10.1016/j.gfs.2020.100488
Gomiero T (2016) Soil degradation, land scarcity and food security: reviewing a complex challenge. Sustainability 8(3):281. https://doi.org/10.3390/su8030281
Gordo O, Jose Sanz J (2010) Impact of climate change on plant phenology in Mediterranean ecosystems. Glob Chang Biol 16:1082. https://doi.org/10.1111/j.1365-2486.2009.020844
Grey S (2014) Indigenous knowledge. In: Michalos AC (ed) Encyclopedia of quality life and well being research. Springer, Dordrecht. https://doi.org/10.1007/978-94-00753-5_1448
Heininen C (2022) Climate change and the great power rivalry in the Arctic. Rethin Environ Secur 24(2):25–38. https://doi.org/10.25253/99.2022242.2
Herbert ER, Boon P, Burgin AJ, Neubauer SC, Franklin RB, Ardón M, Hopfensperger KN, Lamers PM, Gell P (2015) A global perspective on wetland salinization: ecological consequences of a growing threat to freshwater wetlands. Ecosphere 6(10):1–43. https://doi.org/10.1890/es14-00534.1
Huang J, Ji M, Xie Y, Wang S, He Y, Ran J (2015) Global semi-arid climate change over last 60 years. Clim Dyn 46(3–4):1131–1150
Intergovernmental Panel on Climate Change (2021) Climate change: the physical science basis. Cambridge University Press, Cambridge
Intergovernmental Panel on Climate Change (2022) Impacts, adaptation and vulnerability. In: Partner HO et al (eds) Contribution of working group II to the sixth assessment report of the IPCC. Cambridge University Press
Intergovernmental Panel on Climate Change (2023) Climate Change 2023 Synthesis Report. Hoesung L Jose R (eds). IPCC_AR6_SYR. 1–186

IOM & UNCCD (2019) Addressing the Land Degradation – Migration Nexus: The Role of the United Nations Convention to Combat Desertification. International Organization for Migration and United Nations Convention to Combat Desertification, IOM, Geneva

Isaac ME, Dawoe E, Siesiechowiez K (2009) Assessing local knowledge use in agroforestry management with cognitive maps. Environ Manag 43(6):13211329

Iwu MM (2013) Handbook of African medicinal plant, 2nd edn. CCR Press, Taylor and Francis Group

Jiri O, Mafongoya PL, Musundire R (2017) The use of underutilized crops and animal species in managing climate change risks. In: Mafongoya PL, Ajayi OC (eds) Indigenous knowledge systems and climate change management in Africa. CTA, Wageningen

Kusumastutie R, Hidayanto AN et al (2023) Innovation through indigenous knowledge sharing, organisational memory, and indigenous knowledge erosion on indigenous batik enterprise (a structural equation model in action). Int J Innov Learn 33(1):71. https://doi.org/10.1504/ijil.2023.127953

Leal FW, Matandirotya NR, Lütz JM, Alemu EA, Brearley FQ, Baidoo AA, Kateka A, Ogendi GM, Adane GB, Emiru N, Mbih RA (2021) Impacts of climate change to African indigenous communities and examples of adaptation responses. Nat Commun 12(1). https://doi.org/10.1038/s41467-021-26540-0

Li X, Sun J, Zhang H, Tan H, Hui R, Qi J, Ward ND (2023) Warming decreases desert ecosystem functioning by altering biocrusts in drylands. J Appl Ecol 60(12):2676–2687

Mafongoya PL et al (2017) Using indigenous knowledge for seasonal quality prediction in managing climate risk in sub-Saharan Africa. In: Mafongoya PL, Ajayi OC (eds) Indigenous knowledge and climate change management in Africa, vol 3. CTA, Wageningen, pp 48–63

Malferethane IO (2012) The role of indigenous knowledge in disaster and risk reduction: a critical analysis. Dissertation. Northwest University Pocheftsroom, South Africa

Masinde M, Bagula A, Muthama NJ (2012) The role of ICTs in downscaling and up-scaling integrated weather forecasts for farmers in sub-Saharan Africa. In Proceedings of the Fifth International Conference on Information and Communication Technologies and Development (pp. 122–129)

Masinde M, Bagula A (2012) ITIKI bridge between African indigenous knowledge and modern science of drought prediction knowledge. Manag Dev J 7(3):274–290. https://doi.org/10.1080/19474199.2012.683444

Matowayinka JZ (2016) Indigenous knowledge systems, community – based climate observation practices and synergies with climate services and adaptation in Zimbabwe. Paper presented at The Southern Africa regional climate services workshop towards exploiting the full potential of climate services, Victoria Falls, Zimbabwe, Nov. 29-Dec. 2, 2016

Millennium Ecosystem Assessment (2005) Ecosystems and human well-being: desertification synthesis. World Resources Institute, Washington

Ncube B, Largadien A (2015) Insight into indigenous coping to drought for adaptation in agriculture: a Karoo scenario. In: Indigenous knowledge systems for managing climate change in South Africa, vol 12. University of Technology Cape Peninsular, p 256

Nigerian Natural Medicine Development Agencys (2013) In: Okujagu TF, Etutavie SO, Ndukwe M, Salihu T, Mbaoji C, Barnabas O, Chukwubuikem A, Ifeoma C (eds) Medicinal plants of Nigeria south zone, vol 1. Nigerian Natural Medicine Development Agency, Lagos

Nishi M Subramanian SM (2023) Ecosystem restoration through managing socioecological production landscapes and seascapes (SEPLS). https://doi.org/10.1007/978-98199-1292-6

Nyong A, Adesina F, Osman EB (2007) The value of indigenous knowledge in climate change mitigation and adaptation strategies in the African Sahel. Mitigation and Adaptation Strategies for Global Change https://doi.org/10.1007/511027-007-9099-0

Okonya JS, Ajayi OC, Mafongoya PL (2017) The value of Indigenous knowledge in seasonal weather forecasting and planning of farm activities by rural crop farmers in Uganda. In Indigenous knowledge systems and climate change management in Africa. 11:249. Technical Centre for Agricultural and Rural Cooperation, The Netherlands

Okpara UT, Stringer LC, Dougill AJ, Bila MD (2015) Conflicts about water in Lake Chad: are environmental, vulnerability and security issues linked? Prog Dev Stud 15(4):308–325. https://doi.org/10.1177/1464993415592738

Radcliffe C, Raman A, Parissi C (2021) Entwining indigenous knowledge and science knowledge for sustainable agricultural extension: exploring the strengths and challenges. J Agric Educ Ext 27(2):133–151

Rodriguez F, Alsop S (2016) Rethinking climate change education research in a world of shifting climates. ICERI proceedings. https://doi.org/10.21125/iceri.2016.1075

Stavi I, De Pinho JR, Paschalidou AK, Adamo SB, Galvin K, De Sherbinin A, Even T, Heaviside C, Van Der Geest K (2021) Food security among dryland pastoralists and agro-pastoralists: the climate, land-use change, and population dynamics nexus. Anthrop Rev 9(3):299–323. https://doi.org/10.1177/20530196211007512

Steenwerth KL, Hodson AK, Bloom AJ, Carter MR, Cattaneo A, Chartres CJ, Hatfield JL, Henry K, Hopmans JW, Horwath WR, Jenkins BM, Kebreab E, Leemans R, Lipper L, Lubell MN, Msangi S, Prabhu R, Reynolds MP, Solis SS, Jackson LE (2014) Climate-smart agriculture lobal research agenda: scientific basis for action. Agric Food Secur 3(1). https://doi.org/10.1186/2048-7010-3-11

Stringer LC, Mirzabaev A, Benjaminsen TA, Harris RM, Jafari M, Lissner TK, Stevens N, Pahlen CTD (2021) Climate change impacts on water security in global drylands. One Earth 4(6):851–864. https://doi.org/10.1016/j.oneear.2021.05.01

Sukula SK (2006) Developing indigenous knowledge databases in India. Electron Lib 24(1):83–93. https://doi.org/10.1108/02640470610649263

Tuomas JM, Noora V (2024) Agricultural limitations to soil carbon sequestration: plant growth, microbial activity, and carbon stabilization. Agric Ecosyst Environ 367:108986. https://doi.org/10.1016/j.agee.2024.108986

UNEP (2019) Global environment outlook 6. United Nations Environment Program, Nairobi. Available via https://www.unep.org Accessed 10 Jan 2025

United Nations Convention to Combat Desertification (2020) The global land outlook. UNCCD, Bonn. Available via https://www.unccd.org. Accessed 10 Jan 2025

United Nations Convention to Combat Desertification (2022) An introduction to the United Nations Convention to Combat Desertification. Available via https://www.unccd.int Accessed on 10 Jan 2025

United Nations Development Program (2023) Climate change adaptation. Sowing resilience. How Colombia's indigenous peoples adapt to climate change. Available via https://www.undp.org Accessed 10 Jan 2025

United Nations Development Program (2024a) Local communities and indigenous peoples' platform web portal: UN indigenous sociocultural groups: Asia. Available via https://www.undp.org Accessed 10 Jan 2025

United Nations Development Program (2024b) Climate Promise. Indigenous knowledge is crucial in the fight against climate change – here's why. Available via https://www.undp.org Accessed 10 Jan 2025

United Nations Framework Convention for Climate Change (2020) Local communities and indigenous peoples' platform web portal overview. Available via https://icipp.ppunfccc.int//icipp.background/overview. Accessed on 6 Oct 2024

Weiskopf SR, Rubenstein MA, Crozier LG, Gaichas S, Griffis R, Halofsky RE, Kimberly JW, Morelli TL, Morisette JT, Muñoz RC, Pershing RJ, Peterson RL, Poudel R, Staudinger MD, Sutton-Grier AE, Thompson L, Vose J, Weltzin JF, Whyte KP (2020) Climate change effects on biodiversity, ecosystems, ecosystem services, and natural resource management in the United States. Sci Total Environ 733:137782

Wolka K, Mulder J, Biazin B (2018) Effects of soil and water conservation techniques on crop yield, runoff and soil loss in sub-Saharan Africa. A review. Agric Water Manag 207:67–69

World Health Organisation (2020) Climate change: land degradation, desertification. Available via https://www.who.int Accessed on 10 Jan 2025

Part V
Adaptive Policies, Institutional Governance, and Long-Term Sustainability Strategies

Chapter 25
Mainstreaming Circular Ecosystem-Based Approach into National Climate Adaptation Strategies: Enhancing Resilience and Sustainability of Crop-Livestock Systems in Africa's Drylands

Mujeyi Angeline, Musemwa Lovemore, and Munyati Vincent

Abstract Africa's drylands are highly vulnerable to climate change, threatening the livelihoods of millions of smallholder farmers and pastoralists. To enhance resilience and sustainability, there is a need to mainstream circular ecosystem-based approaches into national climate adaptation strategies. This study examines the potential of circular ecosystem-based adaptation (CEBA) to enhance the resilience and sustainability of crop-livestock systems in Africa's drylands. The study reviews successful CEBA practices and policies in Southern Africa, East Africa, and West Africa, highlighting their impacts on ecosystem services, livelihoods, and climate resilience. Case studies from South Africa, Mali, Kenya, Niger, and other countries demonstrate the effectiveness of CEBA approaches in improving soil fertility, increasing crop yields, promoting biodiversity, and enhancing ecosystem services. The study identifies key enabling factors, including policy support, institutional capacity, and community engagement, and highlights the need for a coordinated approach to mainstream CEBA into national climate adaptation strategies. The findings of this study provide valuable insights for policymakers, practitioners, and researchers seeking to promote sustainable and resilient crop-livestock systems in Africa's drylands.

Keywords Climate change resilience · Circular economy · Sustainable agriculture · Ecosystem-based adaptation

M. Angeline (✉) · M. Lovemore · M. Vincent
Department of Agricultural Economics, Education and Extension (AEEE), Faculty of Agriculture and Environmental Sciences, Astra Campus, Bindura University of Science Education (BUSE), Bindura, Zimbabwe

Y. D. I. Khan et al. (eds.), *Land Restoration Through Ecosystem-Based Approach*, Disaster Resilience and Green Growth, https://doi.org/10.1007/978-981-95-0389-6_25

25.1 Introduction

Climate change poses significant threats to dryland ecosystems of Africa, exacerbating existing vulnerabilities. Africa has experienced significant warming, with temperatures rising by over 1 °C since 1901, increasing the frequency and severity of heatwaves and extremely hot days. A comprehensive review of 230 statistical crop modeling studies (Hu et al. 2024) found that global crop yields decline with increasing temperatures, with a 1 °C warming resulting in yield reductions of 7.5% (maize), 6.0% (wheat), 6.8% (soybean), and 1.2% (rice), exhibiting spatial heterogeneity due to nonlinear crop responses to temperature variations. Dryland ecosystems, home to 40% of the world's population, are disproportionately vulnerable to the impacts of climate change and highly susceptible to climate-related shocks, including heat stress, drought, and changing vegetation patterns (Stringer et al. 2021). These changes have devastating consequences for rural livelihoods, exacerbating poverty and food insecurity. The 2024 drought in Southern Africa left a trail of destruction, impacting countries such as Angola, Botswana, Malawi, Mozambique, Namibia, South Africa, Zambia, and Zimbabwe, where over 27 million people were affected. The drought's far-reaching consequences were evident in the reduced water levels at Zambia's Kariba Dam, which led to widespread power outages affecting both Zambia and Zimbabwe (Zhang et al. 2025). In India's drylands, climate-related shocks have reduced agricultural productivity by up to 30% (Kumar et al. 2020). Climate change also threatens biodiversity in dryland ecosystems, with up to 30% of plant and animal species facing extinction (Weiskopf et al. 2020). For example, in Australia's drylands, climate change has altered the distribution of native vegetation, compromising ecosystem services (Eldridge and Beecham 2018). Climate-related shocks have severely impacted Africa's drylands, reducing agricultural productivity by up to 30% and compromising water availability (Fouque and Reeder 2019). These changes have far-reaching consequences, affecting not only agricultural productivity but also human health, biodiversity, and ecosystem services. For instance, altered temperature and precipitation patterns have increased the risk of malaria and trypanosomiasis by changing the distribution of disease-carrying vectors like mosquitoes and tsetse flies (Fouque and Reeder 2019). The resilience of dryland ecosystems is compromised, threatening the livelihoods of over 100 million people who depend on these ecosystems. Consequently, many governments are putting in place climate change adaptation strategies that support farming sectors.

Among these strategies, ecosystem-based adaptation (EBA) has emerged as a promising approach, leveraging the natural resilience of ecosystems to help communities adapt to climate change. By harnessing the benefits of healthy ecosystems, EBA offers a sustainable and cost-effective solution for enhancing the resilience of both dryland ecosystems and the livelihoods they support (Nalau et al. 2018). Ecosystem-based adaptation (EbA) is a nature-based approach to climate change adaptation that harnesses the power of biodiversity and ecosystem services to help communities cope with the impacts of climate change (Zölch et al. 2018; Simonson et al. 2021). By restoring and conserving natural ecosystems, EbA provides cost-effective and sustainable solutions that offer multiple benefits, including carbon sequestration, biodiversity conservation, and the provision of essential ecosystem services such as food, water, and shelter. EbA approaches are highly effective, providing long-term and landscape-scale benefits that enhance the resilience of both ecosystems and human communities. Despite the growing recognition of ecosystem-based adaptation (EbA) as a vital strategy for climate change adaptation, existing approaches often overlook the importance of circularity in enhancing the resilience and sustainability of ecosystems. The majority of EbA frameworks focus on conventional adaptation strategies, neglecting the potential benefits of circular economy principles in promoting resource efficiency, reducing waste, and fostering regenerative ecosystems. This study aims to address this knowledge gap by exploring the integration of circularity with EbA and investigating its potential to enhance the effectiveness of adaptation strategies in dryland ecosystems. The chapter will answer the following research questions:

- How can circular economy principles be integrated with ecosystem-based adaptation to enhance resilience and sustainability in dryland ecosystems?
- What are the key barriers (institutional, financial, knowledge gaps) to implementing circular ecosystem-based adaptation (CEbA) in dryland ecosystems, and how can they be addressed?
- How can sustainable crop-livestock systems be designed and implemented to promote resource efficiency, ecosystem services, and adaptation in dryland ecosystems?
- Why is it important to integrate circular ecosystem-based adaptation into national and regional development plans, and what are the implications for policy, planning, and practice in dryland countries?

25.2 Context and Rationale

African drylands are characterized by savannas and open shrublands, with a dynamic mix of trees and grasses (Wei et al. 2021). The availability of resources like water and soil nutrients, as well as disturbance regimes like fire and herbivory, regulates this vegetation. Research has shown that rainfall patterns are the primary driver of vegetation composition and growth in these ecosystems (Zhang et al. 2019). Additionally, human activities like land conversion, farming, and deforestation are increasingly impacting ecosystem properties. These complex interactions between resources, ecosystems, and disturbances pose challenges for dryland ecosystems, which support a significant portion of Africa's population, who rely on livestock and agriculture for their livelihoods. Dryland ecosystems in Africa are highly vulnerable to socioeconomic factors (population and income growth, trade liberalization, urbanization, and markets), land degradation, water scarcity, and climate change challenges. These impact poverty and food insecurity and cause limited access to basic services such as healthcare and education. Climate change is exacerbating these challenges, with droughts, floods, and heatwaves increasingly frequent and severe.

The numerous challenges facing dryland farm households impact their production, consumption, and investment decisions. To address these challenges, international and national research and development organizations, in partnership with government departments, have developed climate change strategies to transform dryland agriculture. This has led to the promotion and support of various climate-smart technologies in Africa. The adoption of climate-smart agriculture (CSA) practices has shown promising results in enhancing resilience and reducing vulnerability in sub-Saharan Africa. For instance, a review by Thierfelder et al. (2017) found that CSA principles, such as conservation agriculture (CA), have positive effects on adaptation and productivity, although yield benefits may take two to five cropping seasons to materialize. Similarly, Michler et al. (2019) discovered that CA helped farmers in Zimbabwe adapt to rainfall variability, despite not yielding immediate benefits. Furthermore, research from the Climate Change, Agriculture, and Food Security (CCAFS) Kaffrine Climate-Smart Village revealed that farmers who adopted a suite of CSA practices, including inorganic fertilizer use, were able to enhance resilience and reduce vulnerability (Bonilla-Findji et al. 2020). These findings suggest that CSA can be an effective strategy for reducing vulnerability and enhancing resilience in specific contexts.

Climate change poses significant threats to livestock production systems in drylands, leading to crop failures, fodder scarcity, reduced productivity (milk, meat, and eggs), and increased incidence of endemic animal diseases (Ramana 2022). To mitigate these impacts, various adaptation strategies have been promoted, including enhanced fodder production, scientific feeding practices, provision of shelter to protect against extreme weather events, disease surveillance, and timely immunization. Additionally, integrating livestock production systems with agriculture and allied sectors has been encouraged to reduce vulnerability to climate change.

However, these climate change adaptation and mitigation strategies have several weaknesses. Livestock strategies have often focused on short-term gains, neglecting the long-term sustainability of livestock production systems. Moreover, they may not fully account for the complex interdependencies between livestock, agriculture, and ecosystems. Furthermore, the implementation of these strategies can be hindered by limited access to resources, inadequate institutional support, and insufficient community engagement. These weaknesses underscore the need for a more circular and integrated approach to livestock production, one that prioritizes sustainability, resilience, and the well-being of both people and the planet. While crop climate-smart technologies have shown promise in enhancing resilience and productivity, they often fall short in addressing the complex, interconnected challenges facing agricultural ecosystems. These technologies tend to focus on single crops or farming systems, neglecting the broader ecosystem services and synergies that underpin long-term sustainability. Furthermore, they may rely heavily on external inputs, such as synthetic fertilizers and pesticides, which can exacerbate environmental degradation and undermine ecosystem health. As a result, these technologies may not be sufficient to address the scale and complexity of the climate crisis, highlighting the need for a more holistic, circular, ecosystem-based approach that prioritizes the health, resilience, and regeneration of entire ecosystems.

Current climate change adaptation strategies in dryland Africa, such as drought-tolerant crops and conservation agriculture, have shown promise but have limitations. These strategies often require significant investments of time, money, and labor, which can be prohibitive for many dryland communities. Despite these challenges, there are opportunities for innovation and transformation in dryland Africa. By supporting climate-resilient agriculture, improving water management, and promoting ecosystem restoration, it is possible to enhance the livelihoods of dryland communities while protecting the natural resources on which they depend. However, this will require a coordinated effort from governments, civil society, and the private sector, as well as a commitment to supporting the unique needs and priorities of dryland communities.

25.3 Circular Ecosystem-Based Adaptation: Concept and Framework

Ecosystem-based adaptation (EbA) strategies are crucial for enhancing climate change resilience and mitigation efforts in dryland ecosystems. Smallholder farmers can benefit from farm management practices based on agro-biodiversity and ecosystem services, which provide adaptation benefits. Various agricultural practices, such as agroforestry and conservation agriculture, can help farmers improve their farming systems and increase resilience to climate change.

Dryland EbA focuses on restoring, conserving, and sustainably managing ecosystems to enhance resilience to climate change. This involves promoting drought-tolerant vegetation, conserving soil moisture, and protecting biodiversity (UNCCD 2019). EbA prioritizes local knowledge, renewable inputs, and affordable implementation for smallholder farmer livelihoods and food security. Effective dryland EBA strategies integrate ecological, social, and economic considerations, including conservation agriculture practices, agroforestry, and restoring degraded lands (FAO 2017).

Studies have shown the effectiveness of EbA strategies in enhancing ecosystem services and farmer livelihoods. For example, integrating rainwater harvesting, crops, trees, and livestock in smallholder farming systems under rain-fed conditions significantly enhances ecosystem services and farmer livelihoods (Palsaniya et al. 2024). In Togo's Savannah region, EbA strategies have shown promise in addressing climate change impacts on smallholder farming, with practices such as agroforestry, crop rotation, and intercropping reducing vulnerability to climate risks (Kissi et al. 2023). Integrating circular economy (CE) principles into EbA can amplify its effectiveness. CE focuses on waste reduction, resource efficiency, and closed-loop production. By adopting CE principles, EbA can minimize waste, promote sustainable resource use, and foster resilient ecosystems. Successful integration of CE principles in EbA requires policy support, stakeholder engagement, and community participation.

Circular economy principles offer a promising approach to enhance the resilience and sustainability of dryland ecosystems. In the context of agriculture, circular economy principles involve the design and implementation of regenerative and restorative systems that promote resource efficiency, reduce waste, and foster ecosystem services. This approach prioritizes the use of renewable energy, minimizes the use of synthetic fertilizers and pesticides, and promotes the conservation and efficient use of water. Ecosystem-based adaptation (EbA) is a complementary approach that recognizes the importance of ecosystem services in supporting human well-being and adaptation to climate change. EbA involves the conservation, restoration, and sustainable management of ecosystems to enhance their resilience and reduce the vulnerability of human communities. Key components of EbA include the maintenance of ecosystem diversity, the promotion of ecological connectivity, and the support of sustainable livelihoods.

The integration of circular economy principles with EbA offers significant synergies for enhancing ecosystem health and resilience. Circular ecosystem-based adaptation (CEbA) integrates circular economy principles with ecosystem-based adaptation to enhance resilience and mitigate climate-related risks (UNEP 2024). By promoting regenerative and restorative practices, circular economy principles can enhance ecosystem services, such as soil formation, water cycling, and carbon sequestration. Conversely, EbA can provide a framework for scaling up circular economy practices, promoting their adoption and implementation at the landscape scale. By combining these approaches, it is possible to create more resilient, sustainable, and regenerative ecosystems that support both human well-being and ecosystem health.

CEBA approaches prioritize resource efficiency, waste reduction, and closed-loop systems, promoting ecosystem services and biodiversity conservation. CEbA strategies have shown significant benefits in dryland ecosystems, enhancing water use efficiency by up to 30% and reducing greenhouse gas emissions. For example, in Africa's drylands, CEbA has promoted climate-resilient agriculture, improving livelihoods and food security (Thorntorn et al. 2018). CEbA approaches also foster social cohesion, knowledge sharing, and adaptive capacity among rural communities. Community-led CEbA initiatives can enhance resilience, reducing vulnerability to climate-related shocks.

25.4 Integration into National Development Plans

The integration of circular ecosystem-based adaptation into national development plans is crucial for enhancing the resilience and sustainability of dryland ecosystems in Africa. All 54 African countries have ratified the Paris Agreement and submitted nationally determined contributions (NDCs), showing a strong commitment to mitigating global temperature rise (Akeredolu 2024). However, the successful implementation of NDCs requires effective integration of adaptation strategies into national development plans. The importance of integrating adaptation strategies into national development plans is underscored by the Intergovernmental Panel on Climate Change (IPCC) Africa assessment, which highlights the compounding risks faced by African countries due to climate change. These risks include reduced food production, increased heat-related mortality, and flooding from sea level rise. Effective integration of adaptation strategies into national development plans can help mitigate these risks and enhance the resilience of dryland ecosystems.

Circular ecosystem-based adaptation (CEbA) can be effectively integrated into national adaptation plans by incorporating its key principles and strategies into existing policy frameworks. This can be achieved by identifying and prioritizing ecosystem-based adaptation measures that promote circularity, such as agroforestry, conservation agriculture, and integrated water management. These measures can be mainstreamed into national adaptation plans through the development of specific policy and regulatory frameworks, capacity building and training programs for stakeholders, and the allocation of dedicated funding and resources. Furthermore, CEbA can be aligned with the nationally determined contributions (NDCs) under the Paris Agreement by incorporating circular economy principles into greenhouse gas emission reduction targets and adaptation strategies. By integrating CEbA into national adaptation plans, countries can enhance their resilience to climate change, promote sustainable development, and support the achievement of the Sustainable Development Goals (SDGs).

The Green Climate Fund (GCF) plays a critical role in supporting the implementation of National Adaptation Plans (NAPs) and integrating adaptation strategies into national development plans (Chivhenge et al. 2023). As of November 2023, 44 developing countries with submitted NAPs had received approval for 83 adaptation projects from the GCF, totaling USD 18.6 billion. This support can help African countries integrate circular ecosystem-based adaptation into their national development plans, enhancing the resilience and sustainability of dryland ecosystems.

25.5 Sustainable Crop-Livestock Systems in Drylands

Crop-livestock integration is essential for sustainable agriculture in drylands, promoting soil fertility, improving water use efficiency, and enhancing biodiversity. This integration provides livelihood benefits, including food security, income generation, and employment opportunities. Effective crop-livestock integration requires careful planning and management, balancing crop and livestock production, managing water and soil resources, and mitigating climate-related risks (Thornton et al. 2018). Successful integration can lead to increased agricultural productivity, improved ecosystem services, and enhanced resilience to climate change. To enhance resilience and sustainability in dryland agriculture, farmers must adopt climate-resilient agricultural practices, such as agroforestry, conservation agriculture, and integrated water management. These practices can help mitigate the impacts of climate change, including increased temperatures, altered precipitation patterns, and increased frequency of extreme weather events (IPCC 2014). Conservation agriculture has increased crop yields by over 20% in Africa's drylands (Phiri et al. 2025). Agroforestry practices have also shown significant benefits, increasing livestock productivity by up to 15% and reducing greenhouse gas emissions (Kay et al. 2019). In Latin America's drylands, agroforestry has promoted ecosystem services, including pollination and pest control.

Additionally, dryland farmers require access to climate information, financial services, and markets to enhance their resilience and adaptive capacity. By adopting these strategies, dryland farmers can improve their livelihoods, enhance ecosystem services, and promote sustainable agriculture in the face of climate change.

The integration of cropping enterprises with rangelands is crucial for achieving a resilient and sustainable ecosystem in dryland areas. This integration can be achieved through various strategies, including agro-pastoralism, where livestock grazing is combined with crop production. Livestock manure from rangelands can be used as fertilizer in cropping enterprises, reducing the need for synthetic fertilizers and promoting nutrient cycling. Additionally, cropping enterprises can provide feed and fodder for livestock, while livestock can provide draft power and manure for crops. Adopting circular economy principles can further enhance the sustainability of dryland ecosystems. By promoting the reuse and recycling of resources, circularity can help reduce waste and the consumption of new resources (Fig. 25.1). For instance, crop residues can be used as fodder for livestock or as organic amendments for soils. Similarly, livestock manure can be composted and used as fertilizer, reducing the need for synthetic fertilizers. By closing these nutrient loops,

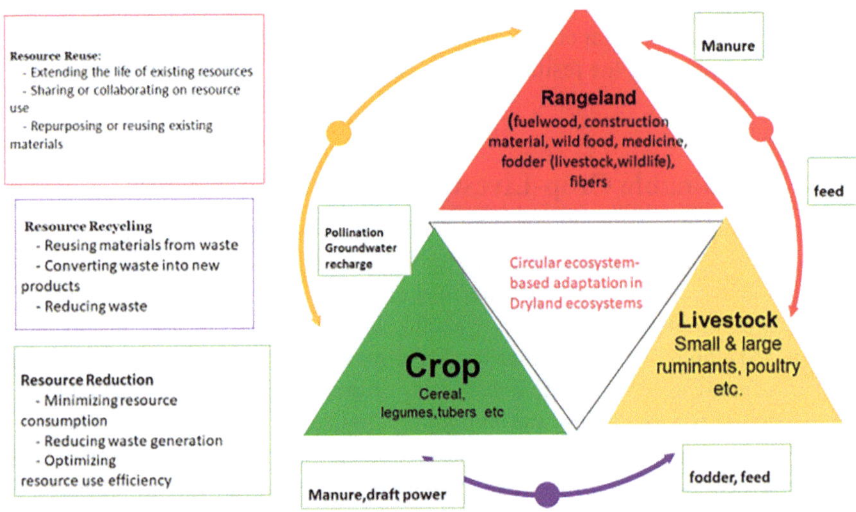

Fig. 25.1 Visualizing the integration and circularity concepts

circularity can help maintain soil health, promote biodiversity, and support ecosystem services.

The benefits of circularity in dryland ecosystems extend beyond environmental sustainability. By promoting the efficient use of resources and reducing waste, circularity can also enhance the economic viability of dryland farming systems. For example, the use of crop residues as fodder or organic amendments can reduce the costs of feed and fertilizer, improving the profitability of farming enterprises. Moreover, adopting circular economy principles can create new business opportunities and income streams, such as the production and sale of compost or organic fertilizers. By promoting economic development and environmental sustainability, circularity can help achieve a more resilient and sustainable ecosystem in dryland areas.

25.6 Dryland Agroforestry Practices

Agroforestry, the integration of trees into agricultural landscapes, is vital for enhancing ecosystem resilience and sustainability in dryland regions. By combining trees with crops and livestock, agroforestry promotes ecological interactions that improve soil health, biodiversity, and ecosystem services. In dryland regions, agroforestry is crucial in boosting community resilience to climate change and land degradation. Research has shown that tree-based agriculture practices have provided a lifeline for millions of Africans, enhancing their capacity to cope with drought and climate shocks (Mbow et al. 2014; Alemu 2016). Agroforestry enables farmers to harvest timber, fodder, and other tree products, which can replace failed crop production and minimize vulnerability.

Agroforestry contributes significantly to ecosystem-based adaptation (EbA) strategies in drylands. Studies in Ethiopian and Malawian drylands demonstrate that on-farm trees provide income, improve livelihood resilience, and increase crop production (Ndalama et al. 2015; Linger 2014). Approximately one-third of household income in Ethiopian drylands comes from trees on farmland, which is less affected by climate hazards like drought. Agroforestry has been shown to enhance farmland ecosystem resilience, improve income and food security, and ultimately enhance livelihood resilience to climate and other shocks (Zada et al. 2022; Nyong et al. 2019; Quandt 2020). The benefits of agroforestry in drylands are clear. For instance, research in northwestern Ethiopia found that home-garden agroforestry significantly increased farmers' cash income (Linger 2014). To scale up agroforestry's impact, policymakers and practitioners should prioritize supportive policies, capacity building, and community engagement. By promoting agroforestry practices, dryland communities can build resilience, improve livelihoods, and adapt to climate change.

Dryland agroforestry and sustainable agriculture practices are critical EbA strategies. Agroforestry involves integrating drought-tolerant trees into farming systems to promote ecological interactions and synergies. Sustainable agriculture practices, such as conservation agriculture and organic farming, reduce soil erosion, improve water retention, and enhance biodiversity. Dryland agroforestry and sustainable agriculture practices have numerous benefits, including improved crop yields, increased livestock productivity, and enhanced ecosystem services. These practices promote climate resilience by reducing soil degradation, improving water use efficiency, and supporting carbon sequestration. Case studies demonstrate the effectiveness of dryland agroforestry and sustainable agriculture practices. For example, farmers in Malawi's drylands have adopted agroforestry practices, improving maize yields by 20% and reducing soil erosion by 30% (Amadu et al., 2020).

25.6.1 Integrated Soil Fertility Management

Soil health is a critical component of sustainable farming practices in dryland crop-livestock systems. Healthy soils support biodiversity, improve water retention, and enhance ecosystem services, ultimately contributing to climate resilience and agricultural productivity. In drylands, soil degradation and fertility decline are major concerns, compromising crop and livestock productivity. Integrated soil fertility management (ISFM) innovations offer a promising solution, promoting soil health through the combined use of organic (e.g., compost and manure) and inorganic fertilizers, crop rotation, and conservation tillage.

ISFM innovations have shown significant benefits in dryland crop-livestock systems. In Africa's drylands, ISFM has increased maize yields by up to 50%, improved soil organic carbon by 20%, and reduced soil erosion by 30% (Vanlauwe et al. 2020). Moreover, ISFM promotes climate resilience by enhancing soil water holding capacity, reducing drought stress, and supporting biodiversity. For example, in Zimbabwe's drylands, ISFM has improved soil fertility, increasing cattle productivity by up to 25% and enhancing rural livelihoods.

Scaling up ISFM innovations requires a holistic approach, addressing policy, institutional, and technical constraints. This includes developing context-specific ISFM practices, enhancing farmer knowledge and skills, and promoting market access for organic fertilizers. Additionally, integrating ISFM with climate-smart agriculture practices, such as conservation agriculture and agroforestry, can enhance synergies and maximize benefits. By prioritizing soil health and ISFM innovations, dryland crop-livestock systems can enhance resilience, improve productivity, and support sustainable agricultural development. Sustainable farming practices also enhance livelihood benefits, improving rural incomes and food security. In Asia's drylands, sustainable agriculture has increased farmer incomes by up to 25% (Kumar et al. 2020).

Microdosing, the strategic application of small amounts of organic fertilizers, offers a vital solution for enhancing soil fertility in dryland regions. By applying precise amounts of nutrients, typically 5–20 kg/ha, microdosing stimulates soil biota, improves nutrient cycling, and boosts water retention. Results from some on-farm trials in Zimbabwe drylands showed that smallholder farmers could increase their yields by 30–100% through application of micro doses, as little as 10 kg Nitrogen ha-1 (Twomlow et al. 2011). This targeted approach helps overcome soil fertility limitations, common in drylands, where nutrient-depleted soils exacerbate drought vulnerability. Microdosing with organic amendments like compost, manure, or green manure increases crop yields, enhances root development, and promotes drought resilience. Moreover, microdosing reduces fertilizer waste, conserves resources, and mitigates environmental pollution. As a low-cost, climate-resilient strategy, microdosing empowers dryland farmers to improve soil health, increase agricultural productivity, and ensure sustainable livelihoods.

Integrated soil fertility management also includes the use of improved germplasm alongside the use of inorganic fertilizer and organic inputs. The development and dissemination of improved germplasm have transformed agricultural productivity in dryland regions. Breeding programs have yielded climate-resilient crop and livestock varieties, tolerant to drought, heat, and soil salinity. These enhanced germplasm, such as drought-tolerant maize, sorghum, and cowpea, and heat-tolerant wheat and chickpea, enable farmers to maintain yields despite harsh environmental conditions. Additionally, researchers have identified and conserved valuable genetic diversity in dryland-adapted crop landraces and livestock breeds, preserving valuable traits for future breeding efforts. The introduction of improved germplasm has increased crop yields by up to 50%, enhanced food security, and empowered dryland communities to cope with climate variability and change.

25.6.2 Climate-Resilient Water Management Strategies in Drylands

Effective water management is vital in dryland regions, where water scarcity compounds agricultural challenges. To build resilience, climate-resilient water management strategies are critical, enabling sustainable agriculture and mitigating the impacts of water scarcity. These strategies involve promoting water conservation, efficient use, and sustainable management. Climate-resilient water management in drylands requires integrating climate information into decision-making processes. Effective climate-resilient water management strategies in drylands include rainwater harvesting, groundwater recharge, and efficient irrigation systems. These strategies reduce the risks associated with droughts, floods, and water scarcity. Examples of climate-resilient water management strategies in drylands include rain water harvesting, greywater reuse, and conservation agriculture, as well as the implementation of efficient irrigation systems, watershed management, and drought-tolerant crop varieties. Climate-Resilient Water Management can enhance water efficiency in agriculture. Adoption of drip irrigation technologies can also enable precise water delivery to crops, thus minimizing water losses through evaporation and runoff (Yang et al. 2023). Use of soil moisture monitoring tools.

Water scheduling tools offer a vital solution, enabling farmers to optimize water use and reduce waste. Advanced tools, such as crop water modeling software and mobile apps, provide real-time data on soil moisture, weather forecasts, and crop water requirements. These tools allow farmers to precisely schedule irrigation, reducing evapotranspiration and runoff. Additionally, decision-support systems integrate satellite imagery, GIS mapping, and weather data to identify areas of high water stress, enabling targeted interventions. By adopting water scheduling tools, dryland farmers can increase water productivity, reduce consumption by up to 30%, and enhance crop yields, ultimately contributing to sustainable agriculture and resilience in water-scarce regions.

In-field water harvesting technologies are pivotal for enhancing water availability in dryland regions. These innovative solutions capture and store rainwater or runoff within the field, reducing evaporation and runoff losses. Techniques such as contour ridging, mulching, potholing, tied ridges, and micro-catchments enable farmers to harvest and conserve water for irrigation, supplementing scarce rainfall. Additionally, technologies like rainwater harvesting ponds, reservoirs, and infiltration trenches help recharge groundwater, providing a reliable source of water for crops. By adopting in-field water harvesting, dryland farmers can increase water productivity, reduce drought risk, and boost crop yields, ultimately enhancing food security and livelihood resilience.

25.6.3 Restoring Ecosystems

Rangelands, a vital source of feed for livestock and wildlife, face severe degradation due to climate change, overgrazing, population pressure, and inadequate rainfall, posing a significant global environmental threat. (Daba and Mammo 2024). These factors culminate in fodder scarcity and range deterioration, exacerbating feed shortages. Bush encroachment is a major form of land degradation in rangelands that diminishes the availability and quality of forage for grazing animals, impacting meat, milk, and leather production (Shikangalah and Mapani 2020). This phenomenon also increases the risk of animal diseases, provides hiding spots for predators, and hinders accessibility to limited grass forage. Bush encroachment has led to significant declines in grazing capacity, with a 1% increase in bush cover resulting in a 26.5 kg/ha decrease in grass biomass in Zimbabwe. In Namibia, bush encroachment decreased the carrying capacity from 10 ha/LSU to 30 ha/LSU, while in South Africa's Molopo region, veld productivity for cattle was suppressed by 970 kg/ha of dry matter (Geißler et al. 2023). The economic implications of bush encroachment are substantial. In Namibia, the decline in cattle numbers resulted in an estimated economic loss of N$700–1.6 billion (US$ 94.8 million–US$ 217 million) per annum. Similarly, in Zimbabwe, woody encroachment caused a loss of £16 million (US$74 million) in beef production in 1948. These declines in livestock production can significantly impact food security, particularly in countries like Namibia, where livestock production contributes up to 75% to the total agricultural output (Shikangalah and Mapani 2020).

Dryland regions face heightened vulnerability due to harsh climate and soil conditions, making livestock production a more feasible agricultural pursuit than cropping. However, rangelands offer a multitude of benefits beyond livestock production. These secondary resources include medicinal plants, firewood, wild foods, and essential nutrients through milk and meat production (Mudau et al. 2022). They play a vital role in supporting livelihoods, ensuring food security, and providing ecosystem services that sustain local communities. Effective rangeland reclamation strategies include practices such as rotational grazing, reseeding with native species, and soil conservation measures. These approaches can lead to significant improvements in range condition, increased forage production, and enhanced ecosystem services. Restoring rangelands is particularly crucial in drylands, where rangelands are highly sensitive to environmental stressors, such as drought, soil erosion, and overexploitation. In dryland areas, rangelands restoration can not only support biodiversity and improve soil fertility but also enhance the resilience of these areas to the impacts of climate change.

Successful rangeland restoration also requires community engagement and involvement. Local communities often possess valuable traditional knowledge and skills essential for effective rangeland management. Collaborative approaches that integrate local knowledge with scientific expertise can ensure sustainable rangeland management and support the livelihoods of pastoralist communities. By restoring degraded rangelands, we can maintain ecosystem services, support biodiversity, and ensure the long-term productivity of these critical ecosystems. Restoring ecosystems is a critical strategy for enhancing climate resilience, mitigating climate change, and improving the livelihoods of people in vulnerable regions such as drylands. By integrating soil conservation, vegetation restoration, water management, biodiversity conservation, and sustainable land use practices, ecosystems can be restored to function effectively and sustainably. Ecosystem restoration not only benefits the environment but also plays a pivotal role in improving food security, providing sustainable livelihoods, and protecting vulnerable communities from the impacts of climate change.

25.6.4 Water Provision in the Dryland Rangeland

Water provision is a critical component of dryland rangeland management, ensuring the sustainability of livestock production in these water-scarce regions. Effective water management strategies, such as range water balance, are essential to optimize water use and balance water supply and livestock needs. Range water balance involves assessing water availability, demand, and runoff to inform grazing management, water harvesting, and storage decisions, reducing the risk of overgrazing and water depletion. Initiatives such as rainwater harvesting, groundwater development, and water conservation infrastructure are vital to enhance water provision in dryland rangelands. These may include installing water tanks and pans, rehabilitating existing water sources, and implementing water-saving technologies like drip irrigation and solar-powered pumps. By improving water accessibility and efficiency, these initiatives can increase livestock productivity, enhance pastoralist resilience to climate change, and support the overall well-being of dryland communities.

25.6.5 Enhancing Dryland Ecosystems Through Climate-Smart Forestry

Afforestation and reforestation are effective, low-cost restoration methods for degraded lands. Climate-adaptive tree species with high water use efficiency and low flammability are essential, as well as fast-growing, pest-resistant species that optimize carbon sequestration, mitigating climate change (Woodcock et al. 2018; Yosef et al. 2018; Cunningham et al. 2015; Berrahmouni et al. 2015). Afforestation efforts in dryland regions have been gaining momentum as a strategy to combat desertification and land degradation. Utilizing native drought-tolerant species is crucial for successful afforestation in these areas. These species, such as Acacia tortilis and Commiphora africana, have adapted to the harsh dryland conditions and require minimal watering (Mengich et al. 2020). By planting native species, afforestation initiatives can promote biodiversity, improve soil health, and provide essential ecosystem services.

Studies have demonstrated the effectiveness of native drought-tolerant species in enhancing ecosystem resilience in drylands. A study in the Sahel highlighted the potential of afforestation to enhance precipitation (Yosef et al. 2018) and carbon sequestration potential. For instance, research in the African Sahel found that planting drought-resistant tree species increased soil organic carbon by 30% and reduced soil erosion by 25% (Diao and Sarjono 2015). Similarly, a study in Australia's drylands showed that native vegetation restoration with shallow-rooted annual agricultural plants improved groundwater recharge and reduced salinity (Clarke et al. 2002). By embracing native drought-tolerant species, afforestation initiatives can contribute to sustainable land management and climate change mitigation in dryland regions. A study in Ghana showed that farmers benefited from ecosystem-based adaptation (EbA) practices such as integrated crops and trees and keeping local breeds of farm animals (Dong-Uuro and Peprah 2023). Benefits of EbA included improvements in soil fertility, leading to high crop yield and availability of food and income. Challenges of adopting EbA were the long timespan for the benefits of most EbA practices to begin to accrue, lack of extension information, and the increasing cost of hired labor. The study recommended the integration of fruit tree crops and food crops to address hunger and increase farmers' incomes, mobile extension technology to bridge the extension services gap, and for government policy objectives to include EbA by farmers under the flagship "planting for food and jobs" program.

25.6.6 Challenges and Barriers

Circular ecosystem-based adaptation (CEbA) has emerged as a promising approach to enhance resilience and mitigate climate-related risks in vulnerable ecosystems. However, despite its potential, CEbA adoption remains limited due to various barriers. The implementation of EbA strategies is constrained by various factors, including economic and financial, governance and institutional, social and cultural, knowledge, and physical and biological constraints (Nalau et al. 2018). One of the primary barriers to CEbA adoption is the lack of policy support and regulatory frameworks, the absence of strong innovative extension for smallholder farmers, and a lack of understanding of the adoption determinants of EbA practices (Vignola et al. 2015). In many countries, policies and regulations often favor traditional linear economy approaches, hindering the transition to circular economy practices.

The adoption of circular ecosystem-based adaptation (CEbA) is hindered by several economic and financial constraints. One major obstacle is the limited availability of financing mechanisms (UNEP 2024). This scarcity of funds is exacerbated by the fact that the bulk of funding is allocated to protecting existing biodiversity rather than restoring degraded lands (Nalau et al. 2018). As a result, international adaptation finance dedicated to biodiversity and ecosystem services is severely limited, creating a significant barrier to CEbA adoption. CEbA initiatives often require significant upfront investments, which can be challenging for small-scale farmers and communities to access. Additionally, the lack of technical capacity and knowledge-sharing platforms hinders the widespread adoption of CEbA practices. Social and cultural barriers also play a significant role in limiting CEbA adoption. In some communities, traditional practices and cultural values may not align with CEBA principles, making it challenging to implement and sustain CEbA initiatives. Furthermore, the lack of market demand and economic incentives for CEbA products and services can discourage entrepreneurs and businesses from investing in CEbA.

Governance and institutional constraints also hinder EbA implementation. Lack of citizen and community participation, co-development of projects, and attention to differing governance arrangements can lead to ineffective EbA strategies (Wamsler and Pauleit 2016). Institutional fragmentation, organizational silos, and problems across vertical and horizontal governance also constrain EbA implementation. Social and cultural constraints are also significant. Top-down management efforts can be perceived as interfering with farmers' land, leading to resistance. Gender disparities also exist, with men often being better informed about climate change and adaptation due to their prominent roles in decision-making. This creates social and cultural constraints on women's access to knowledge.

Infrastructure constraints, such as inadequate water management systems and transportation networks, also hinder CEbA adoption (IPCC 2014). In many dryland regions, inadequate infrastructure limits the availability of resources, such as water and energy, essential for CEBA practices. Addressing these barriers requires a comprehensive approach that involves policy reforms, financial incentives, capacity building, and social and cultural sensitization. By understanding and addressing these barriers, policymakers, practitioners, and researchers can work together to promote CEbA adoption and enhance resilience in vulnerable ecosystems.

25.6.7 Facilitating CEbA Adoption: Strategies for Success

25.6.7.1 Community-Based Adaptation (CBA) Initiatives

Community-based adaptation (CBA) initiatives are crucial for building resilience in climate-vulnerable dryland communities, but their success hinges on effective community engagement and tailored implementation strategies. Despite challenges like power imbalances, community-led approaches have gained traction, fostering local innovation, agency, and sustained resilience. Research highlights the importance of partnerships between community organizations, academia, and international agencies to develop context-specific, culturally sensitive solutions. CBA initiatives have been applied across various climate adaptation programs, including disaster risk reduction, sustainable agriculture, and water management. However, adaptation constraints in developing countries, such as limited access to resources, credit, and technology, must be addressed. Fortunately, recent examples demonstrate that low-cost, domestically resourced CBA initiatives can be effective, emphasizing the value of community-led interventions, international support, and collaborative efforts to achieve Sustainable Development Goals (SDGs) and enhance climate resilience.

25.6.7.2 Digital Agriculture Platforms

Digital agriculture platforms leverage cutting-edge technologies such as mobile apps, satellite imaging, and data analytics to revolutionize decision-making in agriculture. These platforms provide farmers with real-time information on weather patterns, soil health, and crop growth, enabling them to make informed decisions. Additionally, digital platforms facilitate access to markets, extension services, and financial products. By scaling out digital agriculture platforms, farmers can increase productivity, reduce costs, and enhance resilience to climate change.

25.6.7.3 Farmer-Led Innovation Hubs

Farmer-led innovation hubs are community-driven centers that foster knowledge sharing, innovation, and collaboration among farmers. These hubs provide a platform for farmers to share experiences, test new practices, and develop context-specific solutions. By empowering farmers to take ownership of innovation, hubs accelerate the adoption of climate-resilient practices and technologies. Scaling out farmer-led innovation hubs ensures that solutions are tailored to local needs, increasing their effectiveness and sustainability.

25.6.7.4 Public-Private Partnerships (PPPs)

Public-private partnerships (PPPs) bring together governments, private sector companies, NGOs, and research institutions to scale out climate-resilient agriculture. PPPs leverage the strengths of each partner, combining resources, expertise, and networks to achieve greater impact. By collaborating, partners can develop and disseminate climate-smart technologies, provide training and extension services, and facilitate market access. Scaling out PPPs accelerates the transition to climate-resilient agriculture, benefiting farmers, communities, and the environment.

25.6.7.5 Climate-Smart Agriculture (CSA) Financing

Climate-smart agriculture (CSA) financing addresses the financial barriers that farmers face in adopting climate-resilient practices. Specialized financial instruments, such as green bonds, climate-resilient loans, and insurance products, help farmers invest in climate-smart technologies and practices. By scaling out CSA financing, farmers can access affordable capital, manage climate-related risks, and increase resilience. This approach also attracts private sector investment, supporting the transition to a low-carbon agriculture sector.

25.6.7.6 Capacity Building and Training

Vulnerable communities, often situated in remote areas, lack access to modern technology, knowledge, and educational opportunities, underscoring the urgent need for targeted capacity-building initiatives to bridge this gap and enhance their resilience to climate change (Shammin et al. 2022). Capacity building and training enhance farmers' skills and knowledge, empowering them to adopt climate-resilient practices. Workshops, demonstrations, and mentoring programs provide farmers with hands-on experience and technical expertise. By scaling out capacity building and training, farmers can optimize climate-smart technologies, improve productivity, and enhance resilience. This approach also strengthens extension services, ensuring that farmers receive ongoing support and guidance to address emerging challenges. Effective capacity building and training programs are critical to sustaining climate-resilient agriculture.

25.7 Case Studies and Best Practices

25.7.1 Southern Africa

South Africa's Working for Water (WfW) scheme is a pioneering project that showcases circular ecosystem-based adaptation in drylands. Launched in 1995, the project aims to clear invasive alien plants that threaten biodiversity, exacerbate drought, and impede agriculture. By using a payment for ecosystem services (PES) approach, WfW provides jobs and economic empowerment for marginalized communities, while also restoring ecosystem health. The project's impact has been significant, with over 50,000 jobs created annually and a focus on supporting vulnerable groups such as women, youth, and people living with HIV/AIDS. WfW has also expanded its focus to include education, training, and healthcare services for its employees. With a total annual budget of over R400 million, the project has become the largest single poverty relief and public works program in South Africa. The project's environmental impact has also been substantial, with an estimated 46 million m3 per annum increase in streamflow due to the clearing of invasive alien plants from riverbanks between 1997 and 2006. This is critical in addressing South Africa's worsening water scarcity. While challenges remain, including balancing environmental and socioeconomic goals, WfW serves as a model for circular ecosystem-based adaptation in drylands, demonstrating the potential for sustainable livelihoods, ecosystem restoration, and climate resilience.

25.7.2 East Africa

The Green Belt Movement, founded by Wangari Maathai, has been working to rehabilitate degraded forest land through the planting of indigenous tree seedlings. The organization's efforts focus on promoting community-based climate resilience and livelihood improvement through the implementation of food security and water harvesting initiatives. Since 2007, the organization has planted over 11 million trees within the Mt. Kenya forest ecosystem, resulting in increased tree cover, regeneration of new animal habitats, and mitigation of flash floods. The organization's forest restoration project aims to restore degraded forest land in Kenya's water towers, including the Mt. Kenya and Mt. Elgon ecosystems. The project promotes the circular economy concept of the 3Rs (reduce, reuse, recycle) by reducing environmental pollution, reusing natural resources, and recycling materials. In 2018, the Green Belt Movement partnered with ECOSIA GmbH from Germany to restore degraded forest land in Mt. Elgon and Cherangany water towers. The project aimed to reestablish a sustainable water supply through the protection of water catchments within the water towers.

During Phase 2 (2019–2020), the organization planted 374,508 trees in Mt. Elgon and 500,000 trees in the Mau Complex. The project has had a significant impact on the environment and local communities, increasing tree cover and biodiversity, regenerating new animal habitats, mitigating flash floods, improving livelihoods for local communities, and promoting community-based climate resilience. The project's success can be attributed to the organization's community-based approach, which engages local communities in the restoration efforts. The project's focus on promoting sustainable livelihoods and climate resilience has also contributed to its success.

25.7.3 West Africa

Mali's agroforestry program has demonstrated the potential of circular ecosystem-based adaptation in drylands. By integrating trees into agricultural landscapes, farmers have improved soil fertility, increased crop yields, and enhanced food security. The program's focus on agroforestry treatments has shown significant results, with maize grain yields from Gliricidia treatment plots exceeding regional averages by 8%. The program's use of Gliricidia plus Stylosanthes treatments has been particularly successful, providing livestock fodder and enhancing maize productivity without the use of synthetic fertilizer inputs. These treatments have yielded the highest maize grain yield of any treatment, at 2.9 tons/ha, outperforming the use of synthetic fertilizers. The program's emphasis on sustainable agricultural practices has not only improved crop yields but also promoted ecosystem health and resilience. The Mali agroforestry program's success has important implications for dryland agriculture in Africa. By adopting circular ecosystem-based adaptation approaches, farmers can improve their livelihoods, enhance food security, and promote ecosystem resilience. The program's focus on agroforestry treatments has shown that simple, sustainable measures can have profound effects on communities, making it a model for replication in other dryland regions.

In Niger, a 15-year project (1984–1999) aimed at restoring ecological balance and promoting social and economic development yielded impressive results. The project increased incomes by over $6 million per year, with 42% derived from livestock, 25% from vegetables, 22% from cereals, and 11% from wood. Land degradation and desertification were halted, and large surfaces were recovered through the strategic planting of nearly 18 million trees. The project led to a 300% increase in woodlands, a 133% increase in wood production, and a 57% increase in leaf biomass availability. Additionally, interventions such as block plantations, trench planting, and bench terraces enabled farmers to reclaim arable land, reduce water erosion, and increase forage yields by over 90%. The project's overall impact was a net growth of production value of about 700%, with some interventions yielding as high as 1400% growth in yearly production (Oakland 2015).

25.8 Conclusion

This chapter underscores the vital importance of CEbA strategies in fostering resilience and sustainability in dryland agricultural systems. The findings highlight the effectiveness of CEbA approaches, such as agroforestry, conservation agriculture, and climate-resilient water management, in enhancing crop yields, reducing soil degradation, promoting biodiversity, and supporting ecosystem services. Moreover, CEbA strategies can sequester carbon, promote climate resilience, and contribute to sustainable development, poverty reduction, and food security. To address the pressing challenges facing dryland agriculture, policymakers, practitioners, and stakeholders must prioritize CEbA strategies. Future research should focus on scaling up CEbA approaches, developing context-specific solutions, and enhancing stakeholder engagement. Policy initiatives should integrate CEbA principles, promote climate-resilient agriculture, and support sustainable land management. Additionally, research should investigate the economic benefits of CEbA strategies, develop innovative financing mechanisms, and enhance capacity building for farmers and practitioners.

The urgent need to promote CEbA strategies in dryland regions cannot be overstated. Governments, international organizations, civil society, and private sector entities must collaborate to develop and implement policies, programs, and projects that prioritize CEbA principles. Empowering farmers, practitioners, and local communities to adopt CEbA strategies and continuing to develop and disseminate knowledge on CEbA approaches are critical steps toward enhancing resilience, promoting sustainability, and ensuring a food-secure future for dryland communities. By working together, we can unlock the potential of CEbA strategies to transform dryland agriculture and promote a more sustainable and resilient future.

References

Akeredolu F (2024) African climate politics and multilateralism: domestic factors and cross-country variation in climate ambitions. S Afr J Int Aff 31(2):273–292. https://doi.org/10.1080/10220461.2024.2413686

Alemu MM (2016) Indigenous agroforestry practices in Southern Ethiopia: the case of Lante, Arba Minch. Open Access Library J 03(12):1–12. https://doi.org/10.4236/oalib.1103278

Amadu FO, Miller DC and McNamara PE (2020) 'Agroforestry as a pathway to agricultural yield impacts in climate-smart agriculture investments: Evidence from southern Malawi', Ecological Economics 16. https://www.sciencedirect.com/science/article/pii/S0921800918316793

Berrahmouni N, Regato P, Parfondry M (2015) Global guidelines for the restoration of degraded forests and landscapes in drylands: building resilience and benefiting livelihoods. Forestry Paper No. 175, Food and Agriculture Organization of the United Nations, Forestry Paper No. 175, Rome. Available at: https://www.fao.org/policy-support/tools-and-publications/resources-details/en/c/449353/

Bonilla-Findji O, Ortega A, Ouedraogo M, Fall M, Chabi A, Andrieu N, Eitzinger A, Zougmoré R, Läderach P (2020) How are smallholder farmers coping with and adapting to climate-related shocks in Kaffrine Climate-Smart Village, Senegal? https://cgspace.cgiar.org/handle/10568/111561

Chivhenge E, Mabaso A, Museva T, Zingi GK, Manatsa P (2023) Zimbabwe's roadmap for decarbonisation and resilience: an evaluation of policy (in)consistency. Glob Environ Chang 82:102708. https://doi.org/10.1016/j.gloenvcha.2023.102708

Clarke CJ, George RJ and Bell RW (2002) 'Dryland salinity in south-western Australia: its origins, remedies, and future research directions', Soil Research, 40(1):93–113. https://doi.org/10.1071/SR01028

Cunningham SC, Mac Nally R, Baker PJ et al (2015) Balancing the environmental benefits of reforestation in agricultural regions. Perspect Plant Ecol Evol Syst 17:301–317

Daba B, Mammo S (2024) Rangeland degradation and management practice in Ethiopia: a systematic review. Environ Sustain Indic 23:100413

Dong-Uuro PP, Peprah K (2023) Ecosystem-based adaptation practices to stem climate change impacts: smallholder farmers' perspectives. J Environ Manag 323:573–585. https://doi.org/10.1080/14693062.2023.2269121

Eldridge DJ, Beecham G (2018) The impact of climate variability on land use and livelihoods in Australia's rangelands. In: Gaur M, Squires V (eds) Climate variability impacts on land use and livelihoods in drylands. Springer, Cham. https://doi.org/10.1007/978-3-319-56681-8_14

FAO (2017) Ecosystem-based adaptation (EbA) in agricultural sectors: Context, approaches and lessons learned. Rome: FAO

Fouque F, Reeder JC (2019) Impact of past and on-going changes on climate and weather on vector-borne diseases transmission: a look at the evidence. Infect Dis Poverty 8(1):51. https://doi.org/10.1186/s40249-019-0565-1

Geißler K, Blaum N, von Maltitz GP, Smith T, Bookhagen B, Wanke H et al (2023) Biodiversity and ecosystem functions in southern African savanna rangelands: threats, impacts and solutions. In: von Maltitz GP et al (eds) Sustainability of southern African ecosystems under global change, ecological studies 248. Springer, pp 407–435. https://doi.org/10.1007/978-3-031-10948-5_15

Hu T, Zhang X, Khanal S, Wilson R, Leng G, Toman EM, Wang X, Li Y, Zhao K (2024) Climate change impacts on crop yields: a review of empirical findings, statistical crop models, and machine learning methods. Environ Model Softw 179:106119

IPCC (2014) Climate change 2014: impacts, adaptation and vulnerability. Part a: global and sectoral aspects. In: Field CB, Barros VR, Dokken DJ, Mach KJ, Mastrandrea MD, Bilir TE, Chatterjee M, Ebi KL, Estrada YO, Genova RC, Girma B, Kissel ES, Levy AN, MacCracken S, Mastrandrea PR, White LL (eds) Contribution of working group II to the fifth assessment report of the intergovernmental panel on climate change. Cambridge University Press, Cambridge

Kay S, Rega C, Moreno G, den Herder M, Palma JHN, Borek R, Crous-Duran J, Freese D, Giannitsopoulos M, Graves A, Jäger M, Lamersdorf N, Memedemin D, Mosquera-Losada R, Pantera A, Paracchini ML, Paris P, Roces-Díaz JV, Rolo V, Rosati A and Herzog F (2019) 'Agroforestry creates carbon sinks whilst enhancing the environment in agricultural landscapes in Europe', Land Use Policy 83:581–593. https://doi.org/10.1016/j.landusepol.2019.02.025

Kissi AE, Villamor GB, Abbey GA (2023) Ecosystem-based adaptation practices of smallholder farmers in the Oti Basin, Togo: probing their effectiveness and co-benefits. Ecologies 4(3):535–551. https://doi.org/10.3390/ecologies4030035

Kumar S, Mishra AK, Pramanik S, Mamidanna S, & Whitbread A (2020) Climate risk, vulnerability and resilience: Supporting livelihood of smallholders in semiarid India. Land Use Policy 97:104729. https://doi.org/10.1016/j.landusepol.2020.104729

Linger E (2014) Agro-ecosystem and socio-economic role of homegarden agroforestry in Jabithenan District, North- Western Ethiopia: implication for climate change adaptation. Springer Plus 3(1):1–9

Mbow C, Smith P, Skole D, Duguma L, Bustamante M (2014) Achieving mitigation and adaptation to climate change through sustainable agroforestry practices in Africa. Curr Opin Environ Sustain 6:8–14. https://doi.org/10.1016/j.cosust.2013.09.002

Mengich EK, Macharia JM, Mitloehner R, Too DK and Muturi GM (2020) 'Diameter distribution of indigenous trees as indicator of adapted species in semi-arid rangelands of Kenya', Journal of Tropical Forest Science, 32(2):144–153
Michler JD, Baylis K, Arends-Kuenning M, Mazvimavi K (2019) Conservation agriculture and climate resilience. J Environ Econ Manag 93:148–169. https://doi.org/10.1016/j.jeem.2018.11.008
Mudau HS, Msiza NH, Sipango N, Ravhuhali KE, Mokoboki HK, Moyo B (2022) Veld restoration strategies in South African semi-arid rangelands. Are there any successes? A review. Front Environ Sci 10:1–12. https://doi.org/10.3389/fenvs.2022.960345
Nalau J, Becken S, Mackey B (2018) Ecosystem-based adaptation: a review of the constraints. Environ Sci Pol 89:357–364. https://doi.org/10.1016/j.envsci.2018.08.014
Ndalama E, Kamanga-Thole G, Missanjo E (2015) Agroforestry contribution to the improvement of rural community agroforestry contribution to the improvement of rural community livelihoods in Balaka, Malawi. Int J For Hortic 1(1):5–11
Nyong AP, Ngankam TM, Felicite TL (2019) Enhancement of resilience to climate variability and change through agrofor estry practices in smallholder farming systems in Cameroon. Agrofor Syst 94(3):687–705. https://doi.org/10.1007/s10457-019-00435-y
Oakland (2015) Restoring Ecological Balance And Bolstering Social And Economic Development In Niger. https://www.oaklandinstitute.org/sites/default/files/files-archive/Restoring_Ecological_Balance_Niger.pdf
Palsaniya DR, Kumar S, Das MM et al (2024) Ecosystem services from rain water harvesting, agroforestry and livestock based smallholder rain-fed integrated farming system. Agrofor Syst 98:2617. https://doi.org/10.1007/s10457-024-01055-x
Phiri M, Martinsen V, Simusokwe G, Smebye AB, Obia A, Shitumbanuma V, Selby J, Cornelissen G, Makate C and Mulder J (2025) Enhancing yields and climate resilience through conservation agriculture: multiyear regional on-farm trials in Zambia. Plant Soil https://doi.org/10.1007/s11104-024-07191-8
Quandt A (2020) Contribution of agroforestry trees for climate change adaptation: narratives from smallholder farmers in Isiolo, Kenya. Agrofor Syst 94(6):2125–2136. https://doi.org/10.1007/s10457-020-00535-0
Ramana DBV (2022) Livestock based production systems for climate adaptation in dryland areas. In: Poshiwa X, Ravindra Chary G (eds) Climate change adaptations in dryland agriculture in semi-arid areas. Springer, Singapore. https://doi.org/10.1007/978-981-16-7861-5_9
Shammin MR, Haque AKE, Faisal IM (2022) A framework for climate resilient community-based adaptation. In: Haque AKE et al (eds) Climate change and community resilience. Springer Nature Singapore, Singapore, p 15. https://doi.org/10.1007/978-981-16-0680-9_2
Shikangalah RN, Mapani BS (2020) A review of bush encroachment in Namibia: from a problem to an opportunity? J Rangel Sci 10(3)
Simonson D, Miller E, Jones A, García-Rangel S, Thornton H, McOwen C (2021) Enhancing climate change resilience of ecological restoration – a framework for action. Perspect Ecol Conserv 19(3):300–310. https://doi.org/10.1016/j.pecon.2021.05.002
Stringer LC, Mirzabaev A, Benjaminsen TA, Harris RMB, Jafari M, Lissner TK, Stevens N, Tirado-von der Pahlen C (2021) Climate change impacts on water security in global drylands. One Earth 4(6):851–864
Thierfelder C, Chivenge P, Mupangwa W, Rosenstock TS, Lamanna C, Eyre JX (2017) How climate-smart is conservation agriculture (CA)? Its potential to deliver on adaptation, mitigation and productivity on smallholder farms in Southern Africa. Food Secur 9(3):537–560. https://doi.org/10.1007/s12571-017-0665-3
Thornton PK, Whitbread A, Baedeker T, Cairns J, Claessens L, Baethgen W, Bunn C, Friedmann M, Giller K.E, Herrero M, Howden M, Kilcline K, Nangia V, Ramirez-Villegas J, Kumar S, West PC and Keating B (2018) 'A framework for priority-setting in climate smart agriculture research', Agricultural Systems 167:161–175. https://doi.org/10.1016/j.agsy.2018.09.009

Twomlow S et al (2011) Micro-dosing as a pathway to Africa's green revolution: evidence from broad-scale on-farm trials. In: Bationo A, Waswa B, Okeyo J, Maina F, Kihara J (eds) Innovations as key to the green revolution in Africa. Springer, Dordrecht. https://doi.org/10.1007/978-90-481-2543-2_113

United Nations Environment Programme (2024) Adaptation Gap Report 2024: Come hell and high water — As fires and floods hit the poor hardest, it is time for the world to step up adaptation actions. Nairobi. https://doi.org/10.59117/20.500.11822/46497

UNCCD (2019) Drought Resilience, Adaptation and Management Policy Framework: Supporting Technical Guidelines. Bonn, Germany: UNCCD. https://www.unccd.int/sites/default/files/relevant-links/2019-09/190906%20UNCCD%20drought%20resilience%20technical%20guideline%20EN.pdf

Vanlauwe B, Chivenge P & Zingore S (2020) Soil fertility maintenance and nutrient management for agricultural transformation. In R. Sikora, E.R. Terry, P. Vlek and J. Chitja, Transforming agriculture in Southern Africa: constraints, technologies, policies and processes. Oxfordshire, UK: Routledge. (p. 67–76)

Vignola R, Harvey CA, Bautista-Solis P, Avelino J, Rapidel B, Donatti C, Martinez R (2015) Ecosystem-based adaptation for smallholder farmers: definitions, opportunities and constraints. Agric Ecosyst Environ 211:126–132

Wei F, Wang S, Brandt M, Fu B, Meadows ME, Wang L et al (2021) Responses and feedbacks of African dryland ecosystems to environmental changes. Curr Opin Environ Sustain 48:29–35. https://doi.org/10.1016/j.cosust.2020.09.004

Weiskopf SR, Rubenstein MA, Crozier LG, Gaichas S, Griffis R, Halofsky JE, Hyde KJW, Morelli TL, Morisette JT, Muñoz RC, Pershing AJ, Peterson DL, Poudel R, Staudinger MD, Sutton-Grier AE, Thompson L, Vose J, Weltzin JF, Whyte KP (2020) Climate change effects on biodiversity, ecosystems, ecosystem services, and natural resource management in the United States. Sci Total Environ 733:137782

Woodcock P, Cottrell JE, Buggs RJA et al (2018) Mitigating pest and pathogen impacts using resistant trees: a framework and overview to inform development and deployment in Europe and North America. Forestry 91(1):1–16

Yosef G, Walko R, Avisar R et al (2018) Large-scale semi-arid afforestation can enhance precipitation and carbon sequestration potential. Sci Rep 8:996

Zada M, Zada S, Ali M, Zhang Y, Begum A, Han H, Ariza- Montes A, Araya-Castillo L (2022) Contribution of small-scale agroforestry to local economic development and livelihood resilience: evidence from. Land 11(1):71. https://doi.org/10.3390/land11010071

Zhang WM, Brandt M, Penuelas J, Guichard F, Tong XY, Tian F, Fensholt R (2019) Ecosystem structural changes controlled by altered rainfall climatology in tropical savannas. Nat Commun 10

Zhang W, Zhou T, Ye W, Zhang T, Zhang L, Wolski P, Risbey J, Wang Z, Min S-K, Ramsay H, Brody M, Grimm A, Clark R, Ren K, Jiang J, Chen X, Fu S, Li L, Tang S, Hu S (2025) A year marked by extreme precipitation and floods: weather and climate extremes in 2024. Adv Atmos Sci 42(1):1–11. https://doi.org/10.1007/s00376-025-4540-4

Zölch T, Wamsler C, Pauleit S (2018) Integrating the ecosystem-based approach into municipal climate adaptation strategies: the case of Germany. J Clean Prod 170:966–977. https://doi.org/10.1016/j.jclepro.2017.09.14

Chapter 26
Restoring Dryland Ecosystems: Local Knowledge, Native Species and Tree Restoration Policies and Practices in Northern Nigeria

Nicola Favretto, Lindsay C. Stringer, Aliyu S. Barau, Jibrin Mohammed Jibrin, Saminu Ado, Ibrahim B. Yakubu, and Martin Dallimer

Abstract Despite the growing global interest in the role of trees in ecosystem restoration as part of the solution to the global climate crisis, projects and programmes often fail to achieve their restoration targets. Underlying this failure are deficiencies in stakeholder engagement, from the local community to policy levels, and the absence of an enabling policy environment. Communities also often have a keen sense of landscape stewardship, possessing extensive local knowledge and experience. Actively harnessing and integrating this knowledge is vital for restoration success. At the same time, poor coherence between policies regarding tree restoration fail to support one another across sectors, resulting in a lack of understanding of wider benefits and risks. This chapter uses a case study in the drylands of northern Nigeria. It undertakes policy analysis and community engagement activities to address the following research questions: (i) To what extent do policies recognise the role of trees in fostering mitigation, adaptation, and development in Nigeria?, (ii) What is the level of policy coherence across sectors, and what are the barriers and solutions to strengthen the coherence of tree restoration efforts across government levels?, and (iii) How can local communities be engaged in initiatives to revive, plant, and care for trees over the long term? We conclude that although the role of trees in ecosystem restoration is prominent in many policies, there is a lack of coherence across sectors. However, addressing policy coherence issues is only part of any solution. We show that there is considerable local knowledge about, and

N. Favretto · L. C. Stringer
Department of Environment and Geography, University of York, York, UK

A. S. Barau · J. M. Jibrin · S. Ado · I. B. Yakubu
Department of Environmental Management, Bayero University Kano, Kano, Nigeria

M. Dallimer (✉)
Centre for Environmental Policy, Imperial College London, London, UK
e-mail: m.dallimer@imperial.ac.uk

Y. D. I. Khan et al. (eds.), *Land Restoration Through Ecosystem-Based Approach*, Disaster Resilience and Green Growth, https://doi.org/10.1007/978-981-95-0389-6_26

appetite for, tree restoration among smallholder farmers and their associated communities. Ensuring a suitable enabling policy and agriculture extension service environment will help build on the desires of local communities. Doing so will further support countries across dryland Africa to meet their aspirations on ecosystem restoration, tree planting and the associated benefits for climate change mitigation, adaptation and improved livelihoods.

Keywords Ecosystem restoration · Stakeholder engagement · Local knowledge · Sahel

26.1 Introduction

Globally, we have lost 30% of the planet's forest cover over the past century, while an additional 20% has been degraded (Keenan et al. 2015). Such rapid land cover alterations are a major driver of biodiversity declines, and both processes are exacerbated by climate change. Implications extend beyond environmental issues and pose far-reaching economic and societal challenges, often hampering progress towards diverse Sustainable Development Goals (SDGs) (Dannevig et al. 2022). Thus, land-use change, climate change and development are intimately interlinked and cannot be tackled in isolation. As cross-cutting issues, climate mitigation and adaptation and forest restoration need to be mainstreamed into policy, farming practices and the consciousness of communities across a range of sectors and levels (Stringer et al. 2014).

Tackling the loss and degradation of forests and other landscapes with trees has often been central to ecosystem-based approaches and nature-based solutions. These interventions are encouraged by initiatives such as the United Nations Decade on Ecosystem Restoration (2021–2030) (Nelson et al. 2024). Such restoration aims to improve the condition of an ecosystem by reorienting it from its current state towards the trajectory it would have followed in the absence of degradation, while accounting for environmental change. Often, restoration is undertaken to reinstate or augment the livelihood options for natural resource–dependent populations (Favretto et al. 2020). In response to climate crises and biodiversity loss, large-scale restoration projects are expanding globally in scope and financial investment despite evidence suggesting that such investments remain insufficient (Mirzabaev et al. 2022). Many international development initiatives have entered the global restoration arena, with projects largely proposed, funded and delivered by international organisations and large international non-governmental organisations (NGOs), often in partnership with local organisations or governments.

Within these, a range of forest and landscape restoration approaches are promoted to restore the loss of ecological functions and improve human well-being in deforested or degraded landscapes, to generate multiple benefits by actively engaging diverse stakeholders and allowing a suite of possible interventions tailored to local conditions (Djenontin et al. 2020). For example, the Bonn Challenge seeks to restore 350 million hectares of land globally by 2030, under which several sub-Saharan African countries made pledges towards restoring 100 million hectares of degraded landscapes through initiatives such as the African Forest Landscape Restoration Initiative (AFR100) (Djenontin et al. 2020).

Many of these initiatives have targeted desertifying areas and drylands, which include the world's hyper-arid, arid, semi-arid and dry sub-humid areas (Stringer et al. 2021). Aside from their extensive spatial extent, drylands are important carbon stores, host considerable biodiversity and are inhabited by approximately 40% of the global human population (van der Esch et al. 2017). Far from being barren systems, drylands support a range of ecosystem functions and processes that sustain their human inhabitants while providing important ecosystem services at wider scales. The major international agreement for tackling desertification (or dryland degradation) is the United Nations Convention to Combat Desertification (UNCCD), established in 1994. Although the UNCCD is an international treaty, emphasis is on actions at the national level, with the implementation of National Action Programmes and the setting of Land Degradation Neutrality Targets (for more information about Land Degradation Neutrality and associated SDG target 15.3, see Orr et al. (2017) and Cowie et al. 2018)). Despite many years of efforts and ambitious aspirations, many initiatives have failed to deliver on their goals.

Some commentators have highlighted the need for a systems approach that requires restoration activities to operate across sectors and consider multiple issues and variables, supported by a coherent policy environment (Kong et al. 2024). According to the Organisation for Economic Co-operation and Development (OECD), policy coherence is defined as *the systematic promotion of mutually reinforcing policy actions across government departments and agencies creating synergies towards achieving the agreed objectives* (OECD 2005: 4). Analyses of coherence are key to assess where multi-sectoral policies are supporting or conflicting with one another. Coherent policy approaches can improve effectiveness and reduce competition for limited budgets and resources. Without a coherent approach, important trade-offs can occur. For example, restoration in the Loess Plateau in China has been critiqued for using non-native species and focusing on soil erosion only as a single ecosystem service/benefit to wider society. By doing so, efforts neglected to assess the whole system and the impacts of vast-scale tree planting on water and its downstream availability to expanding urban areas (Feng et al. 2016). Similarly, Kong et al. (2024) demonstrate how restoration efforts need to be developed to consider the changing social, economic and technological context and the desired environmental changes.

The Great Green Wall of the Sahara and Sahel (GGW) was endorsed by the African Union in 2007. The GGW offers another example of an ambitious initiative that supports re-greening, ecological restoration, adaptation and resilience-building. Despite making some gains, the GGW has encountered difficulties. The approach relies on implementing a broad range of sustainable land management approaches across its member countries, facilitated and delivered by a large diversity of partners with the involvement of local communities. It was intended to tailor its approaches to meet the local development and environmental contexts, with evidence of success in this respect (Sacande and Berrahmouni 2016); however, progress has been slow. By 2020, only between 4% and 20% of the 100 million ha restoration goal had been realised, primarily due to a gap between the necessary financial investments and those that have materialised, despite the potential for substantial social, economic and environmental returns (Mirzabaev and Wuepper 2023; Mirzabaev et al. 2022).

Other restoration initiatives have been conceived with limited engagement with the local context. Frequently, they rely on species that are not native to a region and require expensive infrastructure, such as nurseries, and expertise, such as irrigation and care, to ensure that planted trees survive. Even where restoration initiatives have demonstrated success in enhancing tree cover, in many cases, this has been at the expense of native non-treed habitats (Parr et al. 2024). Further, establishing large areas of non-native species presents missed opportunities, as these species offer poor habitats for biodiversity. They also typically offer limited benefits to communities, such as livelihood opportunities from non-timber forest products, enhanced food security or improved agricultural yields through soil conservation, water management and nitrogen fixation. Often, these challenges emerge due to a lack of coherence, with the efforts of one ministry, project or department inadvertently competing with, or undermining the efforts of, another.

There is increasing recognition that restoration requires engagement with social sciences disciplines to better understand local perspectives, policy and governance aspects, power dynamics, knowledge and inequalities (Pritchard 2021; Austen et al. 2022). Various frameworks have thus been proposed to guide the capture of these. For instance, in the Dryland Development Paradigm (DDP), Reynolds et al. (2007) recognised drylands as coupled, dynamic and co-adapting systems that are hierarchically nested and networked across multiple scales. Local ecological knowledge was acknowledged as central to co-adaptation, while a small number of 'slow variables' were found to determine dryland socio-ecological system dynamics. Thresholds in key slow variables controlled by different processes were noted to define different system states, with those thresholds changing over time. Stringer et al. (2017) extended this work, identifying how these aspects could be considered in dryland research and development activities. Their findings pointed towards three core principles for dryland research and development initiatives to succeed (1) *unpack* relationships and system interlinkages, livelihood portfolios and value chains; (2) *traverse* scales, sectors, stakeholders and ways of knowing to incorporate the perspectives of all actors and their interactions; and (3) *share* experiences, lessons and knowledge and co-produce actions, taking into account the scalar aspects of knowledge, enabling empowerment and internally driven research in development approach, as opposed to externally driven research for development. Emphasis in each of the principles is placed on working with local communities and stakeholder networks, including policymakers, weaving together local and scientific knowledge through processes of co-development to develop appropriate and feasible options for each context (Sinclair and Coe 2019), supported by coherent policies across sectors. Similarly, the Society for Ecological Restoration's Standards of Practice draws on ten principles (see Gann et al. 2019) that guide restoration efforts towards (1) contributing to global policy frameworks, (2) promoting fair and inclusive engagement, (3) including a continuum of restorative activities, (4) aiming at the highest recovery possible to benefit nature and people, (5) addressing the causes of degradation, (6) incorporating all types of knowledge, (7) setting ecological, cultural and socio-economic goals, (8) tailoring activities to local and land/seascape contexts, (9) measuring results and adapts actions, and (10) integrating policies and measures for lasting impacts. These have been adopted in restoration efforts under the UN International Decade on Ecosystem Restoration (Nelson et al. 2024).

Where tree restoration efforts engage with both the social and natural sciences, several common elements support success. These include an emphasis on working with local communities, policy stakeholders, and diverse forms of knowledge, rather than imposing external ideas and expertise. Success also depends on ensuring that engagement occurs from the conception of a restoration initiative through to its delivery, ongoing monitoring, and integration with wider policy and practice. This chapter presents a case study in dryland northern Nigeria, drawing on core elements from the frameworks outlined above. These frameworks underscore the importance of both understanding the broader policy context and engaging with local communities and individuals throughout the restoration process. We analyse key policies to see how they contribute towards an enabling environment for restoration that considers climate change adaptation and mitigation, as well as the need for development. We also engaged stakeholders in a range of training activities, building on their local knowledge to develop a novel approach to dryland restoration through Irrigation-Free Indigenous Tree Restoration (IFITR). To guide our work, we asked:

1. To what extent do policies recognise the role of trees in fostering mitigation, adaptation and development in Nigeria?
2. What is the level of policy coherence across sectors, and what are the barriers and solutions to strengthen the coherence of tree restoration across government levels?
3. How can local communities be engaged in initiatives to revive, plant, and care for trees over the long term?

26.2 Case Study: Northern Nigeria

Nigeria provides a suitable case study, given it has Africa's largest human population and is tackling multiple challenges found elsewhere on the African continent, offering the potential to deliver important lessons for other locations with some of the same interacting pressures. Dryland Northern Nigeria, which includes the states of Kano and Jigawa (Fig. 26.1), is subject to land degradation and desertification, driven by climate change and unsustainable land use practices. Almost 70% of the population is involved in agriculture, which forms the backbone of the regional economy, making this a key barrier to sustainable development (FAO 2022). Poverty in the region is the highest in Nigeria, with land degradation exacerbated by

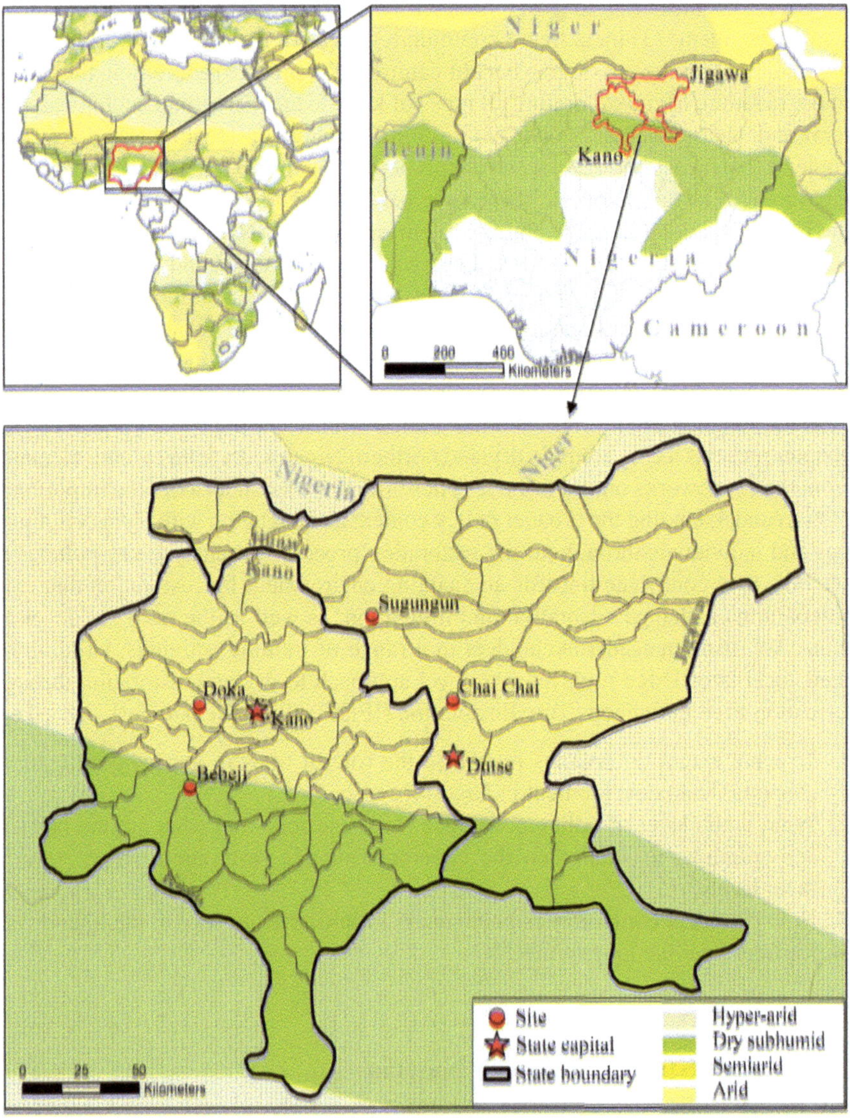

Fig. 26.1 Location of Nigeria, the states of Jigawa and Kano, and four demonstration sites showcasing irrigation-free indigenous tree restoration. Shading indicates aridity status. (FAO 2019)

migration, conflict and low recognition of the rights and roles of women. Women are disproportionately impacted by poverty and land degradation, given their prominent roles in farming households and engaging mainly in unskilled, labour-intensive agricultural work, lacking access to finance, training and decision-making.

Nigeria's federal and state-level government ministries have identified ecosystem restoration initiatives, such as tree planting, agroforestry, improved uptake of farmer-managed natural regeneration of vegetation and IFITR as priorities that can help to mitigate climate change and deliver benefits to farming communities (Nigeria's Federal Ministry of Environment 2021). IFITR is a novel approach that centres on indigenous tree species. Doing so can help harness climate change mitigation benefits through improved tree establishment and improved soil health, and enhanced vegetation and soil carbon sequestration. It also offers wider socio-economic and environmental conservation benefits, such as increased yields from food and fodder production, livelihood diversification for smallholders, provision of habitat and food for wildlife, increased biodiversity and ecosystem function and increased availability of traditional medicines (Yakubu 2020).

As climate change mitigation and development are among Nigeria's most crucial cross-cutting issues, government departments have taken steps to explore synergies and integration among sectors. However, implementation challenges persist due to the tendency of government ministries and departments to operate in silos. Improved communication, information sharing and collaboration can support more effective and coherent operations to reach shared goals across sectors.

26.3 Policy Analysis

26.3.1 Methods

To answer our first two research questions, we applied the intersectionality-based policy analysis framework (Hankivsky et al. 2014) to assess how environmental and development policies across various sectors in Nigeria support IFITR and ecosystem restoration through increased tree cover and planting to achieve mitigation, adaptation and development. We followed a multi-step qualitative document analysis approach (Altheide et al. 2008). First, we identified policy documents and strategies across the climate change, agriculture, forestry, national development and energy sectors publicly available in May 2022 (Table 26.1). We focused on climate mitigation, adaptation and development implications of these documents to identify the role of ecosystem restoration, including increasing tree cover in delivering environmental and societal benefits. We also wished to explore cross-sectoral linkages such as the broader topic of sustainable land management and related areas, including gender and equality and stakeholder engagement. Where revisions had been made to earlier policy documents, we considered the most recent policy document.

An intersectionality-based policy analysis approach was used to understand if the policy and strategic documents support increased tree cover in Nigeria to achieve mitigation, adaptation and development goals (Hankivsky et al. 2014). The policies were analysed through qualitative content analysis (Neuendorf 2017). Cross-sectoral coherence was assessed using a subjective scoring system based on the

Table 26.1 Policies identified and analysed across the climate change, agriculture, forestry, national development and energy sectors in Nigeria

Sector	Policy
Climate change	National Climate Change Policy 2021–2030 Nationally Determined Contribution 2021 National Action Plan on Gender and Climate Change 2020 National Adaptation Plan Framework 2020 National Adaptation Strategy and Plan of Action on Climate Change for Nigeria 2011
Agriculture	Agriculture Promotion Policy 2016–2020
Forestry	National Forest Policy 2020
National development	National Development Plan 2021–2025 Economic Sustainability Plan 2020
Energy	National Energy Policy of Nigeria 2018

Table 26.2 Scoring criteria to assess the coherence of tree restoration for climate mitigation, adaptation and development policy in Nigeria

Coherence level	Description
High	The policy strongly supports climate change mitigation, adaptation and development across sectors in relation to tree planting and ecosystem restoration. It includes detailed complementary activities and plans, such as clear objectives and solutions
Partial	The policy somewhat supports climate change mitigation, adaptation and development across sectors in relation to tree planting and ecosystem restoration. Activities and plans may be included but lack specifics/details
Limited	The policy mentions climate change mitigation, adaptation and development across sectors in relation to tree planting and ecosystem restoration, but no details on activities or plans are provided
None	There is no evidence that the policy supports climate change mitigation, adaptation and development across sectors in relation to tree planting and ecosystem restoration

expert knowledge and understanding of the authors (Table 26.2; adapted from England et al. 2018).

Content analysis was carried out in four steps. First, thematic sets of keywords covering the focus of this analysis on mitigation, adaptation and development were developed, namely: (1) climate change mitigation and greenhouse gas emission(s); (2) climate change adaptation; (3) tree(s) and forest(s); (4) restoration, soil fertility and sustainable land management; (5) farmer managed natural regeneration (FMNR) and agroforestry; (6) poverty, livelihood(s), poor farmer(s) and job creation (7) gender, women and youth; (8) partnership, participation, inclusive and stakeholder engagement; and (9) financing and access to finance. Second, each identified sectoral policy was assessed against the nine thematic keyword sets based on the content of the paragraphs in which they were located. The assessments were compiled into a table summarising the discursive content identified in each document under the thematic keyword sets. The coherence of each policy document with each thematic keyword set was qualitatively scored on a scale from three (high coherence) to zero (no coherence; Table 26.2). Average coherence scores were calculated for each policy and sector. This enabled cross-comparison of the main emphases of each policy on tree restoration for climate mitigation, adaptation and development.

26.3.2 Results

Most policies in Nigeria acknowledge the potential of trees in addressing climate and development challenges, but concrete actions are often absent. Table 26.3 summarises the key objectives for tree restoration in climate mitigation, adaptation and development as identified in the assessed policies, along with the corresponding plans and measures to support implementation. A detailed coherence analysis is available in Appendix. A notable exception is the National Forest Policy 2020, which prioritises indigenous species and improved multipurpose trees, such as gum Arabic, shea butter and locust beans, to tackle deforestation, degradation and desertification. The policy also emphasises non-timber forest products for income generation and soil enhancement, aiming to integrate these resources into the national economy. It calls for capacity-building initiatives to empower stakeholders in non-timber forest products management and large-scale agroforestry systems to reduce rural poverty. Responsibilities are distributed across federal and state governments, focusing on access to markets for forest commodities, and local governments are tasked with land preparation, planting and maintenance. These efforts align with global recommendations for integrated climate and development strategies and provide a way forward for national-level implementation (Donatti et al. 2022).

In contrast, most other policies lack explicit references to indigenous species. The National Climate Change Policy 2021–2030 supports adaptation and mitigation measures, such as agroforestry, reforestation and afforestation, but provides no detailed implementation plans. Similarly, the Nationally Determined Contribution (NDC) 2021 mentions natural forest management, emphasising the need to reduce fuelwood harvesting and to promote agroforestry for a low-carbon transition, but lacks actionable frameworks. Limited focus on trees is also seen in the National Action Plan on Gender and Climate Change 2020, which highlights nursery establishment and tree planting for water resource management in gender-based programmes. This plan advocates for gender-responsive approaches in implementing the Paris Agreement and integrating them into energy and transport sector policies. The National Adaptation Strategy and Plan of Action on Climate Change 2011 emphasises afforestation and reforestation to enhance welfare and reduce climate risks. It promotes sustainable forest management through community-based approaches and encourages business incentives for tree planting investments. Similarly, the Agriculture Promotion Policy 2016–2020 highlights tree planting as a measure for erosion control, land degradation reduction and soil fertility

Table 26.3 Overview of key objectives for tree restoration in climate mitigation, adaptation and development from the assessed policies, along with associated plans and measures for implementation

Policy	Key objectives for tree restoration in climate mitigation, adaptation and development	Associated plans and measures for implementation
National Climate Change Policy 2021–2030	Support agroforestry, reforestation and afforestation	No plans or measures outlined—Advocates for integration of mitigation and adaptation across sectors and highlights the need for private sector engagement for leveraging financial resources
Nationally Determined Contribution (NDC) 2021	Promote natural forest management Reduce fuelwood harvesting and promote agroforestry for a low-carbon transition	No plans or measures outlined
National Action Plan on Gender and Climate Change 2020	Nursery establishment and tree planting for water resource management	No plans or measures outlined—Emphasises building women's capacity in gender-sensitive planning
National Adaptation Plan Framework 2020	Promote a multi-sectoral approach to climate change that also considers forestry	Strengthen the role of the private sector and NGOs in improving access to climate finance Align national policies with global agreements, i.e. Paris Agreement, Sendai Framework for Disaster Risk Reduction and Agenda 2063
National Adaptation Strategy and Plan of Action on Climate Change 2011	Promote afforestation and reforestation to enhance welfare and reduce climate risks	Forest management is to be sustained through community-based approaches and agricultural extension services Provide business incentives for tree planting investments
Agriculture Promotion Policy 2016–2020	Foster tree planting for erosion control, land degradation reduction and soil fertility enhancement in drought-prone areas Invest in long-term soil improvement and water conservation through tree planting	No plans or measures are outlined
National Forest Policy 2020	Prioritise indigenous species and improve multipurpose trees to tackle deforestation, degradation and desertification Use non-timber forest products for income generation and soil enhancement	Participatory forestry and climate change planning, and capacity-building of institutional stakeholders for tree products management Federal and state governments should enable access to markets for forest commodities Local governments will support land preparation, planting and maintenance

(continued)

Table 26.3 (continued)

Policy	Key objectives for tree restoration in climate mitigation, adaptation and development	Associated plans and measures for implementation
National Development Plan 2021–2025	Track tree planting efforts as an indicator for reversing deforestation trends	No plans or measures are outlined
Economic Sustainability Plan 2020	It neglects climate-development goals entirely	N/A
National Energy Policy 2018	Cultivate fast-growing tree species to accelerate forest regeneration for energy production (overlooks native species)	No plans or measures are outlined

enhancement, particularly in drought-prone areas. The document calls for investments in long-term soil improvement and water conservation through tree planting. These policy goals are crucial for Nigeria's progress towards the SDGs and NDCs, but they require comprehensive mainstreaming, implementation and monitoring across sectors (Dannevig et al. 2022). Despite acknowledging the importance of trees, the National Development Plan 2021–2025 provides limited focus, tracking tree planting efforts as an indicator for reversing deforestation trends but lacks substantive measures. Similarly, the Economic Sustainability Plan 2020 neglects climate-development goals entirely. The National Energy Policy 2018 calls for cultivating fast-growing tree species to accelerate forest regeneration for energy production but overlooks native species, implying reliance on exotic varieties. These gaps highlight the need for stronger support measures to align tree-focused policies with broader sustainability goals (Ripple et al. 2022).

Coherence across Nigeria's policies on mitigation, adaptation and development is inconsistent. While many address themes like livelihoods, gender equality and financing, actionable strategies are often lacking. The National Forest Policy 2020 demonstrates the highest coherence, integrating participatory forestry and climate change planning, mirroring practices in sub-Saharan Africa (Mutune and Lund 2016). The National Adaptation Plan Framework 2020, building on the National Adaptation Strategy and Plan of Action on Climate Change 2011, identifies adaptation as integral to development, emphasising its mainstreaming into national agendas. The Federal Government is tasked with promoting poverty reduction through agricultural extension services and integrating adaptation with mitigation efforts. The National Adaptation Plan Framework 2020 stresses a multi-sectoral approach to climate change, recognising its differential impacts across agriculture, forestry, water resources and other sectors. It provides a platform for coordinated national responses and encourages private sector and NGO involvement to improve access to climate finance. The framework calls for alignment with global agreements like the Paris Agreement, the Sendai Framework for Disaster Risk Reduction and Agenda

2063. However, significant barriers persist, including siloed governance, complex bureaucratic processes and low awareness of funding opportunities. Enhancing access to financing mechanisms is essential for fostering cross-sectoral engagement and action (Manuamorn et al. 2020).

The National Climate Change Policy 2021–2030 similarly emphasises integrating climate considerations into national and sectoral planning. It identifies policies and strategies to incorporate environmental management into Nigeria's socio-economic development. The policy also advocates for inclusive governance, integrating mitigation and adaptation across sectors. Private sector engagement is highlighted as vital for leveraging financial resources to support these measures. The National Action Plan on Gender and Climate Change 2020 provides a more integrated perspective, prioritising agriculture, forestry and land use for mitigation. It emphasises building women's capacity in gender-sensitive planning and promoting alternative livelihoods. This aligns with global literature advocating for marginalised groups' inclusion in sustainable tree value chains (Carias et al. 2022). In agriculture, the Agriculture Promotion Policy 2016–2020 emphasises stakeholder participation and financial access but lacks alignment with tree recovery and adaptation strategies. In the energy sector, the National Energy Policy 2018 promotes integrated planning but offers little clarity on achieving its objectives. The National Development Plan 2021–2025 identifies funding shortfalls and low environmental awareness as barriers to cross-sectoral integration. However, it highlights opportunities for creating sustainable supply chains, integrating responsible investments and leveraging private sector funds. The plan also prioritises enacting the Climate Change Bill to support decarbonisation pathways and adaptation strategies.

While Nigeria's policies demonstrate varying degrees of coherence, significant gaps remain in actionable measures and cross-sectoral integration. Addressing these challenges through targeted planning, private sector engagement and streamlined governance is essential for achieving meaningful climate and development outcomes. We next assess how communities can be engaged in efforts to revive, plant and care for trees over the long term through a farmer engagement and training programme.

26.4 Engaging Local Communities in Efforts to Revive, Plant and Care for Trees Over the Long Term

Examining policy content and coherence provides an overview of the extent to which the national and state-level policy landscape may facilitate the wider uptake of efforts to enhance the role of trees in ecosystem restoration. Although our analysis revealed shortcomings in terms of coherence and commitments to concrete actions at the policy level, we were still able to show that the importance of trees for restoration initiatives is widely recognised. Supporting smallholder farmers in adopting new or rediscovering traditional farming practices requires support and information. Farmers and agricultural extension workers tend to get information from various sources, including popular media, such as radio and television. However, even when provided with information, barriers remain to uptake, such as a lack of capacity, technical skills and knowledge of the specifics of practices (Antwi-Agyei and Stringer 2021). Such barriers can be somewhat addressed using workshops to deliver capacity and expertise.

Here, we began with a stakeholder analysis. Following a desk-based review of project documents and initial stakeholder identification, a stakeholder map was created using a bottom-up approach to data collection (Favretto et al. 2021a). An analysis of stakeholders' roles linked to IFITR development was undertaken. Stakeholders were ranked by relevance within four categories using an Influence/Interest matrix to analyse their engagement needs. This allowed us to target farming stakeholders, extension workers and decision-makers in Nigeria at national and state levels, working on sustainable land management, climate change and ecosystem restoration.

Working with representatives of these groups, we co-developed suitable restoration approaches and associated training materials that focused on local priorities and highlighted local expertise and context-specific knowledge (Favretto et al. 2021b). Training manuals were developed by collating best practice guidelines covering best IFITR approaches across northern Nigeria, focused on market-driven, locally led tree restoration, working with native species to produce non-timber forest products for income generation and soil improvement. This drew on a desk-based literature review of material from peer-reviewed literature, alongside analysis of reports and experiences from northern Nigeria. The manuals were developed during two co-creation workshops, attended by different stakeholder groups (smallholder farmer and women's associations, traditional leaders, herbalists, extension workers and policymakers). The manuals were finalised after an expert review workshop, in which the manuals' guidelines were reviewed and further validated across stakeholder groups.

Having developed the manuals, we went on to deliver on-the-ground training at four demonstration sites: Doka and Bebeji in Kano State and Sugungu and Chaichai in Jigawa State (Fig. 26.1). Training at the demonstration sites equipped 200 smallholder farmers and extension workers with the necessary silvicultural and technical skills that enabled them to identify, protect and manage the growth of trees and shrubs that can be restored naturally in their fields. The manuals were employed at the demonstration sites to enable the trainees to harness opportunities for livelihood diversification and empowerment of marginalised groups through new products and markets linked to the showcased practices. Return site visits between 1 and 2 years afterwards allowed an interim evaluation of the implications of our efforts for long-term sustainability. To make this evaluation, we considered a range of different indicators: tree density, ability to coppice, replication and adoption by neighbouring farmers, supported by extension and other stakeholders, and the emergence of income-generating opportunities. Together, these indicators covered environmental, economic and social pillars of sustainability.

As farmers prepared for the clearing and planting season in June 2022, we revisited the villages that hosted the four demonstration plots (Fig. 26.1). Farmers continued to manage the growing trees at all four sites in ways compatible with the training the project had provided. There was clear enthusiasm and commitment to the IFITR approach and ethos. Further return visits were undertaken in July 2023, and a more comprehensive assessment was undertaken. Across all four sites, tree species composition had stabilised (at Doka and Bebeji) or improved. Stem density at Chaichai and Sugunu was 13% and 17% higher, respectively, than in 2022. In all four sites, trees were developing new foliage, and the majority (56%) of stems were suitable for coppicing, which had subsequently shown increased signs of growth and regeneration. Farmers demonstrated that they continued to follow good silvicultural practices. Our project was only able to establish demonstration sites and deliver training. This was done to encourage more farmers to adopt good silvicultural practices that encourage IFITR. In 2023, adoption rates by farmers in the four villages ranged from 97% in Sugungu, 94% in Chaichai, 86% in Bebeji and 77% in Doka. Lower adoption rates were observed in the villages in Kano State, where villages are closer to the major urban centre of Kano; therefore, land is both scarce and more expensive than in Jigawa State. These adoption rates align well with the fact that all farmers that we were able to speak to confirmed that they had implemented IFITR practices on their land and instructed their neighbours on how to encourage tree regrowth as well. We were also able to observe that interactions between extension officers and farmers across the four villages were excellent and that these interactions were clearly important in helping ensure that IFITR practices were still being followed despite challenges such as browsing by stray domestic animals and the associated need for enhanced fencing and tree protection measures.

26.5 Conclusion

We examined two key aspects associated with tree-based ecosystem restoration, namely the policy context and local community engagement, focusing on a case study from northern Nigeria. While many policies lack coherence, the results point towards the need for improved collaboration between policy stakeholders across sectors and governance levels. The National Adaptation Plan Framework 2020 highlights the importance of aligning national policies with global agreements such as the Paris Agreement, the Sendai Framework for Disaster Risk Reduction and Agenda 2063 to improve policy coherence.

The findings suggest partial integration of native trees with climate mitigation, adaptation, development, livelihoods diversification, job creation, gender equality and financing across sectors. The National Forest Policy 2020 explicitly emphasises indigenous species for restoration and poverty alleviation, aligning with objectives in the Agriculture Promotion Policy 2016–2020, the National Action Plan on Gender and Climate Change 2020 and the National Energy Policy 2018, albeit focusing more broadly on tree planting rather than indigenous varieties. More general afforestation and reforestation efforts are advocated by the National Climate Change Policy 2021–2030, the National Adaptation Strategy and Plan of Action on Climate

Change 2011 and the NDC 2021. In contrast, tree restoration is largely overlooked in the Economic Sustainability Plan 2020, the National Development Plan 2021–2025 and the National Adaptation Plan Framework 2020. A key observation is that none of the analysed policies, except the National Forest Policy 2020, provide concrete plans and measures for implementing tree restoration in climate mitigation, adaptation or development. The National Forest Policy details the responsibilities of federal, state and local governments to enable land restoration through planting, maintenance and financing mechanisms. This suggests a considerable gap in creating an enabling environment for tree-based restoration efforts. None of the policies scored higher than two in the coherence analysis, indicating limited capacity to support tree planting initiatives, let alone those focused on native species.

Even when coherence was present, which was most notable between climate change and forestry policies, significant barriers, such as major constraints to private sector investment, limit the achievement of forestry development goals. Similar challenges are reported globally, particularly about inadequate investment in restoration activities like reforestation (Mirzabaev et al. 2022; Favretto et al. 2020). Addressing these barriers entails the promotion of finance-focused measures, including supporting small-scale forest enterprises (as highlighted in the National Adaptation Strategy and Plan of Action on Climate Change 2011), strengthening the role of the private sector and NGOs in mobilising financial resources (as outlined in the National Climate Change Policy 2021–2030 and the National Adaptation Plan Framework 2020) and developing markets and value chain opportunities (as per the National Forest Policy 2020). Training programmes for community-based tree management and skill-building for smallholders are critical to maximise these efforts. Such initiatives would empower local stakeholders to capitalise on emerging opportunities, facilitating more sustainable and impactful tree-based restoration activities.

However, a lack of clear vision for achieving concrete results often leads to unrealistic restoration targets and timelines, frequently accompanied by inconsistent species selection that fails to integrate with local ecosystems. Inadequate community involvement can mean local views and preferences for the target landscapes can be neglected, leading to opposition to restoration activities. Engaging with local communities to set restoration priorities and to ensure a systems-level approach is taken (Dallimer and Stringer 2018; Okpara et al. 2018) is therefore essential. Thus, regardless of the broader policy context, regular engagement with smallholder farmers and local communities remains essential if restoration efforts involving tree care are to be successful. Here, we provided initial training that covered IFITR and coupled this with follow-up visits. While feedback from participants in our training sessions indicated they were happy to be involved, consulted and trained on tree care to support restoration, a stronger enabling environment would increase the durability of these efforts. In our return visits and informal follow-up, the contact between extension officers and farmers was key to success.

Nevertheless, despite the anecdotal evidence from our follow-up visits, unless funding is available over the longer term to carry out monitoring and ensure that capacities and skills are maintained by farmers and extension workers, scalable impacts on the ground are unlikely. Our project was, unfortunately, only funded over a short term. Restoration efforts tend to be short-sighted and limited in scope without adequate funding systems and long-term monitoring, restricting their potential for scaling and hindering the realisation of socio-economic and environmental benefits over the medium to long term. This leads to a more challenging context for other important efforts, such as capacity building and outreach. At a national level, investments in extension services have declined across African nations. Ensuring adequate investment in capacity, expertise and extension services will be key to many countries delivering on commitments to tree planting and ecosystem restoration.

Acknowledgements The team were funded by the UK PACT (Partnering for Accelerated Climate Transitions) Green Recovery Challenge Fund, grant number FPNBS\100004. We thank the support staff of Bayero University Kano for their valuable help provided in the field and during the workshops, as well as all the workshop participants for sharing their knowledge and time. Xueqin Li developed Fig. 26.1.

Appendix

Coherence of policy documents for identified climate mitigation, adaptation and development keywords (given in column headings) for Nigeria. Coherence was scored as: 3 = high coherence; 2 = partial coherence; 1 = limited coherence; 0 = no coherence.

Climate

Policy	Climate change mitigation and greenhouse gas emission(s)	Climate change adaptation	Tree(s) and forest(s)	Restoration, soil fertility and sustainable land management	Farmer Managed Natural Regeneration (FMNR) and agroforestry	Poverty, livelihood(s), poor farmer(s) and job creation	Gender, women and youth	Partnership, participation, inclusiveand stakeholder engagement	Financing and access to finance	Average
National Action Plan on Gender and Climate Change 2020	(3) Promotes multi-sector gender-sensitive mitigation. Identifies action steps, indicators, responsible institutions and outcomes	(3) Promotes multi-sector gender-sensitive adaptation. Identifies action steps, indicators, responsible institutions and outcomes	(1) Mentions the use of trees to improve integrated water resource management. Identifies limited activities and plans	(2) Aims to enhance local participation in forestry and agriculture. Identifies action steps, indicators, responsible institutions and outcomes	(0) No details or evidence provided	(3) Promotes alternative livelihoods and local participation. Identifies action steps, indicators, responsible institutions and outcomes	(3) Aims to mainstream gender considerations across all climate programmes/ policies. Sets specific objectives, action steps, indicators, responsible institutions and outcomes	(3) Identifies national institutions with mandates related to gender and environment. Sets specific objectives, action steps, indicators, responsible institutions and outcomes	(2) Provides list of institutions to fund implementation and calls for partnership building and coordination with general details of the modalities	2.2
National Adaptation Plan Framework 2020	(0) No details of mitigation measures, activities or plans	(1) Supports adaptation and development in the form of general statements. No details on activities or plans	(0) No details or evidence provided	(1) Calls for ecosystem-based adaptation approaches with no details on activities or plans	(0) No details or evidence provided	(1) Aims to enhance rural livelihoods. No details on activities or plans	(1) Calls for a fully gender-responsive National Action Plan approach with no details on activities or plans	(2) Mandates a multi-sectoral approach to managing climate impacts. Some assumptions are included but lacks comprehensive detail	(1) Stresses the need for coherent approaches to fund adaptation through multi-stakeholder engagement with no details on activities or plans	0.7

Climate

Policy	Climate change mitigation and greenhouse gas emission(s)	Climate change adaptation	Tree(s) and forest(s)	Restoration, soil fertility and sustainable land management	Farmer Managed Natural Regeneration (FMNR) and agroforestry	Poverty, livelihood(s), poor farmer(s) and job creation	Gender, women and youth	Partnership, participation, inclusiveand stakeholder engagement	Financing and access to finance	Average
National Adaptation Strategy and Plan of Action on Climate Change for Nigeria 2011	(1) Promotes a low-carbon economy through the integration of adaptation with mitigation. No detailed activities or plans	(3) Sets adaptation as an integrated component of sustainable development. Identifies action steps and responsible institutions	(3) Fosters afforestation and reforestation to enhance human welfare. Sets specific objectives and action steps	(3) Fosters adoption of improved agricultural systems. Sets specific objectives and action steps	(0) No details or evidence provided	(3) Fosters increased adaptation through improved livelihoods. Sets specific objectives and action steps	(3) Supports adaptation by vulnerable groups, including women, children and youth, the elders and people with disabilities. Sets specific objectives and action steps	(3) Lays out partnership roles and responsibilities for national adaptation. Outlines actions and responsibilities.	(2) Calls for increased allocation of financial resources to climate change issues. Outlines guidelines for funding the national strategy	2.3
National Climate Change Policy 2021–2030	(2) Aims for greenhouse gas emissions reduction. Some recommendations are included but lack detail	(2) Supports adaptation measures to promote development. Some activities are included but lack detail	(2) Promotes reforestation for community development. Some activities are included but lack detail	(2) Promotes sustainable land-use systems to foster agriculture and development. Some activities are included but lack detail	(0) No details or evidence provided	(1) Promotes socio-economic development in general. No details on activities or plans are provided	(2) Refers to mainstreaming gender, children, youth and other vulnerable groups into all climate change interventions. Some activities are included but lacks detail	(2) Promotes international partnership and coordinated cooperation. Some activities are included but lacks detail	(2) Mentions finance mobilisation and investment as key enabling conditions. Some activities and recommendations are included but lacks detail	1.7

Climate										
Policy	Climate change mitigation and greenhouse gas emission(s)	Climate change adaptation	Tree(s) and forest(s)	Restoration, soil fertility and sustainable land management	Farmer Managed Natural Regeneration (FMNR) and agroforestry	Poverty, livelihood(s), poor farmer(s) and job creation	Gender, women and youth	Partnership, participation, inclusiveand stakeholder engagement	Financing and access to finance	Average
Nationally Determined Contribution 2021	(2) Promotes agricultural/ forestry measures to move to a low-carbon, climate-resilient status. Lacks detail	(1) Calls for measures in the agriculture, forestry and other land uses sector to foster adaptation by 2050, but no details on activities or plans	(1) Calls for improved natural forest management and forest restoration with no details on activities or plans	(1) Mentions climate-smart agriculture approaches to foster development and reduce emissions with no details on activities or plans	(1) Mentions the use of agroforestry to reduce emissions with no details or evidence	(1) Mentions the role of nature-based solutions to achieve sustainable development and job creation, but no details or evidence provided	(2) Calls for climate mitigation measures that are gender neutral and that enhance social inclusion, but lacks detail	(2) Acknowledges the importance of multi-stakeholder engagement processes in developing and validating the NDC. Lacks detail	(2) Outlines the need for an investment plan to finance low-carbon transition. It provides some figures but lacks detail	1.4
Overall average										1.7

Agriculture										
Policy	Climate change mitigation and greenhouse gas emission(s)	Climate change adaptation	Tree(s) and forest(s)	Restoration, soil fertility and sustainable land management	Farmer Managed Natural Regeneration (FMNR) and agroforestry	Poverty, livelihood(s), poor farmer(s) and job creation	Gender, women and youth	Partnership, participation, inclusive and stakeholder engagement	Financing and access to finance	Average
Agriculture Promotion Policy 2016–2020	(0) Reducing CO_2 emissions is stated as minor goal with no indication of how it could be achieved	(0) Sets the need to improve knowledge for adaptation. No details or evidence were provided	(1) Mentions using trees to reduce erosion and conserve water. Some activities included but lacks detail.	(1) Aims to unlock the potential of agriculture to enhance fertility through improved conservation. Some activities are included but lacks detail	(0) No details or evidence provided	(1) Job creation stated as a federal priority. Some activities are included but lack detail	(2) Fosters full inclusion of youth and women in the agricultural sector. Some activities are included but lack detail	(2) Calls for greater collaboration across federal, state and local levels to maximise engagement. Some activities are included but lacks detail	(3) Promotes enhanced access to finance for agriculture. Some activities are included	1.1
Overall average										1.1

Forestry										
Policy	Climate change mitigation and greenhouse gas emission(s)	Climate change adaptation	Tree(s) and forest(s)	Restoration, soil fertility and sustainable land management	Farmer Managed Natural Regeneration (FMNR) and agroforestry	Poverty, livelihood(s), poor farmer(s) and job creation	Gender, women and youth	Partnership, participation, inclusive and stakeholder engagement	Financing and access to finance	Average
National Forest Policy 2020	(2) Promotes the role of forestry in mitigation. Some activities are included but lack detail	(2) Promotes the role forestry in adaptation. Some activities are included but lack detail	(3) Fosters forest conservation for development. Includes complementary activities and plans	(0) No details or evidence provided	(2) Supports agroforestry to enhance livelihoods some activities included. Lacks detail	(2) Aims to upscale the contribution of forests to economic development. Some activities are included but lack comprehensive detail	(2) Promotes participation of women and vulnerable groups in forest resources management. Some activities are included but lack detail	(2) Promotes stakeholders participation in forest resources management, research and development. Some activities included but lacks detail	(2) Promotes investment of the public and private sectors in forest resource management and utilisation. Some activities included but lacks detail	2
Overall average										2

National development										
Policy	Climate change mitigation and greenhouse gas emission(s)	Climate change adaptation	Tree(s) and forest(s)	Restoration, soil fertility and sustainable land management	Farmer Managed Natural Regeneration (FMNR) and agroforestry	Poverty, livelihood(s), poor farmer(s) and job creation	Gender, women and youth	Partnership, participation, inclusive and stakeholder engagement	Financing and access to finance	Average
National Development Plan 2021–2025	(0) Supports the development of decarbonisation pathways, but no details or evidence provided	(0) Supports climate adaptation in sustainable production practices, but no details or evidence is provided	(0) No details or evidence provided	(0) No details or evidence provided	(0) No details or evidence provided	(1) Promotes inclusive job creation, but limited details are provided	(3) Targets women empowerment and the mainstreaming of gender issues at all policy levels. Sets specific action steps and indicators	(0) Calls for strategic partnerships and collaboration, but limited details are provided	(2) Calls for increased access to finance and support for businesses and projects in environmentally sustainable sectors. Some activities and plans are included	0.7
Economic Sustainability Plan 2020	(0) No details or evidence provided	(0) No details or evidence provided	(0) No details or evidence provided	(0) No details or evidence provided	(0) No details or evidence provided	(0) Supports job creation through the adoption of agricultural programme, but does not relate to achieving climate-development goals	(0) No details or evidence provided	(0) Fosters partnership, but not to achieve any climate-development goal	(0) Aims to engage with multilateral and donor agencies to access funding for crisis response. No action relates to achieving climate-development goals	0
Overall average										0.35

Energy										
Policy	Climate change mitigation and greenhouse gas emission(s)	Climate change adaptation	Tree(s) and forest(s)	Restoration, soil fertility and sustainable land management	Farmer managed natural regeneration (FMNR) and agroforestry	Poverty, livelihood(s), poor farmer(s) and job creation	Gender, women and youth	Partnership, participation, inclusive and stakeholder engagement	Financing and access to finance	Average
National Energy Policy of Nigeria 2018	(1) Promotes conservation and carbon management best practices in the nation's energy supply chain. Limited details are provided	(0) No details or evidence provided	(1) Encourages cultivation of fast-growing tree species to accelerate forest regeneration. Limited details are provided	(0) No details or evidence provided	(0) No details or evidence provided	(0) Mentions job creation, rural and agricultural development. No details or evidence were provided	(2) Promotes gender sensitivity and special attention to rural energy needs. Some activities are included but lack detail	(1) Targets effective coordination of national energy planning, programmes and policy implementation. Limited details are provided	(1) Aims to ensure the availability of adequate funding for the energy sector. Limited details are provided	0.66
Average										0.66

References

Altheide D, Coyle M, DeVriese K et al (2008) Emergent qualitative document analysis. In: Hesse-Biber SN, Leavy P (eds) Handbook of emergent methods. Guilford press, New York, pp 127–151

Antwi-Agyei P, Stringer LC (2021) Improving the effectiveness of agricultural extension services in supporting farmers to adapt to climate change: insights from northeastern Ghana. Clim Risk Manag 32:100304. https://doi.org/10.1016/j.crm.2021.100304

Austen GE, Dallimer M, Irvine KN et al (2022) The diversity of people's relationships with biodiversity should inform forest restoration and creation. Conserv Lett 16:e12930. https://doi.org/10.1111/conl.12930

Carias D, Page T, Smith H et al (2022) Beyond the 'field of dreams' model in smallholder forestry: building viable timber value chains for smallholder tree growers in developing countries. Land Use Policy 120:106227. https://doi.org/10.1016/j.landusepol.2022.106227

Cowie AL, Orr BJ, Castillo Sanchez VM et al (2018) Land in balance: the scientific conceptual framework for land degradation neutrality. Environ Sci Pol 79:25–35

Dallimer M, Stringer LC (2018) Informing investments in land degradation neutrality efforts: a triage approach to decision making. Environ Sci Pol 89:198–205. https://doi.org/10.1016/j.envsci.2018.08.004

Dannevig H, Hanssen Korsbrekke M, Hovelsrud GK (2022) Advancements of sustainable development goals in co-production for climate change adaptation research. Clim Risk Manag 36:100438. https://doi.org/10.1016/j.crm.2022.100438

Djenontin IN, Zulu LC, Etongo D (2020) Ultimately, what is forest landscape restoration in practice? Embodiments in sub-Saharan Africa and implications for future design. Environ Manag 68:619–641. https://doi.org/10.1007/s00267-020-01360-y

Donatti CI, Andrade A et al (2022) Ensuring that nature-based solutions for climate mitigation address multiple global challenges. One Earth 5:493–504. https://doi.org/10.1016/j.oneear.2022.04.010

England MI, Dougill AJ, Stringer LC et al (2018) Climate change adaptation and cross-sectoral policy coherence in southern Africa. Reg Environ Chang 18:2059–2071. https://doi.org/10.1007/s10113-018-1283-0

FAO (2019) Trees, forests and land use in drylands: the first global assessment. FAO Forestry Paper 184. Rome. https://openknowledge.fao.org/server/api/core/bitstreams/4db91cfa-6a0d-4e40-82ce-a7d2297c6a6c/content. Accessed 05 Jan 2025

FAO (2022) Nigeria agriculture at a glance. Rome: FAO. https://www.fao.org/nigeria/fao-in-nigeria/nigeria-at-a-glance/en/. Accessed 05 Jan 2025

Favretto N, Afionis S, Stringer LC et al (2020) Delivering climate-development co-benefits through multi-stakeholder forestry projects in Madagascar: opportunities and challenges. Land 9:157. https://doi.org/10.3390/land9050157

Favretto N, Dallimer M, Stringer LC et al (2021a) Harnessing benefits for climate change mitigation through irrigation-free indigenous tree restoration: sharing knowledge and building capacity. Stakeholder mapping report. Centre for Dryland Agriculture, Kano

Favretto N, Dallimer M, Stringer LC et al (2021b) Harnessing benefits for climate change mitigation through irrigation-free indigenous tree restoration: sharing knowledge and building capacity. Best practice guidelines manual for institutional stakeholders. Centre for Dryland Agriculture, Kano

Feng X, Fu B, Piao S et al (2016) Revegetation in China's loess plateau is approaching sustainable water resource limits. Nat Clim Chang 6:1019–1022. https://doi.org/10.1038/nclimate3092

Gann GD, McDonald T, Walder B et al (2019) International principles and standards for the practice of ecological restoration. Restor Ecol 27:S1–S46. https://doi.org/10.1111/rec.13035

Hankivsky O, Grace D, Hunting G et al (2014) An intersectionality-based policy analysis framework: critical reflections on a methodology for advancing equity. Int J Equity Health 13:119. https://doi.org/10.1186/s12939-014-0119-x

Keenan RJ, Reams GA, Achard F et al (2015) Dynamics of global forest area: results from the FAO global Forest resource assessment 2015. For Ecol Manag 352:9–20. https://doi.org/10.1016/j.foreco.2015.06.014

Kong ZH, Paavola J, Stringer LC (2024) National environmental programs and local social-ecological system change in dryland China: implications for environmental governance. Ecol Soc 12. https://doi.org/10.5751/ES-15330-290312

Manuamorn OP, Biesbroek R, Cebotari V (2020) What makes internationally-financed climate change adaptation projects focus on local communities? A configurational analysis of 30 adaptation fund projects. GEC 61:102035. https://doi.org/10.1016/j.gloenvcha.2020.102035

Mirzabaev A, Wuepper D (2023) Economics of ecosystem restoration. Annu Rev Resour Econ 15:329–350. https://doi.org/10.1146/annurev-resource-101422-085414

Mirzabaev A, Sacande M, Motlagh F et al (2022) Economic efficiency and targeting of the African Great Green Wall. Nat Sustain 5:17–25. https://doi.org/10.1038/s41893-021-00801-8

Mutune JM, Lund JF (2016) Unpacking the impacts of 'participatory' forestry policies: evidence from Kenya. Forest Policy Econ 69:45–52

Nelson CR, Hallett JG, Romero Montoya AE et al (2024) Standards of practice to guide ecosystem restoration – a contribution to the United Nations Decade on Ecosystem Restoration 2021–2030. Rome, FAO, Washington, DC, SER & Gland, Switzerland, IUCN CEM. https://www.fao.org/3/cc9106en/cc9106en.pdf. Accessed 05 Jan 2025

Neuendorf K (2017) The content analysis guidebook. SAGE Publications, Inc. https://doi.org/10.4135/9781071873045

Nigeria's Federal Ministry of Environment. 2021. Nigeria's Nationally Determined Contribution – 2021 update. Abuja: Nigeria's Federal Ministry of Environment. https://unfccc.int/sites/default/files/NDC/2022-06/NDC%20INTERIM%20REPORT%20SUBMISSION%20-%20NIGERIA.pdf. Accessed 05 Jan 2025

OECD (2005) Policy Coherence for development: promoting institutional good practice, the development dimension. OECD Publishing, Paris. https://doi.org/10.1787/9789264108554-en. Accessed 05 Jan 2025

Okpara U, Stringer LC, Akhtar-Schuster M et al (2018) A social-ecological systems approach is necessary to achieve land degradation neutrality. Environ Sci Pol 89:59–66. https://doi.org/10.1016/j.envsci.2018.07.003

Orr BJ, Cowie AL, Castillo Sanchez VM et al (2017) Scientific conceptual framework for land degradation neutrality. a report of the science-policy interface. United Nations Convention to Combat Desertification, Bonn, Germany. https://www.unccd.int/sites/default/files/2018-09/LDN_CF_report_web-english.pdf. Accessed 05 Jan 2025

Parr CL, te Beest M, Stevens N (2024) Conflation of reforestation with restoration is widespread. Science 383:698–701. https://doi.org/10.1126/science.adj0899

Pritchard R (2021) Politics, power and planting trees. Nat Sustain 4:932–932. https://doi.org/10.1038/s41893-021-00769-5

Reynolds JF, Stafford Smith DM, Lambin EF et al (2007) Global desertification: building a science for dryland development. Science 316:847–851. https://doi.org/10.1126/science.1131634

Ripple WJ, Moomaw WR, Wolf C et al (2022) Six steps to integrate climate mitigation with adaptation for social justice. Environ Sci Pol 128:41–44. https://doi.org/10.1016/j.envsci.2021.11.007

Sacande M, Berrahmouni N (2016) Community participation and ecological criteria for selecting species and restoring natural capital with native species in the Sahel. Restor Ecol 24:479–488. https://doi.org/10.1111/rec.12337

Sinclair F, Coe R (2019) The options by context approach: a paradigm shift in agronomy. Exp Agric 55:1–13. https://doi.org/10.1017/S0014479719000139

Stringer LC, Dougill AJ, Dyer JC (2014) Advancing climate compatible development: lessons from southern Africa. Reg Environ Chang 14:713–725. https://doi.org/10.1007/s10113-013-0533-4

Stringer LC, Reed MS, Fleskens L (2017) A new dryland development paradigm grounded in empirical analysis of dryland systems science. Land Degrad Dev 28:1952–1961. https://doi.org/10.1002/ldr.2716

Stringer LC, Mirzabaev A, Benjaminsen TA et al (2021) Climate change impacts on water security in global drylands. One Earth 4:851–864. https://doi.org/10.1016/j.oneear.2021.05.010

van der Esch S, ten Brink B, Stehfest E et al (2017) Exploring future changes in land use and land condition and the impacts on food, water, climate change and biodiversity: Scenarios for the UNCCD Global Land Outlook. PBL Netherlands Environmental Assessment Agency: The Hague, The Netherlands. https://catalogue.unccd.int/1600_Exploring_future_changes_landuse_biodiversity.pdf. Accessed 05 Jan 2025

Yakubu IB (2020) Training manual on Sustainable Land and Water Management (SLWM) Practices. The Nigeria Erosion and Watershed Management Project (NEWMAP), Kano State office, Kano, Nigeria. https://doi.org/10.13140/RG.2.2.18699.41765. Accessed 05 Jan 2025

Chapter 27
Understanding Government Policies on Climate-Smart Agriculture in West Africa: Insights from Burkina Faso

Eveline Sawadogo Compaoré, Roman Hinz, Reginald Tang Guuroh, Jewel Andoh, Constantin Gueswindé Compaoré, Sibiri Birba, and Farid Traoré

Abstract Integrating climate-smart agriculture (CSA) - understood both as a strategic framework and a set of context-specific practices - into national policies remains a critical challenge in West Africa, where agriculture is a cornerstone of rural livelihoods. Achieving sustainable development goals in this region requires addressing a complex mix of political priorities, institutional roles, and socioeconomic realities. This study focuses on Burkina Faso, looking at its policy approaches to CSA and providing insights into the political and institutional contexts that shape CSA strategies at a national level. It also offers lessons for improving policy coherence and scaling such strategies nationally. We examine government policies, programs, strategies, frameworks, and action plans using the National Innovation System framework. Additionally, the research incorporates insights from 27 expert interviews conducted between 2022 and 2023. The findings reveal a strong awareness of climate change impacts and the importance of CSA but also highlight fragmented policies and weak institutional collaboration, which hinder effective implementation. Advocacy and outreach efforts face challenges due to resource constraints, limited coordination, and a lack of locally adapted implementation strategies. In Burkina Faso, the term CSA is mostly not used explicitly but is

E. S. Compaoré (✉) · S. Birba · F. Traoré
Institute for Environment and Agricultural Research, Ouagadougou, Burkina Faso
e-mail: compeve@yahoo.fr; sibiri_birba@yahoo.fr; farid.traore@cnrst.gov.bf

R. Hinz
Kassel Institute for Sustainability, University of Kassel, Kassel, Germany
e-mail: roman.hinz@uni-kassel.de

R. T. Guuroh · J. Andoh
CSIR-Forestry Research Institute of Ghana, Kumasi, Ghana
e-mail: rtguuroh@csir-forig.org.gh; jandoh@csir-forig.org.gh

C. G. Compaoré
Nazi Boni University, Bobo-Dioulasso, Burkina Faso
e-mail: tintincomp@yahoo.com

Y. D. I. Khan et al. (eds.), *Land Restoration Through Ecosystem-Based Approach*, Disaster Resilience and Green Growth, https://doi.org/10.1007/978-981-95-0389-6_27

reflected through CSA-aligned practices and strategic goals. Despite these challenges, certain programs, such as agroforestry initiatives, have shown potential for broader adoption. However, stakeholders frequently cited barriers such as inadequate funding, limited access to technology, and insufficient farmer training. Moreover, CSA represents one approach to climate-resilient agriculture alongside alternatives such as agroecology. Experts highlighted the need for a more integrated policy approach combining multiple concepts to effectively address national priorities.

Keywords Public policy · Climate-smart agriculture · Burkina Faso

27.1 Introduction

Climate-smart agriculture (CSA) is a framework and approach designed to address the challenges posed by climate change and its impact on agriculture. It focuses on three core objectives: increasing agricultural productivity and incomes, enhancing resilience to climate variability, and reducing greenhouse gas emissions (FAO 2013a, b). CSA includes a range of strategies, such as improved water and soil management, sustainable land-use planning, and the adoption of locally relevant technologies (Wreford et al. 2017; Douxchamps et al. 2015). Beyond practices at the farm level, CSA emphasizes the importance of supportive policies, investments, and capacity-building initiatives to create enabling environments for its implementation (Wiederkehr et al. 2018). However, its adoption is highly context-dependent and faces barriers such as policy fragmentation, resource constraints, and limited institutional coordination (FAO 2021; Thorton et al. 2018).

Climate change (CC) has emerged as a critical global challenge, particularly in regions like West Africa, where agriculture forms the backbone of livelihoods. The Sahel region is among the most climate-vulnerable areas globally, grappling with recurrent droughts, floods, and extreme weather events. These crises exacerbate vulnerabilities, including food insecurity, low incomes, and limited institutional capacity. Rural populations, who depend heavily on agriculture, face significant livelihood threats as CC undermines agricultural productivity, depletes natural resources, and drives economic instability (FAO 2013a, b; Lalou et al. 2009). Addressing these interlinked challenges requires holistic approaches integrating resilience, sustainability, and productivity, making CSA a compelling strategy framework (Wreford et al. 2017; World Bank 2022).

Despite growing interest in CSA practices in West Africa, their integration into public policy remains marginal and often unclear within the dominant agricultural paradigms. The region's fragmented nature of agricultural policies reflects fragile compromises and divergent priorities. These are shaped by institutional frameworks, the players involved, the influence of the agribusiness sector, and political objectives, particularly concerning food self-sufficiency and agriculture's role in economic development. This complexity underscores the need for policies that effectively balance short-term productivity goals with the longer-term objectives of sustainability and resilience.

Regional initiatives such as the ECOWAS Agricultural Policy (ECOWAP) and the Comprehensive Africa Agriculture Development Programme (CAADP) have been pivotal in addressing food insecurity and fostering agricultural development, particularly following the 2008 food crisis. These frameworks encourage countries to allocate at least 10% of their national budgets to agriculture, as outlined in the Maputo Declaration of 2003, aiming to improve food security, reduce poverty, and integrate regional agricultural markets. Yet, the tension between rapid production growth and sustainability goals reveals an ongoing challenge in aligning regional policies with the broader principles of CSA, especially in the face of environmental degradation and climate pressures (Fouilleux 2017).

In Burkina Faso, these dynamics are reflected in National Agricultural Investment Plans (NAIPs), which align with regional policies and have achieved progress in enhancing agricultural productivity. Nonetheless, the focus on short-term economic gains occasionally limits the incorporation of climate resilience and environmental conservation. For instance, the prioritization of short-term productivity gains over long-term climate adaptation highlights a gap that needs to be addressed through integrated and forward-looking policy approaches. Similarly, donor-driven initiatives such as the Alliance for a Green Revolution in Africa (AGRA) have contributed to yield increases through input subsidy programs for hybrid seeds and synthetic fertilizers. While these approaches address immediate food security concerns, their growing environmental costs underscore the need to consider more sustainable alternatives.

Intergovernmental meetings like the 2021 United Nations Summit on Food Systems held in Dakar have highlighted the importance of sustainable agricultural frameworks like CSA. These forums provide opportunities to align national policies with global climate and food security goals. However, translating these priorities into coherent and actionable strategies remains a significant challenge, particularly in institutional contexts dominated by productivist approaches.

This chapter examines the institutional and sociopolitical dynamics influencing CSA adoption in Burkina Faso. Using the National Innovation System (NIS) framework, it explores how national policies, local contexts, and international donor priorities shape the uptake of CSA. By analyzing developments in Burkina Faso, the chapter identifies opportunities to align national strategies with CSA principles, address climate challenges, promote food security, and foster sustainable agricultural development.

27.2 Theoretical Framework

Since the 1960s, governments in West Africa have consistently sought to implement development policies to improve their populations' economic and social conditions. Drawing on a broad definition of technology that encompasses social technologies,

Table 27.1 Key variables of the NIS theoretical concept

NIS focus: Enabling environment		NIS focus: Policy implementation	
Structural components	Standards for enabling environment	Institutional performance and functional roles	Standards for policy implementation
Government strategies	- Reform and policies supporting professional education - Technology foresight to guide policy setting - Robust regulatory frameworks - Systemic upgrading of public infrastructure	Organizations strategically positioned to enable the updake and diffusion of CSA principles and practices	Institutionalizing mandatory evaluation and formulating adaptive strategies informed by regional policies

Source: Process by authors based on: OECD (1997). Managing national innovation systems, Paris: Organisation for Economic Co-operation and Development, p. 24

this study employs the NIS analytical framework to examine the institutional and policy dynamics surrounding the adoption of CSA (Table 27.1).

As defined in the OECD's 1999 report, the NIS concept refers to *"the set of institutions that (jointly and individually) contribute to the development and diffusion of new technologies. These institutions provide the framework within which governments form and implement policies to influence the innovation process. As such, it is a system of interconnected institutions to create, store, and transfer the knowledge, skills, and artefacts that define new technologies"* (Metcalfe in OECD 1997: 24). From this perspective, a country's innovation capacity is shaped by the interconnectedness of its firms, science systems, supporting institutions, and other research bodies. These components collectively generate, diffuse, and apply knowledge (OECD 1997).

The NIS framework specifically focuses on the nation-state level, emphasizing that the dynamics of institutions are interlinked with the broader macroeconomic and regulatory context. Key components include communication infrastructures, factor and product market conditions, and the education and training system (OECD 1997). This interrelatedness underscores the importance of examining not just the individual contributions of institutions but also their interactions within a collective system of knowledge creation and use. These interactions are influenced by values, norms, and legal frameworks (Smith in OECD 1997).

Governments play a central role within the NIS by shaping policy options and regulations that guide innovation trajectories. Through their policy interventions, governments influence the behavior and interconnections of other actors in the system, including universities (as knowledge producers) and industries or firms (as users of knowledge). Lundvall (1985, Lundvall 1992, Lundvall 2007) identifies these three actors—government, universities, and firms—as the core components of the NIS. Similarly, Goto (2000: 104) highlights the interdependence of these sectors, stating that *"the national innovation system essentially consists of three sectors: industry, universities, and the government, with each sector interacting with the others while simultaneously fulfilling its distinct role."*

In this framework, the government is tasked with creating an enabling environment to enhance the effectiveness of the interconnected actors. By fostering collaboration, providing funding, and establishing supportive legal and institutional frameworks, governments can significantly influence the innovation processes necessary to adopt CSA-aligned practices.

27.3 Government's Role

NIS scholars consistently emphasize the critical role of government in fostering innovation (Freeman 1995). The acquisition, use, and diffusion of technology require active government involvement, as the government's role is fundamental to the functioning of the innovation system. Through its policymakers, the government shapes strategies at the national level, acting as a key decision-maker and facilitator (Freeman 1987). One important dimension of this role is its influence on Foreign Direct Investment (FDI). While FDI inflows in some contexts may be limited in scale, they have been shown to generate employment in the formal sector and contribute to local value addition, as supported by the government's broader policy framework (United Nations 2009).

Governments also play a significant role in designing intellectual property rights systems, directing scientific research, and financing universities. By supporting business Research and Development (R&D) and contributing to the formation of human and social capital, governments help evaluate, adapt, and implement foreign technologies (Feinson 2003). As Freeman's case study on Japan demonstrated, R&D can serve as a critical tool for knowledge assimilation and domestic technological innovation (Freeman 1987). These government-led incentives provide administrative support and the infrastructure, facilities, and legitimacy needed to enhance networking between knowledge producers and users (Lundvall 1985; Freeman 1995; OECD 1997).

Institutional support from the government creates a favorable social and political context for collaboration between firms and knowledge producers, which is essential for innovation. Founders of the NIS framework reject mainstream economic theories that treat knowledge, technology, and technical change as purely market-driven phenomena. Instead, they emphasize the enabling environment as a key determinant in adopting practices such as CSA. They argue that innovation requires an institutional approach to understand the unique characteristics of technical and knowledge-driven changes.

The term "national" in the National Innovation System (NIS) highlights the institutional dimension of innovation, particularly CSA practices. As Chesnais points out, "national" underscores the importance of systemic relations, linkages, and institutions beyond just markets (Chesnais in Sharif 2006). This perspective reinforces the idea that governments must actively shape policy frameworks to foster innovation.

In this context, the NIS framework offers a dual perspective. On the one hand, it serves as an analytical tool to explore the dynamics of innovation with openness and reflexivity, allowing for a deeper understanding of the processes involved. On the other hand, it acts as a narrative for policy, providing structured definitions and assumptions aimed at achieving specific outputs and outcomes. These policy narratives are particularly relevant in addressing climate change and promoting CSA. Governments must develop coherent policies that align with the institutional realities and systemic relations needed to drive meaningful changes in agricultural practices.

27.4 Methodology

27.4.1 Study Area: Burkina Faso

Burkina Faso is a landlocked country in West Africa bordering Mali, Niger, Benin, Togo, Ghana, and Côte d'Ivoire (Fig. 27.1). It has a surface area of 274,200 km^2 and is ranked 186th out of 193 countries on the Human Development Index (as of year

Fig. 27.1 Map showing Burkina Faso in Bleu

2023). Its economy is based on agriculture, although gold exports are growing. It is a fairly uniform landscape, with soils often characterized by laterite, giving them an ochre coloration. It has a flat relief, which does not retain large quantities of water, hence the water shortage in certain regions. The climatic instability affects the country's agriculture and calls for enhanced efforts to preserve the environment while maintaining agricultural production. Farming requires a more climate-smart approach that supports environmental protection and long-term sustainability, as opposed to conventional agircultural methods.

27.4.2 Research Design and Data Collection

This study primarily utilized qualitative research to gather in-depth insights from policy documents and expert interviews in Burkina Faso, specifically on climate-smart agriculture (CSA). A comprehensive desk review of government policies and programs was conducted to identify explicit and implicit references to CSA practices in the country. While numerous policies were identified that indirectly or broadly supported CSA principles, this study focuses on policies explicitly promoting CSA practices. Additionally, we concentrated on key policies selected based on their relevance, impact, and alignment with CSA principles, ensuring a manageable scope for detailed analysis.

To complement the policy review, 30 professionals in Burkina Faso were interviewed between January 2022 and December 2023. The experts represented various organizations and perspectives, as detailed in Table 27.2.

The data collected from policy reviews and transcribed expert interviews were analyzed using a content analysis approach. The analysis focused on identifying key emerging variables, including:

Policies, strategies, and action plans promoting CSA practices. Challenges in implementing CSA policies and strategies. This methodology ensures a robust understanding of the current CSA policy landscape (Table 27.3).

27.5 Results

Burkina Faso, like many other West African countries, is actively integrating climate-smart agriculture (CSA) into their national development agendas. This effort is part of broader strategies aimed at addressing the impacts of climate change and achieving the sustainable development goals. The focus is on balancing mitigation, adaptation, and productivity objectives within the Agriculture, Forestry, and Other Land Use (AFOLU) sector. Various action plans are being developed to enhance food security while preserving the environment. Additionally, there are several policies that explicitly or implicitly support climate-smart agriculture

Table 27.2 Sectoral distribution of Burkinabe organizations surveyed in 2022 and 2023 with expertise in the fields of agriculture, forestry, and climate change

Stakeholder groups	Interviewed organizations: Burkina Faso
International organizations	FAO Burkina, SOS-Sahel Burkina, Permanent Interstate Committee for Drought Control in the Sahel
Governmental organizations	Ministry of Environment: Directorate of Project and Program Coordination (DPPC), Ministry of Environment: General Directorate of Green Economy and Climate Change (GDGECC), Permanent Secretariat of the National Council for Sustainable Development (PS/NCSD), Ministry of Environment: General Directorate for Environmental Preservation (GDEP), Regional Directorate of the Ministry of Agriculture (RD), Provincial Directorate of the Ministry of Agriculture (PD), Ministry of Agriculture: Technical Support Zones (TSZ), Permanent Secretariat REED+ (PS/REDD+), Ministry of Environment: Directorate of Legal Affairs and Litigation (DLAL), Ministry of Environment: Directorate for Environmental Risk Prevention (DERP), Ministry of Agriculture: General Directorate of Hydraulic Developments and Irrigation Development (GDHDI), Ministry of Agriculture: Technical Animation Units (TAU)
Scientific research institutions	Institute of Environment and Agricultural Research (INERA), National Center for Scientific and Technological Research (NCSTR)
Nongovernmental organization/nonprofit organization	Green Wall Initiative for the Sahara and Sahel (IGMVSS), Action for the Promotion of Local Initiatives (APLI), Green Land (terre verte), Tiipalga, Zoramb Nangtaaba Association (ZNA), and National Council for Organic Agriculture (CNABio) Beog Puuto Project, Agricultural Productivity Improvement Project through Water and Soil Conservation (PACES)
Total	30

mechanisms and infrastructure. Table 27.4 presents a summary of various government policies and action plans promoting climate-smart agriculture practices in Burkina Faso.

27.6 Climate-Smart Agriculture Landscape in Burkina Faso

Burkina Faso has integrated CSA into its national climate change adaptation plans, focusing on improving agricultural productivity while enhancing resilience. The implementation of CSA measures is supported not only by national efforts but also by regional and international strategies and funding mechanisms. For instance, the government allocates specific funds for CSA initiatives within its national budget for sustainable agriculture. This includes financing projects that improve access to sustainable agricultural technologies. The sources of funding that the state relies on are its national budget and those from its financial partners. For instance, the Regional Project for the Promotion of CSA, initiated by ECOWAS with the support of the West African Development Bank, finances subprojects of climate-smart

Table 27.3 Policy documents reviewed in Burkina Faso and their sources

Country	Document	Source
Burkina Faso	Presidential Initiative for Agricultural Production and Food Self-Sufficiency	https://presidential initiative for agriculture bears its first fruit
	Regional Project for the Promotion of Climate-Smart Agriculture (CSA)	Promotion de l'Agriculture Intelligente face au Climat en Afrique de l'Ouest – AIC I Agence Régionale pour l'Agriculture et l'Alimentation de la CEDEAO https://www.araa.org/fr/projets/promotion
	National Climate Change Adaptation Plan (PNA)	Burkina Faso NAP_English.pdf
	Rural Development Strategy (RDS) for 2016–2025	https://www.bing.com/search?q=Rural+Development+Strategy+(RDS)+for+2016-2025
	Laws and decrees	AN, Loi n°003–2011/AN du 05 avril 2011 portant Code forestier au Burkina Faso. AN, Loi n°006–2013/AN du 02 avril 2013 portant Code de l'environnement au Burkina-Faso AN, Loi n° 008–2014/AN du 08 avril 2014 portant loi d'orientation sur le développement durable au Burkina Faso AN, Loi n°034–2009/AN du 16 juin 2009 portant régime foncier rural AN, Loi n°034–2012/AN du juillet 2012 portant réorganisation agraire et foncière AN. 2018 Loi n°034–2018/AN portant pilotage et gestion du développement décret du 20 mai portant organization du ministère de l'environnement, de l'économie verte et du changement climatique

agriculture in several countries, including Burkina Faso. It includes establishing tailored climate services and developing resilient agricultural practices. Regarding policies, Burkina Faso has engaged in several political initiatives and regulations to promote climate-smart agriculture. As such, the National Policy for Food Security and Nutrition integrates climate resilience issues as a priority into its objectives. Among others, we have the Presidential Initiative for Agricultural Production and Food Self-Sufficiency (IP-P3A) and the Regional Project for the Promotion of Climate-Smart Agriculture (CSA). Burkina Faso has developed specific programs to encourage the adoption of CSA practices, such as promoting drought-resistant crops. These include the National Climate Change Adaptation Plan (PNA) 2015. The PNAs aim to (i) reduce vulnerability to the impacts of climate change by developing adaptation and resilience capacities and (ii) facilitate the integration of climate change adaptation coherently into policies, programs, or activities, new or existing, within specific development planning processes and strategies at relevant sectors and levels. In addition, the Rural Development Strategy (RDS) for 2016–2025 aims to contribute sustainably to food and nutrition security, strong economic

Table 27.4 Government policies/programs/strategies promoting climate-smart agriculture practices in Burkina Faso

Country	Policy/strategy/action plan	Year introduced	Aim	Some policy actions	Performance (organizations and standards)
Burkina Faso	Law N° 034–2002/AN on the orientation of pastoralism. Conservation of forests and the environment	2002	It defines pastoral areas and makes their use subject to compliance with environmental regulations, i.e., the protection being considered here	Make sure agro-pastoralists are getting along in planning areas for animals to move. Also, protect the forest resources	The Ministry of Agriculture and Pastoralism in addition to the Ministry of Environment is collaborating to achieving sustainable actions
	National Environmental Policy (Politique Nationale en matière d'Environnement—PNE)	2007	To create a reference framework for considering environmental issues in development policies and strategies, as well as in economic activities	Provide support for planting trees using more ecological practices to protect the environment	The Ministry of Environment made special the need for tree planting at all stages and supporting institutions. The sustainable management of wetlands in Burkina Faso is based on major national, regional, subregional, and international guidelines on the environment and sustainable development
	Regional Project for the Promotion of Climate-Smart Agriculture (CSA)		To produce sufficient and sustainable food	Being able to produce sufficiently to feed its 300 million inhabitants in a context marked by poor yields (less than 3 tons/ha for rice against 6 tons/ha in Asia), a limited use of fertilizers (on average 10 kg/ha against 100 kg/ha in Asia), a high demographic growth rate (between 2.2% and 3.6% per year depending on the country), and a very limited use of irrigation.	With the support of FAO, the 10 principles. Research infrastructures supported by the REDD agency from the Ministry of Environment

	National Climate Change Adaptation Plan (PNA)	2015	To reduce vulnerability to the impacts of climate change by developing adaptation and resilience capacities; to facilitate the integration of climate change adaptation coherently into policies, programs, or activities, new or existing, within specific development planning processes and strategies at relevant sectors and levels	The objectives of the NAPs are to (i) reduce vulnerability to the impacts of climate change by building capacity for adaptation and resilience, (ii) facilitate the integration of adaptation to climate change, coherently, into new or (ii) facilitate the integration of climate change adaptation in a coherent manner	Government agencies that promote existing policies, programs, or activities into specific planning processes and strategies within relevant sectors and at different levels
	The Nationally Determined Expected Contribution (NDEF)	2016	The first Nationally Determined Contribution (NDC) was implemented from 2016 to 2020. This first NDC set GHG emission reduction targets for 2030 of 21,574.63 Gg CO2eq, or 18.2%, compared with the reference scenario for mitigation actions, and 43,707 Gg CO2eq, or 36.95%, through the implementation of adaptation actions	To help meet the global goal of pursuing 1.5 °C, adapt to climate impacts, and ensure sufficient finance to support these efforts	Ministries and other agencies, such as the Ministry of Environment, Science, Technology, and Innovation, are well-positioned to implement the policy. To ensure effectiveness, these ministries and agencies are to be integrated into already existing monitoring and evaluation systems

growth, and poverty reduction. This was supplemented with the five-year agricultural program 2024–2028 called the Presidential Initiative for Agricultural Production and Food Self-Sufficiency (IP-P3A), with a cost of 120,893,950,000 FCFA, including climate-smart agriculture, intensification of rice, wheat, and vegetable production, as well as mobilizing the army for food security purposes. This program is divided into four components, one of which concerns promoting and establishing climate-smart agriculture on 5,000 ha linked to climate change adaptation and mitigation. In the context of the annual implementation of sectoral plans related to climate-smart agriculture (CSA), the activities carried out primarily include raising awareness among farmers about CSA practices, training farmers on CSA practices, identifying and preparing sites for interventions, establishing climate-smart agriculture, disseminating and promoting CSA practices, and monitoring and evaluating implemented practices.

It is important to note that in West Africa, the fragmentation of efforts across ministries and donor-driven initiatives frequently leads to overlap and inefficiency. Many of the challenges government agencies face are linked to uncoordinated actions across different governance levels. One interviewee described the situation as follows:

"It's very difficult. We are making efforts at different levels, but it's not coordinated at some point. The project I'm coordinating involves Burkina and all the neighbouring countries; we usually meet for yearly reports. But here we do our work. Not with any other ministry or something like this... if we have activities like workshops, we do invite the researchers and people from other ministries, that's all."

This example highlights that cross-sectoral collaboration tends to occur only on an ad hoc basis—often limited to single events —rather than being systematically institutionalised. Such fragmentation undermines the development of coherent long-term strategies and shared implementation structures.

In addition to coordination issues, limited or delayed funding often slows down the progress of CSA initiatives. Interviewees repeatedly pointed to unsustainable financing dynamics, which hinder the continuous implementation of national or regional plans. Nonetheless, government agencies do engage with NGOs, financial partners, and other stakeholders to support CSArelated activities, especially at the project level.

These challenges reflect broader structural issues within Burkina Faso's national innovation system (NIS). As with past development efforts since the 1960s, institutions, enterprises, and human actors have often been considered inadequate to promote innovation systems in Burkina Faso. Government support for these systems has also been inconsistent. As a result, linkages between actors - especially between knowledge producers and users - remain weak or entirely absent. The following statement illustrates these systemic difficulties:

"We are working from our side for the Green Wall initiative, but sometimes it's hard due to uncoordinated actions… sometimes because policies do not tell us what to do on these. We usually follow project goals by inviting other ministries and departments that usually deal with the same actions."

Finally, regarding clear policies on CSA, it is important to note that is is merged in the full package of climate and conservation initiatives at the country level, rather than addressed through a dedicated or standalone CSA policy framework.

27.7 Discussion

There seems to be an omission in highlighting the need for a coherent, long-term strategy that fully incorporates agriculture as a critical sector for achieving climate goals. Stakeholders emphasize the importance of regional planning to address Burkina Faso's diverse agro-ecological zones, each with unique challenges and opportunities. This was complemented by a leading actor in government institutions in Burkina Faso who stated that *"policies by themselves do not make any difference. We do not usually have a clear idea of how to operate the existing policies. No matter how good they are..."*

Burkina Faso's CSA efforts highlight a mix of challenges and opportunities. While robust policy frameworks and international partnerships provide a strong foundation, gaps in financing, institutional coordination, and local implementation hinder progress. The international yearly conferences, such as the COP conference recommendations, are yet to generate expected outcomes at international, regional, and national level initiatives. Despite great efforts from FAO, SILSS, WASCAL, funders, and donors, the practical challenge remains high. Addressing these barriers requires a comprehensive approach to integrating CSA into national climate strategies, enhancing resource allocation, and empowering local stakeholders. By leveraging opportunities in agroforestry, traditional knowledge, and carbon markets, Burkina Faso can advance its CSA goals - enhancing livelihoods, strengthening resilience, and contributing to global mitigation efforts..

FAO (2010a, b) proposed to adopt climate-smart agriculture. CSA has three pillars: food security and income, adapting to climate change, and reducing greenhouse gas emissions. Climate change issues, consequences, global concern; the different manifestations of CC impacts on agriculture and farmers' living conditions. Farmers from Burkina Faso perceive the effects of climate change on many factors: temperature, rainfall, wind, agricultural production, epidemic diseases, etc. Many strategies focus on Assisted Natural Regeneration (ANR), such as the so-called Green Belt. Interventions have also looked at improving farmers' agricultural practices through best agronomic practices, agroforestry, and climate-smart agriculture (reduced tillage, use of natural/organic pesticides, improved fallow, soil and water conservation, etc.). It usually includes reports, and the interviews highlight the following:

The findings suggest several policy implications. First, there is an urgent need for targeted interventions to enhance financial, human, and natural capital among farming households. Policymakers should prioritize programs that promote income diversification, improve educational attainment, and increase farm sizes and soil fertility through access to modern inputs and soil improvement practices. Financial

Capital: Policies should facilitate income diversification by promoting off-farm activities and providing financial support for small-scale enterprises. This could include microcredit and microinsurance schemes, training in alternative livelihoods, and support for market access. Human Capital: Enhancing education and training programs for farmers is crucial. Agricultural extension services should be strengthened to provide comprehensive and accessible advice on modern farming techniques, climate information, and adaptive strategies. Programs should be designed to address the specific educational needs of rural populations, emphasizing practical and context-relevant knowledge. Natural Capital: Policies should support soil fertility improvement and access to agricultural inputs. Subsidies for fertilizers and other inputs and training on sustainable farming practices can enhance soil productivity and farm resilience. Additionally, land tenure reforms to secure larger farm sizes for smallholders can facilitate the adoption of resilient agricultural practices. Support for Vulnerable Households: Special attention should be given to households that have experienced natural hazards. Comprehensive support programs, including financial aid, technical assistance, and infrastructure improvements, can help these households recover and build resilience. Addressing the impacts of climate change in Burkina Faso requires a holistic approach that recognizes the differentiated nature of vulnerability and adaptation. Consequently, developing quality programs and policies for climate change adaptation and mitigation must go beyond a neutral approach and prioritize inclusion and empowerment. The interview concluded with a discussion on the need for inclusive, stakeholder-driven strategies and the potential for deeper engagement in carbon markets. State decentralization and the practical implementation of public policies are crucial to supporting this process. Burkina Faso's policies recognize the importance of climate-smart agriculture, but experts highlighted gaps in implementation and enforcement. Policies often focus on high-level goals (FAO and World Bank strategies) without clear guidelines for local implementation. The government's role in fostering collaboration among stakeholders—including researchers, extension services, and farmers—was identified as critical to improving CSA adoption.

Despite these challenges, interviewees acknowledged growing awareness of climate change impacts among farmers and policymakers. The potential for public-private partnerships and the integration of local knowledge into national policies were key opportunities for scaling CSA practices.

27.8 Conclusion

The findings emphasize that financial, human, and natural capital are critical for resilience, with significant policy implications for targeted interventions. In other words, policies have become the primary means of characterizing a state increasingly producing goods and services. In addition to the legal legitimacy that determines the right to command, the exercise of power is manifested in public policies that are both institution-building (they socially define reality and the problems to be solved) and policy-making (they define the state's role in society). In this regard, climate-smart agriculture implementation will be all the more challenging because of limited resources and no clear and enforced operational actions. That is why the World Bank has developed tools and analytical work so that countries can assess their opportunities, find trade-offs, prioritize their investments, and identify the key anchors of their climate-smart agriculture.

How do policies then rely on operational-level actions in a more systematic way? This is a key question that calls for more reflection and future research can build on these findings by considering additional factors and employing dynamic data sets to capture the evolving nature of adaptive capacity over time in terms of low regulations, research, practices, etc. Integrating vulnerability and social justice considerations into adaptation policy can further enhance the effectiveness of these interventions, ensuring that they address the needs of the most vulnerable populations. The results of this study provide a foundation for developing comprehensive, context-specific policies that enhance resilience and well-being. This research underscores the need for policies that promote diverse CSA practices while addressing country-specific challenges. It also calls for targeted, actionable policies to better support collective action and bridge the gap between policy and practice.

Bibliography

Bukina Faso Rural Development Strategy (RDS) for 2016-2025, Report

Douxchamps S et al (2015) Households and food security: lessons from food secure households in East Africa. Agric Food Secur 4:23. https://doi.org/10.1186/s40066-015-0042-4

FAO (2010a) 'Climate-smart' agriculture policies, practices and financing for food security, adaptation and mitigation. Rome: Author. Retrieved from https://www.fao.org/docrep/013/i1881e/i1881e00.htm

FAO (2010b) Climate-Smart Agriculture-Policies, Practices and Financing for Food Security, Adaptation and Mitigation. Available from http://www.fao.org/docrep/013/i1881e/i1881e00.htm

FAO (2013a) Statistical Year Book, World food and agriculture, Food and Agriculture Organization of the United Nations, https://www.fao.org/4/i3107e/i3107e00.pdf

FAO (2013b) Climate-smart agriculture. Source Book, Rome, FAO, http://www.fao.org/docrep/018/i3325e/i3325e00.htm

FAO (2021) FAO Strategy on Climate Change 2022–2031

Feinson S (2003) National Innovation Systems: Overview and Country Cases, Rockefeller Foundation, New York

Fouilleux E (2017) Sécurité Alimentaire, souverainité alimentaire et politiques publiques: de quoi parle-t-on? Cain

Freeman C (1987) Technology policy and economic performance: lessons from Japan. Pinter, London

Freeman C (1995) The 'National system of innovation' from a historical perspective. Camb J Econ 19(1):5–24

FRSIT/CRDI (2008) Compte rendu de l'Atelier de Lancement du Projet FRSIT/CRDI: "Analyse des Acteurs du System d'Innovation et Renforcement des liens entre les Acteurs au Service du Developpement Socio-Economique au Burkina Faso"

FRSIT/CRDI (2009) Analyse des Acteurs du System d'Innovation et Renforcement des liens entre les Acteurs au Service du Developpement Socio-Economique au Burkina Faso Atelier National sur l'Etat des Lieux de la situation des Acteurs du System d'Innovations, du 24 au 27 Juin 2009, Kombissiri, numero 1

Goto (2000) Godin, B. (2007) 'National innovation system: The system approach in historical perspective,' project on the history & sociology of STI statistics working paper, 36

Goto, Akira, (2000) Japan's National Innovation System: Current Status and Problems, Oxford Review of Economic Policy, Oxford University Press and Oxford Review of Economic Policy Limited, vol. 16(2), pages 103–113, Summer

IRD (2021) Contrat d'objectifs, de moyens et de performance 2021–2025. https://openknowledge.fao.org/handle/20.500.14283/cc2274en

Lalou R et al (2009) les sociétés rurales face aux changements climatiques et environnementaux en Afrique de l'Ouest, IRD editions

Lundvall, B.-Å. (1985) Product Innovation and User-Producer Interaction, Aalborg, Aalborg University Press

Lundvall B-A (1992) National systems of innovation: towards a theory of innovation and interactive learning. Pinter, London

Lundvall B-A (2007) national innovation systems-analytical concept and development tool. Ind Innov 14(1):95–119

National Climate-Smart Agriculture and Food Security Action Plan (2016–2020)

National REDD+ Strategy (NRS) 2016

Nations Unies (2009) Programme des Nations Unies pour le développement Rapport annuel 2009. OECD 1999

OECD (1997) National innovation systems. OECD Publication, Paris

OECD (1999) Managing National Innovation Systems, OECD Publishing, Paris, https://doi.org/10.1787/9789264189416-en

Thornton PK, Whitbread A, Baedeker T, Cairns J, Claessens L, Baethgen W, Bunn C, Friedmann M, Giller KE, Herrero M, Howden M, Kilcline K, Nangia V, Ramirez-Villegas J, Kumar S, West PC, Keating B (2018) A framework for priority-setting in climate smart agriculture research. Agricultural Systems, 167. https://doi.org/10.1016/j.agsy.2018.09.009

Van Wijk Ken E et al (2021) Small farms and development in sub-Saharan Africa: farming for food, for income or for lack of better options? Food Secur 13:1431–1454. https://doi.org/10.1007/s12571-021-01209-0

Wiederkehr C, Beckmann M, Hermans K (2018) Environmental change, adaptation strategies and the relevance of migration in Sub-Saharan drylands. In Environmental Research Letters 13(11). https://doi.org/10.1088/1748-9326/aae6de

World Bank Group. (2022) G5 Sahel Region Country Climate and Development Report. CCDR Series. © World Bank. https://hdl.handle.net/10986/37620 License: CC BY 3.0 IGO

Wreford, A., A. Ignaciuk and G. Gruère (2017), "Overcoming barriers to the adoption of climate-friendly practices in agriculture", OECD Food, Agriculture and Fisheries Papers, No. 101, OECD Publishing, Paris. https://doi.org/10.1787/97767de8-en

Normatif Documents

Action-sociale. https://www.action-sociale.gov.bf › ministere › organisa..

AN, Loi n°003-2011/AN du 05 avril 2011 portant Code forestier au Burkina Faso

AN, Loi n°006-2013/AN du 02 avril 2013 portant Code de l'environnement au Burkina-Faso

AN, Loi n° 008-2014/AN du 08 avril 2014 portant loi d'orientation sur le développement durable au Burkina Faso

AN, Loi n°034-2009/AN du 16 juin 2009 portant régime foncier rural

AN, Loi n°034-2012/AN du juillet 2012 portant réorganisation agraire et foncière

AN. 2018 *Loi n°034-2018/AN portant pilotage et gestion du développement*

Article 13 du DECRET N°2013-856/PRES/PM/MPFG du 03 octobre 2013 portant organisation du Ministère de la Promotion de la Femme et du Genre. JO N°05 DU 30 JANVIER 2014

BRIDGET BURNS, 2018. Guide de poche sur l'égalité de genre sous la CCNUCC P.12-14

Burkina Faso 1991. *Constitution*

DECRET N°2008-482/PRES/PM/MPF/MEF portant création, attributions et composition de la Commission nationale de suivi de la mise en œuvre des engagements du Burkina Faso en faveur de la femme. JO N°36 DU 04 SEPTEMBRE 2008

décret du 20 mai portant organisation du ministère de l'environnement, de l'économie verte et du changement climatique

Résolution n°002-2021/AN portant règlement de l'Assemblée Nationale P18

Chapter 28
The Potential and Policy Options for Mainstreaming Ecosystem-Based Adaptation in Africa's Drylands

Leonard Omullo Orondo

Abstract The drylands of Africa are fragile ecosystems with rampant land degradation and desertification arising primarily from unsustainable human land use activities, such as over-grazing and over-cutting of vegetation. Internal factors such as high poverty levels and poor natural resource governance and policies have fueled these stressors, enhancing their impacts. The impacts of climate change further aggravate this situation, increasing the vulnerability of ecosystems and their populations beyond traditional coping strategies. This situation calls for innovative adaptation strategies to improve the resilience of dryland ecosystems and populations. This chapter reviews the various adaptation options implemented for restoration programs in the African drylands to evaluate their successes and challenges. It finds that African governments have implemented continental strategies, such as the Great Green Wall (GGW), the AFRI100 Forest Landscape restoration program, and other national initiatives, with mixed results. It further finds that simple community-led restoration practices have better chances of success and should be up-scaled. It proposes policy reforms to anchor the mainstreaming of such proven processes. Most of the initiatives have adopted sectoral approaches, even though some can be said to be systemic. The chapter recommends the adoption of an ecosystem-based adaptation (EbA) as the most appropriate strategy for restoration in the drylands, due to its ability to holistically deliver ecosystem, community, and climate change resilience. EbA has various approaches that enhance adoption flexibility. It makes good governance, benefit sharing, and stakeholder participation central to restoration processes, thereby supporting sustainability.

Keywords Dryland restoration · EbA · Climate change resilience · Ecosystem services · Adaptive governance

---

L. O. Orondo (✉)
Landscapes Integrity Solutions (LIS), Nairobi, Kenya

Y. D. I. Khan et al. (eds.), *Land Restoration Through Ecosystem-Based Approach*,
Disaster Resilience and Green Growth, https://doi.org/10.1007/978-981-95-0389-6_28

Abbreviations

AFWC	African Forestry and Wildlife Commission
AR4	IPCC Fourth Assessment Report
IGAD	The Intergovernmental Authority on Development (IGAD)
IPCC	International Panel on Climate Change
UNFCCC	United Nations Framework Convention on Climate Change
DAP	Desertification Action Plan
EbA	Ecosystem-based adaptation
Eco-DRR	Ecosystem Disaster Risk Reduction
TWENDE	Towards Ending Drought Emergencies
CIFOR	Center for International Forestry Research
ICRAF	The International Council for Research in Agroforestry
NCCRP	National Climate Change Response Policy
CBD	Convention on Biodiversity
GBF	Global Biodiversity Framework
ASALS	Arid and semiarid areas
NBS	Nature-based solutions
UNEA	United Nations General Assembly
SDGs	Sustainable Development Goals
IUCN	International Union for the Conservation of Nature
NAP	National Action Plan
NBSAP	National Biodiversity Strategy and Action Plan
MEAS	Multilateral Environmental Agreements
NDCs	Nationally Determined Contributions
WWF-UK	World Wide Fund, UK
GBF	Global Biodiversity Framework
IPLC	Indigenous Peoples and Local Communities
GEF	Global Environmental Facility
WEF	World Economic Forum
LDN	Land Degradation Neutrality
FMNR	Farmer Managed Natural Regeneration

28.1 Introduction

Drylands are areas where productivity is limited by water availability (Leiter et al. 2019). The drylands can be categorized into four types, namely arid, semiarid, dry sub-humid, and hyper-arid lands and encompass a variety of ecosystems. The drylands in Africa occupy about 60% of its total surface area (Kolding et al. 2016). The drylands are less arid than deserts and are characterized by low and erratic precipitation, high temperatures and evapotranspiration. These zones have supported livelihoods and economic activities for centuries despite their ecological fragility. They are biophysically and socially diverse. Existing Biomes range from woodlands and savannas on either side of Central Africa's equatorial forests, to Mediterranean shrublands in the North and South, enclosing the hyper-arid Sahara in Northern Africa and the smaller Namib desert in the Southern part.

Many dryland inhabitants are pastoralists (sedentary or nomadic) or agro-pastoralists, combining livestock-rearing and crop production. They also use non-timber forest products such as honey, fodder, medicinal herbs, fuelwood, grass, and trees to support their livelihoods. The inhabitants of the African drylands have adapted to living in conditions of permanent water scarcity, variable rainfall, and the recurrent risks of weather-related shocks (Campbell 1977; IAASTD 2009). However, due to the changing socio-economic conditions such as high population growth, increased poverty, and political circumstances, their traditional coping strategies are becoming increasingly insufficient to deal with the emerging climatic risks and impacts. Incidents of unsustainable land management such as overgrazing, excessive fuelwood collection, over-cultivation, and poor irrigation technologies have led to increased land degradation and desertification. Climate change impacts, superimposed on the ongoing degradation processes and socioeconomic conditions, have led to the vulnerability of the African dryland populations and ecosystems. These processes have been reinforced by inadequate land governance structures and policies (UNEP 2006).

28.1.1 *Characteristics of the African Drylands*

African dryland ecosystems experience high levels of environmental degradation due to land conversion (to cropland), overgrazing, changing fire regimes, introduction of non-native plants, water misuse, overharvesting of wild species, and long-term impacts of climate change. These degradation effects impact ecosystem integrity and functions, compromising their ability to deliver ecosystem services and goods. These also make the impacts of climate change more severe. However, the African drylands communities have become resilient to these conditions by using their indigenous knowledge to adapt to the situations. Indigenous knowledge has been applied to support animal husbandry, biodiversity conservation, disaster early warning, and various livelihood activities in other sectors.

Drylands are generally environmentally fragile ecosystems, characterized by challenging agro-climatic conditions, and having low primary productivity, hence prone to degradation (Noble et al. 2014). The drylands are mainly vulnerable to food insecurity compared to other global drylands (IPCC 2018). Most of these drylands are in remote areas, hence politically and socially marginalized, making them prone to losing resources to influential individuals, further compounding the livelihood issue (Leiter et al. 2019). Most of those living in the African drylands are poor and lack basic amenities. The poverty and food insecurity make the households more vulnerable to climate change impacts, land degradation, desertification, and weak institutions.

Due to the fragility of these ecosystems, the risks of climate change to natural and managed ecosystems are expected to be higher in the drylands than in the humid areas (Leiter et al. 2019). Results show that temperatures in the drylands are increasing by more than 20% as compared to the humid areas (IPCC 2018) and that temperature and rainfall variability are increasing every year, making cropping, pastoralism, and water management even more challenging (Janetschek et al. 2020). Climate change impacts such as a rise in temperature will cause water scarcity, an increase in vector-borne diseases, reduced crop and livestock productivity, and damage to infrastructure in the drylands (McSweeney et al. 2012). The IPCC fourth assessment report (AR4) has concluded that Africa is at the highest risk from climate change impacts given the existing stresses in the Continent (IPCC 2007). It is conceivable, therefore, that in the coming years, most parts of dryland of Africa will witness high variability in rainfall patterns, with increased incidences of extreme events such as floods and droughts.

Climate change will affect the land use and livelihoods of the resident populations, becoming more intense with rising human populations. Climate change is one of the catalysts for ecosystem degradation in the drylands, causing problems to livelihoods and poor ecosystem health. In particular, these zones have heavily felt the impact of droughts and floods, with severe impacts on livestock, infrastructure and loss of life. Climate change and the fragility of the ecological conditions, significantly reduce the coping capacities of ecosystems and communities. Despite a lot of efforts, food insecurity and under-nourishment are on the rise in nearly all countries in sub-Saharan Africa. Consequently, any meaningful development in Africa must be integrated with climate adaptation for successful outcomes to be realized.

28.1.2 Climate Change and Vulnerability in the African Drylands

Dryland ecosystems are continually under stress from a multitude of factors, both internal and external. These include weather variability, recurrent and unpredictable droughts and floods, land degradation and desertification caused by unsustainable land use practices on a fragile resource base. These processes are fueled by local factors such as increased human population pressures and dynamics, poverty, high dependence on subsistence rain-fed agriculture, prevalence of HIV/AIDS, and civil conflicts, among others. These processes are further affected by other external factors, such as poor governance, policies, and ineffective land tenure systems (MA 2005). A common outcome of these threats is a situation of chronic food insecurity in the African drylands, evidenced by under-nourishment, stunting in children, and out-migrations. Most of those affected by food insecurity are the poor populations of the African drylands, whose livelihoods depend on local natural resources. Increased temperatures are expected to lead to water stress by causing soil moisture loss, further reducing rain-fed agricultural yields in the drylands. This will put additional pressure on the already existing food insecurity and undermine the livelihoods of the Pastoralist and agropastoralist inhabitants.

Dryland populations rely on ecosystem services for survival, such as flood control. These services are supported by nature, such as soils, biodiversity, water, and other natural resources, which are in turn threatened by several activities, such as development, urban expansion, and settlements. As a result, these ecosystems are becoming more degraded, unproductive, and unable to sustain the provision of services.

The arid populations have always adapted to these stressors. Still, the impacts of climate change have exacerbated the impacts of these stressors beyond the natural capacity of humans and ecosystems to cope. Consequently, many traditional adaptive knowledge and livelihood strategies used before can no longer hold (Boko et al. 2007). Therefore, attempts to reduce the vulnerabilities of the dryland populations to climate change impacts and other risks must boost their risk management strategies or introduce new ones tailored to their contexts. This will require the adoption of various adaptation strategies.

Several adaptation initiatives have been proposed to restore and sustainably develop the African drylands, with mixed results. Ecosystem restoration and adaptation are two intricately linked terms. Restoration seeks to provide a more accurate understanding of the original state of the ecosystem, and allows adaptation strategies to adopt more informed decisions. Climate adaptation on the other hand, involves the change of behaviours,systems,economies,our environment or lifestyles, in order to protect ourselves from the impacts of climate change. So restoration efforts feeds into adaptation processes,with Mitigation co-benefits. This chapter sets to document the restoration options implemented in the African drylands, their successes and challenges. It further seeks to conduct a comparative analysis of the various options in a bid to identify and recommend better ones. A final objective is to assess policy options for mainstreaming and upscaling proven adaptation strategies. The methodology used in this process employs a systematic review of the restoration and adaptation options that have been used in the African region since the 1970s.

The chapter presents the case of an ecosystem-based adaptation (EbA) as the most superior strategy that can deliver sustainable development in the African drylands by addressing issues of climate resilience, disaster risk reduction, and ecosystem and community resilience to minimize the vulnerability to the impacts of climate change and environmental degradation, and spur sustainable development of the drylands. It concludes by recommending policy options that can be used to mainstream EbA, the potential challenges, and the way forward. In the discussions on policy options, references are made to relevant international policy documents such as the UNCCD, for the simple reason that most domestic policy frameworks are aligned with them (International policies), because environmental issues are global. Additionally, doing so is an attractive strategy to access donor funding, which is a critical element in implementing restoration efforts. It also helps to achieve international collaboration and partnerrhips in addressing global environmental issues, such as climate change.

More examples are from the Sahel region and the Horn of Africa. However, they are realistically up-scalable to other drylands on the continent, with slight modifications, due to similarities in agro-ecological conditions.

28.1.3 The Principles and Concepts of Ecosystem-Based Adaptation (EbA)

An ecosystem-based adaptation (EbA) approach is one of the many aspects of nature-based solutions. It aims to holistically manage our ecosystems to promote conservation, restoration, improved livelihoods, and sustainable use of resources equitably. Central to the idea of ecosystem-based adaptation is the philosophy that we should work with nature to manage environmental challenges rather than against it. This means that we can deal with global challenges such as climate change through adaptation and mitigation strategies, environmental management, disaster risk reduction, while meeting the needs of the people at the same time, by adopting nature-based solutions. It aims to achieve climate resilience by improving community and ecosystem resilience. These are achieved through biodiversity enhancement, improved livelihoods and ecosystem restoration initiatives. The process employs various approaches to achieve each of these objectives. Good governance is a central to theme in the implementation of EbA, which is also multi-sectoral and multi-stakeholder focused.

The implementation of EbA processes is aimed at achieving multiple objectives at the Landscape and ecosystem levels and within associated Communities. The following multiple objectives can simulteneously be achieved by an Eba process.

(i) Improved livelihood.
(ii) Ecosystem and community resilience.
(iii) Climate resilience.
(iv) Disaster risk reduction.
(v) Achievements of international reporting mechanisms for conventions, treaties, and protocols.
(vi) Achievements of various Sustainable Development Goals (SDGs).

The process can employ various nature based tools to realize the above objectives. Some of the common approaches used in the adaptation processes are mentioned later in this chapter.

28.1.4 Some Factors That Influence Adaptation Options

The adaptation strategies will depend on the vulnerability of the population and the opportunities presented, according to each agro-ecological zone. Previous attempts to address the dryland development have focused on a top-down approach and large state-sponsored projects, with minimal involvement of the Communities, such as in mega-irrigation projects and Dams as seen in Kenya. Such programs have largely not been effective (ICSU 2017). Major UN reports show that community-based approaches work better (Wiggins and Keats 2013). Adapting to climate change should focus on improving the resilience of the productive systems in the drylands, such as pastoralism, agro-pastoralism, agro-forestry, and the associated socio-economic and agro-climatic factors. There are also opportunities to promote traditional agro-products and deliver these to the markets for income generation to further address the resilience of the people. Traditional and indigenous knowledge can also be exploited for conservation and other development purposes, such as in livestock, human health and disaster early warning. An elaborate assessment of opportunities for adaptation is a necessary step to profiling adaptation options. Such assessments may include the risks and vulnerabilities occasioned by climatic hazards to the ecosystems and populations. Other internal factors such as policies and governance issues may also pose risks and vulnerabilities.

Generally, the potential for the sustainable development of the drylands will be based on its local natural resources ranging from solar energy, minerals, water resources, wildlife, a rich cultural heritage, biodiversity, and people, including women, youth, and children. It has now been shown that livelihoods based on natural capital can deliver significant economic gains to the drylands and reduce poverty. However, climate change and environmental degradation continue to pose major challenges to the development of these areas; hence, Climate change adaptation is very critical to the development of the drylands. Finally, the funding, cost-effectiveness, and success of the restoration strategies are important.

28.1.5 Current Status of Ecosystem-Based Adaptations and Similar Initiatives in Africa's Drylands

There are various adaptation options that have been used in the drylands of Africa to respond to the impacts of climate change. Each dryland zone may have its own specific options that can be applicable, even though some options cut across all the dryland zones. The socio-economic and agro-climatic characteristics of the zones are important in determining which adaptation options are applicable. Such options include:

(i) Natural resource-based options: These are based on the sustainable management of land, water, soil, and animal resources.
(ii) Market-based options: These focus on improving market access, improving incomes, and reducing population vulnerability. Carbon credits and other recent nature based financial mechanism also fall under this category.
(iii) Institutional options: These aim at improving local level changes such as micro-credit, awareness creation, and nature-based businesses. Institutional changes may also capture policy and governance reforms.

The ecosystem based approach is designed to be a holistic approach that captures all of the above parameters. The following paragraphs give an elaborate view of the various restoration initiatives that have been carried out in various parts of Africa, particularly the Sahel and Horn of Africa. It comprises a few examples which can meet some of the requirements of EbA, and others that may not be so integrated. Some initiatives have been made at a continental level, while others are country-focused such as by the government, farmers, communities, producer organizations, and the private sector. The continental level interventions are those that Africa has agreed with governments, donors, and other partners to implement. They involve a number of countries in the drylands. These include the following.

28.1.5.1 The Great Green Wall Initiative

This initiative was launched by the African Union (AU) in 2007 as a buffer to manage land degradation in the Sahel and Horn of Africa. The AU signed agreements with 11 countries in 2020 to combat climate change and desertification across the Sahel and Horn of Africa. The idea was to plant a belt of trees 15 km wide and 7000 km long, from Senegal to Djibouti, in areas that receive 100 mm to 400 mm mean annual rainfall. But lately, the area of intervention in the 11 countries was increased to about 156.1 million hectares (UNCCD 2016). The goal of the GGR has also shifted from planting trees to supporting an integrated ecosystem management approach in the targeted areas. This would include a mosaic of various land uses and production systems that incorporate sustainable dryland management and restoration, regeneration of natural vegetation, and associated soil and water retention and conservation measures. UNCCD reports that by early 2020, a total of 17.8Mha had been rehabilitated within member states.

The initiative aims to restore biodiversity loss and desertification arising from the encroachment of the Sahara desert in the Sahel by creating a living barrier of trees and vegetation, restoring degraded land along the 8000 Km stretch green wall, enhances food security and creates jobs, and transforms lives in 11 countries from Djibouti in the East to Senegal in the West (Fig. 28.1). The GGWI contributes to the implementation of the Rio conventions while ensuring alignment with the African Union's Agenda 2063, the 2030 Sustainable Development Goals and the United Nations Decade on Ecosystem Restoration (2021–2030). The Great Green Wall Status Report, UNCCD (2016), shows that between 2007 and 2018, 20 million hectares of land were restored, and 350,000 jobs were created through the diversification of economic activities and income regeneration, thus improving the quality of life for millions of people in the Sahel. Revenues from income-generating activities amounted to $90 million across all 11 countries: Burkina Faso, Chad, Djibouti, Eritrea, Ethiopia, Mali, Mauritania, Niger, Nigeria, Senegal, and Sudan.

This is truly an ecosystem-based adaptation approach because the program embraces various approaches and delivers multiple objectives to the ecosystems and communities. With improved funding and upscaling of lessons learnt, the program will have major restoration impacts.

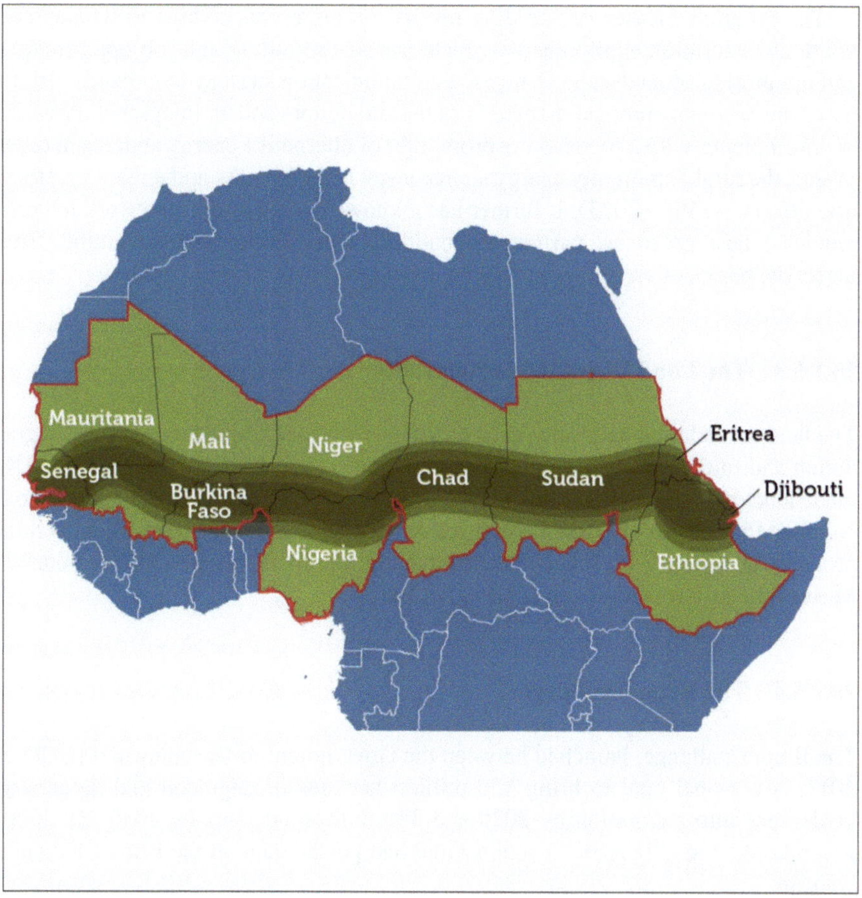

Fig. 28.1 Nairobi Express Highway,Kenya: Underneath a large concrete overpass, multiple support columns are covered withvertical gardens of lush green plants. The scene includes a road with cars and a motorbike passing by. In the background,there are buildings and trees, suggesting an urban setting. The sky is overcast, indicating a cloudy day

28.1.5.2 The African Forest Landscape Restoration Initiative

Ten (10) African countries launched the African Forest Landscape Restoration Initiative (AFR100) in 2015 to restore 100 million ha by 2030. AFRI100 contributes to the Bonn Challenge, the African Agenda 2063 and to the Sustainable Development Goals. In its early years, the AFRI100 has focused on mobilizing partners, developing capacity, piloting projects, and creating an international standing required for its success. The IUCN has supported the program in capacity development through its land restoration accelerator program. A monitoring framework for progress in the implementation of AFRI100 was created but has not given its report due to several factors, like the Covid-19 epidemic.

The program focuses on building restoration entrepreneurships to women and youth as restoration entrepreneurs, create access to funding and job opportunities and monitoring of landscape changes; a communication strategy to enhance visibility of the program amongst member's states and donors and an integrated approach to development with a focus to the promotion of alternative energy sources in order to shift the rural community energy source away from biomass and enhance restoration efforts, AFWC (2023). It further has a knowledge sharing framework to communicate best practices, partnership building, and gender mainstreaming. This meets the basics of an ecosystem-based approach.

28.1.5.3 The Land Degradation Neutrality

The land degradation neutrality aims to achieve a balance between new land degradation and future efforts to improve degradation (UNCCD 2016). In 2015, the UN made land degradation neutrality (LDN), a target of Sustainable Development Goal Number 15 (combat degradation and restore degraded land and soil). There is little progress in Africa, with few Countries making reports to the UNCCD secretariat. Most of the progress has been achieved in Europe.

28.1.5.4 The Bonn Challenge

The Bonn Challenge, launched between the Government of Germany and IUCN in 2011, is a global goal to bring 150 million hectares of degraded and deforested landscapes into restoration by 2020 and 350 million hectares by 2030. By 2020, Rwanda was the only African country that had posted data on the Bonn Challenge website.

The Food and Agricultural organization (FAO) has also launched a program called "Action against desertification." If the other similar initiatives have not had much progress in Africa, their fate is likely similar.

28.1.5.5 Farmer Land Community Managed Restorations (FLCMR)

We have several examples of successful farmer restoration efforts across the Sahel and Ethiopia. Thousands of Niger farmers have protected the woody species that regenerate naturally on their farmlands covering over 5 Million hectares in Africa, making the largest restorative transformation in Africa.

Hundreds of Villages in Senegal's Peanut basin now have more trees than several decades ago. In two villages surveyed, protection and resulting regeneration of community grazing lands have transformed them from degraded shrub to dense wooded savanna with 61% tree cover (Tappan et al. 2020). The lesson here is that locally managed initiatives are very successful.

In Senegal, the World Vision reported that many farmers had been trained on farmer-managed natural regeneration (FMNR). Between 2015 and 2020, the area under FMNR in Kaffrine region (of Senegal) alone increased by almost one-third to 85,000 hectares, as farmers saw the benefits of these practices and were further supported by training and extension services.

Farmer-managed natural regenerations are cheaper to implement than planted trees, and the success rates are also higher. Most of the trees regenerated were also multipurpose, serving various purposes and improving resilience. Trees also improve soil fertility and crop yields. In Mali, landscapes which were treeless in the early 1980s now have a good cover of trees extending thousands of hectares (Allen et al. 2020). The planting of trees was limited by scarcity of water, but farmers who adopted FMNR have a better success rate. Tree cover improved soil fertility, provided fuelwood and fodder for animals, and the millet yield improved on the farms. The trees also reduced the time used by girls and women to look for firewood since they could now use pruned branches.

Similar stories were told in Bukina Faso (Belemviré et al. 2020) and Niger (Hassane and Reij 2020), both of which had faced droughts and food shortages since the 1970s. The improvements in water harvesting techniques, e.g. traditional planting pits (Zai), half-moons, and contour stone bunds, recharged groundwater levels and improved land productivity. Both regions are now green, and the ground wells have water, and food productivity has increased. The farmers have been able to sustain these land management techniques beyond the project cycle, which ended in 1995. This is evidence of ownership, transferred capacity, and ability to sustain the program due to its cost-effectiveness.

Farmers in the Sahel have had to cope with drought and famines. In recent decades, they have also had to cope with extreme weather events. Farmers in Niger have been able to adapt to climate change using a set of simple techniques such as water harvesting, FMNR, mulching, use of early maturing varieties, crop rotation, and developing their own climate smart agriculture. The simple water harvesting techniques and the FMNR introduced to farmers in the 80 s and 90 s, have enabled the farmers to build resilience to climate change and increase food productivity, thereby alleviating poverty and improving incomes through the sale of foodstuffs.

As a result of these innovations, farmers noticed a reduction in conflicts over natural resources and the development of local governance institutions that are effective and respected. In Senegal, researchers have also developed a "climate-smart village" approach. It is based on strengthening local governance of natural resources, in combination with promotion of agroforestry, planting fruit and fodder species, FMNR, and management of inter-village pastoral grazing areas, among other practices of conservation (Sanogo et al. 2020).

The area around Northern Ethiopia represents a large-scale transformation into a green area by trees (Nyssen et al. 2009). The previously degraded watersheds of Abreha We Atsbeha have largely been restored using area enclosures, restoring 1500 ha. Other methods used to restore the watershed include terracing, natural regeneration, and contour bunds. Farmlands have also been restored through the development of check dams and re-greening, which increased the irrigable area of the watersheds, resulting in more food production.

Several rangeland rehabilitation techniques have also been used in Ethiopia, such as enclosures, rotational grazing, reseeding, and building of bunds. These techniques improve rangeland resilience and increase the productivity of the pastoral systems.

Several of the techniques applied by farmers and communities fall within the realm of ecosystem-based adaptation, and deliver multiple benefits to the landscapes and communities.

28.1.5.6 The TWENDE Project in Kenya

The TWENDE (Towards Ending Drought Emergencies) project in Kenya targets 11 counties in two major climate zones in the drylands, which have devolved governance structures under Kenya's new constitution. Building capacity and institutions for the improved implementation of devolution was seen as necessary to enhance the climate resilience in the arid and semiarid counties (*Garissa, Tana River, Isiolo, Marsabit, Samburu, Kajiado, and Taita–Taveta*). The restoration interventions focus on increasing the adaptive capacities of Communities and local institutions to develop evidence-based landscape planning and management. This will be done by increasing accessibility to climate data and information and enhancing the ability of community-based cottage industries to access markets and financial services, thereby improving their resilience.

The TWENDE (Towards Ending Drought Emergencies) project in Kenya's drylands has used an ecosystem-based adaptation in the arid and semiarid Rangelands (ASALS) to reduce the cost of climate change induced drought on the National economy, by increasing resilience of the livestock and other land use sectors through restored and effectively governed rangeland ecosystems and resources.

28.1.5.7 The Private Sector Initiatives

Progress in restoration in the Sahel shows that farmers and community-led initiatives are the main drivers of restoration at scale, at minimal costs. The report notes that large-scale projects have played a role, but that the contribution of the private sector has been largely limited.

However, the informal groups and organizations, largely called producer organizations, can have very significant impacts on restoration efforts. They are usually a good entry point for restoration efforts. However, producer organizations often face challenges such as limited technical knowledge, unfavorable tenure arrangements, limited involvement in landscape planning and decision making, and lack of financial support. These groups have significant local knowledge and should be included in restoration planning and implementation. The exploitation of Gum Arabic by small-holder farmers in Sudan provides small-holder farmers with a significant amount of income. The structuring of the gum Arabic sector pilot project in North Kordofan State (2014–2018), introduced open systems that paid producers competitive prices (Elhassan et al., 2020). This gave rise to the development of a new idea of using Gums for adaptation and mitigation in Sudan. This provides a good example on using nature based businesses to support conservation and improve biodiversity and resilience.

The Sawlog timber production scheme in Uganda is another of the private sector-led initiative projects supporting afforestation in the drylands. It is currently involved in the afforestation program in *Karamoja*, the driest part of the country (400–700 mm mean annual rainfall), through the development of demonstration sites and training courses. It supports small-holder farmers to produce timber through the provision of seedlings and financial support. The success of this program was based on convincing communities of the economic benefits of the program. Giving financial initiatives to farmers through soft loans was another trigger.

28.2 Other Cross-Cutting Factors

Success for restoration in African drylands is enhanced when communities, Governments, local institutions work together in planning and implementing restoration efforts, applying existing policies and bylaws (Haile and Gebregziabher 2020). In Burkina Faso and Niger, 17 local conventions at the commune (rural district) level now support sustainable landscape use, strengthen responsive decentralized governance, and reinforced rights and responsibilities (Johnson and Steinberg 2020). This has led to better resource management, improved FMNR and reduced conflicts, reduced degradation, and improved climate resilience.

In Ghana, it has been shown that achieving equitable resource governance in FMNR requires a shared vision for the restored landscapes, as communities and farmers are able to see mutual benefits (Gebrekirstos et al. 2020). This has also been applied to community-managed forests and pastures. The need for all land users, including pastoralists, to participate in decision-making, has also been recognized. For all these to succeed, there is a need to strengthen community-led resource governance. Inter-community and cross-jurisdictional collaborations, support from Governments, NGOs, and traditional institutions, like the traditional chiefs holding land custody and registries in Ghana are crucial for success.

The involvement of women, the youths, indigenous communities, and other marginalized groups in the community is very important to the success of regeneration processes. For instance, the youth and women can benefit from the reduced time to fetch fuelwood as these are available from the tree branches that have been pruned (Reij et al. 2009), as well as the availability of water due to improved groundwater recharge (Hassane and Reij 2020; Belemviré et al. 2020) as a result of the regenerated trees. Gender is important in restoration efforts because it determines who does what. Usually, Women are also constrained in implementing restoration practices as they do not have the same resource rights and resources as men due to deep-rooted gender norms. Improvement in land tenure security and economic empowerment will therefore enhance the role of women in restoration programs. In Ethiopia, the youth involved in restoration efforts now own tree nurseries and are able to lease and manage nurseries and plantation sites, giving them control over the forests they are planting. It is clear that incentives and benefits from restoration activities enhance participation and outcomes.

28.2.1 Land Tenure Rights in the Drylands

It is widely held by Western scholars that natural resources are more secure under private land ownership. However, experience from the African continent has shown widespread failures in the registration and succession processes of private land, largely due to corruption and inefficient land administration systems. In the case of Kenya, land administration before the 2010 Constitution was affected by many land laws and a lack of a comprehensive land policy, which compromised administration and management, leading to inefficiencies and degradation (Orondo L 2020). Land tenure influences land use resource use and management. A land owner's risk perception of investments in land depends on how long he can hold on to the land, his freedom to use the land, and whether he can sell it or not.

It is commonly assumed that having a private Land title will result in access to bank credits, and an improvement of livelihoods and development of the holder. However, a study by the World Bank in Kenya failed to demonstrate this (Bruce and Migot-Adholla 1993). The study concluded that the transformation of communal land tenure in the drylands would not necessarily lead to better development.

In Kenya, pastoralists move cattle enclosures from one place to another, and there are clear advantages to hold land individually or in small groups, as opposed to being a user of trust land. For the group ranches (under trust tenure), the movement of enclosures created a demand for individual titles. In Kenya, some group ranches agreed on how to use different portions of the ranches (land use planning). The trust land tenure system proved to be an obstacle toward the movement of enclosures, as they were constructed without the consideration of any legal land rights. The County Councils then held the legal rights for all members of a community, leading to poor governance. An individual could therefore not prevent the rest from using it. Further, as members of the council do not use the local land, they had little understanding of the responsibility of holding land in trust for local residents. The trust land in most cases in Kenya was therefore converted into group ranches, in order to make the local communities more responsive to the use and management of land and local resources.

Kenya finally adopted a mix of private land ownership and Communal land tenure in its dryland areas. Certainly, these present two different dimensions to environmental and natural resource management and ecosystem sustainability. Most of the group ranches have today folded up, a number are sub-dividing their lands into individual parcels or groups. But this has not helped slow down environmental degradation in the drylands, suggesting that private land tenure may not necessarily result in better environmental and resource management, even though it may enhance the productivity of the land.

Under the tragedy of the Commons doctrine, it was assumed that a lack of tenure rights contributes to environmental degradation. The situation in Kenya seems neutral between private and communal ownership. An exception to this may be territories such as forests, still under the management of the indigenous communities. Kenya has since developed a Community Land Act of 2017, which recognizes customary tenure rights and gives the communities a central role in the management of community land, through the Community Land Management Committees. The Governor of the County where a community is located is given the mandate of forwarding the names of members to the community land management committees, for purposes of registration in the National Land Registry at the Ministry of Land. But experience shows that a number of governors have forwarded their own lists, not approved by community assemblies, to the registry (pers.comm.)! This is evidence that decentralized land management governance does not necessarily improve land governance, and more checks and balances are still needed at the grassroots. However, when well-managed, the community land management committees stand a better chance to properly administer and manage community lands in Kenya, which are mostly in the drylands of Kenya. The Community Land Management Committee could be a key institution for the entry of EbA-focused restoration programs. Having legal rights over resources will motivate the local communities to take an interest in local initiatives such as EbA, which is crucial for better governance and sustainability.

28.3 Ecosystem-Based Restoration for the Drylands of Africa

Ecosystem-based adaptation also includes reseeding of the rangelands or the use of drought-resistant species. The Center for International Forestry Research and World Agroforestry (CIFOR-*ICRAF*) has used a drought-resistant grass species (*Sporobolus species*) that can be used during severe droughts as a restoration strategy for Rangeland Management of livestock to address the shortage of forage during severe droughts.

An ecosystem-based approach to restoration is a recent strategy still under discussion and refinement globally, even though its benefits have been proven. It employs nature to enhance biodiversity and improve climate resilience while supporting livelihoods too. It is an elaboration of a broad strategy called "Nature Based Solutions (NBS)." The United Nations Environment Assembly (UNEA) delivered a multilaterally agreed-upon definition of NBS at its meeting in Nairobi in March, 2022, which states as follows:

"Nature-based solutions are actions to protect, conserve, restore, sustainably use and manage natural or modified terrestrial, freshwater, coastal and marine ecosystems which address social, economic and environmental challenges effectively and adaptively, while simultaneously providing human well-being, ecosystem services, resilience and biodiversity benefits" UNEA (2022).

Keywords from the strategy are to "protect conserve restore and sustainably manage," "social economic and environmental challenges," "adaptively," "livelihoods" "benefits," "resilient." It also means the approach can be applied to a natural or modified ecosystem achieving all these targets will require a multidimensional approach.

Several strategies or tools have been proposed for the implementation of an ecosystem-based approach, which also includes disaster risk reduction strategies (Eco-DRR). These include:

(i) Integrated water resource management.
(ii) Sustainable land management (SLM).
(iii) Protected areas management.
(iv) Spatial urban/rural planning and management.
(v) Integrated coastal zone management.
(vi) Sustainable natural resources management.

Essentially, these tools are an integration of the earlier sector-based approached, but have multiple objectives and benefits, hence more impactive. The exact mix of the strategies applied will depend on a particular landscape context, and should be targeted to particular restorations. In each EbA process, a disaster risk reduction strategy is built in each tool, in order to minimize the impacts caused by climate related hazards such as droughts,fire or floods. This approach enables the desired multiple benefits to accrue to the ecosystems and communities.

For dryland restoration, the ecosystems are already degraded, and restoration should therefore focus on restoring ecosystem integrity and functions and mitigating the impacts of climate change. To achieve this in practice, the nature-based solutions for the African drylands employ a combination of ecosystem-based adaptation (EbA) strategies and disaster risk reduction strategies simultaneously, achieving the objectives of improving climate resilience and reducing disaster risks (Eco-DRR) from climatic factors. This approach has proven to have multiple benefits such as enhancement of biodiversity, improvement of climate resilience, enhancing ecosystem function and provision of goods, services, and livelihoods. However, the NBS should be designed and implemented across the various sectors, involving all stakeholders and professionals, defining clear participatory governance structures and strategies for scaling up, for it to maximize impact. The governance structure should be devolved (see discussion under governance). Nature-based solutions also support many Sustainable Development Goals (SDGs) because NBS actions normally address multiple objectives and needs spanning ecological, economic, and social spheres, within the context of multidisciplinary processes.

28.4 Design, Planning, and Implementation of Ecosystem-Based Adaptation

The design and implementation of NBS is critical to the delivery of targeted objectives and the realization of desired outcomes at the ecosystem and community level. While this section does not intend to go into the details on the design and implementation of ecosystem-based adaptation initiatives, it gives an overview of some of the approaches. Several conservation bodies, conservation policy organs, and implementers have developed robust guidelines or Standards that can guide the design of ecosystem restoration actions. The IUCN Global standards for NBS, Conservation International Guidelines for Design and Implementation of NBS are examples. We also have best practice compilations such as those found in PANORAMA-Solutions for a Healthy Planet, which contains best practices for different ecosystems and regions globally. However, we still have no unanimity on a common standard that should be used, and this is an ongoing discussion.

In this chapter, I highlight important considerations for the design of NBS according to the IUCN standard for NBS implementation . The Standard enables the translation of NBS concepts into targeted actions for implementation. This chapter only highlights the criteria for design and implementation. I believe a discussion of restoration actions based on EbA would be inadequate without this.

(i) **Criteria 1**

Identifies the societal challenges to be addressed by the NBS.

Currently these include climate change (adaptation and mitigation), disaster risk reduction, ecosystem degradation and biodiversity loss, food security, human health, social and economic development, and water security.

One or more challenges can be the entry point for NBS, with the focus being the potential to deliver multiple benefits.

(ii) **Criteria 2**

Manage the NBS against established baselines, in view of the dynamic nature of the systems that influence the solutions.

(iii) **Criteria 3, 4, 5**

These correspond to the three pillars of Sustainable Development Goals: environmentally sustainable, socially equitable, and economically viable.

(iv) **Criteria 6**

Addresses the balancing of trade-offs and choices that need to be made to achieve short and long-term gains, and how to ensure that there is a transparent, equitable and inclusive process to determine such trade-offs.

(v) **Adaptive Management**

This criterion addresses the need for adaptive management, to continuously learn about system-wide processes and adapt the NBS according to such changes.

A careful consideration of these criteria, using any guidelines may help to deliver NBS actions. It has to be emphasized that the outcomes of an NBS will vary from ecosystem type, location, and even region, reflecting varying contexts and circumstances.

28.4.1 Challenges and Opportunities

Ecosystem-based approaches, being multidisciplinary and cross-sectoral processes, face several challenges. Some of these challenges are used for adaptive management of the processes. Very intractable issues should be a subject of administrative, policy, and legal development or reforms.

EbA processes present opportunities of being used to assist a country meet its international reporting requirements' to conventions, treaties, and protocols and the Sustainable development Goals. This should be appropriately tapped, like in improving country-level international reporting on the same. The lessons learnt in the process of implementation can also be used as synergies and linkages to improve process implementation and outcomes, besides other National development initiatives.

28.5 The Policy, Legal, and Institutional Frameworks for Ecosystem-Based Adaptation in the Drylands

Nature-based adaptations are cross-sectoral, multidisciplinary and involve the participation of various stakeholders. It is almost impossible to manage such a diverse pool of stakeholders, with different governance structures, unless there are policies and laws to support coordination and collaborations. Policies are important for giving directions and regulating how different players can interact at an operational level, including budgetary issues. Further, domestic policies in the environment and natural resources sectors are dependent on global policy frameworks, and therefore, there has been reference to the latter. This linkage should be exploited in developing a policy for Eba processes.

In the environment and natural resources sectors, policy development has tended to take a sectoral dimension in most countries in Africa. This is also the case for Kenya. In themselves, these policies have overlaps, at times conflicting, making mainstreaming a challenge. This is not surprising as there is a suite of land-based natural resources such as water, minerals, wildlife, forests, and land which are siloed in different Ministries. For instance, climate change itself is an environmental resource, yet in circumstances, may be a separate Ministry from Environment! This architecture has resulted in cross-sectoral overlaps not being explicitly addressed at the policy level. Instead, this is left to be dealt with administratively, at least in the case of Kenya, which never happens in practice. The latter has caused conflicts in the administration and management of these resources, affecting projects in the sector such as ecosystem restoration and adaptation projects. Cross-sectoral projects have been the most vulnerable to this gap.

Most countries, particularly those signatory to the UNCCD, have used this global policy framework to develop actions targeted at managing land degradation at the national level. But most of these programs have suffered from shortages of donor funds and poor Coordination across the different sectors. As a consequence, there have been little achievements in reversing land degradation in most African Countries. Uganda and Tanzania went around this challenge by just identifying climate-relevant projects and coordinating already existing programs, instead of looking for funds for new programs.

Several factors pose challenges to the restoration efforts in the African drylands. Lack of land and resource rights, lack of spatial information on degraded areas, emergence of invasive species, such as *Prosopis juliflora*, and poor awareness creation are some of these factors. Other factors include land use activities such as overgrazing, overcutting of bushlands for multiple purposes, conversion of communal rangeland into private lands, poor irrigation-leading to salinity, Mining, insecurity-leading to reduced grazing corridors. A number of efforts have been made by countries in the region to address some of these, like the IGAD Transhumance pastoralist protocol aimed at securing pastoral routes for Livestock. The protocol allows free movement of livestock across borders, and farmers are issued with passports to facilitate seamless movements. In the Western region, ECOWAS has also arranged for their pastoralists to be given certificates of movement to facilitate free cross-border movements of livestock. These are necessary adaptation measures, as they allow for natural pasture regeneration, besides spreading risks to livestock loss due to droughts or disease outbreaks

Inadequate land and land use policies in most countries in Africa are a setback to sustainable land use, natural resource management and restoration actions. Without tenure rights, like in the case of women and youth, the latter are not incentivized to participate in restoration activities. These policies are also critical for mainstreaming actions at the sectoral level. An overarching restoration policy in the environment and natural resources sectors, which cuts across would be ideal for this purpose. Fundamentally, such a policy should address issues of stakeholder coordination and funding, besides other critical aspects of EbA.

Several African Countries have developed policies to address land degradation and environmental conservation. But implementation and progress are slow, as evidenced by reports of progress in actions discussed earlier in this chapter.

28.5.1 Mainstreaming Ecosystem-Based Approaches into Policies and Plans

Due to the absence of comprehensive restoration policies that articulate how cross-sectoral implementation will be managed; most restoration efforts in the African drylands have been fragmented and poorly coordinated, leading to little impact at scale. It is in this context that ecosystem-based approaches (EbA) have been proposed. EbA has traditionally not been included in most policies. Some earlier policy documents such as the CBD at the global level and NBSAPs at the National levels, recommended "the ecosystem approach" for biodiversity conservation. But even then, the understanding was that of biodiversity management from an "ecosystem focus," rather than using an ecosystem-based approach to manage biodiversity. The concept and principles of EbA were also not fully embraced and applied.

Even though mainstreaming of nature-based solutions has not been popular before, recent interests by various policy bodies on nature-based solutions have seen the integration of NBS into policy documents at the national, regional, and global levels. Recently, nature-based solutions have received recognition at many International policy platforms on the environment and natural resources as an approach that can be employed to support sustainable development, e.g., UNFCCC, CBD, NBSAPS, etc. A number of G7 and G20 leaders have pledged enhanced financial contributions to promote NBS with environmental and social safeguards. The UNCCD decision 15/8 of 2022 implores Parties to determine synergies in MEAS in the implementation of sustainable land management, ecosystem-based approaches or nature-based solutions." The mainstreaming of EbA policies at the global level is significant because most countries have aligned their programs to global policies through multilateral agreements (MEAs) or conventions. There are two fundamental motivations for this: one is the professional and technical guidance this provides, and the other is the potential funding opportunities provided through associated funding instruments such as the Global Environmental Facility (GEF), the Adaptation Funds, and the Green Climate Fund. Mainstreaming at the global level therefore, gives countries a leverage to mainstream ecosystem based approaches (EbAs) into their domestic policy documents. African countries should mainstream EbA into their policy documents. This should entail provisions on how ecosystem-based adaptations should be managed at the national and local level, including required coordination, funding, institutions, and legislations.

However, NBS have recently found recognition in many country national policies, strategies, and actions such as the NDCs and NAPs. According to a 2021 review of updated NDCs by WWF-UK9, 105 out of 114 (92%) of NDCs include explicit references to nature and nature-based approaches, and of these, many make specific mention of "nature-based solutions. Several targets of the Kunming-Montreal Global Biodiversity Framework (GBF) explicitly specify inclusion of nature-based solutions in conservation approaches (e.g., targets 2, 8, 11). Similarly, NBS have also been recognized at regional levels. For instance, the European Union's (EU) Green Deal emphasizes the use of NBS on climate adaptations and mitigations.

The Indigenous Peoples and Local Communities (IPLC) through their indigenous knowledge systems and practices have a role to play in EbA mainstreaming. An EbA policy framework should provide for the recognition and use of Indigenous Knowledge in addition to the accommodation of fundamental human and resource rights of the IPLCs and other categories of stakeholders, particularly those directly and indirectly impacted by the ecosystem-based processes. As far as possible, the indigenous knowledge should be integrated with modern science. This means the policy has to recognize the critical role of stakeholder engagement in EbA actions. The policy framework should equally speak to the design, planning, implementation, and scaling of ecosystem-based adaptation processes to ensure that no critical or vulnerable stakeholder's interests are ignored, and that equity is achieved in the accrued benefit-sharing mechanism. Inclusive stakeholder engagements are helpful in identifying synergies and linkages among the various actions for project sustainability. Stakeholder consultations should also be used to support progressive project design and adaptive management.

The NBS processes should be mainstreamed across all sectors beyond the traditional dockets of Environment and Climate Change, e.g., to finance, budgeting, planning and development, procurement, business, etc. Mainstreaming should be done through policies, strategies, plans, and budgetary processes. Mainstreaming will enhance cross-sectoral and multidisciplinary participation by various players and assist in the coordination and realization of impacts. A comprehensive EbA inclusive policy should address the issue of leadership and coordination of EbA processes as well as the need to identify a business case for nature-based solutions in a bid to incentivize the private sector to invest in EbA processes. NBS should be integrated into sectoral planning by assigning it to a high-level governance structure capable of delivering an inter-ministerial and inter-sectoral governance and decision support. This is how the climate change policy has been managed in Kenya, where the president is the chairman of the Climate Change Council, and the vice-president is the deputy chair. Such a high-level governance structure allows quick mobilization of funds and actions within the government. Most countries have inter-ministerial teams which are administratively formed at top management level, but mostly end up being ineffective due to lack of dedicated budgets to operate. There is therefore need to link EbA projects at planning, budgeting and implementation levels.

An NBS inclusive policy should encourage the use of best practices and case studies such as found in compilations like "PANORAMA-Solutions for a heathy Planet," and globally agreed guidelines and standards such as the IUCN Global Standards for nature-based solutions. There is also need for more discussions and consensus-building on the practices and concepts of NBS at all levels to inform policy. This will allow issues like monitoring and management of multiple objectives to be addressed, and agreed on.

In the application of ecosystem-based adaptation at a local or regional level, the National policies should be cascaded accordingly, with regional coordination institutions for maximum impact at scale.

Finally, an EbA inclusive policy should address critical challenges that arise from implementation such as governance and benefit sharing concerns.

28.5.2 Private Sector and Financing NBS

Ecosystem-based adaptations require heavy financial outlays because they quite often mix with grey infrastructure. Such heavy financing is usually beyond the capacities of most public sector funding. The outcomes also take long time and are quite often difficult to quantify for a business case. This makes them less attractive for the private capital. That's also why adaptation, in general, has attracted less private finance than mitigation in climate change financing globally.

Therefore, policies that support private sector investments in nature-based solutions such as EbA, should be developed, in consultation with policy and decision makers, planners, financial institutions, and other relevant players. Recently, the African Developments Bank's African Accelerated Adaptation Program (AAAP) has been launched to address adaptation issues in Africa. It seeks to scale up new technologies, designs, and nature-based solutions to adapt urban and rural infrastructure to Africa's current and future climate, as one of its pillars. This project seeks to mobilize substantial funds to climate-proof such projects and also improve the climate resilience of communities. For instance, the Nairobi Express highway has a green short shrub infrustructure(with live shrubs on highway pillars) in its design in a bid to integrate climate resilience into its design (Fig. 28.2).

The African Adaptation Accelerator Program (AAAP) is a joint program between the African Development Bank (ADP) and the Global Centre for Adaptation (GCA). They aim to mobilize $ 25 billion for adaptation in Africa by 2025 in four pillars: (i) food security, (ii) resilient infrastructure, (iii) youth innovation and job creation, and (iv) innovative finance. To date, AAAP has supported 42 projects, including 20 investment operations and 22 technical assistance projects in 36 countries and has garnered support from over 30 African heads of state and other global leaders, recognizing it as a key vehicle to operationalizing the Africa Adaptation Initiative's mandate (ADB website).

Policy makers should have discussions with the private sector umbrella organizations to come up with workable incentives for the sector to engage in investments in adaptation. The government can also dedicate capital for ecosystem-based adaptation through public–private partnerships (PPPs). Some donors are already allocating

Fig. 28.2 The new Nairobi express highway with live short shrubs on underpass columns that contribute to climate resilience

funds for nature-based solutions. For instance, the IUCN Nature+ accelerator Fund, with initial support from the Global Environmental Facility (GEF), is solely dedicated to the funding of NBS restoration projects.

28.6 Governance and Stakeholder Issues in Ecosystem-Based Approaches

.Governance in ecosystem-based adaptation relates to how the Government interacts with other stakeholders and citizens in project implementation, how decisions are taken on processes, how responsibilities are allocated and benefits shared. Studies of adaptation projects have shown that a major barrier to mainstreaming adaptation actions relates to governance of such projects. The governance requirements for EbA projects are multifaceted and challenging. But governance is underpinned by policies, laws, and institutions, hence the need for mainstreaming EbA into National policies. Weak or a lack of policies will result in poor governance of EbA processes.

The UNFCCC and the Paris Agreements (Article 7) recommend that adaptation should be made a fully participatory and transparent approach, taking into consideration vulnerable communities and ecosystems. Notwithstanding this, the government still remains a dominant player in adaptation projects, partly due to the huge costs involved and the sector not being attractive to private sector investments. Poor governance concerns on such processes do not always deliver equity to communities and stability to ecosystems, necessitating the need for more attention.

EbA is a people-centric approach to adapt to climate change using nature. Climate change is one of the greatest risks for economic development, WEF, (2019). EbA projects should therefore aim at restoring biodiversity and ecosystems, plus improving the resilience of the people and ecosystems. Enhanced biodiversity in terms of quantity and quality will improve ecosystem resilience, while cushioning the Community against climatic hazards will reduce the incidence and impact of disasters on them, thereby improving their resilience.

A governance structure that is capable of delivering these multiple objectives must be participatory, devolved, and polycentric, yet coordinated. Governance is the greatest variable in the successful mainstreaming of EbA into policies and actions for climate change resilience. The focus of governance for EbA is society, and how the power is distributed across society for the purposes of achieving the EbA objectives. These objectives should also be consultatively developed. The challenge in governance is to reconcile competing, at times conflicting stakeholder demands and interests while still keeping track of the multiple process objectives and outcomes. Current multistakeholder engagement strategies should be employed to identify and profile the stakeholders, for meaningful engagement and deployment.

Good governance will facilitate transparency and equity in decision making, ensuring accrual of benefits (ecological, social, and economic) to communities and ecosystems. Poor governance, on the other hand, might result in a lack of sustainability of the EbA processes, or increased risks and unnecessary costs to projects. These may therefore, ultimately compromise the delivery of ecosystem services to the people and compromise overall ecosystem functions, leading to a failure of the EbA processes.

Stakeholder engagements and consultations are an important ingredient of EbA governance, particularly for the drylands due to their ecological fragility. Such consultations will allow the people to participate in identifying critical issues that need to go into EbA project design, address how they prefer them to be tackled, air their grievances, and make suggestions. This also has the additional benefit of ensuring ownership and sustainability of processes.

Besides being adaptive, governance must be polycentric and decentralized and based on the best scientific basis. Adaptive governance means that results from the monitoring framework of the processes are used to improve project design or modify objectives or change benefit-sharing mechanism, particularly for communities. Adaptive governance also provides a better understanding of different aspects of the community, allowing synergies and linkages to be positively exploited for the benefit of the ecosystems and the community.

EbA governance structure suitability will also depend on context and circumstances. The institutions and rules governing EbA projects must be dynamic and be changed to fit changing circumstances. A good governance structure should address some of the following critical issues:

(i) All the issues, concerns, and interests of all stakeholders, including vulnerable groups, minorities, human rights, gender, etc., to ensure inclusivity.
(ii) How decisions on the project processes will be made.
(iii) How grievances will be addressed, including contact points.
(iv) How responsibilities are to be allocated.
(v) How will information about the process flow.
(vi) Provide a leadership structure through a democratic, inclusive, and transparent manner.
(vii) How community benefits are to accrue and be shared, if any.
(viii) The role of communities and government in sustainability.

28.7 Discussion

This chapter demonstrates that Africa has implemented several adaptation initiatives in an attempt to address the persistent issues of biodiversity loss, water scarcity, land degradation, and food insecurity in its drylands, with mixed results. Several countries have developed policy documents to coordinate these processes, but implementation is weak, as evidenced by the reports to the United Nations Convention to Combat Desertification (UNCCD) website. Again, most of the programs have exhibited a lack of a clear strategy for systemic integration and linkages. This is the gap that an ecosystem-based approach is likely to address.

It seems that Continental programs such as the Bonn Challenge, the Great Green Wall (GGW), and the Afri 100 Landscape restoration programs have been well received, with significant success. They were also quite well integrated to deliver multiple benefits. This can be attributed to significant Donor funding and project visibility. Most of the adaptation processes have largely been mapped on the opportunities in the drylands for adaptation and mitigation, and have adopted a systemic or partial-systemic approach. I find that this demonstrates an evolution of an ecosystem-based approach rather than an absence. Most of the projects discussed in this chapter have addressed multiple objectives, with the GGW and the AfRI100 adequately complying with most of the principles of EbA. The major weaknesses in most of these programs were the stark absence of a clear governance and benefit-sharing mechanisms that are a requirement for EbA. Even though a number of projects discussed illustrate the development and successful implementation of local level governance structures, these seem to have been implemented by default, rather than a compliance with any requirements. Equally, the issue of who participates, benefits, and how, was not clear in many programs. These gaps point to the need for and role of policy instruments in regulating restoration programs. In all the cases sampled community-led programs such as the farmer-managed natural regeneration (FMNR) appear to have the greatest success, over large-scale top-down government-led initiatives. Simple, easy-to-implement restoration practices like rangeland enclosures, pit seedling planting, banding, terracing, and mulching received better success in most of the cases reviewed. One reason for their success was cost-effectiveness. However, funding remained a challenge in most of the programs.

One fact that was clear from most of the programs was the luck of collaborations between projects and programs. Most appeared to be stand-alone, and did not have mechanisms for collaborations, and synergies. This underpins the need for comprehensive programs for up-scaling of all restoration efforts. This calls for comprehensive strategies for mainstreaming the processes. We indicate in the chapter the need to develop appropriate policies that can mainstream proven adaptation strategies from the local level to the National and regional levels. Nearly all programs reviewed had strong and elaborate objectives that could respond to Land degradation, livelihoods, and climate resilience; but all were quite weak or had no clear provisions for governance, coordination, stakeholder participation, and benefit sharing. Restoration is a local process, and without the commitment, ownership, and participation by the immediate affected communities and other stakeholders, including the Government, NGOs, producer organizations, and success will be compromised. Restoration efforts should be an all-inclusive process.

The private sector involvement in restoration programs was limited, even though the Continental programs benefited from donor funding. The private sector is risk-averse, but business models based on restoration outcomes may be more attractive and less risky, as shown by the case of Gum Arabic in Sudan. However, the government should develop public-private partnership instruments, integrated with business models to incentivize the private sector. This is besides the development of private sector friendly policies to support restoration interventions.

A challenge faced by nearly all the programs was the lack of inadequate or lack of monitoring and reporting progress. In the course of writing this chapter, progress on most of the initiatives was not easily accessible, and where it was, it was done by a third party, such as Food and Agriculture Organization (FAO). A clear communication strategy of the progress and achievements of these initiatives would not only help mobilize the stakeholders but also enhance the program's visibility. This strategy would benefit greatly from a comprehensive stakeholder mapping and engagement. The various stakeholders involved in restoration, including scientists, governments, NGOs, etc., should also come up with clear mechanisms to monitor the multiple objectives addressed by these restoration programs and also evaluate project impacts. The universal adoption of an ecosystem-based approach in program implementation would be a significant step in achieving significant impacts in restoration efforts in the drylands and deliver resilience. EbA has many benefits and is clearly well structured. The EbA approach is able to deliver both ecosystem and community resilience, necessary for the sustainable development of the African drylands.

In the context of Africa, national and regional collaborations on restoration programs stand a better chance to deliver the restoration objectives at scale, and also address the impacts of climate change. The responsibility of addressing land degradation, desertification and climate change impacts should be given a higher profile in respective Governments, because they demand serious commitments, and are very critical for the sustainable development of the continent.

28.7.1 Way Forward

Despite Africa implementing many restoration programs in its drylands, Land degradation, biodiversity loss, climate change impacts, poverty and food insecurity continue to be some of the greatest challenges to the sustainable development of the drylands and its increasing populations of humans and livestock. Much more efforts in terms of upscaling proven restoration and adaptation strategies, increasing budgetary allocation, community awareness, addressing negating internal and external factors, need to be made. An ecosystem-based adaptation strategy shows greater potential to deliver this promise, due to its multiple benefits. But there is need to coordinate this with all other ongoing efforts at the landscape and regional levels. To achieve impact at scale, policy, legal reforms and the building of coordinating institutions, are recommended. Major challenges facing the implementation of restoration programs such as funding, capacity development, etc; will need to be addressed.

References

AFWC (2023) Report of the 24th. Session of the African Forestry and Wildlife Forest Commission, Arusha, Tanzania 2023

Allen M, Diakhite M, Gana D (2020) Two decades of farmer managed natural regeneration on the Seno plain, Mali

Belemviré A, Ouedraogo M, Reij C, Tappan G (2020) History and impacts of dryland restoration in Yatenga, Burkina Faso. In: Pasiecznik N, Reij C (eds) Restoring African drylands. Tropenbos International, Wageningen, p 266

Boko M, Niang I, Nyong A, Vogel C, Githeko A, Medany M, Osman-Elasha B, Tabo R, Yanda P (2007) Africa. Climate change 2007: impacts, adaptation and vulnerability. In: Parry ML, Canziani OF, Palutikof JP, Van der Linden PJ, Hanson CE (eds) Contribution of working group II to the fourth assessment report. Cambridge University Press, Cambridge

Bruce J, Migot-Adholla S (1993) Searching for land tenure security in Africa. World Bank, Washington

Campbell, DJ. (1977) Strategies for coping with drought in the Sahel: A study of recent population movements in the Department of Maradi, Niger. Unpublished Ph.D. Dissertation, Clark University, Worcester, MA, USA

Gebrekirstos A et al (2020) Context of participation in restoration

Haile M, Gebregziabher D (2020) Successful landscape restoration in Abreha we Atsbeha watershed. Tigray, Ethiopia

Hassane A, Reij C (2020) Post-project impacts of restoring. Two decades of farmer managed natural regeneration on the Seno plain, Mali. In: Pasiecznik N, Reij C (eds) Restoring African drylands. Tropenbos International, Wageningen, p 266

IAASTD (2009) Agriculture at Crossroads: Volume V: Sub –Saharan, Available at: http://islandpress.org/bookstore/details.php?isbn=9781597265409

ICSU (2017) A guide to SDG interactions: from science to implementation. International Council for Science

IPCC (2007) Summary for policymakers. In: Solomon S, Qin D, Manning M, Chen Z, Marquis M, Averyt MB, Tignor M, Miller HL (eds) Climate change 2007: the physical science basis. Contribution of working group I to the fourth assessment report of the intergovernmental panel on climate change. Cambridge University Press, Cambridge and New York

IPCC (2018) Global warming of 1.5°C. Intergovernmental panel on climate change. Geneva, Switzerland

Janetschek H, Clara B, Adis D, Hackmann B (2020) The 2030 agenda and the Paris agreement: voluntary contributions towards thematic policy coherence. Clim Pol 20:430–442. https://doi.org/10.1080/14693062.2019.1677549

Johnson B, Steinberg D (2020). Local land-use plans, bylaws and conventions reduce resource-based conflicts

Kolding J, van Zwieten PA, Marttin F et al (2016) Fisheries in the drylands of sub-Saharan Africa— "fish come with the rains". Building resilience for fisheries-dependent livelihoods to enhance food security and nutrition in the drylands. FAO, Rome

Leiter T et al (2019) Adaptation metrics. Current landscape and evolving practices. UNEP DTU partnership. In: Northrop E, Biru H, Lima S, Bouye M, Song R (eds) Examining the alignment between the intended nationally determined contributions and sustainable development goals. World Resources Institute

MA (2005) Ecosystems and human Well-being: desertification synthesis. World Resources Institute, Washington

McSweeney C, New M, Lizcano G (2012) UNDP climate change country profiles: Morocco

Noble I et al (2014) Adaptation needs and options. In: Climate change 2014: impacts, adaptation, and vulnerability. Part a: global and sectoral aspects. Contribution of working group II to the fifth assessment report of the intergovernmental panel on climate change. Cambridge University Press, Cambridge/New York, pp 659–708

Nyssen J, Haile M, Naudts J, Munro N, Poesen J, Moeyersons J, Frankl A, Deckers J, Pankhurst R (2009) Desertification? Northern Ethiopia re-photographed after 140 years. Sci Total Environ 407:2749–2755. https://doi.org/10.1016/j.scitotenv.2008.12.016

Orondo L (2020) Land rights as an imperative for sustainable land and natural resources management in Kenya. US-China Law Rev 17:103–112. https://doi.org/10.17265/1548-6605/2020.03.003

Reij C,Tappan G, Smale M (2009) Agroenvironmental transformation in the sahel: another kind of green revolution. Washington,D.C: International Food Policy Research Institute

Sanogo D, Sall M, Camara BA, Diop M, Badji M, Ba HS (2020) The climate-smart village approach: communities at the heart of restoration in Senegal. In: Pasiecznik N, Reij C (eds) Restoring African drylands. Wageningen, Tropenbos International, p 266

Tappan G, McGahuey M, Winterbottom R (2020) Restoration of agricultural landscapes and dry forests in Senegal. In: Pasiecznik N, Reij C (eds) Restoring African drylands. Tropenbos International, Wageningen, p 266

UNCCD (2016) Land degradation neutrality target setting – a technical guide draft for consultation during the land degradation neutrality target setting program inception phase

UNEP (2006) Africa environment outlook 2 – our environment, our wealth. UNEP, Nairobi

Wiggins, S. & Keats, S. (2013). Leaping and learning: linking smallholders to markets in Africa. Overseas Development Institute, London

Chapter 29
Probing into Nexus Between Industrial Development and Dryland Degradation in the Global South

K. Vishaal, Aamir Ahmad Teeli, Hanh Pham Thi, and Showkat Anwar Bhat

Abstract In the march toward achieving the "Green Economy (GE)," both developed and developing countries are resorting to a number of strategies. The reason behind such strategies is the disturbing trend observed, mostly in developing countries, called the "industrial pollution and land degradation," which directly threatens the achievement of "Green Economy," and "Sustainable Development Goals (SDGs)." This chapter elucidates the complex interactions between industrial activities and the quality of dryland and environment in order to analyze the pathways to green economy from the point of view of both technological and non-technological IPM strategies. Also, we study current empirical evidences of India's march toward the green economy. The study suggests an integrated, localized, factorial, and institutional approach (especially for a developing country like India) called "Industrial Pollution Mitigation and Sustainable Land Management" as one of the approaches, in order to mitigate industrial pollution and its impacts on drylands. From these conceptual analyses, it has been proposed that India can improve the prospects of sustainable development in the years to come and achieve a green economy in a progressive manner by adopting such a comprehensive framework.

Keywords Industrial pollution mitigation · Technological innovation · Sustainable Dryland Management Framework

---

K. Vishaal · A. A. Teeli (✉)
Department of Economics, Central University of Tamil Nadu, Thiruvarur, Tamil Nadu, India

H. P. Thi
Head of Tourism Department, Faculty of Business Administration, Vietnam Women's Academy, Ha Noi, Vietnam
e-mail: hanhpt_qtkd@vwa.edu.vn

S. A. Bhat
Population Research Centre, Central University of Kashmir,
Srinagar, Jammu and Kashmir, India

Y. D. I. Khan et al. (eds.), *Land Restoration Through Ecosystem-Based Approach*, Disaster Resilience and Green Growth, https://doi.org/10.1007/978-981-95-0389-6_29

29.1 Introduction

Industrial activities are fundamental to modern economies in providing essential goods and services that underpin daily life and achieving higher rates of economic growth and urbanization (Khademi et al. 2019), as advocated by the different growth and development theories (see Smith 1776; Kaldor 1966; Rostow 1960; Solow 1956). Though these industrial activities exert a range of positive externalities like contributing to appreciably high economic growth, generating employment opportunities and exerting positive spillover effects on the rest of the economy, on the other hand, they generate significant levels of negative environmental externalities (Han et al. 2020). Industrial activities emit environmental pollutants, which contribute to air and water pollution, which further leads to land degradation. Land degradation is a serious environmental issue and an immediate public health issue, such as air and water pollution (Wolf et al. 2022). This issue thereby acts as a factor that directly and indirectly threatens "Sustainable Development Goals (SDGs)" (Vlek et al. 2017). Soil erosion and land degradation pose a major threat to global food security and to the achievement of the Sustainable Development Goals (SDGs) on ensuring the well-being of at least 3.2 billion people around the world (FAO 2015). Because of such human activities on land resources, soil erosion (an aspect of land degradation) has increased exorbitantly high in the recent past (See, Yang et al. 2003; Hurni et al. 2015). The above environmental externalities further lead to a string of negative externalities in social and economic dimensions. Since there is an interrelationship among the three dimensions, such negative externalities impact the three-dimensional "Environmental, Social and Economic Sustainability" (ESES) at both local and global levels.

Industrial pollution is releasing hazardous elements into the environment due to industrial processes. At the same time, land degradation involves declining land quality caused by natural or human-induced factors, leading to reduced productivity and ecological and ecosystem health (UNCCD 2014). One of the main risks to humanity's future is the degradation of agricultural areas under drylands, which provide food and raw materials (Oldeman 1997).

The exponential growth of industrialization is consuming large tracts of agricultural land (Ha et al. 2020; Seyam et al. 2023), which also seriously degrades soil fertility (Karthika et al. 2022). Such urban and industrial construction activities further lead to the release of high levels of carbon. Especially, in the past 40 years, a colossal loss of approximately one-third of the world's area of arable agricultural lands has been experienced due to pollution and erosion, caused by urbanization and industrialization (ELD Initiative and UNEP 2015). Not only is the cover and fertility of agricultural drylands lost due to the process of industrialization, but the health of most ecologically and economically productive ecosystems, like the dryland ecosystem, has also been declining over the last three decades. The extent of land degradation is vast, accounting for over 23% of the Earth's landmass, growing at a rate of 5–10 million hectares yearly (Stavi and Lal 2015). Approximately 1.5 billion people worldwide are impacted by land degradation (Gnacadja 2012a; Stavi and Lal 2015) due to the lack of intervention in relevant policies and localized action plans.

29.1.1 Industrialization and Industrial Pollution in the Global South

The industrialization related to manufacturing activities has been rampant in the Global South since the 1990s and especially since 2000 (World Bank 2021; Hira et al. 2022; Kruse et al. 2023). Such expeditious industrial growth has increased production by around three times in South Asian and seven times in Sub-Saharan African countries (World Bank 2021). Though phenomenal industrial growth has been observed in the Global South, it has been associated with higher environmental pollution and illnesses (Babatola 2018; Sumaira and Siddique 2023). These industrial pollution issues have been linked mainly with the energy use sector, extractive industries (International Energy Agency 2022). In the Global South, the energy use sector mostly relies on fossil fuel consumption (Fuhr 2021). These fossil fuels are raw materials and depend on the mining and extractive industries (United Nations 2021). Sixty-three countries of the Global South that have been classified under the low- as well as middle-income categories depend majorly on these extractive industries since 2000 (Roe and Dodd 2016).

Out of the total emissions of greenhouse gases in the world, the Global South alone shares 63% (Fuhr 2021). In terms of total global carbon dioxide (CO_2)emissions of 37,150 $MtCO_2$, the two highest contributors, namely China and India, alone, contribute to around 11,397 and 2830 $MtCO_2$, respectively (Global Carbon Atlas 2021). These two countries are from the Global South and account for around 38.3% of global carbon emissions. In the Global South, the emissions of greenhouse gases, heavy metal pollutants, and particulate matter have led to higher levels of dryland degradation (UNCCD 2024).

29.1.2 Dryland Degradation in the Global South

Drylands are lands that are arid, semiarid, and dry sub-humid, with large water scarcity due to its low cum irregular precipitation, which is mainly unpredictable (UNCCD 2024). Drylands brings forth various ecosystem services, such as provisioning and cultural and regulatory ecosystem services. Drylands occupy 44% of total tilled land, facilitating 69% of livestock worldwide (IPBES 2019). These drylands form 46% of the earth's area of land, of which Africa hosts one-third (Global Centre on Adaptation 2021). The fall in productivity in terms of economic and biological dimensions of these lands is called the "Dryland Degradation," which further leads to desertification (IUCN 2017).

The dryland degradation is driven by anthropogenic and natural factors. The anthropogenic factors are deforestation, degrading agricultural activities, unsustainable grazing, extractive industries (mining), irregular irrigation practices, urban growth and sprawl, soil erosion and unsustainable utilization of chemicals, etc. (IUCN 2017; IPBES 2019; GEF 2024). All these indicate that unsustainable land use change is the predominant dryland degradation driver (Erb et al. 2018). Since Africa and Asia together host almost 70% of the drylands in the world, such degradation/desertification of the drylands of the Global South has become a serious issue.

Especially in the Sub-Saharan African region, 73% (or almost three-fourth of drylands have been degraded (FAO and ITPS 2015). That too, nearly half of those degraded drylands have undergone rigorous degradation (ibid). This has led to a sharp dip in nutrient levels of almost 40% of soils, land compaction, water erosion, and logging in Sub-Saharan Africa (Tully et al. 2015; FAO and ITPS 2015). In case of South Asia, almost 6% of the total dryland area has been affected by desertification in the last 40 years, and approximately 20% are under the purview of speculative risk. In Latin America, 5520 sq. km (31% of total land area) is occupied by drylands (Zhang et al. 2024). Droughts have intensified the degradation of these drylands, further creating a dip in the growth of vegetation (ibid). The above incidents of dryland degradation indicate that the Global South's drylands have been disproportionately affected by industrial pollution and other anthropogenic factors.

Industrial pollution through emissions of greenhouse gases and other pollutants, as well as various pathways such as water, air, and soil, has negatively impacted/degraded the drylands in the Global South to a large extent (UNEP 2015). The above sections also reiterate the same. Though it is possible to identify massive dryland degradation, the relative impact of industrial pollution must be examined to establish a substantial nexus between industrial pollution and dryland degradation in the Global South. IPBES (2019) has empirically mapped and studied the relative impacts of the varied drivers of degradation of lands in the Global South. The normalized difference vegetation index (NDVI) dataset was utilized to study the degradation pattern. IPBES (2019) opined that the extractive industries, energy development, industrialization, infrastructure development, and urbanization degrade the lands in the Global South more than the other anthropogenic drivers. Those anthropogenic drivers are overgrazing, expansion of crops/agricultural land, tree plantations, natural resource extraction other than timber, and invasive species growth. This indicates a substantial nexus between industrial pollution and dryland degradation in the Global South.

If due attention is not paid to dryland management and land degradation is not effectively addressed, global food security targets will be missed (Hossain et al. 2020) since land degradation reduces productivity (Bhattacharyya et al. 2015), and the amount of food provided. Due to this phenomenon, between the 1970s and the early 2000s, the proportion of the world's Dryland area that is said to be vulnerable to severe drought more than doubled, and in the upcoming decades, there is a chance that widespread droughts may occur. In particular, these droughts threaten rural livelihoods in developing nations, such as several in Africa and Central Asia, where poor socioeconomic infrastructures and adverse biophysical circumstances collide

(UNCCD 2011). Africa, in particular, hosts 36 out of the 80 countries that have been greatly affected by land degradation. Moreover, compared to the present scenario, demands for food, water, and energy are expected to increase by 50%, 40%, and 4%, respectively, by 2030 (Gnacadja 2012b). It is also anticipated that the expanding human population and rising needs for energy, water, and food would significantly increase pressure on land (Conacher 2009).To this end, the primary objective of this research paper is to explore the theoretical underpinnings of industrial pollution and the dryland damage caused by the former. Specifically, this chapter aims to analyze the mechanisms through which industrial activities contribute to dryland degradation and assess the ecological, economic, and social impacts caused by industrial pollution, and put forth some suggestions that could mitigate them.

29.2 Review of Literature

Industries such as mining, smelting, oil refineries, ore-dressing refineries, ports, thermal power plants, textiles, and metallurgy discharge vast amounts of pollutants into the soil (Izah and Angaye 2016; Sharma et al. 2021). Industrial pollutants can be broadly categorized into air, soil, and water pollutants (Fig. 29.1). Each type has distinct pathways and impacts on land quality. Industrial emissions, including sulfur dioxide (SO_2), nitrogen oxides (NO_x), volatile organic compounds (VOCs), and particulate matter (PM) like heavy metals, generally settle on land surfaces (Han et al. 2021), contributing to soil acidification and nutrient imbalances leading to significant damage to drylands (Hédl et al. 2011). The greenhouse gas emissions from industries lead to increased global warming (Mohajan 2017; Kumar 2018) and further local- and global-level climate change that exacerbates the following impacts of industrial pollution and land degradation (Stern 2007).

These pollutants discharged from industrial activities can contaminate soil in various ways, reducing land fertility. Direct disposal of industrial waste on land, such as slag, fly ash, and chemical residues, can further degrade soil structure and chemistry (Adriano et al. 2002). These can result in reduced agricultural productivity and ecological function or deteriorated ecosystem health (Oldeman 1997; Vlek et al. 2017).

29.3 Pathways of Land Degradation

Land degradation caused by industrial pollution occurs through numerous pathways, mainly chemical degradation. Chemical degradation caused by the accumulation of noxious materials, such as heavy metals and persistent organic pollutants (POPs), can alter soil chemistry and reduce fertility and biological activity. These changes have been realized especially in South and Southeast Asian countries (Herawati et al. 2000). Physical degradation, which is sourced from Industrial

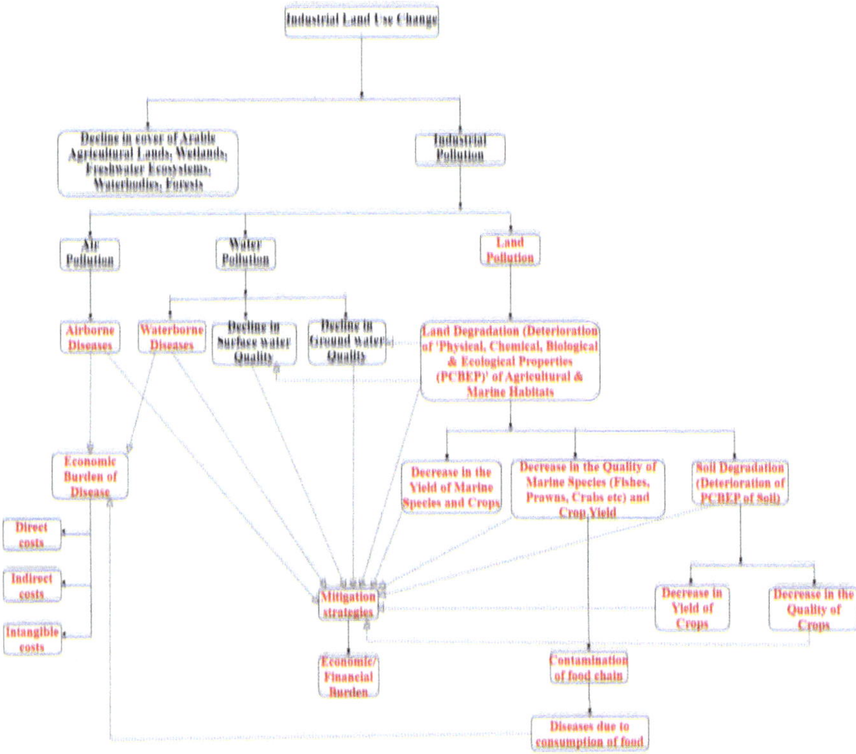

Fig. 29.1 Industrial pollution and land degradation. (Source: Author's Compilation)

activities, can lead to soil erosion, subsoil compaction, and loss of soil structure, thereby damaging drylands (Stolte et al. 2015). These physical changes reduce the land's ability to support vegetation and retain water. The pollution can also disrupt soil microbial communities, which are essential for nutrient cycling and plant growth (Edwards 2002), especially in the case of fly ash pollution (Jambhulkar et al. 2018). Such reduced microbial diversity and activity can further impair soil health and productivity.

29.4 Impacts of Industrial Pollution and Land Degradation

Dryland degradation due to industrial pollution has profound ecological consequences on biodiversity, which can lead to habitat loss and reduced species diversity (Xu et al. 2019). Plants and animals that rely on healthy soil ecosystems are particularly vulnerable to pollution-induced changes (Šiukšta et al. 2022). Soil degradation caused by such pollution then affects critical ecosystem services (Baude et al. 2019), such as water filtration, carbon sequestration, and nutrient cycling (Lal 2015). These changes can disrupt local and global ecological balance (León and Osorio 2014). In addition to this, this phenomenon can have substantial economic costs, including reduced agricultural productivity because degraded soils yield lower crop outputs, affecting food security and farmer livelihoods (Bindraban et al. 2012). The cost of remediation and loss of arable land further exacerbate economic burdens. Lastly, it has an impact on social aspects of economies, including the displacement of Indigenous communities (Hens 2012; Tran 2015), particularly in peri-urban areas of developing regions where land tenure is insecure (Adam 2014). This displacement can result in social unrest and loss of cultural heritage. Marginalized communities generally bear the disproportionate impact of industrial pollution, facing higher levels of exposure and fewer resources for mitigation and adaptation (UNESCAP 2018; Shen et al. 2020).

29.5 Mitigation Strategies

The following first two types of mitigation strategies, such as environmental regulations and technological innovation, do not act only as strategies but also as determinants of environmental performance at the macro/global level (Pei et al. 2019).

29.5.1 Environmental Regulations

National and sub-national governments must enforce stringent pollution control standards, including emission limits, industrial waste disposal regulations, and penalties for non-compliance with industrial norms and pollution regulations. Implementing the polluter pays principle ensures that industries are financially responsible for the environmental damage they cause, incentivizing pollution prevention. Carbon pricing can be an effective way to minimize carbon emissions in the long term. Carbon pricing is an economic strategy to reduce greenhouse gas emissions by assigning a monetary cost to emitting carbon dioxide. This can be implemented through carbon taxes or cap-and-trade systems, where companies pay for their emissions. Unlike carbon credits, which represent a tradable permit to emit a specific amount of carbon, carbon pricing directly incentivizes reductions by making emissions financially costly.

Though carbon pricing is necessary to achieve zero/reduced emissions, the sufficient condition is to achieve "Technological Innovation" and the increased "Deployment Incentives for Low-Emission Technologies" (Stern 2007). Out of these two, technological innovation needs to be focused on first. Not only the renowned Stern (2007) report has advocated the premise, but also the studies such as Ahmadova et al. (2023), Chen et al. (2022), and Satheedevi et al. (2022), regarding the nexus between economic growth and environmental sustainability, have suggested exact "Technological Innovation/Change," to establish parity among the both.

29.5.2 Technological Innovation

Adopting cleaner production techniques, such as waste minimization, resource efficiency, and green chemistry, can significantly reduce pollutant emissions, march toward achieving SDGs, and nullify/reduce the impact of industrial pollution on three-dimensional sustainability. Carbon capture and storage (CCS) technology, which captures carbon dioxide emissions from industrial processes or power plants and stores them underground in geological formations to prevent them from entering the atmosphere, can be used as one of the technological means to mitigate carbon emissions. CCS is a critical long-term carbon mitigation strategy for sectors with hard-to-abate emissions. Other complementary technologies include electric vehicles (EVs), which reduce emissions from transportation, and solar energy, which provides a renewable and emission-free source of electricity. Together, these technologies form a diverse toolkit to combat climate change and transition to a low-carbon economy.

Technologies such as scrubbers, filters, and bioremediation can effectively capture and treat pollutants before they reach the environment. Even the adoption of "Robotics" for autonomous production has decreased the intensity of pollution emissions at the micro/firm level, it may not have been very successful in decreasing the level of total emissions (Song et al. 2023). Even though the "Research & Development" on these innovations is efficiently led and diffused by the private sectors and agencies, the coordination of government (in terms of funding and support) with the latter has been very less and needs to be developed, in order to catapult emissions control (Stern 2007). Since the energy sector contributes to two-thirds of emissions in the world, the innovations focusing on cleaner technologies need to take place in the non-renewable energy use sector on one side and, on the other, must increase the share of renewable energy and its storage potential, etc.

29.5.3 International Organizational Initiatives

Many earnest international-level efforts are being put into combating this issue and restoring as much damage as possible to drylands. An idea of "Net Zero Land Degradation" has been adopted lately by the United Nations to protect drylands and minimize the damage. The United Nations Convention to Combat Desertification (UNCCD) President made the formal introduction of the idea of Zero Net Land Degradation (ZNLD), also known as Land Degradation Neutrality (LDN), in 2011 (UNCCD 2011). According to the idea of net-zero land degradation, the amount of degraded land worldwide will either decline or stay stable (Stavins and Lal 2015).

29.5.4 Nature-Based Solutions (NbS)

Nature-based solutions (NbS) utilize natural processes and ecosystems to address environmental challenges. To mitigate dryland degradation, NbS include reforestation, agroforestry, sustainable grazing practices, and soil restoration techniques like contour farming and biochar. These approaches enhance soil fertility, increase water retention, and promote biodiversity while reducing erosion and desertification. Integrating traditional knowledge with modern science is critical to tailoring solutions to local conditions. NbS not only restore ecosystems but also improve livelihoods and resilience to climate change in vulnerable dryland areas.

Implementing activities such as crop rotation, agroforestry, and no-till farming can improve soil health and resilience to degradation (Kopsidis and Pfister 2013; Fahad et al. 2022; Angon et al. 2023). Agroforestry systems are seen to be effective in restoring degraded drylands (Nair et al. 2010). In the case of soil erosion control, breaking the spatial continuity of slopes is one of the most effective ways. This can be accomplished in two ways: by creating earthworks, including contour-based terraces (Lal 1997), or by preserving buffer strips of vegetation that include woody (Jacinthe et al. 2009). By preventing erosion and storing significant levels of carbon in soils, root systems, and canopies (utilizing the above techniques), the use of trees and shrubs in afforestation and reforestation systems has been shown to effectively restore the productive potential of degraded lands (Behan and Misek 2008). Remediation of contaminated sites through techniques like phytoremediation, soil washing, and chemical stabilization can also be used to restore degraded lands.

29.6 Conclusion

Industrial pollution and dryland degradation are interconnected challenges that require efforts at the local and global levels to be addressed. By understanding the mechanisms and impacts of these issues, we can develop effective strategies for mitigation and sustainable development, but using a localized institutional framework to catapult the efficiency of such strategies is the need of the hour. It is evident from all the above that mere mitigation without efficiency in all the above factors and a localized institutional approach for specific industrial regions, high scores of E-IPM and improvement in environmental and land quality cannot be possible. So, it would be ideal to build a localized institutional "Industrial Pollution Mitigation and Sustainable Land Management Framework" that incorporates efficiency in all these factors to mitigate industrial pollution and land degradation at the point source to build sustainable industrialization. Such a localized approach will only help achieve zero/reduced emissions and land degradation. The future research and policy efforts must continue to prioritize these areas to ensure a healthier and more sustainable environment for future generations.

References

Adam AG (2014) Peri-urban land tenure in Ethiopia. Doctoral dissertation, KTH Royal Institute of Technology https://www.diva-portal.org/smash/record.jsf?pid=diva2%3A773927&dswid=-7062

Adriano DC, Weber J, Bolan NS et al (2002) Effects of high rates of coal fly ash on soil, turfgrass, and groundwater quality. Water Air Soil Pollut 139:365–385

Ahmadova G, Bueno García M, Delgado-Márquez B et al (2023) Firm-and country-specific advantages: towards a better understanding of MNEs' environmental performance in the international arena. Organ Environ 36(3):468–497. https://doi.org/10.1177/10860266221129699

Angon PB, Anjum N, Akter MM et al (2023) An overview of the impact of tillage and cropping systems on soil health in agricultural practices. Adv Agric 1:8861216. https://doi.org/10.1155/2023/8861216

Babatola SS (2018) Global burden of diseases attributable to air pollution. J Publ Health Africa 9(3). https://doi.org/10.4081/jphia.2018.813

Baude M, Meyer BC, Schindewolf M (2019) Land use change in an agricultural landscape causing degradation of soil based ecosystem services. Sci Total Environ 659:1526–1536. https://doi.org/10.1016/j.scitotenv.2018.12.455

Behan J, Misek J (2008) Environmental effects of forest biomass removal. In: State Library of Oregon. Oregon Department of Forestry. Available via Digital Collections. https://digitalcollections.library.oregon.gov/nodes/view/129303. Accessed 1 Dec 2008

Bhattacharyya R, Ghosh BN, Mishra PK et al (2015) Soil degradation in India: challenges and potential solutions. Sustainability 7(4):3528–3570. https://doi.org/10.3390/su7043528

Bindraban PS, van der Velde M, Ye L et al (2012) Assessing the impact of soil degradation on food production. Curr Opin Environ Sustain 4(5):478–488. https://doi.org/10.1016/j.cosust.2012.09.015

Chen Y, Cheng L, Lee CC (2022) How does the use of industrial robots affect the ecological footprint? International evidence. Ecol Econ 198:107483. https://doi.org/10.1016/j.ecolecon.2022.107483

Conacher A (2009) Land degradation: a global perspective. N Z Geogr 65(2):91–94. https://doi.org/10.1111/j.1745-7939.2009.01151.x

Edwards CA (2002) Assessing the effects of environmental pollutants on soil organisms, communities, processes and ecosystems. Eur J Soil Biol 38(3–4):225–231. https://doi.org/10.1016/S1164-5563(02)01150-0

Erb KH, Kastner T, Plutzar C, Bais ALS, Carvalhais N, Fetzel T, Gingrich S, Haberl H, Lauk C, Niedertscheider M, Pongratz J (2018) Unexpectedly large impact of forest management and grazing on global vegetation biomass. Nature 553(7686):73–76. https://doi.org/10.1038/nature25138

Fahad S, Chavan SB, Chichaghare AR, Uthappa AR, Kumar M, Kakade V, Pradhan A, Jinger D, Rawale G, Yadav DK, Kumar V (2022) Agroforestry systems for soil health improvement and maintenance. Sustainability 14(22):14877. https://doi.org/10.3390/su142214877

FAO and ITPS (2015) Status of the World's Soil Resources (SWSR – main report. Food and Agriculture Organization of the United Nations and Intergovernmental Technical Panel on Soils, Rome, Italy. https://openknowledge.fao.org/server/api/core/bitstreams/6ec24d75-19bd-4f1f-b1c5-5becf50d0871/content

Fuhr H (2021) The rise of the global south and the rise in carbon emissions. Third World Q 42(11):2724–2746. https://doi.org/10.1080/01436597.2021.1954901

Global Centre on Adaptation (2021) State and trends report. https://gca.org/wp-content/uploads/2022/07/03_WTW_14855_GCA_2021_Sect2_DRYLANDS_v7.pdf

Gnacadja L (2012a) From combating desertification in drylands to global land degradation neutrality and the zero net land degradation. In: Proceedings of the 4th international conference on drylands, deserts & desertification. Ben Gurion University of the Negev, Sde-Boqer, pp 64–65

Gnacadja L (2012b) Moving to zero-net rate of land degradation. Statement by Executive Secretary. UN convention to combat desertification, Rio de Janeiro

Ha TV, Tuohy M, Irwin M et al (2020) Monitoring and mapping rural urbanization and land use changes using Landsat data in the northeast subtropical region of Vietnam. Egypt J Remote Sens Space Sci 23(1):11–19. https://doi.org/10.1016/j.ejrs.2018.07.001

Han Q, Wang M, Cao J et al (2020) Health risk assessment and bioaccessibilities of heavy metals for children in soil and dust from urban parks and schools of Jiaozuo, China. Ecotoxicol Environ Saf 191:110157. https://doi.org/10.1016/j.ecoenv.2019.110157

Han Q, Liu Y, Feng X et al (2021) Pollution effect assessment of industrial activities on potentially toxic metal distribution in windowsill dust and surface soil in Central China. Sci Total Environ 759:144023. https://doi.org/10.1016/j.scitotenv.2020.144023

Hédl R, Petřík P, Boublík K (2011) Long-term patterns in soil acidification due to pollution in forests of the eastern Sudetes Mountains. Environ Pollut 159(10):2586–2593. https://doi.org/10.1016/j.envpol.2011.06.014

Hens L (2012) Environmentally displaced people. Encycl Life Support Sys (EOLSS) 2:48–51

Herawati N, Suzuki S, Hayashi K et al (2000) Cadmium, copper, and zinc levels in rice and soil of Japan, Indonesia, and China by soil type. Bull Environ Contaminat Toxicol 64(1):33–39

Hira A, Pacini H, Pereira AS et al (2022) Shifting to circular manufacturing in the global south: challenges and pathways. J Dev Soc 38(3):310–335. https://doi.org/10.1177/0169796X221107217

Hossain MA, Amin MN, Sultana J et al (2020) Climate change impact on agriculture and related sustainable land management practices in Bangladesh–a review. Int J Environ Clim Change 10(2):53–69. https://doi.org/10.9734/IJECC/2020/v10i230181

Hurni K, Zeleke G, Kassie M et al (2015) Economics of Land Degradation (ELD) Ethiopia Case Study: Soil degradation and sustainable land management in the rainfed agricultural areas of Ethiopia: An assessment of the economic implications. https://boris.unibe.ch/72796/1/ELD-ethiopia_i_06_72dpi-D.pdf

IUCN (2017) Issues Brief. https://iucn.org/sites/default/files/2022-07/drylands_and_land_degradation_issues_brief_0.pdf. Accessed Jun 2017

Izah SC, Angaye TC (2016) Heavy metal concentration in fishes from surface water in Nigeria: potential sources of pollutants and mitigation measures. Sky J Biochem Res 5(4):31–47

Jacinthe PA, Lal R, Owens LB (2009) Application of stable isotope analysis to quantify the retention of eroded carbon in grass filters at the north Appalachian experimental watersheds. Geoderma 148(3–4):405–412. https://doi.org/10.1016/j.geoderma.2008.11.013

Jambhulkar HP, Shaikh SMS, Kumar MS (2018) Fly ash toxicity, emerging issues and possible implications for its exploitation in agriculture; Indian scenario: a review. Chemosphere 213:333–344. https://doi.org/10.1016/j.chemosphere.2018.09.045

Kaldor N (1966) Marginal productivity and the macro-economic theories of distribution: comment on Samuelson and Modigliani. Rev Econ Stud 33(4):309–319. https://doi.org/10.2307/2974428

Karthika V, Sekaran U, Jainullabudeen GB et al (2022) Advances in bioremediation of industrial wastewater containing metal pollutants. In: Biological approaches to controlling pollutants. Woodhead Publishing, pp 163–177. https://doi.org/10.1016/B978-0-12-824316-9.00001-X

Khademi H, Gabarrón M, Abbaspour A et al (2019) Environmental impact assessment of industrial activities on heavy metals distribution in street dust and soil. Chemosphere 217:695–705. https://doi.org/10.1016/j.chemosphere.2018.11.045

Kopsidis M, Pfister U (2013) Agricultural development during early industrialization in a low-wage economy: Saxony, c. 1790–1830 (No. 39)

Kruse H, Mensah E, Sen K et al (2023) A manufacturing (re) naissance? Industrialization in the developing world. IMF Econ Rev 71(2):439. https://doi.org/10.1057/s41308-022-00183-7

Kumar A (2018) Global warming, climate change and greenhouse gas mitigation. Biofuels: greenhouse gas mitigation and global warming: next generation biofuels and role of biotechnology, 1–16. https://link.springer.com/chapter/10.1007/978-81-322-3763-1_1

Lal R (1997) Soil degradative effects of slope length and tillage methods on alfisols in western Nigeria. I. Runoff, erosion and crop response. Land Degrad Dev 8(3):201–219

Lal R (2015) Restoring soil quality to mitigate soil degradation. Sustainability 7(5):5875–5895. https://doi.org/10.3390/su7055875

León JD, Osorio NW (2014) Role of litter turnover in soil quality in tropical degraded lands of Colombia. Sci World J 2014(1):693981. https://doi.org/10.1155/2014/693981

Mohajan HK (2017) Greenhouse gas emissions, global warming and climate change. In Proceedings of the 15th chittagong conference on mathematical physics, Jamal Nazrul Islam Research Centre for Mathematical and Physical Sciences (JNIRCMPS), Chittagong, Bangladesh (Vol. 16)

Nair PR, Nair VD, Kumar BM et al (2010) Carbon sequestration in agroforestry systems. Adv Agron 108:237–307. https://doi.org/10.1016/S0065-2113(10)08005-3

Oldeman LR (1997) Soil degradation: a threat to food security? In international conference on time ecology: time for soil culture-temporal perspectives on sustainable use of soil, April 6–9, 1997, Tutzing (p. 15). ISRIC

Pei Y, Zhu Y, Liu S et al (2019) Environmental regulation and carbon emission: the mediation effect of technical efficiency. J Clean Prod 236:117599. https://doi.org/10.1016/j.jclepro.2019.07.074

Roe A, Dodd S (2016) Like it or not, poor countries are increasingly dependent on mining and oil & gas. In: UNU-WINDER Blog. https://www.wider.unu.edu/publication/it-or-not-poor-countries-are-increasingly-dependent-mining-and-oil-gas Accessed June 2016

Rostow WW (1960) The five stages of growth – a summary, p 4–16. https://books.google.co.in/books?hl=en&lr=&id=i06C7MgJ84oC&oi=fnd&pg=PA81&dq=Rostow+WW+(1960)+The+Five+Stages+of+Growth%2D%2DA+Summary,+p.+4-16.&ots=Y0yAFg3nak&sig=FqRr93lYfHBBEWgFUMNrdLjcQSg&redir_esc=y#v=onepage&q=Rostow%20WW%20(1960)%20The%20Five%20Stages%20of%20Growth%2D%2DA%20Summary%2C%20p.%204-16.&f=false

Satheedevi AG, Sharma A, Dhar M (2022) How do the anthropogenic factors affect the environment in India? Evidence from the urban provinces. Technol Forecast Soc Chang 183:121950. https://doi.org/10.1016/j.techfore.2022.121950

Seyam MMH, Haque MR, Rahman MM (2023) Identifying the land use land cover (LULC) changes using remote sensing and GIS approach: a case study at Bhaluka in Mymensingh, Bangladesh. Case Stud Chem Environ Eng 7:100293. https://doi.org/10.1016/j.cscee.2022.100293

Sharma KV, Sarvalingam BK, Marigoudar SR (2021) A review of mesocosm experiments on heavy metals in marine environment and related issues of emerging concerns. Environ Sci Pollut Res 28(2):1304–1316. https://doi.org/10.1007/s11356-020-11121-3

Shen W, Srivastava S, Yang L et al (2020) Understanding the impacts of outdoor air pollution on social inequality: advancing a just transition framework. Local Environ 25(1):1–17. https://doi.org/10.1080/13549839.2019.1687431

Šiukšta R, Pukenytė V, Kleizaitė V et al (2022) The butterfly effect: mild soil pollution with heavy metals elicits major biological consequences in cobalt-sensitized broad bean model plants. Antioxidants 11(4):793. https://doi.org/10.3390/antiox11040793

Smith A (1776) An inquiry into the nature and causes of the wealth of nations: Volume One. London: printed for W. Strahan; and T. Cadell, 1776. https://era.ed.ac.uk/handle/1842/1455

Solow RM (1956) A contribution to the theory of economic growth. Q J Econ 70(1):65–94. https://doi.org/10.2307/1884513

Song J, Chen Y, Luan F (2023) Air pollution, water pollution, and robots: is technology the panacea. J Environ Manag 330:117170. https://doi.org/10.1016/j.jenvman.2022.117170

Stavi I, Lal R (2015) Achieving zero net land degradation: challenges and opportunities. J Arid Environ 112:44–51. https://doi.org/10.1016/j.jaridenv.2014.01.016

Stern N (2007) The economics of climate change: the Stern review. HM Treasury. https://doi.org/10.1257/aer.98.2.1

Stolte J, Tesfai M, Øygarden L et al (2015) Soil threats in Europe. Luxembourg: Publications Office. https://op.europa.eu/en/publication-detail/-/publication/0c21db96-a4a6-11e5-b528-01aa75ed71a1/language-en
Sumaira, Siddique HMA (2023) Industrialization, energy consumption, and environmental pollution: evidence from South Asia. Environ Sci Pollut Res 30(2):4094–4102. https://doi.org/10.1007/s11356-022-22317-0
Tully K, Sullivan C, Weil R et al (2015) The state of soil degradation in Sub-Saharan Africa: baselines, trajectories, and solutions. Sustainability 7(6):6523–6552. https://doi.org/10.3390/su7066523
UNCCD (2014) Land degradation neutrality: Resilience at local, national and regional levels. https://catalogue.unccd.int/858_V2_UNCCD_BRO_.pdf
UNCCD Secretariat (2011) Preparatory Process for the Rio+ 20 Conference. https://catalogue.unccd.int/850_Rio_6_pages_english.pdf. Accessed 18 Nov 2011
UNEP (2015) The Economics of Land Degradation in Africa: Benefits of Action Outweigh the Costs. https://wedocs.unep.org/bitstream/handle/20.500.11822/7467/-The_Economics_of_Land_Degradation_in_Africa_Benefits_of_Action_Outweigh_the_Costs;_A_complementary_report_to_the_ELD_Initiative-2015ELD-unep-report_.pdf?sequence=2&amp%3BisAllowed=. Accessed Oct 2015
United Nations Convention to Combat Desertification (UNCCD) (2024) Global Drylands. https://www.unccd.int/cop16#:~:text=The%20sixteenth%20session%20of%20the,desertification%2C%20land%20degradation%20and%20drought
United Nations Environment Programme (2021) Air pollution series: actions on air quality – a global summary of policies and programmes to reduce air pollution. https://www.unep.org/resources/report/actions-air-quality-global-summary-policies-and-programmes-reduce-air-pollution
Vlek PL, Khamzina, A, Tamene LD (2017) Land degradation and the sustainable development goals: threats and potential remedies. http://ciat-library.ciat.cgiar.org/Articulos_Ciat/biblioteca/LAND_DEGRADATION_AND_THE_SDGs-THREATS_AND_POTENTIAL_REMEDIES.pdf
Wolf MJ, Emerson JW, Esty DC et al (2022) 2022 environmental performance index (EPI) results. Yale Center for Environmental Law & Policy, New Haven
Xu L, Chen SS, Xu Y et al (2019) Impacts of land-use change on habitat quality during 1985–2015 in the Taihu Lake Basin. Sustainability 11(13):3513. https://doi.org/10.3390/su11133513
Yang D, Kanae S, Oki T et al (2003) Global potential soil erosion with reference to land use and climate changes. Hydrol Process 17(14):2913–2928. https://doi.org/10.1002/hyp.1441
Zhang Y, Huang K, Zhang J et al (2024) Dryland social-ecological systems in Americas. In: Dryland social-ecological systems in changing environments. Springer Nature, Singapore, pp 325–357. https://doi.org/10.1007/978-981-99-9375-8_10

Chapter 30
Restoring Drylands: Policy Pathways, Governance, and Stakeholder Synergy for Ecosystem Revival

Deepanjana Saha

Abstract Dryland ecosystems, covering 40% of Earth's terrestrial surface and supporting over 2 billion people, face critical threats from climate change, land degradation, and unsustainable practices. Ecosystem-based restoration (EbR) emerges as a promising solution to reverse degradation, bolster ecological resilience, and enhance livelihoods. This chapter critically examines governance structures, stakeholder engagement, and policy frameworks essential for successful dryland restoration. Emphasis is placed on integrated governance models, participatory approaches, and cross-sectoral collaboration to address challenges such as fragmented policies, insufficient funding, and weak enforcement mechanisms. Case studies, including Africa's Great Green Wall and India's community-driven initiatives, demonstrate the potential of coordinated restoration efforts in achieving ecological, social, and economic gains. Key gaps are identified, with recommendations to improve policy coherence, mobilize resources, and adopt innovative technologies like AI and remote sensing. Aligning local actions with global frameworks like the UNCCD and Bonn Challenge, this analysis highlights the need for sustained partnerships to ensure dryland ecosystems' long-term viability.

Keywords Ecosystem-based restoration · Dryland restoration · Governance · Stakeholder engagement · Sustainable land management

D. Saha (✉)
Conservation Genetics lab, Ashoka Trust for Research in Ecology and the Environment (ATREE), Bengaluru, Karnataka, India

Manipal Academy of Higher Education (MAHE), Manipal, Karnataka, India
e-mail: deepanjana.saha@atree.org

Y. D. I. Khan et al. (eds.), *Land Restoration Through Ecosystem-Based Approach*, Disaster Resilience and Green Growth, https://doi.org/10.1007/978-981-95-0389-6_30

30.1 Introduction

Dryland ecosystems, which cover approximately 40% of the Earth's terrestrial surface, are characterized by low and highly variable precipitation, directly impacting nearly 2 billion people worldwide who depend on these areas for their livelihoods and resources (FAO 2017; Davies et al. 2012). These ecosystems span regions in Africa, Asia, Latin America, and parts of North America and are particularly vulnerable due to climatic constraints and ecological fragility (Maestre et al. 2022). While drylands play a critical role in supporting biodiversity and human livelihoods, they are increasingly threatened by climate change, unsustainable land-use practices, and resource over-extraction (Bai et al. 2008; Akhtar-Schuster et al. 2021).

One of the most pressing challenges for dryland ecosystems is the intensifying impact of climate change. Rising global temperatures have led to more frequent and severe droughts, prolonged heat waves, and erratic rainfall patterns, exacerbating water scarcity, decreasing soil fertility, and shifting vegetation patterns. These changes accelerate land degradation and reduce agricultural productivity (IPCC 2021). Additionally, human-induced activities like overgrazing, deforestation, and unsustainable farming practices have further degraded soil health and increased erosion, contributing to biodiversity loss and declining ecosystem services vital to local communities (Reed and Stringer 2016).

Ecosystem-based restoration (EbR) has emerged as a holistic approach to addressing these challenges by reversing dryland degradation, enhancing ecological resilience, and supporting sustainable livelihoods (Abhilash 2021). Unlike conventional restoration approaches, EbR emphasizes restoring ecosystem functions—such as nutrient cycling, water regulation, and carbon sequestration—that underpin long-term sustainability in these regions (Table 30.1). Large-scale initiatives, such as the *Great Green Wall*, which aims to restore 100 million hectares of degraded land across the Sahel by 2030, highlight the potential of coordinated restoration efforts to achieve ecological and socioeconomic benefits (UNCCD 2022). These initiatives also demonstrate the importance of addressing the root causes of degradation by promoting inclusive, community-driven governance and sustainable resource management (Chomba et al. 2016; Girardin et al. 2022).

30.2 Ecosystem-Based Restoration (EbR) in Drylands

30.2.1 Definition and Principles

EbR focuses on leveraging natural processes to restore degraded ecosystems while balancing ecological resilience and socioeconomic integration. The approach is grounded in ecological principles prioritizing biodiversity conservation and landscape functionality, emphasizing that restored ecosystems should be resilient enough to recover from disturbances and continue providing ecosystem services.

Table 30.1 Comparison of EbR practices vs. traditional land restoration approaches

Aspect	Ecosystem-based restoration (EbR)	Traditional land restoration approaches
Primary Goal	Focuses on restoring ecosystem functions and resilience (Cowie et al. 2018)	Emphasis on physical land reclamation and productivity restoration
Approach	Utilizes nature-based solutions, integrating ecological and socioeconomic systems	Often technocratic, focusing on engineering solutions (e.g., soil and water control)
Biodiversity Impact	Prioritizes native species to enhance biodiversity and ecosystem services (Chazdon et al. 2017)	Limited to specific species introduction; often monoculture planting
Soil health improvement	Enhances soil organic matter, fertility, and water retention through natural processes	Relies on synthetic fertilizers, soil amendments, or irrigation
Community involvement	Engages local communities, promoting knowledge sharing and participatory management (Reed et al. 2009)	Limited community input, primarily top-down decision-making
Long-term resilience	Builds ecosystem resilience to withstand climate impacts and other stresses (Abhilash 2021)	Focuses on immediate outcomes, often requiring ongoing interventions
Cost and resource efficiency	Cost-effective over time due to natural regeneration; uses locally available resources	High initial costs; depends on imported materials and heavy machinery
Economic benefits for locals	Supports sustainable livelihoods by enhancing ecosystem services and local economies	Primarily focused on land productivity; limited direct economic benefits to locals
Policy and governance requirements	Requires multilevel governance, adaptive policies, and stakeholder collaboration (Akhtar-Schuster et al. 2017)	Often implemented through rigid policies with limited flexibility
Example projects	Great Green Wall initiative, rangeland restoration projects in Africa and Asia	Reforestation projects, commercial forestry, soil reclamation projects

Source: Author's compilation

EbR also integrates human needs by incorporating sustainable land management practices that support local livelihoods, which is especially critical in dryland regions where communities rely heavily on ecosystem services (FAO 2017).

30.2.2 Benefits of EbR

EbR offers numerous ecological and socioeconomic benefits. By restoring vegetation cover, EbR enhances biodiversity, creating habitats for various species and contributing to ecosystem stability (Chomba et al. 2016). Restorative actions such as reforestation and soil management improve soil structure and fertility, essential for sustained agricultural productivity in drylands (Reed and Stringer 2016).

Furthermore, EbR can significantly boost local livelihoods by fostering sustainable agricultural practices and generating new income sources from ecotourism and non-timber forest products, contributing to poverty reduction (Ferraro and Pattanayak 2006).

30.2.3 Implementation Challenges

Despite its advantages, implementing EbR in drylands presents significant challenges. Funding limitations often hinder large-scale restoration projects, which require sustained financial investment over time (Cowie et al. 2018). Institutional inertia, characterized by existing policies and bureaucratic structures that are slow to adapt, can impede the integration of EbR principles into land management frameworks (Akhtar-Schuster et al. 2017). Additionally, sociopolitical barriers, including conflicts over land use and resource rights, can create resistance among local communities, further complicating EbR initiatives. Addressing these challenges requires coordinated efforts among policymakers, local communities, and international organizations to create enabling environments for EbR.

30.3 Policy Frameworks for Dryland Restoration

30.3.1 International Policy Landscape

Global frameworks play a pivotal role in shaping dryland restoration strategies by providing strategic guidance and fostering international cooperation. One of the primary initiatives in combating land degradation is the United Nations Convention to Combat Desertification (UNCCD). The UNCCD offers a comprehensive international framework to address desertification, land degradation, and drought (DLDD), focusing particularly on the challenges faced by dryland ecosystems. A key component of the UNCCD is the Land Degradation Neutrality (LDN) target, which seeks to balance land degradation and restoration by 2030. This target aligns with Sustainable Development Goal 15, specifically targeting "Life on Land" (Cowie et al. 2018; Akhtar-Schuster et al. 2021). LDN focuses on maintaining or improving land productivity, soil health, and ecosystem resilience, directly addressing the challenges of dryland degradation (Le et al. 2023).

Recent global initiatives have strengthened the international policy landscape for combating land degradation and promoting resilience. For example, the **International Drought Resilience Alliance (IDRA)**, launched at COP27 in 2022, aims to enhance global preparedness for droughts by fostering cooperation, innovation, and knowledge sharing among member countries. Similarly, the **G20 Global Land Initiative**, launched under the G20 framework in 2021, seeks to accelerate sustainable land management practices worldwide, specifically focusing on restoring degraded lands and enhancing food security in vulnerable regions (G20 2022). These initiatives build on existing frameworks, emphasizing global solidarity in addressing the socioeconomic impacts of land degradation.

The Intergovernmental Panel on Climate Change (IPCC) also provides essential insights into the relationship between climate change and land degradation. The IPCC's Sixth Assessment Report highlights the urgency of restoring dryland ecosystems to strengthen resilience against climate impacts, including prolonged droughts and extreme weather events (IPCC 2021). These findings emphasize the need for integrated approaches such as ecosystem-based restoration (EbR), which focus on enhancing ecosystem services like water regulation, carbon sequestration, and biodiversity conservation (Maestre et al. 2022).

Global efforts like the **Great Green Wall Initiative**, which targets the restoration of 100 million hectares of degraded land in the Sahel region by 2030, illustrate how coordinated restoration projects can yield ecological, social, and economic benefits (UNCCD 2022). Similarly, the **UN Decade on Ecosystem Restoration (2021–2030)** promotes large-scale restoration to combat land degradation, improve biodiversity, and mitigate climate change impacts (UNEP 2021).

30.3.2 Regional Initiatives

Several regional initiatives have emerged to tackle dryland degradation and promote larger-scale ecosystem restoration, targeting specific challenges in different regions worldwide.

Africa: One notable example is Africa's **Great Green Wall (GGW)** initiative. The GGW is a massive restoration effort to combat desertification and improve livelihoods across the Sahel region. The GGW also emphasizes cross-border cooperation, involving 20 countries and multiple stakeholders to achieve restoration outcomes. Recently, funding commitments under GGW have exceeded $19 billion, reflecting renewed global interest in addressing land degradation in Africa (Maestre et al. 2022).

In Asia, particularly in China and Central Asia, initiatives like the Three-North Shelterbelt Project (Great Green Wall of China) aim to combat desertification and restore ecosystems across arid regions. This project spans several decades and involves large-scale tree planting and land restoration to reduce soil erosion and improve agricultural productivity. The project has already restored over 30 million hectares of degraded land and significantly reduced sandstorms in northern China. Additionally, Central Asian nations are working under the Central Asian Desert Initiative (CADI) to promote sustainable land use and restore dryland ecosystems across deserts in Uzbekistan, Kazakhstan, and Turkmenistan.

Latin America: In Latin America, various initiatives address the challenges dryland ecosystems face, particularly in countries like Argentina, Chile, and Mexico. Ecosystem-Based Adaptation (EbA) and agroforestry projects are implemented to restore degraded landscapes while enhancing agricultural productivity. For example, the Gran Chaco Initiative in Argentina and Paraguay works to restore one of the largest dry forest ecosystems in the world by promoting sustainable farming practices and conserving biodiversity (IUCN 2022) Similarly, in Mexico, the ProNatura Dryland Restoration Program has focused on restoring arid and semiarid regions by integrating traditional knowledge with scientific techniques to restore water availability and enhance soil fertility (Castellanos and Perez 2021).

North America: In North America, initiatives such as the Conservation Reserve Program (CRP) in the United States aim to combat dryland degradation through sustainable land management. By incentivizing farmers to convert marginal farmland to natural vegetation cover, the CRP has restored millions of hectares of degraded land, improving biodiversity and reducing soil erosion (USDA 2022). In Canada, programs focus on restoring native grasslands in semiarid regions, critical for wildlife habitat and carbon storage (Nature Conservancy Canada 2023).

These regional initiatives reflect a global effort to address the challenges of dryland degradation. They emphasize ecosystem restoration, biodiversity conservation, and sustainable livelihoods, aligning with international frameworks like the UNCCD's LDN targets and the UN Decade on Ecosystem Restoration. By leveraging regional strengths and promoting cross-border collaboration, these initiatives build resilience in some of the world's most vulnerable ecosystems.

30.3.3 National and Local Policies

At the national and local levels, policies aimed at dryland restoration in India integrate regulatory measures, financial incentives, and community participation to address the unique challenges posed by the country's diverse and often arid landscapes. India, with nearly 30% of its land classified as degraded or under desertification risk, has implemented various strategies to restore its drylands.

30.3.3.1 National Policies Supporting Dryland Restoration

- **National Action Programme to Combat Desertification (NAPCD):** India's **NAPCD** developed in alignment with the UNCCD, emphasizes sustainable land management (SLM) practices to combat desertification and drought. It focuses on enhancing soil fertility, water conservation, and biodiversity restoration. The program has been instrumental in implementing reforestation, agroforestry, and climate-resilient cropping systems, particularly in dryland states like Rajasthan and Gujarat (MoEFCC 2022).
- **National Afforestation Programme (NAP):** The **NAP** supports afforestation and eco-restoration of degraded forests and drylands. The program ensures participatory approaches in forest and dryland restoration by engaging local communities through Joint Forest Management Committees (JFMCs). Its initiatives

have been particularly impactful in the semiarid regions of Andhra Pradesh, Tamil Nadu, and Maharashtra, improving soil fertility and vegetation cover (Kumar and Rajagopal 2021).

- **Mahatma Gandhi National Rural Employment Guarantee Act (MGNREGA):** Under **MGNREGA**, dryland restoration projects, including soil conservation, water harvesting, and reforestation activities, are prioritized. For example, in Rajasthan's Thar Desert, MGNREGA has supported the construction of water retention structures and dune stabilization projects, improving local livelihoods and ecological resilience.
- **Watershed Development Programs:** The **Integrated Watershed Management Programme (IWMP)** focuses on soil and water conservation, particularly in dry and semiarid regions. By restoring degraded watersheds, these initiatives enhance agricultural productivity, water availability, and ecosystem services in states like Madhya Pradesh, Telangana, and Odisha (Jain and Sharma 2022).
- **Desert Development Programme (DDP):** The **DDP**, implemented in desert-prone areas like Rajasthan, Gujarat, and Haryana, emphasizes sand dune stabilization, reforestation, and promotion of drought-resistant crops. These efforts aim to reduce desertification while enhancing the livelihoods of rural communities (MoRD 2021).

30.3.3.2 Local-Level Initiatives and Participatory Approaches

Local policies in India often involve direct community engagement to ensure restoration efforts are tailored to the needs of specific regions.

- **Rajasthan:** In Rajasthan, community-led water conservation initiatives like *Johad* construction have restored groundwater levels and revived drylands. Organizations like the Tarun Bharat Sangh have worked with local communities to build water-harvesting structures, benefiting agriculture and biodiversity in arid districts like Alwar and Jodhpur (Singh et al. 2022).
- **Gujarat:** Gujarat's Banni Grasslands Restoration Project adopts a participatory approach to manage and restore semiarid grasslands. The project involves pastoral communities in conserving grassland biodiversity while promoting sustainable livestock practices, ensuring livelihoods and ecosystem health (Patel and Desai 2021).
- **Madhya Pradesh:** The Sukhi Baliraja Initiative in Madhya Pradesh focuses on sustainable watershed management and reforestation in dryland districts. Local self-help groups (SHGs) actively participate in restoring degraded lands and improving water retention and agricultural productivity (Deshpande and Sharma 2022).
- **Tamil Nadu:** Tamil Nadu's Rainwater Harvesting Scheme, implemented across rural dryland regions, emphasizes local participation in constructing rainwater storage systems to mitigate water scarcity. These efforts have enhanced resilience against drought and improved agricultural output.

- **Maharashtra:** The Jalyukt Shivar Abhiyan in Maharashtra is a state-level initiative to make villages drought-free through participatory water conservation techniques like deepening reservoirs and desilting. The program has significantly restored degraded drylands in drought-prone regions like Marathwada .

30.3.3.3 Global Alignment and Future Directions

India's national and local policies often integrate global frameworks like the UNCCD's Land Degradation Neutrality (LDN) goals and the Bonn Challenge. For example, India has pledged to restore 26 million hectares of degraded land by 2030, aligning its policies with international commitments.

Through a combination of national strategies and community-driven local efforts, India demonstrates the importance of participatory governance in achieving dryland restoration. By blending traditional knowledge with modern techniques, these policies restore degraded ecosystems, build resilience against climate change, and support sustainable livelihoods.

30.4 Governance Structures for Effective EbR

The restoration of drylands is an increasingly urgent priority as these ecosystems face challenges from climate change, land degradation, and unsustainable land-use practices. Effective governance structures are essential for facilitating ecosystem-based resilience strategies that can restore ecological functions, improve livelihoods, and enhance resilience in drylands. Governance in this context requires coordination among multiple stakeholders across various levels, from local to international, to ensure the effective implementation of policies and strategies (Barton et al. 2004).

30.4.1 Multilevel Governance: Importance of Coordination Among Local, National, and International Bodies

Multilevel governance is central to the success of EbR in drylands, as it involves coordinating efforts across local, national, and international levels. Local communities are often the primary stewards of dryland ecosystems and hold vital knowledge about local environmental conditions and practices (Barton et al. 2004). However, the broader national and international policy frameworks provide the necessary support, funding, and legal structures that enable local actions to scale up. Coordination across these levels ensures that restoration efforts are not fragmented and that local needs and knowledge are integrated into national and international policies. Such coordination fosters a sense of collective responsibility, ensuring that policy interventions are comprehensive and adaptable to specific ecological and socioeconomic contexts. The role of international organizations, such as the UNCCD, is also crucial in facilitating global cooperation and resource mobilization for dryland restoration. These global institutions can help harmonize efforts and provide financial and technical support for national and local-level initiatives.

30.4.2 Institutional Collaboration: Role of Cross-Sectoral and Cross-Departmental Efforts

Effective ecosystem-based restoration (EbR) in drylands is not just a governmental responsibility; it requires active participation from civil society organizations (CSOs) and the private sector through corporate social responsibility (CSR). These stakeholders bring resources, innovation, and community-driven approaches essential for tackling the complex challenges of dryland restoration.

CSOs work directly with local communities to build awareness and encourage active involvement in restoration efforts. For instance, the Tarun Bharat Sangh in Rajasthan has implemented water-harvesting structures like *johads*, reviving groundwater levels and restoring dryland ecosystems. Similarly, the Foundation for Ecological Security (FES) has collaborated with rural communities in Gujarat and Andhra Pradesh to restore degraded lands and ensure sustainable resource management. CSOs serve as bridges between traditional knowledge and modern scientific practices. They organize training sessions and workshops to empower local stakeholders with restoration techniques, such as sustainable agroforestry and soil conservation. Promoting participatory governance, they help communities co-design restoration strategies tailored to their unique challenges. Civil society is critical in advocating for policy reforms and holding governments accountable for their commitments to combat land degradation. Organizations like the Watershed Organisation Trust (WOTR) in Maharashtra have successfully lobbied to include community-centric approaches in national restoration policies.

Many corporations are leveraging CSR funds to support large-scale restoration programs. For example, the Mahindra Group has initiated projects on afforestation and watershed management in arid regions, contributing to ecosystem restoration and climate mitigation. Companies are integrating advanced technologies like remote sensing, GIS mapping, and artificial intelligence to monitor and assess restoration progress. For instance, ITC Limited's Mission Sunehra Kal incorporates precision agriculture techniques to rehabilitate degraded lands and improve water-use efficiency in dryland farming. CSR programs often collaborate with CSOs and government agencies to scale restoration efforts. For example, the Coca-Cola Foundation has partnered with NGOs like WWF-India to restore water bodies and improve water security in arid regions, directly benefiting local communities.

30.4.3 Challenges in Governance: Identifying Governance Gaps

Despite the importance of multilevel governance and institutional collaboration, several challenges persist in the governance of drylands restoration. One major challenge is insufficient stakeholder engagement, particularly the lack of involvement from marginalized groups, such as women, indigenous communities, and smallholder farmers, who often possess valuable traditional ecological knowledge (Kooiman et al. 2008). Inadequate stakeholder engagement can lead to policies that are not contextually appropriate or fail to address local communities' real needs, ultimately hindering restoration initiatives' success.

Another significant governance gap is the weak enforcement of policies and regulations. While many countries have enacted policies aimed at combating desertification and restoring degraded lands, these policies often lack the mechanisms for effective implementation and monitoring. Weak enforcement can result in the continuation of unsustainable land-use practices and the failure to achieve desired ecological outcomes. As Duit et al. (2010) highlight, the challenges of enforcement are compounded by inadequate resources, political will, and the fragmented nature of governance in drylands.

30.5 Stakeholder Engagement and Community Participation

Successful ecosystem restoration goes beyond technical interventions; it hinges on actively involving stakeholders, especially local communities, whose knowledge, needs, and participation are critical for achieving sustainable results. In dryland restoration, local communities often serve as the primary stewards of the land, making their involvement in planning and decision-making processes indispensable. Their participation contributes valuable local insights, enhances the legitimacy of restoration efforts, and significantly improves the chances of achieving long-term sustainability (Reed et al. 2009)

30.6 Conclusion

Restoration of dryland ecosystems is a pressing global priority given their vulnerability to climate change, desertification, and socioeconomic inequities. Effective restoration requires robust governance frameworks, participatory approaches, and policy alignment across scales. Ecosystem-based restoration (EbR) highlights the importance of integrating ecological, social, and economic dimensions, promoting nature-based solutions that address the root causes of degradation while supporting sustainable livelihoods.

Success stories, such as Africa's Great Green Wall and India's participatory watershed initiatives, exemplify the benefits of multi-stakeholder collaboration, blending traditional knowledge with modern innovations. However, gaps in funding, policy coherence, and enforcement remain significant barriers. Addressing these requires adaptive governance, resource mobilization, and technological integration to enhance monitoring and implementation.

Future strategies must focus on fostering global partnerships and empowering local communities to ensure sustainable outcomes. By aligning efforts with international commitments like the UNCCD's Land Degradation Neutrality targets, the global community can restore drylands to their ecological potential while fostering socioeconomic resilience for millions dependent on these fragile ecosystems.

References

Abhilash PC (2021) Restoring degraded lands for sustainable ecosystem services in drylands. J Environ Manag 287:112287. https://doi.org/10.1016/j.jenvman.2021.112287

Akhtar-Schuster M, Stringer LC, Erlewein A, Metternicht G, Minelli S, Safriel U, Sommer S (2017) Unpacking the concept of land degradation neutrality and addressing its operation through the Rio conventions. J Environ Manag 195:4–15

Akhtar-Schuster M, Thomas RJ, Stringer LC, Chasek P, Seely M (2021) Improving the enabling environment to combat land degradation: institutional, financial, legal, and science-policy challenges and solutions. Land Degrad Dev 32(1):1–15

Bai ZG, Dent DL, Olsson L, Schaepman ME (2008) Proxy global assessment of land degradation. Soil Use Manag 24(3):223–234

Barton DN, Miller FL, Robinson M (2004) Governance for sustainable development: strategic approaches to integrating economic, social, and environmental goals. Routledge

Castellanos E, Perez J (2021) Restoring arid and semi-arid landscapes in Mexico: ProNatura dryland restoration program. J Environ Manag 284:112123

Chazdon RL et al (2017) The role of community-based restoration in ecological sustainability. Restor Ecol 25(1):10–18

Chomba S, Kariuki J, Lund JF, Sinclair F (2016) Roots of inequity: how the implementation of REDD+ reinforces past injustices. Land Use Policy 50:202–213. https://doi.org/10.1016/j.landusepol.2015.09.021

Cowie AL, Orr BJ, Castillo Sanchez VM, Chasek P, Crossman ND, Erlewein A, Welton S (2018) Land in balance: the scientific conceptual framework for land degradation neutrality. Environ Sci Pol 79:25–35. https://doi.org/10.1016/j.envsci.2017.10.011

Davies J, Poulsen L, Schulte-Herbruggen B, Mackinnon K, Crawhall N, Henwood WD et al (2012) Conserving dryland biodiversity. Issues Environ Sci Technol 37:161–184

Deshpande S, Sharma M (2022) Sustainable watershed management in Madhya Pradesh: Sukhi Baliraja initiative. Sustain Sci 17(1):45–60

Duit A, Galaz V, Eckerberg K (2010) Governance, complexity, and resilience. Glob Environ Chang 20(3):363–368

FAO (2017) The future of food and agriculture: trends and challenges. Food and Agriculture Organization of the United Nations. Available at: http://www.fao.org/3/i6583e/i6583e.pdf

Ferraro PJ, Pattanayak SK (2006) Money for nothing? A call for empirical evaluation of biodiversity conservation investments. PLoS Biol 4(4):e105. https://doi.org/10.1371/journal.pbio.0040105

G20 (2022) G20 global land initiative. Available at: https://www.g20.org

Girardin CAJ, Jenkins S, Seddon N, Allen M, Lewis SL, Wheeler CE et al (2022) Nature-based solutions can help cool the planet – if we act now. Nature 603(7899):105–108. https://doi.org/10.1038/s41586-022-04474-x

IPCC (2021) Climate change 2021: the physical science basis. Cambridge University Press

IUCN (2022). Gran Chaco Initiative: Restoring Dry Forest Ecosystems in Latin America. International Union for Conservation of Nature. Available at: https://www.iucn.org

Jain S, Sharma P (2022) Integrated watershed management programme in India: impacts on dryland agriculture and water resources. Agric Water Manag 260:107296

Kooiman J, Bavinck M, Jentoft S, Pullin R (2008) Fish for life: interactive governance for fisheries. Amsterdam University Press

Kumar R, Rajagopal S (2021) National afforestation programme: achievements and future directions. J For Res 35(2):225–233

Le QB, Nkonya E, Mirzabaev A (2023) Biomass productivity-based mapping of global land degradation hotspots. ZEF-Discussion Papers on Development Policy, 193. https://doi.org/10.2139/ssrn.2852404

Maestre FT, Salguero-Gómez R, Quero JL (2022) It's getting hot in here: ecological and economic challenges for dryland restoration under climate change. Nat Rev Earth Environ 3(5):310–320

MoEFCC (2022) National Action Programme to Combat Desertification (NAPCD). Ministry of Environment, Forest and Climate Change, Government of India

MoRD (2021) desert development programme: addressing desertification in India. Ministry of Rural Development, Government of India

Nature Conservancy Canada (2023) Prairie conservation action plan. Available at: https://www.natureconservancy.ca

Patel R, Desai H (2021) Participatory grassland restoration in Gujarat: the Banni grasslands project. Ecol Manage Restor 22(4):345–355

Reed MS, Stringer LC (2016) Land degradation and restoration in East Africa. Land Use Policy 56:85–98

Reed MS, Dougill AJ, Taylor MJ (2009) Integrating local and scientific knowledge for adaptation to climate change. Front Environ Sci 3:11. https://doi.org/10.3389/fenvs.2015.00011

UNEP (2021) Global environment outlook: healthy planet, healthy people. United Nations Environment Programme

Part VI
Epilogue

Chapter 31
Lessons Learned and the Way Forward for Land Restoration Through Ecosystem-Based Approach in the Drylands

Y. D. Imran Khan, Saurabh Purohit, Jibrin Mohammed Jibrin, and Madhav Dwarkanath Gholkar

Global drylands (hyperarid, arid, semiarid, and dry subhumid) cover about 42% of the earth's land surface and support nearly 2.5 billion people (Wang et al. 2023; Lewin et al. 2024). These drylands are home to diverse ecosystems with low vegetation cover, livelihoods, and diversified, rich cultural traditions that have evolved to thrive in harsh and variable climatic conditions (Wang et al. 2024). However, land degradation has become a key challenge due to unsustainable land use, climate change, population pressure, and deforestation (Galvin 2021). This is why urgent and sustainable solutions are needed in response to the effects of climate change, from decreased agricultural productivity to loss of biodiversity and increased risk to local communities. Ecosystem-based approaches (EbA) to land restoration provide a promising framework to address these challenges. Rather than concentrating solely on the technical aspects of restoration as with conventional methods, EbA integrates natural processes, biodiversity, and ecosystem services to restore the altered landscapes, build resilience, and increase adaptive capacity to climate change (United Nations Environment Programme 2021). Such approaches appreciate the complex interactions between ecosystems and peoples, applying science and traditional and local knowledge for sustainability.

Y. D. I. Khan (✉) · S. Purohit · M. D. Gholkar
WOTR Centre for Resilience Studies (W-CReS), Watershed Organisation Trust (WOTR), Pune, Maharashtra, India
e-mail: ikeducationkhan08@gmail.com; saurabhpurohit10@gmail.com; madhav.gholkar@wotr.org.in

J. M. Jibrin
Centre for Dryland Agriculture (CDA), Bayero University, Kano, Nigeria
e-mail: jibrin@buk.edu.ng

Y. D. I. Khan et al. (eds.), *Land Restoration Through Ecosystem-Based Approach*, Disaster Resilience and Green Growth, https://doi.org/10.1007/978-981-95-0389-6_31

With this background, this book explores the role of the EbA in land restoration in the drylands of the global south, where environmental degradation and socio-economic vulnerabilities intersect. Exploring case studies from different countries of the global south emphasizes how community-led initiatives, water conservation techniques, agroecological practices, and policy frameworks contribute to restoring degraded dryland ecosystems. Moreover, it highlights the importance of knowledge co-production, participatory governance, and inclusive decision-making in ensuring long-term success and sustainability.

This book aims to offer researchers, policymakers, practitioners, and development organizations some understanding of the potential benefits and bottlenecks of implementing EbA within the context of dryland restoration. It seeks to resolve the dichotomy between theoretical frameworks and practical realities by providing valuable evidence and experience-based approaches that can be implemented in different ecological, social, and political environments.

In the context of the multiplicative challenges of climate change, biodiversity loss, and land degradation, restoring drylands using ecosystem-based approaches is not only an environmental concern but also an important step toward food and nutritional security, alleviating poverty, and enhancing resilience to climate change. In this regard, this book seeks to engage readers in contemplating land restoration not solely as a technical challenge but as a complex and multidisciplinary problem that needs collaboration and innovation, as well as respect for the people and landscapes within the world's drylands. The chapters of this book provide insights to understand both existing and required actions in the successful implementation of EbA in restoring the dryland ecosystems across the global south. Highlights of the learning drawn from the chapters are discussed below. All the chapters of the book are segregated under the following five parts:

Part I: Restoration Dynamics in Dryland Ecosystems.
Part II: Nature-Based Restoration Strategies for Climate-Resilient Ecosystems.
Part III: Sustainable Land Management and Innovative Agricultural Practices.
Part IV: Cutting-Edge Innovations, Digital Tools, and Co-creation Strategies.
Part V: Adaptive Policies, Institutional Governance, and Long-Term Sustainability Strategies.

Part I, Restoration Dynamics in Dryland Ecosystems, explores sustainable restoration strategies for fragile dryland landscapes, emphasizing climate resilience, ecosystem-based approaches (EbA), and nature-based solutions (NbS). Sagar et al. (Chap. 2) highlight how EbA enhances resilience against climate change through rainfall mapping, soil moisture analysis, and reforestation. Chapter 3 by Gorain et al. discusses NbS, including agroforestry and water harvesting, as cost-effective restoration methods that support biodiversity and local economies. Chapter by Dadas et al. (Chap. 4) examines the undervaluation of rangelands in India, advocating for restoration using the EbA framework. R.J. Ranjeet Daniels focuses on Tamil Nadu's dryland restoration, emphasizing traditional agroecological practices and keystone species like the Palmyra palm (Chap. 5). Chapter 6 by Mhaske et al. assesses irrigation's impact on groundwater quality in Maharashtra, identifying issues like salinization and nitrate contamination. These chapters of Part I emphasize the need for sustainable land management, integrating scientific research with community-driven initiatives to enhance ecological health, climate adaptation, and livelihood sustainability in dryland ecosystems.

Part II, Nature-Based Restoration Strategies for Climate-Resilient Ecosystems, explores how ecosystem-based and nature-based solutions enhance dryland resilience. Verma et al. discuss integrated resource management strategies like agroforestry and sustainable livestock management to improve soil fertility and biodiversity (Chap. 7). Chapter 8 (Kumar et al.) emphasizes nature-based solutions, such as drought-tolerant crops and afforestation, to balance productivity and resilience. Kumar et al. (Chap. 9) highlight ecosystem-based adaptation (EbA) strategies like forest restoration and water flow regulation to combat climate risks. Chapungu and Zhanda review soil restoration techniques in Southern Africa, addressing policy challenges (Chap. 10). In Chap. 11, Saha et al. focus on nature-based disaster risk management in Cameroon through wetland restoration and dike construction. Purohit et al. examine avian diversity in Indian wetlands, advocating for conservation measures (Chap. 12). Mukate et al., in Chap. 13, analyze groundwater pollution in Maharashtra, stressing sustainable management. Together, these chapters of Part II highlight ecosystem-based restoration, sustainable land use, and community-driven strategies as essential for mitigating climate change impacts on dryland ecosystems.

Part III, *Sustainable Land Management and Innovative Agricultural Practices*, explores strategies to enhance dryland resilience through ecosystem-based approaches, organic farming, agroforestry, and plant breeding innovations. Chapter 14 by Omega et al. examines soybean farmers in Northern Ghana, highlighting row planting and weed control as key adaptation strategies while advocating for institutional support. A chapter by Sreekumar et al. discusses organic farming in Maharashtra, emphasizing agroforestry, crop diversification, and soil conservation but noting challenges like market access (Chap. 15). In Chap. 16, Simon CK Rop focused on Kenya, showcasing the role of agroforestry in restoring degraded lands and improving farmer livelihoods. Rawat et al. review a World Bank-funded vermicomposting initiative in Madhya Pradesh and Chhattisgarh, emphasizing its benefits for soil health and economic sustainability in Chap. 17. Chapter 18, Yetgin et al., explores climate-smart plant breeding, emphasizing CRISPR, marker-assisted selection, and drought-tolerant crops to enhance food security. Overall, Part III emphasizes the significance of sustainable agriculture, legislative interventions, and community involvement in alleviating climatic effects on dryland ecosystems.

Part IV, *Cutting-edge innovations, digital tools, and co-creation strategies*, explores sustainable dryland management. MacCarthy et al., in Chap. 19, examine decision-support tools for optimizing resource allocation among different farmer categories in Northern Ghana, emphasizing the need for dynamic agricultural planning. Gaitan et al. present the Innovation Niche Approach in Northern Cameroon, promoting climate-smart agriculture through Agricultural Innovation Platforms and Watershed Approaches in Chap. 20. A review by Mureithi et al. on soil and water conservation technologies, highlighting local participation and watershed management in dryland restoration (Chap. 21). Agboka et al. employ MaxEnt modeling to predict climate change's impact on *Anacardium occidentale* habitats in Togo, urging climate-resilient agricultural planning (Chap. 22). Chapter 23 by Bhuyan et al. integrates remote sensing and land use analysis to identify potential wetland restoration areas in India. Chitongo and Zhanda emphasize the role of multi-stakeholder participatory approaches in Sub-Saharan Africa's dryland restoration (Chap. 24). Afolabi et al. highlight indigenous ecological knowledge as a viable strategy for climate adaptation in Chap. 25. This section on *cutting-edge innovations, digital tools, and co-creation strategies* showcases diverse technological and traditional solutions for enhancing dryland resilience.

Part V, *Adaptive Policies, Institutional Governance, and Long-term Sustainability Strategies,* examines the role of governance, policies, and community engagement in sustainable dryland management. Angeline et al. (Chap. 26) highlight the circular ecosystem-based approach (CEbA) in Africa's crop-livestock systems, emphasizing policy support and institutional capacity. Favretto et al. highlight tree restoration policies in northern Nigeria, advocating for local knowledge integration in Chap. 27. A review by Sawadogo et al. on climate-smart agriculture (CSA) in Burkina Faso and Ghana emphasizes the need for better policy coordination (Chap. 28). Leonard Omullo Orondo evaluates ecosystem-based adaptation (EbA) strategies, finding community-led initiatives more effective than large-scale programs like the Great Green Wall (Chap. 29). Vishaal et al., in Chap. 30, analyze industrial pollution's role in India's dryland degradation, proposing a mitigation framework. Deepanjana Saha (Chap. 31) discusses policy pathways for restoration, emphasizing participatory governance and cross-sector collaboration. The section concludes that adaptive governance, policy coherence, and global-local alignment are crucial for long-term dryland sustainability and climate resilience.

Way Forward

- Implement rainfall mapping and soil moisture analysis to guide climate-resilient dryland reforestation efforts and ensure optimal land use and water retention.
- Expand afforestation efforts and use drought-tolerant crops to enhance dryland resilience while maintaining agricultural productivity.
- Promote agroforestry and water harvesting as cost-effective, nature-based solutions for ecological restoration and economic sustainability.
- Scale up traditional agroecological practices and protect keystone species to restore degraded drylands while preserving indigenous biodiversity.
- Develop sustainable irrigation practices and groundwater monitoring systems to mitigate risks like salinization and nitrate contamination.
- Strengthen wetland restoration initiatives and dike construction to enhance water retention and reduce the impacts of drought and desertification.

- Increase wetland conservation efforts to protect avian biodiversity and improve overall ecosystem health.
- Develop sustainable groundwater management strategies to combat pollution and secure long-term water availability.
- Address policy gaps in soil restoration by integrating scientific research with community-driven conservation initiatives.
- Promote integrated resource management by combining agroforestry with sustainable livestock practices to enhance soil fertility and biodiversity.
- Strengthen community-led soil and water conservation projects to restore degraded drylands effectively.
- Invest in climate-smart plant breeding, including CRISPR and drought-resistant crops, to enhance dryland food security.
- Develop and deploy decision-support tools that provide real-time guidance for farmers on resource allocation and land restoration.
- Scale up agroforestry initiatives to rehabilitate degraded lands while improving farmer livelihoods.
- Expand vermicomposting projects to enhance soil fertility and create economic opportunities for local communities.
- Use modeling and technology to guide climate-adaptive land use planning and reforestation in vulnerable dryland areas.
- Apply remote sensing and GIS to map and prioritize wetland restoration areas, ensuring targeted restoration interventions.
- Expand the Innovation Niche Approach to promote climate-smart agriculture and watershed restoration.
- Recognize rangelands as vital ecosystems and integrate them into national and global restoration plans using the EbA framework.
- Strengthen policy support for circular ecosystem-based approach (CEbA) to improve the sustainability of crop-livestock systems.
- Enhance policy coordination for climate-smart agriculture (CSA) to optimize restoration outcomes.
- Promote tree restoration policies by incorporating local ecological knowledge to enhance dryland resilience.
- Prioritize community-led ecosystem-based adaptation (EbA) initiatives over large scale, top-down programs to ensure effective and inclusive restoration.
- Implement industrial pollution control measures and sustainable land management frameworks to combat land degradation in drylands.
- Strengthen participatory governance and cross-sector collaboration to ensure policy coherence and long-term sustainability in dryland restoration efforts.
- Promote multi-stakeholder participation in dryland restoration to ensure long-term success and community ownership.
- Integrate indigenous ecological knowledge with modern restoration techniques to create culturally relevant and sustainable adaptation strategies.

References

Galvin KA (2021) Transformational adaptation in drylands. Curr Opin Environ Sustain 50:64–71

Lewin A, Murali G, Rachmilevitch S, Roll U (2024) Global evaluation of current and future threats to drylands and their vertebrate biodiversity. Nat Ecol Evol 8:1448–1458. https://doi.org/10.1038/s41559-024-02450-4

United Nations Environment Programme (2021) Guidelines for integrating ecosystem-based adaptation into national adaptation plans: supplement to the UNFCCC NAP technical guidelines. Nairobi

Wang H, Liu Y, Wang Y, Yao Y, Wang C (2023) Land cover change in global drylands: a review. Sci Total Environ 863:160943

Wang Y, Liu Y, Shan L, Du J, Liu Y, Li T, Cui X (2024) Ecosystem management and sustainable livelihoods in drylands. In: Dryland social-ecological Systems in Changing Environments. Springer Nature Singapore, Singapore, pp 139–157

MIX
Papier aus verantwortungsvollen Quellen
Paper from responsible sources
FSC® C105338

FSC
www.fsc.org

If you have any concerns about our products,
you can contact us on
ProductSafety@springernature.com

In case Publisher is established outside the EU,
the EU authorized representative is:
Springer Nature Customer Service Center GmbH
Europaplatz 3, 69115 Heidelberg, Germany

Printed by Libri Plureos GmbH
in Hamburg, Germany